Fundamentals of
PHYSICAL SCIENCE

SECOND EDITION

James T. Shipman
Ohio University

Jerry D. Wilson
Lander University

Aaron W. Todd
Middle Tennessee State University

D. C. HEATH AND COMPANY
Lexington, Massachusetts Toronto

To

The editorial team, Richard, Barbara, and Martha — JTS

All my former students — JDW

Four excellent high school teachers:
Frances Hobgood, Sarah Murray, Lee Pate, and Jere Warner — AWT

Address editorial correspondence to
D. C. Heath and Company
125 Spring Street
Lexington, MA 02173

Acquisitions Editor: Richard Stratton
Developmental Editors: Joanne Williams and Barbara Withington Meglis
Production Editor: Martha Wetherill
Photo Researchers: Judy Mason and Sharon Donahue
Art Editors: Diane Grossman and Prentice Crosier
Production Coordinator: Charles Dutton
Permissions Editor: Margaret Roll

Cover image: Alan Becker/The Image Bank

For permission to use copyrighted materials, grateful acknowledgment is made to the copyright holders listed on pages A19–21, which are hereby considered an extension of this copyright page.

Copyright © 1996 by D. C. Heath and Company.

Previous editions copyright © 1992 by D. C. Heath and Company.

All rights reserved. No part of this publication may be reproduced or transmitted in any form or by any means, electronic or mechanical, including photocopy, recording, or any information storage or retrieval system, without permission in writing from the publisher.

Published simultaneously in Canada.

Printed in the United States of America.

International Standard Book Number: 0-669-39764-4

Library of Congress Catalog Number: 94-74303

10 9 8 7 6 5 4 3 2 1

Preface

In this technological society, it is imperative that we, as educators, stimulate students' interest in the sciences and present the information and skills they need in today's world. In *Fundamentals of Physical Science*, Second Edition, we make concepts easily accessible by developing them in a logical rather than a chronological fashion and by discussing them in the context of everyday experience. We provide real-world examples of physical science throughout the text to enhance the readers' understanding of their natural world.

Fundamentals of Physical Science, Second Edition, is designed for the college nonscience major. The textbook emphasizes the basic concepts in physics, chemistry, and astronomy and presents an overview of meteorology and geology. Because many courses cover only physics, chemistry, and perhaps astronomy, we have concentrated our efforts in those areas of physical science; however, we do provide broad-brush coverage of meteorology and geology for courses that present more of a survey of physical science.

Conceptual Development

One of the outstanding features of *Fundamentals of Physical Science*, Second Edition, is the importance we place on fundamental concepts. The presentation emphasizes the fundamental concepts of each division of physical science by defining, explaining, and giving examples relative to the physical world. We begin with the most basic concepts (length, mass, time, and electric charge) and then proceed to new concepts, building on the basic ones. Our approach is more descriptive than quantitative. We believe that a working knowledge of the concepts and of the relationship of the concepts to one another provides students with a good and lasting comprehension of the physical world.

Chapter 1 begins with the fundamental concepts of measurement. From these fundamentals, we progress to the concepts of motion, force, energy, wave motion, heat, electricity, magnetism, and modern physics. We then use these concepts to explain the principles of chemistry and astronomy.

The structure of *Fundamentals of Physical Science* facilitates student involvement in the material. Care has been taken to move logically through each chapter, example, and worked-out solution without skipping steps and potentially leaving students confused.

Even though we have treated each discipline both descriptively and quantitatively, little emphasis is placed on math. We have provided mathematical assistance for students who may need it, but the relative emphasis, whether descriptive or quantitative, is left to the instructor's discretion. To those who wish to emphasize the descriptive approach in teaching physical science, we recommend using only the Questions at the end of each chapter and omitting the Exercises.

Changes in this Edition

The second edition of *Fundamentals of Physical Science* has been thoroughly revised. We have condensed the Earth Science section from five chapters to three, updated the text, added new photographs, redrawn some line art, and added composite photos and new illustrations to show the comparisons of the planets. We believe that the following changes will be of particular interest to instructors.

- New co-author and chemist, Aaron W. Todd, has rewritten the chemistry chapters completely, updating the discussions and providing real-world examples to explain essential concepts. He has also rewritten the chapter on nuclear physics and given more emphasis to the historical aspects of that field.

- We have reduced the amount and level of mathematics used. We do not exclude math, but we keep it to a minimum. Wherever appropriate, we provide a step-by-step approach for solving problems and highlight the steps for quick reference. The step-by-step solutions to each in-text example show students how the problem-solving approach is applied. Confidence Questions enable students to check their understanding of the preceding material before moving on to the next section. Answers to the Confidence Questions are provided at the end of each chapter.

- "Highlights"—boxed essays covering historical or descriptive material of special interest—are now provided in each chapter. For example, the "Highlight" in the chapter dealing with the Moon presents the latest theory accepted by many astronomers on the Moon's origin.

- More examples relevant to the environment have been added.

- Learning goals are now listed at the beginning of each section to help students study more effectively.

- At the end of each chapter, multiple-choice questions, keyed to the appropriate section of the text, have been added.

- Paired exercises in the end-of-chapter material are new to this edition. In one exercise, we provide the answer; in the other we omit the answer so that students can work out the problem. A third exercise, similar to the paired exercises in the text, is given in the students' *Study Guide*, with a complete solution.

- Thought questions, a new feature at the end of each chapter, stimulate cumulative learning and further appreciation of concepts covered in the text.

- The glossary has been improved by the inclusion of the page number where each glossary term is first defined in the text.

Supplements

The *Study Guide* is written by James T. Shipman of Ohio University, Aaron W. Todd of Middle Tennessee State University, and Clyde D. Baker, also of Ohio University. It features study goals, discussion, review questions, solved problems, and multiple-choice questions with explanations.

The *Instructor's Guide* and *Test Item File*, by James T. Shipman, Jerry D. Wilson, and Aaron W. Todd, have been updated and combined into one supplement for convenience. Each chapter of the *Instructor's Guide* includes a brief discussion, suggested demonstrations, answers to textbook questions, solutions for exercises, and a brief quiz with answers. The *Test Item File* offers a printed version of more than 1200 questions available in completion, multiple-choice, and short-exercise formats. The Appendix includes a Teaching Aids section for each of the five sciences, as well as up-to-date audiovisual resources.

The *Laboratory Guide*, by James T. Shipman and Clyde D. Baker, offers 53 experiments. Each experiment includes an introduction, learning objectives, a list of required apparatus, a detailed procedure for collecting data (requiring students to generate tables and graphs, and to perform calculations), and questions about the experiment.

The *Instructor's Resource Manual for the Laboratory Guide*, also by James T. Shipman and Clyde D. Baker, now includes an integrated equipment list to assist instructors in planning experiments. For most experiments, additional data and calculations are provided, as well as answers to questions, a discussion of each experiment, and additional questions. The *Laboratory Guide* and the *Instructor's Resource Manual for the Laboratory Guide* are appropriate for courses using either *Fundamentals of Physical Science* or the Seventh Edition of *Introduction to Physical Science*.

The *Computerized Testing Program*, Esatest III, provides the *Test Item File*'s questions in electronic format. Instructors can easily produce chapter tests, midterms, and final exams that include graphics. Instructors can edit existing questions or add new ones as desired and can preview questions on screen. The computerized testing program is available for IBM computers.

The Transparencies—more than 80 in one, two-, and four-color—illustrate important concepts from the text. They are available to adopters of either *Fundamentals of Physical Science* or *Introduction to Physical Science*.

Acknowledgments

We wish to thank our colleagues and students for the many contributions they made to this text through correspondence, discussions, questionnaires, and classroom use of the text material. We would also like to thank the following reviewers for their suggestions and comments: Richard M. Bowers, Weatherford College; Talbert Brown, Southeast Oklahoma State; George Canty, Jr., Fort Valley State College; Darry S. Carlstone, University of Central Oklahoma; Robert Carlton, Middle Tennessee State University; Michael W. Castelaz, East Tennessee State; Douglas Chan, Langston University; Jack Couch, Bloomsburg State University; Milton W. Ferguson, Norfolk State University; Jack Gaiser, University of Central Arkansas; Charles Irish, Westark Community College; Michieal L. Jones, University of Georgia; Steven D. Kamm, Oklahoma City Community College; George Knott, Cosumnes River College; Lauree Lane, Tennessee State University; Kenneth LaSota, Robert Morris College; Michael Merchant, University of Arkansas; James F. Nugent, Salve Regina University; John Rives, University of Georgia; R. Neil Rudolph, Adams State College; John J. Stith, Virginia State University; Aleksandar Svager, Central State University; Barbara Z. Thomas, Brevard Community College; Lynn Thompson, Ricks College; Hanno T. Tohver, University of Alabama; Michael Torbett, Macon College; Walter Watson, Alabama A&M University; Jai-Ching Wang, Alabama A&M University; James C. White II, Middle Tennessee State University; Carey Witkov, Broward Community College.

We are grateful to those individuals and organizations who contributed photographs, illustrations, and other information used in this text. We are also indebted to the D. C. Heath staff for their dedicated and conscientious efforts in the production of *Fundamentals of Physical Science,* Second Edition. We especially wish to thank Richard Stratton, Senior Acquisitions Editor; Barbara Withington Meglis, Developmental Editor; Martha Wetherill, Production Editor; and Judy Mason and Sharon Donahue, Photo Researchers. Finally, we acknowledge the contributions of Sarah and Sudie Shipman, Sandy Wilson, and Clara Todd.

We welcome comments from students and instructors of physical science and invite you to send your impressions and suggestions.

J. T. S.
J. D. W.
A. W. T.

Contents

PART ONE PHYSICS

1 MEASUREMENT — 1

1.1 The Senses 2
1.2 The Scientific Method and Fundamental Quantities 3
1.3 Standard Units 5
  Highlight: The Metric System and the SI 6
1.4 Derived Quantities 10
1.5 Conversion Factors 14
1.6 Scientific Notation and Metric Prefixes 16

2 MOTION — 23

2.1 Position and Path 23
2.2 Speed and Velocity 26
2.3 Acceleration 31
Highlight: Galileo and the Leaning Tower of Pisa 35
2.4 Projectile Motion 36

3 FORCE AND MOTION — 43

3.1 Force and Newton's First Law of Motion 43
3.2 Newton's Second Law of Motion 46
  Highlight: Isaac Newton 47
3.3 Newton's Law of Gravitation 53
3.4 Newton's Third Law of Motion 57
3.5 Momentum and Impulse 58
  Highlight: Impulse and the Automobile Air Bag 64

4 WORK AND ENERGY — 69

4.1 Work 69
4.2 Power 72
4.3 Kinetic Energy and Potential Energy 75
4.4 The Conservation of Energy 79
4.5 Forms and Sources of Energy 82
  Highlight: The Conservation of Mass-Energy 84

5 TEMPERATURE AND HEAT — 89

5.1 Temperature 90
Highlight: Freezing from the Top Down 94
5.2 Heat 96
5.3 Specific Heat and Latent Heat 97
5.4 Thermodynamics 101
5.5 Heat Transfer 106
5.6 Phases of Matter 108
Highlight: Kinetic Theory, the Ideal Gas Law, and Absolute Zero 110

6 WAVES — 117

6.1 Wave Properties 118
6.2 Electromagnetic Waves 121
6.3 Sound Waves 122
Highlight: Noise Exposure Limits 127
6.4 The Doppler Effect 128
6.5 Standing Waves and Resonance 130

7 WAVE EFFECTS AND OPTICS — 137

7.1 Reflection 137
7.2 Refraction and Dispersion 139
7.3 Diffraction, Interference, and Polarization 143
Highlight: The Rainbow 145
7.4 Spherical Mirrors 151
Highlight: Liquid Crystal Displays (LCDs) 154
7.5 Spherical Lenses 157

8 ELECTRICITY AND MAGNETISM — 165

8.1 Electric Charge and Current 165
8.2 Voltage and Electrical Power 170
8.3 Simple Electric Circuits and Electrical Safety 174
Highlight: Superconductivity 175
8.4 Magnetism 180
Highlight: Electric Shock 181
8.5 Electromagnetism 186

9 ATOMIC PHYSICS — 197

9.1 The Dual Nature of Light 197
9.2 The Bohr Theory of the Hydrogen Atom 200
9.3 Quantum-Physics Applications 205
9.4 Matter Waves and Quantum Mechanics 209
Highlight: Fluorescence and Phosphorescence 210
9.5 Multielectron Atoms and the Periodic Table 215

10 NUCLEAR PHYSICS — 223

- 10.1 The Atomic Nucleus 223
- 10.2 Radioactivity 227
  - **Highlight:** The Discovery of Radioactivity 230
- 10.3 Half-life and Radiometric Dating 233
- 10.4 Nuclear Reactions 238
- 10.5 Nuclear Fission 241
  - **Highlight:** The Building of the Bomb 244
- 10.6 Nuclear Fusion 247
- 10.7 Biological Effects of Radiation 251

PART TWO CHEMISTRY

11 THE CHEMICAL ELEMENTS — 259

- 11.1 Classification of Matter 259
- 11.2 Names and Symbols of Elements 262
- 11.3 Occurrence of the Elements 264
- 11.4 The Periodic Table 266
  - **Highlight:** Mendeleev and the Periodic Table 268
- 11.5 Naming Compounds 272
- 11.6 Groups of Elements 276

12 CHEMICAL BONDING — 285

- 12.1 Law of Conservation of Mass 285
- 12.2 Law of Definite Proportions 286
  - **Highlight:** The Origins of Chemistry 288
- 12.3 Dalton's Atomic Theory 290
- 12.4 Ionic Bonding 292
- 12.5 Covalent Bonding 298
- 12.6 Metallic Bonding and Hydrogen Bonding 304
- 12.7 The Stock System of Nomenclature 307

13 CHEMICAL REACTIONS — 315

- 13.1 Chemical and Physical Properties and Changes 315
- 13.2 Chemical Equilibrium 316
- 13.3 Balancing Equations 318
- 13.4 Energy and Rate of Reaction 320
- 13.5 Acids and Bases 326
- 13.6 Single-Replacement Reactions 332
  - **Highlight:** The Mole and Avogadro's Number 336
- 13.7 Electrochemistry 337

14 ORGANIC CHEMISTRY — 345

14.1 Bonding in Organic Compounds 345
14.2 Aliphatic Hydrocarbons 347
14.3 Aromatic Hydrocarbons 352
14.4 Derivatives of Hydrocarbons 354
14.5 Synthetic Polymers 361
Highlight: Drugs 364

PART THREE ASTRONOMY

15 THE SOLAR SYSTEM — 373

15.1 The Planet Earth 374
15.2 The Solar System 377
15.3 The Terrestrial Planets 385
15.4 The Jovian Planets and Pluto 391
15.5 Other Solar System Objects 401
Highlight: Space Flights in the Solar System 402
15.6 The Origin of the Solar System 407
15.7 Other Planetary Systems 410

16 PLACE AND TIME — 417

16.1 Cartesian Coordinates 417
16.2 Latitude and Longitude 418
16.3 Time 420
Highlight: The Concept of Time 423
16.4 The Seasons 427
16.5 Precession of Earth's Axis 432
16.6 The Calendar 433

17 THE MOON — 440

17.1 General Physical Properties 440
17.2 Composition and Origin of the Moon 444
Highlight: The Origin of the Moon 445
17.3 Lunar Motions 445
17.4 Phases of the Moon 446
17.5 Solar and Lunar Eclipses 451
17.6 Ocean Tides 454

18 THE UNIVERSE — 462

18.1 The Sun 463
18.2 The Celestial Sphere 466
Highlight: Solar Neutrinos 467
18.3 Stars 469
18.4 Gravitational Collapse and Black Holes 477
18.5 Galaxies 479
18.6 Quasars 486
18.7 Cosmology 487
Highlight: Other Cosmologies 492

PART FOUR   EARTH SCIENCE

19 THE ATMOSPHERE, WEATHER, AND CLIMATE — 498

19.1 Atmospheric Composition and Structure 499
19.2 Atmospheric Energy Content and Measurements 501
Highlight: Why the Sky is Blue and Sunsets are Red 504
Highlight: The Greenhouse Effect 506
19.3 Air Motion and Clouds 508
19.4 Air Masses and Storms 517
19.5 Atmospheric Pollution and Climate 526
Highlight: The Ozone Hole 532

20 MINERALS, ROCKS, AND GEOLOGIC EVENTS — 539

20.1 Minerals 540
20.2 Rocks 545
Highlight: Gems 546
20.3 Volcanoes 556
20.4 Earthquakes 562

21 STRUCTURAL GEOLOGY AND GEOLOGIC TIME — 572

21.1 Earth's Structure 573
21.2 Continental Drift and Seafloor Spreading 575
21.3 Plate Tectonics 581
21.4 Geologic Time 584
Highlight: The Origin of the Appalachian Mountains 587

APPENDIXES

Appendix I   The Seven Base Units of the International System of Units (SI)   A1
Appendix II   Solving Mathematical Problems in Science   A1
Appendix III   Equation Rearrangement   A2
Appendix IV   Analysis of Units   A5
Appendix V   Positive and Negative Numbers   A7
Appendix VI   Powers-of-10 Notation   A8
Appendix VII   Significant Figures   A9
Appendix VIII   Length Contraction, Time Dilation, and Relativistic Mass Increase   A11
Appendix IX   Alphabetical List of the Elements   A13
Appendix X   Psychrometric Tables (Pressure: 30 in. of Hg)   A14
Appendix XI   Seasonal Star Charts   A15

Photo Credits   A19
Glossary   A22
Index   A37

Guide to Features

Guide to Features

New Chemistry Section

The chemistry chapters, written by our new coauthor Aaron W. Todd, include real-world examples to generate interest in chemistry and thorough explanations of chemical concepts to help students understand the chemistry around them.

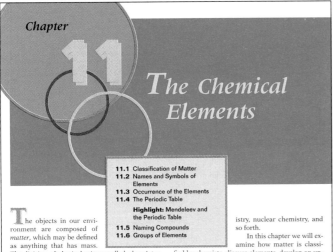

Chapter 11: The Chemical Elements

11.1 Classification of Matter
11.2 Names and Symbols of Elements
11.3 Occurrence of the Elements
11.4 The Periodic Table
Highlight: Mendeleev and the Periodic Table
11.5 Naming Compounds
11.6 Groups of Elements

The objects in our environment are composed of *matter*, which may be defined as anything that has mass. The division of physical science called **chemistry** deals with the composition and structure of matter, and the reactions by which substances are changed into other substances. The modern understanding of chemical reactions, based on the electron configuration of atoms, is relatively recent. However, chemistry had its beginnings early in history (see the Chapter 12 Highlight).

Chemistry is separated into five major divisions. *Physical chemistry*, the most fundamental of the five divisions, applies the theories of physics (especially thermodynamics) to the study of chemical reactions. *Analytical chemistry* identifies what substances are present in a material and determines how much of each substance is present. *Organic chemistry* is the study of compounds that contain carbon (see Chapter 14). The study of all other chemical compounds is called *inorganic chemistry*. *Biochemistry*, where chemistry and biology meet, deals with the chemical reactions that occur in living organisms. As you might expect, chemistry's five major divisions overlap, and there are other, smaller divisions such as polymer chemistry, nuclear chemistry, and so forth.

In this chapter we will examine how matter is classified by chemists, discuss elements, develop an understanding of the periodic table, and learn how compounds are named.

11.1 Classification of Matter

Learning Goals:
- To explain how chemists classify matter.
- To identify three types of solutions.

In Chapter 5 we saw that matter by its physical phase: solid, liquid However, chemists find the schem Fig. 11.1 particularly useful. It ter into pure substances and mixt

A **pure substance** is a type o all samples have fixed composit properties. Pure substances are compounds. An **element** is a su all the atoms have the same num same atomic number (Section 10

Figure 11.21 Emerald, the green gem variety of beryl.
Beryllium, an alkaline earth metal, comes primarily from the mineral named beryl, $Be_3Al_2Si_6O_{18}$.

Hydrogen

Although hydrogen is commonly listed in Group 1A, it is not considered an alkali metal. Hydrogen is a nonmetal that usually reacts like an alkali metal, forming HCl, H_2S, and so forth (similar to NaCl, Na_2S, etc.). Yet sometimes hydrogen reacts like a halogen, forming NaH (sodium hydride), CaH_2, and so forth (compare with NaCl and $CaCl_2$). It is the least dense of the elements and so was used in early dirigibles (Fig. 11.23). At room temperature hydrogen is a colorless, odorless gas and consists of diatomic molecules (H_2). It does not liquefy at normal atmospheric pressure until cooled to $-253°C$ (20 K).

Although hydrogen burns in air, it is actually a safer fuel than gasoline. It will probably be the main fuel used in transportation in the future. Since it burns to form only water, it is apparent how its use would help the pollution problem.

Applications

Expanded coverage of current issues and new environmental examples emphasize the connection between physical science and the world around us.

to build roads and to neutralize acid soils. Tums tablets are mainly calcium carbonate. Also composed of calcium carbonate are the impressive formations of stalactites and stalagmites in underground caverns.

Strontium compounds produce the red colors in fireworks, and barium compounds produce the green. When radioactive strontium-90 (from fallout) is ingested, it goes into the bone marrow because of its similarity to calcium. If enough strontium-90 is ingested, it can destroy the bone marrow or cause cancer. Fortunately, radioactive fallout is minimal now. A barium compound encountered all too frequently is barium sulfate ($BaSO_4$), the white material in the thick solutions a patient ingests prior to having X-rays taken of his or her gastrointestinal tract (Fig. 11.22).

Radium is an intensely radioactive element that glows in the dark. Its radioactivity is a deterrent to practical uses, although a few decades ago watch dials were painted with radium chloride ($RaCl_2$) so they could be read in darkness.

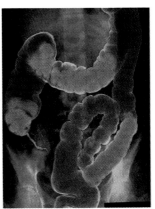

Figure 11.22 A common use of barium sulfate.
The contrast in this radiograph of a patient's large intestine is produced by the introduction of $BaSO_4$ which absorbs X-rays much better than do body tissue and bone.

Guide to Features

Thought Questions

Thought Questions in the end-of-chapter section allow students to integrate the material learned in the chapter with their own knowledge of the world.

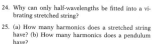

longer wavelengths, (c) a shift toward increased frequency, (d) a sonic boom.

19. How is the wavelength of sound affected when (a) a source moves toward a stationary observer, and (b) an observer moves away from a stationary source?

20. What would be the situation for a sound (a) "blue shift" and (b) "red shift"?

21. Compare the crack of a whip with a sonic boom.

6.5 Standing Waves and Resonance

22. Stationary points in a standing wave are called (a) normal modes, (b) zero points, (c) nodes, (d) antinodes.

23. A stretched string will be driven in resonance when driven at the frequency of its (a) first overtone, (b) second overtone, (c) first harmonic, (d) all of these.

24. Why can only half-wavelengths be fitted into a vibrating stretched string?

25. (a) How many harmonics does a stretched string have? (b) How many harmonics does a pendulum have?

26. What is the effect if a system is driven in resonance? Is a particular frequency required?

27. What determines the pitch or frequency of a string on a violin? How does the violinist get a variety of notes from one string?

FOOD FOR THOUGHT

1. Here's an old one: If a tree falls in the forest and no one is present to hear it, is there sound?

2. If an astronaut on the Moon dropped a hammer, would there be sound? Explain. (*Follow-up:* How do astronauts communicate with each other and mission control?)

3. Discuss the effects if (a) a sound source and an observer were both moving with the same velocity, and (b) a sound source approached a stationary observer going faster than the speed of sound.

4. If a jet pilot is flying faster than the speed of sound, will he or she be able to hear any sound?

5. When one sings in the shower, the tones sound full and rich. Why is this?

EXERCISES

6.1 Wave Properties

1. A periodic wave has a period of 3.0 s. What is the frequency?

   Answer: (a) 0.40 Hz

   ... has a frequency of 200 Hz. What is the ... en adjacent crests or compressions of ... ume an air temperature of 20°C.)

   ... und in water is 1530 m/s. What is the ... a 2000-Hz sound wave in water?

6.3 Sound Waves

5. Compute the wavelength of ultrasound with a frequency of 30 kHz if the speed of sound in water is 1500 m/s.

   Answer: 0.050 m

6. During a thunderstorm, 4 s elapse between observing a lightning flash and hearing the resulting thunder. Approximately how far away (in km and mi) was the lightning flash? (*Hint:* Use the fractional approximations.)

   Answer: 4/3 km or 4/5 mi

7. A subway train has a sound intensity level of 90 dB, and a rock band has a sound intensity level of 110 dB. How many times greater is the sound intensity of the band than the subway train?

   Answer: 100 times

HIGHLIGHT

Impulse and the Automobile Air Bag

A relatively new, major automobile safety feature is the air bag. As learned earlier, seatbelts restrain you so you don't follow along with Newton's first law when the car comes to a sudden stop. But, what is the principle underlying the action of the air bag?

When a car has a head-on collision with another vehicle, or hits an immovable object such as a tree, it stops almost instantaneously. If not buckled up, the driver could be seriously injured in hitting the steering wheel and column, and you can imagine what might happen to a passenger in the front seat. Even with seatbelts, the impact of a head-on collision may be such that seatbelts do not restrain you completely and injuries could occur. Enter the air bag. This balloon-like bag inflates automatically on hard impact and cushions the driver (Fig. 1). Passenger-side air bags are becoming more common, and back-seat air bags are also available.

If we look at the air bag in terms of impulse, the bag increases the contact time in stopping a person, thereby reducing the impact force. Also, the impact force is spread over a large general area and not applied to just certain parts of the body, as in the case of seatbelts.

Being inquisitive, you might wonder what causes an air bag to inflate and what inflates it. Keep in mind that inflation must occur in a fraction of a second to do any good. (How much time would there be between the initial collision contact and a driver hitting the steering wheel column?) In current designs the air bag's inflation is initiated by an electronic sensing unit. This unit contains sensors that detect rapid decelerations, such as those that occur in a high-impact collision. The sensors have threshold settings so that normal hard braking does not activate them.

Sensing an impact, a control unit sends an electric current to an igniter in the air-bag system that sets off a chemical explosion. The gases (mostly nitrogen) rapidly inflate the thin nylon bag. The total process from sensing to complete inflation takes only about 25 thousandths of a second (0.025 s)!

The sensing unit is equipped with its own electrical power source. In a front-end collision, a car's battery and alternator are among the first to go. The currently installed automobile air bags offer protection for only front-end collisions, in which the car's occupants are thrown forward (more accurately, continue to travel forward—Newton's first law). No protection is offered for side-impact collisions. However, a foreign manufacturer has announced that its cars will be equipped with side air bags, as well as front air bags. Most cars will probably have this extended protection in the future. But, always remember to buckle up—even if the vehicle is equipped with air bags. (Maybe we should make this Newton's fourth law of motion.)

Figure 1 Life-saving impulse. An illustration of an inflated air bag protecting a driver. The bag increases the contact time, thereby reducing the impact force (which would be quite large if the driver hit the steering column). The seat belt slows the forward motion of the body and adds extra protection. Always wear them.

Highlight Boxes

Interest boxes, covering historical or descriptive material of particular interest to the student of physical science, now appear in each chapter.

---

have two oppositely rotating rotors (Fig. 3.24a). Smaller helicopters have a small "antitorque" rotor on the tail, which counteracts the rotation of the helicopter body.

CONFIDENCE QUESTION 3.4

Take a piece of chalk (or similar object) and tie a string about one-half meter long securely to the chalk. Hold the string in your hand, securing it firmly between the thumb and forefinger, and whirl the chalk around in a vertical circle. Upon extending the forefinger so the string winds itself around this finger, what happens to the motion of the chalk and why? (Try this simple experiment yourself.)

Guide to Features

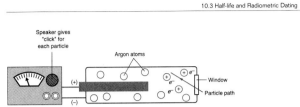

10.3 Half-life and Radiometric Dating   235

Figure 10.9 A schematic representation of a Geiger counter.
A high-energy particle from a radioactive source enters the window and ionizes argon atoms along its path. The ions and electrons formed produce a pulse of current, which is amplified and counted.

higher the ratio of radiogenic ^{206}Pb to ^{238}U, the older the rock (Fig. 10.10).

It is possible to tell how much ^{206}Pb in a rock is radiogenic by determining how much ^{204}Pb is present, because all ^{204}Pb is primordial and the ratio in which primordial ^{206}Pb and ^{204}Pb occurs is known. If a rock contains ^{238}U, then ^{235}U will also be present, and this radionuclide decays with a half-life of 7.04×10^8 y to ^{207}Pb. Thus the ratio of ^{207}Pb to ^{235}U can be used as a check on the age obtained by the ^{206}Pb/^{238}U ratio. Such checks substantiate the validity of the method. Dating of some types of rocks is achieved by use of other radionuclides, such as potassium-40 (half-life = 1.25×10^9 y), which decays to argon-40.

The oldest rocks on Earth have been dated at 3.8 billion y, which provides a minimum possible age of Earth—the time since the solid crust first formed. For meteorites, which are assumed to have solidified at the same time as other solid objects in the solar system including Earth, ages have been determined to be 4.4 billion to 4.6 billion y. The oldest Moon rocks returned to Earth by the Apollo missions were found to be 4.5 billion y old. These measurements and other evidence indicate that the age of Earth is about 4.6 billion y.

EXAMPLE 10.6 Dating a rock by use of ^{235}U

Uranium-235 has a half-life = 7.04×10^8 y and decays to ^{207}Pb. If analysis of a rock shows that enough radiogenic ^{207}Pb is present to indicate that only one-eighth of the original ^{235}U is undecayed, how old is

the rock? (This is an example of finding the elapsed time.)

Solution

STEP 1
Start with $\frac{1}{1}$ and see how many times you must halve to get to $\frac{1}{8}$.

$$\frac{1}{1} \rightarrow \frac{1}{2} \rightarrow \frac{1}{4} \rightarrow \frac{1}{8}$$

If only one-eighth of the ^{235}U remains, the rock has existed for three half-lives. (Count the arrows in the sequence above.)

STEP 2
To find the rock's age, multiply the number of half-lives by the half-life.

(3 half-lives)(7.04×10^8 y/half-life) = 2.11×10^9 y

So the rock is about 2.1 billion years old.

CONFIDENCE QUESTION 10.6
Potassium-40 has a half-life of 1.25×10^9 y and decays to argon-40. If analysis of a rock shows that only one-fourth of the original ^{40}K is undecayed, about how old is the rock?

To find the age of organic (once-living) remains, such as charcoal, parchment, or bones, scientists use **carbon dating**, developed in 1950 by Willard F. Libby (Nobel Prize in chemistry, 1960). This procedure measures the activity of ^{14}C in the

Confidence Questions

Following each worked-out example in the chapter is a Confidence Question that provides immediate reinforcement of students' understanding of the material just covered. Students can check their solutions to the Confidence Questions at the end of each chapter.

The Universe

has received broad acceptance
the observed cosmological red shift,
mic microwave background, and the
of hydrogen to helium in stars.
predicts an open universe.
provides information about the
ck to the very beginning.
and (b) are true.
al principle
(a) is based on the assumption that the universe is homogeneous.
(b) implies that the universe has no edge.

(c) is based on the assumption that the universe is isotropic.
(d) implies that the universe has no center.
(e) all of the above
62. Describe the Big Bang model of the universe.
63. State three experimental facts that support the Big Bang model.
64. What property of the universe can be determined by taking the reciprocal of Hubble's constant?
65. How old is (a) the Sun and (b) the universe?
66. Why is the universe thought to be expanding?

FOOD FOR THOUGHT

1. Where on Earth's surface must an observer be in order to see all visible stars over a period of one year?
2. For an Earth observer, is it conceivable that the universe is both finite and infinite? (Does Hubble's law set a limit on time?)
3. Do you think our universe is one of a kind, or is it possible that others exist? Why or why not.
4. What is your mental concept of the universe? Has your personal view changed since entering college?

EXERCISES

18.2 The Celestial Sphere
1. Find the distance in light-years to a star with a parallax of 0.20 s.
   Answer: 16 ly
2. Find the distance in parsecs to a star with a parallax of 0.20 s.
3. Calculate the number of miles in a light-year, using 186,000 mi/s as the speed of light.
   Answer: 5.87×10^{12} mi
4. Calculate the number of meters in a light-year, using 3.00×10^8 m/s as the speed of light.
5. How long in years does it take light to reach us from Alpha Centauri, which is about 4.3 ly away?
   Answer: about 4.3 ly
6. How many parsecs away is Alpha Centauri, which is at a distance of about 4.3 ly?

18.3 Stars
7. The Crab Nebula is expanding at a rate of 70 million mi/day. If it is the remnant of a supernova of A.D. 1054, how many miles in diameter will it be in A.D. 2000?

18.4 Gravitational Collapse and Black Holes
8. Determine the radial distance of the event horizon of a black hole formed from the gravitational collapse of a star having a mass of 15×10^{30} kg.
   Answer: 22 km
9. Calculate the distance between the singularity and the event horizon for a gravitationally collapsed star having a mass of 2.0×10^{32} kg.

Color-coded, paired exercises

The end-of-chapter material contains paired exercises. One exercise has the answer provided, and the other is left for the student to work out. A third exercise with its complete solution is in the student Study Guide.

Worked-Out Examples

Worked-out examples give the students clear, step-by-step solutions to solving problems in physical science using minimal mathematics.

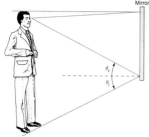

Figure 7.4 Complete figure.
For a person to see his or her complete figure in a plane mirror, the height of the mirror must be at least one-half of the height of the person, as can be easily shown by ray tracing.

The image is located by drawing two (or more) rays emanating from the object and applying the law of reflection. The image is located where the rays intersect or appear to intersect. Notice that for a plane mirror the image is located "inside" or behind the mirror at the same distance the object is in front of the mirror.

Figure 7.4 shows a ray diagram for the light rays involved when a person sees a complete or head-to-toe image. How big (tall) a mirror is needed for this? Applying the law of reflection reveals that one can see one's complete image in a plane mirror that is only one-half of one's height. Also, the distance one stands from the mirror is not a factor.

It is the reflection of light that allows us to see things. Look around you. What you see in general is light reflected from the walls, the ceiling, the floor, and other objects. Of course, there must be one or more sources of light present, such as the Sun or lamps. If you are in a completely dark room, then there is no reflected light and you can't see anything.

You have probably noticed that at night in a lighted room, a transparent glass windowpane reflects light and acts as a mirror. Yet during the day,

Figure 7.5 Natural reflection.
Beautiful reflections, such as shown here for a water surface, are often seen in nature.

we see through it. Why is this? The glass itself doesn't act any differently night or day. Light is still reflected back into the room during the day. However, the large amount of transmitted light from the outside masks the reflected light during the day, whereas during the night the masking effect of transmitted light is absent.

We often see beautiful reflections in nature, as shown in Fig. 7.5. Is the picture really right-side-up? Turn the book over and see.

7.2 Refraction and Dispersion

Learning Goals:

To explain the phenomenon of refraction and how this leads to the dispersion of light.

To explain how the boundary of transparent media can be used as a mirror through total internal reflection, and how this is applied in fiber optics.

To define dispersion and describe some of its effects.

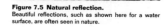

Learning Goals

Each section of the chapter begins with a list of Learning Goals that will help students study and synthesize the material in that section.

HIGHLIGHT

Noise Exposure Limits

Sounds with intensities of 120 dB and higher can be painfully loud to the ear. Brief exposures to even higher sound intensity levels can rupture eardrums and cause permanent hearing loss. However, long exposure to relatively lower sound (noise) levels can also cause hearing problems. (Noise is defined as unwanted sound.) Such exposures may be an occupational hazard, and in some jobs ear protectors must be worn (see the accompanying figure). You may have experienced a temporary hearing loss after being exposed to a loud band for a long time or a loud bang for a short time.

Federal standards now set permissible noise exposure limits for occupational loudness. These limits are listed in Table 1. Notice that a person can work on a subway train (90 dB, Fig. 6.13) for 8 h, but a person should only play in (or listen to) an amplified rock band (110 dB) continuously for $\frac{1}{2}$ h.

Table 1 Permissible Noise Exposure Limits

Maximum Duration per Day (h)	Sound Level Intensity (dB)
8	90
6	92
4	95
3	97
2	100
1½	102
1	105
½	110
¼ or less	115

Figure 1 Sound-intensity safety. An airport worker wears ear protectors to prevent ear damage from the high sound-intensity levels of jet engines.

In general, as the density of the medium increases, the speed of sound therein increases. The speed of sound is about 4 times faster in water than in air and, in general, about 15 times faster in solids.

Using the speed of sound and the frequency, we can easily compute the wavelength of a sound wave.

EXAMPLE 6.2 Finding Audible and Ultrasound Wavelengths

What is the wavelength of a sound wave in air with a frequency of (a) 2200 Hz and (b) 22 MHz?

Solution

We have

Given: (a) f = 2200 Hz
(b) f = 22 MHz = 22,000,000 Hz
Find: λ (wavelength)

Notice in (b) the frequency was given in megahertz (MHz). *Mega-* denotes million, and we write the frequency directly in hertz.

The wavelength and frequency are related by Eq. 6.3 ($v = \lambda f$), which we write in the form $\lambda = v_{sound}/f$. To calculate λ, a value for the speed of sound must be obtained, and we will assume this to be its 20°C value, v_{sound} = 344 m/s (known). Then, we can put in the numbers:

(a) For f = 2200 Hz, which is in the audible range,

$$\lambda = \frac{v_{sound}}{f} = \frac{344 \text{ m/s}}{2200 \text{ Hz}} = 0.16 \text{ m}$$

This is a wavelength of about $\frac{1}{2}$ ft.

(b) For f = 22 MHz, or 22,000,000 Hz, which is in the ultrasonic region,

$$\lambda = \frac{v_{sound}}{f} = \frac{344 \text{ m/s}}{22,000,000 \text{ Hz}} = 0.000016 \text{ m}$$

Chapter 1

Measurement

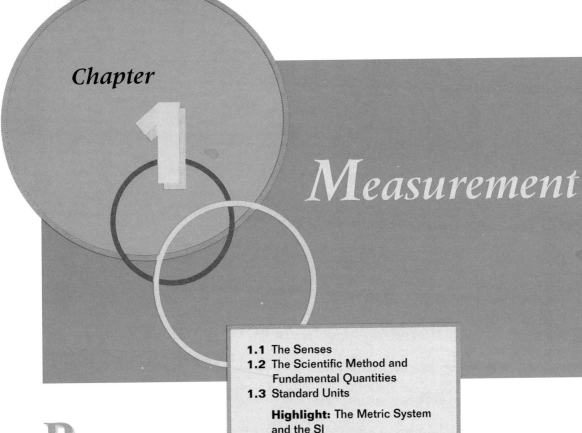

1.1 The Senses
1.2 The Scientific Method and Fundamental Quantities
1.3 Standard Units

Highlight: The Metric System and the SI

1.4 Derived Quantities
1.5 Conversion Factors
1.6 Scientific Notation and Metric Prefixes

Physical Science is concerned with the understanding and description of the physical world around us. We see things that occur and want to know why. To understand and talk about something in science, we must first describe it, and this is done through **measurement,** which involves comparing a physical quantity to some standard. Over the years, humans have developed better and better methods of measurement; and scientists make use of the most advanced, as well as the simplest of these. A general knowledge of measurement methods and an awareness of their limitations help one better understand the information that measurements convey.

We are continually making measurements in our daily lives. Each day we plan our activities as a function of time. With watches and clocks, we measure the times for events to take place. Every 10 years we take a census and determine (measure) the country's population. We count our money, the amount of precipitation, the minutes, days, and years of our lives. Some of us keep accurate measurements of food and drugs taken into the body because of illness. Many lives depend on accurate measurements being made by medical doctors, laboratory technicians, and pharmacists in the diagnosis and treatment of disease.

Meteorologists measure the many elements that make up the weather, such as temperature, pressure, humidity, and wind speed. This information, along with weather forecasts, is relayed to millions of people by the communications media, which operate on monitored (measured) assigned frequencies of transmission.

The ability to know and predict phenomena is a function of accurate measurements. Scientists must measure the very large and the very small, exploring outward to discover an ever-larger universe and probing inward to examine small particles. At one time, it was thought that all things could be measured with exact certainty. However, as smaller and smaller objects were measured, it was learned that the very act of measuring distorted the measurement. This uncertainty in making measurements is very small and is discussed in Chapter 9.

The preceding examples show the relevance of measurement and underscore the need to know the concepts of measurement. Understanding measurements provides the first step in the understanding of our physical environment.

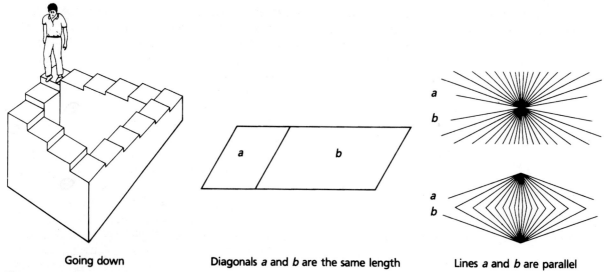

Figure 1.1 Some optical illusions.
We can be deceived by what we see—or what we think we see.

1.1 The Senses

Learning Goal*:
To identify some of the limitations of our senses and how they can be overcome.

Our environment stimulates our senses, either directly or indirectly. The five human senses—sight, hearing, smell, touch, and taste—are our link to the world. Most information about our environment comes through sight. Visual information is not always a true representation of the facts because our eyes, and therefore the mind, can be fooled. There are many well-known optical illusions, such as those in Fig. 1.1. Unknowingly, people may be convinced that what they see in such a drawing actually exists as they perceive it.

Hearing ranks second to sight in supplying the brain with information about the external world. The senses of touch, taste, and smell, although very important for good health and happiness, rank well below sight and hearing in providing environmental information.

All the senses can be deceived and thus can provide false information. Anyone who has gone to the beach to swim during the early morning hours when the air is cool knows how warm the water feels to the body. Yet, later in the afternoon when the air has become warm, the water, which has remained practically the same temperature, feels cool. Thus if we were asked to judge the temperature of the water, our answers would probably vary according to the change in temperature of the air.

Not only can our senses be deceived, they also have limitations. For example, if two distant stars are close together, the unaided eye cannot distinguish them and sees what appears to be a single bright spot or star. Also, the unaided eye is unable to distinguish easily between the stars of our galaxy and the planets of our solar system. However, by making observations over long periods of time, we observe that planets move relative to the more distant stars. (In fact, the word *planet* comes from the Greek word meaning "wanderer.") The unaided eye is also limited from seeing all the planets of the solar system because it is not sensitive enough to see the dim light reflected from the outer three planets. (Do you know the names of these planets?) Similarly, other senses have their own limitations.

The handicaps of the senses can be overcome by using instruments to make measurements and observations. For example, the planets Uranus,

---

*Students should be able to achieve this goal after reading and studying this section.

Neptune, and Pluto were discovered after the invention of the telescope. Or, in the diagram of the length optical illusion in Fig. 1.1, you might want to use a ruler to measure the diagonals (a) and (b) to verify that they are equal. We extend our ability to measure various things by using instruments or tools. However, even the most precise instruments have limitations. We can measure time with a wristwatch, but most watches cannot be used to measure time intervals of less than $\frac{1}{10}$ second.

1.2 The Scientific Method and Fundamental Quantities

Learning Goals:

To explain how theories are tested by the scientific method.

To list the fundamental quantities that are used in measurements and physical descriptions.

How, then, do we go about scientifically describing our physical environment? What is generally done is to observe a *phenomenon* (something that happens in the environment) and then try to formulate a fundamental concept. A **concept** is a meaningful idea that can be used to describe a phenomenon. Then, a *hypothesis* explaining the phenomenon is developed. This is a guess or possible explanation of why or how something happens. But is the hypothesis accurate? To determine this, it must be tested by experiment and measurement.

A hypothesis predicts or describes what should happen in an associated experimental situation. If the experimental results in such a situation are consistent with predictions, then the hypothesis has merit. However, a hypothesis must be tested again and again for a variety of aspects and situations. If, as a result of extensive testing, a hypothesis describes what is observed, it is then commonly referred to as a *theory*. Basically, this process of investigating nature is known as the **scientific method,** which holds that no theory or model of nature is valid unless the results are in accord with experiment; or more briefly, a theory or hypothesis must be substantiated by experiment. The development of the scientific method is generally attributed to Galileo, an Italian scientist (Chapter 2) and Francis Bacon, an English philosopher. The scientific method puts science on a firm basis, requiring that speculations and opinions be put to the test.

A theory may lead to the establishment of a scientific *"law,"* which is a description, often mathematical, of *what* is observed. For example, in Chapter 3 you will study Newton's laws of motion (three of them) and his law of gravitation. A law describes what is observed, while a theory or hypothesis explains *why* something occurs.

So, in applying the scientific method, what do we measure in our experiments? Certainly we should make things as simple and basic as possible. The physical characteristics of phenomena can be expressed in what are called **fundamental quantities,** and our comprehension of the physical world is based on these. Scientists generally identify four quantities—**length, mass, time,** and **electric charge**—as being *fundamental*. These quantities are fundamental in the sense that we cannot think of any that are more basic in describing nature. They form the foundation for other quantities needed to understand the physical sciences. That is, more complex quantities are really combinations of the fundamental quantities.

What quantities do you use to describe your environment? Think about questions you might ask about it when going grocery shopping (particularly the first week you were a freshman). "Where is the store?" "When does it open?" "Should I buy the amount I have on the scales or double it?" These questions of "Where?" "When?" and "How much?" refer to the basic concepts of space, time, and matter.

Length

The description of space might refer to a location or to the size of an object. To measure locations and sizes, we use the fundamental quantity of length, which is defined as the measurement of space in any direction.

Space has three dimensions, each of which can be measured by a length (Fig. 1.2). The three dimensions can easily be seen by considering a rectangular object. It has length, width, and height, but each of these dimensions is actually a *length*. A sphere, such as a ball, has a radius, a diameter, and

CHAPTER 1 Measurement

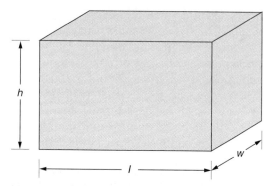

Figure 1.2 Dimensions and length.
The dimensions of a rectangular box are commonly given in terms of length (*l*), width, (*w*), and height (*h*), but all are length measurements in different directions.

a circumference. Again, all of these dimensions are easily described by length measurements.

Time

Once we know where something is located, we are frequently interested in what is happening to it. "Is the car moving?" "When will the next plane leave?" "What day will you be going home?" All of these questions can be answered using the fundamental quantity of time. Each of us has an idea of what time is, but you may find it difficult to define or explain it. Some terms that are often used in referring to time are duration, period, or interval. Time is sometimes described as the continuous, forward flowing of events. Without events or happenings of some sort, there would be no perceived time (● Fig. 1.3). The mind has no innate awareness of time, merely the awareness of events taking place in time. That is, we do not perceive time as such, only events that mark locations in time, similar to the marks on a meterstick marking length intervals. Note that time has only one direction—forward. That is, time has never been observed to run backwards, as it might appear to do so when a film is run backwards in a projector.

Time and space are linked together. In fact, time is sometimes added as a fourth dimension to the three dimensions of space. For the most part, however, we tend to regard space and time as separate fundamental quantities. But keep in mind that if something exists in space, it must also exist in time.

Figure 1.3 Time and events.
A runner crosses the finish line in a New York City marathon. The time for the event is seen to be 2 h, 9 min, and 40 s.

Mass

Because we are often asked questions about amounts of matter, we need a third fundamental quantity known as *mass*. To define mass precisely, we need to understand the concepts of force and acceleration, and/or gravity, which are discussed in Chapters 2 and 3. For now, let us simply say that **mass** refers to the amount of matter an object contains.

Living on Earth, many of us tend to measure matter in terms of *weight*. And indeed, the greater the weight of an object, the more matter it contains. However, weight is *not* a fundamental quantity. An astronaut with a backpack who weighs 300 pounds on Earth would weigh $\frac{1}{6}$ of that amount, or 50 pounds, on the surface of the Moon (● Fig. 1.4).

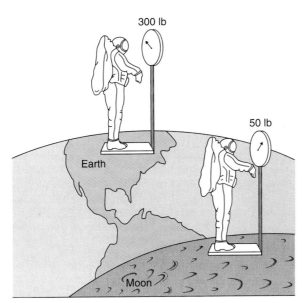

Figure 1.4 Mass and weight.
The weight of an astronaut on the Moon is $\frac{1}{6}$ of that on Earth. For example, the astronaut weighs 300 lb (with backpack) on Earth and 50 lb on the Moon. The astronaut's mass or quantity of matter, however, remains the same. Mass is the fundamental quantity.

But, the astronaut's mass or quantity of matter would be the same on Earth and on the Moon.

Weight is related to the force of gravity, which is the attractive force a celestial body has for an object. Your weight on Earth is the gravitational attraction Earth has on you. On the Moon, the gravitational attraction is only $\frac{1}{6}$ of that on Earth, and hence the weight of an object would be different. Weight changes depending on where you are in the universe. On the other hand, the mass of an object would be the same throughout the universe, so mass is the *fundamental* quantity. The relationship between mass and weight is discussed in more detail in Chapter 3.

Electric Charge

A fourth fundamental quantity is **electric charge**, the property associated with some particles that gives rise to electrical forces and electrical phenomena (electricity). There are two kinds of electric charge, which are referred to as positive (+) and negative (−). Two particles with positive charges or two particles with negative charges repel each other, whereas opposite charges attract (Fig. 1.5).

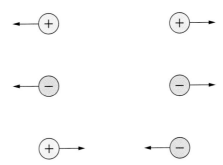

Figure 1.5 Two types of electric charge.
Electric charges are designated as positive (+) and negative (−). Two like charges (two positive or two negative) repel (top). Unlike charges (positive and negative) attract (bottom).

Electric charge is an important property of matter, because all atoms contain electrically charged particles, and most matter is composed of atoms. Electric current is simply the flow of electric charge. The concepts of electric charge and electric force are discussed in more detail in Chapter 8.

1.3 Standard Units

Learning Goal:

To define the standard units of the metric system and the British system of units.

In order to measure fundamental quantities and their various combinations, we need to compare them with standards. As an example, we use a ruler to measure length; that is, we compare the unknown length with a standard length—the ruler, or more accurately, a foot (or 12 in.). An adopted measurement standard is called a standard unit. A standard unit is a fixed and reproducible value for the purpose of taking accurate measurements. A set or group of standard units is referred to as a *system of units*. Two major systems of measurement are used around the world, each of which uses different standard units.

The United States is one of the few nations that still uses the British engineering system of measurement. The **British system** uses the foot

HIGHLIGHT

The Metric System and the SI

Historians generally agree that the metric system originated with Gabriel Moulton, a French mathematician, when in 1670 he proposed a comprehensive decimal system based upon a physical quantity of nature and not the human anatomy. Moulton proposed that the fundamental base unit of length be equal to one minute of arc of a great circle of Earth, and a pendulum constructed with this length be used to define the unit of time.

Over one hundred twenty years later, in the 1790s during the French Revolution (1789–1799), the French Academy of Science recommended the adoption of a decimal system with a unit of length equal to one ten-millionth the distance on the surface of Earth from the equator to the North Pole. The unit of length was named the "metre" from the Greek *metron*, meaning "to measure." A cubing of one-tenth of this length was proposed to be a unit of volume, and a unit of mass was proposed to be equal to the amount of pure water needed to fill the cube. Thus an easy usable system based on multiples or submultiples of 10 and defined on a single base unit related to a physical quantity of nature was established.

The metric system as originally conceived had problems, especially with standard units. To solve these problems, the French government in 1870 legalized a conference to work out standards for a unified measurement system. Five years later on May 20, 1875, the Treaty of the Meter was signed in Paris by 17 nations, including the United States.

The Treaty established a General Conference on Weights and Measures as the supreme authority for all actions. The treaty also established an International Committee of Weights and Measures with the responsibility for the supervision of the International Bureau of Weights and Measures—a permanent laboratory and world center of scientific metrology (science of measurement).

The United States officially adopted the metric system in 1893, but there were no mandatory requirements, and the British units have continued to be used. The United States Congress enacted Public Law 94-168, the Metric Conversion Act of 1975, that stated that "the policy of the United States shall be to coordinate and plan the increasing use of the metric system in the United States and to establish a United States Metric Board to coordinate the voluntary conversion to the metric system." However, no mandatory requirements were made, and the United States continues to use the British units of measurement.

In 1960 a modernized metric system consisting of six basic standard units (meter, kilogram, second, ampere, kelvin, and candela) was established by the 11th General Conference on Weights and Measures. This system is known as the International System of Units, or SI for short (SI is an abbreviation for the French Le Système Internationale d'Unités).

The metric mass unit (kilogram) adopted in 1960 did not meet the needs of chemistry, so the 14th General Conference meeting in 1971 established the seventh base unit, the mole, for the amount of substance. See Appendix I for the seven base units and their definitions.

---

for the standard unit of length. Quantities of matter are expressed in weight units (pounds) rather than mass units. Hence, we say the British system is a gravitational system. (Mass and weight are directly related, as discussed in Chapter 3.) There is a British system unit of mass, the slug, but it is rarely used. One slug of mass weighs approximately 32 pounds on Earth's surface.

All systems of measurement use the *second* as the standard unit of time, and the *coulomb* is usually the standard unit of electric charge.

The **metric system** of units, which is used by most countries of the world, is much simpler than the British system. One great advantage of the metric system is that converting to larger or smaller units is accomplished by using factors of 10. For example, in the metric system 1 kilometer is 1000 meters, whereas in the British system 1 mile is 5280 feet. Thus if you were asked how many meters there are in 2 kilometers, you could immediately answer 2000. But, if you were asked how many feet there are in 2 miles, you would probably need a pencil and paper or a calculator to get the answer.

Another advantage of the metric system is the consistent use of multiple and submultiple prefixes. For example, knowing that a kilometer is 1000 m allows the accurate prediction that 1 kilo-

1.3 Standard Units

Table 1.1 Standard Units for the Metric and British Systems of Measurement

Fundamental Quantity	Absolute Systems		Gravitational System
	Metric (mks)	Metric (cgs)	British
Length	meter (m)	centimeter (cm)	foot (ft)
Mass	kilogram (kg)	gram (g)	slug
Time	second (s)	second (s)	second (s)

gram is 1000 grams (*kilo* means 1000). Similarly, knowing that 1 centimeter is $\frac{1}{100}$ of a meter (*centi* means $\frac{1}{100}$ or 0.01), you know that 1 centigram is $\frac{1}{100}$ of a gram. Inversely, there are 100 centimeters in a meter and 100 centigrams in a gram. In the British system, on the other hand, knowing that a foot has 12 inches gives you no clue at all that a pound has 16 ounces. (Metric prefixes are discussed more fully in Section 1.6.) In summary, the metric system is (1) easier to *use,* and (2) easier to *learn.*

The standard units of the metric system for length, mass, and time are the *m*eter, *k*ilogram, and *s*econd, respectively. Using the first letter of these units, we sometimes refer to the **mks system.** These units are part of a modernized metric system called the **International System of Units,** officially abbreviated **SI** from the French spelling. (See the chapter Highlight.)

A system of metric units that is sometimes used for smaller quantities is the **cgs system** (*c*entimeter, *g*ram, *s*econd). Table 1.1 lists the standard units for the SI (mks), cgs, and British systems for the fundamental quantities of length, mass, and time. Let's take a closer look at the metric units.

The Meter

The standard length in the mks system is the **meter** (m, from the Greek *metron,* "to measure"), which was originally intended to be one ten-millionth of the distance from Earth's geographic north pole to the equator, along a meridian (Fig. 1.6). The unit was first adopted by the French in the 1790s, and it is now used in scientific measurements of length throughout the world. Note in Fig. 1.6 that the meter is slightly longer than the British yard.

From 1889 to 1960 the standard meter was a platinum-iridium bar kept in the vaults at the International Bureau of Weights and Measures near Paris, France. However, the accuracy of this metal standard could vary. For example, metals expand

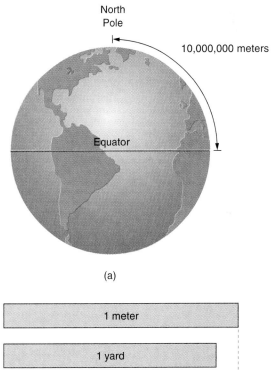

Figure 1.6 Metric length unit—the meter.
(a) The meter was originally defined as one ten-millionth of the distance from the North Pole to the equator. Hence, the distance between these locations would be 10,000,000 meters. (b) The meter is a bit longer than the yard—3.37 inches longer. (Not to scale.)

Figure 1.7 Kilogram standard.
Prototype kilogram number 20 is the United States standard of mass. This prototype is a platinum-iridium cylinder, 39 mm in diameter and 39 mm tall.

and contract with temperature changes, so more invariable standards were adopted in 1960 and 1983. The 1960 standard was referenced to the wavelength of light, but now the meter is defined in the 1983 reference to the standard unit of time. One meter is the length of the path traveled by light in a vacuum during the time interval of $1/299{,}792{,}458$ of a second. That is, the speed of light in vacuum is defined to be 299,792,458 meters per second.

From the basic meter standard, other units of length are defined. For example, the millimeter (mm) is $\frac{1}{1000}$ of a meter, the centimeter (cm) is $\frac{1}{100}$ of a meter, and a kilometer (km) is 1000 meters.

The Kilogram

The SI or mks standard unit of mass is the **kilogram** (kg). This unit was originally defined in terms of a specific volume of water at its maximum density (4°C).* This volume was taken to be that of a cube 10 cm ($\frac{1}{10}$ meter) on a side. Notice that this is a volume of 1000 cm³ (10 cm × 10 cm × 10 cm). As stated earlier, the metric prefix *kilo* means one thousand, and 1 kilogram is equivalent to 1000 grams. One **gram** (g), then, is the mass of 1 cm³ of water at its maximum density.

However, the water standard for the kilogram was not a convenient reference because of handling and cooling, so a comparable solid standard was made. Currently, the kilogram is defined to be the mass of a cylinder of platinum-iridium kept at the International Bureau of Weights and Measures. The United States prototype is kept at the National Institute of Standards and Technology (NIST, formerly the National Bureau of Standards) in Washington, D.C. (Fig. 1.7). Of length, mass, and time, the kilogram standard unit of mass is the only one based on an artifact (an object). As was learned, the meter is defined as a fraction of the distance light travels in one second in vacuum; and as will be discussed shortly, the standard unit of time is defined in terms of the radiation from an excited atom.

CONFIDENCE QUESTION 1.1
The metal standard for the meter was replaced in part because it could vary in length with temperature changes. Shouldn't the metal standard for the kilogram also be replaced for this reason? Explain. Would wear from usage make a difference for either standard?

When giving examples of mass, be careful to use mass units and not weight units. Here are some examples of mass: 1000 cm³ of water has a mass of 1 kg, and 1 cm³ of water has a mass of 1 g; the mass of a textbook is 1.4 kg; the student has a mass of 55 kg. In case you're wondering how kilograms and pounds compare, on Earth 1 kg has a weight of 2.2 lb. (How much does the 55-kg student weigh in pounds?)

The Second

The standard unit of time in both the metric and British systems is the **second** (s). For many years, the second was defined in terms of a mean (average) solar day. (A solar day is measured by two suc-

---

*The density of water varies with temperature.

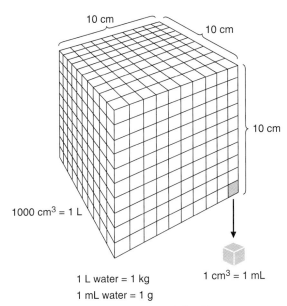

Figure 1.8 The kilogram and the liter.
A kilogram of mass was originally defined as that of a volume of water 10 cm on a side, or a volume of 1000 cm³. This volume is taken to be a unit of capacity, the liter (L). Hence, 1 cubic centimeter (cm³) is equivalent to 1 milliliter (mL). Since 1 L of water effectively has a mass of 1 kg, then 1 cm³ or 1 mL of water has a mass of 1 gram (g).

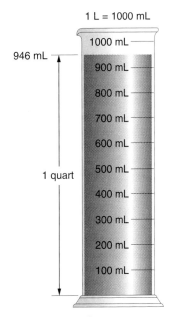

Figure 1.9 The liter and quart.
One liter is slightly larger than one quart—54 mL larger. That is, 1 L = 1.06 qt.

cessive daily positions of the Sun.) With 24 hours in a day, 60 minutes in an hour, and 60 seconds in a minute, there are 86,400 seconds in a day; or 1 second is equal to 1/86,400 of a day. A mean solar day was taken because solar days are not of equal duration, because Earth's path around the Sun is not a perfect circle.

Today, scientists use an atomic definition of the second, based on the vibrations of a cesium atom as it radiates a certain wavelength of light. The second is defined as the duration of 9,129,631,770 cycles of the radiation associated with a specific transition of the cesium-133 atom.

Here are some examples of time: The science class met for 50 minutes; the period of a pendulum is 2.8 seconds; the student drove back to school after spring break in 4 hours.

The Liter

The standard unit of volume in the SI (mks system) is the cubic meter (m³). Since this is a rather large unit (larger than a cubic yard), it is common to use a smaller, nonstandard unit called the liter. The liter (L) is a volume of 1000 cm³, which was the volume used to define a kilogram mass (Fig. 1.8). Note that since this volume of water was originally used to define the kilogram, a liter of water has a mass of one kilogram.*

The liter can be divided into 1000 parts and the prefix *milli*, which means $\frac{1}{1000}$, is applied. Thus, 1 L is equal to 1000 mL (milliliters). Since this volume is also equal to 1000 cm³, then 1000 cm³ = 1000 mL; or

$$1 \text{ cm}^3 = 1 \text{ mL}$$

The cubic centimeter (cm³) is sometimes abbreviated cc, particularly in medicine. As illustrated in Fig. 1.9, a liter is slightly larger than a quart—54 mL larger, or 1 L = 1.06 qt.†

Most countries now use the metric system. In fact, the United States is the only major country not officially using it. However, as you know from

---

*Interestingly enough, the SI standard unit of volume of m³ is associated with a mass unit. A cubic meter is 1000 L, so the mass of 1 m³ of water is 1000 kg. This is a *metric ton* or *tonne*.

†A lower case "ell" (l) is sometimes used to abbreviate the liter, e.g., 1.5 l or 20 ml. However, a capital "ell" (L) is generally preferred so as to avoid confusion with the numeral one (1).

Figure 1.10 Examples of metric conversions.
(a) Soft drinks are commonly sold in one-liter bottles, as well as two- and three-liter bottles. (b) A highway sign showing the speed limit in both British and metric units.

(a) (b)

experience, the United States is gradually using more and more metric units—food items have the equivalent masses in grams and kilograms, metric wrenches are needed to work on automobiles and other machinery, and soft drinks are now commonly purchased in liter bottles (Fig. 1.10). Because most soft drinks have about the same density as water, a liter of any soft drink will have a mass of approximately one kilogram. Since a liter and a quart are about the same volume, we may one day buy liters of milk instead of quarts of milk. In a few places in the United States, gasoline is sold in liters.

Figure 1.10b shows a highway sign with equivalent British and metric units. In metric countries, highway distances are measured in kilometers (1 km = 0.62 mi, or 1 mi = 1.6 km). Speeds are then expressed in km/h (kilometer per hour), whereas we use mi/h (often abbreviated mph).

The conversion to a new and different system of units may sound overwhelming, but it isn't. Many units in both systems are very similar in value and only a few will be needed in everyday life, as the flyer in Fig. 1.11 points out.

1.4 Derived Quantities

Learning Goals:

To explain derived quantities in terms of fundamental quantities.

To analyze how simple mathematical relationships can be used to derive information.

Most of the phenomena that we observe can be described in terms of the four fundamental quantities. However, because of the many interactions in nature, combinations of one or more of these quantities must be used. Combinations of one or more fundamental quantities are called **derived quantities**. For example, area, volume, speed, and density are derived quantities, and they are defined as follows:

$$\text{area} = (\text{length})^2$$
$$\text{volume} = (\text{length})^3$$
$$\text{speed} = \frac{\text{length}}{\text{time}}$$
$$\text{density} = \frac{\text{mass}}{\text{volume}} = \frac{\text{mass}}{(\text{length})^3}$$

All You Will Need to Know About Metric
(For Your Everyday Life)

10

Metric is based on Decimal system

The metric system is simple to learn. For use in your everyday life you will need to know only ten units. You will also need to get used to a few new temperatures. Of course, there are other units which most persons will not need to learn. There are even some metric units with which you are already familiar: those for time and electricity are the same as you use now.

BASIC UNITS

(comparative sizes are shown)

METER: a little longer than a yard (about 1.1 yards)
LITER: a little larger than a quart (about 1.06 quarts)
GRAM: a little more than the weight of a paper clip

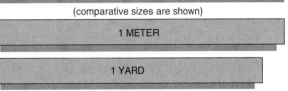

25 DEGREES FAHRENHEIT

COMMON PREFIXES
(to be used with basic units)

milli: one-thousandth (0.001)
centi: one-hundredth (0.01)
kilo: one-thousand times (1000)

For example:
1000 millimeters = 1 meter
 100 centimeters = 1 meter
1000 meters = 1 kilometer

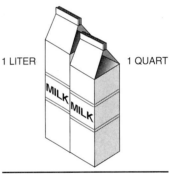

1 LITER   1 QUART

OTHER COMMONLY USED UNITS

millimeter:	0.001 meter	diameter of paper clip wire
centimeter:	0.01 meter	a little more than the width of a paper clip (about 0.4 inch)
kilometer:	1000 meters	somewhat further than $1/2$ mile (about 0.6 mile)
kilogram:	1000 grams	a little more than 2 pounds (about 2.2 pounds)
milliliter:	0.001 liter	five of them make a teaspoon

25 DEGREES CELSIUS

OTHER USEFUL UNITS

hectare: about $2\,1/2$ acres
metric ton: about one ton

WEATHER UNITS: **FOR TEMPERATURE**   **FOR PRESSURE**
                    degrees Celsius   kilopascals are used
                                             100 kilopascals = 29.5 inches of Hg (14.5 psi)

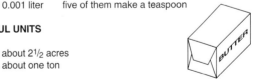

−40 −20   0   20   37   60   80   100   °C
−40   0   32   80 98.6   160   212 °F
water freezes body temperature   water boils

1 POUND

1 KILOGRAM

Figure 1.11 A government flyer describing what you need to know about going metric.

Table 1.2 Derived Quantities and Units

Derived Quantity	Unit
Area (length)^2	m^2, cm^2, ft^2, etc.
Volume (length)^3	m^3, cm^3, ft^3, etc.
Density (mass per volume)	kg/m^3, g/cm^3, etc.

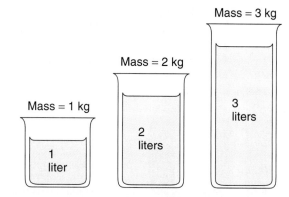

Figure 1.12 Equal densities.
Whether you have one, two, or three liters of pure water (with masses of one, two, and three kilograms, respectively), the density is the same. The density of water in each of the three containers is 1 kg/L, which is 1 g/cm^3 or 1000 kg/m^3.

The fundamental quantities are combined to derive most of the terms used in science. Some derived quantities and associated units have already been used in our discussion, and more such combinations will become evident in the course of our study. The common examples of derived quantities given previously are shown in Table 1.2, along with their associated units.

Density refers to how compact matter is in a substance or object. In more formal language, **density** (ρ, the Greek letter rho) is the amount of mass in a definite volume, or simply the mass per volume (mass/volume), or in equation form:

$$\rho = \frac{m}{V} \quad (1.1)$$

For example, suppose an object has a mass of 20 kg and occupies a volume of 5 m^3. Its density is then $\rho = m/V = 20$ kg/5 m^3 = 4 kg/m^3. Note that density expresses the mass per *unit volume*. The unit volume is one m^3, so each cubic meter (m^3) of the substance of which the object is composed has a mass of 4 kg. The units of kg/m^3 are standard SI units; however, when you are working with smaller amounts of substances, the smaller units of g/cm^3 (grams/cubic centimeter) may be more convenient.

If mass is uniformly, or evenly, distributed throughout the volume, then we say the density is uniform, and each unit volume contains the same amount of matter. Another way of saying this is that the density is constant. Fig. 1.12 shows that if you have a uniform substance, such as water, the density remains the same no matter how much of the substance you have. However, if the mass is not distributed uniformly throughout a volume or object, then the calculation of density gives an *average* density. For example, if you computed the density of Earth, this result would be an average density since Earth's mass is not uniformly distributed throughout its nearly spherical volume.

From our discussion of the kilogram and liter, we know that the density of water is 1 g/cm^3 or 1 kg/L. (In SI units, water's density is 1000 kg/m^3.) We can compare the densities of other substances with that of water. This type of comparison is particularly easy in g/cm^3 units. For example, a rock might have an average density of 3.3 g/cm^3, and so is 3.3 times as dense as water. Pure iron has a density of 7.9 g/cm^3, and Earth has an average density of 5.5 g/cm^3.

Scientists use mathematical relationships such as Eq. 1.1 for density, to investigate and predict. These relationships may be thought of as the "language" of science. Equation relationships provide a concise way of communicating information. To paraphrase an old saying, an equation is worth a thousand words. Mathematics provides the "tools" of science. In general, the more tools one has, the more and better jobs one can do. Similarly, with more and advanced mathematical methods, scientists can continually probe nature through investigation and predictions. (The match of results with predictions must be verified—the scientific method.)

In the course of our study, you will be given the explanations needed for an *overview* of mathematical procedures and presentations. That overview is meant to provide you with a basic understanding and appreciation of the workings of science. Density is a good example of how one can use scientific relationships to obtain information. Suppose you had a spherical steel ball bearing with a volume of 100 cm³ and you wanted to know its mass. You could go to the lab and put it on a scale to find out, but using the density equation, you would not have to get up from your desk. Since density equals mass/volume, with slight manipulation (see Appendix III), we have

$$\text{mass} = \text{density} \times \text{volume}$$

or $m = \rho V$. The densities of various common substances are listed in tables; if you look up the density of steel, it is found to be 7.8 g/cm³. Then, the mass of the ball bearing is

$$\text{mass} = \text{density} \times \text{volume} = 7.8 \text{ g/cm}^3 \times 100 \text{ cm}^3$$
$$= 780 \text{ g}$$

Similarly, you might be considering a given mass of a substance, say, 400 g of mercury, which has a density of 13.6 g/cm³, and would like to know if a 25-mL beaker would hold this quantity of the liquid metal. To find out, you can compute the volume of the 400 g of mercury. From the density relationship, we have the volume equal to mass/density ($V = m/\rho$). You should be able to show that the mercury would occupy a volume of 29.4 cm³ (or 29.4 mL—why?), so using a 25-mL beaker would not be a good idea because of overflow and spillage. (Mercury is a very toxic substance. Extreme care should be taken in handling it.)

The densities of liquids such as blood or alcohol may be measured by means of a device called a *hydrometer*. A hydrometer commonly consists of a weighted glass bulb with a stem that floats in the liquid. The higher the hydrometer floats, the greater is the liquid's density, which can be read on a calibrated scale on the stem (Fig. 1.13).

When a medical technologist checks a sample of urine, one test that is run is for density. Urine of a healthy person has a density range from about 1.015 to 1.030 g/cm³, and consists of mostly water and dissolved salts. When the density is out of the normal range, the urine may have an excess or deficiency of dissolved salts, perhaps caused by an illness.

The hydrometer is commonly used in testing automobile battery acid and antifreeze. In a car battery, the liquid is a solution of sulfuric acid, which is more dense than pure water. As a battery cell supplies electric current, sulfuric acid is chem-

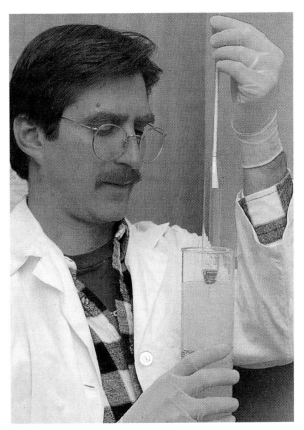

Figure 1.13 Liquid density.
A hydrometer is used to measure the density of a liquid.

ically converted to water. When fully charged, the battery liquid has a density of about 1.3 g/cm³. If, after charging, the liquid is found to have a lower density close to that of water (1.0 g/cm³), this indicates that something is wrong with the cell. The hydrometer in a battery tester is inside a glass tube into which the battery liquid is drawn up using a rubber bulb. This design prevents burns of the skin from contact with the sulfuric acid solution during testing.

The testing of automobile radiator antifreeze is similar. The closer the density of the antifreeze and water solution is to 1.0 g/cm³, the closer it is to being pure water. For this test, the hydrometer is calibrated directly in temperature degrees rather than density units. A density of 1.0 g/cm³ corresponds to a hydrometer reading of 0°C or 32°F, the freezing point of water. The further the mixture is from being pure water, the lower the temperature readings. Some newer antifreeze testers have a pivoted plastic indicator instead of a glass bulb hydrometer. The indicator pivots upward in the solution and points to the appropriate temperature on a scale on the side of the plastic body of the tester.

As we have seen, quantities are expressed with numerical values *and units*. Measured quantities are often used in mathematical equations. In an equation, not only must the numerical values on both sides of the equation be equal (as indicated by the equals sign), but also the units on both sides of the equation. That is, we cannot equate a quantity with units of speed (such as m/s) to a quantity with units of area (such as m^2)—we cannot equate apples to oranges, so to speak.

When a combination of units gets bulky and complicated, we frequently give it a name of its own for convenience. Here are a few examples that will be applied in later chapters:

$$\text{newton (N)} = \text{kg} \times \text{m/s}^2$$
$$\text{joule (J)} = \text{kg} \times \text{m}^2/\text{s}^2$$
$$\text{watt (W)} = \text{kg} \times \text{m}^2/\text{s}^3$$

Note that even if we give a derived quantity another name, it is still made up of a combination of fundamental units. The names of the units given here are in honor of famous scientists. In general, when a unit is named after a person, the abbreviation is capitalized; for example, N for newton (after Isaac Newton). You may have noticed this difference after working with meter (m), kilogram (kg), and second (s). [The only common exception is L for liter, because a lowercase "ell" (l) can be confused with the numeral one (1), as noted previously.]

1.5 Conversion Factors

Learning Goal:

To explain how to use conversion factors in converting quantities from one unit to another, for example, inches to centimeters.

Frequently, we want to convert from one unit to another. Using wrong units has resulted in some costly errors. Once, the robot arm of the space shuttle was locked in the wrong position because of an error in units. Another time, a river boat was sunk when it was loaded with a number of kilograms instead of the same number of pounds.

A unit conversion may be within a particular system of units or from one system to another. For example, if you were asked to convert 6 yards to feet, you'd know that this was 18 ft, since you know there are 3 feet per yard, or 3 ft/yd. This quantity (3 ft/yd) is called a *conversion factor*.

With our gradual conversion to the metric system, we often need to convert from one system of units to another (British to metric, or vice versa) in order to make comparisons. To do this, one must know the relationship or equivalence between the different units. For example,

$$1 \text{ in.} = 2.54 \text{ cm}$$

is an *equivalence statement*. As learned previously, the numerical values and units on both sides of an equation must be the same. In the instance of an equivalence statement, we use an equals sign to indicate that 1 in. and 2.54 cm represent *the same length* in their respective units. The numbers are different because they refer to different units of length.

The previous equivalence statement can yield two conversion factors:

$$\frac{1 \text{ in.}}{2.54 \text{ cm}} \quad \text{or} \quad \frac{2.54 \text{ cm}}{1 \text{ in.}}$$

Hence, we see that **conversion factors** are equivalence statements expressed as ratios. Various equivalence statements are listed on the inside of the back cover of this book.

To convert the units of a quantity to other units, we simply use the appropriate conversion factor as *determined by the units*. For example, suppose we want to know how many centimeters there are in 100 inches. To determine this, we multiply 100 in. by a conversion factor:

$$100 \text{ in.} \times \frac{2.54 \text{ cm}}{1 \text{ in.}} = 254 \text{ cm}$$

cancel inches

Notice how the units (in.) cancel, much like numerical fractions, and we are left with the proper equation: 100×2.54 cm = 254 cm. (Both sides of the equation are equal in magnitude and units, just as an equation must be.) As long as we use a conversion factor that allows the units to cancel properly, we know that we are correct. Notice that using 100 in. × (1 in./2.54 cm) would give some rather weird units (in²/cm)! Here's what you did when you earlier converted 6 yd to feet in your head: 6 yd × (3 ft/yd) = 18 ft. So, to do a unit conversion is quite easy. Basically, you first must know or be given the conversion factor and then let the units do the rest.

Here are the general steps for doing unit conversions:

STEP 1

To convert a quantity from one unit to another, use a conversion factor—that is, a ratio that may be obtained from an equivalence statement. (Often these factors or statements are not known offhand and must be given or looked up in a table in order for you to do the conversion.)

STEP 2

Choose the appropriate form of a conversion factor (or factors) so that the unwanted units cancel.

STEP 3

Check to see that the units cancel and that you are left with desired unit. Then perform the multiplication and/or division of the numerical quantities.

Here are a couple of examples to illustrate this procedure:

EXAMPLE 1.1 Converting Quantities

A student weighs 132 lb. What is the student's mass in kilograms?

Solution

One might represent this problem by an equivalence statement:

$$132 \text{ lb} = ? \text{ kg}$$

(That is, 132 lb is equivalent to how many kilograms?) Then, we carry out the conversion using an appropriate conversion factor:

STEP 1

$$1 \text{ kg} = 2.2 \text{ lb} \quad \text{(equivalence statement)}$$

(1 kg of mass is equivalent to or weighs 2.2 lb),

and $\dfrac{1 \text{ kg}}{2.2 \text{ lb}}$ or $\dfrac{2.2 \text{ lb}}{\text{kg}}$ (conversion factors)

(Mass and weight are different quantities, but a given mass has an equivalent weight on Earth. Notice how the number 1 is commonly left out of the denominator of a conversion factor, that is, 2.2 lb/kg, not 2.2 lb/1 kg.)

STEP 2

The first form of conversion factor, 1 kg/2.2 lb, would allow the lb unit to be canceled. (The lb unit is the unit in the denominator of the ratio.)

STEP 3

Checking this and performing the operation:

$$132 \text{ l\!\!\!b} \times \left(\frac{1 \text{ kg}}{2.2 \text{ l\!\!\!b}} \right) = 60 \text{ kg}$$

cancel lb

Can you find your mass in kilograms? In metric countries, people express their "weights" in kilograms, or "kilos."

EXAMPLE 1.2 Conversion Factors

A cafeteria tray has a length of 1.50 ft. What is this length in meters?

Solution

In equivalence statement form we have:

$$1.50 \text{ ft} = ? \text{ m}$$

Looking at the equivalence statements involving feet and meters (inside back cover), we see that 1 ft = 0.3048 m and we have the conversion factor 0.3048 m/ft. Also, we see that 1 m = 3.28 ft, giving a conversion factor of 3.28 ft/m. Notice that we have m/ft and ft/m. So, which conversion factor do we use? Wanting to "get rid of" the ft unit, it should be evident that this unit will cancel if the 1.50 ft is multiplied by the first conversion factor:

$$1.50 \text{ ft} \times \left(\frac{0.3048 \text{ m}}{\text{ft}}\right) = 0.457 \text{ m}$$

cancel ft

[Multiplication is usually more convenient in unit cancellation, but division will work so long as you use the right conversion factor. Note that ft/(ft/m) = m.]

CONFIDENCE QUESTION 1.2

(a) Soft drinks are sold in an economy 3-L size. What is the volume of such a container in quarts? (Recall that 1 L = 1.06 qt.)

(b) Show that the conversion factor 3.28 ft/m can also be used to convert 1.50 ft to meters. (*Hint:* Think of unit cancellation in division.)

We see how easy it is to convert from one unit to another by just letting the units help you do the "setting up." If you simply use the conversion factors that make the unwanted units cancel, the conversion will come out right.

With the use of the metric system becoming increasingly common, the proper use of conversion factors will become increasingly important. As was shown in Fig. 1.10b, speed limit signs may one day be in km/h (kilometer/hour). For example, the 65 mi/h interstate speed limit would become

$$65 \text{ mi/h} \times \left(\frac{1.609 \text{ km/h}}{\text{mi/h}}\right) = 105 \text{ km/h}$$

Here and in later conversions, the canceling of the units (mi/h) is not shown explicitly with slashes because it can be done mentally. Note that this equation has the same units on both sides (km/h). You may have noticed that most newer cars have speedometers calibrated in both mi/h (commonly abbreviated mph) and km/h (Fig. 1.14).

CONFIDENCE QUESTION 1.3

In a European country, the school zone speed limit is posted as 40 km/h. What is this in mi/h?

Figure 1.14 Miles per hour and kilometers per hour.
The speedometers of many automobiles are calibrated in mph (mi/h) and km/h.

1.6 Scientific Notation and Metric Prefixes

Learning Goals:

To explain how very large and very small numbers are expressed in the shorthand scientific (or powers-of-10) notation.

To define metric system prefixes that indicate multiples and submultiples of 10.

In physical science many numbers are very large or very small. In order to express such numbers conveniently, scientists use **scientific notation** (sometimes called *powers-of-10 notation*). When the number 10 is squared or cubed, we get

$$10^2 = 10 \times 10 = 100$$
$$10^3 = 10 \times 10 \times 10 = 1000$$

You can see that the number of zeros is just equal to the superscript number of the 10, which is called the *exponent,* or *power.* For example, 10^{23} has an exponent or power of 23 and is equivalent to 1 fol-

1.6 Scientific Notation and Metric Prefixes 17

Table 1.3 Numbers Expressed in Scientific or Powers-of-10 Notation

Number	Powers-of-10 Notation
0.025	2.5×10^{-2}
0.0000408	4.08×10^{-5}
0.0000001	1×10^{-7}
0.00000000000000000016	1.6×10^{-19}
247	2.47×10^{2}
186,000	1.86×10^{5}
4,705,000	4.705×10^{6}
9,000,000,000	9×10^{9}
30,000,000,000	3×10^{10}
602,300,000,000,000,000,000,000	6.023×10^{23}

Table 1.4 Prefixes Representing Powers of 10

Multiple	Prefix (Abbr.)	Pronunciation
10^{24}	yotta- (Y)	yot′ta (*a* as in *a*bout)
10^{21}	zetta- (Z)	zet′ta (*a* as in *a*bout)
10^{18}	exa- (E)	ex′a (*a* as in *a*bout)
10^{15}	peta- (P)	pet′a (as in *pet*al)
10^{12}	tera- (T)	ter′a (as in *terra*ce)
10^{9}	giga- (G)	ji′ga (*ji* as in *ji*ggle, *a* as in *a*bout)
10^{6}	mega- (M)	meg′a (as in *mega*phone)
10^{3}	kilo- (k)	kil′o (as in *kilo*watt)
10^{2}	hecto- (h)	hek′to (*heck-toe*)
10	deka- (da)	dek′a (*deck* plus *a* as in *a*bout)
10^{-1}	deci- (d)	des′i (as in *deci*mal)
10^{-2}	centi- (c)	sen′ti (as in *senti*mental)
10^{-3}	milli- (m)	mil′li (as in *mili*tary)
10^{-6}	micro- (μ)	mi′kro (as in *micro*phone)
10^{-9}	nano- (n)	nan′oh (*an* as in *an*nual)
10^{-12}	pico- (p)	pe′ko (*peek-oh*)
10^{-15}	femto- (f)	fem′toe (*fem* as in *fem*inine)
10^{-18}	atto- (a)	at′toe (as in an*ato*my)
10^{-21}	zepto- (z)	zep′toe (as in *zep*pelin)
10^{-24}	yocto- (y)	yock′toe (as in *so*ck)

lowed by 23 zeros. Negative powers of 10 can also be used. For example,

$$10^{-2} = \frac{1}{10^2} = \frac{1}{100} = 0.01$$

Numbers written just as a power or multiple of 10, like 10^6, may be written more fully in the extended form 1×10^6. This is simply 1 multiplied by 10 six times, or 1,000,000, where the decimal point has been moved to the right six places. Another example is 2.5×10^2 equals 250 (as you can see by expanding the power of 10; that is, $2.5 \times 100 = 250$).

If a power of ten is a negative exponent, to write the number in decimal form we shift the decimal place to the left the number of places in the exponent number. Thus, one micrometer, which is 1×10^{-6} m, equals 0.000001 m. (The understood decimal point after the 1 is shifted six places to the left.) Table 1.3 shows examples of large and small numbers expressed in scientific notation.

In the metric system, there are standard prefixes that are used to represent powers of 10 (Table 1.4). The list is extensive, however, only a few have common usage. These are **kilo, centi,** and **milli.** Some examples are

1 *kilo*meter $= 10^3$ m $= 1000$ m (1 *thousand* meters)
1 *centi*meter $= 10^{-2}$ m $= 0.01$ m (1 *one-hundredth* of a meter)
1 *milli*gram $= 10^{-3}$ g $= 0.001$ g (1 *one-thousandth* of a gram)

Notice that a centimeter is analogous to one cent of a dollar (a centidollar). There are 100 cents in one dollar, and one cent is $\frac{1}{100}$ (or 0.01) of a dollar. Similarly, there are 100 centimeters in one meter, and one centimeter is $\frac{1}{100}$ (or 0.01) of a meter.

CONFIDENCE QUESTION 1.4

Property taxes and school bond levies are expressed in terms of mils. A mil is $\frac{1}{10}$ of a cent. Using a metric prefix, tell what part of a meter is analogous to a monetary mil and explain why. (How about a dime?)

You may also experience the *mega* (M), *micro* (μ), and *nano* (n) prefixes:

$$mega\text{ton} = 10^6 \text{ tons (1 million tons)}$$

$$1 \text{ microsecond} = 10^{-6} \text{s}$$
(1 *one-millionth* of a second)

$$1 \text{ nanometer} = 10^{-9} \text{ m}$$
(1 *one-billionth* of a meter)

It is quite easy to represent large and small numbers in scientific notation. For example, Earth is about 93 million miles from the Sun, or 93,000,000 miles. Expressed using powers of 10, this is 93×10^6 miles (93 megamiles). As an example of a small

Table 1.5 Some Values of Length, Mass, and Time (Values Are Approximate and Rounded to Nearest Power of 10)

Length (meters)		Mass (kilograms)		Time (seconds)	
Radius of known universe	10^{25}	Known universe	10^{51}	Time for light waves to reach Earth from most distant quasar	10^{17}
Diameter of Milky Way	10^{21}	Milky Way Galaxy	10^{41}		
One light year	10^{16}	Sun	10^{30}	Half-life of uranium-235	10^{16}
Distance from Earth to Sun	10^{11}	Earth	10^{25}	Half-life of carbon-14	10^{11}
Distance light travels in one second	10^{8}	Man	10^{2}	One day	10^{5}
		One liter of water	10^{0}	One class session	10^{3}
Diameter of Earth	10^{7}	One dime	10^{-3}	One minute	10^{2}
Length of football field	10^{2}	Postage stamp	10^{-5}	Time between heartbeats	10^{0}
Width of hand	10^{-2}	Red blood cell	10^{-12}	Time a discharged bullet travels the barrel of a rifle	10^{-3}
Thickness of paper	10^{-4}	Iron atom	10^{-25}		
Diameter of hydrogen atom	10^{-10}	Proton	10^{-27}	Time for a beam of light to travel the length of a football field	10^{-6}
Diameter of proton	10^{-15}	Electron	10^{-30}		
				Half-life of polonium-212	10^{-7}
				Time for the electron to revolve once around nucleus of the hydrogen atom	10^{-15}

number, let's say a sheet of paper is 0.00045 m thick. This may be written 4.5×10^{-4} m. Thus it can be seen that the exponent, or power of 10, changes when the decimal point of a number is shifted. The general rules for this are as follows:

If the decimal point is shifted so the prefix number becomes smaller, the exponent, or power of 10, becomes correspondingly larger. Similarly, if the decimal point is shifted so the prefix number becomes larger, the exponent becomes correspondingly smaller.

The rules also apply to numbers already expressed in powers of 10. For example, the Earth's distance from the Sun could equivalently be expressed as 93×10^6 mi = 9.3×10^7 mi (prefix number made smaller, exponent becomes correspondingly larger—one decimal point shift, six increased by one to seven) or 93×10^6 mi = 930×10^5 mi (prefix number made larger, exponent becomes correspondingly smaller—one decimal shift, six decreased by one to five).* An example with a negative exponent is $4.5 \times 10^{-4} = 0.45 \times 10^{-3}$ (prefix number made smaller, exponent becomes correspondingly larger—one decimal shift, -4 increased by one to -3) and $4.5 \times 10^{-4} = 45 \times 10^{-5}$ (prefix number made larger, exponent becomes correspondingly smaller—one decimal shift, -4 decreased by one to -5).

Only single decimal point shifts are shown here, but the rule applies to any number of shifts. What would be the exponent for the solar distance expressed as $9300 \times 10^?$? (Right! Sounds just like a warning they call out in golf.)

CONFIDENCE QUESTION 1.5

Express the measurements of 695,000 kg and 0.000024 s in scientific notation in three forms as in the preceding examples.

Not only may quantities be written in scientific notation, but mathematical operations, such as addition, subtraction, multiplication and division, may be done with numbers in this form. The rules for these operations and examples are given in Appendix VI. Your instructor may wish you to learn these so as to be able to handle the very large numbers associated with topics in the chemistry and astronomy sections.

We end this chapter by showing in Table 1.5 the vast range of values of length, mass, and time measurements expressed in powers of 10. They illustrate the convenience of expressing very large and very small numbers in scientific notation. (You wouldn't want to have to write out, say, the diameter of our Milky Way Galaxy or the mass of an electron with zeros, would you?)

---

*When we are not told otherwise, it is conventional or standard to express numbers in scientific notation with one digit to the left of the decimal point, as 9.3×10^7 mi, for example.

IMPORTANT TERMS

These important terms are for review. After reading and studying the chapter, you should be able to define and/or explain each of them.

concept
scientific method
fundamental quantities
length
mass
time
electric charge
standard unit
British system
metric system
mks system
SI
cgs system
meter (m)
kilogram (kg)
gram (g)
second (s)
liter (L)
derived quantities
density
conversion factors
scientific notation (powers-of-10)
kilo
centi
milli

IMPORTANT EQUATIONS

$$\text{density} = \frac{\text{mass}}{\text{volume}} \quad \left(\rho = \frac{m}{V}\right)$$

or $\quad \text{mass} = \text{density} \times \text{volume} \ (m = \rho V)$

or $\quad \text{volume} = \dfrac{\text{mass}}{\text{density}} \ \left(V = \dfrac{m}{\rho}\right)$

QUESTIONS

1.1 The Senses

1. Which of the following senses provides us with the most knowledge about our environment: (a) touch, (b) taste, (c) sight, (d) hearing?
2. In which order does hearing rank in supplying us with information about the external world: (a) first, (b) second, (c) third, (d) fifth?
3. Do all measurements ultimately depend on our senses? Explain. p. 2+3
4. Discuss the limitations of our senses other than sight. p. 2+3

1.2 The Scientific Method and Fundamental Quantities

5. A scientific hypothesis (a) is a physical law, (b) is a guess, (c) need not be tested by the scientific method, (d) none of these.
6. Which of the following is *not* a fundamental quantity: (a) weight, (b) mass, (c) length, (d) time?
7. If two electrically charged particles repel each other, then we know (a) they have opposite charges, (b) both have positive (+) charges, (c) both have negative (−) charges, (d) only that they have like charges.
8. Define and give the standard metric units of (a) length, (b) mass, and (c) time. length - m, cm   mass - kg, g   Time = sec
9. (a) Does a 70-kg astronaut have the same mass on the Moon as on Earth? Yes
   (b) Does the astronaut have the same weight on the Moon and Earth? No
10. Explain why time is sometimes considered to be a fourth dimension along with the three dimensions of space. pg. 4

1.3 Standard Units

11. The standard unit of mass in the SI is the (a) slug, (b) kilogram, (c) gram, (d) pound.
12. In the following pairs, which metric unit is smaller than its British counterpart on a one-to-one basis: (a) meter-yard, (b) kilogram-pound, (c) liter-quart, (d) kilometer-mile?

13. Which of the following is longer: (a) 1 mi, (b) 1 km, (c) 1000 yd?
14. Which of the following is the largest volume: (a) 1 qt, (b) 1 L, (c) 1 m^3, (d) 1 ft^3?
15. A common abbreviation for mile per hour is mph. Explain why this abbreviation might confuse a foreign visitor.
16. Give the original and current definitions of the (a) meter, (b) kilogram, and (c) second.
17. Give the standard units of length, mass, and time in three unit systems.
18. Considering the SI and British system of units, discuss any advantages of one over the other.

1.4 Derived Quantities

19. A derived quantity (a) only applies to the British gravitational system, (b) is a mathematical derivation, (c) describes nonfundamental phenomena, (d) is a combination of fundamental quantities.
20. The derived unit of density in the SI is (a) kg/cm^3, (b) g/cm^3, (c) kg/m^3, (d) g/m^3.
21. (a) Which is more dense, a kilogram of iron or a kilogram of feathers?
    (b) Which has more mass?
22. Express in equation form (a) the volume of an object in terms of its density and mass, and (b) the mass of an object in terms of its density and volume.
23. Explain the principle of checking (a) the battery liquid, and (b) the antifreeze of an automobile. Speculate why glass instruments are used for the former and plastic instruments are commonly used for the latter.
24. Suppose you would like to compute the average density of your body. Describe how you might do this. (*Hint:* See Exercise 3.)

1.5 Conversion Factors

25. The form 1 in. = 2.54 cm is (a) an equation, (b) an equivalence statement, (c) a conversion factor, (d) a conversion statement.
26. In converting inches to centimeters, an appropriate conversion factor to multiply by would be (a) 1in./2.54cm, (b) 2.54 cm/in., (c) either of these.

1.6 Scientific Notation and Metric Prefixes

27. For a number expressed in scientific notation, when the decimal point is shifted to the right, the exponent in the power of ten (a) increases, (b) decreases, (c) sometimes increases, sometimes decreases.
28. How many millimeters are there in 100 cm: (a) 10, (b) 100, (c) 1000, (d) 10,000?
29. The number 2,340,000 may be expressed in scientific notation as (a) 0.234×10^7, (b) 234×10^4, (c) 2.34×10^6, (d) all of the preceding.
30. The number 1×10^7 is how many times larger than 1×10^4?
31. Taking the national debt to be 5 trillion dollars, express this amount in scientific notation.

Food for Thought

1. We sometimes hear that people have a "sixth sense." Do you think humans could have a sixth sense? If so, speculate what this might be.
2. Legal laws are sometimes repealed. Could a scientific law ever be repealed? Explain.
3. Would time exist if there were no motion? Explain.
4. If one object is heavier than another, does this mean that the heavier object has a greater density? Explain.
5. As pointed out in the chapter, a penny or one cent (or centidollar) is analogous to a centimeter. Are there any metric prefix analogies for a nickel, a quarter, or a half-dollar? Explain.

Exercises

1.1 The Senses

1. Answer the questions in Fig. 1.15 then verify your answers.

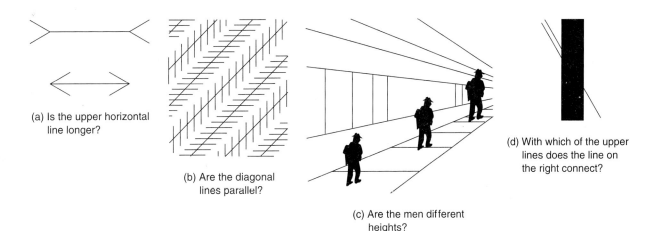

Figure 1.15 Do you believe what you're seeing?
Answer the questions below the figures. (A ruler might help you convince yourself.)

1.4 Derived Quantities

2. A volume of a particular liquid is 1.2 times heavier than an equal volume of water. What is the density of the liquid?

   Answer: 1.2 g/cm^3

3. A geologist finds a rock, and its mass measures 1100 g on a scale. By immersing it in a cylinder of water, the volume is found to be 215 cm^3. What is the average density of the rock?

   Answer: 5.12 g/cm^3

4. *A piece of pure iron has a mass of 2500 g. If the density of iron is 7.86 g/cm^3, what is the volume of the iron piece in cm^3?

   Answer: 318 cm^3

5. If a uniform iron ball has a volume of 100 cm^3, what is its mass in grams?

6. A 154-lb student immerses himself in a tub and finds that he has a body volume of 0.175 m^3. He then computes his average density. What is it?

   Answer: 400 kg/m^3

1.5 Conversion Factors

7. A computer disk has 1.4 MB (megabytes) of memory. (a) How many bytes is this? (b) How many kilobytes?

   Answer: (a) 1.4 million bytes
   (b) 1400 kilobytes

8. A person is measured to have a height of 178 cm. What is the height of the person in inches?

   Answer: 70.1 in.

9. A woman has a height of 5 ft 2 in. She must list this on a passport in centimeters. What would this be?

10. What is the mass of a 240-lb football player in kilograms? (And what is your mass in "kilos"?)

    Answer: 109 kg

11. Which has more mass, a 2.0-kg package or a 2.0-lb package, and how much more?

    Answer: 2.0-kg has 1.09 kg more

12. A car travels in a school zone at 15 mph. What is this speed in km/h?

    Answer: 24 km/h

13. A driver in a car with a metric speedometer observes that the car is traveling 95 km/h in a 55 mi/h speed zone. Is the speed limit being exceeded?

14. A motorist wants to buy 10 gallons of gasoline, but the pump dispenses gas in liters. About how many liters would this be in round numbers?

    Answer: 38 L

1.6 Scientific Notation and Metric Prefixes

15. Write each of the following quantities in scientific notation.
    (a) 16 kilotons   (b) 50 micrograms
    (c) 33 megawatts   (d) 33 milliwatts

    Answer: (a) 16×10^3 tons
    (b) 50×10^{-6} µg

*Red exercise numbers indicate two similar exercises with the answer given for the first one. When two or more sets of paired exercises follow one another, each set of exercise numbers is printed in different colors to help you easily distinguish between pairs. Worked examples similar to the paired text exercises may be found for each chapter in the student *Study Guide*.

16. Fill in the blank with the correct power of ten:
    (a) 2.41 kilograms = 2.41 × _____ grams
    (b) 3.54 milliseconds = 3.54 × _____ seconds
    [*Hint:* For (c), (d), and (e) write the metric prefix as a power of 10, and then shift the decimal point.]
    (c) 150 kilovolts = 1.50 × _____ volts
    (d) 0.65 megatons = 6.5 × _____ tons
    (e) 0.0087 microseconds = 8.7 × _____ seconds
    Answer: (a) 2.41×10^3 g
    (c) 1.5×10^5 volts

17. Write the following numbers as a series of digits.
    (a) 134.6×10^4 (b) 0.0234×10^{-5}
    (c) 2477.35×10^{-2} (d) 0.00999×10^3
    Answer: (a) 1,346,000
    (b) 0.000000234

18. Write each of the following in conventional, or standard, scientific notation (one digit to the left of the decimal point).
    (a) 115   (b) 0.00045
    (c) 5280   (d) 0.007030
    Answer: (a) 1.15×10^2
    (b) 4.5×10^{-4}

19. Express the following numbers in conventional, or standard, scientific notation. (*Remember:* If the prefix number gets larger, the exponent gets smaller, and vice versa.)
    (a) 25.0×10^6 (b) 458×10^{-4}
    (c) 0.314×10^3 (d) 0.00107×10^{-8}

20. Using conversion factors and scientific notation, calculate (a) the lifetime of a 70-year-old person in seconds, and (b) your own lifetime in seconds. (How accurate can you make the latter calculation?)
    Answer: (a) 2.2×10^9 s

Answers to Multiple-Choice Questions

1. c   5. b   11. b   19. d   25. b   27. b
2. b   6. a   12. d   20. c   26. b   28. c
      7. d   13. a   29. d
             14. c

Solutions to Confidence Questions

1.1 Thermal expansion and contraction (see Chapter 5) would affect the length of the meter standard and make it variable. Thermal expansion of the metal kilogram standard would cause it to change size (affecting density); but even so, the amount of matter would be the same and the mass invariable. Wear would cause both standards to have variable values. The rounding off of edges of the meter standard would cause error, and wear (loss of mass) on the kilogram standard would cause it to have a smaller value, however slight.

1.2 (a) Using the given conversion factor:

$$3 \text{ L} \times \left(\frac{1.06 \text{ qt}}{\text{L}}\right) = 3.18 \text{ qt}$$

So you get almost 0.2 quart more in a 3-L bottle. A good buy—if the 3-qt and 3-L prices are the same.

(b) Here, to cancel units, the 1.50 ft must be *divided* by the conversion factor:

$$\frac{1.50 \text{ ft}}{(3.28 \text{ ft/m})} = 0.457 \text{ m}$$

Recall that in the division of fractions (and conversion factors), you invert and multiply: 1.50 ft × (1 m/3.28 ft) = 0.457 m.

1.3 $40 \text{ km/h} \times \left(\frac{0.62 \text{ mi/h}}{\text{km/h}}\right) = 25 \text{ mi/h}$

1.4 With a mil being 1/10 (0.1) of a cent, there are 10 mils per cent or *1000 mils per dollar* (10 mils/cent × 100 cents/dollar = 1000 mils/dollar). Hence, the mil is analogous to a millimeter (1000 millimeters per meter). A dime would be a decidollar.

1.5 $659{,}000 \text{ kg} = 65.9 \times 10^4 \text{ kg} = 6.59 \times 10^5 \text{ kg}$
$= 659 \times 10^3 \text{ kg}$
$0.000024 \text{ s} = 0.24 \times 10^{-4} \text{ s} = 2.4 \times 10^{-5} \text{ s}$
$= 24 \times 10^{-6} \text{ s}$

Chapter 2

Motion

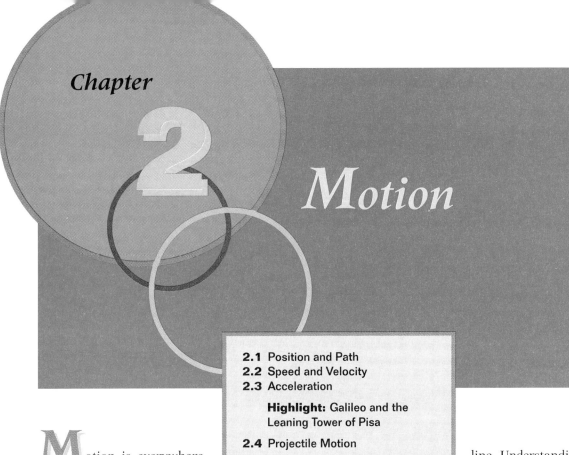

2.1 Position and Path
2.2 Speed and Velocity
2.3 Acceleration

Highlight: Galileo and the Leaning Tower of Pisa

2.4 Projectile Motion

Motion is everywhere. We walk to class. We drive to the store. Birds fly. The wind blows the trees. Rivers flow. Even the continents drift. And in the larger environment, Earth rotates, the Moon revolves about Earth, and Earth revolves around the Sun. On a larger scale, the Sun moves in the galaxy (the Milky Way), and the galaxies move with respect to one another. And even in the microcosm, we picture molecules of matter to be continuously in motion, and the electrons of atoms to be orbiting their nuclei.

This chapter focuses on the description of motion in our environment, with definitions and discussion of terms such as *speed*, *velocity*, and *acceleration*. (We all have a general understanding of their meaning, but can you define them?) We will examine these concepts without considering any forces involved, reserving that discussion for Chapter 3.

Although motion will be discussed in general, two basic types will be considered in detail—straight-line motion (the simplest) and circular motion (a very common form). For instance, we drive in straight lines on highways and go around circular curves. When going around a curve, there is a different sensation from traveling in a straight line. Understanding acceleration is the key to understanding this type of motion, as we will see. Also, we will touch on projectile motion. The motion of a thrown ball is an example of projectile motion. Here again we have an acceleration, but in one direction and not in another.

2.1 Position and Path

Learning Goals:

To define motion.

To contrast the designations of path lengths through the concepts of scalar and vector quantities.

The term **position** refers to the location of an object. To designate a position, we must give or imply a reference point. For example, the entrance to campus is located 1.6 km (1 mi) from the shopping mall. My friend is standing 1 m (3.28 ft) from me. Or, as illustrated in Fig. 2.1a, my school (point B) is located 100 km (62 mi, since 1 km = 0.62 mi) from my hometown (point A) along a straight line drawn on the map. Note that a position is designated by a length from the reference point.

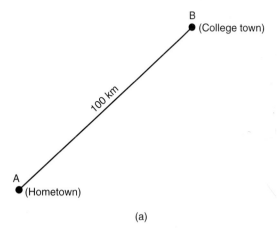

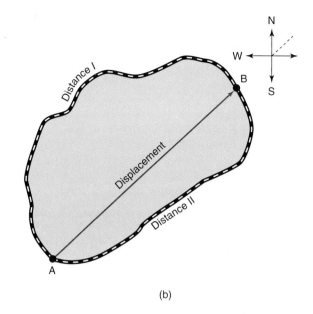

Figure 2.1 Position and path.
(a) Position is designated by a length from a reference point. In this case, the college town is located 100 km from the hometown. (b) Even so, a *change* of position may occur by different routes. Here are examples of two different routes from points A to B, whose distances are the lengths of the actual paths traveled. However, there is only one straight-line displacement route—the straight-line distance plus direction (northeast in this case).

With an understanding of what is meant by position, we can now ask: What is motion? We all know that when something is in motion it is moving; but more specifically, it is changing its position. Hence, when an object is undergoing a continuous change of position, we say that the object is in **motion**.

However, a change in position may occur along different routes. For example, in Fig. 2.1, in traveling between home and school, there is only one straight-line route, but many indirect ones (two are shown in Fig. 2.1b).

In science, quantities are classified as being either scalar or vector. A **scalar** quantity is one that has magnitude or size only. For example, you may have driven 160 km or 100 mi (the magnitude includes units). In length measurements, this is know as **distance**, which is simply the actual path length traveled. In Fig. 2.1b, the two routes shown obviously have different lengths and hence different distances.

We are quite familiar with scalar quantities, but a vector quantity adds another helpful dimension—direction. A **vector** quantity, is one with both magnitude and direction. In length measurements, the straight-line distance between two points, with a direction, is called **displacement**.

This is the shortest route between the initial and final points, with the direction of measurement specified. For example, in Fig. 2.1b, the displacement from point A to point B would be 100 km (magnitude) to the northeast (designated NE, using compass directions). Notice that the magnitude of the vector displacement is simply the straight-line scalar distance between the two points.

So, keep in mind,

Distance is a scalar quantity—magnitude (with units) only.

Displacement is a vector quantity—magnitude (with units) *and direction*.

CONFIDENCE QUESTION 2.1

What would be the displacement from point B to point A in Fig. 2.1b?

Straight-line motion is conveniently described using the Cartesian coordinate system, with which you are familiar—*x*- and *y*-axes. Let's consider the straight-line route in Fig. 2.1 to be along the *x*-axis, as illustrated in Fig. 2.2a. It is represented horizontally for convenience with the +*x* direction being taken as northeast. You may have noticed that we represent the displacement with an arrow. This is a convenient way to represent vectors in general, with the length of the arrow indicating magnitude,

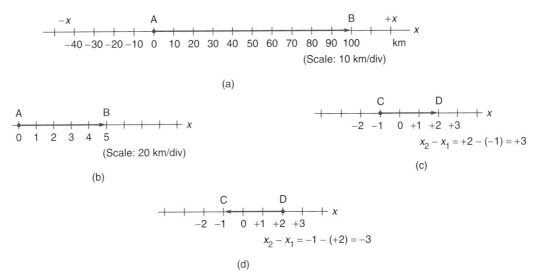

Figure 2.2 Displacement vectors to scale.
Vectors may be represented by arrows drawn to scale, with the length of an arrow proportional to the magnitude of the vector and the arrowhead indicating its direction. (a and b) A displacement in the x direction with different scales. (c and d) Even if a vector does not start at the origin, the value of the vector may be obtained by subtracting the final and initial coordinates. In (c) the vector has a value of +3 (arbitrary units), and in (d) a value of −3 (arbitrary units). The signs indicate the directions of the vectors.

and the arrowhead, the direction. To make vector arrows more manageable, they may be drawn to scale to make them shorter. As illustrated in Fig. 2.2b, the arrow is shortened using a scale of 20 km/division (km/div).

You may have also noticed that the starting point was taken to be at the origin (0 km). This starting location is convenient because in finding the length of a vector, we really subtract the beginning and end coordinate positions, $x_2 - x_1$, and x_1 is zero. However, the end coordinate may not always be zero, as illustrated in Figs. 2.2c and 2.2d. Note how the plus and minus signs indicate the directions: $+x$ and $-x$. (The plus sign is often omitted as being understood.) This process is analogous to finding a "length" of time. Say your starting point was 1 P.M. and the stopping point 4 P.M. Then, $\Delta t = t_2 - t_1 = 4 - 1 = 3$ hours. [The Δ (Greek delta) is used to indicate an interval difference, or a "change" in a quantity or "difference" in a quantity. Similarly, $\Delta x = x_2 - x_1$.] Incidentally, time is a scalar quantity, even though we talk about a forward flow of events.

We need length measurements to describe motion, but there is another ingredient, as you may have surmised. Motion involves a change of position, but it takes time to make such changes, and some changes may occur faster than others. That length and time are used to describe motion is evident in racing. For example, as shown in Fig. 2.3, runners try to run a certain length or distance in the shortest possible time. Combining length and time to give the time rate of change of position is the basis of describing motion in terms of speed and velocity, as discussed in the following section.

Figure 2.3 Motion.
Motion involves a change of position with time. By combining length and time to give the time rate of change of position, we can tell which runner is fastest or covers a given distance in the shortest time.

2.2 Speed and Velocity

Learning Goals:

To distinguish between speed and velocity.

To explain how to interpret the information expressed by position-versus-time graphs.

How fast is that car going? You may respond by estimating its speed or velocity. The terms *speed* and *velocity* are often used interchangeably; however, these terms have distinct meanings. Basically, speed is a scalar quantity and velocity is a vector quantity. For example, the speedometer on your car tells you how fast you are going, but not the direction. (Hence, we don't have a velocity-meter.)

How fast an object is moving is expressed in terms of length and time, or the *time rate of change of position;* for example, in units of km/h or mi/h (as on a speedometer). *Speed* describes motion in general, and is associated with distance (a scalar). The **average speed** of an object is the total distance traveled divided by the time spent in traveling the total distance. That is,

$$\text{average speed} = \frac{\text{distance traveled}}{\text{time to travel distance}}$$

An example of average speed might be that calculated for a trip, say, for your drive home from school. If the distance is 160 km (100 mi) and it takes four hours, then

$$\text{average speed} = \frac{160 \text{ km}}{4 \text{ h}} = 40 \text{ km/h}$$

$$\left(\text{or } \frac{100 \text{ mi}}{4 \text{ h}} = 25 \text{ mi/h}\right)$$

In equation form, we may write the average speed ($\bar{v}$) as

$$\bar{v} = \frac{d}{t} \quad (2.1)$$

where the bar over the symbol indicates that it is an average value. Note that d and t are length and time intervals. They are sometimes written as Δd and Δt to indicate this explicitly. However, we will use d and t for convenience, remembering that they are *intervals*. (As measurements, distance and time are always intervals, aren't they?)

Taken over a time interval, speed is an average. (As an analogy, think of an average class grade.) So

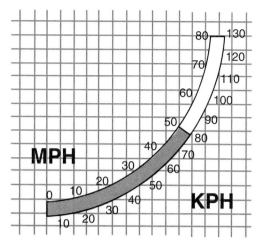

Figure 2.4 Instantaneous speed.
The speed indicated on an automobile speedometer is an example of nearly instantaneous speed—the speed the car is going at a particular instant. Notice that the speedometer is calibrated in both mi/h and km/h. (*Note:* mph and kph are not accepted abbreviations.)

average speed gives only a general description of motion. For example, during a time interval in which you are making a long trip by car, you speed up, slow down, and even stop. The average speed, however, is a single value that represents the average rate for the entire trip.

The description of motion can usually be made more specific by taking a smaller time interval; for example, a few seconds or even an instant. The speed of an object at any instant of time may be quite different from its average speed, and it gives a more accurate description of the motion. In this case, we have an instantaneous speed.

The **instantaneous speed** of an object is the speed at that instant of time (Δt being extremely small). A common example of nearly instantaneous speed is the speed registered on an automobile speedometer (Fig. 2.4). This value is the speed at which the automobile is traveling right then, or instantaneously.

Now let's look at describing motion with velocity. *Velocity* is similar to speed, but there is a direction involved. So instead of using the scalar distance to indicate change in position, for velocity we use the vector displacement. **Average velocity** is the displacement divided by the total travel time. (Recall that displacement is the straight-line distance between the initial and final positions, with the direction toward the final position—a vector

Figure 2.5 Constant velocity.
The car travels equal distances in equal periods of time in straight-line motion. With a constant speed and a constant direction, the velocity of the car is constant.

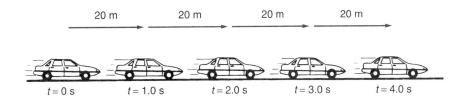

quantity). For straight-line motion in one direction, speed and velocity are very similar. Their magnitudes are the same because the lengths of the distance and displacement are the same. The distinction between them in this case is that a direction must be specified for the velocity.

The directions for straight-line motion are commonly designated by signs (+ and −). A plus sign ($+\bar{v}$) indicates a velocity in one direction; for example in the $+x$ direction on a graph. A negative sign ($-\bar{v}$) indicates a velocity in the opposite direction, for example, in the $-x$ direction.

As you might guess, there is also **instantaneous velocity (v),** which is the velocity at any instant of time. For example, a car's instantaneous speedometer reading plus the direction the car is traveling at that instant gives its instantaneous velocity. Of course, the speed and/or direction of the car may, and usually do, change. This motion is then accelerated motion, which is discussed in the following section.

If the velocity is constant or uniform, then we don't have to worry about changes. Suppose an airplane is flying at a constant speed of 320 km/h (200 mi/h) directly eastward. Then, the airplane has a constant velocity and flies in a straight line. (Why?) Also for this special case, you should be able to convince yourself that the constant instantaneous velocity and the average velocity are the same ($v = \bar{v}$). By analogy, think about everyone in your class getting the same (constant) test score. How do the class average and individual scores compare?

A car traveling with a constant velocity is illustrated in Fig. 2.5. Let's see how we can describe its motion with what we know so far.

EXAMPLE 2.1 Describing Motion—Speed and Velocity

Describe the motion of the car in Fig. 2.5 in terms of speed and velocity.

Solution

(In solving example exercises, we will follow the general steps noted here. You will find it helpful to use such an approach. Also, see Appendix II.)

First, read the problem, identifying what chapter principle(s) apply to it. (Think about what you are being asked to find.) Then, write down what is given in symbol notation. From Fig. 2.5, we have

Given: $d = 80$ m
$t = 4.0$ s

Then, determine what is wanted or what we are to find.

Find: Speed and velocity

After surveying the relationships we have for these quantities, it should be evident that Eq. 2.1 ($v = d/t$) applies. And further analyzing the situation, you should notice that the car has a constant or uniform motion; that is, it travels 20 m each second. Then we can use Eq. 2.1 to find the speed with $\bar{v} = v$. (Why?) But first we look at the units of the given quantities to make sure they are appropriate (usually standard) and in the same system. These are both in the SI system. (Suppose the time had been given in minutes or hours. What would you do?)

Performing the simple math gives

$$v = \frac{d}{t} = \frac{80 \text{ m}}{4.0 \text{ s}} = 20 \text{ m/s}$$

The constant speed of the car is 20 m/s. You may have expected this result since the car travels 20 m each second, or 20 m/s.

The motion is in one direction, or in a straight line, so the constant speed is the magnitude of the car's velocity, which is also constant (no change in magnitude and/or direction). We may write the velocity as +20 m/s, where the + sign indicates the direction of the car (taken here to be to the right in the figure). For such horizontal motion in a straight line, we sometimes take this direction to be along the x-axis (Section 2.1), and Eq. 2.1 is written $v = x/t$.

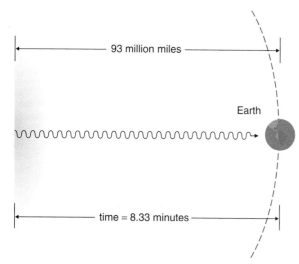

Figure 2.6 The speed of light.
Although light travels about 186,000 miles per second, it still takes over 8 minutes for light from the Sun to reach us. See Example 2.2.

The units are all in the same system; the unit of distance, miles in both given quantities, can be canceled. (Note that we could have worked in the metric system since v was also given in m/s, and the table inside the back cover lists the distance in km. If we had chosen to work in the metric system, a unit change to meters would have been necessary.)

Then, rearranging our speed equation, we get $t = d/v$ (see Appendix III if you have trouble doing this), and

$$t = d/v = \frac{93,000,000 \text{ mi}}{186,000 \text{ mi/s}} = 500 \text{ s}$$

From this example we realize that although light travels very fast, it still takes 500 s, or about 8.3 min, to arrive at Earth after leaving the Sun (Fig. 2.6). This means that when we look at the Sun, we see it as it was 8.3 min ago. Here again, we are working with a constant speed and velocity.

Note that the simple equation $v = d/t$ can easily be rearranged, similar to density ($\rho = m/V$) in Chapter 1, so as to find distance or time. That is, given v and t, the distance is given by $d = vt$. And now let's look at an example in which we find time.

EXAMPLE 2.2 Calculating Travel Time

The speed of light in space or vacuum is about 186,000 mi/s (3.00×10^8 m/s). How long does it take the Sun's rays to reach Earth?

Solution

Here, you are given speed and asked to find time. A distance is involved—the distance the light travels, which is the distance of Earth from the Sun. The distance is a known quantity that you are expected to look up (like the numerical value of π if you don't know it). From the solar system data (inside the back cover), we see that Earth is 93×10^6 miles, or 93 million miles, from the Sun. Then, we have

Given: $d = 93,000,000$ mi (known)
$v = 186,000$ mi/s

We want the time it takes for light to travel from the Sun, so

Find: t (time)

QUESTION

If an object has a constant speed, does it also have a constant velocity?

Answer

Not always. If an object travels with a constant speed in a straight line, then it also has a constant velocity—that is, constant speed and constant direction. However, an object may move with a constant speed in a curved path. In this case the velocity is not constant because the direction of motion is continuously changing. Keep in mind that a vector has magnitude and direction, and can be changed by changing either or both of these features.

CONFIDENCE QUESTION 2.2

The Olympic 100-m dash was run under 10 s for the first time in 1968 by Jim Hines (USA) in a time of 9.92 s. What was his average speed in m/s?

Graphs

We often represent situations graphically, particularly as functions of time. For example, a weight-conscious person may plot weight versus time, or a good student who is value-conscious may plot tu-

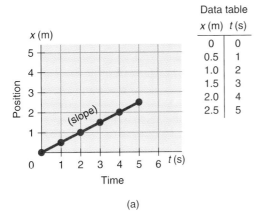

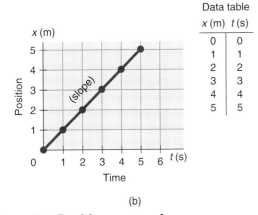

Figure 2.7 Position versus time.
Graphical representations help analyze motions. Here we have two instances of uniform motions—equal distances covered in equal times. The inclinations or slopes of the lines indicate the velocities of the motions. See text for description.

ition cost versus time. Such graphs allow one to "see what is happening," so to speak.

Graphs may also be used in describing or illustrating motion. Suppose you are given graphs of position versus time as shown in Fig. 2.7. Here we will assume that the motion is along the x-axis, so our general equation for speed and velocity is $v = x/t$ or $x = vt$. What do the graphs in Fig. 2.7 tell you? First we see from the accompanying data tables that the motion is uniform in each case—equal distances are covered in equal time periods. The object represented in graph (a) moves 0.5 m each second, and the object represented in graph (b) moves 1 m each second. As a result, the plots are straight lines. The inclination of such a straight line is called the **slope**. This slope has the value of the constant speed, or the magnitude of the velocity, as can be seen by putting x and t values from the data table into $v = x/t$. (They are all the same for a given graph. Try it.) If $x = 0$ at $t = 0$, this means that the object was at the origin when the measurement started at $t = 0$.

Each graph gives you another bit of information. Note that the position (x coordinate) increases with time. This means that the motion is in the positive x direction. For similar straight-line x-versus-t graphs for which the line "slopes" upward, we say that the slope is positive, since $v = x/t$ has positive values. A positive slope means that the car is moving in the positive direction. But, what is the difference between the situations represented in Figs. 2.7a and 2.7b? They both represent motions in the positive x direction, but the slopes or inclinations of the lines are different. What does this mean?

As seen from the data tables, the motion in Fig. 2.7b has a greater speed or magnitude of velocity than that in Fig. 2.7a. Hence, the greater the slope of the line, the greater the speed and magnitude of the velocity in the positive x direction. Slopes should be visually compared only when the graph scales are the same. Different scales can make things look differently, but the numerical slope value remains the same no matter how you plot the data. To avoid any confusion, it is sometimes convenient to plot two or more sets of data on the same graph.

So in a sense, graphs are pictures of equations; that is, graphical mathematics, so to speak. We will use this graphical approach whenever possible to illustrate equations and principles.

For example, what does the graph in Fig. 2.8 tell you? It shows that the object, whatever it is, is

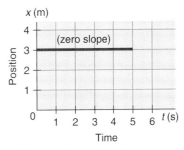

Figure 2.8 Not much motion!
A zero (horizontal) slope indicates no change of position with time; the represented object is at rest.

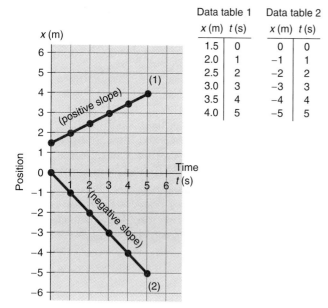

Data table 1		Data table 2	
x (m)	t (s)	x (m)	t (s)
1.5	0	0	0
2.0	1	−1	1
2.5	2	−2	2
3.0	3	−3	3
3.5	4	−4	4
4.0	5	−5	5

Figure 2.9 Positive and negative slopes.
See Example 2.3.

at the same position or location as time goes on, so it is not moving. The slope of a horizontal line is zero (no inclination), and so the speed and velocity of the represented object are zero.

EXAMPLE 2.3 Graphical Representations

What do the plots on the graph in ● Fig. 2.9 tell you?

Solution

Here we have two situations plotted on the same graph. In plot (1), we see the slope of the straight line is positive, so the motion represented is uniform along the +x-axis. But what about the vertical intercept (the point where the line touches the vertical axis)? Unlike the situation in Fig. 2.7, this graph is telling you that at $t = 0$ when time started to be measured, the object was at $x = 1.5$ m, and as time went on, the object moved along the x-axis with a constant speed (and velocity—why?).

Now let's look at plot (2). The data for the plot starts at $t = 0$, so whatever was moving was at the origin at the time it was moving along with a constant speed and velocity (as shown by the straight line). But here, the slope or inclination of the line is downward. (In this case, we say the slope of the line is negative.) Note that as time goes on, the position of the moving object is farther along the negative x-axis; hence the object is moving in the −x direction.

To make sure you understand graphical analysis, let's do a comprehensive example on slopes.

EXAMPLE 2.4 Graphical Analysis

Analyze the general motions of an object in the different regions indicated on the graph in ● Fig. 2.10.

Solution

Keep in mind what slopes mean. The observation of the motion began at $t = 0$, and the motion had a constant rate (velocity) as indicated by the straight line in region (1). Also the motion was in the +x direction. (Why?) In region (2) the motion began to slow. (The instantaneous velocity is indicated by a line drawn *tangent* to the curve at a particular point. *Tangent* describes a straight line that just touches the curve at a particular point.) Note that in region (2) the tangent line has a smaller slope or is less inclined compared to (1), meaning the motion is slowing. At the top of the plot, the slope of the tangent line is horizontal (compare Fig. 2.8), and the object has slowed to a stop.

Then, the slope turns negative (region 4), and the object starts to move in the −x direction. In the case of a car, it would have come to a stop and now is backing up. (Notice that the x position values decrease with time, which means the object is moving in the −x direction.) Going from region (4) to region (5), the slope becomes more negative (greater negative incline), meaning the speed of the object increases. Comparing region (5) to region (1), we can see that the constant speed in the −x direction is less than that in the +x direction. (Why?)

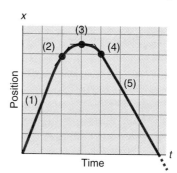

Figure 2.10 Changing slopes.
What does this mean? See Example 2.4.

CONFIDENCE QUESTION 2.3

What would it mean if the plot in Fig. 2.10 were continued for a time so that the position of the object was below the x-axis?

2.3 Acceleration

> **Learning Goal:**
> To define acceleration and its relationship to velocity-versus-time graphs.

When you drive down a straight interstate highway and suddenly increase your speed, say, from 20 m/s (45 mi/h) to 29 m/s (65 mi/h), you feel as though you are being forced back against your seat. When driving fast on a circular cloverleaf, you feel forced to the outside of the circle. These experiences result from changes in velocity.

There are two ways that you can change the velocity of an object. You can (1) change its *magnitude* when traveling in a straight line (which changes the length of the velocity vector), and/or (2) you can change the *direction* of motion (which changes the direction of the velocity vector). When any of these changes occur, we say that the object is *accelerating*. The faster the change in the velocity, the greater the acceleration.

Acceleration is defined as the time rate of change of velocity. If we use the Δ symbol to mean "change in," the equation for acceleration can be written as

$$\text{acceleration} = \frac{\text{change in velocity}}{\text{time for change to occur}} = \frac{\Delta v}{t}$$

The change in velocity is just the final velocity v_f minus the original velocity v_o. (Note that these are instantaneous velocities.) Thus, in symbolic form, we can define acceleration as

$$\bar{a} = \frac{\Delta v}{t} = \frac{v_f - v_o}{t} \tag{2.2}$$

Keep in mind that an acceleration is a measure of a change in velocity during a given time period. Eq. 2.2 then gives an *average* acceleration, as indicated by the overbar ($\bar{a}$).

That acceleration is a change in velocity can be easily remembered from a familiar application: The common name for the gas pedal of a car is the *accelerator*. As you push down on the accelerator, you speed up (increase the magnitude of the velocity), and when you let up on the accelerator, you slow down (decrease the magnitude of the velocity). Since velocity is a vector quantity, acceleration is also a vector quantity. For an object in straight-line motion, the acceleration vector may be in the same direction as the velocity vector or the acceleration may be in the opposite direction of the velocity (Fig. 2.11). In the first instance, the acceleration describes the object as speeding up, and the velocity is increasing. If the velocity and acceleration are in opposite directions (have opposite signs), the acceleration slows down the object, a change that is sometimes called a *deceleration*. Since we call the gas pedal the accelerator, maybe we should call the brake pedal the "decelerator."

How about the units of acceleration? Looking at Eq. 2.2, we see that acceleration is velocity divided by time. So the units of acceleration in the SI are (m/s) divided by (s); that is, (m/s)/s or m/s² (pronounced "meter per second squared"). These units may be confusing at first, but just remember that

Figure 2.11 Acceleration and deceleration.
For an object in straight-line motion, when the acceleration is in the same direction as the velocity, the object's speed increases. If the acceleration is in the opposite direction of the velocity, there is deceleration and the speed decreases.

the units of acceleration are based on those of velocity—meter per second (velocity unit) in a given time unit (per second)—so the acceleration units become (meter per second) per second, (m/s)/s = m/s².

EXAMPLE 2.5 Calculating Acceleration

A drag racer starting from rest achieves a speed of 30 m/s in 4.0 s while traveling in a straight line. What is the average acceleration of the dragster?

Solution

Starting from rest, the original or initial velocity is zero. (This is an example of how some information not given specifically in numerical form must be extracted from the problem.) So, we have

Given: $v_o = 0$ $t = 4.0$ s
$v_f = 30$ m/s

and

Find: $\bar{a}$ (average acceleration)

We will use Eq. 2.2, and noting that the units of the given quantities are all right, we have

$$\bar{a} = \frac{v_f - v_o}{t} = \frac{30 \text{ m/s} - 0}{4.0 \text{ s}} = 7.5 \text{ m/s}^2$$

Here the velocity is taken to be in the positive direction ($+v_f$) and the acceleration, also a vector quantity, is in the same direction.

CONFIDENCE QUESTION 2.4

A speedboat traveling in a straight line at 40 m/s slows down at an average rate of 2.0 m/s². What is the speed of the boat at the end of 5.0 s? (*Hint:* Velocity and acceleration are in opposite directions. Why?)

An acceleration may also be constant (and $\bar{a} = a$). To help understand what this means, consider a constant acceleration of 9.8 m/s² for an object initially at rest. This value means that the velocity changes magnitude by 9.8 m/s each second. (Also, the motion is in a straight line. Why?) Thus, as the number of seconds increases, the velocity goes from 0 to 9.8 m/s during the first second, to 19.6 m/s (9.8 m/s + 9.8 m/s) at the end of the second second, to 29.4 m/s (19.6 m/s + 9.8 m/s) at the end

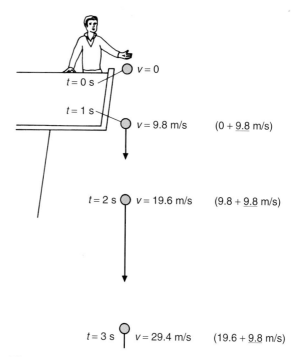

Figure 2.12 Constant acceleration.
For an object with a constant downward acceleration of 9.8 m/s², its velocity increases 9.8 m/s each second. The increasing lengths of the velocity arrows indicate the increasing velocity of the object. (*Note:* The distances of fall are not to scale.)

of the third second, and so on, with an additional 9.8 m/s each second. This sequence is illustrated in Fig. 2.12 for an object that is dropped and falls with an acceleration of 9.8 m/s².

In general, we will work with constant accelerations during the course of our study and so omit the bar over the acceleration a.

We can rearrange Eq. 2.2 to give an equation for the final velocity of an object: $v_f - v_o = at$, and

$$v_f = v_o + at \qquad (2.3)$$

This equation is useful in analyzing accelerated motion graphically with velocity-versus-time (v-versus-t) graphs. Two such graphs are shown in Fig. 2.13. Here, the values of the velocity on the vertical axis are those of v_f, the "final" velocity at the end of a given time. With a constant acceleration, Eq. 2.3 is the equation of a straight line, and the graphs are analyzed as was done previously for the x-versus-t graphs.

Let's see what the graphs tell us. In Fig. 2.13a, the slopes of the straight lines are the values or

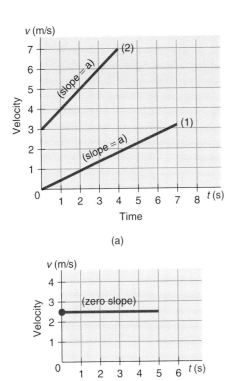

Figure 2.13 Velocity versus time.
For an object with constant acceleration, the graph of v-versus-t is a straight line. (a) The greater the slope, the greater the acceleration or increase in velocity. Note that the accelerations in these cases are positive or in the same direction as the initial velocities, the magnitudes of which are indicated by the y intercepts of the lines. (b) Here, the velocity does not change with time, so the horizontal line with zero slope indicates a zero acceleration.

with a velocity of $v_o = +3$ m/s (a speed of 3 m/s in the positive direction). Fig. 2.13b is an easy one to read. A horizontal line has a slope of zero, and the acceleration is zero since the velocity does not change with time. The velocity read from the graph is +2.5 m/s.

CONFIDENCE QUESTION 2.5

Describe the motion of the object represented in the v-versus-t graph in Fig. 2.14.

As pointed out earlier, we will consider primarily constant or uniform accelerations. There is one very special constant acceleration associated with falling objects. The **acceleration due to gravity** at Earth's surface is directed downward and is denoted by the letter g. Its magnitude in the SI is about

$$g = 9.80 \text{ m/s}^2$$

This value corresponds to 980 cm/s² or approximately 32 ft/s².

The acceleration due to gravity varies slightly depending on such factors as how far you are from the equator and how high up you are. However, the variations are small, and for our purposes we will take g to be the same everywhere on Earth's surface (9.80 m/s²).

The renown Italian physicist Galileo Galilei (1564–1642, known universally by his first name, Galileo) was one of the first scientists to assert that all objects fall with the same acceleration. Of magnitudes of the constant accelerations. Note that the steeper the slope, the greater the change in velocity with time, hence, the greater the acceleration. That is, plot (2) shows a greater acceleration than plot (1). Also, the acceleration and velocity are in the same (positive) direction in each case. (Why?)

There is some more information on the graph: Note where the lines intercept the vertical axis. The intercepts are at $t = 0$. The intercept values are the initial velocity (v_o) values. As you can see, the initial velocity of the object in plot (1) was zero ($v_o = 0$), and the object in plot (2) was initially moving

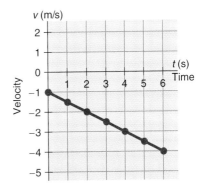

Figure 2.14 A negative slope and a negative y intercept.
What does this mean? See Confidence Exercise 2.5.

course, this assertion assumes that frictional effects are negligible. We can state Galileo's principle as follows:

> **If frictional effects can be neglected, every freely falling object near Earth's surface accelerates at the same rate, regardless of the mass of the object.**

One can illustrate this experimentally by dropping a small mass, such as a coin, and a larger mass, such as a ball, at the same time from the same height. They will hit the floor, as best as can be judged, at the same time. (Assume negligible air resistance.) Legend has it that Galileo himself performed such experiments. See the chapter Highlight.

The effect of air resistance or friction can be demonstrated by dropping a piece of tissue paper and a coin. The air resistance will prevent the tissue paper from falling as fast as the coin. However, if the tissue paper is wadded up into a small ball to minimize air resistance, it will fall with about the same acceleration as the coin.

On the Moon there is no atmosphere, so there is no air resistance. One of the astronauts dropped a feather and a hammer simultaneously from the same height (Fig. 2.15). They hit the surface of the Moon at the same time because neither was slowed by air resistance. Of course, objects fall on the Moon at a slower rate than those on Earth, because the acceleration due to gravity on the Moon is considerably less than on Earth. (Recall that it is only $\frac{1}{6}$ as much.)

The velocity of a freely falling object increases 9.8 m/s each second; that is, it increases uniformly with time. But how about the distance covered each second? Distance is not covered uniformly because the object speeds up. How the distance and velocity for a dropped object vary with time are illustrated in Fig. 2.16. From the figure, we can see that at the end of the first second the object has fallen 4.9 m. At the end of the second second, the total distance fallen is 19.6 m. At the end of the

Figure 2.15 No air resistance and equal accelerations.
Astronaut Scott demonstrated that a feather and a hammer fall at the same rate on the Moon. There is no atmosphere on the Moon and hence no air resistance or friction to slow the feather, so the feather is accelerated at the same rate as the hammer.

third second, the object has fallen 44.1 m, which is about the height of a 10- or 11-story building, and it is falling at a speed of 29.4 m/s, which is about 65 mi/h—pretty fast!*

If we throw an object straight upward, it slows down; that is, the velocity decreases 9.8 m/s each second. In this case, the velocity and acceleration are in opposite directions and there is a deceleration (Fig. 2.17). The object slows down, until it stops instantaneously at its maximum height. Then it starts to fall downward as though it were dropped from this height. The travel time upward is the same as the travel time downward to the original starting position. Also, the object returns with same speed as it had initially.

---

*The magnitude of the velocity of a dropped object is computed from Eq. 2.3 with $v_o = 0$; that is, $v = at$; and the distance traveled is given by $d = \frac{1}{2}at^2$.

HIGHLIGHT

Galileo and the Leaning Tower of Pisa

There is a popular story that Galileo dropped stones or cannonballs of different masses from the top of the Tower of Pisa to experimentally determine whether objects fall at the same rate. Whether this legend is true is questionable (Fig. 1).

Galileo did indeed question Aristotle's view that objects fell because of their "earthiness"; and the heavier or more earthy an object, the faster it would fall in seeking its "natural" place at the center of Earth. His ideas are evident in the following excerpts from his writings.*

How ridiculous is this opinion of Aristotle is clearer than light. Who ever would believe, for example, that . . . if two stones were flung at the same moment from a high tower, one stone twice the size of the other, . . . that when the smaller was halfway down the larger had already reached the ground?

And:

Aristotle says that "an iron ball of one hundred pounds falling a height of one hundred cubits reaches the ground before a one-pound ball has fallen a single cubit." I say that they arrive at the same time.

―――――――――――――
*From L. Cooper, *Aristotle, Galileo, and the Tower of Pisa* (Ithaca, N.Y.: Cornell University Press, 1935).

Although Galileo mentions a *high tower*, the Tower of Pisa is not mentioned in his writing, and there is no independent record of such an experiment. The first account of a Tower of Pisa experiment is found in a biography written by one of his pupils over ten years after Galileo's death. Fact or fiction? No one really knows. What we do know is that all freely falling objects near Earth's surface fall with the same acceleration.

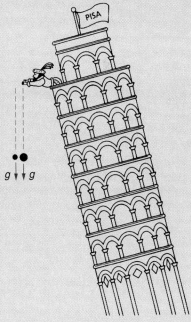

Figure 1 (a) Galileo Galilei (1564–1642). Understanding the motion of objects was one of Galileo's many scientific pursuits. **(b) Free fall.** All freely falling objects near Earth's surface accelerate downward at $g = 9.80$ m/s^2. Galileo is alleged to have dropped cannonballs or stones of different masses (weights) simultaneously from the Leaning Tower of Pisa. Over short distances, air resistance can be neglected, so the objects would have struck the ground at the same time.

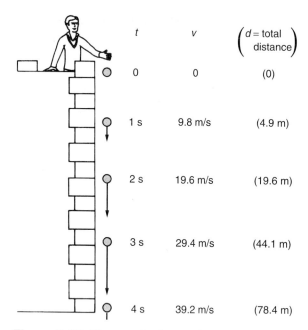

Figure 2.16 Time, velocity, and distance.
Listed are the magnitudes of the velocities and the distances traveled by a freely falling object at the ends of the first few seconds. The magnitude of the velocity increases by 9.8 m/s each second. The total distance traveled is computed by using the equation $d = \frac{1}{2}gt^2$ for the case when the object is dropped from rest. (The distances of fall are obviously not to scale in the figure.)

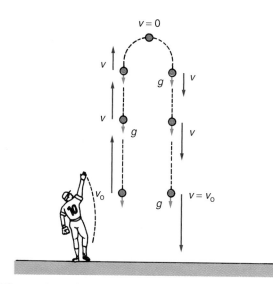

Figure 2.17 Up and down.
An object projected vertically upward initially slows down because the acceleration due to gravity g is in the direction opposite to that of the velocity, and the object stops ($v = 0$) for an instant at its maximum height. Then it accelerates downward and returns to its starting point with a velocity equal and opposite to the initial upward velocity. (The downward path is displaced in the figure for clarity.)

2.4 Projectile Motion

Learning Goals:

To explain how two-dimensional projectile motion can be described in terms of its one-dimensional components.

To identify what factors affect projectile range.

We frequently throw or project objects. A vertical projection, as just discussed, is in a straight line (one dimension). However, in general, projectile motion is in two dimensions. To analyze such motion it is convenient to look at the *components* of motion which means investigating the motion of the object in each dimension. In so doing, we analyze straight-line motions in each dimension. The combination of these is the total motion.

Consider, for example, a horizontal projection, say, a ball thrown horizontally (Fig. 2.18). The ball initially has a horizontal velocity and so travels in a horizontal direction. But, as the ball travels outward, gravity acts on it, and with a downward acceleration (due to gravity), the ball travels downward while moving horizontally at the same time.

So, if air resistance is neglected, a horizontally projected object essentially travels in a horizontal direction with a constant velocity (no acceleration in that direction) while falling under the influence of gravity. The resulting path of the combination of these motions is the curved arc we observe. The motion continues as described until the projectile hits the ground. The horizontal distance a projectile travels is called its **range.**

Note that the downward motion of a horizontally projected object is the same as that of a dropped object. This can be demonstrated as shown in Fig. 2.19. An object thrown horizontally will fall at the same rate as an object that is dropped.

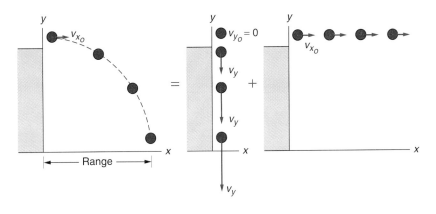

Figure 2.18 Horizontal projection.
A horizontal projection is a combination of two components of motion—free-fall in the downward *y* direction and uniform motion in the *x* direction. The distance the object travels in the *x* direction is called its *range*.

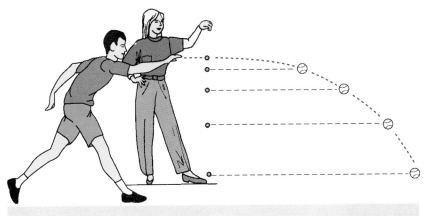

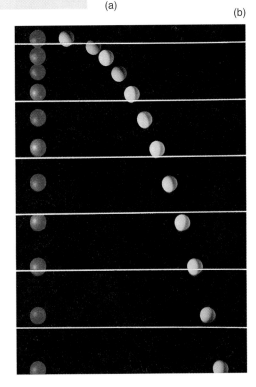

Figure 2.19 Same vertical motions.
(a) If a ball is thrown horizontally and a ball dropped simultaneously from the same height, both will hit the ground at the same time (if air resistance is neglected) because the vertical motions are the same. (Diagram not to scale.) (b) A multiflash photograph of two balls—one projected horizontally at the same time the other was dropped. Note from the horizontal lines that they fall vertically at the same rate.

Occasionally, a sports announcer claims that a hard throwing quarterback can throw a football so many yards "on line," meaning in a straight line. This cannot be true, as you can see. All objects thrown horizontally begin falling as soon as they are released.

If an object is projected at an angle θ (Greek theta) to the horizontal, it will follow a symmetric curved path such as illustrated in Fig. 2.20a, where air resistance is again neglected. The curved path is essentially the result of the combined motions in the vertical and horizontal directions. The

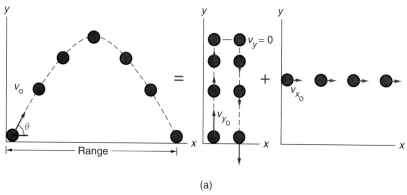

(a)

Figure 2.20 Projectile motion.
(a) The curved path of a projectile is the result of combined motions in the vertical horizontal directions. (b) A basketball player seems to "hang" in the air. See text for description.

(b)

projectile goes up and down vertically, while at the same time traveling horizontally with a constant velocity.

Now you can understand why basketball players when jumping to score seem to be suspended momentarily, or "hang" in midair (Fig. 2.20b). Notice that for the vertical motion near the maximum height, the velocity is quite small—going to zero and then increasing from zero (see Fig. 2.17). During this time, the combination of the slow vertical motion and the constant horizontal motion gives the illusion that the player hangs in the air.

If a ball or other object is thrown at various angles, the path that it takes will depend on the angle at which it is thrown. If air resistance is neglected, each path will resemble one of those shown in Fig. 2.21. As shown, the range is maximum when the object is projected at an angle of 45° relative to level ground. Notice in the figure that for a given initial speed, projections at complementary angles (angles that add up to 90°—for example, 30° and 60°) have the same range so long as there is no air resistance.

CONFIDENCE QUESTION 2.6

Two projectiles have the same initial speed. (a) If air resistance is neglected, which projectile would have the greater range—one with a projection angle of 55° or one with a projection angle of 65°? (b) What other projection angle will give this greater range?

With little or no air friction, projectiles have symmetric paths. (These paths are called *parabolas*.) However, when a ball or object is thrown or hit hard, air resistance comes into effect. In this case the projectile path resembles one of those shown in Fig. 2.22 and is no longer symmetric. Air resistance reduces the velocity of the projectile, particularly in the horizontal direction. As a result, the maximum range now occurs for a projection angle less than 45°.

Athletes such as football quarterbacks and baseball players are aware of the best angles at

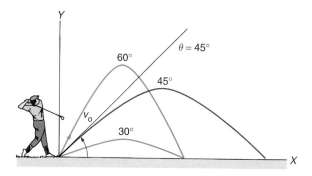

Figure 2.21 Maximum range.
A projectile's maximum range on a horizontal plane with any given initial speed can be achieved with a projection angle of 45°. This assumes there is no air resistance.

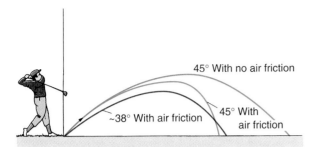

Figure 2.22 Effects of air resistance.
Long football passes or hard-hit baseballs follow trajectories similar to those shown here for a golf ball. Air resistance reduces the range. (Diagram not to scale.)

which to throw in order to get the maximum distance. A good golfing drive also depends on the angle at which the ball is hit. (Of course, in most of these instances there are other considerations, such as spin.) In track and field events such as discus and javelin throwing, both spin and the projection angle are considerations. Figure 2.23 shows an athlete hurling a javelin at an angle of less than 45° in order to get the maximum distance.

Figure 2.23 Going for the distance.
With air resistance, the angle for maximum range is less than 45°.

Important Terms

These important terms are for review. After reading and studying the chapter, you should be able to define and/or explain each of them.

position
motion
scalar
distance
vector

displacement
average speed
instantaneous speed
average velocity
instantaneous velocity

slope (of graph)
acceleration
acceleration due to gravity
range (of projectile)

Important Equations

Average speed: $\bar{v} = \dfrac{d}{t}$ $\left(\dfrac{\text{distance traveled}}{\text{time to travel distance}} \right)$

Average acceleration: $\bar{a} = \dfrac{\Delta v}{t} = \dfrac{v_f - v_o}{t}$

Acceleration due to gravity: $g = 9.80 \text{ m/s}^2$ ($\approx 32 \text{ ft/s}^2$)

Final velocity: $v_f = v_o + at$
(constant acceleration)

Questions

2.1 Path and Position

1. A vector quantity has (a) direction, (b) magnitude, (c) both of the preceding.
2. The magnitude of the displacement of an object is always (a) greater than the distance traveled, (b) equal to the distance traveled, (c) less than the distance traveled, (d) less than or equal to the distance traveled.
3. What is needed to designate the position of an object? Relate your answer to designating a position on a Cartesian graph.
4. What is the meaning of the Greek symbol Δ (delta) as used in Δx or Δt?
5. Are only the quantities of length and time needed to describe motion? Explain.

2.2 Speed and Velocity

6. The units of m/s apply to (a) speed, (b) average velocity, (c) instantaneous velocity, (d) all of the preceding.
7. On an x-versus-t graph, the magnitude of the slope of a straight line is the object's (a) displacement, (b) distance, (c) speed, (d) acceleration.
8. If a straight line on an x-versus-t graph has a negative slope or inclination, this means that the object (a) has a negative speed, (b) is moving in the $-x$ direction with a constant speed, (c) is accelerating in the $-x$ direction, (d) is coming to a stop.
9. Explain how distance and displacement are associated with speed and velocity, and why.
10. Can the average speed and instantaneous speed of an object be the same; that is, have the same value? How about the average velocity and instantaneous velocity?
11. For straight-line motion, is it possible to have a positive (+) displacement and a negative (−) velocity? Explain.
12. A jogger jogs two blocks directly north.
    (a) How do the jogger's average speed and magnitude of the average velocity compare?
    (b) If the return jog is over the same path, how do the average speed and average velocity magnitude compare for the total trip?
13. On an x-versus-t graph, a straight line starts with a negative x intercept and crosses the t-axis. Generally describe the represented motion.

2.3 Acceleration

14. Acceleration is the time rate of change of (a) speed, (b) velocity, (c) displacement, (d) distance.
15. The SI units of acceleration are (a) (m/s)/s, (b) m²/s, (c) m/s, (d) m-s/s.
16. For straight-line motion, a positive (+) acceleration would produce a deceleration for an object (a) with $+v$, (b) with $-v$, (c) with $v_o = 0$, (d) in no instance.
17. Can an object have an instantaneous velocity of 9.8 m/s in one direction and simultaneously have an acceleration of 9.8 m/s² in the same or opposite direction? Explain.
18. When Galileo dropped two objects from a high place, they hit the ground at almost exactly the same time. Why was there a slight difference in when they hit?
19. Would it be appropriate to call the steering wheel of an automobile an "accelerator"? Explain.
20. Can an object have an instantaneous velocity of zero and still be accelerating? (*Hint:* Think about an object projected vertically upward at its maximum height.)

2.4 Projectile Motion

21. If air resistance is neglected, a ball thrown at an angle θ to the horizontal has (a) a constant velocity in the x direction, (b) a constant acceleration in the $-y$ direction, (c) a changing velocity in the $+y$ direction, (d) all of the preceding.
22. In the absence of air resistance, a projection at an angle of 33° above the horizontal will have the same range as a projection at (a) 45°, (b) 52°, (c) 57°, (d) 66°.
23. For projectile motion, what quantities are constant? (Neglect air resistance.)
24. On what does the range of a projectile depend?
25. Can a baseball pitcher throw a fastball on a straight, horizontal line? Why or why not?
26. Taking into account air resistance or friction, how would you throw a ball to get the maximum range?

Food for Thought

1. We have highway speed limits, but not velocity limits. What would a velocity limit imply?
2. A fellow student tells you that a positive velocity ($+v$) means that there is an acceleration and that a negative velocity ($-v$) means there is a slowing down or a deceleration. Do you agree? That is, is this always the case? Also, is it correct to associate $+v$ and $-v$ with forward and backward motion, respectively?
3. Two projectiles have projection angles of 50° and 35°, respectively, and they both have the same range (if air resistance is neglected). Explain how this is possible.
4. Why is projectile range reduced with air resistance?

Exercises

2.1 Path and Position

1. In a 100-m swimming competition, a swimmer swims in a straight line down and back in a 50-m length pool. What are the swimmer's distance and displacement (a) when at the opposite end of the pool, and (b) on returning to the starting point?

   Answer: (b) 100 m and 0 m

2. To get to a shopping mall, a lady drives 3 km north and then 4 km east. What are (a) the distance and (b) the magnitude of the displacement in this case? (*Hint:* Recall the Pythagorean theorem.)

   Answer: (a) 7 km; (b) 5 km

3. An airplane flies with a constant speed of 200 km/h on a northwest heading. Draw the plane's velocity vector on a Cartesian graph with a scale of 50 km/h per division.

2.2 Speed and Velocity

4. At a track and field meet, a runner runs the 100-m dash in 15 s. What is his average speed?

   Answer: 6.7 m/s

5. An airplane flying directly eastward at a constant rate travels 300 km in 2.0 h.
   (a) What is the average velocity of the plane in km/h and m/s?
   (b) What is its instantaneous velocity?

   Answer: (a) 150 km/h or 42 m/s, east; (b) same

6. Prove that the plot in Fig. 2.8 has a zero slope.
7. Describe the motion of the object represented in Fig. 2.24. Give the velocity for each segment. (Is the represented motion physically possible? Explain.)

   Answer: seg 1, $v = +1.2$ m/s; seg 2, $v = 0$; seg 3, $v = -0.88$ m/s

8. Describe the motion of the object represented in Fig. 2.25. Give the velocity for each segment.

2.3 Acceleration

9. An airplane on takeoff achieves a speed of 100 km/h in 6.0 s. What is the plane's average acceleration?

   Answer: 4.6 m/s², down runway

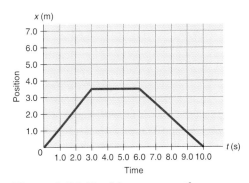

Figure 2.24 Position versus time. See Exercise 7.

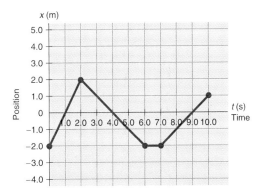

Figure 2.25 Position versus time. A whole lot of changes. See Exercise 8.

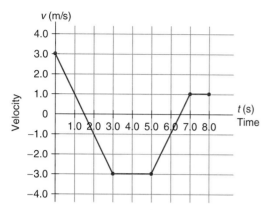

Figure 2.26 Velocity versus time.
What have we here? See Exercise 14.

10. A ball is dropped from the top of a tall building with a height of 100 m (32 floors).
    (a) What is the ball's velocity at the end of the first second of fall?
    (b) At the end of the second second? (*Hint:* This can be done mentally.)

    Answer: (b) 19.6 m/s, downward

11. A sprinter starting from rest on a straight and level track is able to achieve a speed of 12 m/s in a time of 4.0 s. What is the sprinter's average acceleration?

    Answer: +3.0 m/s^2

12. A race car traveling at 180 km/h ejects a parachute and slows down uniformly to 10 km/h in a time of 4.5 s. What was the racer's acceleration in SI units?

13. Discuss Fig. 2.10 in terms of acceleration.

14. Describe the motion represented in Fig. 2.26. Are the sharp changes realistic? If not, how would a more realistic graph look in these regions?

    Answer: seg 1, $a = -2.0$ m/s^2

15. Sketch (a) v-versus-t and (b) a-versus-t graphs for an object thrown downward with an initial speed of 4.0 m/s.

2.4 Projectile Motion

16. An airplane flies horizontally west at a speed of 200 m/s (448 mi/h). A package of supplies is dropped from the plane to some stranded campers.
    (a) What is the package's initial velocity? Give both speed and direction.
    (b) What are the magnitude and direction of the package's acceleration?

    Answer: (a) 200 m/s, west

17. Neglecting air resistance, at what projection angle would an object have the same range as one projected at 49° if both had the same initial speed?

18. Two projectiles with the same initial speed are projected at angles of 50° and 39°, respectively. Which has the greater range? (Neglect air resistance.)

19. Sketch graphs of (a) x versus t, and (b) a versus t for a horizontal projectile.

20. Sketch graphs of (a) x versus t and (b) a versus t for an object projected at an angle θ above the horizontal. (Neglect air resistance.)

ANSWERS TO MULTIPLE-CHOICE QUESTIONS

1. c   6. d   8. b   15. a   21. d
2. d   7. c   14. b   16. b   22. c

SOLUTIONS TO CONFIDENCE QUESTIONS

2.1 100 km, southwest

2.2 $v = \dfrac{d}{t} = \dfrac{100 \text{ km}}{9.92 \text{ s}} = 10.1$ m/s

2.3 That the object had passed through the origin and the motion continues in the $-x$ direction.

2.4 $v_f = v_o + at = 40$ m/s $+ (-2.0$ m/s$^2)(5.0$ s$) = 30$ m/s, where v is taken as positive, and a is taken as negative ($-a$).

2.5 The object is initially ($t = 0$) in motion with a velocity of $v_o = -1$ m/s (in the $-x$ direction) and accelerates in that direction at a constant rate (straight line with negative slope). Note that the magnitude of the velocity increases by $\Delta v = 0.5$ m/s each second, so the magnitude of the acceleration is $a = 0.5$ m/s per second or 0.5 (m/s)/s $= 0.5$ m/s^2.

2.6 (a) 55° (closer to 45°)
    (b) 35°, since 55° + 35° = 90°. Another way of looking at this is that 55° is 10° above 45° (maximum range), so the other angle is 10° below 45°, or at 35°.

Chapter 3

Force and Motion

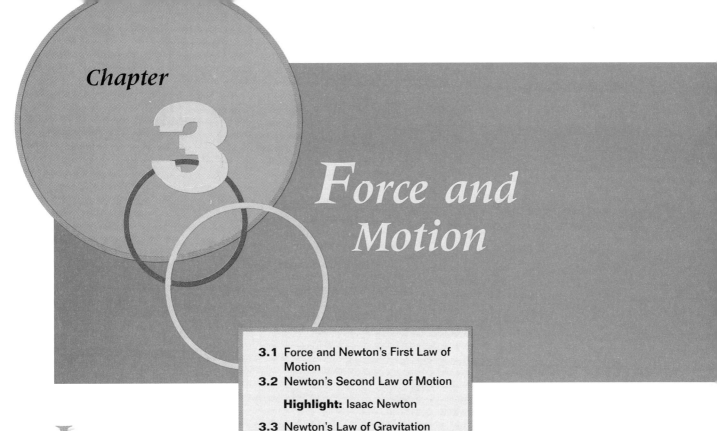

3.1 Force and Newton's First Law of Motion
3.2 Newton's Second Law of Motion

Highlight: Isaac Newton

3.3 Newton's Law of Gravitation
3.4 Newton's Third Law of Motion
3.5 Momentum and Impulse

Highlight: Impulse and the Automobile Air Bag

In Chapter 2 we talked about motion without regard to what causes it. You know how to set something in motion—give it a push or a pull. That is, you apply a force. Force and motion are intimately related, a sort of cause-and-effect relationship. However, if you stop and think about it, the effect of a force is to cause a *change* in motion of an object, whether it is initially at rest or in motion. In this chapter you will gain an insight into the nature of forces and their effects on the motions of objects on which they are applied.

As we saw in Chapter 2, Galileo was interested in the motion of falling objects. He was also interested in other types of motions, such as the motion of planets that he observed with one of the first telescopes to be constructed. Galileo was one of the first scientists to make a formal statement concerning objects at rest and in motion. But it remained for Isaac Newton (1642–1727), who was born the year Galileo died, to actually formulate a set of laws of motion. In addition, Newton formulated his law of gravitation and described the motions of planets and other celestial objects. Most of this was done when Newton was only 25 years of age. His extensive works have clearly established Newton as one of the greatest scientists of all time. Newton's life is considered in more detail in a chapter Highlight.

Now let's get on with our study of force and motion.

3.1 Force and Newton's First Law of Motion

Learning Goals:

To define the concepts of force and inertia.

To explain the principle of Newton's first law of motion (or law of inertia).

We all have an intuitive idea of what is meant by a force, but let's be more specific. Pushes and pulls are examples of forces. We might say that these are examples of *contact forces*, a term that describes how the forces are applied. Other types of forces are sometimes described as *action-at-a-distance* forces. An example is gravity, a force that will be described later in the chapter. It is the force of

Figure 3.1 Balanced and unbalanced forces.
(a) When two forces are equal and opposite, they balance each other and there is no net force (and no motion if the system is initially at rest). (b) When F_2 is greater than F_1, there is an unbalanced or net force equal to the vector difference, and motion occurs.

gravity that keeps the Moon in orbit about Earth and the planets in orbit about the Sun.

Let's develop a general definition of a force. Basically, a **force** is a quantity that is capable of producing motion or a change in motion. A force applied to an object at rest may cause it to be set into motion, or applied to an object in motion may cause a change in its motion. In either case, there may be an acceleration (Chapter 2).

Here's another important characteristic: *Force is a vector quantity and therefore has magnitude and direction.* Also, more than one force may act on an object. As a result, a force may act on a body with no resulting motion. This is because its effect is canceled out by other forces. Notice in the above definition, a force does not necessarily have to actually produce a change in motion, as long as the capability is there. For example, in the tug-of-war shown in • Fig. 3.1a, forces are being applied, but there is no motion. The forces in this instance are balanced, that is, equal and opposite, and in effect cancel each other. A change in motion occurs only when the applied forces are unbalanced, or don't cancel each other. The forces may act on an object in different directions; but for a change in motion or an acceleration to occur, there must be an **unbalanced force** (sometimes called a *net force*) for a net change or effect. The effect is as though there were only a single unbalanced or net force (Fig. 3.1b).

Similarly, there may be an unbalanced or net force for three or more applied forces. However, if the net force is zero, then the forces are balanced and there is zero net effect. (The net force is the vector sum of the forces; a discussion of vector computation is beyond the scope of this text.)

Another force description we will use is *external force*. An external force is an applied force—that is, one applied on or to an object or system. There are also *internal forces*. For example, suppose the object is an automobile and you are a passenger traveling inside. You can push (apply a force) on the floor or dashboard, but there is no effect on the car's motion because this is an internal force. Another car hitting yours in a "fender bender" would be an example of an (unpleasant) external force.

Before the time of Galileo and Newton, scientists had asked themselves, "What is the natural state of motion?" According to the theory of

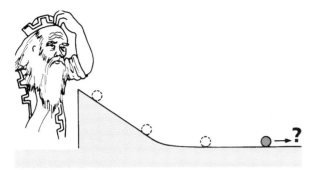

Figure 3.2 Motion without resistance.
If the level surface could be made perfectly smooth, how far would the ball travel?

motion of the Greek philosopher Aristotle, which had prevailed some twenty centuries after his death, an object required a force to keep it in motion. That is, the natural state of an object was one of rest, with the exception of celestial bodies, which were naturally in motion. It was readily observed that moving objects naturally tended to slow down and come to rest, so a natural state of rest must have seemed logical to Aristotle.

Galileo studied motion using a ball rolling onto a level surface from an inclined plane. The smoother he made the surface, the farther the ball would roll (Fig. 3.2). He reasoned that if he could make a very long surface perfectly smooth, there would be nothing to stop the ball, and it would continue in motion indefinitely or until something stopped it. Thus, contrary to Aristotle's theory, Galileo concluded that objects could naturally remain in motion.

Newton also recognized this phenomenon and incorporated Galileo's result in his first law of motion. This law can be stated several ways. One statement of **Newton's first law of motion** is as follows:

> An object will remain at rest or in uniform motion in a straight line unless acted upon by an external, unbalanced force.

"Uniform motion in a straight line" means that the velocity is constant. Thus, another way to state Newton's first law is to say that the natural state of motion is at a constant velocity. If the constant velocity is equal to zero, we say that an object is at rest.

Because of the ever-present forces of friction and gravity on Earth, it is difficult to observe a moving object in a natural state with a nonzero constant velocity. However, in free space where there is no friction and negligible gravitational attraction, an object initially in motion maintains a constant velocity. For example, after being projected on its way, an interplanetary satellite approximates this condition. Some *Pioneer* and *Voyager* spacecraft have left the solar system, and such a spacecraft will keep on going with a constant velocity until an appreciable external, unbalanced force alters this velocity.

Motion and Inertia

Galileo also introduced another concept relating to motion. It appeared that objects have a tendency to maintain a state of motion. That is, there seemed to be a resistance to changes in motion. Similarly, if an object was at rest, it seemed to "want" to remain so. Galileo called this behavior inertia, and we say that **inertia** is the tendency of an object to resist changes in motion.

Newton went one step further and related the concept of inertia to something that could be measured—mass. That is, **mass** is a measure of inertia. The greater the mass of an object, the greater its inertia, and vice versa.

As an example of the relationship of mass and inertia, suppose you push two different people on swings initially at rest, one a very large man and the other a small child (Fig. 3.3a). It is quickly evident that it is more difficult to get the adult moving. That is, there would be a noticeable difference in the resistance to motion between the man and the child. Also, once you got them swinging and then tried to stop the motions, you'd notice a difference in the resistance to a change in motion again. Being more massive, the man has the greater inertia and greater resistance to change.

Another example or demonstration of inertia is shown in Fig. 3.3b. The stack of coins has inertia and resists being moved when at rest. If the paper is quickly jerked, the inertia of the coins will prevent them from toppling. You may have pulled a magazine from the bottom of a stack with a similar action.

Because of the relationship of motion and inertia, Newton's first law is sometimes called the *law*

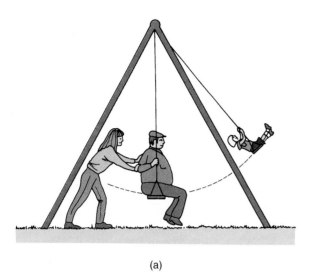 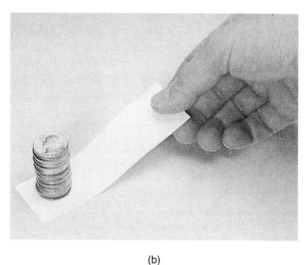

(a) (b)

Figure 3.3 Mass and inertia.
(a) An external, applied force is needed to put an object in motion, and the motion depends on the mass and inertia of the object. (b) Because of inertia, the paper can be removed from beneath the stack of quarters without toppling the stack, by giving it a quick jerk.

of inertia. It can be applied in various situations. For example, suppose you were in the front seat of a car traveling at a high speed down a straight road and the driver suddenly put on the brakes for an emergency stop. What would happen, according to Newton's first law, if you were not wearing a seat belt and the car was not equipped with a passenger-side airbag? Well, certainly the friction on the seat of your clothing would not be enough to appreciably change your motion, so you'd keep on moving in accordance with the first law while the car was coming to a stop. The next external, unbalanced force you might experience would not be pleasant. Newton's first law makes a good case for buckling up and also for airbags. (The principle of the automobile airbag will be discussed in a later chapter Highlight.)

3.2 Newton's Second Law of Motion

Learning Goals:

To explain the relationships of force, mass, and acceleration as expressed in Newton's second law of motion.

To clearly distinguish between mass and weight using the second law.

To define centripetal force.

In our initial study of motion (Chapter 2), acceleration was defined as the time rate of change of velocity ($\Delta v/t$). Nothing was said about what causes acceleration, only that a change in velocity resulted. Now we ask: What causes an acceleration? Newton's first law answers the question. If an external, unbalanced force is required to produce a change in velocity, then it follows that an unbalanced force produces an acceleration.

Newton was aware of this, but he went a step further and also related acceleration to inertia or mass. Because inertia is the resistance to a change in motion, a reasonable assumption is that the greater the inertia or mass of an object, the smaller the change in motion or velocity for a given applied force. Such insight was typical of Newton in his many contributions to science. (See the chapter Highlight.) Hence we have in summary:

1. The acceleration produced by an unbalanced force acting on an object (or mass) is directly proportional to the magnitude of the force ($a \propto F$) and in the direction of the force. ($\propto$ is a proportionality sign, and in a direct proportion, when one quantity becomes larger, so does the other.)

2. The acceleration of an object being acted upon by an unbalanced force is inversely proportional to the mass of the object, $a \propto 1/m$. (In an inverse proportion, when one quantity gets larger, the other gets smaller.)

HIGHLIGHT

Isaac Newton

Isaac Newton (Fig. 1) and Albert Einstein are usually considered to be the two greatest scientists in history. Newton's laws of motion were one of many contributions he made to a variety of subjects in physics. Newton was born on Christmas day in 1642 in the village of Woolsthorpe in Lincolnshire, England. He showed no particular genius in his early schooling, but fortunately a teacher encouraged him to pursue his education, and in 1661 he entered Trinity College at Cambridge.

Four years later he received his degree and planned to continue studying for an advanced degree. But an epidemic of the bubonic plague broke out and the university was closed. Newton returned to Woolsthorpe, and in the next two years he laid the groundwork for many of his contributions in physics, mathematics, and astronomy. In Newton's own words, "I was in the prime of my age for invention, and minded mathematics and philosophy [science] more than any time since."

Over the next 20 years Newton was very productive, and at the age of 45 he published his famous treatise, *Philosophiae Naturalis Principia Mathematica [Mathematical Principles of Natural Philosophy]** or *Principia* for short. In this book he set forth his laws of motion, along with the theory of gravitation. The publication of the *Principia* was financed by a friend, Edmund Halley, who used Newton's theories to predict the return of the comet that bears his name.

---

*Physics was once called natural philosophy.

(a)   (b)

Figure 1 The man and the book.
(a) Sir Isaac Newton (1642–1727). **(b)** The title page of the *Principia*. Can you read the Roman numerals at the bottom of the page that give the year of publication?

Newton was reportedly a shy man, but he often got into disputes about his theories and achievements. A famous one is his dispute with Gottfried Leibniz about who first developed calculus. Newton was elected to Parliament and later appointed Master of the Mint, where he supervised the task of recoining the English currency. He was knighted by Queen Anne in 1705.

Sir Isaac Newton died in 1727 at the age of 85 and was buried with honor in Westminster Abbey. Some insight about this austere bachelor and giant of science may be gleaned from the following excerpts from his writings:

I seem to have been only a boy playing on the seashore and diverting myself in now and then finding a smoother pebble or a prettier shell than ordinary, whilst the great ocean of truth lay all undiscovered before me.

If I have been able to see farther than some, it is because I have stood on the shoulders of giants.

About Newton, the poet Alexander Pope wrote:

Nature and Nature's laws lay hid in night God said, Let Newton be! and all was light.

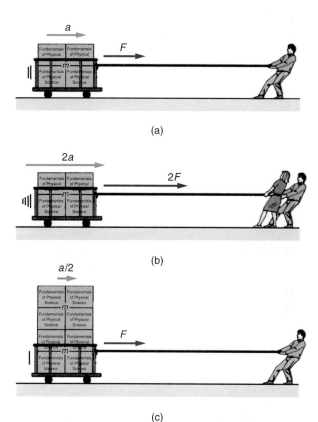

Figure 3.4 Force, mass, and acceleration.
(a) An unbalanced force *F* acting on a mass *m* produces an acceleration *a*. (b) If the mass remains the same and the force is doubled (2*F*), the acceleration is doubled. (c) If the mass is doubled (2*m*) and the force remains the same (*F*), the acceleration is half as much.

Combining these effects of force and mass on acceleration, we have a statement of **Newton's second law of motion,** which may be expressed simply and conveniently as

$$\text{acceleration} \propto \frac{\text{unbalanced force}}{\text{mass}}$$

or

$$a \propto \frac{F}{m}$$

These relationships are illustrated in ● Fig. 3.4. Notice that if you double the force acting on a mass, the acceleration doubles (direct proportion, $a \propto F$, and $2a \propto 2F$). If, however, the same force is applied to twice as much mass, the acceleration is one-half as much [inverse proportion, $a \propto F/m$, and $F/2m = (1/2) F/m = a/2$].

When appropriate units are used, we can write Newton's second law in equation form as

$$a = \frac{F}{m}$$

which is commonly rearranged as

$$F = ma \quad (3.1)$$

(in words)   force = mass × acceleration

This equation is a mathematical statement of Newton's second law. Note from Fig. 3.4 that the mass *m* is the *total mass* of the system or all of the mass that is accelerated. A system may consist of two or more separate masses. Also, *F* is the net or unbalanced force, which may result from two or more applied forces. The direction of a net force is generally taken to be along a coordinate axis and so is indicated by a + or − sign. Unless otherwise stated, we will take a general reference to a force in our discussion as meaning an external unbalanced force.

Hence, we see that Newton's second law is somewhat of a cause (force) and effect (acceleration) relationship. We often observe an object speeding up, slowing down, and/or changing direction. These changes indicate an acceleration. From Newton's second law we know that an acceleration is evidence of a (net) force.

In the metric system (SI), the units of mass and acceleration are kg and m/s², respectively, so force ($F = ma$) has the basic units of kg-m/s². This derived unit is given the special name of **newton** (abbreviated N), for obvious reasons, and force is expressed in units of newtons.* One newton (1 N) is the magnitude of the force necessary to accelerate 1 kg at a rate of 1 m/s². [From $F = ma$ we get (1 kg)(1 m/s²) = 1 N.]

In the British system the unit of force is the pound (lb). This unit too has derived units of mass times acceleration, but recall that the British system is a gravitational or force system and mass is rarely used. Instead, quantities of matter are expressed in pounds (force). As will be seen shortly, weight is a force and is expressed in newtons in the SI. Equivalent weight units are shown in ● Fig. 3.5. If you are familiar with the story that Newton gained insight by observing (or being struck) by a falling apple while meditating on the concept of gravity, you can easily remember that an average apple weighs about one newton.

---

*Another metric unit of force is the *dyne*. Based on $F = ma$, 1 dyne (abbreviated dyn) is the cgs unit g-cm/s². The dyne is a relatively small unit, only 1/100,000 as large as a newton.

Weight
one newton
or ≈ 0.225 lb

Figure 3.5 Approximately 1 newton.
An average-sized apple weighs approximately 1 N (or 3.6 oz or 0.225 lb).

Here's an example using $F = ma$ that illustrates that F is the net or unbalanced force and m is the total mass.

EXAMPLE 3.1 Applying Newton's Second Law

A horizontal force of 6.0 N is applied to a block at rest on a frictionless surface by means of a string, as illustrated in Fig. 3.6. (a) If the mass of the block is 2.0 kg and the mass of the string is negligible, what is the acceleration of the block? (b) What would be the acceleration of the block if the magnitude of the applied force were doubled? (c) What would be the acceleration if there were two blocks of 2.0 kg mass with an applied force of 6.0 N?

Solution

Summarizing what is given and what we are to find, we have:

Given: (a) $F = 6.0$ N $m = 2.0$ kg
(b) $F = 12$ N $m = 2.0$ kg
(c) $F = 6.0$ N $m = 4.0$ kg

Find: a (acceleration) for all cases

(Since the surface is frictionless, we don't have to worry about an opposing frictional force. Basically, the force of friction opposes motion between surfaces in contact when there is relative motion or a tendency for relative motion. Keep this in mind; however, we will generally ignore friction in our discussion.) The units are all in the SI, so no conversion is necessary. Now that we are getting multiple units, keep in mind that kg-m/s^2 is a newton (N). For simplicity, the cancelation of units won't be shown explicitly.

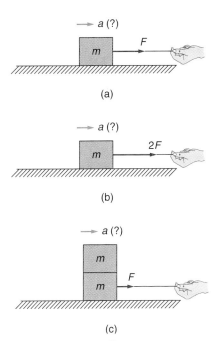

Figure 3.6 Cause and effect.
See Example 3.1.

(a) To find acceleration, we use Eq. 3.1 in the form $a = F/m$, and putting in the numbers, we have

$$a = \frac{F}{m} = \frac{6.0 \text{ N}}{2.0 \text{ kg}} = 3.0 \text{ m/s}^2$$

in the direction of the force, taken here as positive. (Remember that force and acceleration are vector quantities and the direction of the acceleration is the same as that of the net applied force.)

(b) Here, the force is doubled. If you understand Newton's second law, you know what the answer is already. But, let's work it out.

$$a = \frac{F}{m} = \frac{12 \text{ N}}{2.0 \text{ kg}} = 6.0 \text{ m/s}^2$$

in the direction of the force. (Double the force and you double the acceleration, $a \propto F$.)

(c) Here, the mass is doubled, and with the original applied force we have

$$a = \frac{F}{m} = \frac{6.0 \text{ N}}{4.0 \text{ kg}} = 1.5 \text{ m/s}^2$$

in the direction of the force, and the acceleration is one-half of that in part (a).

CONFIDENCE QUESTION 3.1

(a) In part (b) of Exercise 3.1, what would be the result if an additional force *F* were applied in the opposite direction?

(b) Suppose the 2.0-kg mass in Fig. 3.6a accelerated to the left at a rate of 1.5 m/s². What is the net force acting to produce this acceleration?

The symbol *F* is used to represent any force, and as you see, Newton's second law of $F = ma$ describes what a force does, not what the force is. In virtually all of our everyday situations, all forces are of two fundamental types: the *gravitational force* and the *electromagnetic force*. An example of the gravitational force is weight, and the description of this force will be considered in detail in the next section. Examples of electromagnetic force include electrostatic effects, electrostatic cling, and picking up pieces of metal with a magnet. We will study the electric force and the magnetic force separately in Chapter 8, but they are two aspects of the same electromagnetic force. On the submicroscopic level, electromagnetic forces hold atoms and molecules together, and exert forces on other atoms and molecules. Although it often is not apparent, electromagnetic forces through these interactions are at work in all normal situations that are not gravitational—for example, friction.

The only two other known fundamental forces are associated with the nucleus of the atom, and we cannot sense them directly as we do gravitational and electromagnetic forces. There is the *strong nuclear force* (or simply *strong force*) that binds or holds together the particles of the nucleus—the nuclear "glue," so to speak. The other is the *weak nuclear force* (or simply *weak force*), which is involved in certain kinds of radioactive decay. These forces will be considered in Chapter 10.

The electric and magnetic forces were once thought to be independent. However, it was shown that they are different aspects of the same (electromagnetic) force. Scientists, in trying to simplify nature, would like to show that the four fundamental forces are related. This unification has been done in part. It has been shown that the weak force and the electromagnetic force are different aspects of the same force—the *electroweak force*. Perhaps someday we will be able to show that all forces are aspects of a single superforce.

Mass and Weight

Now that we are familiar with Newton's second law, this is a good place to make a firm distinction between mass and weight. These quantities were generally defined previously: *Mass* refers to the amount of matter an object contains (it is also a measure of inertia), and *weight* is related to the force of gravity; that is, the gravitational attraction on an object by a celestial body (such as Earth). However, mass and weight are related, and Newton's second law clearly shows this relationship.

On the surface of Earth, where the acceleration due to gravity is relatively constant (9.8 m/s²), the weight force *w* on an object with mass *m* is given by the equation

weight = mass × acceleration (due to gravity)

$$w = mg \qquad (3.2)$$

Notice that this is a special case of $F = ma$, in which different symbols have been used to distinguish particular quantities.

Now you can easily see why mass is the fundamental quantity. Generally, a physical object has the same amount of matter (*m*), and so it has the same mass. But the weight of an object may differ, depending on the value of *g*. For example, on the surface of the Moon the acceleration due to gravity (g_m) is one-sixth of what it is on the surface of Earth [$g_m = g/6 = (9.8$ m/s²$)/6 = 1.6$ m/s²]. Similarly on the surface of Mars, the acceleration due to gravity is 0.38 times that of Earth, or $g_M = 3.7$ m/s². As you can see from Eq. 3.2 with different *g*'s, an object will have the same mass on the Moon, Earth, or Mars, but its weight will be different on each.

Also notice that weight has the unit of newtons in the SI; however, in metric countries, people express their "weight" in terms of mass in "kilos" (kilograms). Since weight and mass are related by the constant *g*, scales may be easily calibrated in mass units (kilograms or grams) or in weight units (newtons or pounds). As shown in Fig. 3.7, 1 kg has a weight of 9.8 N, as you can readily determine from Eq. 3.2.

In Chapter 2, it was learned that all freely falling bodies fall at the same rate (with an acceleration *g*). Galileo verified this result, but he could not explain why it was so. Looking at Eq. 3.2 we

Figure 3.7 Mass and weight.
A mass of 1 kg is suspended on the scale, which has a weight reading of 9.8 N. This is equivalent to a weight of 2.2 lb.

see that an object in free-fall has an unbalanced force acting on it of $F = w = mg$ (downward). Different masses would have different forces, so why do objects in free-fall descend at the same rate? Newton's laws explain the observation.

Suppose there were two objects in free-fall, one with twice as much mass as the other, as illustrated in Fig. 3.8. According to Newton's second law, the more massive object would have twice the weight, or gravitational attraction. However, by Newton's first law, the more massive object has twice the inertia, so it needs twice the force to fall (accelerate) at the same rate. The fact that they both fall with the same acceleration (g) is shown mathematically in the figure. $F = ma$ is a powerful equation, isn't it?

Centripetal Force

Newton's second law is a general relationship that can be applied to many situations. Let's consider a special, but common, case of an object in uniform circular motion. In uniform circular motion, an object has a constant speed. For example, a car going around a circular track at a uniform rate of 90 km/h (55 mi/h) has a constant speed. However, the velocity is not constant, because there is a continuous change of direction. (Recall that velocity is a vector quantity with magnitude and direction, and a change in velocity can result from a change in either or both.) So, the velocity of an object in uniform circular motion is continuously changing; and with a change in velocity, there must be an acceleration (and net force), by Newton's second law.

This acceleration cannot be in the direction of the instantaneous motion or velocity; otherwise,

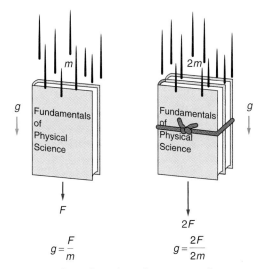

Figure 3.8 Acceleration due to gravity.
The acceleration due to gravity is independent of the mass of a falling object; thus it is the same for all freely falling objects. An object with twice the mass of another has twice the gravitational force acting on it, but it also has twice the inertia and thus falls at the same rate.

the object would speed up and the circular motion would not be uniform. So where are the acceleration and force? Because they cause a change in direction that keeps the object in a circular path, the acceleration and force are actually perpendicular, or at a right (90°) angle, to the velocity vector.

Consider a car traveling in uniform circular motion as illustrated in Fig. 3.9. At any point the instantaneous velocity is tangent to the curve (at an angle of 90° to a radial line at that point). After a short time, the velocity vector has changed (in direction). The change in velocity Δv is given by a vector triangle, as illustrated in the figure.

This change is an average over a time interval Δt, but notice how the Δv vector generally points inward toward the center of the circular path. For instantaneous measurements, this generalization is true; so for an object in uniform circular motion, the acceleration and force is toward the center of the circle. This acceleration is called **centripetal acceleration** (*centripetal* means "center seeking"). When an object moves in a circle of radius r with a constant speed v, the magnitude of the centripetal acceleration a_c is given by the equation $a_c = v^2/r$. Relating this to force through Newton's second

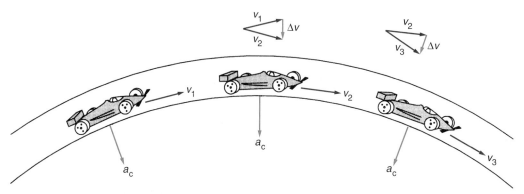

Figure 3.9 Centripetal acceleration.
A car traveling with a constant speed on a circular track is accelerating because its velocity is changing (direction change). The acceleration is toward the center of the circular path and is called *centripetal (center-seeking) acceleration.*

law, we have an expression for the magnitude of the **centripetal force**:

$$F = ma_c = \frac{mv^2}{r} \quad (3.3)$$

Notice that the centripetal force varies with the square of the speed; that is, if you double the speed, the centripetal force would increase by a factor of four. Also, the smaller the radius of the circular path, the greater the centripetal force needed to keep an object in circular motion for a given speed.

So, even though traveling with a constant speed, an object in uniform circular motion must have an inward force acting on it to maintain it in a circular path. Let's look at what actually supplies the centripetal forces in some real situations. For a car going around a circular curve, the centripetal force is supplied by friction on the tires. Should the car hit an icy spot on the road, it would slide outward because the centripetal force is not great enough to keep it in a circular path. Similarly, if you swing a ball on a string in a horizontal circle around your head, you must pull inward on the string to supply the necessary centripetal force in order to change the direction of the ball to keep it in a circular path. If you let go of the string or it breaks, the ball will have an initial tangential velocity and fly off in that direction (and downward with an acceleration g—like a horizontal projectile). Our Moon goes around Earth in nearly a circular orbit. Do you know what supplies the centripetal force to keep it in orbit? There's a gravitational "string" or attraction. We'll take a closer look at such gravitational attractions in the next section.

You have no doubt experienced a lack of centripetal force when riding in a car. In traveling on a straight road and then entering a circular curve at a fast speed, you have a sensation of being thrown outward or away from the center of the curve. From what we've learned about force and motion, you might think that there is an outward force acting on you—a centrifugal (center-fleeing) force. However, this force doesn't exist. Riding in the car before entering the curve, you tend to go in a straight line, in accordance with Newton's first law. As the car makes the turn, you continue to maintain your state of motion, until the car turns in front of you and the door pushes against you (Fig. 3.10a). You may feel that you are being thrown outward toward the door, but actually the door is coming toward you because the car is turning; and when it gets to you, it pushes on you to supply the inward centripetal force you need to go around the curve with the car. The force could be supplied by a seat belt if you are buckled up.

CONFIDENCE QUESTION 3.2

Explain the principle of the practical application of separating water from clothes in a washing machine, as illustrated in Fig. 3.10b.

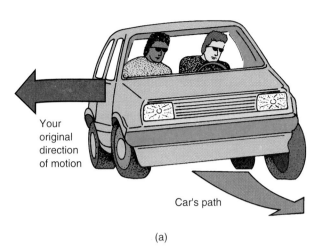 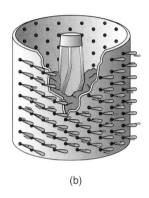

(a) (b)

Figure 3.10 Newton's first law in action.
(a) When a car initially traveling along a straight road goes around a sharp curve or corner, the people in it tend to continue in the original direction, according to Newton's first law. This gives them the sensation of being thrown outward. The door, seat belt, and/or steering wheel provides the necessary centripetal force and acceleration needed to go around the curve with the car. (b) In a washing machine, water is separated from clothes by rotational motion. See Confidence Question 3.2.

3.3 Newton's Law of Gravitation

Learning Goals:

To define Newton's law of gravitation.

To explain how to apply the law in understanding the acceleration due to gravity, tides, and the dynamics of orbiting satellites.

The law governing the gravitational force of attraction between two particles was formulated by Newton from his studies of planetary motion. Known as the law of universal gravitation, it may be stated as follows:

> **Every particle in the universe attracts every other particle with a force that is directly proportional to the product of their masses and inversely proportional to the square of the distance between them.**

Suppose the masses of the two particles are designated as m_1 and m_2, and the distance between them is r (Fig. 3.11a). Then we may write this statement in symbol form as

$$F \propto \frac{m_1 m_2}{r^2}$$

An object is made up of a lot of particles, so the gravitational force on an object is the vector sum of all the particle forces. This computation can be quite complicated, but one simple and convenient case is a homogeneous sphere. (*Homogeneous* means that the mass is distributed uniformly throughout the object.) In this case, the net force acts as though all the mass were concentrated at the center of the sphere, and the separation distance for the mutual interaction with another such spherical object is measured between their centers (Fig. 3.11a). If we assume that Earth, other planets, and the Sun are spheres with uniform mass distributions, we can apply the law of gravitation to such bodies. Also, assuming Earth to be a homogeneous sphere, the separation distance for computing the gravitational attraction for a person

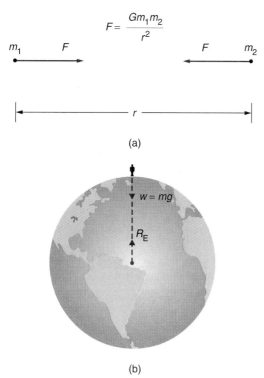

Figure 3.11 Law of gravitation.
(a) Two particles gravitationally attract each other, and the magnitude of the force is given by Newton's law of gravitation. (b) For a homogeneous or uniform sphere, the force acts as though all of the mass of the sphere were concentrated at its center. To a good approximation, a person's weight is determined as though all of the mass of Earth were concentrated at its center and the distance between the "masses" were the radius of Earth.

Table 3.1 What Is G?

$G = 6.67 \times 10^{-11}$ N-m²/kg²
G is a very small quantity.
G is a universal constant. It is thought to have the same value throughout the universe.
G is not g (the acceleration due to gravity).
G is not a force.
G is not an acceleration.

(particle) on Earth's surface is the radius of Earth (Fig. 3.11b). This gravitational force is the person's weight.

Newton's law of universal gravitation may be written in equation form by putting in an appropriate constant (of proportionality):

$$F = \frac{Gm_1m_2}{r^2} \quad (3.4)$$

where G is a constant called the universal gravitational constant—universal because we believe it to be the same throughout the universe, as is the law of gravitation.

Newton used his law of gravitation and his second law to show that gravity supplied the necessary centripetal force on the Moon for it to move in its nearly circular orbit about Earth. However, he did the derivation in ratio form because he did not know or experimentally measure the value of G. Its value was not determined until about seventy years after Newton's death by an English scientist, Henry Cavendish, who used a very delicate balance to measure the force between two masses.

The value of G is very small; it is given in Table 3.1, along with a description of what G is and is not. (The units of G are such that when used in Eq. 3.4, the gravitational force is in newtons. Try it and see.) It is common to distinguish between G and g, the acceleration due to gravity, by referring to "big G" and "little g," respectively. The "big" and "little" here refers to the letter size, not values.

When we drop something, the force of gravity, and the acceleration due to gravity, is evident. Yet, if there is a gravitational force between every two objects, why don't you feel an attraction between you and this textbook? (No pun intended.) Indeed, there is a force of attraction between you and this book, but it is so small that you don't notice it. To illustrate this, let's assume that you and the book are particles and are 1 m apart ($r = 1$ m), that the book has a mass of 1 kg ($m_1 = 1$ kg), and you have a mass of 100 kg (m_2—a little heavy, 220 lb, but it makes the mathematics easy). Then, putting these values into Eq. 3.4 along with the value of G, we find the force of attraction between you and the book to be $F = 6.67 \times 10^{-9}$ N = 0.00000000667 N. This is a very small force. A grain of sand would weigh more. For an appreciable gravitational force to exist between two masses, at least one (or both) of the masses must be relatively large (Fig. 3.12). Your weight is the gravitational attraction between you and Earth. This is appreciable because Earth has a very large mass. (However, you can't blame Earth if you're overweight. Your mass is in Eq. 3.4 too.)

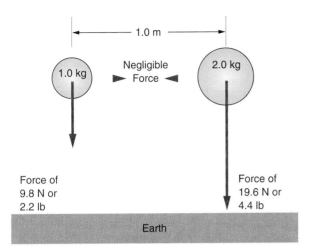

Figure 3.12 The amount of mass makes a difference.
A 1-kg mass and a 2-kg mass a distance of 1 m apart have a negligible mutual gravitational attraction (≈ 10^{-10} N or 10^{-11} lb). However, because Earth's mass is quite large, the masses are attracted toward Earth with forces of 9.8 N (or 2.2 lb) and 19.6 N (or 4.4 lb), respectively. These forces due to Earth's gravitational attraction are the weights of the masses.

Newton's law of gravitation gives us an insight about why the acceleration due to gravity at Earth's surface is relatively constant, and why this acceleration does not depend on the mass of an object. (Recall that all objects in free-fall have the same acceleration, or fall at the same rate.) Let's generalize first and consider the weight of a mass m due to a large spherical uniform object of mass M. (We'll specify this object to be Earth later.) Then, the force on the smaller mass, or its weight, is $F = mg = GmM/r^2$. Thus, we see that in general,

$$g = \frac{GM}{r^2} \quad (3.5)$$

This equation can be used to find the acceleration due to gravity on the Moon and on the planets. For example, if the radius and mass of the Moon are used, we get the acceleration due to gravity (g_m) on the Moon. As noted earlier, this acceleration turns out to be one-sixth of that on Earth ($g_m = g/6$), basically because the mass of the Moon is much smaller than that of Earth. If you want to lose weight fast, go to the Moon (Fig. 3.13).

Now let's look at Eq. 3.5 specifically for Earth. In this case, we have an object of mass m, the mass of Earth M_E, and the radius of Earth R_E. Then, $w = mg = GmM_E/R_E^2$, and on Earth's surface the acceleration due to gravity is given by

$$g = \frac{GM_E}{R_E^2} \quad \left(= \frac{9.8 \text{ m}}{\text{s}^2} \right) \quad (3.6)$$

Notice that since the mass of the object m cancels out of the equation, it does not appear in the expression for g. Thus, g is independent of the mass of the object, or is the same for all objects. Hence, all objects in free-fall have the same acceleration (resulting from the force of gravity).

Also from Eq. 3.6, we can see why g is relatively constant near Earth's surface. The term G is

Figure 3.13 Mass versus weight.
An astronaut on the Moon carries equipment that may have a mass of about 136 kg (or weigh 300 lb) on Earth. On the Moon, where the acceleration due to gravity is $\frac{1}{6}$ of that on Earth, the mass of the equipment is still the same, but it has a weight of only 50 lb. The astronaut also experiences a weight reduction. For example, an 82-kg (180-lb) astronaut on Earth would weigh only 30 lb on the Moon, but would still have 82 kg of mass.

a constant, and in general, so are the mass and radius of Earth (M_E and R_E). Hence, to a good approximation, g is constant near Earth's surface or over the normal heights of falling objects, since they don't add much to the value of R_E in Eq. 3.6.

The force of gravity and g do decrease with increasing height or altitude. For some height h above Earth's surface we would have in Eq. 3.5, $r = R_E + h$. (As h and r become larger, g becomes smaller—they're inversely proportional.) However, at an altitude of 1.6 km (1 mi), the acceleration due to gravity and hence an object's weight ($w = mg$) is only 0.05% less than g at Earth's surface. At an altitude of 160 km (100 mi), g is still 90% of its surface value.

Notice that the gravitational force varies by the inverse square of the distance ($F \propto 1/r^2$). We refer to such a relationship as an inverse-square law. Because of the inverse-square relationship, the gravitational force falls off rather rapidly with increasing distance. For example, if an object were moved twice the distance from the center of Earth ($r_2 = 2r_1$), then the gravitational attraction would be only $\frac{1}{4}$ as great, since $F_2 \propto 1/r_2^2 = 1/(2r_1)^2 = \frac{1}{4}(1/r_1^2) = \frac{1}{4}F_1$. How about if the distance were tripled? Could you show that the force would be only $\frac{1}{9}$ as great?

This variation of gravitational attraction with distance explains why we have two high and low ocean tides daily. You are probably aware that the major cause of tides is the Moon's gravitational attraction on Earth's oceans. The Moon attracts the water on the nearer side of Earth, and as it orbits, it carries this bulge with it, bringing high tides to land areas. (The depression on the opposite side of Earth where the water is low gives a low tide.) However, this explanation would account for only one high tide and one low tide. Two high and low tides arise because the water on the near side of Earth has the greatest attraction. Being farther away Earth itself has less attraction, and the water on the far side has even less. This differential attraction gives rise to two bulges on opposite sides of Earth—two daily high tides, with intervening low tides. See Chapter 17 where this is illustrated and tides are discussed more fully. You'll find that even the gravitational attraction of the Sun gets into the act.

The gravitational force is said to be an infinite-range force. That is, the only way to get it to approach zero is to separate the masses by a distance

Figure 3.14 Floating around.
Toothbrushes and toothpaste float around in a cabinet in an Earth-orbiting spaceship because of "zero g" (zero gravity) or "weightlessness." Actually, the weights or gravitational attractions of Earth on the objects keep them in orbit with the ship. (Note the Velcro strips on the sides of the cabinet that keep the objects from floating.)

approaching infinity ($r \to \infty$). So, no matter what the altitude is or how far an object is from Earth, there is still a gravitational force.

Gravity supplies the centripetal force that keeps satellites in orbit. This applies to Earth's artificial satellites and to its only natural satellite, the Moon. In effect, the continuous gravitational pull on a satellite causes it to deviate from straight-line motion as prescribed by Newton's first law and keeps it in orbit. In spacecraft orbiting Earth, we see astronauts and articles floating around and hear space-age terms such as zero g and weightlessness (Fig. 3.14). These terms are not really true descriptions. Gravity certainly acts on an astronaut in an orbiting spacecraft. Otherwise, the astronaut (and the spacecraft) would not remain in orbit. Since gravity is acting, the astronaut by definition has weight (and g is not zero).

The reason an astronaut floats in a spacecraft and feels "weightless" is that both the spacecraft and the astronaut are "falling" toward Earth. Keep in mind that the orbiting spacecraft has a centripetal force (gravity) acting on it and therefore is accelerating toward Earth. However, it is also moving tangentially, and the combination of these motions produces a circular (or perhaps an elliptical) orbit. So, the spacecraft and its contents are continuously "falling" toward Earth, and this gives the floating effect. To help see this, imagine yourself in a freely falling elevator standing on a scale. The scale would read zero because it is falling as fast as you are. However, you are not weightless and g is not zero (as you would find out on reaching the bottom of the elevator shaft).

Finally, there is another interesting aspect of the law of gravitation, in particular, as expressed in Eq. 3.6. Suppose you wanted to find your mass. How would you do this? Of course, you would weigh yourself and convert the weight in pounds to equivalent kilograms (or convert pounds to newtons and divide by g). Suppose you wanted to find the mass of Earth. That would take a pretty big scale, but Eq. 3.6 provides an easier way by computing it. The values of g and G have been measured, and the radius of Earth was determined about 250 B.C. (Chapter 15). Putting the values of g, G, and R_E into Eq. 3.6 and solving for M_E, the mass of Earth turns out to be about 6×10^{24} kg.

3.4 Newton's Third Law of Motion

Learning Goals:

To identify the action-reaction force pairs described by Newton's third law.

To explain how the third-law force pair differs from balanced forces.

Newton's third law of motion is sometimes called the law of action and reaction. A common statement of **Newton's third law of motion** is:

> For every action, there is an equal and opposite reaction.

In this statement, the words *action* and *reaction* refer to forces, and Newton's third law can also be stated as:

> For every force, there is an equal and opposite force.

What Newton recognized is that forces occur in pairs. Whenever one object exerts a force on a second object, the second object exerts an equal and opposite force on the first object. Newton gave an illustrative example: "If you press on a stone with your finger, the finger is also pressed upon by the stone." That is, you can't touch without being touched, so to speak.

An important thing to remember is that the two forces of Newton's third-law force pair act on *different objects*. Expressing the law mathematically,

$$F_1 = -F_2$$

where F_1 is the force exerted on object 1 by object 2,

and F_2 is the force exerted on object 2 by object 1.

The minus sign indicates that the forces are in opposite directions ($+F_1$ and $-F_2$). Either force may be called the action or reaction; this is an arbitrary designation. Using Newton's second law, we may write

$$m_1 a_1 = -m_2 a_2$$

From this equation we see that if m_2 is much larger than m_1, then a_1 is much larger than a_2.

Jet propulsion and rockets provide examples of Newton's third law in action. The exhaust gases accelerate out the back of the rocket, and the rocket accelerates in the opposite direction. The molecules of the exhaust gases have very small masses but large accelerations, whereas the rocket has a very large mass and accelerates relatively slowly (Fig. 3.15).

A common misconception is that the exhaust gases push against the launch pad to accelerate the rocket. This is nonsense. If this were true, there would be no space travel, since there is nothing to push against in space. The correct explanation is one of action (gases going out the back) and reaction (rocket propelled forward). The gases (or gas particles) exert a force on the rocket (many tiny

Figure 3.15 Newton's third law of motion in action.
The rockets and exhaust gases exert forces on each other and they are accelerated in opposite directions.

forces, but there are many gas particles), and the rocket exerts a force on the gas particles.

As another example of equal and opposite forces, but of differences in masses and accelerations, consider what occurs in dropping a book. As the book falls, it has a force acting on it (gravity) that causes it to accelerate. What is the opposite reaction? The equal and opposite force is the force of the book's gravitational attraction on Earth. Technically, Earth accelerates upward to meet the book! However, since Earth's mass is so huge compared to that of the book, the acceleration of Earth is minuscule and cannot be detected.

There are many other examples of action and reaction in everyday life. In fact, from Newton's third law it is clear that every time there is a force (action), there must also be a reaction force.

Understanding the difference between the force pair of Newton's third law and balanced forces is important. If a book weighing 15 N lies on a table, the book exerts a force of 15 N downward on the table (Fig. 3.16a). The table, in turn, exerts an upward force of 15 N on the book. This is the force pair of Newton's third law.

Now focus on the book alone (isolated in Fig. 3.16b). There are two forces acting on it. There is the 15-N force of gravity acting downward. Also, there is the 15-N upward force that the table exerts on the book. Note that these forces act on the *same* body and are not a third-law force pair. From Newton's second law, the forces acting on the book are balanced; and with no unbalanced force, the acceleration of the book is zero. (It is at rest on the table.)

If the book is dropped, it falls to the floor (Fig. 3.16c). As it falls, only one force acts on it—the 15-N force of gravity. By Newton's second law, this single, unbalanced force causes the book to accelerate downward.

Thus, Newton's third law relates two equal and opposite forces that act on two separate objects. Newton's second law concerns how forces acting on a single object or system cause an acceleration. If two forces acting on a single object are equal and opposite, then there will be no acceleration.

In some instances the reaction force of Newton's third-law force pair may not be immediately recognized, but it's there. As illustrated in Fig. 3.17, a convenient technique is to consider the forces in analogous situations. We don't usually think of a ceiling exerting a force on a suspended ball or a wall exerting a force on a car, but those forces are there.

3.5 Momentum and Impulse

Learning Goals:

To define linear momentum, impulse, and angular momentum.

To explain and give examples of each of the momentum conservation laws.

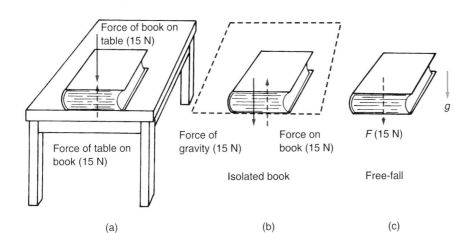

Figure 3.16 Action and reaction.
(a) Equal and opposite forces on *different* objects. The book exerts a force on the table, and the table exerts a force on the book. (b) Two equal and opposite forces act on the isolated book on the table if it is not moving. These are *not* the force pair of Newton's third law. Why? (c) In free-fall, only the force of gravity acts on the book, causing it to accelerate.

Linear Momentum

Stopping a speeding bullet is difficult because it has a high velocity. Stopping a slowly moving oil tanker is also difficult because it has a large mass. In general, the combination or product of mass and velocity is called the **linear momentum** (p):

$$p = mv \qquad (3.7)$$

linear momentum = mass × velocity

Linear momentum is a vector quantity, since velocity is a vector, and is in the same direction as the velocity. Both a speeding bullet and a slowly moving oil tanker have a large linear momentum. We can compare them by actually calculating the linear momentum of each. Indeed, the bullet may have more momentum than the tanker. (It is common to refer to *linear momentum* simply as *momentum*.)

Both mass and velocity are involved in momentum. A small car and a large truck both traveling at 50 km/h may have the same velocity, but the truck has more momentum because it has a much larger mass. If we have a system of two or more masses, the total linear momentum of the system is the vector sum of the linear momenta (plural of momentum) of the individual masses.

The linear momentum of a system is important because if there is no external unbalanced force, it is conserved; it is constant and does not change with time. This property makes it extremely important in analyzing the motions of various systems. The condition for the law of **conservation of linear momentum** can be stated as follows:

> The total momentum of an isolated system remains the same if no external, unbalanced force acts on the system.

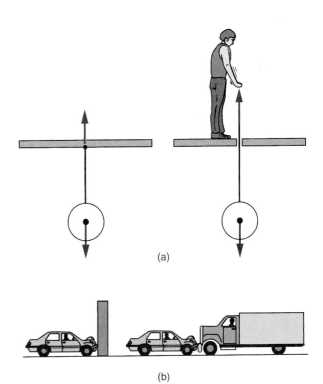

Figure 3.17 Find the forces.
(a) The ceiling exerts an upward force on the mass via the cord, as can be shown by letting someone support the mass. (b) The wall exerts a force on the car. A truck could be substituted for the wall with the same effect.

Figure 3.18 Conservation of linear momentum. With the system initially at rest, the total momentum of the system is zero. When the man jumps toward the shore (an internal force), the boat moves in the opposite direction so as to conserve linear momentum.

Suppose you are standing in a boat near the shore; you and the boat are the system (Fig. 3.18). Let the boat be stationary, so the total linear momentum of the system is zero. (No motion, no linear momentum.) If you jumped toward the shore, you would notice something immediately—namely, the boat would move in the opposite direction. The boat moves because the force you exerted in jumping is an *internal* force, so the total linear momentum of the system is conserved and remains zero. You have momentum in one direction, and to cancel this, the boat must have an equal and opposite momentum. Because momentum is a vector quantity, the momentum vectors can add to zero (as can force vectors).

We looked at jet propulsion or rockets in terms of Newton's third law. Jet propulsion can also be explained in terms of linear momentum. The interaction between the exhaust gas from burnt fuel and the rocket is an internal one, and involves internal forces. Hence, the linear momentum is conserved. The exhaust gases go out the back of the rocket with momentum in that direction, and the rocket goes in the opposite direction to conserve linear momentum. You probably have demonstrated this rocket effect by blowing up a balloon and letting it go. Without a guidance system, the balloon zigzags wildly, but the air comes out of the back and the balloon is "jet" propelled.

The molecules of a rocket's exhaust gases have small masses and large velocities, while the rocket itself has a large mass and a relatively small velocity. Also, there are many, many gas molecules. In a jet engine, the molecules of the exhaust gases have small masses but large velocities, while the rocket, or perhaps the jet aircraft, has a large mass and relatively small velocity. However, jet engines are very powerful, and the continuous internal force accelerates the rocket or jet aircraft to very high speeds. As you are no doubt aware, jet aircraft can "break the sound barrier" or fly faster than the speed of sound (about 340 m/s or 760 mi/h).

If you have ever flown on a large, commercial jet aircraft, you have experienced the force or thrust the jet engines develop as the plane accelerates down the runway for takeoff. The "revved-up" jet engines roar, with gas molecules going in one direction and the plane in the other.

But wait a moment. On landing and after touch-down, the pilot revs-up the engines, as can be heard by the roar. It would seem that the plane should accelerate down the runway, but no, you can feel a deceleration or a braking action. Why is this?

The mechanical brakes of a large aircraft would be burned up in trying to stop the plane alone—it has a lot of momentum. The major braking action is done by the engines through what is called *reverse thrust*. After landing, door arrangements in the engines are positioned to deflect the exhaust gases in the forward direction (the direction in which the plane is moving). The force on the plane is in the opposite direction of the gases, and this force provides the reverse thrust that brakes the plane. If you look closely, you can see small flaps on the outside of an engine that open so the exhaust gases may exit in the forward direction. (Also on landing, metal flaps called ground spoilers pop up on the top of the wings. The spoilers reduce the lift of the wings and provide some braking action.)

CONFIDENCE QUESTION 3.3

Suppose you are standing in the middle of a frozen lake whose smooth, icy surface is nearly frictionless. How could you reach the shore? (Keep in mind that friction is necessary for us to walk. Also, if you fell down and there were no friction, you wouldn't be able to get up.)

Impulse = $F\Delta t$

Figure 3.19 Increasing impulse.
The impulse of a collision may be increased by increasing the contact time Δt. This can be accomplished by "following through" in the case of hitting a ball. Greater impulse gives a greater initial velocity and a longer range or distance.

Impulse

Momentum, or the change in momentum, is related to another interesting quantity called *impulse*. Newton introduced the concept of momentum, calling it a "quantity of motion," and expressed his second law in terms of momentum:

$$F = ma = \frac{m\Delta v}{\Delta t} = \frac{\Delta(mv)}{\Delta t} = \frac{\Delta p}{\Delta t}$$

or force is equal to the time rate of change of momentum. (The mass m is assumed to be constant.) Rearranging, we have

$$F\Delta t = \Delta p \qquad (3.8)$$

(in words)   impulse = change in momentum

where F is the average (or constant) force.

As noted under the equation, the quantity $F\Delta t$ is called **impulse**. The impulse is equal to the product of the applied force times the time interval the force is applied. The force involved in an impulse is usually not constant, but varies with time. Most commonly we think of an impulse force as being associated with a collision, as in billiards when a ball is hit with a cue stick or when a golf club hits a golf ball. Here, an impulse force is applied for only a short time (small Δt). This change in t represents the contact or collision time.

In the case of driving a golf ball as far as one can, the applied force is maximum. So as to increase the impulse, the golfer increases the contact time by "following through" (Fig. 3.19). With $F\Delta t$ being larger, so is Δp, and $\Delta p = m\Delta v = mv_f - mv_o = mv_f$ (with $v_o = 0$, at rest). Hence, by the golfer following through, the velocity immediately after collision is greater, and the ball has a greater range (see Section 2.5).

You have probably learned to manipulate the F and Δt parts of impulse yourself. Have you ever caught a fast-moving, hard ball with your arms extended, as illustrated in Fig. 3.20a? It "stings" your hands. The change in momentum has a constant value of $\Delta p = mv_o$, since the ball comes to a stop ($v_f = 0$). Then,

$$F\Delta t = mv_o$$

where the small Δt indicates a small contact time, giving a large, stinging impulse force.

You quickly learn to move your hands backwards when catching the ball, or to manipulate the impulse (Fig. 3.20b). Doing this, you increase the contact time and reduce the impulse force, since the product is still equal to mv_o (assuming the same initial velocity). In symbol form,

$$F\Delta t = mv_o$$

Another important application of impulse manipulation is in the principle of the automobile air bag. See the chapter Highlight.

Angular Momentum

Another important quantity that Newton found to be conserved is angular momentum. *Angular momentum* arises when objects have angular or rotational motion about an axis of rotation, such as when an object is in uniform circular motion. The magnitude of the angular momentum is given by the product

$$L = mvr \qquad (3.9)$$

(a)

Figure 3.20 Adjusting the impulse.
(a) The change in momentum in catching a ball is constant (mv_o). If the ball is stopped quickly (small Δt), then the impulse force is large and the ball "stings" the hands when caught.

(b)

(b) By increasing Δt by moving the hands as the ball is caught, the impulse force is reduced and catching is made more enjoyable.

where m is the mass, v is the instantaneous velocity, and r is the distance of the object from the axis of rotation.

For example, in swinging a ball on a string in a horizontal path around your head, the axis of rotation would be through your hand.

The linear momentum of a system can be changed by the introduction of an external, unbalanced force. Similarly, the angular momentum of a system can be changed by an external, unbalanced torque (pronounced "tork"). A **torque** is the rotational analog of a force, and gives rise to rotational motion. It is a twisting effect caused by one or more forces. For example, in ● Fig. 3.21 a torque on a steering wheel is caused by two equal and opposite forces acting on different parts of the steering wheel. The forces produce a torque or twist, and cause the steering wheel to turn or rotate.

Angular momentum may also be conserved. The condition for the law of **conservation of angular momentum** is:

> The angular momentum of an object remains constant if there is no external, unbalanced torque acting on it.

[That is,]

angular momentum = angular momentum
at one time   at another time

or, in symbol form:

$$L_1 = L_2$$

or

$$mv_1r_1 = mv_2r_2$$

As an example, consider a planet going around the Sun as illustrated in ● Fig. 3.22. The angular momentum, mvr, of the planet is conserved or remains constant because the mutual gravitational attraction is internal to the system. As the planet gets closer to the Sun, r decreases, so the speed v increases. For this reason, a planet moves faster when it is closer to the Sun. (Although approximated as circles, planetary orbits are slightly elliptical. Planets do move at different speeds in different parts of their orbits, as do comets in elliptical orbits about the Sun.)

Another example of the conservation of angular momentum is illustrated in ● Fig. 3.23. Ice skaters use this principle to spin faster on the ice. The skater extends both arms and perhaps one leg in obtaining a slow rotation in a sweeping manner. Then, the arms and leg are drawn inward and a rapid angular velocity (spin) is gained because of

3.5 Momentum and Impulse 63

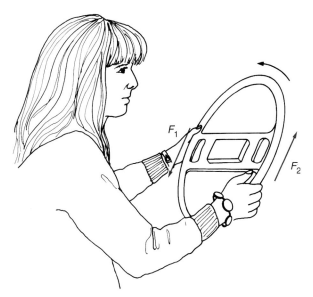

Figure 3.21 Torque.
A torque is a twisting action that produces rotational motion. This is analogous to a force producing linear motion.

(a)

(b)

the decrease in the average radial distance of the mass. [Note that if angular momentum ($L = mvr$) is a constant, then whenever r becomes smaller, v must become larger.]

The conservation of angular momentum explains why helicopters have more than one rotor. With only one rotor, after lifting off, the body of the helicopter would rotate in the direction opposite that of the rotor in order to conserve angular momentum. To prevent rotation, large helicopters

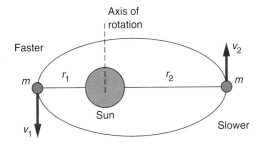

Figure 3.22 Angular momentum.
The angular momentum of a planet going around the Sun is given by mv_1r_1 and mv_2r_2 at different points in the orbit. Because the angular momentum is conserved, $mv_1r_1 = mv_2r_2$. As the planet comes closer to the Sun in its elliptical orbit, the radial distance r decreases, so v must increase. Similarly, the speed v decreases as r increases. Thus, the planet moves fastest when closest to the Sun and slowest when farthest away from the Sun.

Figure 3.23 Conservation of angular momentum.
(a) The skater starts his spin with arms outstretched. (b) When the arms are drawn inward and the average radial distance of the mass decreases, the angular velocity (spin) increases to conserve angular momentum.

HIGHLIGHT

Impulse and the Automobile Air Bag

A relatively new, major automobile safety feature is the air bag. As learned earlier, seatbelts restrain you so you don't follow along with Newton's first law when the car comes to a sudden stop. But, what is the principle underlying the action of the air bag?

When a car has a head-on collision with another vehicle, or hits an immovable object such as a tree, it stops almost instantaneously. If not buckled up, the driver could be seriously injured in hitting the steering wheel and column, and you can imagine what might happen to a passenger in the front seat. Even with seatbelts, the impact of a head-on collision may be such that seatbelts do not restrain you completely and injuries could occur. Enter the air bag. This balloon-like bag inflates automatically on hard impact and cushions the driver (Fig. 1). Passenger-side air bags are becoming more common, and backseat air bags are also available.

If we look at the air bag in terms of impulse, the bag increases the contact time in stopping a person, thereby reducing the impact force. Also, the impact force is spread over a large general area and not applied to just certain parts of the body, as in the case of seatbelts.

Being inquisitive, you might wonder what causes an air bag to inflate and what inflates it. Keep in mind that inflation must occur in a fraction of a second to do any good. (How much time would there be between the initial collision contact and a driver hitting the steering wheel column?) In current designs the air bag's inflation is initiated by an electronic sensing unit. This unit contains sensors that detect rapid decelerations, such as those that occur in a high-impact collision. The sensors have threshold settings so that normal hard braking does not activate them.

Sensing an impact, a control unit sends an electric current to an igniter in the air-bag system that sets off a chemical explosion. The gases (mostly nitrogen) rapidly inflate the thin nylon bag. The total process from sensing to complete inflation takes only about 25 thousandths of a second (0.025 s)!

The sensing unit is equipped with its own electrical power source. In a front-end collision, a car's battery and alternator are among the first to go. The currently installed automobile air bags offer protection for only front-end collisions, in which the car's occupants are thrown forward (more accurately, continue to travel forward—Newton's first law). No protection is offered for side-impact collisions. However, a foreign manufacturer has announced that its cars will be equipped with side air bags, as well as front air bags. Most cars will

Figure 1 Life-saving impulse. An illustration of an inflated air bag protecting a driver. The bag increases the contact time, thereby reducing the impact force (which would be quite large if the driver hit the steering column). The seat belt slows the forward motion of the body and adds extra protection. Always wear them.

probably have this extended protection in the future. But, always remember to buckle up—even if the vehicle is equipped with air bags. (Maybe we should make this Newton's fourth law of motion.)

have two oppositely rotating rotors (Fig. 3.24a). Smaller helicopters have a small "antitorque" rotor on the tail, which counteracts the rotation of the helicopter body.

CONFIDENCE QUESTION 3.4

Take a piece of chalk (or similar object) and tie a string about one-half meter long securely to the chalk. Hold the string in your hand, securing it firmly between the thumb and forefinger, and whirl the chalk around in a vertical circle. Upon extending the forefinger so the string winds itself around this finger, what happens to the motion of the chalk and why? (Try this simple experiment yourself.)

(a)   (b)

Figure 3.24 Conservation of angular momentum.
(a) Large helicopters have two overhead rotors that rotate in opposite directions so as to balance the angular momentum. (b) Small helicopters with one overhead rotor have an "antitorque" tail rotor to balance the angular momentum and prevent the rotation of the helicopter body.

Important Terms

These important terms are for review. After reading and studying the chapter, you should be able to define and/or explain each of them.

- force
- unbalanced (net) force
- Newton's first law of motion
- inertia
- mass
- Newton's second law of motion
- newton (unit)
- centripetal acceleration
- centripetal force
- Newton's law of universal gravitation
- Newton's third law of motion
- linear momentum (conservation of)
- impulse
- torque
- angular momentum (conservation of)

Important Equations

Newton's second law: $F = ma$

Weight: $w = mg$

Newton's law of gravitation: $F = \dfrac{Gm_1 m_2}{r^2}$

Linear momentum: $p = mv$

Impulse: $F\Delta t = \Delta p$

Angular momentum: $L = mvr$

Questions

3.1 Force and Newton's First Law of Motion

1. Which of the following describes the natural state of motion: (a) at a constant velocity, (b) in a circular orbit, (c) accelerated, (d) in free fall?
2. An unbalanced force is necessary for an object to be (a) at rest, (b) in motion with a constant velocity, (c) accelerated, (d) all of the preceding.
3. Inertia (a) sets objects in motion, (b) causes changes in motion, (c) is the result of a net force, (d) is directly proportional to the mass of an object.
4. What is meant when we say that a person has a lot of inertia?
5. An old party trick is to pull a tablecloth from beneath dishes and glasses on a table. Explain how this trick can be done without upsetting or pulling the dishes and glasses with the cloth.
6. Explain the principle of automobile seatbelts in terms of Newton's first law.

3.2 Newton's Second Law of Motion

7. A change in velocity (a) results from inertia, (b) requires an unbalanced force, (c) results from a zero net force, (d) is the natural state of motion.
8. The acceleration of an object is (a) inversely proportional to its mass, (b) directly proportional to the applied force, (c) resisted by inertia, (d) all of the preceding.
9. Give the type of proportionality between each of the following: (a) grades and study time, (b) grades and time spent watching the soaps, (c) grades and extracurricular activities, (d) grades and class attention, (e) grades and class absences.
10. Explain the relationships between (a) force and acceleration and (b) mass and acceleration.
11. Could the second law relationship between force and acceleration be derived from Newton's first law? Explain.
12. (a) Can an object be at rest if forces are being applied to it? Explain.
    (b) If no forces are acting on an object, can the object be in motion? Explain.
13. A 10-lb rock and a 1-lb rock are dropped simultaneously from the same height.
    (a) Some say that because the 10-lb rock has 10 times as much weight or force acting on it, it should reach the ground first. Do you agree? Explain.
    (b) Describe the situation if the rocks were dropped by an astronaut on the Moon.
14. Explain in terms of chapter concepts why the fellow in Fig. 3.25 is getting muddy.

3.3 Newton's Law of Gravitation

15. The acceleration due to gravity (a) is a universal constant, (b) is a fundamental property, (c) decreases with increasing altitude, (d) is different for different objects in free fall.
16. The constant G (a) is a very small quantity, (b) a force, (c) the same as g, (d) decreases with increasing altitude.
17. Discuss the quantities "big G" and "little g" and tell how they are different.
18. In an application of $F = Gm_1m_2/r^2$ to a mass on Earth's surface, what is r? What is r in terms of an Earth-orbiting satellite?
19. The gravitational force is said to have an infinite range. What does this mean?
20. How can the acceleration due to gravity be taken to be constant near Earth's surface, when g varies with altitude?
21. (a) Are astronauts in a spacecraft orbiting Earth "weightless"?
    (b) Is "zero g" possible? Explain.
22. If the Moon is "falling" or accelerating toward Earth, why doesn't it ever get here?

3.4 Newton's Third Law of Motion

23. The force pair of Newton's third law (a) can never produce an acceleration, (b) act on different objects, (c) cancel each other, (d) only exist for internal forces.
24. Two books, one on top of the other, lie on a table. How many forces act on the bottom book: (a) 1, (b) 2, (c) 3, (d) 4?
25. (a) Use Newton's third law to explain how a rocket blasts off.

Figure 3.25 A dirty business.
What doesn't this fellow know? See Question 14.

(b) When approaching the Moon for a landing, the spacecraft is turned around and the rockets fired (the rockets are called retrorockets when fired like this). What is the purpose of this maneuver?

26. If there is an equal and opposite reaction for every force, how can an object be accelerated when the vector sum of these forces is zero?
27. If an object is accelerating, does Newton's third law apply? Explain.
28. Explain the kick of a rifle or shotgun in terms of Newton's third law. Do the masses of the gun and the bullet or shot make a difference?

3.5 Momentum and Impulse

29. A change in linear momentum requires (a) an impulse, (b) a force, (c) an acceleration, (d) all of the preceding.
30. Angular momentum is conserved in the absence of (a) inertia, (b) gravity, (c) a net torque, (d) linear momentum.
31. (a) What are the units of linear momentum?
    (b) What are the units of impulse? Are they the same as those of linear momentum? Explain.
32. Does the conservation of linear momentum follow directly from Newton's first law of motion? Explain.
33. (a) What is the advantage of having a padded dashboard in a car?
    (b) Why are fragile objects often packed in plastic foam?
34. After diving from a high board, divers "tuck in" and spin before straightening out to cleave the water. How does tucking in accomplish spinning?
35. Would you take a ride in a helicopter with only one rotor? If you did, what would happen after lift-off?

Food for Thought

1. The abbreviation for a unit generally follows the unit name; for example, meter (m) and newton (N). Yet, the abbreviation for pound is lb. Why is this?
2. A falling object experiences air resistance. On what factors does air resistance depend?
3. Jupiter, the largest planet, has a surface gravity of (2.53)g, or 2.53 times Earth's surface acceleration due to gravity. Uranus has a surface gravity of (0.89)g, yet it too is larger than Earth. Explain how this can be.
4. What is necessary to put a satellite in an Earth orbit? (Discuss in terms of motions.)
5. Speculate on what would happen if we could "turn off" gravity for a while.

Exercises

3.2 Newton's Second Law of Motion

1. Determine the net force necessary to give an object with a mass of 4.0 kg an acceleration of 2.5 m/s^2.
   Answer: 10 N

2. A force of 2800 N is exerted on a rifle bullet with a mass of 0.014 kg. What will be the bullet's acceleration?

3. What is the unbalanced force on a car moving with a constant velocity of 20 m/s (45 mi/h) northward?

4. What is the weight in newtons of a 6-kg package of nails?
   Answer: 58.8 N

5. What is the force in newtons acting on a 6-kg package of nails that falls off a roof and is on its way to the ground?

6. (a) What is the mass in "kilos" and the weight in newtons of a person who weighs 132 lb.
   Answer: (a) 60.0 kg and 588 N
   (b) What is your weight in newtons?

7. Compute the weight in newtons of a 500-g package of cereal.
   Answer: 4.90 N

8. A bobsled team has a total mass of 310 kg and their bob sled a mass of 40 kg. What is the magnitude of the down-slope force that would accelerate the loaded sled at a rate of 6.0 m/s^2? (What supplies this force?)
   Answer: 2100 N

9. A 2.0-kg block of material sits fixed on top of a 3.0-kg block on a level surface.

(a) If the surface is frictionless and a horizontal force of 10 N is applied to the bottom block, what is the acceleration of the system?
(b) If there were a constant frictional force of 4.0 N between the bottom block and the surface, what would be the acceleration in this case?

3.3 Newton's Law of Gravitation

10. (a) Determine the weight on the Moon of a person whose weight on Earth is 180 lb.
    (b) What would be your weight on the Moon?
    Answer: (a) 30 lb

11. An astronaut on the Moon places a package on a scale and finds its weight to be 18 N.
    (a) What would be its weight on Earth?
    (b) What is the mass of the package on the Moon?

12. If the separation distance between two mass particles (a) tripled and (b) decreased by one-half, how would the force of gravity be affected in each case?
    Answer: (a) $\frac{1}{9}$ as great

13. What would be the acceleration due to gravity at an altitude of (a) R_E and (b) $2R_E$? (*Hint:* Use Eq. 3.5 with $r = R_E + h$ and refer to Eq. 3.6.)

3.5 Momentum and Impulse

14. Calculate the linear momentum for a truck with a mass of 15,000 kg that is traveling at 20 m/s (45 mi/h) eastward.
    Answer: (a) 300,000 kg-m/s, east

15. What is the linear momentum of a small car with a mass of 900 kg traveling at 30 m/s (67 mi/h) northward?

16. What is the magnitude of the angular momentum of a 1000-kg car going at a constant speed of 90 km/h (25 m/s) around a circular track with a radius of 100 m?
    Answer: 2,500,000 kg-m²/s

17. A comet goes around the Sun in an elliptical orbit. At the farthest point, 600 million miles from the Sun, it travels at a speed of 15,000 mi/h. How fast does it travel at its point of closest approach to the Sun at a distance of 100 million miles?
    Answer: 90,000 mi/h

Answers to Multiple-Choice Questions

1. a   7. b   15. c   23. b   29. d
2. c   8. d   16. a   24. c   30. c
3. d

Solutions to Confidence Questions

3.1 (a) With the applied forces equal and opposite, the net force would be zero and the block would remain at rest ($a = 0$).
(b) $F = ma = (2.0 \text{ kg})(-1.5 \text{ m/s}^2) = -3.0$ N. This net force to the left (negative direction) must be the result of an applied force (F_2) to the left that more than balances the applied force to the right ($F_1 = 6.0$ N). You should be able to prove to yourself that $F_2 = -9.0$ N.

3.2 The water is separated from the clothes by spinning action. Inside the rapidly rotating drum, the (centripetal) force exerted on the water by the clothes is not great enough to make it travel in a circular path, and the water flies off through the holes in the drum.

3.3 Your initial linear momentum is zero; since you are isolated, the linear momentum is conserved. If you throw something (a shoe, watch, or article of clothing), you will move in the opposite direction to conserve momentum. Newton's first law then applies and you move toward the shore with a constant velocity. (Also, you could blow puffs of air from your mouth—jet propulsion.)

3.4 There would be an increase in the speed v of the chalk as the distance r between the chalk and the finger becomes smaller. This is an example of the conservation of angular momentum ($L = mvr = $ a constant). (The chalk does experience torques, but the torques on opposite sides of the circle cancel each other much the same as equal and opposite applied forces cancel. The net torque is approximately zero.)

Chapter 4

Work and Energy

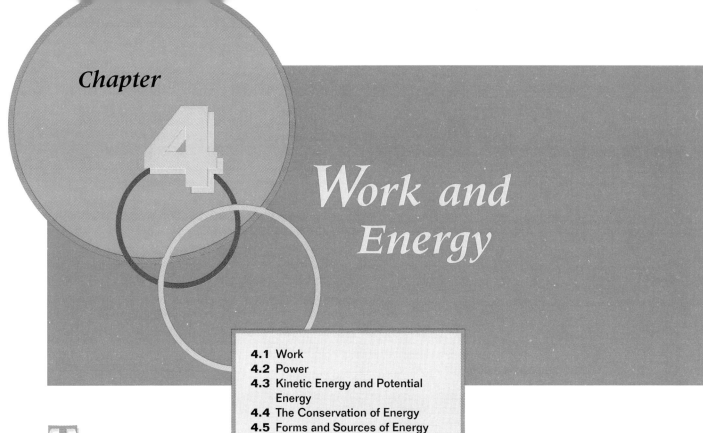

4.1 Work
4.2 Power
4.3 Kinetic Energy and Potential Energy
4.4 The Conservation of Energy
4.5 Forms and Sources of Energy

Highlight: The Conservation of Mass-Energy

The common meaning of the word *work* refers to performing (and hopefully accomplishing) some task or job. When work is done, energy is expended. Hence, work and energy are intimately related.

A person doing work becomes tired. He or she must obtain rest and food in order to continue working efficiently. We know that rest alone is not sufficient to keep a person going; thus food must serve as fuel to supply the necessary energy to do work.

The technical meaning of work is quite different from the common meaning. A student standing at rest and holding several books is doing no work, although he or she may feel tired after a time. Technically speaking, work is accomplished only when a force acts through a distance; that is, when motion is involved. What gives someone or something the ability to do work is energy.

Energy is sometimes referred to as stored work, and our mastery of energy has produced today's modern civilization. From the control of fire to the control of nuclear energy, we have advanced our standard of living through our ability to use and control the flow of energy. Energy takes many forms, such as thermal, chemical, electrical, radiant, nuclear, and gravitational energies. These forms, as will be learned, are classified in the more general categories of kinetic energy and potential energy.

Our main source of energy is the Sun, which each day radiates enormous amounts of energy into space. Only a small portion of this radiant energy is received by Earth. Solar energy supports nearly all life on our planet and provides us with the means for making nature work for Earth's inhabitants.

All of this, and our growing dependence on energy, underscores the importance of having a firm understanding of what work and energy are all about.

4.1 Work

Learning Goal:

To explain the physical definition of work, its units, and applications.

As has been noted, work involves a force. This is reasonable, since you know when you perform a task you generally apply a force. However, there's something more. When a force does work on an object, the force must act through a distance; that is, the object must move. A person just standing and holding a heavy load does no mechanical work, even though the person may become very tired. This may seem unfair, but as a physical concept we define the work done by a constant force as follows:

> The work W done by a constant force F acting on an object is the product of the magnitude of the force (or component of force) and the parallel distance d through which the object moves while the force is applied.

In equation form, we write

$$\text{work} = \text{force} \times \text{parallel distance}$$
$$W = Fd \qquad (4.1)$$

In this form it is easy to see that work involves motion. If $d = 0$, the object has not moved, and no work has been done.

Figures 4.1, 4.2, and 4.3 illustrate the concept of work, and the notion of a component of force and parallel distance. In ● Fig. 4.1, a force is being applied to the wall, but no work is done since the wall does not move through any distance. ● Figure 4.2 shows an object being moved through a distance d by an applied force F. Note that the force and the directed distance are parallel to each other, and the force F acts through the parallel distance d. The work done is then the product of the force and distance, $W = Fd$.

As shown in ● Fig. 4.3, when the force and distance are not parallel to one another, only a component or part of the force acts through a parallel distance. When the lawn mower is pushed on at an angle to the horizontal, only the component of the force parallel to the level lawn (horizontal component F_h) moves through a parallel distance and does work ($W = F_h d$). The vertical component of force (F_v) does no work, since this part of the force does not act through a distance. It only tends to push the lawn mower into the ground.

An important property of work is that it is a scalar quantity. Both the force and the parallel distance (actually displacement) have directions associated with them, but work does not. Work is expressed only as a magnitude (a number with proper units). It has no direction associated with it.

Because work is the product of a force and distance, the units of work are those of force times length ($W = Fd$). The SI units of work are then newton-meter (N-m, force × length). This unit combination is given the special name of **joule** (abbreviated J and pronounced "jool"), in honor of an early English scientist James Prescott Joule.

One joule is the amount of work done by a force of 1 N acting through a distance of 1 m.* Similarly the units of force times length in the British system are pound-foot. For some reason, however, the units are commonly listed in reverse, and we express work in **foot-pound** (ft-lb) units. A force of 1 lb acting through a distance of 1 ft does 1 foot-pound of work. The units of work are summarized in Table 4.1.

Work can be described or classified in various ways. When doing work, we apply a force, and we feel the other part of Newton's third-law force pair acting against us. So, we may say that we are doing work against something; for example, work against inertia, against gravity, or against friction.

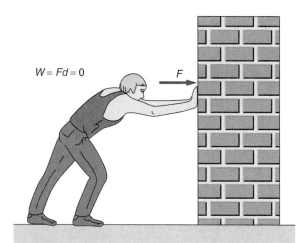

Figure 4.1 No work done.
A force is applied to the wall, but no work is done because there is no movement ($d = 0$).

---

*For the smaller metric force unit of the dyne, the amount of work done by a force of 1 dyn acting through a distance of 1 cm is 1 dyn-cm, or 1 *erg*. One erg is 10^{-7} J.

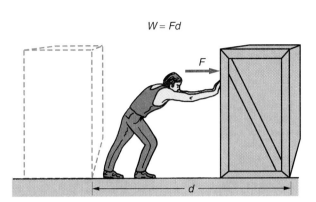

Figure 4.2 Work being done.
An applied force F acts through a parallel distance d. The work equals the force times the distance through which the force is applied when the force and the displacement are in the same direction.

Figure 4.3 Work and no work.
Only the horizontal component F_h of the applied force F does work, because it is parallel to the motion. The vertical component F_v does no work, because $d = 0$ in that direction.

(a) *Work against inertia.* We have learned that any object remains at rest unless acted upon by some external force; or if the object is in motion, it will remain in motion with a constant velocity unless acted upon by some external force (Newton's first law). We call this tendency *inertia*. So, when a force is applied to change the velocity (which may be zero), we say that work is done against inertia. For example, in Fig. 4.4 a mass is shown resting on a frictionless air table. If the velocity of the mass is increased from 0 to 1 m/s, a force must be applied through the distance d, and thus performs work—against inertia.

(b) *Work against gravity.* If we lift an object at a (slow) constant velocity, there is no net force on it, since the object is not accelerating. The weight (mg) of the object presses downward, and we push upward with an equal and opposite force. The distance parallel to the applied upward force is the height h that we lift the object (Fig. 4.5).

Thus, the work done against gravity is

$$W = Fd$$
$$W = mgh \qquad (4.2)$$

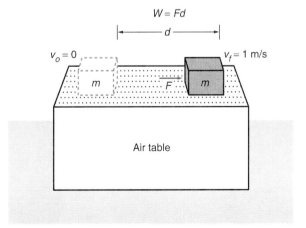

Figure 4.4 Work done against inertia.
A force F is applied to move the mass from rest to a velocity with a magnitude of 1 m/s. The air table provides an almost frictionless surface for the mass.

Table 4.1 Work (and Energy) Units

System	Force × Distance Units $W = F \times d$	Special or Common Name
SI	newton × meter (N-m)	joule (J)
British	pound × foot	foot-pound (ft-lb)

Figure 4.5 Work done against gravity.
A total lifting force at least equal and opposite to the weight mg must be applied to lift the weights. In lifting them a distance or height h, the work done is mgh.

Figure 4.6 Work and gravity.
Walking up stairs requires more work than walking on a level surface. When you go up a distance h, the work you do is increased by an amount mgh over what you would do if you walked on a level surface.

For example, if we lift a 4.0-kg mass a vertical distance of 1.5 m, the work done is

$$W = mgh$$
$$= (4.0 \text{ kg})(9.8 \text{ m/s}^2)(1.5 \text{ m}) = 59 \text{ J}$$

The concept of work against gravity helps to explain why it is much more tiring to walk up stairs than on a level surface. When you walk on level ground, you do little lifting of your body (Fig. 4.6). Of course, there is some work done by your muscles in lifting your legs as you walk, but this is not overly tiring. In fact, most healthy people can walk several miles on a level surface before having to rest. When you walk up stairs, however, you are lifting your whole body. You do an additional amount of work equal to mgh, where mg is your weight and h is the height you go up.

(c) *Work against friction.* Friction always opposes motion. Hence, to move something on a surface in a real situation, we have to apply a force. In doing so, work is done against friction. As illustrated in Fig 4.7a, if an object is moved at a constant velocity, the applied force F is equal and opposite to the frictional force f. The work done by the applied force against friction is $W = Fd = fd$.

When we walk, we must have friction between our feet and the floor; otherwise, we would slip. However, in this case, we do no work against friction (Fig. 4.7b), because the frictional force prevents the foot from moving (slipping). Hence, no motion of the foot, no work. Of course, other (muscle) forces do work, since when walking you are in motion. It is interesting to note in Fig. 4.7b, however, that to move (walk) forward, we actually exert a backward force on the floor.

4.2 Power

Learning Goals:

To define power and its units.

To contrast power and work.

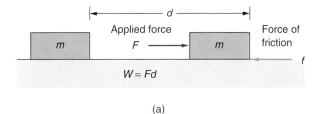

Figure 4.7 Work done against friction.
(a) The mass is being moved at a constant velocity by the force *F*, which is equal and opposite to the force of friction *f*. (b) When you are walking, there is friction between your feet and the floor. This is a static case; the frictional force prevents the foot from moving or slipping, and with no motion, there is no work against friction.

When a family moves into a second-floor apartment, a lot of work must be done to carry all the belongings up the stairs. In fact, each time the steps are climbed, the movers must carry not only the furniture, books, and so on, but also their own weight up the stairs. If the movers do all the work in three hours, they have not worked as rapidly as if the work had been done in two hours. The same amount of work would have to be done in each case, but there's something different—the rate at which work is done.

To express how fast work is done, we use power. **Power** is the time rate of doing work. Power is calculated by dividing the work done by the time required to do the work. In equation form, we have

$$\text{power} = \frac{\text{work}}{\text{time}}$$

$$P = \frac{W}{t} \quad (4.3)$$

Because work is the product of force and distance ($W = Fd$), power may also be written in terms of these quantities: $P = W/t = Fd/t$.

In the SI, work is measured in joules, and power has the units of joule/second (J/s). This unit is given the special name of **watt** (W), after James Watt, a Scottish engineer who developed an improved steam engine; and 1 W = 1 J/s. We rate our light bulbs in watts, for example, a 100-W light bulb. What this means is that such a bulb uses 100 joules of electrical energy each second (100 W = 100 J/s).

One should be careful not to confuse the meaning of the letter W. In the equation $P = W/t$, the *W* stands for "work." In the statement "$P = 25$ W," the W stands for "watts." Notice that when used as variables in equations, the letters are in italic; when used for units, the letters are in regular (roman) type.

In the British system, where the unit of work is the foot-pound, the units of power are foot-pound per second (ft-lb/s). However, a larger unit, the **horsepower** (hp), is commonly used to rate the power of motors and engines, and

1 horsepower (hp) = 550 ft-lb/s = 746 J

The horsepower unit was originated by James Watt, after whom the SI unit of power is named. In the 1700s, horses were used in coal mines to bring coal to the surface and to power water pumps. In trying to sell his improved steam engine to replace horses, Watt cleverly rated the engines in horsepower so as to compare the rates work would be done by an engine and an average horse.

The greater the power of an engine or motor, the faster it can do work—that is, it can do more work in a given time. For instance, a 2-hp motor can do twice as much work as a 1-hp motor in the same time. Or, the 2-hp motor can do the same amount of work as a 1-hp motor in half the time.

The following example shows how power is calculated in a typical situation.

EXAMPLE 4.1 Calculating Power

A force of 150 N is used to push a student's stalled motorcycle 10 m along a flat road in 20 s. Calculate the power, in watts.

Solution

Listing the given data and what we want to find in symbols form:

Given: $F = 150$ N
$d = 10$ m
$t = 20$ s
Find: P (power)

We note that the units are basic and all in the SI, so we can use Eq. 4.3 with the work expressed explicitly as Fd:

$$P = \frac{W}{t} = \frac{Fd}{t} = \frac{150 \text{ N} \times 10 \text{ m}}{20 \text{ s}} = 75 \text{ W}$$

Note that the units are consistent: N-m/s = J/s = W. Here, as stated before, all the units are in the SI, so we know the answer will have the SI unit of power—the watt.

CONFIDENCE QUESTION 4.1

A student carries her 5.0-kg book bag up two flights of stairs (vertical distance 8.0 m) in a time of 1.0 min. How much power was expended in carrying only the book bag?

In the next section, we will see that work produces a change in energy. Thus, power may be thought of as the energy produced or consumed divided by the time taken, and we can write

$$\text{power } (P) = \frac{\text{energy } (E) \text{ produced or consumed}}{\text{time } (t) \text{ taken}}$$

From this equation, we can see that

$$E = P \times t$$

This equation is useful in computing the amount of electrical energy consumed in the home. In particular, since energy is power times time ($P \times t$), it has units of watt-second (W-s). To give a bigger unit of energy, we could use a kilowatt (kW) and an hour (h), which would be a kilowatt-hour (kWh).

Keeping in mind that a kWh is an energy unit, when you pay the power company for electricity, in what units are you charged? If you check an electric bill, you will find that the bill is for so many kilowatt-hours (kWh). Hence, we really pay the power company for the amount of energy consumed (which we use to do work). Maybe we

Figure 4.8 Energy consumption.
Electrical energy is consumed as the motor of the grinder does work and turns the grinding wheel. Notice the flying sparks and that the operator is safely wearing a face shield rather than just goggles as the sign suggests. An electric *power* company is really charging us for *energy* in units of kilowatt-hours (kWh).

should call them the energy company. See Fig. 4.8. The following example illustrates how the energy consumed may be calculated.

EXAMPLE 4.2 Computing Energy Consumed

A 1.0-hp electrical motor runs for 10 hours. How much energy is consumed, in kWh?

Solution

Given: $P = 1.0$ hp
$t = 10$ h
Find: E (in kWh)

We note that the time is in hours, which is what we want. However, the power needs to be converted to kW. We have that 1 hp = 746 W, so converting to kW, we have

1.0 hp = 746 W $\left(\dfrac{1 \text{ kW}}{1000 \text{ W}}\right)$ = 0.746 kW = 0.75 kW

(rounding off).

Then, using the equation

$E = P \times t$
$= 0.75 \text{ kW} \times 10 \text{ h} = 7.5 \text{ kWh}$

This answer is the amount of electrical energy consumed while the motor is running or doing work.

We often complain about our electric bills. In the United States the cost of electricity ranges from about 7¢ to 14¢ per kWh, depending on the location. So to run the motor for 10 h at a rate of 10¢/kWh costs 75¢. That's pretty inexpensive for 10 hours of work output. (Electrical energy is discussed further in Chapter 8.)

4.3 Kinetic Energy and Potential Energy

Learning Goals:

To explain the relationship of work and energy.

To define and distinguish between kinetic energy and potential energy.

When work is done on an object, what happens? When work is done against inertia, the object's speed is changed. When work is done against gravity, the object's height is changed; when it is done against friction, heat is produced, usually with an accompanying change in temperature. In all these examples, some physical quantity changes when work is done.

The concept of energy helps us unify all the possible changes when work is done. Basically, when work is done, there is a change in energy; the amount of work done is equal to the change in energy. But, just what is energy? You may find it somewhat difficult to define exactly. Like force, it is a concept, and easier to describe in terms of what it can do, rather than what it is.

Energy is one of the most fundamental concepts in science. We may describe it as a quantity possessed by an object or system (a group of objects). A common description is that **energy** is the ability to do work. That is, an object or system that possesses energy has the ability or capability to do work. From this, we see where the notion of energy being *stored work* arises. We say that when work is done by a system, the amount of energy of the system decreases. Conversely, when work is done on a system, the system gains energy.

Hence, we see that work is the process by which energy is transferred from one object to another. An object with energy can do work on another object and give it energy. The amount of energy expended is equal to the work done, although this might not be all gained by the other object. In all energy-transfer processes, some of the energy is normally lost to the surroundings, but the total amounts of work and energy are still equal. Therefore, it should not surprise you to learn that work and energy have the same units. In the SI, energy is measured in joules, as is work.

Work and energy are both scalar quantities. That is, they have no direction associated with them. Thus, various amounts of energy can be added and subtracted numerically. By contrast, combining forces can become complicated because of their vector directional nature. But because energy is a scalar quantity, it is easy to use in calculations.

Energy occurs in many forms. We will focus here on mechanical energy, which has two fundamental forms—kinetic and potential. Let's take a look at them.

Kinetic Energy

Kinetic energy is the energy a body possesses because of its motion, or simply, is the energy of motion. As we learned, work requires motion, so when work causes a change in motion, there is a corresponding change in kinetic energy. The kinetic energy of an object can be written as

kinetic energy = $\frac{1}{2} \times$ mass $\times$ (velocity)^2

or in equation form

$$E_k = \tfrac{1}{2} m v^2 \qquad (4.4)$$

Figure 4.9 Work and energy.
The work necessary to increase the velocity of a mass is equal to the increase in the kinetic energy. (We assume no energy is lost.)

This equation gives the amount of kinetic energy an object with mass m has when traveling with a velocity v,* or the change in the kinetic energy when the object is accelerated from rest ($E_k = 0$) to a velocity v. It would be equal to the amount of work done on the object if all the work became kinetic energy. Also, the kinetic energy of an object is equal to the amount of work that would have to be done to bring it to rest.

As an example of doing work to create kinetic energy, consider a pitcher throwing a baseball (● Fig. 4.9). The amount of work needed to accelerate a baseball from rest to a speed v is just equal to the baseball's kinetic energy, $\frac{1}{2}mv^2$.

Work done on a moving body changes the kinetic energy of the body. For this, we have the equation

work = change in kinetic energy
$$W = \Delta E_k = E_{k_2} - E_{k_1} = \tfrac{1}{2}mv_2^2 - \tfrac{1}{2}mv_1^2$$

This is an important equation. It shows the relationship between work and energy. Note that to find the change in kinetic energy, the value must be calculated for *each* instance and then subtracted. It is not correct to subtract the v's and then use the Δv to compute the change in kinetic energy.

---

*Although velocity is a vector, the product $\mathbf{v} \cdot \mathbf{v}$, or v^2, gives a scalar, so kinetic energy is a scalar quantity. Either the magnitude of the instantaneous velocity, or speed, may be used to determine kinetic energy.

CONFIDENCE QUESTION 4.2

The speed of a 1000-kg car traveling on a straight, level road increases from 15 m/s (34 mi/h) to 25 m/s (56 mi/h). What is the change in the kinetic energy of the car?

Because the kinetic energy is proportional to the square of the velocity, doubling the velocity will cause a fourfold increase in the kinetic energy. This relationship can be seen by forming the ratio $E_{k_2}/E_{k_1} = v_2^2/v_1^2 = (v_2/v_1)^2$, and with $v_2 = 2v_1$ or $v_2/v_1 = 2$, we have $E_{k_2}/E_{k_1} = (2)^2 = 4$, or $E_{k_2} = 4E_{k_1}$, or a factor of 4 increase.

EXAMPLE 4.3 Speed and Kinetic Energy

By what factor is the kinetic energy increased when the speed of a car is increased from 30 km/h (about 20 mi/h) to 50 km/h (about 30 mi/h)?

Solution

A factor implies a ratio of the kinetic energies; for example, an increase by a factor of 4 as was just seen. Thus,

Given: $v_1 = 30$ km/h
$v_2 = 50$ km/h
Find: E_{k_2}/E_{k_1} (factor)

From the previous discussion, we have

$$\frac{E_{k_2}}{E_{k_1}} = \left(\frac{v_2}{v_1}\right)^2$$
$$= \left(\frac{50 \text{ km/h}}{30 \text{ km/h}}\right)^2 = 2.8$$

or $E_{k_2} = (2.8)E_{k_1}$

Hence, the final kinetic energy is 2.8 times greater than the initial kinetic energy. Notice that the speed increases only by a factor of

$$\frac{v_2}{v_1} = \frac{50 \text{ km/h}}{30 \text{ km/h}} = 1.7$$

The braking distance of a car is generally defined as the distance a car travels once the brakes are applied for a quick stop. Significantly, an amount of work equal to the car's kinetic energy is needed to cause the stop, and this is equal to the

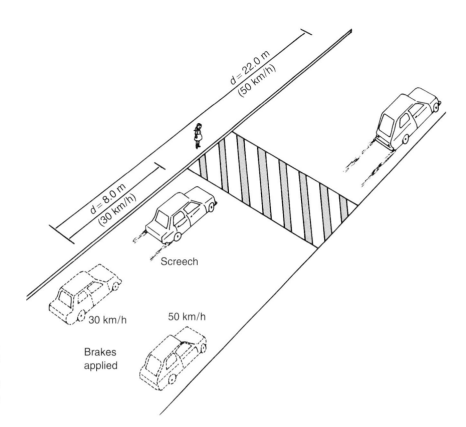

Figure 4.10 Energy and braking distance.
For a constant breaking force, if the braking distance for a car traveling 30 km/h is 8.0 m, then for a car traveling 50 km/h the braking distance is 2.8 times greater, or 22 m.

braking force times the braking distance. If the braking force is assumed to be constant, then the braking distance is directly proportional to the initial kinetic energy of the car, $E_{k_2} = Fd$ and $E_{k_1} \propto d$ (where $E_{k_2} = 0$ since the car is brought to a stop). Thus, from Example 4.3 we see that the braking distance for a car going 50 km/h is almost three times that for a car going 30 km/h.

This braking-distance concept explains why school zones have relatively low speed limits, commonly 20 mi/h, or about 30 km/h as in the preceding example. Suppose the braking distance for a car traveling at 30 km/h (about 20 mi/h) is 8.0 m. Then at 50 km/h (about 30 mi/h) the braking distance will be 2.8 × 8.0 m = 22 m (Fig. 4.10). As this simple example shows, if a driver exceeds the speed limit in a school zone, he or she may not be able to stop for a child darting into the street.

Potential Energy

Potential energy is the energy a body has because of its position or location, or simply, the energy of position. Work is done in changing the position of an object, and hence there is a change in energy.

For example, if a book with a mass of 1.0 kg at rest on the floor is lifted a height of 1.0 m to the top of a table, work is done. The work is done against gravity; and as shown previously (Eq. 4.2), this quantity is equal to $W = mgh$, where mg is the weight of the book, and h is the height lifted. The amount of work done in lifting the book is then

$$W = mgh$$
$$= (1.0 \text{ kg})(9.8 \text{ m/s}^2)(1.0 \text{ m}) = 9.8 \text{ J}$$

While work is being done, the energy of the book changes (increases), and the book on the table then has more energy and more capability to do work because of its height or position. This energy is called *gravitational potential energy*. If the book were allowed to fall back to the floor, it could do work, such as crushing something.

The gravitational potential energy E_p is equal to the work done, so we may write

gravitational potential energy = weight × height

or
$$E_p = mgh \qquad (4.5)$$

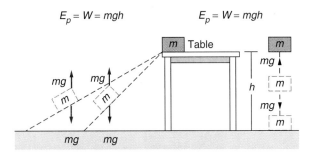

Figure 4.11 Independent of path.
The work done in placing a mass on the table is independent of the path taken and depends only on the initial and final positions, or the difference in height.

When work is done by or against gravity, the potential energy changes, and

$$\begin{aligned}\text{work} &= \text{change in potential energy} \\ &= E_{p_2} - E_{p_1} \\ &= mgh_2 - mgh_1 \\ &= mg(h_2 - h_1) \\ &= mg\Delta h\end{aligned}$$

An object has a particular value of potential energy at a particular height or position; for example, mgh_2 or mgh_1. Work is done when there is a change in position, so keep in mind that the h in Eq. 4.5 is really a height difference (Δh).

Also, the work done is independent of path and depends only on the initial and final positions, h_1 and h_2. That is, the value of the work done is the same no matter how you get between these points (● Fig. 4.11). The force necessary to lift an object from the floor to the tabletop height is the force needed to overcome gravity, which acts downward. Therefore, the applied force used to lift the object must be upward, no matter what path is taken to arrive at the tabletop. (Notice in Fig. 4.11 that h is really a height difference.)

The value of the gravitational potential energy at a particular position depends on the reference point; that is, on the zero point from which the height is measured. Near the surface of Earth where the acceleration due to gravity is relatively constant, the designation of a zero position or height is arbitrary. Any point will do. Using an arbitrary zero point for the gravitational potential energy is like using a point other than the zero mark on the end of a meterstick to measure a length (● Fig. 4.12). This practice gives rise to negative positions and displacements, such as + and − positions on the axis of a Cartesian graph (Fig. 4.12).

Height displacements may be positive or negative relative to the zero reference point, but notice that the height difference or change in potential energy between two positions is the same in both cases in the figure (and in any case). A negative (−) h gives a negative potential energy. Such a potential energy is somewhat like being in a potential "well," or like being in a hole or well shaft, since we usually designate $h = 0$ at Earth's surface.

There are other types of potential energy besides gravitational potential energy. For example, when a spring is compressed or stretched, work is done (against the spring force). The spring has potential energy, or more correctly, a change in potential energy as a result of a change in length (position). Also, work is done when a bowstring is drawn back. The bow and the bowstring bend and acquire potential energy. This potential energy is then capable of doing work on the arrow, which produces motion or kinetic energy. Note again how work is the process of transferring energy.

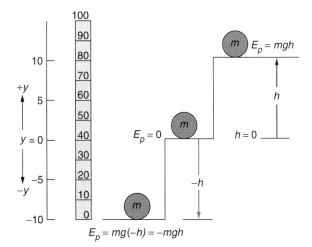

Figure 4.12 Reference point.
The reference point for measuring heights is arbitrary. For example, the zero reference point may be that on a Cartesian axis or at one end of a meterstick (left). For positions below a chosen zero reference such as on the Cartesian y-axis, the potential energy is negative because of the negative displacement. Referenced to the zero end of the meterstick, the potential energy values would be positive. The important idea is that the energy differences are the same for any reference.

4.4 The Conservation of Energy

Learning Goals:

To describe the principle of the conservation of energy.

To distinguish between the conservation of total energy and the conservation of mechanical energy.

To explain why such conservation laws are so important.

Although the meaning is the same, the law of conservation of energy (or simply, the conservation of energy) can be stated in a variety of ways. Examples are:

1. The total energy of an isolated system is conserved or remains constant.
2. Energy can be neither created nor destroyed.
3. In changing from one form to another, energy is always conserved.

For the **conservation of total energy,** we will use the statement:

The total energy of an isolated system is conserved (remains constant).

Thus, although energy may be changed from one form to another, energy is not lost from the system.

A *system* is something enclosed by boundaries, which may be real or imaginary, and *isolated* means that nothing from the outside affects the system (and vice versa). For example, the students in a classroom might be considered a system. They may move around in the room, but if the room were isolated and no one left or entered, then the number of students would be conserved ("conservation of students"). We sometimes say that the total energy of the universe is conserved. This is true, as the universe is the largest isolated system we can think of, and all the energy in the universe is in it somewhere, in some form.

In equation form we can state the conservation of total energy as

$$(\text{total energy})_{\text{time 1}} = (\text{total energy})_{\text{time 2}}$$

That is, the total energy does not change with time.

To simplify the understanding of the conservation of energy, we often use ideal systems in which the energy is only in two forms—kinetic energy and potential energy—which are together called **mechanical energy.** In this case, we talk about the **conservation of mechanical energy,** which may be written in equation form as

$$(E_k + E_p)_{t_1} = (E_k + E_p)_{t_2}$$

$$\left(\tfrac{1}{2}mv^2 + mgh\right)_{t_1} = \left(\tfrac{1}{2}mv^2 + mgh\right)_{t_2} \quad (4.6)$$

where the potential energy is taken to be gravitational (Eq. 4.5), and t_1 and t_2 indicate the energy at different times; that is, at initial and final energies.

In writing this equation, we are assuming an ideal or conservative system in which no energy is lost in the form of heat energy due to frictional effects. Actually, these heat effects are often important and are discussed in Chapter 5. However, in first introducing the conservation of energy, for convenience we consider frictional effects to be negligible.

EXAMPLE 4.4 Finding Potential and Kinetic Energies

A girl on a sled starts from rest down a snowy slope from a vertical height of 25 m above level ground (Fig. 4.13). What will be the potential and kinetic energies of the girl and the sled at the heights indicated in the figure if she and the sled have a total mass of 50 kg? (Assume no losses due to friction.)

Solution

In this case the girl and the sled are the system. According to the law of conservation of energy, the total energy (E_T) will be the same everywhere along the slope. When girl and sled are at rest at the top of the hill, the total energy of the system is all potential energy $(E_T = E_p)$, since $v = 0$ and $E_k = 0$.

At any point on the slope the potential energy will be $E_p = mgh$, where $mg = 50$ kg $\times$ 9.8 m/s^2, or

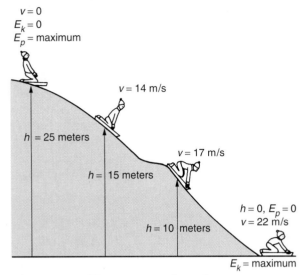

Figure 4.13 The conservation of energy.
The total energy of the girl and the sled at the top of the hill is potential energy, since they are at rest. As the sled and rider move down the hill, the potential energy decreases and the kinetic energy increases (height decreases and speed increases). If we assume no loss to friction, at the bottom of the hill the total energy is all kinetic with the velocity being a maximum, and the potential energy is zero. See Example 4.4.

Table 4.2 Energy Summary for Example 4.4

Height (m)	E_T (J)	E_p (J)	E_k (J)	v (m/s)
25	12,250	12,250	0	(0)
15	12,250	7,350	4,900	(14)
10	12,250	4,900	7,350	(17)
0	12,250	0	12,250	(22)

Note: Significant figures assumed.

490 N, and h is the height. We can find the potential energy at heights 25 m, 15 m, 10 m, and 0 m.

$$E_p = mgh$$

At $h = 25$ m: $E_p = 490$ N $\times$ 25 m $= 12{,}250$ J
At $h = 15$ m: $E_p = 490$ N $\times$ 15 m $= 7350$ J
At $h = 10$ m: $E_p = 490$ N $\times$ 10 m $= 4900$ J
At $h = 0$ m: $E_p = 490$ N $\times$ 0 m $= 0$ J

Because the total energy is conserved or constant, the kinetic energy (E_k) at any point can be found from the equation $E_T = E_k + E_p$. Therefore, $E_k = E_T - E_p = 12{,}250$ J $- E_p$, because we know that $E_T = E_p$ at the top of the hill. (Check and see if the equation gives $E_k = 0$ at the top of the hill.) A summary of the results is given in Table 4.2.

From Table 4.2 or Fig. 4.13 notice that as the sled travels down the hill (decreasing h), the potential energy of the sled and rider becomes less and the kinetic energy becomes greater (increasing v). That is, potential energy is converted into kinetic energy. (Note that the shape of the slope makes no difference in these calculations; only the height at a given point is important. Recall that gravitational potential energy is independent of path.)

At the bottom of the slope, where h is zero, all the energy of the system is then in the form of kinetic energy, and the velocity of the sled is a maximum. This can be seen mathematically using Eq. 4.6 with the total energies (E_T) at the top and bottom of the slope:

total energy at top = total energy at bottom

or
$$(E_T)_{\text{top}} = (E_T)_{\text{bottom}}$$

or
$$\left(\tfrac{1}{2}mv^2 + mgh\right)_{\text{top}} = \left(\tfrac{1}{2}mv^2 + mgh\right)_{\text{bottom}}$$

but, $v = 0$ at the top and $h = 0$ at the bottom, so we have

$$(0 + mgh)_{\text{top}} = \left(\tfrac{1}{2}mv^2 + 0\right)_{\text{bottom}}$$

or
$$mgh_{\text{top}} = \tfrac{1}{2}mv^2_{\text{bottom}}$$

Hence, the total energy was all potential at the top, and is all kinetic at the bottom. The kinetic energy at the bottom is equal to what the potential energy was at the top, or the total (mechanical) energy is conserved.

CONFIDENCE QUESTION 4.3

Two students throw identical snowballs as shown in Fig. 4.14, with both snowballs having the same initial speed v_o. Which ball will have the greater speed on striking the level ground at the bottom of the slope? Justify your answer using the energy considerations discussed earlier.

4.4 The Conservation of Energy

Figure 4.14 Away they go.
The identical snowballs are thrown from the same height and with the same initial speed, but in different directions. Which will have greater speed on striking the ground?

The simple pendulum (■ Fig. 4.15) also illustrates many of the features of the conservation of energy. A *simple pendulum* consists of a mass (called a *bob*) attached to a light string and supported so the bob can swing back and forth, or oscillate, freely. The pendulum is called simple because most of the pendulum's swinging mass is in the bob and the mass of the string is negligible.

Note that as the pendulum swings back and forth, the height of the bob changes, with $h = 0$ being taken at the point of lowest swing. With continuous increases and decreases in height between $h = 0$ and the maximum height h, there are continuous changes in the potential energy.

CONFIDENCE QUESTION 4.4

A simple pendulum as illustrated in Fig. 4.15 oscillates back and forth. Use the letter designations to identify the pendulum's position(s) described in the following list. (*Position(s)* is used to imply there may be more than one answer. Consider the pendulum to be ideal with no frictional losses.)

1. Position(s) of instantaneous rest _____
2. Position(s) of maximum velocity _____
3. Position(s) of maximum E_p _____
4. Position(s) of maximum E_k _____
5. Position(s) of minimum E_p _____
6. Position(s) of minimum E_k _____
7. Position(s) after which E_p increases _____
8. Position(s) after which E_k increases _____
9. Position(s) after which E_p decreases _____
10. Position(s) after which E_k decreases _____

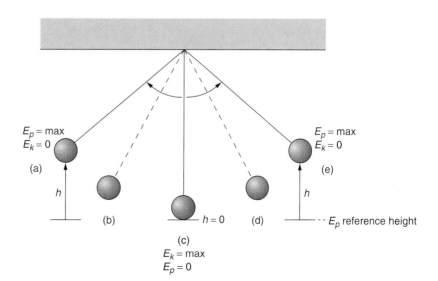

Figure 4.15 The simple pendulum and energy.
The pendulum swinging back and forth involves transformations between kinetic and potential energies. What can you say about the total mechanical energy of the pendulum at points (b) and (d)?

4.5 Forms and Sources of Energy

Learning Goals:

To identify some common forms of energy.

To compare the major sources of energy and the main sectors of energy consumption.

To list some "alternative" energy sources and explain their pros and cons.

We commonly talk about various *forms* of energy, such as chemical energy and electrical energy. Many forms of energy exist; however, the main unifying concept is the *conservation of energy*. We cannot create or destroy energy, but we can change it from one form to another.

For example, if we consider the conservation of energy to its fullest, we have to look at the examples in Section 4.4 in more detail. In actuality, we know that both the sled and the pendulum will come to a stop. Then, we might ask the question: Where did the energy go? Of course, friction is involved. In most practical situations, the kinetic and/or potential energy of objects eventually end up as heat. *Heat energy* will be studied at some length in Chapter 5, but for now let's say that heat is transferred energy that becomes associated with kinetic and potential energies on the molecular level. As things get hot, the motions of the atoms and molecules that make up different substances increase—they vibrate faster. So, there is more kinetic energy *and* potential energy (stored in the chemical bonds, which simplistically may be thought of as little springs).

We have already studied *gravitational energy*. We see its application in hydroelectric plants where the gravitational energy of water is used to generate electricity. Electricity may be described in terms of electrical force and *electrical energy* (Chapter 8). This energy is associated with the motions of electric charges or electric currents. It is electrical energy that runs numerous appliances and machines that do work for us.

Electrical forces hold or bond atoms and molecules together, and there is potential energy in these bonds. When fuel is burned (a chemical reaction), a rearrangement of electrons and atoms in the fuel occurs and energy is released. We refer to this as *chemical energy*. Our main fuels—wood, coal, petroleum, and natural gas—are indirectly the result of the Sun's energy. This *radiant energy* or light from the Sun is electromagnetic radiation. When electrically charged particles are accelerated, they "radiate" electromagnetic waves (Chapter 6). Visible light, radio waves, TV waves, and microwaves are examples of electromagnetic waves.

A more recent entry into the energy sweepstakes is *nuclear energy*. Nuclear energy is the source of the Sun's energy. Fundamental nuclear forces are involved, and the rearrangement of nuclear particles to form different nuclei results in the release of energy. In energy-releasing nuclear processes, we find that some of the mass of nuclei is converted to energy. Thus, we now consider mass to be a form of energy. See the chapter Highlight.

We have learned to control one type of nuclear process that releases energy (called *fission*, Chapter 10), and use it to generate electricity. This process is different from that of the Sun's (called fusion). The Sun gives off vast amounts of energy in burning huge amounts of nuclear fuel (hydrogen); however, it should be with us for at least another couple of billion years. Meanwhile on Earth, we have not yet learned to control nuclear fusion. Research goes on, but it is estimated that nuclear energy from this source will not be available until well into the twenty-first century. (See Chapter 10.)

As we go about our daily lives, each of us is constantly using and giving off energy in the form of body heat. The source of this energy is our food. An average adult radiates heat energy at about the same rate as a 100-W light bulb. This heat radiation explains why a crowded room can soon get hot. In winter, extra clothing helps to keep our body heat from escaping. In the summer, the evaporation of perspiration helps remove heat and cool our bodies.

The commercial sources of energy on a national scale are mainly coal, oil (petroleum), and natural gas. Figure 4.16 shows the percentage consumption of each of these resources over several decades. Nuclear and hydroelectric energies are the only other significant commercial sources. About one-half of our oil consumption comes from imported oil. The United States does have large re-

4.5 Forms and Sources of Energy 83

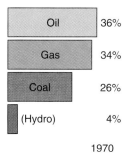

1970

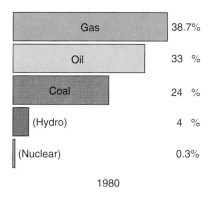

1980

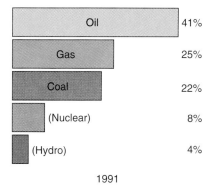

1991

Figure 4.16 Comparative fuel consumption.
The bar graphs show the relative percentages of fuel consumption in the United States for various years. Notice the increased relative use of nuclear energy and the changes in gas and oil use.

serves of coal, but there are some pollution problems with this resource (Chapter 19).

Perhaps you're wondering where all of this energy goes and who consumes it. Figure 4.17 gives a general breakdown of use by sector. We are no doubt becoming an increasingly electrical-dependent society. How do you think the sector breakdown would have looked for the other years in Fig. 4.16?

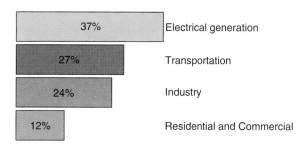

Figure 4.17 Energy consumption.
The bar graph shows the relative consumption of energy by various sectors of the economy (1991).

All of these forms of energy go into satisfying a growing demand. With only about 4.7% of the world's population, the United States accounts for approximately 25% of the world's annual energy consumption of fossil fuels—coal, oil, and natural gas. With increasing world population and development in the "third world" countries, there is an ever-increasing demand for energy. Where will it come from?

Of course, fossil fuels will continue to be used, but their increasing use gives rise to pollution and environmental concerns (Chapter 19). Research is being done on so-called *alternative fuels* and energy sources, which would be nonpolluting supplements to our energy supply. In closing this chapter, let's consider a few of these.

Hydropower is widely used to produce electricity (Fig. 4.18). We would like to increase hydropower production because falling water is nonpolluting. However, most of the best sites for dams have already been developed. Also, the damming of rivers usually results in the loss of agricultural land and alters ecosystems.

Because of our large agriculture capabilities, we can produce large amounts of corn from which ethanol (an alcohol) can be made. Mixed with gasoline, we have "gasohol," a fuel that can be used to run our cars. Ethanol has been advertised as reducing air pollution when mixed and burned with gasoline. Some pollutants are reduced, but others are added or increased. Also, there is the disposal of waste by-products from ethanol production to consider; unfortunately, twice as much fossil energy is used in ethanol production than the ethanol produces.

HIGHLIGHT

The Conservation of Mass-Energy

For many years there were independent conservation laws for energy and mass. They held that the total energy of an isolated system remains constant, as does the total mass of such a system. Both might change forms—for example, energy from kinetic to potential, and mass through a phase change or chemical reaction. Neither could be created nor destroyed, so to speak.

These concepts were combined and extended by Einstein's theory of relativity, which he developed in the early 1900s. This theory essentially predicted that *mass is a form of energy*, as given by the simple relationship.

energy = mass × (speed of light)^2

or $\qquad E = mc^2$

This mass-energy relationship has been verified experimentally many times—that is, the conversion of mass to energy *and* vice versa.

If there are mass-energy conversions, you may wonder why they are not readily observed in everyday life. Even when something like a piece of paper is burned, no mass loss is detected by delicate scales when the combustion gases are taken into account. What must be realized is that mass-energy conversion is appreciable only in special cases. The theory of relativity predicts that the mass of a particle will increase when it is accelerated (given energy) to a speed that is an appreciable fraction of the speed of light. Ordinary, everyday motions are nowhere near the speed of light (3.0×10^8 m/s, or about 186,000 mi/s). However, such speeds can be obtained in particle accelerators (Chapter 10), and the particles exhibit the behavior of increased mass. Also, in some special reactions energy is converted directly to mass, and mass converted directly to energy.

The mass-energy equivalence does not mean that we can convert just any mass to energy at will. If we could, our energy problems would be solved. For example, if one gram of mass could be converted completely to energy, according to the $E = mc^2$ equation this would be equivalent to the average electrical energy used monthly in 20,000 homes.

Mass-energy conversion does take place on a limited scale in nuclear reactions (Chapter 10). The conversion of mass to energy is used in nuclear reactors, which generate about 20% of the electricity used in the United States.

Also, a huge mass-energy conversion goes on continuously in our Sun (and other stars). To produce the Sun's luminosity, about 300 billion kilograms (600 million tons) of hydrogen must be converted to helium each second, with the simultaneous conversion of about 2 billion kilograms (4 million tons) of matter into radiant energy each second. (In a nuclear fusion reaction, only a small fraction of the reactant mass is converted directly into energy. The fusion of hydrogen nuclei produces helium nuclei. See Chapter 10.)

The conversion of 600 million tons of hydrogen per second may seem like a drain on the Sun; but don't worry. The Sun is so massive that at this rate only about $\frac{1}{1000}$ of 1% of its mass will be converted into energy in a billion years.

As a result of mass-energy conversion, we now generally refer to the conservation of mass-energy.

If you drive from Los Angeles to Palm Springs, California, in the desert you will suddenly come upon acres and acres of windmills (see cover photo). Wind power has been used for centuries. In fact, windmills for pumping water were once common on American farms. There have been significant advances in wind technology, and the wind turbines shown in the photo generate electricity di-

Figure 4.18 Hoover Dam.
The potential energy of dammed water can be used to do work. Here, Lake Meade stretches into the distance behind Hoover Dam, the major dam on the Colorado River.

Figure 4.19 Solar energy generating systems. (a) Distributed receivers use rows of trough mirrors that focus sunlight on a central-pipe receiver. (b) Central receivers use computerized, sun-tracking mirrors, called heliostats, to collect and concentrate the sunlight on a receiver on a central tower.

(a)

(b)

rectly. The wind is free and nonpolluting, but the availability of sites with sufficient wind (at least 20 km/h) limits widespread development of wind power. And, the wind does not blow continuously.

The Sun is our major source of energy, and we can put this to more use. Solar heating and cooling systems are used in homes and businesses. Also, there are technologies that focus on concentrating solar radiation for large-scale energy production. Distributed receivers use rows of trough mirrors that focus sunlight on a central-pipe receiver (Fig. 4.19a). Fluid in the pipe is heated and used to generate steam for electrical production. Central receivers use computerized, sun-tracking mirrors, called *heliostats,* to collect and concentrate the sunlight on a receiver on a central tower (Fig. 4.19b).

However, the most environmentally promising solar application is the photovoltaic cell (or photocell, for short). These cells convert sunlight directly to electricity. The light meter used in photography is a photocell. A problem with photocells has been efficiency, which was originally only about 4%. With advanced technology, the efficiency is now over 20%. Even so, electricity from photocells costs approximately 30¢ per kWh, which is not economically competitive with that produced from fossil fuels (on the order of 10 to 15¢ per kWh). Photocell arrays could be put on the roofs of buildings to reduce the need for additional land. However, electrical backup systems would be needed, since the photocells could be used only during the daylight hours. Also, clouds would reduce the efficiency.

Hopefully we will one day learn to mimic the Sun and produce energy by nuclear fusion. You'll learn more about this in Chapter 10.

Important Terms

These important terms are for review. After reading and studying the chapter, you should be able to define and/or explain each of them.

work
joule (J)
foot-pound (ft-lb)
power
watt (W)
horsepower (hp)
energy
kinetic energy
potential energy
conservation of total energy
mechanical energy (conservation of)

Important Equations

Work: $W = Fd$

Work against gravity: $W = mgh$

Power: $P = \dfrac{W}{t} = \dfrac{Fd}{t}$

Kinetic energy: $E_k = \tfrac{1}{2}mv^2$

Potential energy (gravitational): $E_p = mgh$

Conservation of mechanical energy:

$$\left(\tfrac{1}{2}mv^2 + mgh\right)_{t_1} = \left(\tfrac{1}{2}mv^2 + mgh\right)_{t_2}$$

Questions

4.1 Work

1. Work may be done against (a) inertia, (b) friction, (c) gravity, (d) all of the preceding.
2. The SI unit of work is the (a) ft-lb, (b) watt, (c) newton, (d) joule.
3. Do all forces do work? Explain.
4. Show that in terms of fundamental units, those of work are kg-m^2/s^2.
5. A weight lifter holds 900 N (about 200 lb) over his head. Is he doing any work on the weights? Did he do any work on the weights? Explain.
6. What is meant by doing work *against* (a) inertia, (b) gravity, and (c) friction?

4.2 Power

7. With an increase in power, (a) more work must be done, (b) a greater force must be applied, (c) a task will take longer to do, (d) there is an increase in the time rate of doing work.
8. Which one of the following is a unit of power: (a) joule-m, (b) N-m/s, (c) kg-m^2/s^2, (d) watt/s?
9. Persons A and B do the same job, but B takes a longer time. Who does the greater amount of work? Who is more powerful?
10. What is a watt in terms of fundamental SI units?
11. What does a greater power rating mean in terms of (a) the amount of work that can be done in a given time, and (b) how fast a given amount of work can be done?
12. What do we pay the electric power company for?

4.3 Kinetic Energy and Potential Energy

13. Car B is traveling twice as fast as car A, but car A has three times more mass than car B. Does (a) car A or (b) car B have the greater kinetic energy, or (c) do they both have the same kinetic energy?
14. Two identical cars, A and B, start from the same point and travel to the top of a hill by different routes. If the distance travelled by car A is 1.5 times that of car B, does (a) car A or (b) car B have the greater change in potential energy, or (c) do they both have the same change?
15. Define energy in general terms.
16. Explain how the braking distance of an automobile is related to kinetic energy.
17. Can you have energy without motion? Explain.
18. An object is said to have negative potential energy. If you do not like to work with negative numbers, can you change this value without moving the object? Explain.
19. Why do pole vaulters run fast before vaulting? How about high jumpers?

4.4 The Conservation of Energy

20. Which of the following is always conserved: (a) power, (b) mechanical energy, (c) kinetic energy, (d) total energy of the universe?
21. An object in an isolated system has kinetic energy and gravitational potential energy. If the speed of the object doubles, then the height of the object must decrease (a) twofold, (b) threefold, (c) fourfold, (d) by half as much.
22. What is meant by a *system*?
23. When is the total energy conserved, and what does this mean?
24. A mass suspended on a spring is pulled down and released. It oscillates up and down as illustrated in Fig. 4.20. Assuming the total energy to be conserved, use the letter designations in the figure to identify the spring's position(s) described in the following list.
    1. Position(s) of maximum velocity _____
    2. Position(s) of instantaneous rest _____
    3. Position(s) of minimum E_k _____
    4. Position(s) of minimum E_p _____
    5. Position(s) of maximum E_k _____
    6. Position(s) of maximum E_p _____
    7. Position(s) after which E_k decreases _____
    8. Position(s) after which E_p decreases _____
    9. Position(s) after which E_k increases _____
    10. Position(s) after which E_p increases _____

4.5 Forms and Sources of Energy

25. Which of the following fuels has the greatest consumption in the United States: (a) natural gas, (b) oil, (c) coal, (d) nuclear?
26. Which sector of the economy consumes the most energy: (a) electric utilities, (b) residential and commercial, (c) industry, (d) transportation?
27. On the average, how much energy do you radiate each second?
28. List five different general forms of energy (other than kinetic and potential energy).
29. Discuss some of the problems, limitations, and advantages of various energy sources.

FOOD FOR THOUGHT

1. We say we do work *against* something, and the work is taken to be positive (+). Is it possible for a force to do negative (−) work? Explain in terms of energy transfer. (*Hint:* Consider the work done by the force of gravity when an object is lifted, and when it falls to the ground.)
2. Suppose the professor asks you to go out and bring back a bucket of energy. Could you do it? Explain.
3. A book lies on a shelf three meters above the floor. A friend tells you that the book's total mechanical energy is zero. Is that possible? Explain.
4. In Confidence Question 4.3 (Fig. 4.14), would it make any difference if the snowballs had different masses? (*Hint:* Write the conservation of energy in equation form.)
5. Someone tells you that the Sun is indirectly the source of hydropower. Could this be true? Explain.

EXERCISES

4.1 Work

1. A worker pushes with a force of 50 N against an upright piece of plywood on the back of a truck to keep it from falling over while the truck moves 2.0 m. How much work is done by the worker?
2. A 5.0-"kilo" bag of sugar is on the counter. (a) How much work is required to put the bag on a shelf a distance of 0.60 m above the counter? (b) What would be the work if this were done on a space station on the Moon?

   Answer: (b) 4.9 J
3. A 4.0-kg box sits on a tabletop. How much work is required to move the box 0.75 m along the table if the force of friction is 20 N while the box is moving?

   Answer: 15 J
4. A man pushes a lawn mower on a level lawn with a force of 80 N. If 40% of this force is directed vertically downward, how much work is done by the man in pushing the mower a distance of 10 m?

   Answer: 480 J

4.2 Power

5. If the man in Exercise 4 pushes the mower 10 m in 45 s, how much power was expended?
6. A student weighing 600 N climbs a stairway (vertical height of 4.0 m) in 20 s.
   (a) How much work was done?
   (b) What is the power output of the student?
   Answer: (b) 120 W
7. How many kilowatt-hours of energy are consumed by a 1200-W hair dryer that is run for 5.0 min?
   Answer: 0.10 kWh

4.3 Kinetic Energy and Potential Energy

8. (a) What is the kinetic energy of a 20-kg dog that is running with a speed of 9.0 m/s (about 20 mi/h)?
   (b) How much work would have to be done to stop the dog?
   Answer: (a) 810 J
9. (a) By what factor is the kinetic energy increased when the speed of a car is increased from 30 mi/h to 40 mi/h?
   (b) What is the factor of decrease in going from 40 mi/h back to 30 mi/h?
   Answer: (a) $1.8\,K_1$

10. What is the potential energy (relative to ground level) of a 3.00-kg object at the bottom of a well 10.0 m deep? Explain the sign of the answer.
    Answer: -294 J
11. How much work is required to lift a 3.00-kg mass from the bottom of a 10.0-m-deep well?

4.4 The Conservation of Energy

12. A sled and rider with a combined weight of 700 N are at the top of a hill that is 5.0 m high.
    (a) What is their total energy at the top of the hill? (Use the bottom of the hill as a reference.)
    (b) Assuming there is no friction, what is the total energy halfway down the hill?
    Answer: (a) 3500 J
13. A 1.0-kg rock is dropped from a height of 6.0 m. At what height is the kinetic energy twice its potential energy?
    Answer: 2.0 m
14. How many joules of energy are given off each second in a crowded room with four 100-W light bulbs and 30 people?

Answers to Multiple-Choice Questions

1. d 7. d 13. b 20. d 25. b
2. d 8. b 14. c 21. c 26. a

Solutions to Confidence Questions

4.1 $P = \dfrac{W}{t} = \dfrac{Fd}{t} = \dfrac{mgh}{t} = \dfrac{(5.0\text{ kg})(9.8\text{ m/s}^2)(8.0\text{ m})}{60\text{ s}} = 6.5\text{ W}$

4.2 $\Delta E_k = \tfrac{1}{2}mv_2^2 - \tfrac{1}{2}mv_1^2 = \tfrac{1}{2}(1000\text{ kg})(25\text{ m/s})^2 - \tfrac{1}{2}(1000\text{ kg})(15\text{ m/s})^2$
$= 312{,}500\text{ J} - 112{,}500\text{ J} = 200{,}000\text{ J}$

4.3 Both will have the same speed. Initially, both have the same kinetic energy and potential energy. Then, just before hitting the ground ($h = 0$), they will have the same kinetic energy and speed.

4.4 1) a and e   2) c   3) a and e   4) c   5) c
   6) a and e   7) c, and b and d on upswings
   8) a and e, and b and d on downswings
   9) a and e, and b and d on downswings
   10) c, and b and d on upswings

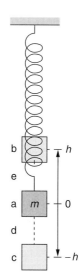

Figure 4.20 An oscillating mass on a spring.
See Question 24, page 87.

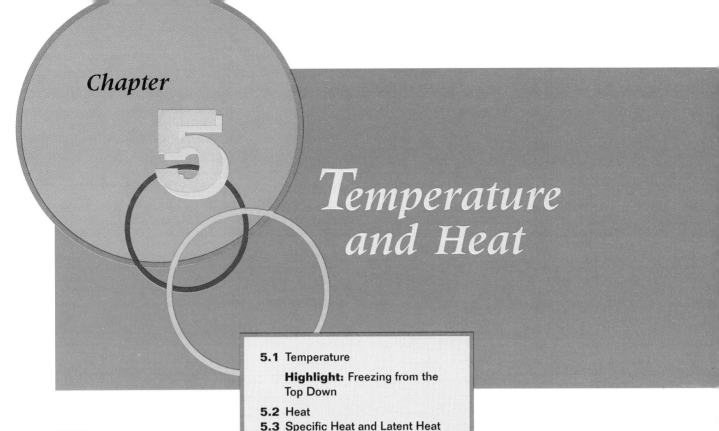

Chapter 5

Temperature and Heat

5.1 Temperature
Highlight: Freezing from the Top Down
5.2 Heat
5.3 Specific Heat and Latent Heat
5.4 Thermodynamics
5.5 Heat Transfer
5.6 Phases of Matter
Highlight: Kinetic Theory, the Ideal Gas Law, and Absolute Zero

The heating effects produced by fire, the sensation received when a piece of ice is held, and the warmth produced when hands are rubbed together are well known. Explaining what takes place in each of these cases, however, is not easy. Both *temperature* and *heat* are terms commonly used when referring to such cases involving hotness and coldness, often interchangeably. However, temperature and heat are not the same thing. They are related, but have different and distinct meanings, as we shall see.

Count Rumford (1753–1814), an American (born Benjamin Thompson), was one of the first to recognize the relationship between mechanical work and heat. In 1789 he published the results of an experiment in which he used a blunt boring tool to drill a cannon barrel filled with water. The temperature of the water rose to the boiling point in two and one-half hours of drilling. This result convinced Rumford that large quantities of heat were produced by friction. He concluded that heat was a form of energy and appeared to be due to the motion of the drill.

Later, James Prescott Joule (1818–1889), an English scientist, determined the quantitative relationship between mechanical energy and heat. He also established the important concept of conservation of energy with respect to heat, and originated many of the basic ideas for the kinetic theory of gases.

The concepts of temperature and heat play an important role in our daily lives. We like our coffee hot and our ice cream cold. The temperature of our living and working quarters is carefully adjusted. The daily temperature reading is perhaps the most important part of a weather report. How hot or how cold it will be affects the clothes we wear and the plans we make.

We will learn in Chapter 19 how the Sun provides heat to our planet. The heat balance between various parts of Earth and its atmosphere gives rise to wind, rain, and other weather phenomena. The thermal pollution of rivers that can be caused by hot water from power plants is a concern because of the effect on the ecology of rivers. On a cosmic scale, the temperature of various stars gives clues to their ages and the origin of the universe.

What is temperature? What is heat? What causes heat? How is heat transferred? The answers to these questions will be examined in this chapter. Knowing the answers, we can explain many of the phenomena occurring around us.

5.1 Temperature

Learning Goals:

To define temperature.

To distinguish the common scales used to express temperature values or readings.

To be able to convert the temperature on one scale to another.

Temperature tells us whether something is hot or cold. In fact, we say that temperature is a relative measure of hotness or coldness. For example, if the water in one cup is at a higher temperature than the water in another cup, then we know that the water in the first cup is hotter. But this cup of hot water would be colder than a cup of water with an even higher temperature. Hot and cold are *relative* terms.

On the atomic level, we find that temperature depends on the kinetic energy or motion of atoms or molecules of a substance. [In general, some substances (elements) are made up of atoms, and others (compounds) are made up of molecules, which are combinations of atoms. See Chapter 11.] The atoms of all substances are in constant motion. This motion occurs even within solids. The atoms or molecules in solids are held together by interatomic forces, which are sometimes likened to springs holding the molecules together, but allowing movement. The atoms oscillate back and forth about their equilibrium positions.

In general, the greater the temperature of a substance, the greater the motion of its atoms. On this basis we say that **temperature** is a measure of the average kinetic energy of the atoms or molecules of a substance.

Temperature is a measure of hotness and coldness, but is not necessarily a measure of heat, which is generally defined as a form of energy. (More on this later.) A simple experiment will illustrate: Place ice cubes and water in a pan and measure the temperature with a thermometer after the mixture has come to equilibrium (constant temperature). You will find this to be the melting point of ice. Then, slowly apply heat to the pan and stir. As heat is applied, the ice melts. However, when we check the temperature of the stirred ice and water mixture again, we find that the temperature has remained the same. Even with continued heating (and stirring) there is no change in temperature until all the ice is melted. Obviously heat was added, but it didn't change the temperature. We see, then, that temperature does not necessarily measure heat.

As another example, suppose the air temperature of your classroom and the outdoor air temperature are the same. Certainly there is much more heat outside than in the classroom, since there is much more air.

Humans have temperature perception in the sense of touch. However, this perception is unreliable and may vary a great deal with different people. Our sense of touch does not allow us to measure temperature accurately or quantitatively. Nor would you want to use the sense of touch to measure very high or very low temperatures. The quantitative measure of temperature is accomplished through the use of a thermometer. A **thermometer** is an instrument that utilizes the physical properties of materials for the purpose of accurately determining temperature. The temperature-dependent material property most used to measure temperature is **thermal expansion.** Nearly all substances expand with increasing temperature (and contract with decreasing temperature).

The heat-induced change in the linear dimensions or volume of a substance is quite small, but the effects of thermal expansion can be made evident using special arrangements. For example, a bimetallic strip is made of pieces of two different metals bonded together (Fig. 5.1). When the strip is heated, one metal expands more than the other, and the strip bends toward the metal with the smaller thermal expansion. As illustrated in Fig. 5.1a, the strip can be calibrated with a measurement scale. Bimetallic strips in the form of a coil or helix are used for dial-type thermometers. A refrigerator-freezer thermometer is a common example (Fig. 5.1b).

The most common type of thermometer is the liquid-in-glass type, with which you are no doubt familiar. It consists of a glass bulb connected to a glass stem with a capillary bore that is sealed at the upper end. A liquid in the bulb (usually mercury, or

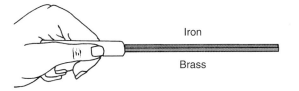
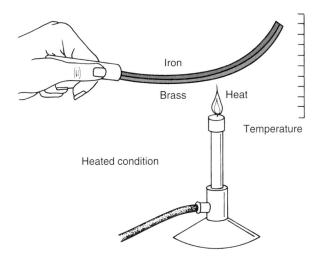

(a)

(b)

Figure 5.1 Bimetallic strip.
(a) Because of different degrees of thermal expansion, a bimetallic strip of two different metals bends when heated. The amount of deflection is proportional to the temperature, and a calibrated scale could be added for temperature readings. (b) Bimetallic coils are not only used in oven thermometers, but also in refrigerator-freezer thermometers as shown here. Note the indicator arrow is attached directly to the coil.

alcohol colored with a red dye to make it visible) expands on heating, and a column of liquid is forced up the capillary. The glass also expands slightly, but the liquid expands much more.

Thermometers are calibrated so that numerical values can be assigned to different temperatures. The calibration requires two reference points and a choice of units. By analogy, think of constructing a stick to measure length. You need two marks or reference points, and then you can divide the interval between the marks into sections or units. If the marks were for a standard meter,

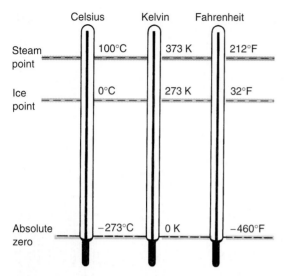

Figure 5.2 Temperature scales.
The common temperature scales are the Fahrenheit and Celsius scales, which have 180- and 100-degree intervals, respectively, between the ice and steam points. A third scale, the Kelvin scale, is used in scientific work and takes zero as the lower limit of temperature—absolute zero. The unit on the absolute or Kelvin scale is the kelvin (K). Absolute zero on the Celsius and Fahrenheit scales are −273°C and −460°F, respectively.

you could divide the interval into 100 sections, or 100 centimeters.

Two common reference points for a temperature scale are the ice (freezing) and steam (boiling) points of water. The **ice point** is the temperature of a mixture of ice and water at one atmosphere of pressure. The **steam point** is the temperature at which pure water boils at one atmosphere of pressure. Common names for the ice and steam points are the *freezing* and *boiling points*, respectively.

The two most common temperature scales are the Fahrenheit and Celsius scales (● Fig. 5.2). The **Fahrenheit scale** has an ice point of 32° (read "32 degrees") and a steam point of 212°. The interval between these points is evenly divided into 180 units. Each unit is called a *degree*. Thus a degree Fahrenheit, abbreviated °F, is 1/180 of the temperature change between the ice and steam points.*

The **Celsius scale** is based on an ice point of 0° and a steam point of 100°. There are 100 equal units or divisions between these points. A degree Celsius, which is abbreviated °C, is thus 1/100 of the temperature change between the ice and steam points. So, a degree Celsius is almost twice as large as a degree Fahrenheit (that is, 100 degrees Celsius and 180 degrees Fahrenheit for the same temperature interval means 1 Celsius degree equals 1.8 Fahrenheit degrees). At one time, the Celsius scale was commonly called the "centigrade" scale. Centigrade referred to the 100 degrees between the ice and steam points (*cent*, "one hundred," as in *century*, and *grade*, German for "degree"). The scale is now correctly referred to as the Celsius scale.†

There is no known upper limit of temperature; however, there is a lower limit. The lower limit occurs at about −273°C or −460°F, and is called **absolute zero**. Another temperature scale, called the **Kelvin scale,** has its zero temperature at this absolute limit (see Fig. 5.2).‡ This scale is sometimes called the *absolute temperature scale*. The unit of the Kelvin scale is the **kelvin,** abbreviated K (not °K), and it has the same interval or unit magnitude as a Celsius degree. Notice that since the Kelvin scale has absolute zero as its lowest reading, there can be no negative Kelvin or absolute temperatures.

Because the kelvin and degree Celsius are equal intervals, we can easily convert from the Celsius scale to the Kelvin scale. We simply add 273 to the Celsius temperature. In equation form, we have

$$T_K = T_C + 273 \qquad (5.1)$$

(Celsius T_C to Kelvin T_K)

For example, a temperature of 0°C equals 273 K (read "273 kelvins"), and a temperature of 27°C is equal to 300 K ($T_K = T_C + 273 = 27 + 273 = 300$ K). Of course, to change a Kelvin temperature to a Celsius temperature, just subtract 273 from the Kelvin value ($T_C = T_K - 273$).

---

*The Fahrenheit scale is named after Daniel Fahrenheit (1686–1736), a German scientist who developed one of the first mercury thermometers. In case you're wondering why he chose 32° and 212° for the ice and steam points on his scale, he didn't. Fahrenheit used two other reference points, and the ice and steam points of water came out to be 32° and 212° on his scale.

†Named after its originator Anders Celsius (1701–1744), a Swedish astronomer. Oddly enough, Celsius originally chose the ice point as 100° and the steam point as 0°, but these references were later changed.

‡Named after Lord Kelvin (William Thomson, 1824–1907), a British physicist who developed it.

Converting from Fahrenheit to Celsius and vice versa is also quite easy. The equations for these conversions are

$$T_F = \tfrac{9}{5} T_C + 32 \quad \text{(Celsius } T_C \text{ to Fahrenheit } T_F\text{)} \quad (5.2a)$$

or

$$= 1.8 T_C + 32$$

and

$$T_C = \tfrac{5}{9}(T_F - 32) \quad \text{(Fahrenheit } T_F \text{ to Celsius } T_C\text{)} \quad (5.2b)$$

or

$$= \frac{T_F - 32}{1.8}$$

These are both the same equation solved for the different temperatures. As examples, try converting 100°C and 32°F to their equivalent temperatures on the other scales. (You already know the answers.)

EXAMPLE 5.1 Changing Fahrenheit to Celsius, and Vice Versa.

What is the equivalent temperature of 5°F on the Celsius scale?

Solution

We are given that $T_F = 5°F$, and the conversion to the Celsius temperature T_C may be made directly using equation Eq. 5.2b:

$$T_C = \tfrac{5}{9}(T_F - 32) = \tfrac{5}{9}(5 - 32) = \tfrac{5}{9}(-27) = 15°C$$

To show that the equations are equivalent in Eq. 5.2, let's convert back the other way; that is, $T_C = -15°C$ to Fahrenheit T_F:

$$T_F = \tfrac{9}{5} T_C + 32 = \tfrac{9}{5}(-15) + 32 = -27 + 32 = 5°F$$

CONFIDENCE QUESTION 5.1

Room temperature is generally taken to be 68°F. What is this temperature on the Celsius and Kelvin scales?

Figure 5.3 Thermal expansion.
Expansion joints are built into bridges and elevated roadways to allow for the expansion of steel girders that is caused by varying temperatures.

As the Confidence Question shows, room temperature on the Fahrenheit and Celsius scales is quite different numerically. Do you suppose that the temperature reading on the Fahrenheit and Celsius scales might ever be equal? The answer is yes, −40°F = −40°C. (See Exercise 6.)

Thermal expansion is not only important in thermometers, but is also a major factor in the design and construction of items ranging from steel bridges and automobiles to watches and dental cements. The cracks in a highway are designed so that in summer the concrete will not buckle and crack as it expands due to heat. Expansion joints are designed into bridges for the same reason (● Fig. 5.3). Thermal expansion characteristics are also used to control the flow of water in car radiators and the flow of heat in homes through the operation of thermostats. An important environmental effect is discussed in the chapter Highlight.

HIGHLIGHT

Freezing from the Top Down

As a general rule, a substance expands when heated and contracts when cooled. An important exception to this rule is water. A volume of water does contract when cooled up to a point—about 4°C. When cooled from 4°C to 0°C, a volume of water expands. This behavior is illustrated in the graph in Fig. 1. Another way of looking at this is in terms of density ($\rho = m/V$, Chapter 1). As a volume of water is cooled to 4°C its density increases (volume decreases), and from 4°C to 0°C, the density decreases (volume increases). Hence we say that water has its maximum density at 4°C (actually 3.98°C).

This rather unique behavior occurs because of water's molecular structure. When water freezes, the water molecules go together in an open hexagonal (six-sided) structure as illustrated in Fig. 2, and as is evident in snowflakes (also six-sided). The open space in the molecular structure explains why ice is less dense than water and floats. Again, this is a rather unusual feature. The solids of most substances are more dense than their liquids.

That water has its maximum density at 4°C explains why lakes freeze at the top first. Most of the cooling takes place at the open surface. As the temperature of the top layer of water is cooled toward 4°C, the cooler water at the surface is more dense than the wa-

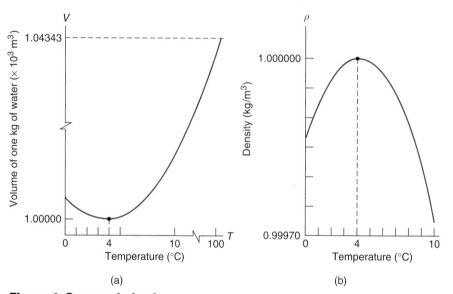

Figure 1 Strange behavior.
(a) Like most substances, the volume of a quantity of water decreases with decreasing temperature—but only down to 4°C. Below this temperature the volume increases slightly. (b) With a minimum volume at 4°C, the density of the water is a maximum at this temperature and decreases with lower temperatures.

ter below, and it sinks to the bottom. This sinking takes place until 4°C is reached. Below 4°C, however, the surface layer is less dense than the water below, and the cooler water remains on top where it freezes at 0°C.

So because of water's unique changes in density with temperature, lakes freeze from the top down. If the water thermally contracted and the density increased all the way to 0°C, the coldest layer would be on the bottom and freezing would begin there first. Think of what this would mean to aquatic life, let alone to ice skating.

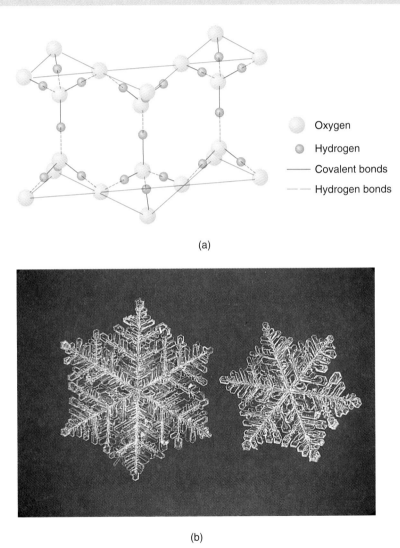

Figure 2 Structure of ice.
(a) An illustration of the open, hexagonal (six-sided) molecular structure in ice. (b) This hexagonal pattern is evident in snowflakes.

5.2 Heat

Learning Goals:

To define heat.

To compare the common units in which heat is measured.

Heat is a form of energy. Basically, we have two kinds of energy—kinetic energy and potential energy. It might be thought that heat is another kind of energy. However, if we were to examine the submicroscopic sources of heat energy, we would discover that the kinetic and potential energies of atoms and molecules are the ultimate sources of heat. Thus, heat can be viewed as either another form of energy or as the manifestation of atomic energy on a macroscopic scale. In any event, *heat is energy*.

We commonly say that an object contains heat, or that we add or remove heat. In the strictest sense, an object contains internal energy. So we define **heat** as the net energy transferred from or to an object as a result of a temperature difference, which results in a change in internal energy. When heat is added to an object, its temperature usually goes up. All the added heat becomes part of the object's internal energy. However, some goes into the kinetic energy of the atoms or molecules, and some into potential energy in the molecular bonds. The temperature of the object is proportional to the average kinetic energy of the object, so for a given amount of heat, increases in temperature will be different for different substances, as will be discussed in Section 5.3.

As a form of energy, heat has the unit of joule (J), as do all forms of energy. A traditional unit for measuring heat energy is the calorie.

A calorie (cal) is defined as the amount of heat necessary to raise the temperature of one gram of water one Celsius degree (technically from 14.5°C to 15.5°C) at normal atmospheric pressure.

In terms of the SI energy unit,

$$1 \text{ cal} = 4.186 \text{ J}$$

This relationship is called *the mechanical equivalent of heat,* and was determined by James Prescott Joule, after whom the joule unit is named. Heat measurements have been commonly made in calories rather than joules. Such measurements are easily converted to joules by using the mechanical equivalent of heat.

The calorie that we have defined is not the same as the one used when discussing diets and nutrition. A kilocalorie is 1000 calories as we have defined it; that is:

A kilocalorie (kcal) is the amount of heat necessary to raise the temperature of one kilogram of water one Celsius degree at normal atmospheric pressure.

A food calorie (Cal) is equal to one kilocalorie and is commonly written with a capital C to avoid confusion. We sometimes refer to a "big" Calorie and a "little" calorie.

$$1 \text{ food Calorie (1 kcal)} = 1000 \text{ calories}$$

$$1 \text{ food Calorie (1 kcal)} = 4186 \text{ J}$$

The Caloric value of food indicates the amount of intrinsic energy it contains and can supply when burned completely by the body.

The unit of heat in the British system is the British thermal unit, or Btu.

One Btu is the amount of heat necessary to raise the temperature of one pound of water one Fahrenheit degree at normal atmospheric pressure.

Air conditioners and heaters (furnaces) are commonly rated in Btu. For example, a window air conditioner may be rated at 25,000 Btu. This is really an hourly rate and indicates that the air conditioner can transfer or remove 25,000 Btu per hour. The Btu is a relatively large unit.

$$1 \text{ Btu} = 0.25 \text{ kcal} = 1055 \text{ J}$$

5.3 Specific Heat and Latent Heat

Learning Goals:

To distinguish between specific heat and latent heat.

To explain how these quantities are applied in heat measurements.

Specific Heat

Heat and temperature, although different, are intimately related. In general, when heat is added to a substance, its temperature increases. For example, suppose you added equal amounts of heat to equal masses of iron and aluminum. How do you think their temperatures would change? You might be surprised to find that if the temperature of the iron increased 100 C°, the corresponding temperature change in the aluminum would be 48 C°. You would have to add more than twice the amount of heat to the aluminum to get the same temperature change as for an equal mass of iron.*

This result reflects the fact that the internal forces of materials are different (different intermolecular "springs," so to speak). In aluminum more energy goes into internal potential energy rather than into kinetic energy, and potential energy is not manifested as a temperature change.

We express this intrinsic difference in terms of specific heat.

> The specific heat of a substance is the amount of heat necessary to raise the temperature of one kilogram of the substance one Celsius degree.

---

*A particular temperature such as $T = 48°C$ is written with °C (48 degrees Celsius), whereas a temperature interval such as $\Delta T = T_2 - T_1 = 100°C - 52°C = 48$ C°, is written with C° (48 Celsius degrees).

By definition, one kilocalorie is the amount of heat that raises the temperature of one kilogram of water one Celsius degree, so it follows that water has a specific heat of 1.0 kcal/kg-C°. For ice and steam, the specific heats are nearly the same, 0.50 kcal/kg-C°. Other substances require different amounts of heat to raise the temperature of one kilogram of a substance by one degree. That is, a specific heat is specific for a particular substance. The specific heats of a few common substances are given in Table 5.1.

Notice that the units of specific heat using kilocalories are kcal/kg-C°. The SI units are J/kg-C°. The greater the specific heat of a substance, the greater is the amount of heat required to raise the temperature of a unit mass. Put another way, the greater the specific heat, the greater is its capacity for heat (given equal masses and temperature change). In fact, the full technical name for specific heat is *specific heat capacity*.

Water has one of the highest specific heats, and so can store more heat energy for a given temperature change. Because of this, water is used in solar energy applications. Solar energy is collected during the day and used to heat water, which can store more energy than most liquids without getting as hot. At night, the warm water may be pumped through a home to heat it.

The amount of heat (H) necessary to change the temperature of a given amount of a substance depends on three factors: the mass (m), the specific

Table 5.1 Specific Heat

Substance	Specific Heat (20°C)	
	kcal/kg-C°	J/kg-C°
Aluminum	0.22	920
Copper	0.092	385
Glass	0.16	670
Human body (average)	0.83	3470
Ice	0.50	2100
Iron	0.105	440
Silver	0.056	230
Soil (average)	0.25	1050
Steam (at constant volume)	0.50	2100
Water	1.000	4186
Wood	0.40	1700

heat (designated by the letter c), and the temperature change (ΔT). In equation form, we write

$$H = mc\Delta T \quad (5.3)$$

(in words)

amount of heat to
change temperature = mass × specific heat × temperature change

This equation applies to a substance that does not undergo a phase change when heat is added (for example, in changing from a solid to a liquid—such as ice to water). As mentioned earlier and as will be considered in more detail shortly, when a substance changes phase, the temperature of the substance does not change when heat is added (or removed).

EXAMPLE 5.2 Calculating Heat Needed

How much heat does it take to heat a bathtub full of water (80 kg) from 12°C to 42°C (about 54°F to 108°F)? Assume no heat is lost to the surroundings.

Solution

As usual, we first write down what is given and what we are to find. The temperature change is calculated directly.

Given: m = 80 kg
c = 1.0 kcal/kg-C° (known)
$\Delta T = T_f - T_i$ = 42°C − 12°C = 30 C°
Find: H (heat)

We see that the units are OK and that the answer will come out in kilocalories. The units could later be changed to joules if so requested. It should be obvious that we use Eq. 5.3, and

$H = mc\Delta T$ = (80 kg)(1.0 kcal/kg-C°)(30 C°)
= 2400 kcal

Notice that this value is positive (+). This simply means that heat was added to the water. If heat had been removed (T_i greater than T_f), then $\Delta T = T_f - T_i$ would be negative, as would the heat (−H). The negative result would indicate that heat was removed.

You might be interested in the cost of heating the water. Recall that the electric company charges for energy in kilowatt-hours (kWh). Looking in the conversion table inside the back cover, we see that each kilocalorie corresponds to 0.00116 kWh, so 2400 kcal is

H = 2400 kcal × 0.00116 kWh/kcal = 2.8 kWh

Assuming the cost of electricity is 10¢ per kWh, the equivalent electrical energy to heat the water would cost 2.8 kWh × 10¢/kWh = 28¢. For a person taking one bath a day for 30 days, it would cost 30 × $0.28 = $8.40 to heat the water (assuming no losses).

CONFIDENCE QUESTION 5.2

Eleven kilocalories of heat are added to each of the following: (a) 1.0 kg of water, and (b) 1.0 kg of aluminum. What is the temperature change for each? (Neglect any losses.)

Latent Heat

Substances found in the environment are usually solids, liquids, or gases. These forms are called *phases* of matter. When heat is added (or removed) from a substance, it may undergo a change of phase. For example, when water is heated sufficiently, it changes to steam (or when enough heat is removed, it changes to ice).

As we know, water changes to steam at a temperature of 100°C (or 212°F) under normal atmospheric pressure. If we keep adding heat to a quantity of water at 100°C, it continues to boil with the conversion of liquid to gas, but the temperature of the water remains the same. Here is a case of adding heat to a substance without a resulting temperature change. Where then does the energy go?

On a molecular level, when a substances goes from a liquid to a gas, work must be done to break the molecular bonds and to separate the molecules. The molecules of a gas are relatively farther apart than the molecules of a liquid. Hence, during a phase change the heat energy must go into the work of separating the molecules, and not into increasing the average molecular kinetic energy, which would increase the temperature. (Phase changes are discussed more fully in Section 5.6.) The heat associated with a phase change is called **latent heat** (*latent* means "hidden").

Referring to Fig. 5.4, let's go through the process of heating a typical substance such as wa-

Figure 5.4 Graph of temperature versus heat energy for a typical substance.
If the typical substance were water, the solid, liquid, and gas phases would be ice, water, and steam, as indicated. Notice that during phase changes heat is added but the temperature does not change. See text for detailed description.

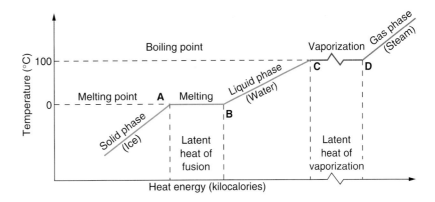

ter, and changing phases from solid to liquid to gas. In the lower left-hand corner, the substance is represented in the solid phase. As heat is applied, the temperature rises. When point A is reached, adding more heat does not change the temperature. Instead, the heat energy goes into doing the work of changing the solid into a liquid. The amount of heat necessary to change one kilogram of a solid into a liquid at the melting point is called the **latent heat of fusion** of the substance. In Fig. 5.4, the latent heat of fusion is simply the amount of heat necessary to go from A to B. This horizontal line indicates the melting point (temperature) on the vertical axis—0°C for water. (For lead, this would be 328°C.) In the region between A and B, the substance is both solid and liquid.

When point B is reached, the substance is all liquid. When point C is reached, adding more heat does not change the temperature. The added heat now goes into the work of changing the liquid into a gas. The amount of heat necessary to change one kilogram of a liquid into a gas at the boiling point is called the **latent heat of vaporization** of the substance. In Fig. 5.4, this is simply the amount of heat necessary to go from C to D. This horizontal line indicates the boiling-point temperature on the vertical axis—100°C for water (and 1744°C for lead). The substance is both liquid and gas in this region.

When point D is reached, the substance is all in the gaseous phase. Further heating again causes a rise in the temperature. Steam with a temperature above 100°C is commonly called *superheated steam*.

CONFIDENCE QUESTION 5.3

Why are the melting point and boiling point of lead so much higher than those of water?

In some instances, a substance can change directly from the solid phase to the gaseous phase. This change is called **sublimation.** Examples are dry ice (solid carbon dioxide), mothballs, and solid air fresheners.

From Fig. 5.4, we get a better understanding of the difference between temperature and heat. Also, we see that the temperature of a substance rises as heat is added only if the substance is not undergoing a change of phase.

When heat is added to ice at 0°C, the ice melts without changing temperature. The more ice there is, the more heat needed to melt it. The heat needed to change a solid completely to a liquid at the freezing point is just the mass of the substance times the latent heat of fusion. So we may write

heat needed to melt a quantity of substance (H) = mass (m) × latent heat of fusion (L_f)

or

$$H = mL_f \tag{5.4}$$

Similarly, if we want to change a liquid into a gas at the boiling point, the amount of heat needed can be written as

heat needed to vaporize a quantity of substance (H) = mass (m) × latent heat of vaporization (L_v)

or

$$H = mL_v \tag{5.5}$$

For water, the latent heats are

L_f = 80 kcal/kg = 80 cal/g = 335,000 J/kg

L_v = 540 kcal/kg = 540 cal/g = 2,260,000 J/kg

Notice that it takes almost seven times as much energy (L_v/L_f = 6.75) to change a kilogram of water to steam than it takes to change a kilogram of ice to water.

and

$H_{melt\ ice} = mL_f$

= (0.200 kg)(80 kcal/kg)

= 16 kcal

so

$H = H_{\Delta T} + H_{melt\ ice}$ = 16.5 kcal

CONFIDENCE QUESTION 5.4

Why does it take so much more heat to change equal amounts of water to steam than ice to water?

EXAMPLE 5.3 Calculating Heat Transfer

Calculate the total amount of heat to change 0.200 kg of ice at −5°C to water at 0°C. (Assume no heat losses.)

Solution

Reading the problem carefully, you should note that there are two parts to the procedure. First, the ice is warmed to the melting point and then additional heat changes the phase from solid to liquid. This means that the total heat (H_t) is composed of two quantities: H_t = H (of temperature change) + H (to melt ice). You cannot mix the two, because they are separate and distinct processes. We will use two equations, the specific heat of ice (c_{ice}) and the other using the latent heat of fusion (L_f), and both of these are known.

Given: m = 0.200 kg

$\Delta T = T_f - T_i = 0°C - (-5°C) = 5\ C°$

L_f = 80 kcal/kg (known)

c_{ice} = 0.50 kcal/kg-C° (known)

Find: $H_t = H_{\Delta T} + H_{melt\ ice}$ (total heat)

A quick check shows that the units are all right, with the answer coming out in kilocalories. We first compute the heat needed to warm the ice to 0°C ($H_{\Delta T}$) and then the heat needed to melt it ($H_{melt\ ice}$):

$H_{\Delta T} = mc\Delta T$

= (0.200 kg)(0.50 kcal/kg-C°)(5 C°)

= 0.5 kcal

CONFIDENCE QUESTION 5.5

Given 0.50 kg of water at room temperature (20°C), how much heat would be needed to bring the water to boiling and to convert it completely to steam? (Assume no heat losses.)

Changing phase is a two-way street. For example, just as 540 kcal of heat is needed to convert 1 kg of water to steam at 100°C, 540 kcal are given up in the reverse process (because of the conservation of energy). For this reason, burns from steam are generally more severe than those from boiling water. When steam condenses, it gives up 540 kcal/kg of latent or "hidden" heat. Steam condensing on skin in quantity, therefore, can cause extra-severe burns.

The freezing and boiling points of water are 0°C and 100°C, respectively, under normal atmospheric pressure (one atmosphere). As a final topic in this section, let's see how pressure affects these temperatures. It is found that the freezing point of water decreases slightly with increasing pressure. This is a small effect. It was once thought that ice skating was possible because the great pressure of the blades on the ice lowered the melting point and a thin film of water below the blades lubricated the gliding action. However, the lowering of the melting point is too small a change for this to be the explanation. Friction between a blade and the ice does contribute to melting, but this friction is relatively small for skates on ice (as compared to skis on snow). It has now been shown that there is normally a thin layer of liquid on the surface of a solid, even at temperatures below the solid's freezing point. This is called *surface melting*.

Pressure has a much greater effect on the boiling point of water than on the freezing point. As ex-

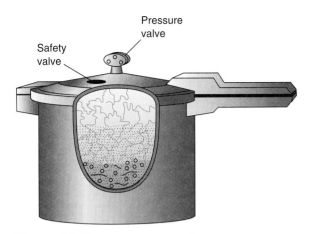

Figure 5.5 The pressure cooker.
Because of increased pressure in the cooker, the boiling point of water is raised and foods cook faster at higher temperatures.

pected, the boiling point of water increases with increasing pressure. Boiling is the process in which energetic molecules escape from the liquid. This energy is gained from heating. If the pressure were greater above the liquid, the molecules would have to have more energy, and the water would have be heated to a higher temperature for boiling to take place.

The increase of the boiling point of water with increasing pressure is the principle on which the pressure cooker is based (Fig. 5.5). Normally, when a heated liquid approaches the boiling point in an open container, pockets of energetic molecules form gas bubbles. When the pressure due to the molecular activity in the bubbles is great enough or greater than the pressure on the surface of the liquid, they rise and break the surface. We then say the liquid is boiling. In the closed container of a pressure cooker, the pressure above the liquid is increased, causing the boiling point to increase. The extra pressure is regulated by a pressure valve, which allows vapor to escape. (There is also a safety valve in the lid in case the pressure valve gets stuck.) Now the contents are cooking at a temperature greater than 100°C, so the cooking time is less. At mountain altitudes, the boiling point of water may be several degrees less than at sea level. For example, at the top of Pike's Peak (elevation 4300 m or 14,000 ft), the atmospheric pressure is reduced to where water boils at about 86°C rather than at 100°C. Pressure cookers come in handy there, if you want to eat on time.

5.4 Thermodynamics

Learning Goals:

To explain the laws of thermodynamics.

To distinguish between heat engines and heat pumps.

As the name implies, **thermodynamics** is generally concerned with the dynamics of heat (thermos)—that is, the production and flow of heat, and the conversion of heat into work. We use heat energy, either directly or indirectly, to do most of the work that is done in everyday life. The operation of heat engines, such as internal-combustion engines, and of refrigerators is based on the laws of thermodynamics. These laws are important because they give relationships among heat energy, work, and the directions in which the thermodynamic process may occur.

The First Law of Thermodynamics

Because one aspect of thermodynamics is concerned with heat-energy transfer, accounting for the energy involved in a process is important. As you may recall, the conservation of energy states that energy can neither be created nor destroyed. The first law of thermodynamics is simply the principle of the conservation of energy applied to thermodynamic processes. For example, consider the heating of an inflated balloon. As energy is added to the system (the balloon and the air inside), the temperature increases so some of the heat goes into the internal energy of the air. However, the balloon also expands and some of the energy goes into the work of expanding the balloon.

Keeping account of the energy, for the **first law of thermodynamics**, we may write in general,

work added to the system = increase in internal energy of the system + work done by the system

or

$$H = \Delta E_i + W$$

So, according to the first law of thermodynamics:

> **Heat added to a closed system goes into the internal energy of the system and/or into doing work.**

To help understand this concept, think of adding heat to a gas in a rigid container. In this case, the energy all goes into the internal energy of the system, with some corresponding increase in temperature. Because the container is rigid, unlike the balloon it does not expand, and no work is done ($W = 0$). By the equation form of the first law then, we have $H = \Delta E_i$, or the added heat causes a change only in the internal energy.

CONFIDENCE QUESTION 5.6

Suppose an inflated balloon is put into a refrigerator freezer and it shrinks. How does the first law of thermodynamics apply here?

Another good example of the first law is the heat engine. A **heat engine** is a device that converts heat energy into work. There are many different types of heat engines: gasoline engines on lawn mowers and in cars, diesel engines in trucks, and steam engines in old-time locomotives. They all operate on the same principle. There is heat input (for example, from the combustion of fuel) and some of the energy (but not all) goes into doing useful work.

In thermodynamics, we are not concerned with the components of an engine but, rather, its principle of operation. We may represent a heat engine schematically, as illustrated in Fig. 5.6. A heat engine is pictured as operating between a high-temperature reservoir and a low-temperature reservoir. These reservoirs are systems from which heat may be readily absorbed and to which heat may be readily expelled at given temperatures. (For an automobile, these would be a cylinder in which the air-fuel mixture is exploded and the environment to which heat is expelled.) In the process, the engine uses some of the input energy to do work. Notice that the widths of the heat and work paths in the figure are in keeping with the conservation of energy:

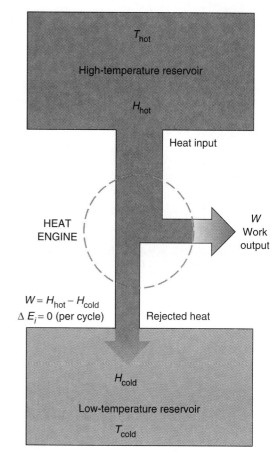

Figure 5.6 Schematic diagram for a heat engine.
A heat engine takes heat, H_{hot}, from a high-temperature reservoir, converts some of it to useful work, W, and rejects the remainder, H_{cold}, to a low-temperature reservoir.

$$\text{heat in} = \text{work} + \text{heat out}$$

or

$$\text{work} = \text{heat in} - \text{heat out}$$

or in terms of the figure labels, $W = H_{hot} - H_{cold}$. For practicality, a heat engine operates in a cycle for continuous energy output. For example, the cylinder process in an automobile continuously goes through one cycle after another.

The successful conversion of heat energy into work by a heat engine is expressed in terms of **thermal efficiency.** Similar to mechanical efficiency (work out/work in—what you get out for what you put in), thermal efficiency is the ratio of the work output to the heat energy input (usually expressed as a percentage). That is,

$$\text{thermal efficiency} = \frac{\text{work output}}{\text{heat input}} \; (\times 100\%)$$

or

$$\text{thermal eff} = \frac{W}{H_{\text{hot}}} \; (\times 100\%) \quad (5.6)$$

For example, if a heat engine absorbs 1000 J each cycle and rejects 400 J and does 600 J of work, then it has an efficiency of 600 J/1000 J = 0.60, or 60%. This efficiency is quite high. The overall efficiency of an automobile is on the order of 15%. That is, 85% of the energy from fuel combustion is lost to friction, and perhaps goes into doing nonessential work not associated with moving the car, for example, running a car air conditioner.

The Second Law of Thermodynamics

As we have seen, the first law of thermodynamics is concerned with the conservation of energy. As long as the energy check-sheet is balanced the first law is satisfied. Suppose that a heat engine operated so that all the heat input was converted to work. This engine doesn't violate the first law, but something is wrong. Such a condition would give a thermal efficiency of 1.0, or 100%, which has never been observed. In real situations, some energy is always lost.

The **second law of thermodynamics** solves this problem and tells us what can and cannot happen thermodynamically. This law can be stated in several ways. One statement of the second law is in terms of heat engines:

> **No heat engine operating in a cycle can convert heat energy completely into work.**

Another way of saying this is that no heat engine operating in a cycle can have a thermal efficiency of 100%. (In mechanical terms, a machine with 100% efficiency would be a perpetual motion machine—you could use the work output to run the machine itself. That is, with no losses, the machine would run forever or perpetually. Hence, in a sense, the second law precludes a perpetual motion machine.)

A heat engine must lose some heat, but what is the best or maximum efficiency that can be obtained? We know it isn't 100%. It can be shown that the maximum ideal efficiency is determined only by the high and low temperatures of the reservoirs (Fig. 5.6). This **ideal efficiency** is given by

$$\text{ideal eff} = \frac{T_{\text{hot}} - T_{\text{cold}}}{T_{\text{hot}}}$$

or

$$\text{ideal eff} = 1 - \frac{T_{\text{cold}}}{T_{\text{hot}}} \quad (5.7)$$

where T_{hot} and T_{cold} are the Kelvin (absolute) temperatures of the high-temperature reservoir and low-temperature reservoir, respectively.

Don't forget this Kelvin temperature requirement. It is a common mistake to use Celsius temperatures.

Keep in mind that the actual thermal efficiency of a heat engine will always be less than its ideal efficiency. The ideal efficiency sets an upper limit, which can only be approached and never achieved.

EXAMPLE 5.4 Calculating Ideal Efficiency

What is the ideal efficiency of a coal-fired steam power plant operating between temperatures of 300°C and 100°C?

Solution

First write down what is given and what we are to find:

Given: $T_{\text{hot}} = 300°C$

$T_{\text{cold}} = 100°C$

Find: ideal efficiency

You should be quick to note that the temperatures are given for the Celsius scale, and that the ideal efficiency is computed with Kelvin temperatures. So, using $T_K = T_C + 273$ (Eq. 5.1) to convert the given temperatures we get:

$T_{\text{hot}} = 300 + 273 = 573$ K
$T_{\text{cold}} = 100 + 273 = 373$ K

(Recall kelvins is abbreviated K, not °K.) Then, simply computing the ideal efficiency using Eq. 5.7 gives:

$$\text{ideal eff} = 1 - \frac{T_{\text{cold}}}{T_{\text{hot}}} \; (\times 100\%)$$

$$= 1 - \frac{373 \text{ K}}{573 \text{ K}} \; (\times 100\%) = 35\%$$

When other losses are taken into account, a typical power plant has an efficiency of about 32%.

Notice from Eq. 5.7 that the second law forbids a cold reservoir of $T_{\text{cold}} = 0$ K. That is, if we could have a reservoir with a temperature of absolute zero, then we could have an ideal efficiency of 100%. Since an ideal efficiency (or real efficiency) of 100% cannot be attained, this statement implies that a temperature of absolute zero cannot be attained. That is, thermodynamically:

> **It is impossible to attain a temperature of absolute zero.**

This result is sometimes called **the third law of thermodynamics**.

Experimentally, absolute zero has never been observed. In low-temperature experiments, scientists have come close to absolute zero—within about 0.000000001 (one-billionth) of a kelvin. As absolute zero is approached, it becomes increasingly difficult to lower the temperature of a material (pump heat from it). Presumably, an infinite amount of work would be required to reach the very bottom of the temperature scale.

Heat Pumps

Here's another statement of the second law:

> **It is impossible for heat to flow spontaneously from a colder body to a hotter body.**

This observation is well known. If a cold object and hot object are placed in contact, the hot object cools and the cold object warms up (Fig. 5.7). The reverse has never been observed to happen, even though there is nothing in the first law that indicates that it could not happen. Notice how the

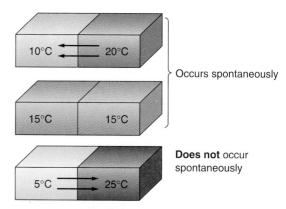

Figure 5.7 Heat flow.
When in thermal contact, heat flows spontaneously from a hotter object to a colder object until they are in thermal equilibrium, or at the same temperature. Heat never flows spontaneously from a colder object to a hotter object. That is, a cold object never gets colder when placed in thermal contact with a warm object. (Assume no heat loss from the system.)

second law tells in which "direction" the process occurs. If heat did flow spontaneously from a colder object to a hotter one, it would be like a ball rolling up a hill to a higher position of its own accord.

Of course, we can get a ball to roll up a hill by applying a force and doing work on it. Similarly, we can get heat to flow up the "temperature hill," so to speak, by doing work. This is the principle of the **heat pump**. A schematic diagram of a heat pump is shown in Fig. 5.8. Work input is required to "pump" heat energy from a low-temperature reservoir to a high-temperature reservoir. Essentially, it is the reverse process of a heat engine.

Refrigerators and air conditioners are examples of heat pumps. Heat is transferred from the inside volume of a refrigerator to the outside through the compressor doing work on a gas (and by the expenditure of electrical energy). The heat output transferred to the room (high-temperature reservoir) is equal to the sum of the heat taken from the refrigerator and the work input. Similarly, an air conditioner transfers heat from the inside of a home or car to the outside (high-temperature reservoir).

The heat pumps used for home heating and cooling have a descriptive name, in that they pump heat—one way in the summer and the other way in the winter. In the summer they operate as air conditioners, pumping heat from inside the house to

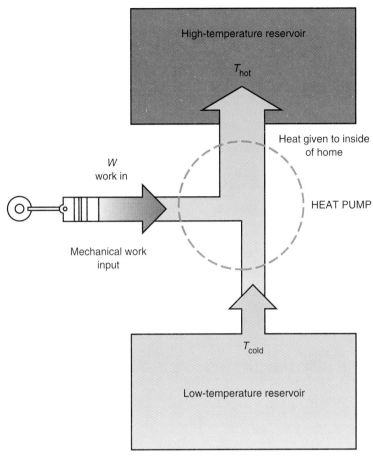

Figure 5.8 Schematic diagram for a heat pump.
Work is required to "pump" heat from a low-temperature reservoir to a high-temperature reservoir.

the outside environment. In the winter they extract heat from the outside air or from a water source, and pump it inside the home for heating. Heat pumps are used extensively in the South, where the climate is mild. In places with very cold winter months, auxiliary heating units (usually electric heaters) are used to supply extra heat.

Entropy

The second law of thermodynamics can also be expressed in terms of entropy. This is a mathematical concept, but we can say that **entropy** is a measure of the disorder of a system. When heat is added to an object or system, its entropy (disorder) increases because the added energy increases the disordered motion of the molecules. As a natural process takes place, the disorder increases. For example, when a solid melts, the molecules are freer to move in random motion in the liquid phase than in the solid phase. Likewise, when vaporization takes place, there is greater disorder and an increase in entropy.

In terms of entropy, the second law can be stated:

> **The entropy of an isolated system never decreases.**

Processes that are left to themselves tend to become more and more disordered, never the reverse. A student's dormitory room naturally becomes disordered, never the reverse. Of course, the room may be cleaned and items put in order, and the entropy of the room system decreases. But to put things back in order, someone must expend energy, with a greater entropy increase than the entropy decrease of the room. Because of this, we sometimes say that the total entropy of the universe increases in every natural process.

Here's another long-term implication of the second law. Heat naturally flows from a region of higher temperature to one of lower temperature. In terms of order, then, we might say that heat energy is more "orderly" when it is concentrated. When transferred naturally to a region of lower temperature, it is "spread out" or more "disorderly," and the entropy increases. Hence, the universe—the stars and galaxies—should eventually cool down to a final common temperature when the entropy of the universe has reached a maximum. This possible fate of the universe, billions of years from now, is sometimes referred to as the "heat death" of the universe.

Table 5.2 Thermal Conductivities of Some Common Substances

Substance	W/C°·m*	Substance	W/C°·m*
Silver	425	Glass	0.4
Copper	390	Wood	0.2
Iron	80	Cotton	0.08
Brick	3.5	Styrofoam	0.033
Floor tile	0.7	Air	0.026
Water	0.6	Vacuum	0

*Note that W/C° = (J/s)/C° is the rate of heat flow per temperature difference ($\Delta H/\Delta t$)/ΔT, where W represents watt. The length unit (m) arises from dimensional considerations of the conductor (area and thickness).

5.5 Heat Transfer

Learning Goals:

To define the three methods of heat transfer.

To identify common examples and applications of these methods.

Because heat is energy in transit, how the transfer is done is an important consideration. Heat transfer is accomplished by three methods: conduction, convection, and radiation.

Conduction is the transfer of heat by molecular activity. The kinetic energy of molecules is transferred from one molecule to another through collision or intermolecular interaction. For example, if you put a metal spoon in a hot liquid, heat is conducted to the handle of the spoon, as you would soon sense when holding the spoon.

How well a substance conducts heat depends on the molecular bonding. Solids are generally the best thermal conductors, with metals being some of the best. In addition to molecular interactions, there are a large number of "free" electrons (not permanently bound) in a metal that can move around. These electrons contribute significantly to heat transfer, or thermal conductivity (as well as to electrical currents, Chapter 8). The **thermal conductivity** of a substance is a measure of its ability to conduct heat. As listed in Table 5.2, metals have relative high thermal conductivities.

Liquids and gases, in general, are relatively poor thermal conductors. Liquids are better thermal conductors than gases because their molecules are closer together and interact more frequently. Gases are poor thermal conductors because their molecules are relatively far apart and conductive collisions do not occur as often. Substances that are poor thermal conductors are referred to as **thermal insulators.**

We make cooking pots and pans out of metals so that heat will be readily conducted to the foods inside. Pot holders, on the other hand, are made out of cloth, a poor thermal conductor or good thermal insulator, for obvious reasons (● Fig. 5.9). Many solids, such as cloth, wood, and plastic

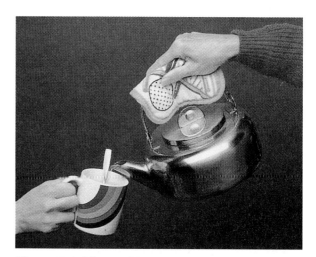

Figure 5.9 Thermal insulator.
Pot holders are made of cloth, a poor thermal conductor, thus preventing heat from being quickly conducted to the hand and causing a burn. Pots and pans, on the other hand, are made of metals so as to promote heat conduction.

5.5 Heat Transfer

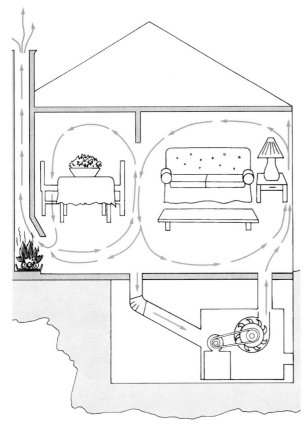

Figure 5.10 Convection cycles.
In a forced-air heating system, warm air is blown into a room. The warm air rises, cold air descends, and a convection cycle is set up that promotes heat distribution. Some of the cold air returns to the furnace for heating. Notice how a great deal of heat energy is lost up the chimney of a fireplace.

"lost its heat," it passes through a cold-air return on its way back to the furnace to be reheated and recirculated (Fig. 5.10).

The warm-air vents are usually in the floor, as are the cold-air ducts, but on opposite sides of the room. The warm air entering the room rises (being "lighter," really less dense, and therefore buoyant). As a result, cooler air sinks toward the floor, and convection cycles that promote even heating are set up in the room. Some of the cooler air returns to the furnace for reheating and recirculation. The transfer of heat by convection currents in a room is similar to the way heat is distributed in Earth's atmosphere (Chapter 19).

The transfer of heat by convection or conduction requires matter or a material medium for the process to take place. However, heat from the Sun comes to us through the void of space by means of electromagnetic waves. The process of transferring heat via electromagnetic waves is known as **radiation.** Electromagnetic waves (Chapter 6), such as light, radio and TV waves, and microwaves, carry energy. So we have energy transfer by transmitted radiation.

A common example of heat transfer by radiation is heat from an open fire or a fireplace. We can readily feel the warmth of the fire on our exposed skin. Yet, the air is a poor conductor; moreover, the air warmed by the fire is rising (up the chimney in a fireplace). Therefore, the only mechanism for appreciable heat transfer here is radiation (Fig. 5.11). All objects emit radiation, and the characteristics of the radiation depend on temperature. This property is used to make thermographs, which are pictures that show different temperatures as different colors. Using the emitted radiations, a "temperature map" of a house can be used to find where heat is escaping.

In general, we find that dark objects are good absorbers of radiation, whereas light-colored objects are poor absorbers. For this reason, we commonly wear light-colored clothing (white) in the summer so as to be cooler. In the winter, we generally wear dark-colored clothes to take advantage of the absorption of the reduced solar radiation.

An application using all three methods of heat transfer is the vacuum or Thermos bottle, which is used to keep liquids either hot or cold. (Fig. 5.12). How does the bottle know which to do? Actually, the three methods of heat transfer are incorporated in the design to prevent heat transfer and

foam (Styrofoam), are porous and have large numbers of air (gas) spaces that contribute to their poor conductivity. For example, Styrofoam coolers depend on this property, as does fiberglass insulation used in the walls and attics of our homes.

The transfer of heat by **convection** requires the mass or bulk movement of a substance from one position to another. Here, energy is transferred with the moving mass. The movement of hot air and hot water are examples of heat transfer by convection.

Most of our homes are heated by convection (movement of hot air). The air is heated in the furnace (or heat pump), then circulated throughout the house by way of metal ducts. When the air has

Figure 5.11 Radiation heat transfer.
Hands are warmed by radiation. Most of the heat is transferred upward by convection.

to keep cold things cold and hot things hot. The sealed, double-walled glass bottle is evacuated to produce a partial vacuum. That is, some of the air is taken (pumped) out from between the glass walls before this space is sealed. The nipple at the bottom of the bottle is the seal.

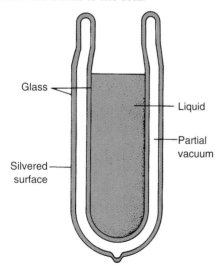

Figure 5.12 A vacuum bottle.
For a vacuum Thermos bottle, glass is a poor conductor, the partial vacuum between the glass walls prevents convection cycles therein, and, by reflection, the silvered surfaces prevent radiation losses. Thus, a vacuum bottle serves to keep hot foods hot and cold foods cold.

Glass is a relatively poor conductor of heat, and any heat conducted through a wall (from the outside in or the inside out) will find the partial vacuum an even greater thermal insulator. Also, in the partial vacuum, heat cannot be transferred very well from one glass wall to the other by convection, as it would if the space contained air. (When there is not much air, there is little convection.) Finally, the surfaces of the glass bottle are silvered to prevent heat transfer by radiation. These silvered walls serve as mirrored surfaces that reflect the radiation to keep it in or out. Thus, hot coffee (or a cold drink) in the bottle remains hot (or cold) for some time.

5.6 Phases of Matter

Learning Goals:

To analyze the molecular structure of the solid, liquid, and gaseous phases of matter.

To contrast what happens in the phase changes of matter.

As we saw in Section 5.3, the addition (or removal) of heat can cause a substance to change phase. The three common **phases of matter** are the solid

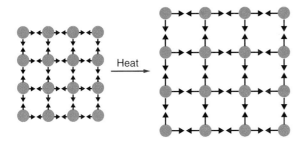

Figure 5.13 Crystalline lattice.
A schematic diagram of a crystal lattice of a solid in two dimensions. Heating causes the molecules to vibrate with greater amplitudes in the lattice, thus increasing the volume of the solid.

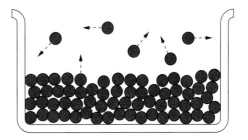

Figure 5.14 Liquid molecules.
A diagram illustrating the arrangement of molecules (small circles) in a liquid. The molecules are packed closely together, form only a slight lattice structure, and are relatively free to move. Some molecules may acquire enough energy to break free of the liquid. This process is called vaporization.

phase, the liquid phase, and the gaseous phase.* At normal room temperature and atmospheric pressure, a substance will be in the phase with which we are most familiar. For instance, at normal temperature and pressure, oxygen is a gas, water a liquid, and copper a solid. However, at other temperature and pressure combinations, a given substance can exist in any of the three phases.

The principal distinguishing features of solids, liquids, and gases can be generally understood if we look at the various phases from a molecular point of view. Most substances are made up of very small particles called molecules; that is, a molecule is the smallest division of a substance.

In many pure **solids,** the molecules are arranged in a specific three-dimensional manner. This orderly arrangement of molecules is called a *lattice*. Fig. 5.13 shows an example of a lattice in only two dimensions. The molecules (represented in the figure by small circles) are bound to each other by electrical forces.

Upon heating, the molecules gain kinetic energy and vibrate. The more heat that is added, the stronger the vibrations become. The vibrating molecules move farther apart, and as shown diagrammatically in Fig. 5.13, the solid expands.

When the melting point of a solid is reached, additional energy (the heat of fusion) causes the bonds that hold the molecules in place to break. As bonds break, holes are produced in the lattice, and nearby molecules can move toward the holes. As more and more holes are produced, the lattice becomes significantly distorted.

Figure 5.14 shows an arrangement of the molecules in a **liquid.** There are many "holes" in the liquid, which has little orderly arrangement. Molecules can easily move to new spots because there are so many holes. With the molecules relatively free to move, a liquid assumes the shape of its container. Upon the heating of a liquid, the individual molecules gain kinetic energy, and even more holes are produced as the liquid expands.

When the boiling point is reached, additional heat energy causes the molecular bonds to be completely broken. The *heat of vaporization* is the heat per kilogram necessary to free the molecules completely from each other. Because the electric forces holding different molecules together are quite strong, the heat of vaporization is fairly large. When the molecules are essentially free from each other, the substance is in the gaseous phase.

Evaporation is another, but relatively slow, process of changing from a liquid phase to a gaseous phase. The molecules in a liquid are in motion at different speeds. A faster-moving molecule near the surface of a liquid may have enough energy to escape the liquid and become part of the air. This process happens occasionally, and we say the liquid evaporates over a period of time. Evaporation is promoted with increasing liquid temperature. (Why?)

---

*States of matter is a term sometimes used; however, in physics a state of matter (or a system) refers to its particular characteristics, such as temperature and pressure. A phase of matter can have many states, so the term *phase* is preferred.

HIGHLIGHT

Kinetic Theory, the Ideal Gas Law, and Absolute Zero

Molecular kinetic theory views the molecules of a gas to be moving randomly at high speeds. The molecules collide with each other and the walls of the container. Each collision on a wall exerts a tiny force on the wall, and the many and continual collisions give rise to a pressure on the wall (Fig. 1). Pressure is defined as the force per unit area ($p = F/A$). The SI units of pressure are N/m², which is called a *pascal* (Pa). We commonly apply a force to a point, but in pressure, the force is spread over an area.

The pressure of a given quantity of gas depends on its volume and temperature. Suppose the temperature of a gas is kept constant (in what is called an *isothermal* process, *iso-* meaning same). And suppose that its volume is decreased by one-half (for example, in a piston-cylinder arrangement). The pressure would then increase—in fact, it would double. Because the gas occupies half the space, the molecules make twice as many collisions with the container walls as before, and hence exert twice the pressure. This inverse relationship between pressure (p) and volume (V) may be expressed as

$$p_1 V_1 = p_2 V_2$$

This expression is known as Boyle's law, after Robert Boyle (1627–1691), an English chemist who discovered it. The law applies to all gases at normal pressures.

Using statistical kinetic theory, one can show that the product of pV is proportional to the average kinetic energy of the gas molecules.

$pV \propto$ average kinetic energy of gas molecules

The greater the kinetic energy of the molecules, the faster they move and the more frequently they collide with the walls. The greater the number of collisions with the walls, the greater the pressure. Also, the greater the molecular energy, the greater is the ability of the molecules to do work and to increase the volume of the container (if not rigid). Note the units of the product of p and V are: $(N/m^2)(m^3) = N \cdot m = J$, the unit of work and energy.

The temperature (T) of a gas is a relative indication of the internal energy of a gas, so we may write

$$pV \propto T$$

Comparing the two proportionalities, you can see that temperature is a measure of the average kinetic energy of the gas molecules.

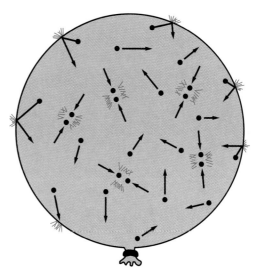

Figure 1 Model of a gas.
Gas molecules are far apart, on the average. They move randomly at high speeds, colliding with each other and the walls of the container. The average force of their collisions with the container walls causes pressure on the walls.

For dilute gases, it is found that pV/T = a constant, so we may write

$$\frac{p_1 V_1}{T_1} = \frac{p_2 V_2}{T_2}$$

where the subscripts indicate the values at different times. This relationship is known as the **ideal (or perfect) gas law**. The temperature in this equation is the Kelvin (absolute) temperature.

An ideal gas is a theoretical system in which there are only collision interactions between gas molecules. The ideal gas law applies to dilute real gases over normal temperature ranges.

Another iso-process for a gas is called an *isovolumetric process*, or more commonly, an *isometric process*. This is one in which the volume of the gas

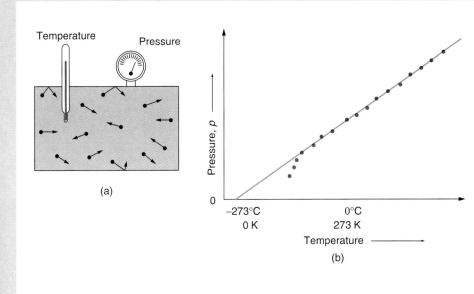

Figure 2 Constant-volume gas thermometer. (a) The volume of a quantity of gas in a rigid container is constant, and the pressure is directly proportional to the temperature ($p = kT$). (b) After an initial graph of pressure versus temperature has been made, temperature may be determined directly from the pressure reading of the gas thermometer. At low temperatures, gases do not follow straight-line relationships. However, if we extend the straight-line segment to the horizontal axis (zero pressure) on the graph, a value may be obtained for the absolute zero of temperature.

doesn't change; for example, a quantity of gas in a rigid container.

In the case of an isometric process, $V_1 = V_2$, and the previous equation becomes

$$\frac{p_1}{T_1} = \frac{p_2}{T_2}$$

If the ratio p/T is the same (constant) at any time, we may write $p/T = k$, where k is a constant.

Hence, for an isometric process, $p = kT$. Because the pressure varies directly with temperature, we have another way to measure temperature: with a constant volume gas thermometer. (See Fig. 2a.) Dilute real gases follow the $p = kT$ relationship. Taking corresponding pressure and temperature readings and plotting them, we get a straight line over the normal temperature range. (Why?) This process calibrates the thermometer. From the graph, we may find the temperature of the gas directly from the pressure reading of the gas thermometer.

At very low temperatures, real gases liquefy and the straight-line relationship no longer holds, making the temperature range of a gas thermometer limited. However, if the straight line of the p-versus-T graph is extrapolated to zero pressure, we obtain a value for the absolute zero temperature (Fig. 2b). This value turns out to be about $-273°C$, which is taken to be 0 K on the Kelvin scale.

The absolute temperature of a gas is directly proportional to the average kinetic energy of its molecules. In a real gas, there are intermolecular forces, and some of the internal energy of the gas is potential energy. However, in an ideal gas, there is only interaction by molecular collision, and so absolute temperature is a relative measure of the total internal energy. (An ideal gas is ideal in the sense that it would remain a gas at any temperature because there are no intermolecular forces to become dominant and liquefy the gas at very low temperatures when the kinetic energy of the molecules becomes small.)

Because the kinetic energy of an ideal gas is its internal energy, it follows that when the temperature of an ideal gas is doubled, its internal energy is also doubled. For example, to double the internal energy of an ideal gas at 200 K, the temperature of the gas would have to be raised to 400 K (with constant volume, so that no work would be done; recall the first law). This relationship is only true on the absolute temperature scale. What would it mean to double the temperature of a gas at 0°C or 0°F?

The escaping molecules take energy with them, and the temperature of the liquid is reduced. Hence, evaporation is a cooling process. In fact, it is a major cooling process of our bodies. When we're hot, we perspire. It's not the perspiring that cools us, but the evaporation of the perspiration. This can be easily demonstrated by standing in front of a blowing fan. The air movement promotes evaporation, and we feel quite cool.

A **gas** is made up of molecules that exert little or no force on one another, except when they collide. The distance between molecules in a gas is quite large compared to the size of the molecules. The molecules in a gas are moving rapidly. As a result, a gas has no definite shape, and assumes the size and shape of its container.

Thus we have:

A *solid* has definite size and shape.

A *liquid* has definite size, but no definite shape and assumes the shape of its container.

A *gas* has no definite size or shape, and assumes the size and shape of its container.

Continued heating of a gas causes the molecules to move faster and faster. Eventually, at high temperatures, molecules and atoms are ripped apart by collisions with one another. Inside hot stars, such as our Sun, atoms and molecules do not exist, and another phase of matter called **plasma** occurs (no relation to blood plasma). Referred to as the fourth phase of matter, a plasma is a gas of electrically charged particles. On Earth we have plasmas in gas discharge tubes, such as fluorescent and neon lamps.

We have seen that when a substance is in a single phase, heating increases the kinetic energy of the molecules of which it is composed. When a substance is changing phases, heating supplies the energy to overcome the attractive forces holding the different molecules together. Since the temperature rises only when the substance is not changing phase, we say that the temperature of a substance can be defined as a measure of the average kinetic energy of its molecules.

As in a gas, not all of the molecules are moving at the same speed. For this reason, we define absolute temperature as a measure of the average kinetic energy of all of the molecules. Some molecules are moving faster and some are moving slower than the average, and the speeds change with the transfer of energy in collisions. A liquid-in-glass thermometer measures the temperature of a gas or other substance by measuring the energy transferred to the thermometer bulb by molecular collisions. In the chapter Highlight, another type of thermometer—a gas thermometer—is described, along with how absolute zero is determined.

Important Terms

These important terms are for review. After reading and studying the chapter, you should be able to define and/or explain each of them.

temperature	Btu	heat pump
thermometer	specific heat	entropy
thermal expansion	latent heat	conduction
ice point	latent heat of fusion	thermal conductivity
steam point	latent heat of vaporization	thermal insulator
Fahrenheit scale	sublimation	convection
Celsius scale	thermodynamics	radiation
absolute zero	first law of thermodynamics	phases of matter
Kelvin scale	second law of thermodynamics	solid
kelvin (unit)	third law of thermodynamics	liquid
heat	heat engine	gas
calorie	thermal efficiency	plasma
kilocalorie	ideal efficiency	

Important Equations

Celsius T_C to Kelvin T_K: $T_K = T_C + 273$

Celsius T_C to Fahrenheit T_F: $T_F = \frac{9}{5}T_C + 32$

$$= 1.8 T_C + 32$$

Fahrenheit T_F to Celsius T_C: $T_C = \frac{5}{9}(T_F - 32)$

$$= \frac{(T_F - 32)}{1.8}$$

Specific heat: $H = mc\Delta T$

(water: $c = 1.0$ kcal/kg-C°)

(ice and steam: $c = 0.50$ kcal/kg-C°)

Latent heat of fusion: $H = mL_f$

Latent heat of vaporization: $H = mL_v$

(water: fusion $L_f = 80$ kcal/kg $= 80$ cal/g)

water: vaporization $L_v = 540$ kcal/kg $= 540$ cal/g)

Thermal efficiency: thermal eff $= \dfrac{W}{H_{\text{hot}}}$ ($\times 100\%$)

Ideal efficiency: ideal eff $= \dfrac{T_{\text{hot}} - T_{\text{cold}}}{T_{\text{hot}}}$

$$= 1 - \frac{T_{\text{cold}}}{T_{\text{hot}}}$$

Questions

5.1 Temperature

1. Temperature is (a) the same as heat, (b) always measured with a liquid-in-glass thermometer, (c) commonly expressed in kelvins, (d) a measure of the average kinetic energy of the molecules of a substance.
2. Which temperature scale has the smallest degree interval: (a) Celsius, (b) Fahrenheit, (c) Kelvin, (d) absolute?
3. Why is temperature a *relative* measurement of hotness and coldness?
4. Would it be possible to use liquids other than alcohol and mercury in thermometers? Explain.
5. What are the ice and steam points on three temperature scales: Celsius, Fahrenheit, and Kelvin?
6. A thermostat of the type used in furnace (or heat pump) controls is shown in Fig. 5.15. The glass vial tilts back and forth so electrical contacts are made via the mercury (an electrically conducting liquid metal) so as to turn the furnace off and on at a set temperature. Explain why the vial tilts back and forth.
7. A graph of Fahrenheit temperature (T_F) versus Celsius temperature (T_C) is shown in Fig. 5.16. Use your knowledge of graphs from Chapter 2 (see Fig. 2.9) to show that this graph reflects Eq. 5.2b ($T_F = \frac{9}{5}T_C + 32$).

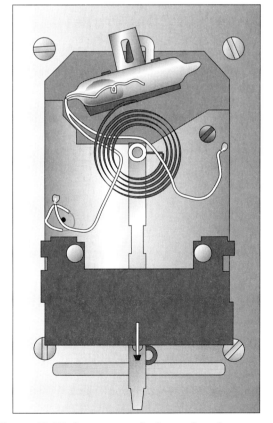

Figure 5.15 An exposed view of a thermostat used to control heating and cooling systems. See Question 6.

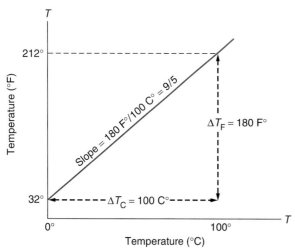

Figure 5.16 A graph of Fahrenheit temperature versus Celsius temperature.
See Question 7.

5.2 Heat

8. Heat is (a) energy being transferred, (b) the same as temperature, (c) the internal energy of a substance, (d) unrelated to energy.
9. The largest unit of heat energy is the (a) kilocalorie, (b) calorie, (c) joule, (d) Btu.
10. Explain what is meant by the statement, "Heat is energy in transit."
11. For practical purposes, what is the difference between (a) the freezing point and melting point, and (b) the steam point and condensation point?
12. Why do sidewalks have segments with grooves or spaces in between, rather than having one single piece?

5.3 Specific Heat and Latent Heat

13. From the equation $H = mc\Delta T$ (Eq. 5.3), we see that the units of specific heat are (a) kg/kcal, (b) kg/J-°C, (c) kcal-kg/J, (d) kcal/kg-C°.
14. Sublimation refers to a phase change of (a) solid to liquid, (b) solid to a gas, (c) gas to liquid, (d) none of these.
15. What is specific about specific heat?
16. Water's high specific heat is one advantage for its use in solar heat storage. What is another?
17. Why does it take a long time for lakes to freeze when the air temperature is already below freezing?
18. When does the addition of heat not result in a temperature change, and why?

5.4 Thermodynamics

19. When heat is added to a system, it goes into (a) doing work, (b) the internal energy, (c) doing work and/or internal energy.
20. The direction of a natural process is indicated by the (a) conservation of energy, (b) thermal efficiency, (c) change in entropy, (d) specific heat.
21. What do the first and second laws of thermodynamics tell you?
22. What is $H_{hot} - H_{cold}$ for (a) a heat engine and (b) a heat pump?
23. Distinguish between thermal efficiency and ideal efficiency.
24. Why is approximately two-thirds of the heat generated by a coal-fired or nuclear power plant wasted? What happens to this wasted heat?
25. In a phase change of a gas to a liquid, or a liquid to a solid, the system becomes more orderly—a decrease in entropy. Doesn't this violate the second law? Explain.
26. Compare and explain the following statements in terms of the laws of thermodynamics: (a) "Energy can neither be created or destroyed," and (b) "Entropy can be created but not destroyed."

5.5 Heat Transfer

27. In metals, much of the thermal conductivity is contributed by (a) radiation, (b) entropy, (c) latent heat, (d) electrons.
28. The method of heat transfer generally involving mass movement is (a) conduction, (b) convection, (c) radiation.
29. What are several examples of good thermal conductors and good thermal insulators? In general, what makes a substance a good conductor or insulator?
30. Why does a vinyl floor feel colder than a rug, even though they are both at the same temperature?
31. If air is a poor conductor, why do we bother putting insulation between the walls of our homes?
32. Some pots and pans have copper bottoms. Is there any practical reason for this or is it purely decorative? (*Hint:* See Table 5.1.)
33. Thermal underwear is knitted with large holes. Wouldn't this defeat the purpose? Explain.

5.6 Phases of Matter

34. Intermolecular bonding is greatest in (a) solids, (b) liquids, (c) gases, (d) plasmas.
35. According to the ideal gas law, if the temperature of a quantity of gas is increased, then (a) only the pressure increases, (b) only the volume increases, (c) both the pressure and volume increase, (d) either or both the pressure and volume may increase.
36. Give descriptions of a solid, a liquid, and a gas in terms of shape and volume.
37. What is an isobaric process? [*Hint:* Part of the word *isobaric* (*-baric*) is associated with the word *barometer.*]

Food for Thought

1. Are the degree intervals the same for mercury and alcohol thermometers? Explain.
2. When a thermometer at room temperature is placed in a boiling liquid, it is observed that the liquid column falls slightly before beginning to rise. Why is this? (Try it yourself.)
3. All snow flakes are hexagonal. Is this a true statement?
4. An automobile radiator is used to cool the engine. Is "radiator" really a good descriptive name?
5. What would it mean to have an efficiency greater than 100%?
6. Water-exchange heat pumps use water from wells or in-ground reservoirs, whereas air-exchange heat pumps simply use the ambient air. Water-exchange heat pump systems are therefore much more expensive. What would be the advantage of using them?

Exercises

5.1 Temperature

1. Normal body temperature is 98.6°F. What is the equivalent temperature on the Celsius scale?
   Answer: 37°C
2. Normal room temperature is about 68°F. What is the equivalent temperature on the Celsius scale? On the Kelvin scale?
3. While in Europe, a tourist hears on the radio that the temperature that day will have a high of 15°C. What is this temperature on the Fahrenheit scale?
4. The temperature of outer space is measured to be 3 K. What is the equivalent temperature on the (a) Celsius scale and (b) Fahrenheit scale?
   Answer: (a) −270°C (b) −454°F
5. Show that absolute zero is equivalent to a temperature of about −460°F.
6. Are the temperature readings on the Fahrenheit and Celsius scales ever equal? The answer is yes. Using Eq. 5.2, show that −40°F = −40°C.

5.3 Specific Heat and Latent Heat

7. How much energy is required to raise the temperature of 1 L of water from 20° to 30°C?
   Answer: 10 kcal
8. (a) How much heat does it take to warm 1.0 kg of water from room temperature (20°C) to the boiling point?
   (b) At 11¢ per kWh, how much does it cost to heat 1.0 kg of water at 20°C for instant coffee?
   Answer: (b) about a penny
9. How many kcal of heat is required to change 500 g of ice at −10°C to water at 20°C?
   Answer: 52.5 kcal
10. A quantity of steam (200 g) at 110°C is cooled to 90°C with a phase change in the process. How much heat was removed?

11. In making iced tea, you cool 1.6 kg of hot tea at 100°C to 0°C. What is the minimum amount of ice necessary to make the iced tea? (Neglect heat losses.)

Answer: 2.0 kg

5.4 Thermodynamics

12. If a heat engine has a thermal efficiency of 40%, how much work is obtained from 50 kcal of heat input?

Answer: 20 kcal

13. Researchers have proposed using the temperature difference between the top and lower portions of the ocean to run a heat-engine power plant. If the surface temperature of the ocean is at 15°C and the lower depth is at 4°C, what is the maximum possible theoretical efficiency?

Answer: 3.8%

14. A coal-fired power plant has operating temperatures of T_{hot} = 320°C and T_{cold} = 100°C. What can you say about the efficiency of the power plant?

Answer: less than 37%

15. A proposed nuclear power plant is said to have T_{hot} = 350°C and T_{cold} = 100°C, with an efficiency of 42%. Would you support this proposal and buy stock in this electric company? (Justify your answer.)

Answers to Multiple-Choice Questions

1. d   8. a   13. d   19. c   27. d   34. a
2. b   9. a   14. b   20. c   28. b   35. d

Solutions to Confidence Questions

5.1 $T_C = \frac{5}{9}T_F - 32) = \frac{5}{9}(68 - 32) = 20°C$

5.2 (a) $\Delta T = \frac{H}{mc} = \frac{11 \text{ kcal}}{(1.0 \text{ kg})(1.0 \text{ kcal/kg-C°})} = 11 \text{ C°}$

   (b) $\Delta T = \frac{H}{mc} = \frac{11 \text{ kcal}}{(1.0 \text{ kg})(0.22 \text{ kcal/kg-C°})} = 50 \text{ C°}$

5.3 The lead atoms are bound much more tightly than the water molecules.

5.4 More work is required to completely separate the molecules in going from a liquid to a gas.

5.5 $H_t = H_{\Delta T} + H_{vap} = mc\Delta T + mL_v$
   $= (0.50 \text{ kg})(1.0 \text{ kcal/kg-C°})(80 \text{ C°})$
   $+ (0.50 \text{ kg})(540 \text{ kcal/kg})$
   $= 310 \text{ kcal}$

5.6 $-H = -\Delta E_i - W$, where $-H$ indicates heat was removed. In the process, internal energy was removed $(-\Delta E_i)$ and work was done on the system $(-W)$ by external pressure.

Chapter 6

Waves

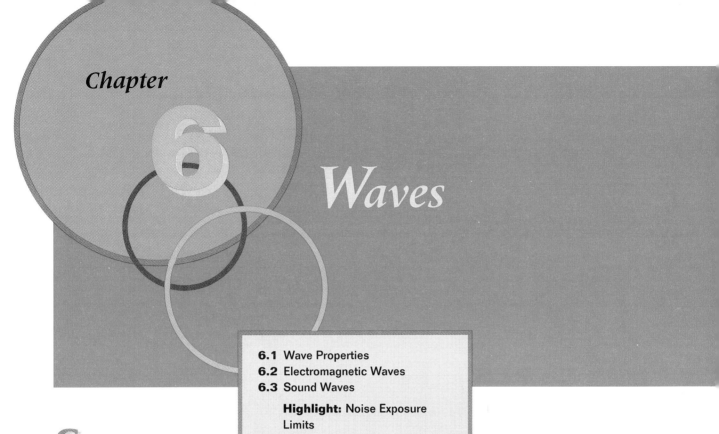

6.1 Wave Properties
6.2 Electromagnetic Waves
6.3 Sound Waves
   Highlight: Noise Exposure Limits
6.4 The Doppler Effect
6.5 Standing Waves and Resonance

Since we began our study of energy, much has been said about the forms of energy, the relationship of energy to work, the conservation of energy, and so on, with many questions raised and answered. Other interesting questions remain, however. For example, how is energy changed from one form to another? How is energy mechanically transferred?

A partial answer to these questions is found in terms of collisions and wave motion, two important ways for transferring energy (Fig. 6.1). If a particle or object applies a force through a distance, then a transfer of energy takes place with something gaining energy and something losing energy. Overall, we find that a transfer of energy takes place when two or more particles collide. Matter is made up of particles. "Collisions" effect energy transfer through the bonding interactions that hold the particles together.

When matter is disturbed, energy emanates from the disturbance through particle interaction. This propagation of energy is called **wave motion**. For example, when a stone is dropped in a pool of water, the water is disturbed and energy is transferred outward from the disturbance by moving waves. Only energy is transferred, not matter (the water), as can be noted by observing a floating fishing bobber that goes up and down with the water.

A similar situation can occur in a solid. For example, when an earthquake occurs, a disturbance takes place because of slippage along a fault; this disturbance is transmitted far and wide by waves. Again, such a disturbance is a transfer of energy, not matter.

A transfer of energy may take place with or without a medium (matter). Sound waves in air and waves on stretched strings or wires as on musical instruments are examples of energy transmissions through media. The neighboring particles of the medium act on one another to transfer a disturbance. However, electromagnetic waves, which include radio, TV, infrared radiation, visible light, and X-rays, can be transferred without a medium. We say these disturbances are radiated through space; for example, light coming from the Sun is radiated through space.

Our knowledge about the planets, the Sun, and other stars comes to us by means of electromagnetic radiation, which we generally refer to as light. In Chapter 9 when quantum mechanics is discussed, we shall study the dual nature of light.

117

6.1 Wave Properties

Figure 6.1 Energy transfer.
Some samples of transferring energy.

Learning Goals:

To distinguish between transverse and longitudinal waves, and give an example of each.

To explain the properties used to describe waves.

The disturbance that generates a wave may be a sudden pulse or force, such as the striking of a desk with the flat of your hand. However, we will be interested in continuous waves that result from periodic disturbances; that is, from disturbances that are repeated again and again at regular intervals. Such waves can be set up by plucking a guitar string or blowing a whistle (Fig. 6.2).

We can set up a periodic wave in a stretched string or cord by shaking (disturbing) one end as shown in Fig. 6.3. The waves travel down the string with a certain speed, or velocity (if direction is given). The **wave velocity** depends on the particle interactions in the medium, and therefore varies from one substance to another. In general, the wave speed is greater in solids than in liquids or gases.

Note that the particles in the stretched string in Fig. 6.3 oscillate up and down (or side to side). This effect can be seen by tying a piece of ribbon or cloth to the string, which would follow the string "particle" motion. The wave velocity is perpendicular to the particle motion, and we characterize this condition as a type of wave. The term **transverse wave** is used to denote wave motion in which the

Strangely enough, light, which we consider to be a wave, has some particle properties.

Our eyes and ears are two wave-detecting devices that serve to link us with the environment. Think of what our lives might be like without light and sound waves. An understanding of waves is critical in understanding many scientific principles and the world around us.

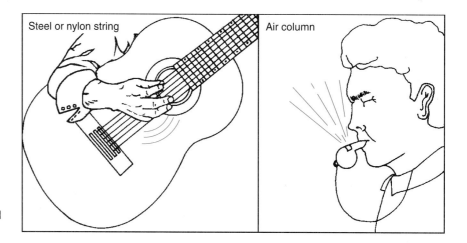

Figure 6.2 Wave motion.
Some common methods used to generate waves.

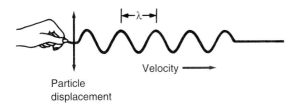

Figure 6.3 Transverse wave.
In a transverse wave, as illustrated here in a stretched cord, the wave velocity (vector) or the direction of wave propagation is perpendicular to the (cord's) particle displacement.

individual particles are displaced perpendicular to the direction of the wave velocity. All electromagnetic waves (Section 6.2), which do not require a medium for propagation, are transverse waves. In a material medium, transverse waves propagate only in solids. Liquids and gases do not have the transverse restoring forces among their molecules for transverse particle oscillations.

We can also set up a wave in a stretched spring (Fig. 6.4). When one end of the spring is shaken back and forth, the disturbance is propagated along the length of the spring. In this case, the spring "particles" oscillate parallel to the wave velocity direction, as can again be seen by tying a piece of ribbon or cloth to the spring. When the particle displacement and the wave velocity are parallel, the wave is called a **longitudinal wave.** Longitudinal waves will propagate in material media—solids, liquids, and gases. There is sufficient molecular bonding in liquids for longitudinal oscillations, and in gases there are compressional effects. Sound waves are longitudinal waves; in air, it is compressional effects that set them up, as we shall see in Section 6.3.

A common wave is that seen on the surface of water. From the profile, it might be thought that these are transverse waves. However, they are really a combination of transverse and longitudinal motions. The water "particles" move in more or less circular paths (Fig. 6.5). The water itself does not move in the wave direction, but merely passes the energy on in wave motion.

The diameters of the circular paths decrease quite rapidly with depth. There may be a fierce storm with big waves on the surface of the ocean. However, a submerged submarine 100 meters below the surface is undisturbed by what's going on above. When an ocean wave approaches the shore, the water particles are forced into steeper paths. Eventually, with decreasing depth, the particles cannot go through the lower part of their paths, and the wave "breaks," with the crest falling forward to form a surf.

Both transverse and longitudinal waves can be described by certain characteristics. As we have seen, wave velocity describes the direction and magnitude of the wave motion (Fig. 6.6), and the wave speed depends on the properties of the medium. The particles of the medium oscillate back and forth, and we call the magnitude of the maximum particle displacement the *amplitude* of the wave. For a transverse wave, this is how "tall" the wave is, whereas for a longitudinal wave, this is the maximum side distance. The amplitude does not affect the wave speed. However, the energy transmitted by the wave is related to its amplitude—a big (tall) wave has more energy than a little wave. Actually, the energy of a wave is proportional to the square of its amplitude.

In periodic wave motion, the particles oscillate back and forth, going through cycles of motion. We characterize this wave property in terms of frequency. The wave **frequency** is the number of oscillations that occur in a given time period. Frequency is generally expressed in cycles per second (cps). This unit is given the name of hertz, after Heinrich Hertz, an early investigator of electromagnetic waves. One **hertz** (Hz) is one cycle per second. For example, if five complete wave crests pass a given spot in one second, the frequency is

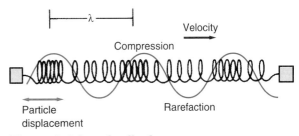

Figure 6.4 Longitudinal wave.
In a longitudinal wave, as illustrated here in a stretched spring, the wave velocity (vector) or direction of propagation is parallel to the (spring's) particle displacement.

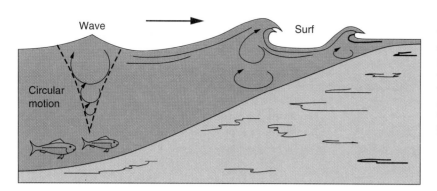

Figure 6.5 Water waves.
The water "particles" of a deep-water wave generally are in circular motion, or in a combination of transverse and longitudinal motions. When the incoming wave can no longer complete this motion in shallow water, it breaks and forms a surf.

five cycles per second, or 5 Hz. Look on your radio. You will see the ranges of the AM and FM bands given in kHz (kilohertz) and MHz (megahertz). Electromagnetic radio waves oscillate pretty fast.

The **period** of a wave is the time it takes for one complete wave oscillation. For example, if five crests of a wave pass by a given point in one second, one crest or complete cycle passes in one-fifth of a second, and the period is one-fifth second. The frequency is five cycles per second, and a simple relationship exists between the frequency (f) and period (T):

or

$$\text{frequency} = \frac{1}{\text{period}}$$

$$f = \frac{1}{T} \quad (6.1)$$

Frequency is expressed generally in cycles per second, and the period is the inverse, with units of seconds per cycle, so to speak. However, *cycle* is a descriptive term and not a unit. The period has the unit of time (second), and from Eq. 6.1 we see that frequency has the unit of inverse time, $1/T$. Hence, the hertz frequency unit expressed in standard units is 1/second, or Hz = 1/s.

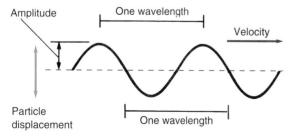

Figure 6.6 Wave description.
Some terms used to describe wave characteristics. See text for description.

The spatial dimensions of a wave are measured in wavelengths. One **wavelength** is the distance from any point on a wave to the same relative point on the next oscillation. Wavelength is commonly measured from crest to crest or some other convenient wave point (Fig. 6.6).

A simple relationship exits among wave speed, wavelength, and period (or frequency). Since speed is distance divided by time ($v = d/t$), we may write in terms of wave parameters,

$$v = \frac{\lambda}{T} \quad (6.2)$$

or by Eq. 6.1

$$v = \lambda f \quad (6.3)$$

where, with the wavelength λ in meters and the period T in seconds (and frequency 1/s), the wave speed v is in m/s.

EXAMPLE 6.1 Finding the Wavelengths of Sound

Consider sound waves in air with a speed of 344 m/s and a frequency of (a) 20 Hz and (b) 20 kHz. Find the wavelength of each of these sound waves.

Solution

Listing the data, we have

Given: $v = 344$ m/s
Find: λ (wavelength)
(a) $f = 20$ Hz
(b) $f = 20$ kHz = 20,000 Hz

The units look consistent, but keep in mind that 1 Hz = 1/s. It should be obvious that we use Eq. 6.3, which gives the wavelength in terms of the speed and frequency:

$$v = \lambda f$$

or

$$\lambda = \frac{v}{f}$$

(a) $$= \frac{344 \text{ m/s}}{20 \text{ Hz}} = \frac{344 \text{ m/s}}{20 \text{ 1/s}} = 17 \text{ m}$$

Here we write Hz = 1/s to show how the units cancel: $\frac{\text{m/s}}{1/\text{s}} = \text{m}$.

(b) $$\lambda = \frac{v}{f} = \frac{344 \text{ m/s}}{20,000 \text{ Hz}} = 0.017 \text{ m}$$

Notice that the higher the frequency f, the shorter the wavelength λ.

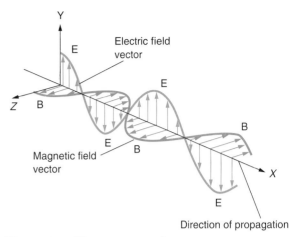

Figure 6.7 Electromagnetic wave.
An illustration of the vector components of an electromagnetic wave. The wave consists of two force fields (electric E and magnetic B) perpendicular to each other and perpendicular to the direction of wave propagation (as shown by the velocity vector).

CONFIDENCE QUESTION 6.1

What is the frequency of a sound wave that has a wavelength of 1.00 m?

The frequencies of 20 Hz and 20 kHz given in Example 6.1 define the general range of audible sound-wave frequencies. (We hear only sounds with frequencies between these extremes.) Thus, the wavelengths of audible sounds cover the range from about 1.7 cm for the highest-sound frequency we can hear to about 17 m for the lowest frequency we can hear. In British units, sound waves have a wavelength range of approximately $\frac{1}{2}$ inch to about 50 feet.

6.2 Electromagnetic Waves

Learning Goals:

To describe electromagnetic waves.

To explain the various regions of the electromagnetic spectrum.

When charged particles such as electrons vibrate (accelerate), energy is radiated from them in the form of **electromagnetic waves**. Electromagnetic waves consist of vibrating electric and magnetic fields, which will be discussed in Chapter 8. They are vectors; we represent these oscillating fields as shown in Fig. 6.7. All electromagnetic (EM) waves travel at "the speed of light," which is 3.00×10^8 m/s (300,000,000 m/s) in vacuum.

EM waves are transverse waves. Note in the figure that for a wave traveling (having a velocity vector) in the x direction, the electric and magnetic field vectors are perpendicular (90°) to this direction. Also, the electric and magnetic field vectors are perpendicular to each other.

Charged particles are accelerated in many different ways to produce electromagnetic waves of various frequencies. Waves with relatively low frequencies (or long wavelengths), known as radio waves, are produced by causing electrons to oscillate, or vibrate, in an antenna. The frequency of oscillation is controlled by the physical dimensions and other properties of the driving circuit. Radio waves have frequencies in the kHz (AM band) and MHz (FM band) ranges.

The production of EM waves with frequencies greater than radio waves is accomplished by molecular excitation. In such cases, radiation occurs from the collision of molecules in hot gases and solids. Because the molecules carry charged particles that are greatly accelerated as the molecules vibrate, the particles radiate EM waves ranging

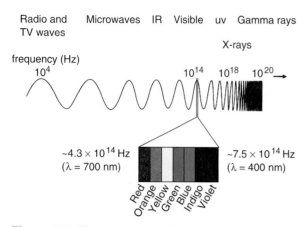

Figure 6.8 Electromagnetic spectrum.
Different frequency (or wavelength) regions are given different names. Notice how the visible spectrum forms only a very small part of the EM spectrum.

from 10^{10} Hz to about 10^{14} Hz. EM waves are distinguished by their frequency (or wavelength), with all of the EM waves making up an electromagnetic spectrum (Fig. 6.8). The radiations from molecular excitation are in the *microwave* and *infrared regions* of the spectrum.

As the temperature of gases and solids is increased to higher values, the atoms comprising the molecules become more excited, and electromagnetic radiation in the *visible* and *ultraviolet regions* of the spectrum is emitted. Notice that the small portion of the spectrum visible to the human eye (Fig. 6.8) lies between the infrared and ultraviolet regions.

High-energy electrons interacting with the atoms of a target material rapidly decelerate, and EM waves of higher frequencies, called *X-rays*, are emitted (Chapter 9). The X-ray region of the spectrum ranges from 10^{17} Hz to 10^{19} Hz. Even higher frequency EM waves, called *gamma rays*, are emitted from nuclear processes (Chapter 10).

The term *light* is commonly used for electromagnetic radiations in and near the visible region; for example, we say *ultraviolet light*. Only the frequency (or wavelength) distinguishes visible EM radiation from the other portions of the spectrum. Our human eyes are only sensitive to certain frequencies or wavelengths, but other instruments can detect other portions of the spectrum. For example, a radio receiver can detect or "pick up" radio waves.

Radio waves are not sound waves. They are electromagnetic waves that are detected and then amplified by the radio circuits. The radio frequency signal is then demodulated—that is, the audio signal is separated from the radio frequency carrier, the assigned frequency of the radio station. It is the audio signal applied to the speaker system that produces sound waves.

All electromagnetic radiation travels at the same speed in a vacuum. This speed is called the speed of light. The **speed of light** in vacuum is designated by the letter c and has a value of

$$c = 3.00 \times 10^8 \text{ m/s} = 300,000,000 \text{ m/s}$$

or

$$c = 186,000 \text{ mi/s}$$

(To a good approximation, this value is also the speed of light in air.)

We can use the equation $c = \lambda f$ to find the wavelength of light or any EM radiation in a vacuum. For example, the wavelength of an AM radio wave with a frequency of 600 Hz comes out to be 500 m—quite long. FM radio waves with higher frequencies have shorter wavelengths. Visible light with frequencies on the order of 10^{14} Hz has relatively short wavelengths—on the order of a millionth of a meter (10^{-6} m). To express such small numbers, the metric prefix of *nano-* (10^{-9}) is commonly used. The visible region of the electromagnetic spectrum corresponds to a wavelength range of 400 nm to 700 nm (Fig. 6.8).

6.3 Sound Waves

Learning Goals:

To define sound.

To explain sound waves, their propagation, and the components of the sound spectrum.

To distinguish sound intensity levels reported in decibels.

In general, **sound** is defined as the propagation of longitudinal waves through matter. Sound waves involve particle displacement in any kind of matter—solid, liquid, or gas.

We are most familiar with sound waves in air, which affect our sense of hearing. However, sound also travels in liquids and solids. When you are swimming underwater and someone clicks two rocks together, you can hear this disturbance. Also, we can hear sound through thin (but solid) walls.

The wave motion of sound depends on the elasticity of the medium. A longitudinal disturbance produces varying pressures and stresses in the medium. For example, consider a vibrating tuning fork, as illustrated in Fig. 6.9. As an end of the fork moves outward, it compresses the air in front of it, and a *compression* is propagated outward. When the fork end moves back, it leaves a region of decreased air pressure and density called a *rarefaction*. With continual vibrations, a series of high-and-low pressure regions travel outward, forming a longitudinal sound wave. The waveform may be displayed electronically on an oscilloscope, as shown in Fig. 6.10.

Sound waves may have different frequencies and so form a spectrum similar to the electromagnetic spectrum (Fig. 6.11). However, the **sound spectrum** has much lower frequencies and is much simpler, with only three frequency regions. These regions are based on the audible range of human hearing, which is about 20 Hz to 20 kHz (20,000 Hz) and defines the *audible region* of the spectrum. Below this (< 20 Hz) is the *infrasonic region*, and above (> 20 kHz) is the *ultrasonic region*. (Note the analogy to infrared and ultraviolet light.) The sound spectrum has an upper limit of about a billion hertz (1 GHz, gigahertz) because of the elastic limitations of materials.

Because of their long wavelengths, infrasonic waves are associated with the movement or slow vibrations of large objects such as portions of the

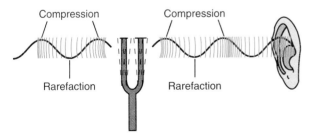

Figure 6.9 Sound waves.
Sound waves consist of a series of compressions (high-pressure regions) and rarefactions (low-pressure regions) as illustrated here being produced in air by a tuning fork. Note how the regions can be described by a waveform.

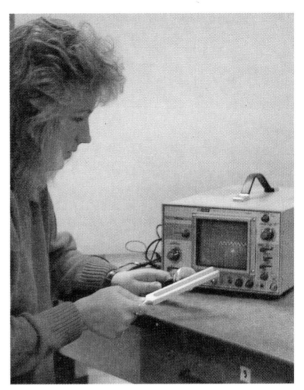

Figure 6.10 Waveform.
The waveform of the tone from a tuning fork can be displayed on an oscilloscope by using a microphone to convert the sound wave into an electrical signal.

Earth; that is, with earth tremors or earthquakes. Ultrasonic waves, on the other hand, have short wavelengths. Because of their rapid or high vibrational frequencies, they are limited to small or less massive objects. This distinction has resulted in a number of technological applications, some of which will be discussed shortly. While sound is sometimes thought of as being those disturbances perceived by the human ear, it is clear from Fig. 6.11 that this definition of sound would omit a vast majority of the sound spectrum.

We hear sound because the propagating disturbance causes the eardrum to vibrate, and sensations are transmitted to the auditory nerve through the fluid and bones of the inner ear. The characteristics we associate with human hearing are physiological and can differ from their physical counterparts. For example, **loudness** is a relative term. One sound may be louder than another, and as you might guess, this property is associated with the energy of the wave.

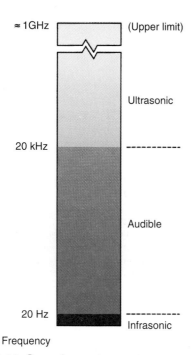

Figure 6.11 Sound spectrum.
The sound spectrum consists of three regions: the infrasonic region ($f < 20$ Hz), the audible region (20 Hz $< f <$ 20 kHz), and the ultrasonic region ($f > 20$ kHz).

The measurable physical quantity is **intensity** (I), which is the rate of energy transfer through a given area. For example, intensity may be given as so many joules per second (J/s) through a square meter (m²). But recall that a joule per second is a watt (W), so intensity has units of W/m².

The loudness or intensity of a sound decreases the farther one is from the source. As the sound is propagated outward, it is "spread" over a greater area, and so has less energy per unit area. This characteristic is illustrated for a point source in Fig. 6.12. In this case, the intensity is inversely proportional to the square of the distance from the source ($I \propto 1/r^2$), that is, it is an inverse square relationship. This relationship is analogous to painting a larger room with the same amount of paint (energy) as used for a smaller room. The paint must be spread thinner and so is less "intense."

The minimum sound intensity detectable by the human ear (called the *threshold of hearing*) is about 10^{-12} W/m². At a much greater intensity (about 1 W/m²), sound becomes painful to the ear (see the chapter Highlight). Because of the wide range, intensity is commonly measured on a more convenient logarithmic scale. This change collapses the scale, making it more manageable. The sound-intensity level is measured on this scale in decibels (dB), as illustrated in Fig. 6.13.

A **decibel** is one-tenth of a bel (B), a unit named in honor of Alexander Graham Bell, the inventor of the telephone. Because the decibel scale is not linear with intensity, when the sound intensity is doubled, the dB level is not doubled. Instead, the intensity level increases by only 3 dB. That is, a sound with an intensity level of 63 dB has twice the intensity of a sound with an intensity level of 60 dB.

Comparisons are made on the dB scale in terms of decibel differences and factors of 10:

An increase of 10 dB increases the sound intensity by a factor of 10.

An increase of 20 dB increases the sound intensity by a factor of 100.

An increase of 30 dB increases the sound intensity by a factor of 1000; and so on.

(For each increment of 10 dB, add another zero.)

CONFIDENCE QUESTION 6.2

A band playing a piece of music at an average intensity level of 60 dB builds up to an ending with a sound intensity of 100 dB (pretty loud). By what factor was the intensity increased?

Loudness is related to intensity, but loudness is subjective and estimates can differ from person to person. Also, the ear does not respond equally to all frequencies. For example, two sounds with different frequencies, but the same intensity level, may be judged by the ear to have different loudnesses. Some occupational problems with loudness and intensity are discussed in the chapter Highlight.

The frequency of a sound wave may be physically measured, whereas **pitch** is the perceived highness or lowness of a sound. For example, a soprano has a high-pitched voice as compared with a baritone. Pitch is related to frequency. But, if a sound with a single frequency is heard at two intensity levels, nearly all listeners will agree that the more intense sound has a lower pitch.

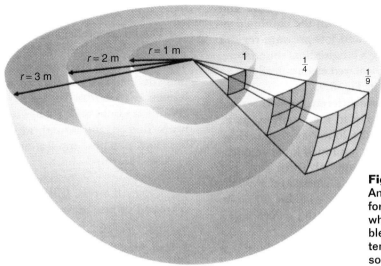

Figure 6.12 Sound intensity.
An illustration of the inverse-square law for a point source ($I \propto 1/r^2$). Notice that when the distance from the source doubles, for example, from 1m to 2m, the intensity is one-fourth as much, because the sound passes through four times the area.

Another characteristic of sound is **quality** (or *timbre*). This refers how a tone sounds. Two people can sing the same note (same frequency), but they sound different. Basically, there is a difference in the waveform that is made up by a number of waves or overtones (Section 6.5). These are set up in our nasal cavities, and because we are all generally built differently, we sound different, or have a different voice quality.

Ultrasound is the term used for sound waves with frequencies greater than 20 kHz. These waves cannot be detected by the human ear, but the hearing frequency range of other animals includes ultrasound frequencies. For example, dogs can hear ultrasound, and ultrasonic whistles used to call dogs don't disturb humans.

An important use of ultrasound is in examining internal parts of the body. Thus, ultrasound is an alternative to X-rays, which can be harmful in some cases. The ultrasonic waves allow different materials, such as tissue and bone, to be "seen" or distinguished by bouncing waves off the object examined. The waves are detected, analyzed, and stored in a computer. An echogram or sonogram, such as the one of an unborn fetus shown in Fig. 6.14 on page 128, is then reconstructed. X-rays might harm the fetus and cause birth defects, but ultrasonic waves have less energetic vibrations and have given no evidence of harming a fetus.

Ultrasound can also be used in a cleaning technique. Minute foreign particles can be removed from an object placed in a liquid bath through which ultrasound is passed. The wavelength of ultrasound is on the same order of magnitude as the particle size, and the wave vibrations can get into small crevices and "scrub" particles free. Ultrasound is especially useful in cleaning objects with hard-to-reach recesses, such as rings and other jewelry. Ultrasonic cleaning baths for dentures are also commercially available.

The speed of sound in a particular medium depends on the makeup of the material. In general, the **speed of sound** in air at 20°C is

$$v_{\text{sound}} = 344 \text{ m/s} \quad (770 \text{ mi/h})$$

or approximately $\frac{1}{3}$ km/s or $\frac{1}{5}$ mi/s.

Over the normal temperature range the speed of sound varies directly with temperature. That is, as the temperature increases, the speed of sound increases. Note that the speed of sound in air is much less than the speed of light.

The relatively slow speed of sound in air may be observed at a baseball game. A spectator may see a batter hit the ball but hear the "crack" of the bat slightly later if he or she is sitting far from home plate. Similarly, one may see the smoke or flash from a fired rifle, but hear the report later, because the sound comes to an observer much slower than the visual signal, which travels at the speed of light.

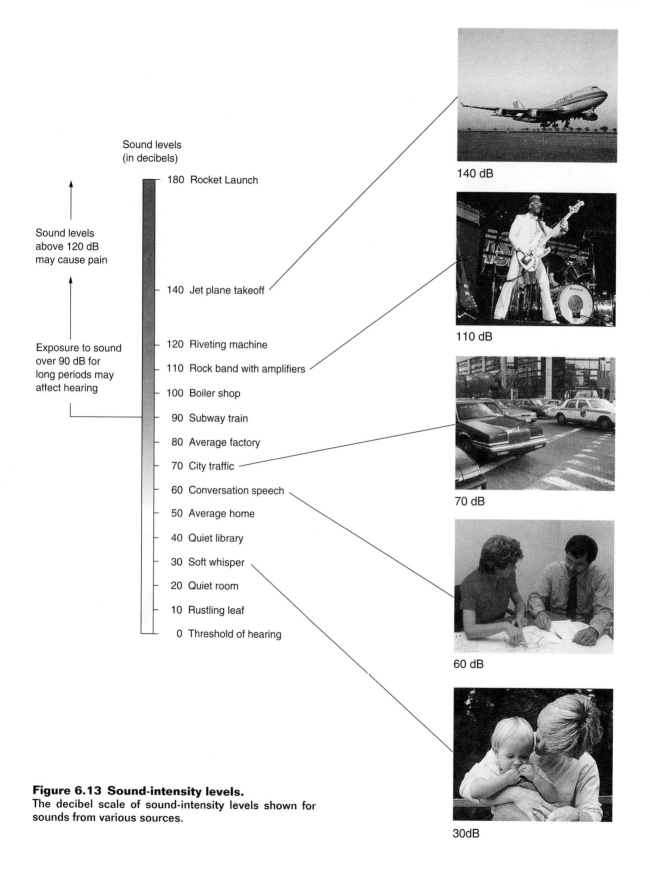

Figure 6.13 Sound-intensity levels.
The decibel scale of sound-intensity levels shown for sounds from various sources.

HIGHLIGHT

Noise Exposure Limits

Sounds with intensities of 120 dB and higher can be painfully loud to the ear. Brief exposures to even higher sound intensity levels can rupture eardrums and cause permanent hearing loss. However, long exposure to relatively lower sound (noise) levels can also cause hearing problems. (Noise is defined as unwanted sound.) Such exposures may be an occupational hazard, and in some jobs ear protectors must be worn (see the accompanying figure). You may have experienced a temporary hearing loss after being exposed to a loud band for a long time or a loud bang for a short time.

Federal standards now set permissible noise exposure limits for occupational loudness. These limits are listed in Table 1. Notice that a person can work on a subway train (90 dB, Fig. 6.13) for 8 h, but a person should only play in (or listen to) an amplified rock band (110 dB) continuously for $\tfrac{1}{2}$ h.

Table 1 Permissible Noise Exposure Limits

Maximum Duration per Day (h)	Sound Level Intensity (dB)
8	90
6	92
4	95
3	97
2	100
1½	102
1	105
½	110
¼ or less	115

Figure 1 Sound-intensity safety. An airport worker wears ear protectors to prevent ear damage from the high sound-intensity levels of jet engines.

In general, as the density of the medium increases, the speed of sound therein increases. The speed of sound is about 4 times faster in water than in air and, in general, about 15 times faster in solids.

Using the speed of sound and the frequency, we can easily compute the wavelength of a sound wave.

EXAMPLE 6.2 Finding Audible and Ultrasound Wavelengths

What is the wavelength of a sound wave in air with a frequency of (a) 2200 Hz and (b) 22 MHz?

Solution

We have

Given: (a) $f = 2200$ Hz
(b) $f = 22$ MHz $= 22,000,000$ Hz
Find: λ (wavelength)

Notice in (b) the frequency was given in megahertz (MHz). *Mega-* denotes million, and we write the frequency directly in hertz.

The wavelength and frequency are related by Eq. 6.3 ($v = \lambda f$), which we write in the form $\lambda = v_{sound}/f$. To calculate λ, a value for the speed of sound must be obtained, and we will assume this to be its 20°C value, $v_{sound} = 344$ m/s (known). Then, we can put in the numbers:

(a) For $f = 2200$ Hz, which is in the audible range,

$$\lambda = \frac{v_{sound}}{f} = \frac{344 \text{ m/s}}{2200 \text{ Hz}} = 0.16 \text{ m}$$

This is a wavelength of about $\tfrac{1}{2}$ ft.

(b) For $f = 22$ MHz, or 22,000,000 Hz, which is in the ultrasonic region,

$$\lambda = \frac{v_{sound}}{f} = \frac{344 \text{ m/s}}{22,000,000 \text{ Hz}} = 0.000016 \text{ m}$$

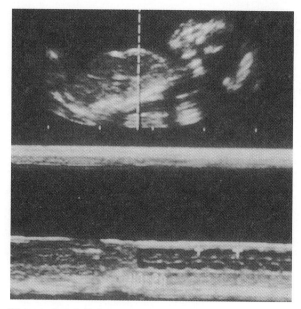

Figure 6.14 Echogram
A fetal echogram, or sonogram, in which the outline of the baby's face can be seen in the top right yellow region. There is no evidence that ultrasound scans can harm a fetus, as X-rays can.

or 16 micrometers. Hence, the wavelength of ultrasound is on the order of particle size and can be used in a cleaning bath, as described earlier.

6.4 The Doppler Effect

Learning Goals:

To explain the Doppler effect.

To identify some Doppler effects and applications.

When we watch a race and a racing car with a loud engine approaches, we hear a higher-than-usual sound frequency. When the car passes by, the frequency or pitch suddenly shifts lower and a low-pitched "whoom" sound is heard. Similar frequency changes may be heard when a large truck passes by.

The reason for the observed change in frequency (and wavelength) of a sound from a moving source is illustrated in ● Fig. 6.15a. As a moving sound source approaches an observer, the waves are "bunched up" in front of the source. With the waves closer together (shorter wavelength), an observer perceives a higher frequency. Behind the source, the waves are "spread out," and with an increase in wavelength, a lower frequency is heard ($f = v/\lambda$). This apparent change in the frequency of the source is called the **Doppler effect.***

If the source is stationary and the observer moves toward and passes the source, similar frequency shifts are also observed. Hence, the Doppler effect depends on the relative motion of the source and the observer.

Waves propagate outward in front of a source as long as the speed of sound is greater than the speed of the source (● Fig. 6.16). However, as the speed of a source approaches the speed of sound in the medium, the waves begin to bunch up. When the speed of the source exceeds the speed of sound in the medium, a V-shaped bow wave is formed. This wave is readily observed for a motor boat traveling faster than the wave speed in water.

In air, when a jet aircraft travels at a supersonic speed (a speed greater than the speed of sound in air), the bow wave is in the form of a shock wave that trails out and downward from the aircraft. When this high-pressure, compressed wavefront passes over an observer, he or she hears a sonic boom. The high-pressure bow wave travels with the supersonic aircraft, and the sonic boom does not occur at the instant the aircraft "breaks the sound barrier" or first exceeds the speed of sound. (There are actually two shock waves, one formed at the front and one at the back of the airplane.)

The Doppler effect is a general effect that occurs for all kinds of waves, such as water waves, sound waves, and light (electromagnetic) waves. In the Doppler effect for visible light, the frequency is shifted toward the blue end of the spectrum when the light source (such as a star or galaxy) is approaching. (Blue light has a shorter wavelength or higher frequency.) In this case, we say a Doppler *blue shift* has occurred. When a stellar light source moves away from us and the frequency is shifted toward the red (longer wavelength) end of the spectrum, we call this a Doppler **red shift**.

---

*After Christian Doppler (1803–1853), an Austrian physicist who first described the effect.

6.4 The Doppler Effect

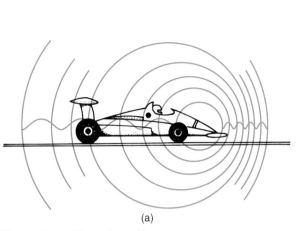

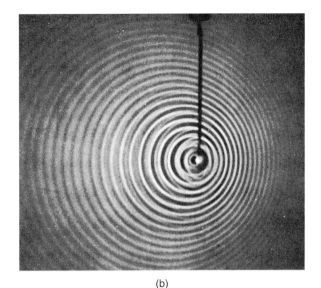

Figure 6.15 Doppler effect.
(a) Because of the motion of the source, illustrated here as a racing car, sound waves are "bunched up" in front and "spread out" in back. This results in shorter wavelengths (and higher frequencies) in front of the source and longer wavelengths (and lower frequencies) behind the source. (b) Doppler effect in water waves in a ripple tank. The source of the disturbance is moving to the right.

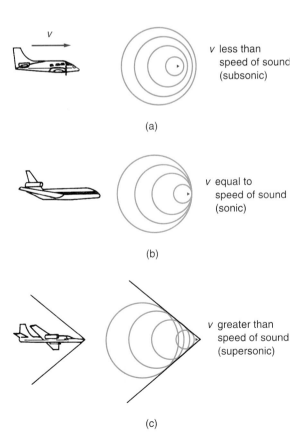

The magnitude of the frequency shift is related to the speed of the source. As we shall see in Chapter 18, light from galaxies shows red shifts, which indicate that they are moving away from us and that the universe is expanding. The degree or magnitude of the red shift is related to the relative recessional speed of a galaxy, and so it can be determined how far receding galaxies are from our own galaxy.

The Doppler shift of electromagnetic waves can be used to determine the speed of an astronomical object. It can also be used to determine the speeds of objects here on Earth. You may have encountered an example in terms of radar (radio

Figure 6.16 Bow waves and sonic booms.
Just as a fast-moving boat forms a bow wave in water, a moving aircraft forms a bow wave in air. The sound waves "bunch up" more in front of an airplane for increasing subsonic speeds [(a) and (b)]. A plane traveling at supersonic speeds forms a high-pressure shock wave in the air (c) that is heard as a sonic boom when it passes over an observer. There are actually two shock waves, one formed at the front and one formed at the back of the aircraft.

waves). *Radar* stands for *radio detecting and ranging*. The detecting and ranging (determining how far an object is away) is done by sending out radio waves, which bounce off an object. The returning waves are detected, and the elapsed time between the emission and receipt of the reflected waves allows the distance d of the object to be computed from $2d = c\Delta t$. (Sonar, which uses sound waves in water, operates on a similar principle.)

However, if the object (a vehicle, for instance) is moving, the Doppler effect comes into play and the returning waves have a frequency shift. This shift is proportional to the speed of the moving vehicle. When the data is fed into a computer in the trunk of a patrol car, your speed pops up on display almost instantaneously. Hopefully, you will not then see a red or blue light that has nothing to do with the Doppler effect.

6.5 Standing Waves and Resonance

Learning Goals:

To analyze standing waves and what produces them.

To explain what is meant by natural frequencies and resonance.

Most of us have shaken one end of a stretched rope and have observed wave patterns that seem to "stand" along the rope when we shake it just right. We refer to these waveforms as **standing waves** (sometimes called *stationary waves*). They are caused by the interference of waves traveling down and back along the rope. When two waves meet, they interfere, and the combined waveform of the superimposed waves is the sum of the waveforms or particle displacement of the medium.

Consider waves traveling in a rope in opposite directions, as illustrated in Fig. 6.17a. These waves may result from shaking one end of the rope, with the reflections returning from the fixed end. Notice that as the waves move, the crests and troughs periodically add and subtract. The zero points, or points that remain stationary, are called **nodes**. The points of maximum displacement are

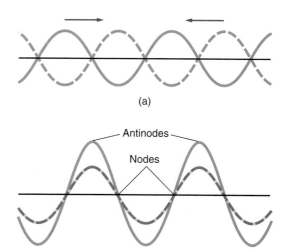
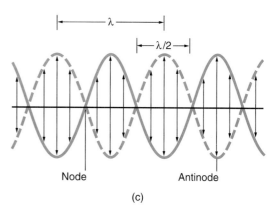

Figure 6.17 Standing waves.
Waves traveling in opposite directions in a stretched rope continually interfere so as to produce a standing wave with nodes and antinodes, as illustrated here.

called **antinodes** (Fig 6.17b). Notice that the distance between two nodes (or two antinodes) is one-half of a wavelength ($\lambda/2$) of the standing wave (Fig. 6.17c). That is, each standing wave "loop" is half a wavelength

Suppose a vibrating string of length L were fixed at both ends (Fig. 6.18). In this situation, there is a node at each end of the string. As a result, only particular numbers of wave loops can "fit in" the string; that is, $L = \lambda/2$, $L = 2(\lambda/2)$, $L = 3(\lambda/2)$, and so on. Hence a series of possible wavelengths is given generally by

$$\lambda_n = \frac{2L}{n} \qquad n = 1, 2, 3, 4, \ldots \qquad (6.4)$$

6.5 Standing Waves and Resonance

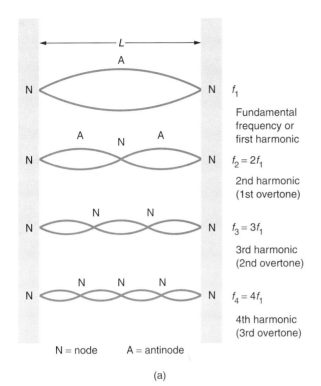

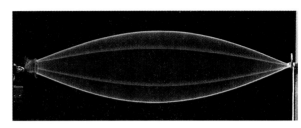

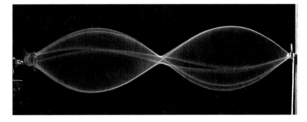

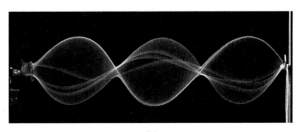

Figure 6.18 Natural frequencies.
(a) An illustration of a series of allowed vibrations, or standing waves, for a string fixed at each end. Notice that only half-wavelength segments can "fit in" the string because of the node conditions at each end. (b) Actual standing waves in a vibrating cord.

For example, for $n = 1$, we have $\lambda_1 = 2L$ or $L = \lambda_1/2$, which corresponds to the first case in Fig. 6.18.

The frequencies of the standing waves may be determined by using Eq. 6.1, $f = v/\lambda$. Substituting λ_n for the general wavelength (Eq. 6.4), we have

$$f_n = \frac{nv}{2L} \quad n = 1, 2, 3, 4, \ldots \quad (6.5)$$

These frequencies are referred to as the **characteristic** or **natural frequencies** of the string. The lowest frequency corresponds to $n = 1$, and $f_1 = v/2L$, which is called the **fundamental frequency** or the **first harmonic**. The higher frequencies, f_2, $f_3, \ldots$, are multiples of the fundamental frequency. Note that for $n > 1$, we have $f_n = nf_1$. Hence, $f_2 = 2f_1$, $f_3 = 3f_1$, etc., and these frequencies are called the *second harmonic, third harmonic,* and so on. Taken together, the harmonics higher than the first harmonic are called *overtones*.

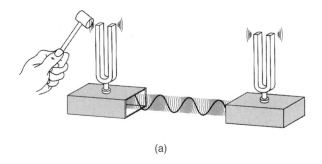

(a)

(b)

Figure 6.19 Resonance.
(a) When one tuning fork is activated, the other tuning fork of the same frequency will be driven in resonance and start to vibrate. (b) Unwanted resonance. The famous Tacoma Narrows Bridge, which collapsed after the wind drove the bridge into resonant vibrations.

CONFIDENCE QUESTION 6.3

If the frequency of the first harmonic of a stretched violin string is 440 Hz, what are the frequencies of the second harmonic (first overtone) and the third harmonic (second overtone)?

When a stretched string or object is acted upon by a periodic driving force with a frequency equal to one of its natural frequencies, a standing wave is formed and the oscillations have large amplitudes. This phenomena is called **resonance,** and when the system is driven in resonance, there is maximum energy transfer to the system.

A common example of driving a system in resonance is pushing a swing. A swing is essentially a pendulum with only one natural frequency, which depends on the length of the rope. When a swing is pushed periodically with a period of $T = 1/f$, energy is transferred to the swing and its amplitude gets larger (higher swings). If the swing is not pushed at its natural frequency, the pushing force may be applied to the swing as it approaches or after it has reached its maximum amplitude and is swinging back. In either case, the swing is not driven in resonance and the energy transfer is not a maximum.

A stretched string, as discussed above, has many natural frequencies—not just one, as a pendulum has. Hence, the string may be driven in resonance at various frequencies.

There are many examples of resonance. The structure of the throat and nasal cavities gives the human voice a particular tone due to resonances. Tuning forks of the same frequency can be made to resonate, as illustrated in Fig. 6.19a. A steel bridge or any elastic structure is capable of vibrating at natural frequencies, sometimes with dire consequences (Fig. 6.19b). Soldiers marching in columns across bridges are told to "break step" and not march at a periodic cadence, which might correspond to a natural frequency of the bridge and result in resonance and large oscillations that could cause structural damage.

Musical instruments use standing waves and resonance to produce different tones. Standing waves are formed on strings fixed at both ends on stringed instruments such as the guitar, violin, and piano. When a stringed instrument is tuned, a string is tightened or loosened, which adjusts the tension and the wave speed (v) in the string. This adjustment changes the frequency or pitch, because the length (and wavelength) of the string is fixed ($\lambda f = v$).

A vibrating string does not produce a great disturbance in air, but the body of a stringed instrument such as a violin acts as a sounding board that disturbs the air and produces more sound. The body of such an instrument acts as a resonant cavity, and sound comes out through holes in the top surface.

Similarly, standing waves are set up in wind instruments in air columns. Organ pipes have fixed lengths similar to fixed strings, so only a certain number of loops or wavelengths can be fitted in. However, the frequency or tone can be varied in some instruments, such as a trombone or trumpet, by varying the length of the air column.

Important Terms

These important terms are for review. After reading and studying the chapter, you should be able to define and/or explain each of them.

wave motion	intensity
wave velocity	decibel
transverse wave	pitch
longitudinal wave	quality
frequency	ultrasound
hertz	Doppler effect
period	red shift
wavelength	standing wave
electromagnetic waves	node
speed of light	antinode
sound	characteristic or natural frequencies
sound spectrum	fundamental frequency (first harmonic)
loudness	resonance

Important Equations

Wave frequency and period: $f = \dfrac{1}{T}$

Wave speed: $v = \dfrac{\lambda}{T} = \lambda f$

Speed of light (in vacuum): $c = 3.00 \times 10^8$ m/s $= 300{,}000{,}000$ m/s

Speed of sound (in air, 20°C): $v_{\text{sound}} = 344$ m/s (770 mi/h)

$(\approx \tfrac{1}{3}$ km/s or $\tfrac{1}{5}$ mi/s$)$

Characteristic frequencies (stretched string): $f_n = \dfrac{v}{\lambda_n} = \dfrac{nv}{2L} \quad n = 1, 2, 3, 4, \ldots$

Questions

6.1 Wave Properties

1. If a piece of ribbon were tied to a stretched string carrying a transverse wave, the ribbon might be observed to oscillate (a) up and down, (b) side-to-side, (c) either (a) or (b), (d) neither (a) nor (b).

2. The energy of a wave is related to its (a) frequency, (b) period, (c) amplitude, (d) wavelength.

3. What is the difference between transverse and longitudinal waves? Give an example of each. Do they propagate in all media? Explain.

4. What are the SI units of (a) wavelength, (b) frequency, and (c) period?

5. Does the equation $v = \lambda f$ give the correct units for speed? Explain.

6.2 Electromagnetic Waves

6. Electromagnetic waves (a) have different speeds in vacuum for different frequencies, (b) require a medium for propagation, (c) are longitudinal waves, (d) are transverse waves.

7. Which one of the following regions lies above the visible region in the EM frequency spectrum: (a) radio wave, (b) ultraviolet, (c) microwave, (d) infrared?

8. Which end of the visible spectrum has the longer wavelength? The higher frequency?

9. Are radio waves sound waves? Explain.

10. What is the range of the wavelengths of visible light? How do these wavelengths compare to those of audible sound?

6.3 Sound Waves

11. Sound waves propagate in (a) solids, (b) liquids, (c) gases, (d) all of these.

12. The upper frequency limit of the audible range of human hearing is about (a) 20 kHz, (b) 2000 Hz, (c) 2 kHz, (d) 2,000,000 Hz.

13. What is the chief physical wave property that describes (a) pitch, (b) loudness, and (c) quality?

14. Can ultrasound be heard? Give some applications of ultrasound.

15. Why does the music from a marching band in a spread out formation on a football field sometimes sound discordant?

16. Does doubling the decibels of a sound level double the intensity? Explain.

6.4 The Doppler Effect

17. A moving observer approaches a stationary sound source. The observer would hear (a) an increase in frequency, (b) a decrease in frequency, (c) the same frequency of the source.

18. If an astronomical light source is moving away from us, we observe (a) a blue shift, (b) a shift toward

longer wavelengths, (c) a shift toward increased frequency, (d) a sonic boom.

19. How is the wavelength of sound affected when (a) a source moves toward a stationary observer, and (b) an observer moves away from a stationary source?

20. What would be the situation for a sound (a) "blue shift" and (b) "red shift"?

21. Compare the crack of a whip with a sonic boom.

6.5 Standing Waves and Resonance

22. Stationary points in a standing wave are called (a) normal modes, (b) zero points, (c) nodes, (d) antinodes.

23. A stretched string will be driven in resonance when driven at the frequency of its (a) first overtone, (b) second overtone, (c) first harmonic, (d) all of these.

24. Why can only half-wavelengths be fitted into a vibrating stretched string?

25. (a) How many harmonics does a stretched string have? (b) How many harmonics does a pendulum have?

26. What is the effect if a system is driven in resonance? Is a particular frequency required?

27. What determines the pitch or frequency of a string on a violin? How does the violinist get a variety of notes from one string?

Food for Thought

1. Here's an old one: If a tree falls in the forest and no one is present to hear it, is there sound?

2. If an astronaut on the Moon dropped a hammer, would there be sound? Explain. (*Follow-up:* How do astronauts communicate with each other and mission control?)

3. Discuss the effects if (a) a sound source and an observer were both moving with the same velocity, and (b) a sound source approached a stationary observer going faster than the speed of sound.

4. If a jet pilot is flying faster than the speed of sound, will he or she be able to hear any sound?

5. When one sings in the shower, the tones sound full and rich. Why is this?

Exercises

6.1 Wave Properties

1. A periodic wave has a period of 3.0 s. What is the wave frequency?

2. Waves moving on a lake have a speed of 2.0 m/s, with a distance of 5.0 m between adjacent crests.
   (a) Determine the frequency of the waves.
   (b) Find the period of the wave motion.
   Answer: (a) 0.40 Hz

3. A sound wave has a frequency of 200 Hz. What is the distance between adjacent crests or compressions of the wave? (Assume an air temperature of 20°C.)

4. The speed of sound in water is 1530 m/s. What is the wavelength of a 2000-Hz sound wave in water?

6.3 Sound Waves

5. Compute the wavelength of ultrasound with a frequency of 30 kHz if the speed of sound in water is 1500 m/s.
   Answer: 0.050 m

6. During a thunderstorm, 4 s elapse between observing a lightning flash and hearing the resulting thunder. Approximately how far away (in km and mi) was the lightning flash? (*Hint:* Use the fractional approximations.)
   Answer: 4/3 km or 4/5 mi

7. A subway train has a sound intensity level of 90 dB, and a rock band has a sound intensity level of 110 dB. How many times greater is the sound intensity of the band than the subway train?
   Answer: 100 times

8. In Table 1 of the chapter Highlight, the maximum exposure limit decreases from 6 to 4 hours when the sound intensity level goes from 92 to 95 dB. By what factor does the intensity increase for this dB increase?

Answer: doubles

9. A rock band with a sound intensity level of 123 dB turns its speakers down so that the intensity is reduced by a factor of 100. What is the new intensity level in dB?

10. A sound speaker is playing with an output of 80 dB. If the volume is turned up so that the output intensity is 10,000 times greater, what will be the new intensity level?

Answer: 120 dB

6.5 Standing Waves and Resonance

11. (a) What is the fundamental frequency of a stretched string 1.0 m long if the wave speed in the string is 240 m/s? (b) What is the frequency of the second harmonic (first overtone?)

Answer: (b) 240 Hz

12. If the frequency of the first harmonic of a stretched string is 256 Hz, what are the frequencies of the first and second overtones?

Answer: 512 Hz and 768 Hz

13. The frequency of the third harmonic of a stretched string is 660 Hz. What is the frequency of the fourth harmonic of the string?

Answers to Multiple-Choice Questions

1. c   6. d   11. d   17. a   22. c
2. c   7. b   12. a   18. b   23. d

Solutions to Confidence Questions

6.1 $f = \dfrac{v}{\lambda} = \dfrac{344 \text{ m/s}}{1.00 \text{ m}} = 344 \text{ Hz}$

6.2 $\Delta dB = 100 \text{ dB} - 60 \text{ dB} = 40 \text{ dB}$, so sound intensity increases by a factor of 10,000.

6.3 $f_2 = 2f_1 = 2(440 \text{ Hz}) = 880 \text{ Hz}$
$f_3 = 3f_1 = 3(440 \text{ Hz}) = 1320 \text{ Hz}$

Chapter 7

Wave Effects and Optics

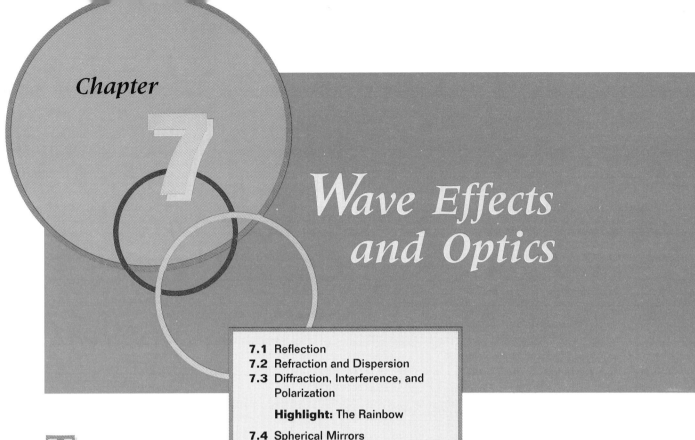

7.1 Reflection
7.2 Refraction and Dispersion
7.3 Diffraction, Interference, and Polarization

Highlight: The Rainbow

7.4 Spherical Mirrors

Highlight: Liquid Crystal Displays (LCDs)

7.5 Spherical Lenses

The effects of waves—particularly sound and light waves—are all around us. We are aware of many of these effects, but they are such a part of our experience that we take them for granted and rarely try to analyze them. For example, if you speak loudly, a person around the corner in another room can hear you, indicating that sound waves are "bent" around corners. But light waves don't appear to bend. That is, you can be heard in the next room but not seen.

When light waves are incident on a soap bubble or an oil slick, we see brilliant displays of colors. Similarly, we sometimes see a colorful rainbow. These phenomena can be described and explained through the effects and interactions of light (electromagnetic) waves.

Mirrors and lenses are commonplace; we look into mirrors daily and many of us wear lenses (glasses). The description of how mirrors and lenses work involves the concepts of reflection and refraction. We will discuss these two effects and then describe the basic principles of mirrors and lenses. From this discussion you will gain an understanding of many optical devices such as the human eye, slide projectors, and eyeglasses.

There are many wave effects. In fact, two of them—the Doppler effect and resonance—have already been discussed (Chapter 6). In this chapter we will consider other important wave phenomena that affect us all the time.

7.1 Reflection

Learning Goals:

To explain the law of reflection.

To distinguish between regular and diffuse reflections.

Waves travel through space in a straight line and will continue to do so unless by some means deviated from their original direction. When waves strike a material surface or boundary, they may rebound with a change in direction. A change in direction by this method is called **reflection**. A common example is the reflection of sound, which we call an *echo*.

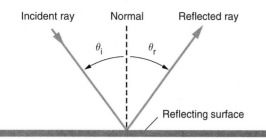

Figure 7.1 Law of reflection.
The angle of incidence (θ_i) equals the angle of reflection (θ_r) relative to the normal (a line perpendicular) to the reflecting surface. The ray and the normal line lie in the same plane.

For light, reflection may be thought of as light "bouncing" off a surface. However, it is really much more complicated and involves the absorption and emission of light by the atoms of the reflecting medium. To describe reflection in a simple manner, we consider the reflection of light rays. A **ray** is a straight line that represents the path of light. A beam of light may be thought of as a group of parallel rays.

The reflection of a light ray is illustrated in Fig. 7.1. The reflection takes place in a particular way. The directions of the incident and reflected rays are indicated by angles that are measured relative to the *normal;* that is, relative to a line perpendicular to the reflecting surface. A special relationship of the incident and reflected rays is described by the **law of reflection:**

> The angle of reflection θ_r is equal to the angle of incidence θ_i.

That is, $\theta_i = \theta_r$. Also, the incident and reflected rays are in the same plane.

CONFIDENCE QUESTION 7.1

If a light ray were incident on a plane mirror surface at an angle of 50° relative to the surface, what would be the angle of reflection?

The reflection from very smooth or mirror surfaces is called **regular (or specular) reflection** (Fig. 7.2). In regular reflection, incident parallel rays are parallel on reflection. However, rays reflected from relatively rough surfaces are not parallel; this is called **irregular (or diffuse) reflec-**

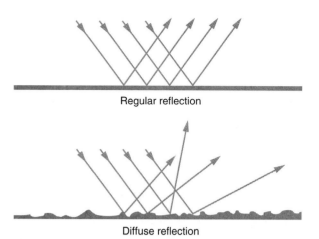

Figure 7.2 Reflection.
A smooth (mirror) surface produces regular or specular reflection. A rough surface produces irregular or diffuse reflection.

tion. The reflection from the page of this book is diffuse reflection. The law of reflection applies to both types of reflections, but the rough surface causes the light rays to be reflected in different directions, thereby diffusing it. Regular reflections from smooth or mirror surfaces, on the other hand, produces images that we can see.

The reflection of rays from a mirror surface may be used to determine the characteristics of the image formed by a mirror. A ray diagram for determining the apparent location of an image formed by a flat or plane mirror is shown in Fig. 7.3.

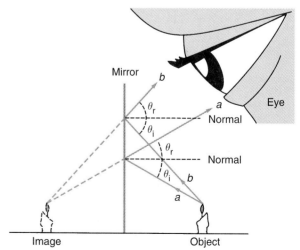

Figure 7.3 Ray diagram.
By tracing the reflected rays, you can locate mirror images where the rays intersect or appear to intersect, behind the mirror.

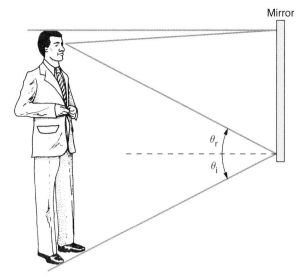

Figure 7.4 Complete figure.
For a person to see his or her complete figure in a plane mirror, the height of the mirror must be at least one-half of the height of the person, as can be easily shown by ray tracing.

Figure 7.5 Natural reflection.
Beautiful reflections, such as shown here for a water surface, are often seen in nature.

The image is located by drawing two (or more) rays emanating from the object and applying the law of reflection. The image is located where the rays intersect or appear to intersect. Notice that for a plane mirror the image is located "inside" or behind the mirror at the same distance the object is in front of the mirror.

Figure 7.4 shows a ray diagram for the light rays involved when a person sees a complete or head-to-toe image. How big (tall) a mirror is needed for this? Applying the law of reflection reveals that one can see one's complete image in a plane mirror that is only one-half of one's height. Also, the distance one stands from the mirror is not a factor.

It is the reflection of light that allows us to see things. Look around you. What you see in general is light reflected from the walls, the ceiling, the floor, and other objects. Of course, there must be one or more sources of light present, such as the Sun or lamps. If you are in a completely dark room, then there is no reflected light and you can't see anything.

You have probably noticed that at night in a lighted room, a transparent glass windowpane reflects light and acts as a mirror. Yet during the day, we see through it. Why is this? The glass itself doesn't act any differently night or day. Light is still reflected back into the room during the day. However, the large amount of transmitted light from the outside masks the reflected light during the day, whereas during the night the masking effect of transmitted light is absent.

We often see beautiful reflections in nature, as shown in Fig. 7.5. Is the picture really right-side-up? Turn the book over and see.

7.2 Refraction and Dispersion

Learning Goals:

To explain the phenomenon of refraction and how this leads to the dispersion of light.

To explain how the boundary of transparent media can be used as a mirror through total internal reflection, and how this is applied in fiber optics.

To define dispersion and describe some of its effects.

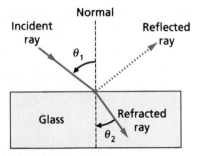

Figure 7.6 Refraction.
When light enters a transparent medium at other than normal incidence, it is deviated, or refracted, from its original path. As illustrated here for light passing from air into a denser medium such as glass, the rays are refracted or "bent" toward the normal ($\theta_2 < \theta_1$, that is, the angle of refraction is less than the angle of incidence). Some light is also reflected at the surface.

When light strikes a transparent medium, such as glass or water, some light is reflected and some is transmitted, as illustrated in Fig. 7.6. On investigation, we find that for other than normal incidence, the transmitted light has changed direction. The deviation of waves from an original path in going from one medium to another is called **refraction**. The refraction of a beam of light in water is illustrated in Fig. 7.7. You have probably observed refraction effects for an object in a glass of water. For example, a fork or pencil in a glass of water will appear to be displaced and perhaps severed. Try it.

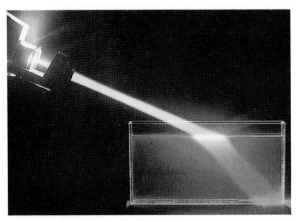

Figure 7.7 Refraction in action.
A beam of light is refracted; that is, its direction is changed on entering the water tank.

The directions of the incident and refracted rays are expressed in terms of the *angle of incidence* θ_1 and the *angle of refraction* θ_2, which are measured relative to a normal (line perpendicular) to the surface boundary of the media (see Fig. 7.6). When light passes obliquely ($\theta_1 > 0°$) into a denser medium—for example, from air into water or glass—the light rays are refracted or "bent" toward the normal ($\theta_2 < \theta_1$, that is, θ_2 is less than θ_1). Complex processes are involved. However, we might intuitively expect that the passage of light by atomic absorption and emission through a denser medium to take a longer time than through a less dense material of equal thickness. Therefore, the speed of light would be different in the different media, with the speed being less in the denser medium. This prediction is indeed the case, and we find that the speed of light in water is about 75% of that in air or vacuum. (Technically, we should talk in terms of *optical* density. This is different than mass density; however, in general the greater the mass density, the greater the optical density.)

CONFIDENCE QUESTION 7.2

When light passes from a denser medium into a less dense medium, for example, from glass to air, how are the rays deviated?

To help understand how light is bent or diffracted when it passes into another medium, consider a band marching across a field and entering a wet, muddy region obliquely (at an angle), as illustrated in Fig. 7.8a. As the marchers enter the muddy region, they keep marching at the same frequency (cadence). But slipping in the muddy ground, the marchers don't cover as much ground and are slowed down.

The marchers in the same row on solid ground continue on with the same stride, and as a result, the direction of the marching column is changed as it enters the muddy region. This change in direction with change in marching speed is also seen when a marching band turns a corner and the inner members of a row mark time.

We may think of the wave fronts as analogous to marching rows (Fig. 7.8b). In the case of light, the wave frequency (cadence) remains the same, but the wave speed and wavelength change ($c_m = \lambda_m f$).

7.2 Refraction and Dispersion

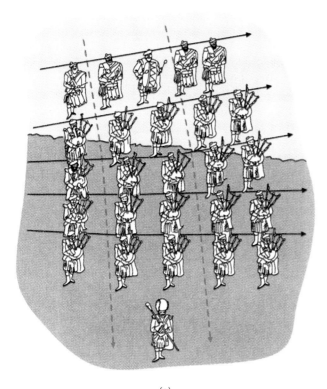

(a)

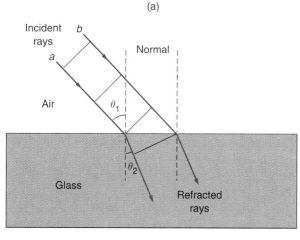

(b)

Figure 7.8 Refraction analogy.
(a) Marching obliquely into a muddy field causes a band to change direction. The cadence or frequency remains the same, but the marchers in the mud slip and travel a shorter distance (shorter wavelength). (b) This is analogous to the refraction of a wave front.

The different speeds in the different media are expressed in terms of the ratio of the speeds. This ratio is known as the **index of refraction** n:

Table 7.1 Indices of Refraction of Some Common Substances

Substance	$n = \dfrac{c}{c_m}$
Water	1.33
Crown glass	1.52
Diamond	2.42
Air (0°C, 1 atm)	1.00029
Vacuum	1.00000

$$\text{index of refraction} = \frac{\text{speed of light in vacuum}}{\text{speed of light in medium}}$$

or

$$n = \frac{c}{c_m} \quad (7.1)$$

The index of refraction is a pure number with no units, since c and c_m have the same units, which cancel. The indices of refraction of some common substances are given in Table 7.1. Notice that the index value for air is close to that for vacuum, so we can use air when determining the index of refraction of a medium.

In general, the greater the index of refraction for a material, the greater its density and the slower the speed of light in it. Also, when a light ray enters a medium with a greater index of refraction, it is bent toward the normal. Why?

EXAMPLE 7.1 Finding the Percentage of Speed

What percentage of the speed of light in vacuum is the speed of light in water?

Solution

To find the percentage, one puts the speed of light in water (c_m, the medium) over the speed of light in vacuum (c), that is c_m/c, and then changes to a percent. This is similar to finding what percentage 40 cents is of a dollar. You would form the ratio $40/100 = 0.40$ ($\times 100\%) = 40\%$.

Noting that the index of refraction, $n = c/c_m$, is the inverse ratio we want, we may write

$$\frac{c_m}{c} = \frac{1}{n} = \frac{1}{1.33} = 0.75 \; (\times 100\%) = 75\%$$

where the index of refraction for water, $n = 1.33$, was obtained from Table 7.1.

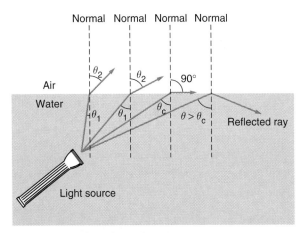

Figure 7.9 Internal reflection.
When light goes from a denser medium (such as water) into a less dense medium (such as air), it is refracted away from the normal. At a certain critical angle, θ_c, the angle of refraction is 90°. For incidence above the critical angle, the light is internally reflected and the surface acts as a mirror.

CONFIDENCE QUESTION 7.3

What percentage of the speed of light in vacuum is the speed of light in air?

When light goes from a denser medium into a less dense medium—for example, from water into air—the rays are refracted, and bent away from the normal. We may see this refraction in Fig. 7.6 by tracing the ray in reverse, from glass into air (denser medium into a less dense one).

This type of refraction is also shown in Fig. 7.9 on the left. But notice that an interesting thing happens as the angle of incidence becomes larger. The refracted ray is bent farther from the normal; and at a particular *critical angle* θ_c, the refracted ray is along the boundary of the two media. For angles greater than θ_c, the light is reflected and none is refracted. This phenomenon is called **total internal reflection.** An illustration of refraction and total internal reflection is shown in Fig. 7.10a. With total internal reflection, a prism can be used as a mirror.

Internal reflection is used to enhance the brilliance of diamonds. In the so-called brilliant cut, a diamond is cut so the light entering the diamond is internally reflected (Fig. 7.10b). The light emerging from the upper portion gives the diamond its beautiful sparkle.

Another example of total internal reflection occurs when a fountain of water is illuminated from below. The light is totally reflected within the streams of water, providing a spectacular effect. Similarly, light can travel along transparent plastic tubes called "light pipes." When the incident angle is greater than the critical angle, the light undergoes a series of internal reflections down the tube (Fig. 7.11a).

Light can also travel along thin fibers, and bundles of such fibers are used in the relatively new field of fiber optics (Fig. 7.11b). You have probably seen fiber optics used in decorative lamps. An important use of the flexible fiber bundle is to pipe light to hard-to-reach places by use of a fiberscope (Fig. 7.11c). Light may also be transmitted down one set of fibers and reflected back through another so that an image of the illuminated area may be seen. In medical applications, fiber optics can illuminate a person's stomach or heart; in industrial applications, it can allow a view of some recessed region in a machine that cannot easily be seen. Fiber optics are also used in telephone communications. In this application, electronic signals in wires are replaced with light signals in fibers.

Dispersion

The index of refraction for a material actually varies slightly with wavelength. This means that when light is refracted, the different wavelengths of light are bent at slightly different angles. This phenomenon is called **dispersion.** When white light (light containing all wavelengths of the visible spectrum) passes through a glass prism, as shown in Fig. 7.12 on page 146, the light rays are refracted on entering the glass; and with different wavelengths refracted at slightly different angles, the light is dispersed into a spectrum of colors (wavelengths). The amount of refraction is a function of the wavelength, with the shortest wavelengths being deviated from their path by the greatest amount. Violet light has shorter wavelength than red light, so it has the greater deviation.

A diamond is said to have "fire" because of colorful dispersion. It also has brilliance due to internal reflection.

Every substance, when sufficiently heated, emits a spectrum of characteristic frequencies or

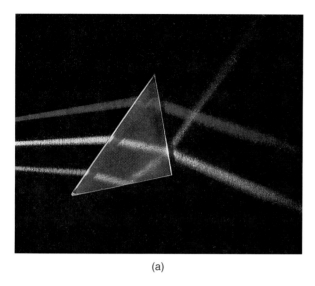

(a)

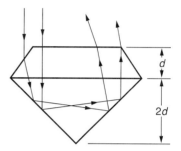

(b)

Figure 7.10 Refraction and total internal reflection.
(a) Beams of colored light from the left are incident on a triangular piece of glass. As the light passes from air into glass, it is refracted, or bent, toward the normal. As the light passes from glass into air, it is refracted away from the normal. The incident angle of the blue beam at the first glass-air interface exceeds the critical angle θ_c and is internally reflected. (b) Refraction and internal reflection give rise to the "brilliance" of a diamond. In the so-called brilliant cut, a diamond is cut with a certain number of faces or facets, along with the proper depths to give the proper refraction and internal reflection.

wavelengths. By dispersing the light into a spectrum and analyzing its characteristic frequencies, scientists can identify a substance. This study of spectra is called *spectroscopy*. A great deal of basic information has been gained by astronomers, chemists, physicists, and other scientists who use the spectra of emitted light and other electromagnetic radiation to study the universe. A prism spectrometer and some spectra are shown in Fig. 7.13 on page 147.

The topics of this section are applied to the rainbow in the chapter Highlight.

7.3 Diffraction, Interference, and Polarization

Learning Goals:

To explain the phenomena of diffraction, interference, and polarization.

To identify some applications of these phenomena.

Diffraction

Water waves passing through a small slit are shown in Fig. 7.14 on page 148. Notice how the waves are bent, or deviated, around the corners of the slit as they pass through. All waves—sound, light, and so on—show this type of bending as they go through relatively small slits or pass by corners. The deviation of waves in such cases is referred to as **diffraction.**

For diffraction to occur effectively, the size of the slit (or object) must be on the same order or smaller than the wavelength of the wave ($d \leq \lambda$,

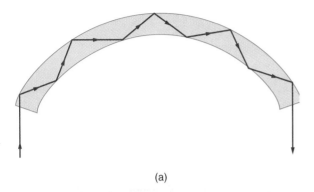

(a)

(b)

(c)

Figure 7.11 Fiber optics.
(a) When light is incident on a fiber above the critical angle, it is internally reflected along the fiber, which acts as a light pipe. (b) A fiber-optic bundle held between a person's fingers. Notice how the ends of the fibers are lit because of the transmission of light by multiple internal reflections. (c) A fiber-optics application. A doctor looks through an endoscope used for digestive tract operations. Light can be reflected down some of the fibers and back through others, allowing a view of otherwise inaccessible places.

where d is the width of the slit). As you might visualize, with a very large slit and relatively short wavelength (distance between adjacent crests), the wave would pass through the slit with little effect. We are well aware that sound waves bend around everyday objects, while light waves do not. For instance, if we hold a newspaper in front of our face and speak, the sound of our voice can be heard by someone in front of the paper, but our face cannot be seen. The slit (or object)-to-wavelength condition for diffraction explains why.

We know that audible sound waves have wavelengths of centimeters to meters (Chapter 6), while visible light waves have wavelengths of around 10^{-6} m (a millionth of a meter). Ordinary objects (and slits) have dimensions of centimeters to meters. Thus, with the wavelengths on the same order as the dimensions of objects, diffraction readily occurs for audible sound. However, the dimensions of ordinary objects or slits are much greater than the wavelengths of visible light, so the diffraction of light is not commonly observed.

HIGHLIGHT

The Rainbow

A beautiful atmospheric phenomenon commonly seen after rain is the rainbow. The colorful are of a rainbow across the sky is the result of several optical effects: refraction, internal reflection, and dispersion. But the conditions must be just right. As we all know, a rainbow is seen after a rain but not after *every* rain.

Following a rain, there are many tiny water droplets in the air. Sunlight incident on the droplets produces a rainbow. But whether a rainbow is seen depends on the relative positions of the Sun and the observer. As you may have noticed, the Sun is generally behind you when you see a rainbow.

To understand the formation and observation of a rainbow, consider what happens when sunlight is incident on a water droplet. On entering the droplet, the light is refracted and then dispersed into component colors as it travels in the water (see Fig. 1a).

When the light enters the droplet above the critical angle, it is internally reflected and the color components emerge from the droplet at slightly different angles. Because of the conditions for refraction and internal reflection, the component colors lie in a narrow range of 40° to 42° for an observer on the ground.

Thus you see the display of colors only when the Sun is positioned so that the dispersed light is reflected to you through these angles. With this condition satisfied and an abundance of water droplets in the air, you see the colorful arc of a *primary rainbow* with colors running vertically from blue to red (see the figure, parts a and c).

Occasionally, conditions are such that you see sunlight that has undergone two internal reflections in water droplets. The result is a vertical inversion of colors in a higher, fainter, and less frequently seen *secondary rainbow* (see Fig. 1b and 1c). Note the bright region below the primary rainbow. Light from the rainbows combines to form this illuminated region. (See the accompanying photo of primary and secondary rainbows.)

The arc length of a rainbow that you see depends on the altitude (angle above the horizon) of the Sun. As the Sun's altitude increases, you see less of the rainbow. On the ground you cannot see a (primary) rainbow if the altitude of the Sun is greater than 42°. The rainbow is below the horizon in this case. However, if your elevation for viewing a rainbow is increased, you see more of the rainbow arc. For instance, airplane passengers commonly view a completely circular rainbow, similar to the miniature "rainbow" that can be seen in the mist produced by a lawn sprayer.

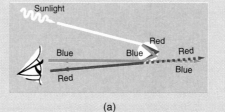

(a)

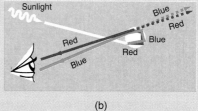

(b)

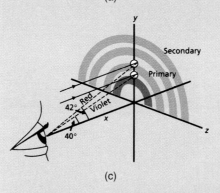

(c)

Figure 1 Rainbow formation. Sunlight may be internally reflected once (a) or twice (b) in a water droplet. The dispersion of the sunlight in the droplets produces the separation of colors, and an observer may see arcs or bows of colors in particular angular regions (c). Both the Sun and the observer must be in the proper positions for the observer to be able to see a rainbow, as in the photo below. Notice the fainter secondary rainbow over the primary rainbow.

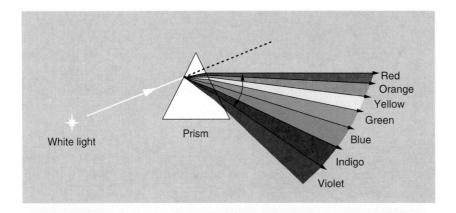

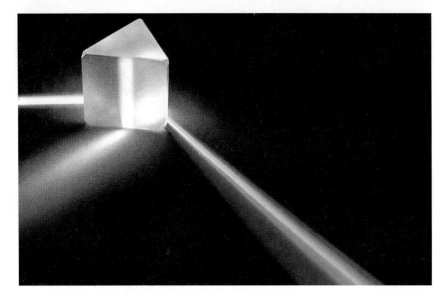

Figure 7.12 Dispersion.
Above: An illustration of white light being dispersed into a spectrum of colors by a prism because the shorter wavelengths toward the blue end of the spectrum are refracted more than the longer wavelengths toward the red end. *Below:* An actual spectrum produced by dispersion in a prism.

Some light diffraction does occur at corners, but this goes largely unnoticed because it is difficult to see. When we speak to someone around the corner in another room, they hear you because of diffraction, even though you are in an acoustical shadow, so to speak. However, when light goes by the corner, a sharp shadow boundary is produced and there is no light in the shadow area such that you may see around the corner. On very close inspection, you would find that the shadow boundary is blurred or fuzzy and that there is actually a pattern of bright and dark regions. This is an indication that some diffraction has occurred, as will be learned later in the section.

As another example, radio waves are electromagnetic waves of very long wavelength, some being hundreds of meters. In this case, ordinary objects and slits are much smaller than the wavelength, and thus radio waves are easily diffracted around buildings, trees, and so on, making radio reception quite efficient.

Interference

Interference occurs when waves meet and interact. Water waves meeting on the surface of a lake interfere with each other. When this occurs, changes in the combined waveform are produced by amplitude and phase relations of the two or more waves interfering. The amplitude of a wave is the magnitude of the maximum displacement of the wave disturbance (Section 6.1). When two or

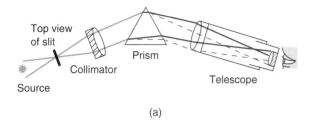

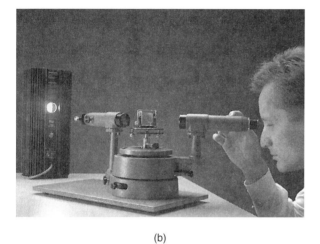

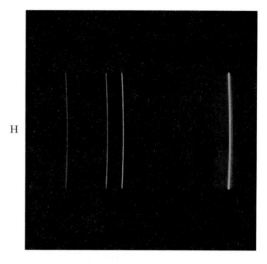

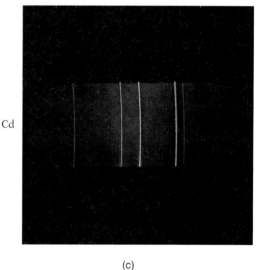

Figure 7.13 Spectroscope and spectrometer. (a) A spectroscope is an instrument used to separate light into its respective component wavelengths or colors. (b) Equipped with a scale to measure the angular deviations from which the wavelengths may be calculated, the instrument becomes a spectrometer. (c) Examples of line spectra whose wavelengths can be measured with a spectrometer.

more similar waves interfere, the combined waveform may be greater than any individual wave if the particle displacements of the waves are changing in the same direction. Or, the combined waveform may be smaller than any individual wave if the particle displacements are changing in opposite directions.

When waves have the same particle displacements (in the same direction at the same time), we say that the waves are *in phase*. If their displacements at all times are in opposite directions, we say that the waves are completely *out of phase*. For these conditions, the waves must have the same frequency. (Why?) Figure 7.15 illustrates this concept, as well as that of interference.

Suppose two waves in phase [(1) and (2) in figure] meet traveling in opposite directions. When exactly superimposed, the particle displacements add, and a waveform with twice the amplitude is observed (4). This is called *complete* **constructive interference.** How about when two waves are completely out of phase [(1) and (3) in figure)]? The displacements still add when exactly superimposed, but in this case the particle of one wave is going up and the other going down, so they cancel and the waveform is flat (5). The crests fill in the troughs, so to speak. This is called *complete* **destructive interference.**

For complete destructive interference, the two waves must have the same amplitude as well as the

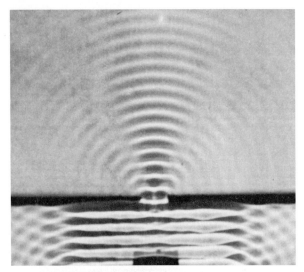

Figure 7.14 Diffraction.
Plane water waves are diffracted as they pass through a slit. Notice how the waves bend or curve around the corners of the slit.

face of water or on a wet road (Fig. 7.16). Part of the light is reflected at the air-oil surface and part is transmitted. Part of the light that travels through the oil film is then reflected at the oil-water surface.

The two reflected waves may be in phase, totally out of phase, or somewhere in between. In Fig. 7.16, the waves are shown in phase, but this result will occur only for certain angles of observation, wavelengths of light (colors), and thicknesses

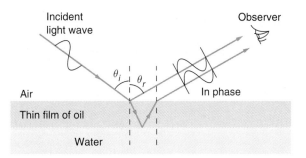

same wavelength. This way, a crest exactly oppositely matches an inverted trough of the other wave.

The colorful displays seen in oil films and soap bubbles can be explained by interference. Consider light waves incident on a thin film of oil on the sur-

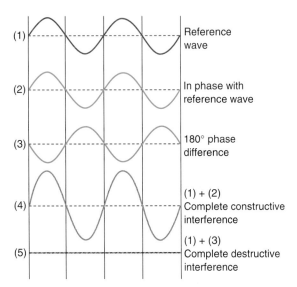

Figure 7.15 Phase differences.
Waves (1) and (2) are in phase. If the waves overlap, complete constructive interference occurs. Waves (1) and (3) are 180° or completely out of phase. If these waves overlap, complete destructive interference occurs.

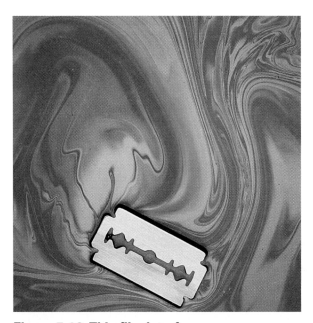

Figure 7.16 Thin-film interference.
When the reflected rays from the bottom and top surfaces of the oil film are in phase, constructive interference occurs and the observer sees only one color of light for a certain angle and film thickness. If the reflected waves are out of phase, destructive interference occurs, which physically means that the light is transmitted at the bottom of the film surface rather than reflected. Because such an oil film can vary in thickness, a colorful display is seen for the different wavelengths of light.

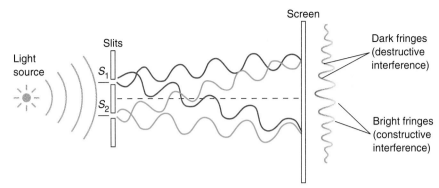

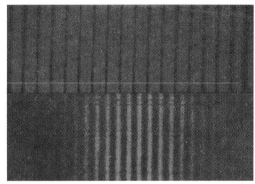

Figure 7.17 Double-slit interference.
Above: Diffracted light through two narrow slits, which act as point sources, interferes, giving rise to regions of constructive interference, or bright fringes, and regions of destructive interference, or dark fringes. *Right:* Actual double-slit interference patterns with different colors of light.

of oil film. At certain angles and oil thicknesses, only one wavelength of light shows constructive interference. The other visible wavelengths interfere destructively. This means that these wavelengths are actually transmitted and not reflected.

Hence, different wavelengths interfere constructively for different oil film thicknesses, and an array of colors is seen. In soap bubbles, where the thickness of the soap film moves and changes with time, so does the array of colors.

Diffraction can also give rise to interference. This interference can arise from the bending of light around the corners of a single slit (see the regions of constructive and destructive interference in Fig. 7.14). An instructive technique employs two narrow double slits that can be considered point sources of light, as illustrated in Fig. 7.17. When the slits are illuminated with monochromatic (single wavelength) light, the diffracted light through the slits spreads out and interferes constructively and destructively at different points where crest meets crest and crest meets trough, respectively. By placing a screen a distance from the slits, an observer can see a pattern of alternate bright and dark slit images or fringes.

This experiment was done in 1801 by the English scientist Thomas Young. It demonstrated the wave nature of light, and also allows the computation of the wavelength of the light from the geometry of the experiment. This is quite an accomplishment, since, as you recall, the wavelength of visible light is on the order of 10^{-6} m.

This interference principle may be extended and applied by adding more slits in the form of a diffraction grating. A **diffraction grating** consists of many narrow, parallel slits spaced very closely together. There may be 10,000 lines or more per centimeter on a diffraction grating. Lines may be ruled on glass, or a thin film of aluminum may be deposited on a glass plate, then some of the metal removed to make regularly spaced, parallel lines. With the advent of modern technology, lasers are used to produce diffraction gratings. The interference patterns produced by diffraction gratings are very sharp compared with patterns for only two slits. Certain wavelengths are seen at certain angles (Fig. 7.18a), and diffraction gratings can be used to separate the wavelengths of light, similar to the dispersion of a prism. For example, the grooves of a compact disc (CD) act as a diffraction grating and the separation of colors is evident (Fig. 7.18b).

Diffraction gratings are used in spectroscopy to analyze light from various sources. Different elements and compounds, when heated and made

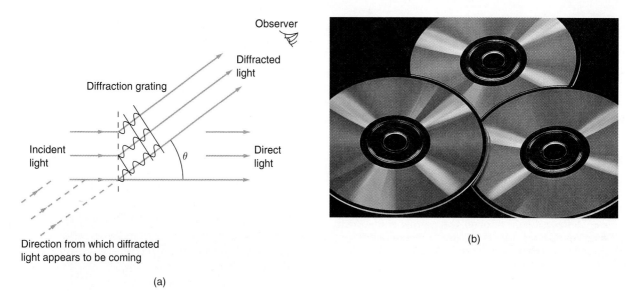

Figure 7.18 Diffraction-grating interference pattern.
(a) An illustration of interference by a diffraction grating. See text for description.
(b) The grooves of a compact disc (CD) act as a diffraction grating and give rise to colorful displays.

incandescent, give off certain characteristic spectra. From the spectra, the composition of a light source can be analyzed and identified. For example, the gaseous element helium was first identified in the solar spectrum—hence the name *helium* (from *helios*, the Greek word for the Sun).

Polarization

Light waves are transverse electromagnetic waves with electric and magnetic field vectors oscillating perpendicular to the direction of propagation (see Fig. 6.7). The atoms of a light source generally emit light waves that are randomly oriented, and a beam of light has transverse field vectors in all directions. When we see a beam of light from the front, the transverse electric field vectors may be depicted as shown in Fig. 7.19a. (Magnetic field vectors are not shown, for simplicity.)

In the figure, the electric field vectors are in a plane perpendicular to the direction of propagation. Such light is said to be unpolarized—that is, the field vectors are randomly oriented. **Polarization** refers to the preferential orientation of the electric field vectors. If there is some partial preferential orientation of the field vectors, the light is said to be partially polarized (Fig. 7.19b). If the vectors are in a single plane, the light is said to be **linearly (or plane) polarized** (Fig. 7.19c).

A light wave may be polarized by several means. A common method is the use of a Polaroid film. Polaroid film has a polarization direction associated with the long molecular chains of the polymer film. The polarizer allows only the components in a plane to pass, as illustrated in Fig.

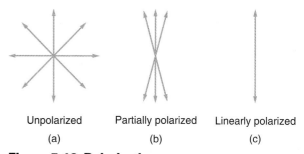

Figure 7.19 Polarization.
(a) When the electric field vectors are randomly oriented as viewed along the direction of propagation, the light is unpolarized. (b) With preferential orientation, the light is partially polarized. (c) If the field vectors lie in a plane, the light is linearly polarized.

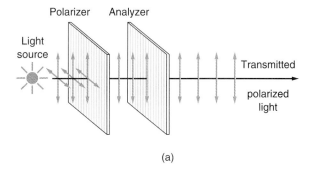

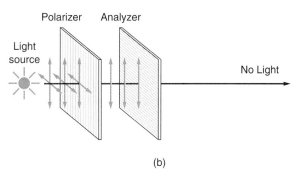

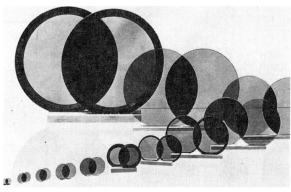

Figure 7.20 Polarized light.
(a) Light is polarized when it passes through a polarizer, and it also passes through the analyzer, if it is similarly oriented. (b) When the polarization direction of the analyzer is perpendicular to that of the polarizer ("crossed Polaroids"), little or no light is transmitted. (c) The actual conditions of (a) and (b).

7.20a. The other field vectors are absorbed and do not pass through the polarizer.

The human eye cannot generally detect polarized light, so an analyzer, perhaps another polarizing sheet, is needed. If a second polarizer is placed in front of another polarizer with the polarization directions at 90° as illustrated in Fig. 7.20b, then little or no light is transmitted. Hence we know that the light coming through the first polarizer was polarized light. In this orientation, the Polaroid sheets are said to be "crossed" (Fig. 7.20c). The polarization of light is experimental proof that light is a transverse wave. Longitudinal waves, such as sound, cannot be polarized.

A common application of polarization is polarizing sunglasses. The lens of these glasses are polarizing sheets oriented so that the polarization direction is vertical. When sunlight is reflected from a surface, such as water or a road, the light is partially polarized in the horizontal direction. The reflections increase the intensity, which an observer sees as a glare. Polarizing sunglasses allow only the vertical component of light to pass, blocking the horizontal component, which reduces the glare (Fig. 7.21).

Another common, but not so well known, application of polarized light is discussed in the chapter Highlight on liquid crystal displays.

7.4 Spherical Mirrors

Learning Goals:

To distinguish between converging and diverging spherical mirrors.

To explain image formation and distinguish between real and virtual images.

To analyze image formation and characteristics through ray diagrams.

Spherical surfaces can be used to make practical mirrors. The geometry of a spherical mirror is shown in Fig. 7.22. A **spherical mirror** is a mirror with the shape of a section of a sphere of radius R. A line drawn through the center of curvature C perpendicular to the mirror surface is called the *principal axis*. The point where the principal axis meets the mirror surface is called the *vertex* (V in the figure).

Another important point in spherical mirror geometry is the *focal point F*. The distance from the vertex to the focal point is called the **focal length** f. (What is "focal" about the focal point and focal length will become evident shortly.) For a spherical mirror, the focal length is one-half the value of the radius of curvature of the spherical surface.

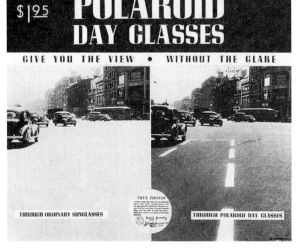

Figure 7.21 Polarizing sunglasses.
Above: Light reflected from surfaces is generally polarized in the horizontal direction. By orienting the polarizing direction of the sunglasses vertically, the horizontal component is blocked, thereby reducing the intensity and glare. *Below:* Old-time glare-reduction advertisement from the Polaroid Corporation archives. Notice the vintage of the cars and the price of the glasses.

Expressed in symbols, the focal length is

$$f = \frac{R}{2} \tag{7.2}$$

where f = the focal length
and R = the radius of curvature.

We can make either side of a spherical section the mirror surface. The inside surface of a spherical section is said to be *concave* (as though looking into a cave or recess); and when light is reflected from this surface, we have a **concave mirror**. A concave mirror is commonly called a **converging mirror**, for the reason illustrated in Fig. 7.23a. Light rays parallel to the principal axis on reflection converge and pass through the focal point. Off-axis parallel rays converge in the focal plane.

Concave mirrors are used to produce parallel beams by putting a light source at the focal point. Reverse the rays in Fig. 7.23 to see this. Flashlights and spotlights have spherical reflectors; with the light source at the focal point, a reasonably parallel beam is obtained.

Similarly, the outside surface of a spherical section is said to be *convex;* and when light is reflected from this surface, we have a **convex mirror**. A convex mirror is commonly called a **diverging mirror**. Parallel rays along the principal axis are reflected so that they appear to diverge from the focal point (Fig. 7.23b).

Reverse the ray tracing (reverse the direction of the light rays). Light rays coming to the mirror from the surroundings are made parallel, and an expanded field is seen in the diverging mirror. Diverging mirrors are used to monitor store aisles (Fig. 7.23c) and to give truck drivers a wide rear view of traffic.

The images formed by spherical mirrors may be found using ray diagrams. We commonly use an

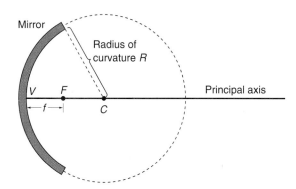

Figure 7.22 Spherical-mirror geometry.
A spherical mirror is a section of a sphere with a center curvature *C*. The focal point *F* is midway between *C* and the vertex *V*. The radius of curvature *R* then is twice the focal length *f*. Hence, *R* = 2*f*, or *f* = *R*/2.

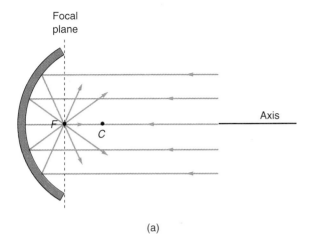

(a)

(c)

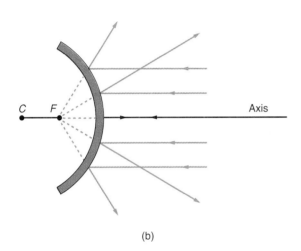

(b)

Figure 7.23 Spherical mirrors.
(a) Rays parallel to the principal axis of a concave, or converging, spherical mirror converge at the focal point. Rays not parallel to the axis converge in the focal plane. Extended images are formed under these conditions. (b) Rays parallel to the axis of a convex, or diverging, spherical mirror are reflected so as to appear to diverge from the focal point inside the mirror. (c) The divergent property of convex mirrors is used to give an expanded field view, as shown here. [Consider reverse ray in diagram (b).]

arrow as an object, and we determine the location and size of the image by drawing two rays, which follow the law of reflection:

STEP 1

Draw a ray from the tip of the object arrow parallel to the principal axis that is reflected through the focal point. (This is called the *parallel ray*.)

STEP 2

Draw a ray from the tip of the object arrow through the center of curvature C that is perpendicular to the mirror surface and is reflected back along the incident path. (This is called the *chief ray*.)

The intersection of these rays locates the position of the tip of the image arrow, the base of which is on the principal axis like the object arrow. These rays are shown in Fig. 7.24 for an object at various positions in front of a concave mirror. The characteristics of an image are described as being (1) real or virtual, (2) upright (erect) or inverted, and (3) larger than (magnified), smaller than (reduced), or the same size as the object. The object distance (D_o) and image distance (D_i) from the mirror are measured from the vertex. Note that these are usually different.

A *real image* **is one for which the reflected light rays converge so that an image can be formed on a screen. A** *virtual image* **is one for which the reflected light rays diverge; it cannot be formed on a screen.**

HIGHLIGHT

Liquid Crystal Displays (LCDs)

When a crystalline solid melts, the resulting liquid generally has no orderly arrangement of atoms or molecules. However, some organic compounds have an intermediate stage in which the liquid still retains some orderly molecular arrangement. Hence the name *liquid crystal* (LC).

Some liquid crystals are transparent, but have an interesting property. When an electrical voltage is applied, the liquid crystal becomes opaque. The applied voltage upsets the orderly arrangement of the molecules and light is scattered, making the LC opaque.

Another property of some liquid crystals can be seen in the way they affect linearly polarized light—they "twist" or rotate the polarization direction by 90°. However, this rotation does not occur if a voltage is applied; instead, molecular disorder results, as mentioned previously. A common application of these properties is in liquid crystal displays (LCDs), which are found on wristwatches, calculators, and small TV and laptop computer screens. How these liquid crystal displays work is illustrated in Fig. 1.

Trace the incident light in Fig. 1a. The unpolarized light is linearly polarized by the first polarizer. The LC rotates the polarization direction, and the polarized light then passes through the second polarizer (which is "crossed" with the first), and is reflected by the

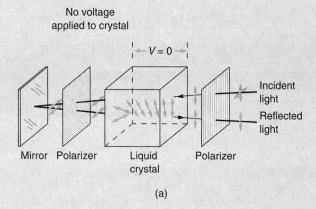

(a)

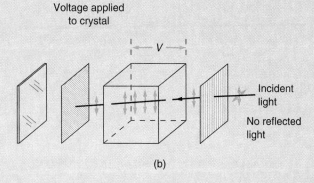

(b)

Figure 1 Liquid crystal display (LCD).
(a) An illustration of how a liquid crystal "twists" the light polarization through 90°. The light passes through the other polarizer and is reflected back and out the crystal with another twist. (b) When a voltage is applied to the crystal, there is no twisting, and light does not pass through the second polarizer. In this case, the light is absorbed and the crystal appears dark.

mirror. On the reverse path, the polarization direction rotation in the LC allows the light to emerge from the LCD, which would appear bright or white.

However, if voltage is applied to the LC (Fig. 1b) so it loses its rotational property, then light is not passed by the second polarizer. With no reflected light, the display would appear dark.

So, by applying voltages to segments of numeral and letter displays, we can form dark regions on a white background (Fig. 2). The white background is the reflected, polarized light. This can be demonstrated by using an analyzer as shown in the figure.

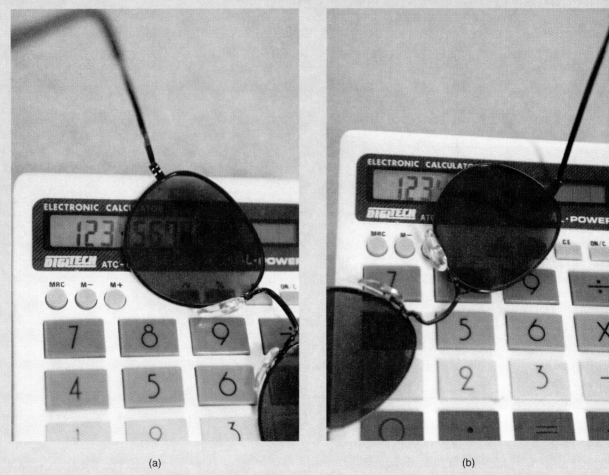

(a)   (b)

Figure 2 LCDs and Polarization.
(a) The light coming from the bright regions of an LCD is polarized, as can be shown by using polarizing sunglasses as an analyzer. (b) Notice that the glasses were rotated 90°.

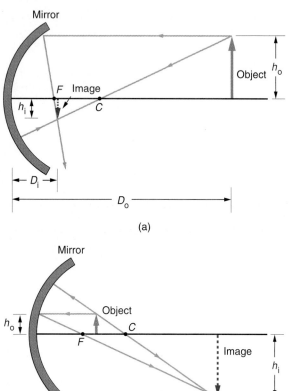

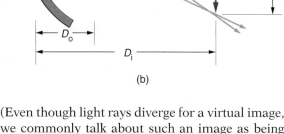

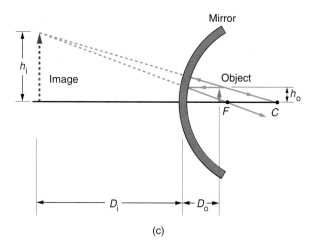

Figure 7.24 Ray diagrams.
(a) The ray diagram for an object beyond the center of curvature C for a concave spherical mirror showing where the image is formed. (b) Ray diagram for the mirror with an object between F and C. The images in these two cases are real and could be seen on screens placed at the image distances. Notice how the image moves out and gets bigger as the object moves toward the mirror. (c) Ray diagram for the mirror with the object inside the focal point. In this case, the image is virtual and is formed behind or "inside" the mirror.

(Even though light rays diverge for a virtual image, we commonly talk about such an image as being "formed.")

Both real and virtual images are formed by concave (converging) spherical mirrors. The images in Figs. 7.24a and b are real; and in (a) the image is inverted and smaller. In Fig. 7.24b, the image is inverted and larger. The image in Fig. 7.24c is virtual, upright, and larger. A practical example of the last case is shown in ■ Fig. 7.25. Concave makeup mirrors magnify so as to get a better view of facial features. The mirror is held so the face object is inside the focal point, and a virtual, upright, magnified image is seen in the mirror. Note the image in the plane mirror behind.

Ray diagram sketches with the approximate dimensions can give the general characteristics of an image. However, ray diagrams can also be drawn to scale. That is, given the location of the object (D_o) and its height (h_o), if the object is drawn to a particular scale, then the location of the image (D_i) and its height (h_i) will be to that scale and can be measured. From the heights you can find the *magnification factor* (M), where $M = h_i/h_o$. For $h_i = h_o$, we have $M = 1$, or the object and the image are the same size. If M is greater than one ($M > 1$), the image is magnified; and if less than one ($M < 1$), the image is smaller or reduced.

Notice that the real images are formed in front of the mirror where a screen could be positioned. Virtual images are formed "inside" or behind the mirror where the light rays appear to converge. For a converging mirror, a virtual image results when the object is inside the focal point. A virtual image always results for a diverging mirror, wherever the object is located.

CONFIDENCE QUESTION 7.4

What are the characteristics of an image formed by a plane mirror?

Figure 7.25 Magnification.
Concave makeup mirrors give moderate magnification so facial features can be better seen.

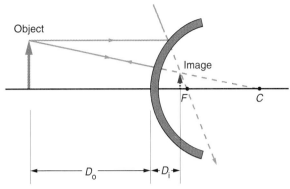

Figure 7.26 Diverging-mirror ray diagram.
The ray diagram for a convex spherical mirror with an object in front of the mirror. A convex mirror always has a virtual image.

Convex mirrors may also be treated by ray diagrams. In this case, the rays of the diagram are drawn using the law of reflection, but the rays are extended through the focal point and center of curvature inside or behind the mirror, as shown in Fig. 7.26. A virtual image is formed where these extended rays intersect. As we may gather from the figure, even though the object distance from the mirror may vary, the image of a convex mirror is always virtual, upright, and smaller than the object.

7.5 Spherical Lenses

Learning Goals:

To distinguish between converging and diverging lenses.

To explain image formation and distinguish between real and virtual images.

To analyze image formation and characteristics through ray diagrams.

A lens consists of material such as a transparent piece of glass or plastic that refracts light waves to give an image of an object.

Lenses are extremely useful and are found in eyeglasses, telescopes, magnifying glasses, cameras, and many other optical devices. The word *lens* comes from the Latin word for lentil, a curved seed with the shape similar to that of a common lens.

In general, there are two main classes of lenses. A **converging lens** or **convex lens** is thicker at the center than at the edge. A **diverging lens** or **concave lens** is thicker at the edge than at the center. These two classes and some possible shapes for each are illustrated in Fig. 7.27.

Light passing through a lens is refracted twice—once at each surface. We will consider what are known as thin lenses. When constructing ray diagrams, we can thus neglect the thickness of the lens, and we can assume that the two surfaces that cause refraction are essentially in the same place.

Also, we will consider only **spherical lenses**—*biconvex (converging) spherical lenses* and *biconcave (diverging) spherical lenses*. These have similar

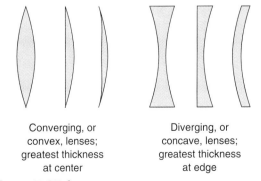

Converging, or convex, lenses; greatest thickness at center

Diverging, or concave, lenses; greatest thickness at edge

Figure 7.27 Lenses.
Different types of converging and diverging lenses.

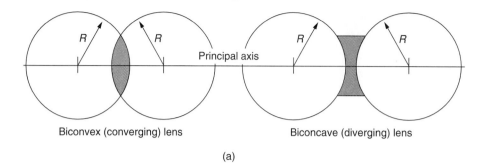

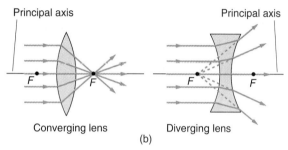

Figure 7.28 Spherical lenses.
(a) Biconvex (converging) and biconcave (diverging) spherical lenses are sections of two spherical surfaces as shown here. (b) For a converging spherical lens, rays parallel to the principal axis and incident on the lens converge at the focal point on the opposite side of the lens. Rays parallel to the axis of a diverging lens appear to diverge from a focal point on the incident side of the lens.

surfaces on both sides and are parts of spherical sections, as shown in ● Fig. 7.28. The principal axis for a lens goes through the center of the lens, as shown in the figure. Rays coming in parallel to the principal axis are refracted toward the principal axis by a converging lens. For a converging lens, the rays are focused at point F, the focal point. (The distance from the center of the lens to the focal point is the focal length f.) You may have used a converging lens or magnifying glass to focus the Sun's rays on a spot to burn a hole in a piece of paper.

For a diverging lens, the rays are refracted away from the principal axis and appear to emanate from the focal point. Since the lenses are symmetric and light can pass through either way, there is a focal point on each side of a lens, one for each lens surface.

We can show that lenses refract light to form images by a ray diagram procedure similar to what we did with mirrors. As before, two rays are drawn from the tip of the object arrow:

> 1. Draw a ray parallel to the principal axis; the ray is then refracted by the lens through the focal point on the opposite side of the lens. (This ray is called the *parallel ray*.)
> 2. Draw an undeviated ray straight through the center of the lens. (This ray is called the *chief ray*.)

The image of the tip of the arrow is located where these two rays intersect, with the base of the image arrow on the principal axis, as in the case of mirrors.

An example of this procedure is shown in ● Fig. 7.29 for a converging lens. Notice that only the focal points for the respective surfaces are shown; these points are all that are needed. The lenses do have radii of curvature, but for spherical lens $f \neq R/2$, in contrast to $f = R/2$ for spherical mirrors.

The image characteristics for lenses are described the same as those for spherical mirrors. However, there is one difference, which should be obvious. In the lens case, a *real image* is formed on the opposite side from the object and can be seen on a screen (Fig. 7.29b). A *virtual image* is formed on the same side of the lens as the object, or object side. For a converging lens, this occurs when the object is inside the focal point, as for a converging mirror.

A converging lens is a simple magnifying glass. Magnifying glasses are used to make things "larger" for better viewing. To do this, the object is brought inside the focal point of the lens, and an enlarged, virtual image is seen through the glass (● Fig. 7.30a). You also have a converging lens in

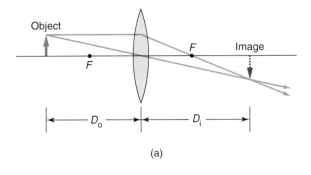

Figure 7.29 Ray diagram.
(a) Ray diagram for a converging lens with the object outside the focal point. A real image is formed on the opposite side of the lens. (b) The real image can be viewed on a screen, as shown here, with a candle for an object.

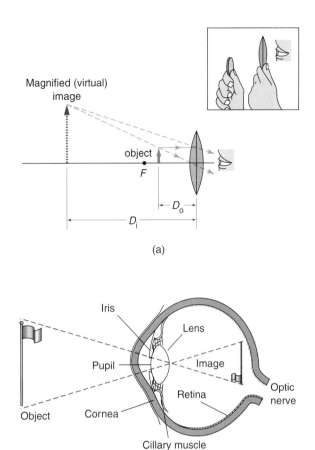

Figure 7.30 Viewing with converging lenses.
(a) A converging lens acts as a simple magnifying glass. With an object inside the focal point, the converging lens forms an enlarged, virtual image that is viewed by the eye, and the object appears magnified. (b) The human eye has a converging lens that is quite amazing. See text for description.

your eye, called the *crystalline lens* (Fig. 7.30b). The lens and other media in the eye refract the incoming light so it is focused on the retina, which contains the photosensitive cells. (The retina contains *cones* for color vision and *rods* for twilight or black-and-white vision.) The crystalline lens is quite amazing. Unlike the magnifying glass, where the object distance is varied by moving the object or the lens, the shape of the crystalline lens is varied so as to change the radius of curvature and the focal length. The crystalline lens is made up of glassy fibers. Changes in tension by the cillary

muscle, which is attached to the lens, cause the fibers to slide over each other and change the shape of the lens so the image is formed on the retina. This change takes place quite rapidly. Stop reading and look at an object across the room. See how fast your lens changed its focal length.

Also, you might note that the image formed on the retina by the converging crystalline lens is upside down (see ray diagram in Fig. 7.29a). It would seem then that we should see the world upside down. However, for some reason, which is quite nice, our brain interprets things right-side-up.

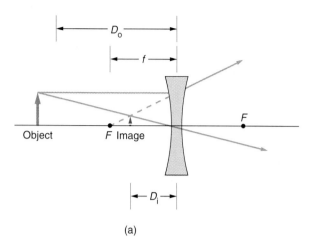

Figure 7.31 Diverging lens.
(a) A ray diagram for a diverging lens. A virtual image is always formed on the object side of the lens. (b) Like a diverging mirror, a diverging lens gives an expanded field of view. See text for discussion.

For a diverging or concave lens, the image is always virtual, upright, and smaller than the object (● Fig. 7.31). When viewing through a concave lens, we see images as shown in the figure. The lens in the photograph is not a biconcave one, but rather is flat on one side and has circular grooves on the other. Such flat plastic lenses are used on the back windows of vans to get a wide-angle view of traffic (similar to a diverging side mirror). The grooves are made as shown in ● Fig. 7.32. Such a lens is called a *Fresnel lens*, after Augustin Fresnel (pronounced free-NEL, 1788–1827), who first realized that all of the refraction of a lens takes place at its surfaces, so the interior part is not really needed. A similar grooving process can be used to make a flat-grooved converging, or plano-convex, lens.

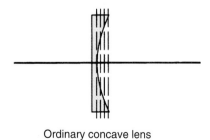

Ordinary concave lens

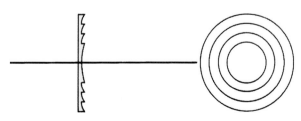

Flat-grooved concave lens   Head-on view of grooves

Figure 7.32 Fresnel lens.
A flat, grooved concave lens is made with the same curvature as an ordinary concave lens.

Important Terms

These important terms are for review. After reading and studying the chapter, you should be able to define and/or explain each of them.

reflection
ray
law of reflection
regular (specular) reflection
irregular (diffuse) reflection
refraction
index of refraction
total internal reflection

dispersion
diffraction
interference
constructive interference
destructive interference
diffraction grating
polarization
linearly (plane) polarized light

spherical mirror
focal length
concave (converging) mirror
convex (diverging) mirror
spherical lenses
converging (convex) lens
diverging (concave) lens

Important Equations

Index of refraction: $n = \dfrac{c}{c_m}$

Focal length (spherical mirror): $f = \dfrac{R}{2}$

Questions

7.1 Reflection

1. The angle of incidence and angle of reflection (a) can never be equal, (b) add up to 90°, (c) are not related, (d) are measured from a line perpendicular to the reflecting surface.
2. The law of reflection applies to (a) regular reflection, (b) irregular reflection, (c) diffuse reflection, (d) all of these.
3. What is a light ray?
4. Explain how the law of reflection applies to diffuse reflection, and give an example of this type of reflection.
5. If you run toward a plane mirror at a given speed, how does your image approach you?

7.2 Refraction and Dispersion

6. When the angle of refraction is greater than the angle of incidence (a) the critical angle is exceeded, (b) the first medium is less dense, (c) the second medium has a smaller index of refraction, (d) the speed of light is greater in the second medium.
7. Dispersion (a) occurs above a critical angle, (b) is the principle of internal reflection, (c) is responsible in part for the rainbow, (d) occurs for diffuse reflection.
8. What is refraction, and what causes it?
9. Explain how the refraction of light on entering a denser medium is analogous to the situation of a marching band.
10. How is light bent or refracted on entering (a) a denser medium and (b) a less dense medium?
11. What is the index of refraction for a vacuum? What is c_m in this case?
12. Draw a diagram to show how total internal reflection occurs.
13. What causes dispersion, and is it of any practical importance?
14. Explain why diamonds have brilliance and "fire."

7.3 Diffraction, Interference, and Polarization

15. Diffraction (a) occurs best when a slit width is less than or equal to the wavelength of the wave, (b) depends on refraction, (c) is caused by interference, (d) does not occur for light.
16. Total destructive interference occurs (a) when the waves are in phase, (b) at the same time as complete constructive interference, (c) when the waves have equal amplitudes and are completely out of phase, (d) for total internal reflection.
17. If the polarization directions of two polarizing sheets are at an angle of 45° to each other, (a) no light gets through, (b) there is maximum transmission, (c) maximum transmission is reduced.
18. A slit opening in a water tank is 0.50 cm wide. Determine above what wavelength the diffraction of water waves on the surface of the water passing through the slit will be easily observed.
19. Describe the constructive and destructive interference for two similar waves with the same frequency, but different amplitudes.
20. What will be the waveform resulting from the interference of a reference wave and a similar wave 90° out of phase (displaced one-quarter wavelength). See Fig. 7.16 for reference. (*Hint:* Make a sketch.)
21. While an observer looks through two polarizing sheets, one of the sheets is rotated through 360°. Describe what would be observed.
22. Explain the principle of polarizing sunglasses.

7.4 Spherical Mirrors

23. A concave mirror (a) has a radius of curvature equal to $2f$, (b) is a diverging mirror, (c) forms only virtual images, (d) forms only magnified images.
24. A real image is one (a) that is always magnified, (b) formed by a converging light ray, (c) formed behind the mirror, (d) that occurs only when $D_i = D_o$.
25. What are the relationships between the center of curvature, the focal point, and the vertex for spherical mirrors?
26. Distinguish between real images and virtual images for spherical mirrors.
27. Explain when real and virtual images are formed by a (a) concave mirror and (b) a convex mirror.

7.5 Spherical Lenses

28. A biconvex lens (a) is a converging lens, (b) is thicker at the edge than at the center, (c) forms virtual images for $D_o > f$, (d) is a Fresnel lens.
29. A virtual image (a) is always formed by a biconvex lens, (b) can be formed on a screen, (c) is formed on the object side of the lens, (d) cannot be formed by a biconcave lens.
30. Distinguish between real images and virtual images for convex and concave lenses.
31. Explain when real and virtual images are formed by (a) a convex lens and (b) a concave lens.
32. How does a magnifying glass magnify?
33. Why are slides put into a slide projector upside down?

Food for Thought

1. How would light be refracted if the two media had the same optical density?
2. Would the marching analogy of refraction apply to going from a denser medium to a less dense one? Explain.
3. The energy of a wave is proportional to its amplitude (actually, amplitude squared). In total destructive interference, the amplitude is zero. Does this mean that the energy is zero, or has been "destroyed" in the destructive interference? Explain.
4. (a) Is a satellite TV dish a spherical mirror? Where is the receiver placed?
   (b) Some dishes are made of open-wire mesh. How can this be a reflecting mirror?
5. For converging mirrors and lenses, we say that an image is formed at infinity when the object is at the focal point. Explain what this means. (*Hint:* Draw a ray diagram.)
6. On automobile passenger-side rearview mirrors, a warning is printed such as: "Objects in mirror are closer than they appear." Why is this and what makes the difference? (*Hint:* The mirrors are convex mirrors.)

EXERCISES

7.1 Reflection

1. Light is incident on a flat mirror surface at an angle of 30° relative to the surface. What is the angle of reflection?

2. How much longer must the minimum length of a plane mirror be for a 6-ft 4-in. man to see his complete (head-to-toe) image than for a 5-ft 2-in. woman?

   Answer: 7 in.

3. Show that for a person to see his or her complete (head-to-toe) image in a plane mirror, the mirror must have a length (height) of at least one-half of the person's height (see Fig. 7.4). Does the person's distance from the mirror make a difference? Explain.

7.2 Refraction and Dispersion

4. What percentage of the speed of light in vacuum is the speed of light in crown glass?

   Answer: 66%

5. What percentage of the speed of light in vacuum is the speed of light in diamond?

   Answer: 41%

6. The speed of light in a certain transparent material is 55% of the speed of light in vacuum. What is the index of refraction of the material?

7. The index of refraction is always greater than one ($n > 1$). Show why this is so.

7.4 Spherical Mirrors

8. Sketch ray diagrams for a concave mirror for objects at (a) $D_o > R$, (b) $R > D_o > f$, and (c) $D_o < f$. Describe how the image changes as the object is moved toward the mirror.

9. A woman holds a makeup mirror with a radius of curvature of 120 cm a distance of 20 cm from her face. Sketch a ray diagram and determine the general characteristics of her image.

   Answer: virtual, upright, and enlarged

10. An object 6 cm tall is placed 30 cm from a concave spherical mirror with a focal length of 20 cm. Draw a ray diagram to scale and determine the image distance, the magnification factor, and the height of the image.

11. An object is placed 10 cm from a convex spherical mirror with a focal length of 4 cm. Sketch a ray diagram and find the general characteristics of the image.

12. By drawing a ray diagram to scale, find the magnification for an object located at the center of curvature (or $D_o = R$) for a concave mirror.

7.5 Spherical Lenses

13. Sketch ray diagrams for a spherical biconvex lens with objects at (a) $D_o > 2f$, (b) $2f > D_o > f$, and (c) $D_o < f$. Describe how the image changes as the object is moved toward the lens.

14. An object is placed 45 cm in front of a converging lens with a focal length of 20 cm. Sketch a ray diagram and determine the general characteristics of the image.

    Answer: real, inverted, and reduced

15. An object with a height of 6 cm is placed in front of a simple magnifying glass that has a focal length of 8 cm. Draw a ray diagram to scale and determine the magnification factor of the image.

16. Draw a ray diagram to scale and find the magnification factor of an object at a distance of $2f$ from a general convex lens.

Answers to Multiple-Choice Questions

1. d   6. c   15. a   23. a   28. a
2. d   7. c   16. c   24. b   29. c
                    17. c

Solutions to Confidence Questions

7.1 $\theta_i = 90° - 50° = 40° = \theta_r$

7.2   Away from the normal. Reverse ray tracing in Fig. 7.6.

7.3 $\dfrac{c_m}{c} = \dfrac{1}{n} = \dfrac{1}{1.00029}$ ($\times\,100\%$) $= 99.97\%$
                             (n value from Table 7.1)

7.4   Virtual, upright, and same size (with $D_i = D_o$).

Chapter 8

Electricity and Magnetism

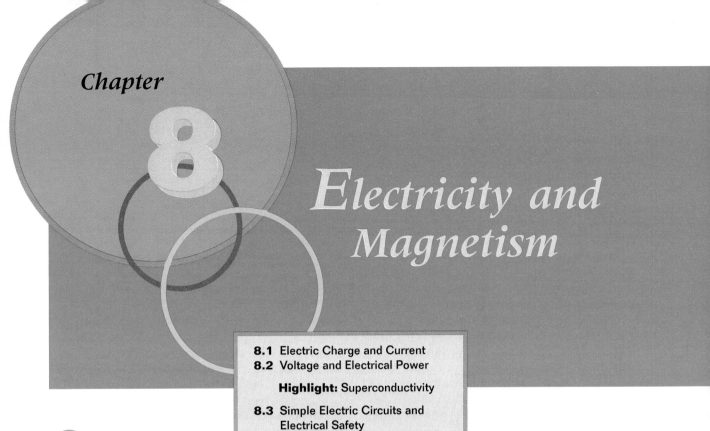

8.1 Electric Charge and Current
8.2 Voltage and Electrical Power

Highlight: Superconductivity

8.3 Simple Electric Circuits and Electrical Safety

Highlight: Electric Shock

8.4 Magnetism
8.5 Electromagnetism

Ours is indeed an electrical society. Think how your life might be without electricity. (We get some idea occasionally when there is a power outage.) Yet, when asked to define electricity, many people have difficulty. The terms *electric current* and *electric charge* come to mind, but what are these?

You may recall that electric charge is listed as a fundamental quantity. That is, we really don't know what it is, so our chief concern is what it does, or the description of electrical phenomena.

As you will learn in this chapter, we know that electric charge is associated with certain particles and that there is an interacting force. Beginning with a force, we can go on to discuss the motion of electric charges (current), and then to electrical energy and power. When we apply these principles, we have the benefits of "electricity" at our disposal. Electricity runs motors, heats homes, gives us lighting, powers our appliances, and on and on.

But the electric force is even more basic than electricity. It keeps atoms and molecules together—even the ones that make up our bodies. It may be said that the electric force holds matter together on a small scale, whereas the gravitational force (Chapter 3) holds our solar system and galaxies together.

Closely associated with electricity is magnetism. In fact, we refer to *electromagnetism* because these phenomena are basically inseparable. For example, without magnetism we would not be able to generate electrical power. As children (and perhaps adults), most of us have been fascinated with the properties of small magnets. Have you ever wondered what causes magnets to attract and repel?

In this chapter you will be introduced to the basic properties of electricity and magnetism—exciting phenomena that are everywhere around you.

8.1 Electric Charge and Current

Learning Goals:

To define electric charge and electric current.

To distinguish between electrical conductors and insulators.

To explain the law of charges, Coulomb's law, and some of the effects and applications of electrostatic charging.

Electric charge is a fundamental quantity; that is, it is the most basic electrical quantity. Electric charge is associated with certain subatomic particles, and experimental evidence leads us to the conclusion that there are two types of charge. They are designated positive (+) and negative (−) for distinction.

All matter, according to modern theory, is made up of small particles called atoms, which are composed of negatively charged particles called **electrons,** positively charged particles called **protons,** and neutral particles that show no electric charge called *neutrons*. As you may be aware, our simplistic model of the atom pictures the protons and neutrons to be in the central nucleus of an atom, with the electrons orbiting the nucleus, similar to the way that planets orbit the Sun in our solar system. Table 8.1 summarizes the fundamental properties of these atomic particles. The unit of electric charge is the coulomb, which will be discussed shortly.

As the table indicates, all three particles have a certain mass, but only the electron and the proton possess electric charge. The magnitudes of the electric charge on the electron and proton are equal, but they are different in nature, as expressed by plus and minus signs. When we have the same number of electrons and protons, the total charge is zero (same number of positive and negative charges of equal magnitude), and we have an electrically neutral condition.

Also, scientists believe that the electronic charge is the fundamental or smallest unit of free charge in nature. Evidence indicates that subatomic particles called **quarks** have fractional electronic charges—for example, charges of $\pm \frac{2}{3}$ and $\pm \frac{1}{3}$ of the electronic charge of 1.6×10^{-19} coulomb. According to this theory, quarks combine to form protons and neutrons. However, quarks are believed to exist only in the nucleus of the atom and not in a free state. Therefore, the fractionally charged quarks are not a consideration in our study of electricity.

The unit of electric charge is the **coulomb** (C), named after Charles A. de Coulomb (1736–1806), a French scientist who studied electrical effects. A net electric charge is usually indicated by the letter q. A $+q$ indicates that an object has more protons than electrons. A $-q$ indicates an excess of negative charge, or more electrons than protons. Often, we simply refer just to an electric charge with the understanding it is associated with some object.

The motion of charge is characterized by electric current. **Current** is defined as the time rate of flow of electric charge. It is measured in units of amperes (or amps, for short), named after André Ampere (1775–1836), another early French investigator of electricity. One **ampere** (A) is equal to the flow of one coulomb of charge per second. (Like heat, electricity was once thought to be a fluidlike substance that was transferred from one body to another, giving rise to a net charge. This is probably why we still talk about the "flow" of electric charge, as we do the "flow" of heat.)

In symbol notation, the electric current, commonly designated by the letter I, and electric charge (q) are related by the equation

$$I = \frac{q}{t} \qquad (8.1)$$

(in words)

$$\text{current} = \frac{\text{charge}}{\text{time}}$$

where I = electric current, measured in amperes (amps),

q = electric charge moving past a given point, measured in coulombs,

and t = time for a given amount of charge to move past the point, in seconds.

So, 1 amp = 1 coulomb/second.

Electrical **conductors** are materials in which an electrical current readily flows. Metals are good electrical conductors. We commonly use metal wires to conduct electrical currents. This conduction is primarily due to the outer, loosely bound electrons of the metal atoms. (Recall from Chapter 5 that electrons also contribute significantly to thermal conduction.)

Table 8.1 Some Properties of Atomic Particles

Particle	Symbol	Mass	Charge
Electron	e⁻	9.11×10^{-31} kg	-1.60×10^{-19} C
Proton	p⁺	1.67×10^{-27} kg	$+1.60 \times 10^{-19}$ C
Neutron	n	1.67×10^{-27} kg	0

Materials in which electrons are more tightly bound do not conduct electricity very well, and such materials are referred to as electrical **insulators**. Examples are wood, glass, and plastics. We coat our electric cords with rubber or plastic so we can safely handle them. Materials that are neither good conductors nor good insulators are called *semiconductors*. Graphite (carbon) is an example of a semiconductor.

In the definition of current in Eq. 8.1, we speak of an amount of charge flowing past a given point. This is not a flow of charge in a manner similar to fluid flow. In a metal wire, for example, the electrons move randomly and chaotically. Some go in one direction past the observation point, and others go in the opposite direction. However, with a current, more electrons go in one direction than the other, so we are really talking about a *net* charge q in the equation (analogous to a net force in Chapter 3).

EXAMPLE 8.1 Finding Electrical Current

In a copper conductor, there is a net flow of 0.50 C of charge past a given point in 3.0 s. What is the current in the conductor?

Solution

We have

Given: $q = 0.50$ C
$t = 3.0$ s
Find: I (current)

It should be obvious that we substitute into Eq. 8.1:

$$I = \frac{q}{t} = \frac{0.50 \text{ C}}{3.0 \text{ s}} = 0.17 \text{ A}$$

CONFIDENCE QUESTION 8.1

A conducting wire carries a current of 1.5 A. How much charge flows past a point in the wire in 2.0 s?

One well-known observation about electric charges can be described by the **law of charges:**

Like charges repel and unlike charges attract.

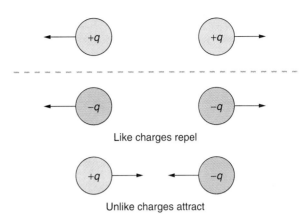

Like charges repel

Unlike charges attract

Figure 8.1 Law of charges.
Like charges repel and unlike charges attract. The magnitude of the force of repulsion or attraction is given by Coulomb's law, $F = kq_1q_2/r^2$, where r is the distance between the point charges.

That is, there is a repulsive force between two like charges ($+q$ and $+q$, or $-q$ and $-q$), and an attractive force between two unlike charges ($+q$ and $-q$). See Fig. 8.1.

Charles de Coulomb developed an expression to describe the electrical force, which is known as *Coulomb's law of electric charges*. **Coulomb's law** states:

> The force of attraction or repulsion between two charges is directly proportional to the product of the two charges and inversely proportional to the square of the distance between them.

In equation form, this is expressed

$$F = \frac{kq_1q_2}{r^2} \qquad (8.2)$$

where F = magnitude of the force of attraction or repulsion, in newtons,
 q_1 = magnitude of first charge, in coulombs,
 q_2 = magnitude of second charge, in coulombs,
and r = the distance between the charges.

The k is a proportionality constant with a value of

$$k = 9.0 \times 10^9 \ \frac{\text{N}-\text{m}^2}{\text{C}^2}$$

The direction of the force on a particular charge is given by the law of charges. Note that for two charges, there are equal and opposite forces (Fig. 8.1). Does this remind you of any of Newton's laws?

Notice that Coulomb's law is similar in form to Newton's law of gravitation (Chapter 3, $F = Gm_1m_2/r^2$). Both forces depend on the square of the separation distance. However, there are important differences. One is that Coulomb's law applies to either an attractive force or a repulsive force, depending on whether the two charges are unlike or like. (This is how we know there are two types of electric charges.) The force of gravity, on the other hand, is always attractive—at least we haven't observed otherwise.

Another important difference is that electric forces are relatively stronger than gravitational forces. An electron and proton attract each other both electrically and gravitationally. However, in dealing with atoms, the electric force is so much stronger than the gravitational forces can be generally ignored; we consider only the electric forces of attraction and repulsion.

An object with an excess of electrons is said to be *negatively charged*, and an object with a deficiency of electrons is said to be *positively charged*. A negative charge can be put on a rubber rod by stroking the rod with fur. This is called *charging by friction*, although it has little to do with friction. In Fig. 8.2, a rubber rod that has been rubbed with

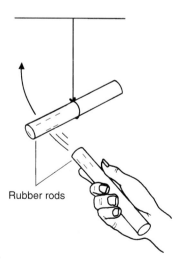

Figure 8.2 Repulsive electric forces.
Two negatively charged objects repel one another.

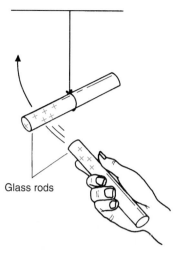

Figure 8.3 Repulsive electric forces.
Two positively charged objects repel one another.

fur is shown suspended by a thin thread that allows the rod to rotate. When a similar rubber rod is brought close to the suspended one, it will rotate away from the second similarly charged rod. That is, the charged rod is repelled.

The same procedure using two glass rods that have been stroked with silk cloth will show similar results (Fig. 8.3). Here electrons are transferred from the rod to the silk.

Although the experiments show repulsion in both cases, the charge on the glass rods is different from the charge on the rubber rods. As shown in Fig. 8.4, a charged rubber rod will be attracted to a charged glass rod. Hence, to have both forces of attraction and repulsion, you must have two types of charges.

Electrostatics deals with the effects of stationary charges like these. The important principle—that unlike charges attract—is the basis of photocopying. In photocopy machines or electrostatic copiers, the material to be copied is placed face down on a glass plate, and a copy is produced. How is this done? As you know, when a page is copied, it is illuminated by a bright light. The light is reflected off the blank parts of the page onto a positively charged cylinder or plate made of a material such as the element selenium (Se). Selenium is a photoconductor; that is, it conducts when exposed to light. Charge leaks away from the areas exposed to light, which leaves an electrostatic "im-

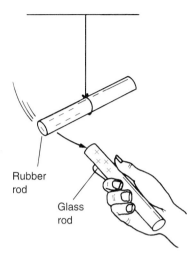

Figure 8.4 Attractive electrical forces.
A negatively charged object and a positively charged object attract one another.

A   Surface of a plate coated with a photo-conductive metal is electrically charged as it passes under wires.

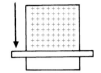

B   Original document is projected through a lens. Plus marks represent latent image retaining positive charge. Charge drained in areas exposed to light.

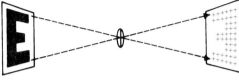

C   Negatively charged powder (toner or "dry ink") is applied to the latent image, which now becomes visible.

D   Positively charged paper attracts dry ink from plate, forming direct positive image.

E   Image is fused into the surface of the paper or other material by heat for permanency.

Figure 8.5 Electrostatic copying.
The steps involved in making an electrostatic photocopy. See text for description.

age" of the dark regions or print of the copy page (● Fig. 8.5).

Next, negatively charged dry ink powder is dusted on the cylinder, and the powder is attracted and sticks to the positively charged regions. A positively charged piece of paper is then passed over the cylinder, and the ink is attracted to the paper. Then the ink is fused onto the paper by heating. The photocopy then pops out of the machine.

Although the inner workings of a photocopier are fairly complicated, the main principle is simply the attraction of unlike charges. Perhaps you have noticed the static charge on photocopies when they are ejected from the machine.

Static charge can also be a problem. After walking across a carpet, you have probably been annoyingly zapped by a spark when you reached for a door knob. You were charged by friction in crossing the carpet, and when reaching for the door, the electric force was strong enough to cause the air to ionize and conduct charge to the metal door knob. This effect occurs best on a dry day. With high humidity, there is a thin film of moisture on objects and charge is conducted away before it can build up. Even so, such sparks are not wanted when working around flammable materials; for example, in an operating room with explosive gases or around gasoline.

Also, static charge gives rise to the problem of the "static cling" of clothes. A lot of money is spent on products that prevent this charge build-up in clothes dryers.

From Coulomb's law, we see that as two charges are brought closer together, the force of attraction or repulsion increases. This effect can give rise to charged regions in a material, as illustrated in ● Fig. 8.6a. When a negatively charged rubber comb is brought near small pieces of paper, the charges in the paper molecules are acted upon by the electrical forces—positive charges attracted

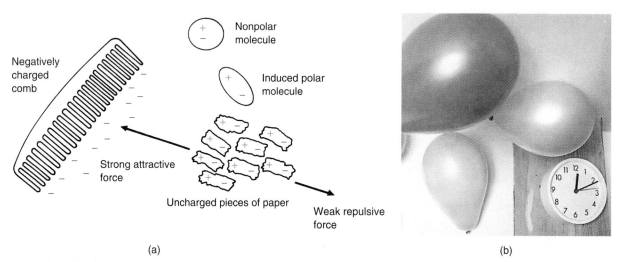

Figure 8.6 Polarization of charge.
(a) When a negatively charged rubber comb is brought near small pieces of paper, the molecules are polarized with definite regions of charge, which give rise to a net attractive force. (b) Charged balloons cling to the ceiling and wall because of the attractive forces from molecular polarization.

and negative charges repelled—and there is an effective separation of charge. The molecules are then said to be *polarized* with definite regions of charge.

Because the positive-charge regions are closer to the comb, the attractive forces are stronger than the repulsive forces due to negative-charge regions, which are farther away. Thus there is a net attraction between the comb and a piece of paper. Small bits of paper may be picked up by the comb, which indicates that the attractive electrical force is stronger than the paper's gravitational weight force. Keep in mind, however, that overall the paper is electrically uncharged; only molecular regions are charged. This process is called *charging by induction*.

Now you know why a balloon will stick to the ceiling or wall after being rubbed on a person's hair or clothing. The balloon is charged by friction through a transfer of charge. When the balloon is placed on a wall, the charge on the balloon induces regions of charge in the molecules of the wall material, and the balloon is attracted to the wall (Fig. 8.6b).

Another demonstration of electrical force is shown in Fig. 8.7. When a charged rubber rod is brought close to a thin stream of water, the water is attracted toward the rod and the stream is "bent." Here the charging is not by induction, but rather because water molecules have a permanent charge separation, or regions of different charges. They are called *polar molecules*.

8.2 Voltage and Electrical Power

Learning Goals:

To distinguish between electric potential energy and voltage.

To analyze Ohm's law and the expressions for electric power in terms of voltage, current, and resistance.

Voltage and Resistance

The effects produced by moving charges give rise to what we generally call electricity. For charges to move, they must be acted upon by forces resulting

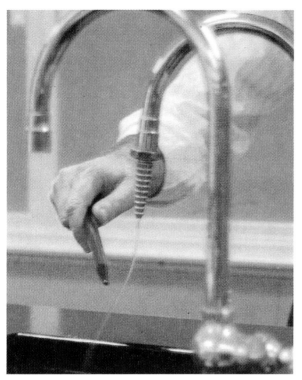

Figure 8.7 Bending water.
A charged rod near a thin stream of water will bend the stream because of the polarization of water molecules.

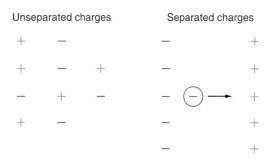

Figure 8.8 Electric potential energy.
Work must be done in separating positive and negative charges. When separated, the charges have electric potential energy and will move if free to do so.

from other positive or negative charges. And, with forces, we have the potential to store electrical energy.

Consider the situation shown in Fig. 8.8. We start out with some unseparated charges and then begin to separate them. It takes relatively little work to pull the first negative charge to the left and the first positive charge to the right. When the next negative charge is moved to the left, it is repelled by the negative charge already there, so more work is needed. Similarly, it takes more work to move the positive charge to the right. As we separate more and more charges, it takes more and more work.

Because work is done in separating the charges, we have **electric potential energy.** If a charge were free to move, it would move toward the opposite charges; for example, a negative charge, as in the separated charges in Fig. 8.8, would move toward the positive charges. Electric potential energy would be converted into kinetic energy, as required by the conservation of energy.

Instead of speaking of electric potential energy, we usually talk about electric potential, or voltage. **Voltage** is defined as the amount of work it would take to move a charge between two points, divided by the value of the charge. That is:

> **Voltage is the work per unit charge, or the electric potential energy per unit charge.**

Or, $$\text{voltage} = \frac{\text{work}}{\text{charge}}$$

$$V = \frac{W}{q} \qquad (8.3)$$

The **volt** (V) is the unit of voltage and is equal to one joule per coulomb. Voltage is caused by a separation of charge. Once the charges are separated, a current can be set up, and we get electricity. A voltage is actually a voltage difference or potential difference. Similar to gravitational potential energy, $E_p = mgh$, where h is really a height difference (Δh), the voltage V is really a voltage difference, ΔV—an electric potential "hill," so to speak.

When we have a current, it meets with some opposition because of collisions within the conducting material. This opposition to the flow of charge is called **resistance.** The unit of resistance is the **ohm** (Ω, Greek letter capital omega, instead of a capital O, so as to distinguish from the numeral zero). This is named after Georg Ohm (1787–1845), a German physicist who formulated a simple relationship between voltage, current,

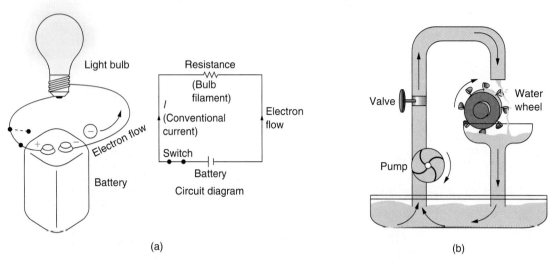

Figure 8.9 Electric circuit and water analogy.
(a) A simple electric circuit in which the battery supplies the voltage and the light bulb the resistance. When the switch is closed, electrons flow from the negative terminal of the battery toward the positive terminal. Electrical energy is expended in heating the bulb filament. A circuit diagram is shown to the right. (b) In the water "circuit," the pump is analogous to the battery, the valve analogous to the switch, and the water wheel analogous to the resistance. Energy is expended and work is done in turning the water wheel.

and resistance. Known as **Ohm's law,** in equation form it is written

$$V = IR \qquad (8.4)$$

(in words)

voltage = current × resistance

A resistance that conforms to this relationship is called an *ohmic resistance*. The resistance of many materials, but not all, is described by Ohm's law.

An example of a simple electric circuit is shown in ● Fig. 8.9, along with a water analogy. The battery provides the voltage to drive the circuit through chemical activity (chemical energy). This device is similar to the pump in the water circuit. When the switch is closed (valve opened in water circuit), there is a current in the circuit. Electrons move away from the negative terminal of the battery toward the positive terminal. The light bulb in the circuit provides resistance, and work is done in lighting it, with electrical energy being converted to heat and radiant energy. The water wheel in the water circuit provides analogous resistance to the water flow and uses potential energy to do work. Notice there is a voltage or potential difference (drop) across the bulb, similar to a gravitational potential difference across the water wheel. The components of an electrical circuit are represented by symbols in a circuit diagram as shown in the figure.

The switch in the circuit allows the path of the electrons to be open or closed. When the switch is open, there is not a complete path or circuit through which charge can flow, and there is no current. (This is called an "open" circuit.) When the switch is closed, the circuit is completed, and there is a current. (The circuit is then said to be "closed.") An electric current requires a closed path or circuit.

Power

When current exists in a circuit, work is done to overcome the resistance of the circuit and energy is expended. Or, over a period of time, we say that power is expended. Recall that the definition of power (*P*) is work per unit time:

$$P = \frac{W}{t}$$

Figure 8.10 Wattage (power) ratings. (a) A 60-W light bulb dissipates 60 J of electrical energy each second. (b) A curling iron uses 13 W at 120 V. Given the wattage and voltage ratings, one can find the current drawn by $I = P/V$.

From Eq. 8.3, $W = qV$, and substituting for W, we have

$$P = \frac{qV}{t}$$

From Eq. 8.1, $q = It$, so $P = \frac{ItV}{t}$, so electric **power** is given by

$$P = IV \qquad (8.5)$$

(in words)   power = current × voltage

Another useful form of this equation may be obtained using Ohm's law ($V = IR$) for V. That is, $P = IV = I(IR)$, and

$$P = I^2 R \qquad (8.6)$$

The unit of power is the watt (W), and electrical appliances are rated in watts (Fig. 8.10).

The power that is dissipated in an electric circuit is frequently dissipated as heat. (For the common incandescent light bulb, over 95% of the electrical energy is dissipated as heat, rather than as visible light.) This heat is called **joule heat** or I^2R **losses** (I squared R losses), as given by Eq. 8.6. This heating effect is used in electric stoves, heaters, hair dryers, and so on. Hair dryers have heating coils of low resistance so as to get a large current for the I^2R losses.

EXAMPLE 8.2 Finding Current and Resistance

Find the current and resistance of a 60-W, 120-V light bulb in operation.

Solution

Following our usual procedure,

Given: $P = 60$ W
$V = 120$ V
Find: I (current)
R (resistance)

The units are standard. Notice that our common electrical units are metric. It should be evident that the current can be found directly from Eq. 8.5, $P = IV$. Rearranging,

$$I = \frac{P}{V} = \frac{60 \text{ W}}{120 \text{ V}} = 0.50 \text{ A}$$

To find the resistance, we can use this current and rearrange Eq. 8.4 ($V = IR$, Ohm's law) to solve for resistance:

$$R = \frac{V}{I} = \frac{120 \text{ V}}{0.50 \text{ A}} = 240 \, \Omega$$

Notice we could also use Eq. 8.6 to solve for R. Rearranging this equation, we have

$$R = \frac{P}{I^2} = \frac{60 \text{ W}}{(0.50 \text{ A})^2} = 240 \, \Omega$$

CONFIDENCE QUESTION 8.2

A coffee maker draws 10 A of current operating at 120 V. How much electrical energy is used by the coffee maker each second?

In metals, electrical resistance varies with temperature, with the resistance generally decreasing with decreasing temperature. If you lowered the temperature enough, could you get rid of resistance completely? In some cases, the answer is yes. See the chapter Highlight.

8.3 Simple Electric Circuits and Electrical Safety

Learning Goals:

To explain the difference between series and parallel circuits.

To describe the aspects and procedures of electrical safety.

There are two principal forms of electric current. In a battery circuit, such as illustrated in Fig. 8.9, the electron flow is always in one direction, from the negative terminal to the positive terminal. This type of current is called **direct current** or **dc.** We use direct current in battery-powered devices such as flashlights, portable radios, and automobiles.

The other common type of current is **alternating current,** or **ac,** which is produced by constantly changing the voltage from positive to negative to positive and so on. (Although the usage is redundant, we commonly say *ac current* and *ac voltage.*) Alternating current is produced by electric companies and is used in the home. The frequency of changing from positive to negative voltages is usually at the rate of 60 cycles per second (cps), or 60 Hz (see Fig. 8.10). The average voltage varies between 110 V and 120 V, and household ac voltage is commonly listed as 110 V, 115 V, or 120 V. The equations for Ohm's law (Eq. 8.4) and power (Eqs. 8.5 and 8.6) apply to both dc and ac circuits containing only resistances.

Once the electricity enters the home or business, it is used in circuits to power (energize) various appliances and other items. Plugging appliances, lamps, and other electrical applications into a wall outlet places them in a circuit. There are two basic ways of connecting elements in a circuit: in series or in parallel.

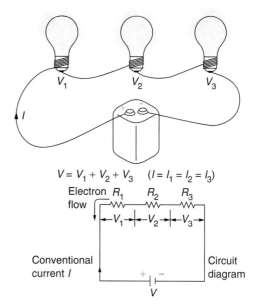

Figure 8.11 Series circuit.
The light bulbs are connected in series, and there is the same current through each of the bulbs. In the circuit diagram, the direction of the current I is taken to be away from the positive terminal or in the direction that positive charges would flow. This convention is used for historical reasons; the current is called the *conventional current.* In actuality, the electrons move in the opposite direction.

An example of connecting elements in a series circuit is shown in Fig. 8.11. The lamps are conveniently represented as resistances in the circuit diagram. In a **series circuit,** the same current passes through all of the resistances. This is analogous to a liquid circuit with a single line connecting several components. The total resistance is simply the sum of the individual resistances. As with different height potentials, the total voltage is the sum of the individual voltage drops, and

$$V = V_1 + V_2 + V_3 + \cdots$$
$$V = IR_1 + IR_2 + IR_3 + \cdots$$

or

$$V = I(R_1 + R_2 + R_3 + \cdots)$$

where the equation is written for three or more resistances. And, if we write $V = IR_s$, where R_s is the total equivalent series resistance, then by comparison

$$R_s = R_1 + R_2 + R_3 + \cdots \quad (8.7)$$

resistances in series

HIGHLIGHT

Superconductivity

In conducting materials, such as metal wires, electrons traveling through a material encounter opposition or *resistance*. On an atomic level this arises from collisions (interactions) with impurities and the lattice atoms or ions of a material. In general, the electrical resistance of most metallic conductors increases as the temperature increases. This is because the increased motions (or vibrations) of the lattice atoms or ions increase the number of collisions, making it more difficult for the electrons to be conducted along the wire.

Conversely, the resistance of metallic conductors generally decreases with decreasing temperature. Therefore, at one time scientists wondered how far the electrical resistance could be reduced. The answer is: to zero!—in certain materials, below certain temperatures, in a phenomenon commonly called *superconductivity*.

In 1908 a Dutch scientist, H. Kamerling Onnes, developed a technique to liquefy helium, which has a boiling point of 4.2 K. This technique allowed him to cool conductors to very low temperatures. He investigated the resistance versus temperature of some metals with a simple experimental technique: a voltage source (V) was used to establish a current (I) in a conductor. From the measured values, the resistance (R) could be obtained from Ohm's law ($R = V/I$).

Onnes investigated the conduction properties of mercury at low temperatures, and much to his surprise, the resistance fell to zero at a temperature near that of liquid helium, about 4 K or −296°C or −452°F (see the accompanying figure). He named the effect superconductivity.

Onnes' result set off a wave of investigations on elements and compounds; materials with higher critical temperatures (T_c, temperature below which superconductivity occurs) were found. In 1973 the highest critical temperature for a material was about 23 K. Then in 1986 came a breakthrough. Superconduction was observed in a metallic ceramic material at 35 K. By 1987 a similar material was found with a T_c of 93 K, which is above the boiling point of liquid nitrogen (77 K or −196°C). The current record-high for T_c is about 125 K. Liquid nitrogen is relatively inexpensive, and magnetic levitation experiments using ceramic superconducting materials are now quite common (see Fig. 1b).

Why electrons or currents encounter no resistance in some materials is not well understood. Current theories are complex and require quantum mechanics (Chapter 9). More readily understood are the potential uses for high-temperature superconductors. Magnets are used in motors, and the greater the current (smaller the resistance), the greater the strength of an electromagnet (Section 8.4) and the more powerful the motor. Also, magnetic levitation may lead to improved "MagLev" trains, which are repelled off the track at high speeds. Electrical energy may be transmitted through superconducting electrical lines with no losses. There are undoubtedly many other uses for superconductivity that scientists and engineers have not even thought of yet!

However, a great many technological problems must be overcome. For example, ceramic materials are brittle and difficult to fabricate into usable wires. Probably the most likely application that will be realized soon will be in the use of fast computer circuitry. Without resistance, electronic signals would travel faster, allowing the computer to operate faster. In any case, you are likely to see some exciting applications of superconductors in the not-too-distant future. Also, scientists would like to develop a room-temperature superconductor. Think how that would revolutionize electrical applications!

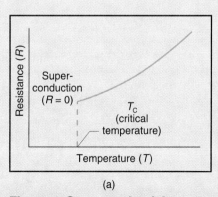

(a)

(b)

Figure 1 Superconductivity.
(a) Resistance versus temperature. The resistance decreases with decreasing temperature, and at the critical temperature (T_c) the resistance drops to zero. At this and lower temperatures, the material is superconducting. (b) Magnetic levitation. The lower superconducting material is cooled with liquid nitrogen, and when the material becomes cold enough to superconduct, the upper magnet is repelled and levitates.

What this equation means is that all of the resistances in series could be replaced with a single resistance R_s, and the same current would flow and the same power would be dissipated. For example, $P = I^2R_s$ is the power used in the whole circuit. (The resistances of the connecting wires are considered negligible.)

The example of lamps or resistances in series in Fig. 8.11 could as easily have been a string of Christmas tree lights connected in series. Christmas tree lights were once connected in this manner. When a bulb burned out, the whole string of lights went out, because there was no longer a completed path for the current, so the circuit was "open." Having a burned-out bulb is like opening a switch in the circuit to turn off the lights.

However, in most strings of lights purchased today, one light can burn out and the others remain lit. Why? Today's lights have a parallel aspect, which will be explained in our discussion of parallel circuits.

The other basic type of simple circuit is called a parallel circuit, as illustrated in Fig. 8.12. In a **parallel circuit,** the voltage across each resistance is the same, but the current through each resistance may vary. Notice that the current from the voltage source (battery) divides at the junction where all the resistances are connected together. This arrangement is analogous to liquid flow in a large pipe coming into a junction where it divides into several smaller pipes.

Because there is no build-up of charge at the junction, the charge leaving the junction must equal the charge entering the junction (conservation of charge), and we may write in terms of current,

$$I = I_1 + I_2 + I_3 + \cdots$$

Using Ohm's law (in the form $I = V/R$), we may write for the different resistances,

$$I = \frac{V}{R_1} + \frac{V}{R_2} + \frac{V}{R_3} + \cdots$$

or

$$I = V\left(\frac{1}{R_1} + \frac{1}{R_2} + \frac{1}{R_3} + \cdots\right)$$

The voltage V is the same across each R, because the voltage source is effectively connected "across"

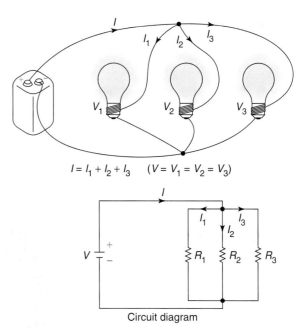

$I = I_1 + I_2 + I_3 \quad (V = V_1 = V_2 = V_3)$

Circuit diagram

Figure 8.12 Parallel circuit.
The light bulbs are connected in parallel, and the current from the battery divides at the junctions (black dots where two or more wires meet). The amount of current through each branch is determined by the relative resistances (of the light bulbs) in the branches; the greatest current is in the path or branch of least resistance.

each resistance and each gets the same voltage effect or drop.

Writing Ohm's law in terms of the total equivalent resistance R_p of the parallel circuit, $I = V/R_p$, by comparison, we have

$$\frac{1}{R_p} = \frac{1}{R_1} + \frac{1}{R_2} + \frac{1}{R_3} + \cdots \quad (8.8)$$

resistances in parallel

For a circuit with only *two resistances in parallel,* this equation can be conveniently written as

$$R_p = \frac{R_1 R_2}{R_1 + R_2} \quad (8.9)$$

As in the case of the series circuit, all of the resistors could be replaced by a single resistance R_p without affecting the current from the battery or the power dissipation.

EXAMPLE 8.3 Finding the Current in a Parallel Circuit

Three resistors have values of $R_1 = 6\ \Omega$, $R_2 = 6\ \Omega$, and $R_3 = 3\ \Omega$. How much current would be drawn from a 12-V battery if they are connected to the battery in a parallel circuit? (See Fig. 8.12.)

Solution

Given: $R_1 = R_2 = 6\ \Omega$
$R_3 = 3\ \Omega$
Find: I (current from battery)

The current drawn from the battery is given by Ohm's law (Eq. 8.4) using the total (parallel) resistance, $I = V/R_p$. Hence, we must first find the total equivalent resistance. For illustration, let's first combine R_1 and R_2 into a single equivalent resistance by using Eq. 8.9:

$$R_{p_1} = \frac{R_1 R_2}{R_1 + R_2} = \frac{(6\ \Omega)(6\ \Omega)}{6\ \Omega + 6\ \Omega} = 3\ \Omega$$

Thus, an equivalent circuit is a resistance R_{p_1} connected in parallel with R_3. Applying Eq. 8.9 again to these resistances to find the total resistance,

$$R_p = \frac{R_{p_1} R_3}{R_{p_1} + R_3} = \frac{(3\ \Omega)(3\ \Omega)}{3\ \Omega + 3\ \Omega} = 1.5\ \Omega$$

That is, the same current would be drawn from the battery for a 1.5-Ω resistance as would be for the three resistances in parallel.

The total resistance could also have been found by using Eq. 8.8, which is solved by using the lowest common denominator for the fractions. Omitting the intermediate ohm symbols for clarity, we have

$$\frac{1}{R_p} = \frac{1}{R_1} + \frac{1}{R_2} + \frac{1}{R_3} = \frac{1}{6} + \frac{1}{6} + \frac{1}{3}$$

$$= \frac{1}{6} + \frac{1}{6} + \frac{2}{6} = \frac{4}{6}\ \Omega$$

Then, inverting the fraction to find R_p,

$$R_p = \frac{4\ \Omega}{6} = 1.5\ \Omega$$

Either procedure to find R_p may be used.

Now, back to finding what was wanted. As noted earlier, we use Ohm's law to find the current,

$$I = \frac{V}{R_p} = \frac{12\ V}{1.5\ \Omega} = 8.0\ A$$

Note that this is not the current through the resistances, because it divides at the junction. The two 6-Ω resistances would have the same current, but the 3-Ω resistance would have twice the amount as through one of the 6-Ω resistances—more current through the "path of least resistance."

An interesting fact about resistances connected in parallel is that the total resistance is always less than the smallest parallel resistance. Such is the case in Example 8.3. You'd be unable to find a parallel circuit that proves otherwise.

CONFIDENCE QUESTION 8.3

Suppose the resistances in Example 8.3 were wired in series and connected to the 12-V battery. Would the battery supply more or less current than for the parallel arrangement? What would be the current in the circuit in this case?

Home appliances are wired in parallel (Fig. 8.13). There are two major advantages to the parallel circuit.

1. The same voltage (110–120 V) is available throughout the house. This makes it much easier to design appliances. (The 110–120 V voltage is obtained by connecting across the hot, or high-voltage, side of the line to ground, or zero potential. This gives a voltage difference of 120 V, even if one of the "high" sides is at a potential of -120 V. The voltage for large appliances, such as central air conditioners and heaters, is 220–240 V, which is available by connecting across the two incoming potentials, as shown in Fig. 8.13. This potential is analogous to a height difference between two positions for gravitational potential energy, one positive and one negative.)

2. If one appliance fails to operate, the others in the circuit are not affected because their circuits are still complete. In a series circuit, if one component fails, none of the others will operate because the circuit is incomplete or "open."

Question

Consider Christmas tree lights that remain on when one bulb burns out. How are they connected?

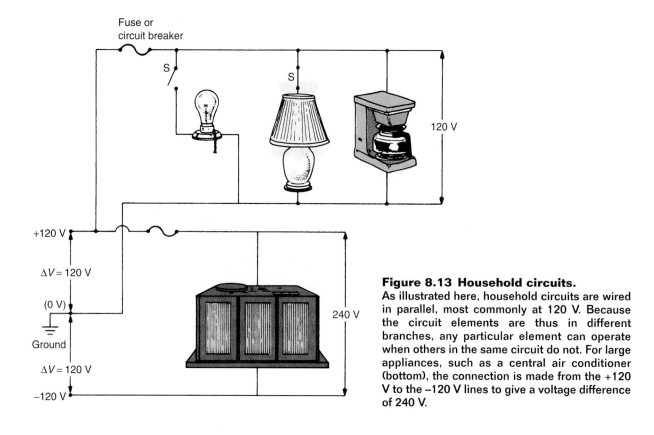

Figure 8.13 Household circuits.
As illustrated here, household circuits are wired in parallel, most commonly at 120 V. Because the circuit elements are thus in different branches, any particular element can operate when others in the same circuit do not. For large appliances, such as a central air conditioner (bottom), the connection is made from the +120 V to the −120 V lines to give a voltage difference of 240 V.

Answer

The bulbs could be wired in parallel. As shown in the household circuit in Fig. 8.13, if a bulb burns out, the other components in the parallel circuit continue to operate. However, the total resistance of a string of lights wired in parallel would be small, and an undesirably and dangerously large current would flow in the circuit. Also, parallel wiring would require a lot of additional wire, which is expensive.

For cost-effectiveness, each bulb is wired in parallel with a "shunt" resistor, as illustrated in Fig. 8.14. When the filament of a bulb burns out, the resistor provides a path for the current, shunting it around the defective bulb, and the other bulbs remain lit. Actually, the shunt resistor is insulated and not part of the circuit when its bulb is lit. However, when the bulb burns out, this places the voltage across the resistor. The resultant sparking then burns off the shunt insulation material so that the resistor makes contact and completes the circuit.

We can also have series-parallel combinations, which give intermediate equivalent total resistances, but we will not go into these.

Electrical Safety

Electrical safety for both people and property is an important consideration in using electricity. For example, in household circuits, as more and more appliances are turned on, there is more and more current and the wires get hotter and hotter. The *fuse* shown in the circuit diagram in Fig. 8.13 is a safety device that prevents the wires from getting too hot and possibly starting a fire. When the current reaches a preset amount, the fuse filament gets so hot that it melts and opens the circuit.

There are several types of fuses commonly used in household circuits. The so-called Edison-base fuse has a base with threads similar to those

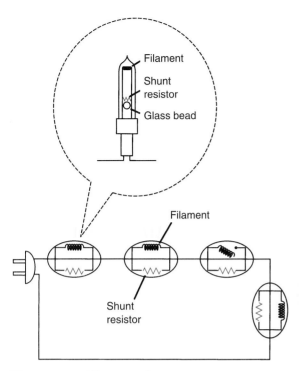

Figure 8.14 Shunt resistors.
Modern Christmas tree lights are wired with an insulated resistor in parallel with each bulb filament. When the bulb filament burns out, the voltage across the shunt burns away the insulation and the resistor becomes part of the circuit, shunting the burned-out bulb and allowing the other bulbs in series to operate.

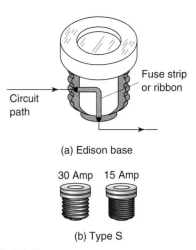

Figure 8.15 Fuses.
(a) An Edison-base fuse. If the fuse exceeds its rating, joule heat causes the fuse strip or ribbon to burn out and the circuit opens. (b) Type-S fuses. Edison-base fuses have the same threads for different ratings, and so a 30-A fuse could be put into a 15-A circuit, which would be dangerous. Type-S fuses have different threads for different fuse ratings and cannot be interchanged.

on a light bulb (Fig. 8.15a). As such, they are replaceable with a fuse of any amp rating. That is, a 30-A fuse can be screwed into the socket which should have a 15-A fuse. Such a mix-up could be dangerous, and so Type-S fuses are often used. Here, a threaded adapter is put in the socket that is specific for a particular fuse. Different-rated fuses have different threads, and a 30-A fuse cannot be screwed into a 15-A socket.

The more popular *circuit breaker* is generally replacing fuses. It serves the same function. When the current in a circuit reaches a preset amperage, the circuit breaker triggers a switch that opens or "breaks" the circuit, halting the current. Circuit opening is accomplished magnetically or by thermal expansion. When the trouble is corrected, the circuit breaker can be reset to close the circuit.

For safety, switches, fuses, and circuit breakers are always placed in the "hot," or high-voltage, side of the line. If they were placed in the ground side, when a circuit was opened, there would be no current; but there would still be 120-V potential to the appliance that could be dangerous to contact.

However, even when circuits are wired properly, fuses and circuit breakers may not always give protection from electrical shock. A "hot" wire inside an appliance or power tool may break loose and come into contact with its housing or casing, putting it at a high voltage. The fuse does not blow if there is not a large current. If the casing is conductive and a person touches it, as shown in Fig. 8.16a, a path is provided to ground and the person would receive a shock.

This condition is prevented by grounding the casing, as shown in Fig. 8.16b. Then, if a hot wire touches the casing, a current flows and the fuse blows. This grounding process explains the use of three-prong plugs on many electrical tools and appliances (Fig. 8.17a)

A polarized plug is shown in Fig. 8.17b. You have probably noticed that some plugs have one blade or prong larger than the other and will only

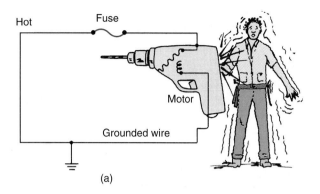

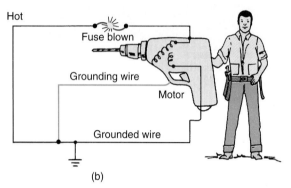

Figure 8.16 Electrical safety.
(a) Without a dedicated ground wire, the casing of an appliance may be at a high potential without blowing a fuse or tripping a circuit breaker. If someone touches the casing, a dangerous shock could result. (b) With a dedicated ground wire through a third prong on the plug, the casing would be grounded. The fuse would be blown, and the casing would be at zero potential.

fit into a wall outlet one way. Polarized plugs are an older type of safety feature. Because the plug is polarized or directional, one side of the plug is always connected to the ground side of the line. The casing of an appliance can be connected to ground by this means, with a similar effect as the three-wire system. However, a dedicated grounding wire is better because the polarized system is dependent on the circuit and the appliance being wired properly, and there is a chance of error. Also, even though connected to the ground side of the line, this is still a current-carrying wire, whereas the dedicated ground is not.

An electric shock can be very dangerous, and touching exposed electric wires should always be avoided. Many people are killed each year through electric shock. The danger is proportional to the

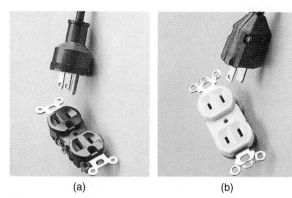

Figure 8.17 Electrical plugs.
(a) A three-prong plug and socket. The third, rounded prong is a dedicated ground used for safety purposes. (b) A two-prong polarized plug and socket. Note that one blade or prong is larger or wider than the other. The ground or neutral (zero-potential) side of the line is wired to the large-prong side of the socket. This distinction or polarization permits paths to ground for safety connections.

amount of electric current that goes through the body. The effects are discussed in the chapter Highlight.

8.4 Magnetism

Learning Goals:

To explain the law of poles, and the magnetic field.

To identify the cause of magnetism, and explain why some materials can be magnetized and some cannot.

To analyze some aspects of Earth's magnetic field.

One of the first things one notices in examining a bar magnet is that it has two regions of magnetic strength or concentration, one at each end of the magnet, that we call *poles*. We designate one as the north pole, N, and the other as the south pole, S. This is because the N pole of a magnet, when used as a compass, is the north-seeking pole or points north, and the S pole is the south-seeking pole.

When examining two magnets, we find that there are attractive and repulsive forces between

HIGHLIGHT

Electric Shock

In an electric shock, the amount of current through the body depends on its resistance, and is given by Ohm's law, $I = V/R_{body}$, where R_{body} is the body resistance. This resistance varies considerably, mainly due to the dryness of the skin. Because our bodies are mostly water, the skin resistance makes up most of the body's resistance.

A dry body can have a resistance as high as 500,000 Ω, and a current from a 110-V source would be 0.00022 A. The danger occurs when the skin is moist or wet. The resistance of the body can go as low as 100 Ω, and the current would rise to 1.1 A.

Because injuries and death from shocks can occur when the skin is wet, appliances such as radios should not be used near a bathtub. Should a plugged-in radio or hair dryer happen to fall into a filled tub, then the whole tub, including a person in it, may be plugged into 110 V.

The general effects of electrical currents on the human body are shown in Table 1. The seriousness of a shock also depends on the path of the current. Through the fingers of the hand would not be as serious as from hand-to-hand through the chest. As can be seen from the table, a current of only 10 mA causes muscle paralysis. Normal muscle reactions are caused by electrical impulses through nerves, and these are influenced by electric shock. A person may not be able to let go if holding on to a conductor when receiving the 10-mA current.

As can be seen, the greater the current, the more serious the effect. With 100 mA or 0.1 A, there can be interference with normal heart muscle action. Resulting uncoordinated movements (ventricular fibrillation) can prevent proper blood circulation, which can then be fatal. With one amp and over, a fatality is likely.

Table 1 General Effects of Electric Shock on the Human Body

Electric Current	Consequences
1.0 mA (0.001 A)	Mild shock
10 mA (0.01 A)	Muscle paralysis
20 mA (0.02 A)	Chest muscle paralysis, respiratory arrest, fatal in a short time
100 mA (0.1 A)	Ventricular fibrillation, irregular heartbeat, fatal in seconds
1000 mA (1.0 A)	Serious burns, almost instantly fatal

---

them, specific to the poles. The forces are described by the **law of poles:**

Like poles repel and unlike poles attract.

That is, N and S poles, N-S, attract, and N-N and S-S poles repel each other (Fig. 8.18a). The strength of the attraction or repulsion depends on the strength of the magnetic poles. Also, in a manner similar to Coulomb's law, the strength of the magnetic force is proportional to the square of the distance between the poles. Figure 8.18b shows some toy magnets that seem to defy gravity due to their magnetic repulsion.

All magnets occur with two poles—that is, as dipoles. Unlike electric charge, which occurs in single charges, magnets are always dipoles. A magnetic monopole would consist of a single N or S pole without the other. There is no known physical reason for magnetic monopoles not existing, but thus far their existence has not been confirmed experimentally. The discovery of a magnetic monopole would be an important fundamental development.

Every magnet produces a force on every other magnet. In order to discuss these effects, we introduce the concept of a magnetic field. A **magnetic field** (B) is a force field characterized by a set of imaginary lines that indicate the direction a small compass needle would point if it were placed near

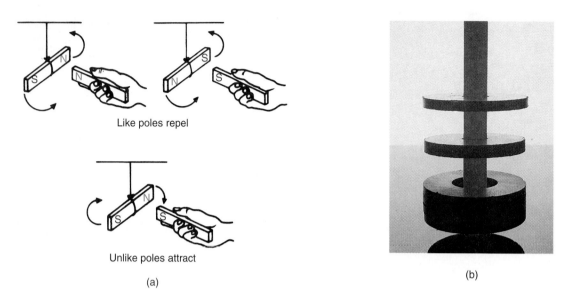

Figure 8.18 Law of poles.
(a) Like poles repel, and unlike poles attract. (b) The adjacent poles of the magnets in the photo must be like poles. Why?

a magnet. Hence, the field lines are indications of the magnetic force—a force field, so to speak.

Figure 8.19a shows the magnetic field lines around a simple bar magnet. The arrows in the field lines indicate the direction in which the north pole of a compass would point. The closer together the field lines, the stronger the magnetic force.

Magnetic field patterns can be "seen" by using small iron filings. The iron filings are magnetized and act as small compass needles. The outline of the magnetic field produced in this manner is shown for two bar magnets in Fig. 8.19b. The field concept can also be used for an electric field around charges, but this force field is not so easily visualized. The electric field is the electric force per unit charge. The electric and magnetic fields are vector quantities, and electromagnetic waves are made up of electric and magnetic fields that vary with time (Chapter 6).

Electricity and magnetism are discussed together in this chapter because they are linked. In fact, the source of magnetism is moving and "spinning" electrons. Hans Oersted, a Danish physicist, first discovered in 1820 that a compass needle is deflected by an electric current–carrying wire. When a compass is placed near a wire in a simple battery circuit and the circuit is closed, there is current in the wire and the compass needle is deflected from its north-seeking direction. When the circuit is opened, the compass needle goes back to pointing north again. Also, it is found that the strength of the magnetic field is proportional to the magnitude of the current—the greater the current, the greater the strength of the magnetic field. Hence, a current produces a magnetic field that can be turned off and on at will. We can investigate such fields by using iron filings.

Different configurations of current-carrying wires give different magnetic field configurations. Some of them are shown in Fig. 8.20. A straight wire produces a field in a circular pattern around the wire. A single loop of wire gives a field not unlike that of a small bar magnet, and the field of a coil of wire with several loops is very similar to that of a bar magnet.

But what produces the magnetic field of a permanent magnet such as a bar magnet? In our simplistic model of the atom, electrons are pictured as going around the nucleus. This is electric charge in motion, or a current loop, so to speak, and it might be expected that this would be a source of a magnetic field. However, it is found that the magnetic field produced by orbiting atomic electrons is very small. Also, the atoms of material are distributed

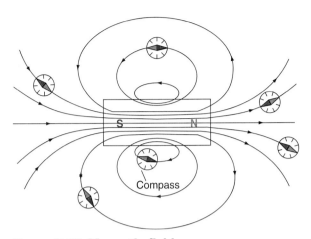

Figure 8.19 Magnetic field.
Magnetic field lines may be mapped using a small compass. The north (N) pole of the compass indicates the direction of the field at any point. As shown in the photo, iron filings become induced magnets and conveniently outline the pattern of the magnetic field.

such that the magnetic fields would be in various directions and generally cancel each other, giving a zero net effect.

Modern theory predicts the magnetic field to be associated with electron "spin." This effect is pictured as an electron spinning on its axis, in the same manner as Earth rotates on its axis. As such, we would have charge in motion. A material has many atoms and electrons, and the magnetic-spin effects of all of these electrons usually cancel out. So most materials are not magnetic (do not become magnetized), or are only slightly magnetic. In some instances, however, the magnetic effect can be quite strong.

Materials that are highly magnetic are called **ferromagnetic.** Ferromagnetic materials include the elements iron, nickel, cobalt, and certain alloys of these and a few other elements. In ferromagnetic materials, the magnetic fields of many atoms combine to give rise to **magnetic domains,** or local regions of alignment. A single magnetic domain acts like a tiny bar magnet.

In iron, the domains can be aligned or nonaligned. A piece of iron with the domains randomly oriented is not magnetic. This effect is illustrated in ● Fig. 8.21. When the iron is placed in a magnetic field, such as that produced by a current-carrying wire, the domains line up or those parallel to the field grow at the expense of other domains, and the iron is magnetized.

(a)

(b)

(c)

Figure 8.20 Magnetic field patterns.
Iron-filing patterns near current-carrying wires outline the magnetic fields for (a) a long, straight wire, (b) a single wire loop, and (c) a coil of wire, or solenoid.

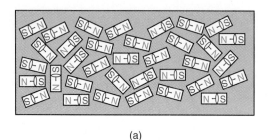

(a) Unmagnetized material

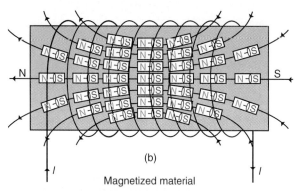

(b) Magnetized material

Figure 8.21 Magnetization.
(a) In a ferromagnetic substance, dipole moments or magnetic domains (groups of magnetic dipoles) are generally unaligned, so there is no net magnetic field. (b) In a magnetic field induced by a current-carrying wire, the domains may grow and become aligned with the magnetic field and the substance becomes magnetized. (Bar magnet representations are used here for simplicity.)

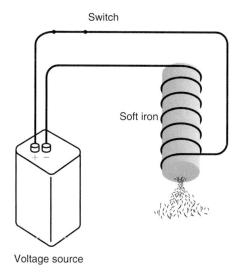

Figure 8.22 Electromagnet.
An electromagnet consists of a coil of wire wrapped around a piece of soft iron. When the switch is closed, charge flows in the wire, giving rise to a magnetic field that magnetizes the iron, thus creating a magnet. When the switch is open, no charge flows and the iron is not magnetized.

When the magnetic field is removed, the domains tend to return to a mostly random arrangement due to heat effects that cause disordering. The amount of domain alignment remaining after the field is removed depends on the strength of the applied magnetic field.

An application of this effect is an electromagnet, which basically consists of a coil of insulated wire wrapped around a piece of iron (● Fig. 8.22). Since a magnetic field can be turned on and off by turning an electric current on and off, we can control whether or not the iron will be a magnet. When the current is turned on, the magnetic field of the coil magnetizes the iron. The aligned domains add to the field, making it about 2000 times stronger.

Electromagnets have many applications. Large ones are routinely used to pick up and transfer scrap iron (Fig. 8.22), and small ones are used in magnetic relays and solenoids, which act as magnetic switches. Solenoids are used in automobiles to engage the starting motor when starting a car.

One type of circuit breaker uses an electromagnetic switch. The strength of an electromagnet is directly proportional to the current in its coils. When there is a certain amount of current in the breaker circuit, an electromagnet becomes strong enough to attract and "trip" a metallic conductor and open the circuit.

The iron used in electromagnets is called "soft" iron. This does not refer to its mechanical hardness, but rather to the ability of this type of iron to be magnetized and quickly become demagnetized. Certain types of iron, along with nickel, cobalt, and a few other elements, are known as "hard" magnetic materials. Once magnetized, they retain their magnetic properties for a long time, and so "hard" iron is used for permanent magnets. When permanent magnets are heated or struck, the domains are shaken from their alignment, and the magnet becomes weaker. In fact, above a certain temperature, called the **Curie temperature,** a material ceases to be ferromagnetic. The Curie temperature of iron is 770°C.*

A permanent magnet is made by "permanently" aligning the domains inside the material. One way to do this is by heating a piece of hard ferromagnetic material above its Curie temperature and then applying a strong magnetic field. The domains line up with the field, and as the material cools, the domain alignment is frozen in, so to speak, producing a permanent magnet.

Earth's Magnetic Field

At the beginning of the seventeenth century, William Gilbert, an English scientist, suggested that Earth acted as a huge magnet. Today, we know that such a magnetic effect does exist for our planet. It is Earth's magnetic field that causes compasses to point north. Experiments have shown that a magnetic field exists within Earth and many hundreds of miles out into space. The aurora borealis and aurora australis, or northern lights and southern lights, common sights in higher latitudes near the poles, are associated with Earth's magnetic field. This effect will be discussed in Section 19.1.

The origin of Earth's magnetic field is not known, but probably the most acceptable theory is that the field is caused by internal currents of electrically charged particles deep within Earth associated with its rotation. It is not due to some huge mass of magnetized iron compound within Earth. Earth's interior is quite hot and above the Curie temperature, so materials would not be ferromagnetic. Also, the magnetic poles slowly change their positions, which suggests changing currents.

Earth's magnetic field does approximate that of a current loop or a huge imaginary bar magnet, as illustrated in ● Fig. 8.23. Notice that the south magnetic pole is near the geographic north pole. It is for this reason that the north pole of a compass needle is attracted northward (law of poles). Even

---

*The Curie temperature is named after Pierre Curie (1859–1906), the French scientist who discovered the effect. Pierre Curie was the husband of Madame Marie Curie. They both did pioneering work in radioactivity (Chapter 10).

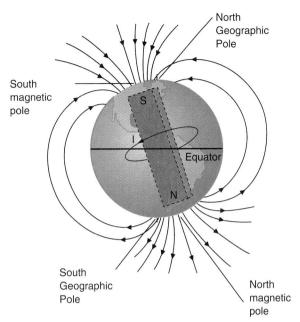

Figure 8.23 Earth's magnetic field.
The magnetic field of Earth is probably caused by internal currents in the liquid outer core in association with the planet's rotation. The magnetic field is similar to that of a giant bar magnet within Earth (but no such bar exists). Note that the magnetic south pole is near the geographic north pole. The north pole of a compass needle points northward because it is attracted toward a magnetic south pole.

so, we commonly refer to a magnetic "north" pole in the direction of *magnetic north* or near the geographic north pole.

The magnetic and geographic poles do not coincide. Presently, the north magnetic pole is some 13° or about 1500 km (930 mi) south of the geographic north pole of Earth's spin axis. The south magnetic pole is displaced even more from its respective geographic pole.

Hence, the compass does not point toward the geographic, or true, north pole, but toward the magnetic north pole. The variation between the two is expressed in terms of **magnetic declination,** which is the angle between geographic (true) north and magnetic north (● Fig. 8.24). The declination may vary east or west of a geographic north-south meridian (an imaginary line running from the geographic north pole to the geographic south pole).

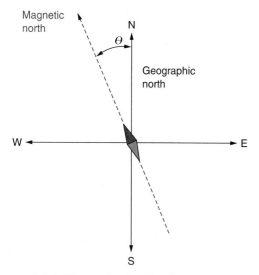

Figure 8.24 Magnetic declination.
The angle θ of the magnetic declination is the angle between geographic (true) north and magnetic north, and is measured in degrees east or west of geographic north.

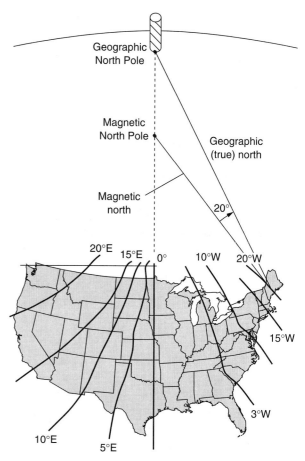

Figure 8.25 Magnetic declinations.
The map shows some isogonic (same magnetic declination) lines for the United States. The 0° line means that magnetic north is in the same direction as true or geographic north.

It is important in navigation to know the magnetic declination at a particular location so the magnetic compass direction can be corrected for true north. This is provided on navigational maps, which show lines of declination expressed in degrees east and west (Fig. 8.25).

The magnetic field of Earth is relatively weak compared with that of magnets used in the laboratory. However, the field is strong enough to be used for orientation by certain animals (including ourselves). For instance, migratory birds and homing pigeons are believed to use Earth's magnetic field to aid them in their homeward flights. Iron compounds have been found in their brains.

8.5 Electromagnetism

Learning Goals:

To identify some electromagnetic interactions and applications.

To distinguish between motors and generators.

To explain the principle of transformers, and how and why they are used in electrical power transmission.

The interaction of electrical and magnetic effects is known as **electromagnetism.** Electromagnetism is one of the most important aspects in physical science, and most of our current technology is directly related to this crucial interaction. Two basic principles of this interaction are as follows:

1. Moving electric charges (current) give rise to magnetic fields.
2. A magnetic field may deflect a moving electric charge.

The first principle forms the basis of an electromagnet, which was considered previously. Electromagnets are found in a variety of applications, such as doorbells, telephones, and devices used to

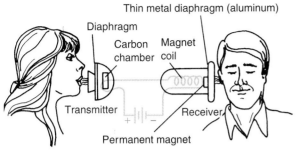

Figure 8.26 Telephone and electromagnetism.
A simplified diagram of a (one-directional) telephone circuit. Sound waves are converted into electrical pulses in a transmitter. The pulses travel along the phone lines to the receiver. There they are converted back to sound waves by the actions they have on an electromagnet that in turn drives a diaphragm.

move magnetic materials (see Fig. 8.22). Let's look at the mechanism in a telephone receiver as another common practical example. A simplified diagram of a telephone circuit is shown in Fig. 8.26.

When a telephone number is dialed, a circuit is completed with another telephone's bell. When the receiver of the ringing phone is lifted, the circuit between the speakers and receivers of the two telephones is completed. In a telephone conversation, the sound waves are converted to varying electric current. This varying current travels along the wire to the other telephone's receiver, where it is converted back into sound.

Here's how this is done. The transmitter of a phone consists of a diaphragm that vibrates in response to the spoken sound waves. The diaphragm vibrates against a chamber that contains carbon granules. As the diaphragm vibrates, the pressure on the carbon granules varies, causing more or less electrical resistance in the circuit. The resistance is low when the granules are pressed together and increases as they spread apart. By Ohm's law, this varying resistance gives rise to a varying electric current in the telephone circuit.

At the receiver end, there is an electromagnet or magnetic coil, and a permanent magnet, which is attached to a disk. The varying electric current gives the electromagnet varying magnetic strengths. The activated electromagnet attracts the permanent magnet with varying force, and the force variation causes the attached disk to vibrate. The vibrations set up sound waves in the air that closely resemble the original sound waves, and the voice is heard at the other end of the telephone.

Magnetic Force on Moving Electric Charge

The second previously mentioned electromagnetic principle may be stated in a qualitative way: A magnetic field can be used to deflect moving electric charges. A stationary electric charge in a magnetic field experiences no force, but when a moving charge enters a magnetic field as shown in Fig. 8.27, it experiences a force. It is found that the magnetic force (F_{mag}) is perpendicular to the plane formed by the velocity vector (v) and the magnetic field (B). In the figure, the force would initially be out of the page, and with an extended field, the negatively charged particle would follow a circular arc path. If the moving charge were positive, it would be deflected in the opposite direction, or into the page. Also, if a charge, positive or negative, is moving parallel to a magnetic field, there is no force on the charge.

This effect can be demonstrated experimentally as shown in Fig. 8.28. A beam of electrons is traveling in the tube from left to right, and is made visible by a piece of fluorescent paper in the tube. In the upper photo, the beam is undeflected

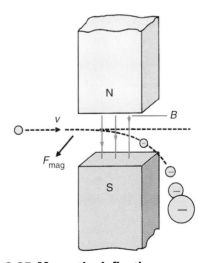

Figure 8.27 Magnetic deflection.
Electrons traveling horizontally in a vertical magnetic field, as shown, experience a force F_{mag} that would deflect them out of the page. See text for description.

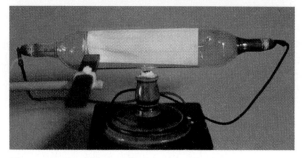

Figure 8.28 Magnetic force on moving charges. *Top:* The presence of a beam of electrons is made evident by a fluorescent strip in the tube that allows the path of the electrons to be seen. *Bottom:* The magnetic field of a magnet gives rise to a force on the electrons and the beam is deflected.

in the absence of a magnetic field. In the lower photo, the magnetic field of a bar magnet causes the beam to be deflected downward. This means the magnetic field is generally directed into the strip of paper. (If you rotated Fig. 8.27 toward you 90°, you'd have the same effect.)

Motors and Generators

The electrons in a conducting wire also experience such force effects caused by magnetic fields. In Fig. 8.29a, a nonconducting wire is shown in a magnetic field. Because there is no current or no net motion of the electrons, there is no force on the wire. However, with a current (moving here to the right), the wire experiences a force out of the page. This situation is similar to the one in Fig. 8.27, except that here the force causes the whole wire to be forced out of the page toward the viewer.

Hence, a current-carrying wire in a magnetic field can experience a force. With a force present, it might quickly come to mind that we could use it to do work, and this is what's done in electric motors.

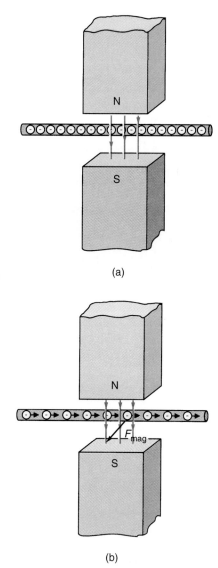

Figure 8.29 Magnetic field and current.
(a) A stationary wire with no current does not experience a force. (b) A current-carrying wire in a magnetic field experiences a force. With a negative electron current to the right, the magnetic force F_{mag} would be out of the page, as in the case illustrated here. The force on a current-carrying wire is the basic principle of an electric motor.

Basically, a **motor** is a device that converts electrical energy into mechanical energy. We plug motors in, and we use the mechanical rotations of their shafts to do work. To help understand the electromagnetic-mechanical interaction of motors (of which there are many types), the diagram of a simple dc motor is shown in Fig. 8.30.

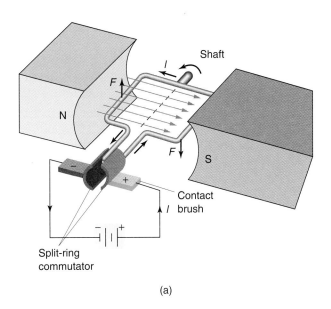

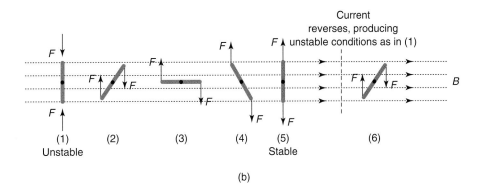

Figure 8.30 A dc motor.
(a) An illustration of a coil in a dc motor. As a current-carrying loop in a magnetic field, the coil experiences a torque and rotates the attached shaft. The split-ring commutator effectively reverses the loop current each half-cycle, so the coil will rotate continuously. (b) The forces on the coil show why the current reversal is necessary.

Real motors have many loops or windings, but only one is shown here for simplicity. The battery supplies current to the loop, which is free to rotate in the magnetic field between the pole faces. The force on the current-carrying loop produces a torque, causing it to rotate.

Continuous rotation requires a split-ring commutator. In contact with the brushes, one part of the ring is positive and the other negative. In the end view sequence in the figure, note that when the plane of the loop is vertical, the forces are such that the loop would stop after a half-cycle. However, because of the split ring, the current in the loop reverses at this time. Note that the wire in contact with the positive part of the ring will be in contact with the negative part of the ring during the next half-cycle, hence the current in the loop reverses each half-cycle. This takes the force situation back to the unstable position (both forces acting inward), but the inertia of the loop carries it through this position and the loop rotates continuously.

Motors generally have many windings (loops) on a rotating armature to enhance the effect. The rotating loops cause a connected shaft to rotate, and this rotation is used to do mechanical work. The conversion of electrical energy to mechanical energy is enhanced by many loops of wire and stronger magnetic fields.

One might ask if the reverse is possible. That is, is the conversion of mechanical energy into electrical energy possible? Indeed it is, and this principle is the basis of electrical generation. Have you ever wondered how electricity is generated?

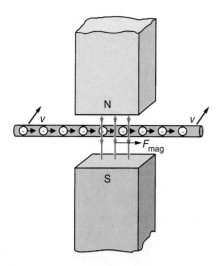

Figure 8.31 Motion and current.
If a wire is moved perpendicularly to a magnetic field as illustrated here, the magnetic force causes the electrons in the wire to move, setting up a current in the wire. This is the basic principle of an electrical generator.

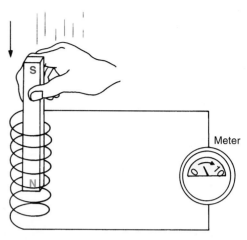

Figure 8.32 Electromagnetic induction.
An illustration of Faraday's experiment showing electromagnetic induction. The reading of the meter indicates a current in the circuit as the magnet moves through the coil.

As illustrated in Fig. 8.31, suppose that an applied (mechanical) force sets a wire in a magnetic field in motion. The electrons in the wire are charges moving in a magnetic field and hence experience a magnetic force (F_{mag}) to the right. The electrons move in the conductor, and an electrical current is set up, as is a voltage, since Ohm's law applies to the wire. Note that the current is generated without batteries, plugs, or other external voltage sources. This illustrates the basic principle of electrical generation.

A **generator** converts mechanical work or energy into electrical energy. A generator operates on what is called *electromagnetic induction*. This principle was discovered by in 1831 by Michael Faraday, an English scientist. An illustration of his experiment is shown in Fig. 8.32. When a magnet is moved toward a loop of wire (or a coil, for enhancement), it is observed that a current is induced in the wire, as indicated on the meter. Investigation shows that this is caused by a time-varying magnetic field through the loop.

The same effect is obtained by using a stationary magnetic field and rotating the loop in the field. The magnetic field through the loop varies with time and a current is induced. A simple ac generator is illustrated in Fig. 8.33. When the loop is mechanically rotated, a voltage and current are induced in the loop that vary in magnitude and alternate back and forth, changing direction each half-cycle. Hence, we have alternating current (ac). There are also dc generators, which are essentially dc motors operated in reverse. However, most electricity is generated as ac, and converted, or rectified, to dc.

Generators are used in power plants to convert other forms of energy to electrical energy. For the most part, fossil fuels and nuclear energy are used to heat water to generate steam that is used to turn turbines that supply mechanical energy in the generation process. The electricity is carried to homes and businesses, where it is either converted back to mechanical energy to do work or converted to heat energy.

Let's end this chapter by discussing how electrical energy or power is transmitted. We have all seen the high-voltage (or high-tension) transmission lines running across the land. The voltage for transmission is stepped-up by means of **transformers.** A transformer is a simple device based on electromagnetic induction (Fig. 8.34). Basically, it consists of two insulated coils of wire wrapped around an iron core, which concentrates the magnetic field when there is current in the input coil. With an ac current in the input or primary coil, there is a time-changing magnetic field as a re-

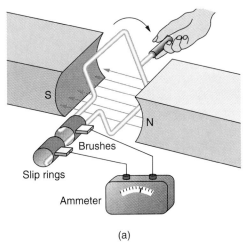
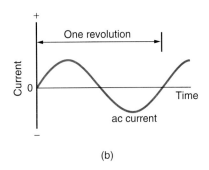

(a)　　　　　　　　　　　　　　　　(b)

Figure 8.33 An ac generator.
(a) An illustration of a coil in an ac generator. If the coil is mechanically turned, a current is induced as indicated by the ammeter. (b) The current varies in direction each half-cycle and hence is called ac or alternating current.

sult of the current going back and forth. The magnetic field goes through the secondary coil, and induces a voltage and current.

Because the secondary coil has more windings than the primary coil, the induced ac voltage is greater than the input voltage (similar to enhancing the effect in a motor or generator), and we call

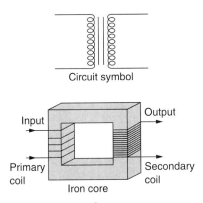

Figure 8.34 The transformer.
The circuit symbol for and the basic features of a transformer. It consists of two insulated coils wrapped around a piece of iron. Alternating current in the primary coil creates a varying magnetic field, which is concentrated by the iron through the secondary coil. This in turn induces a voltage and a current in the secondary coil.

this type of transformer a *step-up transformer*. However, when the voltage is stepped up, the secondary current is stepped down (by the conservation of energy, since $P = IV$). The factor of voltage step-up depends on the ratio of the number of windings on the coils, which we can easily control.

So why do we want to step up the voltage? Actually, its the step down in current that we really are interested in. Transmission lines have resistance, and therefore I^2R losses. So, by stepping down the current, we reduce these losses and save energy that would be lost as joule heat. If you step up the voltage by a factor of 2, the current is reduced by a factor of 2; with one-half the current, you would have only one-fourth of the I^2R losses. (Why?) So for power transmission, the voltage is stepped up to a very high voltage to get the corresponding current step-down, so as to avoid joule heat losses.

Of course, we can't use such high voltages in our homes, which in general have 220–240 V service entries. So how do we get the voltage back down (and the current up)? We use a *step-down transformer*, which steps down the voltage and steps up the current. This is done by simply reversing the input and output coils. If the primary coil has more windings than the secondary, the voltage

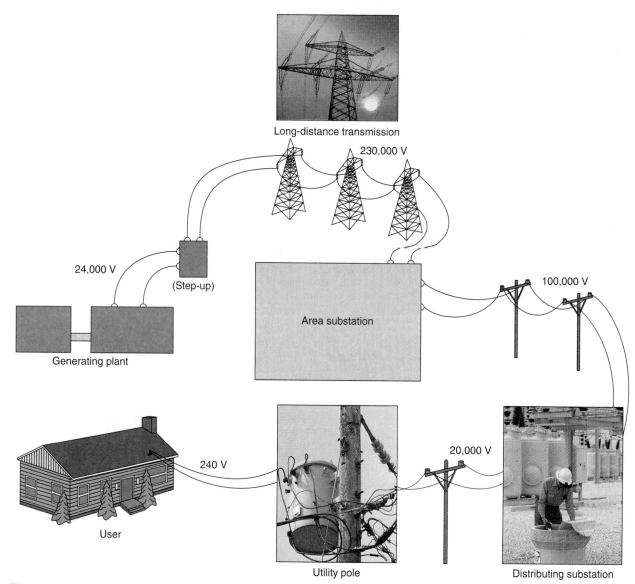

Figure 8.35 Electrical power transmission system.
At the generating plant, the voltage is stepped up with a corresponding current step-down, so as to reduce I^2R losses in the lines for long-distance transmission. The high voltage is then stepped down in distributing substations, and finally to 240 V by the common utility-pole transformer (sometimes on ground) for household usage.

is stepped down. An illustration of this stepping up and stepping down is shown in Fig. 8.35.

This step-up and step-down advantage in reducing joule heat loss in power transmission is a major reason why we use ac electricity in our homes and businesses. Stepping up and down can't be done with dc; that is, a transformer will not work on dc. (Why?) The first commercial electric company in this country was started by Thomas Edison in New York City and did use dc; but with the long distances of transmission we have today, its use would not be practical.

Important Terms

These important terms are for review. After reading and studying the chapter, you should be able to define and/or explain each of them.

electric charge
electrons
protons
coulomb (C)
current
ampere (A)
conductor
insulator
law of charges
Coulomb's law
electric potential energy

voltage
volt (V)
resistance
ohm (Ω)
Ohm's law
power
joule heat (I^2R losses)
direct current (dc)
alternating current (ac)
series circuit
parallel circuit

law of poles
magnetic field
ferromagnetic materials
magnetic domains
Curie temperature
magnetic declination
electromagnetism
motor
generator
transformer

Important Equations

Current: $I = \dfrac{q}{t}$ $\quad \left(\text{current} = \dfrac{\text{charge}}{\text{time}}\right)$

Coulomb's law: $F = \dfrac{kq_1q_2}{r^2}$

$\left(k = 9.0 \times 10^9 \, \dfrac{\text{N-m2}}{\text{C}^2}\right)$

Voltage: $V = \dfrac{W}{q}$

Ohm's law: $V = IR$

Power: $P = IV = I^2R$ $\quad$ (joule heat)

Resistances in series: $R_s = R_1 + R_2 + R_3 + \cdots$

Resistances in parallel: $\dfrac{1}{R_p} = \dfrac{1}{R_1} + \dfrac{1}{R_2} + \dfrac{1}{R_3} + \cdots$

(two resistances): $R_p = \dfrac{R_1 R_2}{R_1 + R_2}$

Questions

8.1 Electric Charge and Current

1. The unit of electric charge is the (a) newton, (b) coulomb, (c) ampere, (d) volt.
2. The unit of electric current is the (a) newton, (b) coulomb, (c) ampere, (d) volt.
3. If two equal negative charges were placed equidistant on either side of a positive charge, the positive charge would (a) experience a net force to the right, (b) experience a net force to the left, (c) have a zero net force.

4. Describe the three basic subatomic particles.
5. What is a quark and where can one be located?
6. What is an electric current, and what are the units of electric current? (Give two equivalent units.)
7. Explain how a charged rubber comb attracts bits of paper and how a charged balloon sticks to a wall or ceiling.
8. Why are some materials good electrical conductors and others are not?

8.2 Voltage and Electrical Power

9. The unit of voltage is the (a) joule, (b) joule/coulomb, (c) amp-coulomb, (d) amp/coulomb.
10. In electrical terms, power has the unit of (a) joule/coulomb, (b) amp/ohm (c) amp-ohm, (d) amp-volt.
11. Distinguish between electrical potential energy and voltage.
12. Explain Ohm's law and give the unit of each term in this law.
13. How does electrical power depend on (a) current and voltage and (b) current and resistance?
14. Explain what is meant by I^2R losses and joule heat.

8.3 Simple Electric Circuits and Electrical Safety

15. Given three resistances, the greatest current would flow in a battery circuit if the resistances were connected in (a) series, (b) parallel, (c) series-parallel.
16. Automobile headlights are wired in (a) series, (b) parallel.
17. Distinguish between alternating and direct currents.
18. Why are home appliances connected in parallel rather than in series?
19. Discuss the safety features of (a) fuses, (b) circuit breakers, (c) three-prong grounded plugs, and (d) polarized plugs.
20. What causes electrical shocks, what are some consequences, and what precautions should be taken to avoid them?

8.4 Magnetism

21. A magnetic field (a) is determined by the law of poles, (b) is in the direction indicated by the north pole of a compass, (c) is initiated above the Curie temperature, (d) is the source of magnetism.
22. The variation of Earth's magnetic north pole from true north is given by (a) a magnetic field, (b) the law of poles, (c) magnetic domains, (d) magnetic declination.
23. Compare (a) the law of charges and the law of poles, and (b) electric and magnetic fields.
24. What is a ferromagnetic material, and what are some examples?
25. Why does a permanent magnet attract pieces of iron? What would happen if the pieces of iron were above the Curie temperature?
26. (a) What does Earth's magnetic field resemble, and where are its poles? (b) What is magnetic declination?

8.5 Electromagnetism

27. A motor is a device that converts (a) chemical energy into mechanical energy, (b) mechanical energy into electrical energy, (c) electrical energy into mechanical energy, (d) mechanical energy into chemical energy.
28. A generator is a device that converts (a) chemical energy into mechanical energy, (b) mechanical energy into electrical energy, (c) electrical energy into mechanical energy, (d) mechanical energy into chemical energy.
29. How do telephone transmitters and receivers work?
30. Describe the basic principle of a dc electric motor.
31. Describe the basic principle of an ac generator.
32. What is the principle of a transformer, and how are transformers used?
33. What is a major reason for using ac electricity in our homes and businesses?
34. What would happen if electric power were transmitted from the generating plant to your home at 120 V?

Food for Thought

1. What is electricity?
2. Speculate about how our environment would be if the gravitational force in the atom were not negligible compared to the electrical force.
3. An old saying about electrical safety states that you should keep one hand in your pocket when working on electricity. What does this mean?
4. What happens if a bar magnet is cut in half? Can it be continually cut in half to finally get two magnetic monopoles? Explain.
5. Why will transformers not operate on dc?

EXERCISES

8.1 Electric Charge and Current

1. A charge of 2.0 C passes by a point in a conductor in 4.0 s. What is the current in the conductor?

   Answer: 0.50 A

2. A current of 1.5 A persists in a conductor for 6.0 s. How much charge passes by a given point in the conductor during this time?

   Answer: 9.0 C

3. A conductor has a current of 0.50 A. How long will it take for 4.0 C of charge to pass a given point in the conductor?

4. What happens to the force between an electron and a proton if the distance between them is (a) doubled, and (b) tripled?

   Answer: (a) reduced one-fourth of original value.

8.2 Voltage and Electrical Power

5. To separate a 0.25-C charge from another charge, 30 J of work is done. What is the voltage between the charges? How is this voltage related to energy?

   Answer: 120 V

6. If an electrical component with a resistance of 40 Ω is connected to a 120-V source, how much current will flow in the circuit?

   Answer: 3.0 A

7. What battery voltage is necessary to supply 0.50 A of current to circuit with a resistance of 3.0 Ω?

8. (a) How much current is drawn by a flashlight using batteries that add to 3.0 V to produce a power output of 0.50 W?
   (b) What is the resistance of the flashlight bulb?

   Answer: (a) 0.17A; (b) 18 Ω

9. Show that electric power is given by $P = \dfrac{V^2}{R}$.

10. How much current is drawn by each of the items shown in Fig. 8.10?

    Answer: bulb, 0.50 A

11. A 100-W light bulb is turned on in the home.
    (a) How much current flows through the bulb?
    (b) What is the resistance of the bulb?
    (c) How much electrical energy is used each second? (Assume a voltage of 120 V.)

    Answer: (a) 0.83 A

8.3 Simple Electric Circuits

12. Show that for two resistances in parallel the total equivalent resistance is given by $R_p = \dfrac{R_1 R_2}{R_1 + R_2}$.

13. A resistance and a shunt resistance are 0.35 Ω and 0.50 Ω, respectively. What is the total equivalent resistance?

    Answer: 0.21 Ω

14. Three resistors with values of 20 Ω, 30 Ω, and 40 Ω, respectively, are connected (a) in series, and (b) in parallel. What is the total equivalent resistance in each case?

    Answer: (b) 9.2 Ω

15. If the arrangements in Exercise 14 are connected to a 90-V battery, what is the current in the circuit in each case?

16. A person with dry skin has a hand-to-hand resistance of 50,000 Ω.
    (a) If the person gets across the terminals of a 12-V battery, what is the current through the body?
    (b) If the person's body is wet with perspiration and the body resistance drops to 2000 Ω, what is the current?

    Answer: (a) 0.24 mA

8.4 Magnetism

17. Sketch the magnetic fields between (a) a north pole and a south pole, and (b) two north poles.

18. Using Fig. 8.25, estimate the magnetic declination where you live and explain what this means. (You must live in the conterminous United States to use the figure.)

Answers to Multiple-Choice Questions

1. b   9. b   15. b   21. b   27. c
2. c   10. d   16. b   22. d   28. b
3. c

Solutions to Confidence Questions

8.1 $q = It = (1.5 \text{ A})(2.0 \text{ s}) = 3.0 \text{ C}$

8.2 $P = IV = (10 \text{ A})(120 \text{ V}) = 1200 \text{ W or J/s}$

8.3 Greater resistance, less current.

$R_s = R_1 + R_2 + R_3$
$= 6 \, \Omega + 6 \, \Omega + 3 \, \Omega = 15 \, \Omega$

$I = \dfrac{V}{R_s} = \dfrac{12 \text{ V}}{15 \, \Omega} = 0.80 \text{ A}$

Chapter 9

Atomic Physics

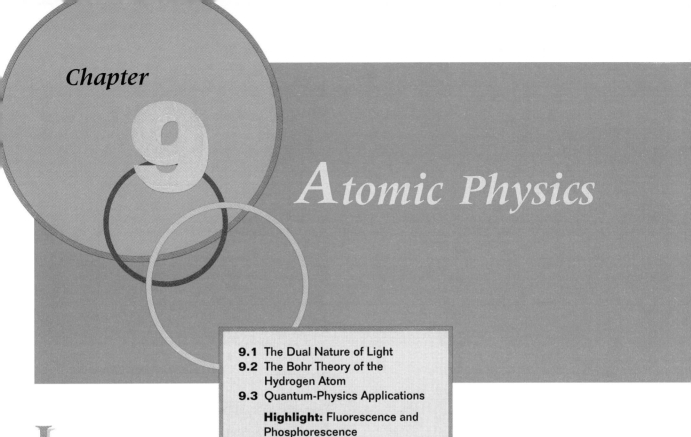

- **9.1** The Dual Nature of Light
- **9.2** The Bohr Theory of the Hydrogen Atom
- **9.3** Quantum-Physics Applications

Highlight: Fluorescence and Phosphorescence

- **9.4** Matter Waves and Quantum Mechanics
- **9.5** Multielectron Atoms and the Periodic Table

In the latter part of the nineteenth century, scientists thought physics was in fairly good order. The principles of mechanics, wave motion, sound, and optics were reasonably well understood. Electricity and magnetism had been combined into electromagnetism, and light had been shown to be a wave (electromagnetic wave). There were some rough edges, but it seemed that only a few refinements were needed.

However, as scientists probed deeper into the microscopic and submicroscopic world of the atom, they observed some strange things—strange in the sense that the phenomena could not be explained by proven classical principles. These discoveries were unsettling, and as time and investigations went on, scientists became aware that not everything about nature was known. In fact, they had to use new approaches in the description of nature that seemed radical at that time.

The development of physics since about 1900 is commonly termed *modern physics*. The preceding classical or Newtonian physics was generally concerned with the macrocosm—the universe, the motions of planets, the description of observable phenomena, and so on—whereas modern physics generally considers the microcosm—the atom and its structure. This chapter and Chapter 10 will be concerned with some of the topics of modern physics. In this chapter we will look chiefly at atomic physics, which deals with the atom as a whole and its electronic structure. In Chapter 10 we will go to the heart of the matter and look at the central core or nucleus of the atom—nuclear physics.

9.1 The Dual Nature of Light

Learning Goals:

To state Planck's hypothesis and explain why it is so radical.

To explain what is meant by the dual nature of light.

To analyze the photoelectric effect and the quantum theory that explains it.

Even before the turn of the century, scientists knew that light of all frequencies was emitted by the atoms of an incandescent (glowing hot) solid, such as the filament of a light bulb. As the temperature

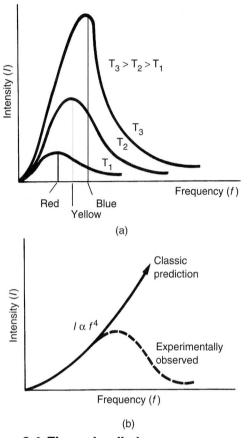

Figure 9.1 Thermal radiation.
(a) As the temperature of an incandescent solid increases, the component of maximum intensity shifts to a higher frequency. (b) Classically, it is predicted that the intensity should be proportional to f^4, which indicates a much greater energy than actually observed. (c) The radiation component of maximum intensity of a hot solid determines its color, as shown here by hot steel coming out of a furnace.

is increased, more radiation is emitted and the component of maximum intensity is shifted to a higher frequency. As a result, a very hot solid appears to go from a dull red to a bluish white (● Fig. 9.1a).

This outcome is expected, because the hotter the solid, the greater the electron vibrations in the atoms and the higher the frequency of the emitted radiation. However, according to classical wave theory, the intensity or energy of the emitted radiation should be proportional to the fourth power of the frequency ($I \propto f^4$), which means that the intensity should increase quite rapidly with frequency (Fig. 9.1b). But this is not what is observed. This discrepancy was termed the **ultraviolet catastrophe**—"ultraviolet" because the difficulty occurred at high frequencies beyond the violet end of the visible spectrum, and "catastrophe" because it predicted the energy intensity to be very much greater than actually observed.

The dilemma was resolved in 1900 by the German physicist Max Planck (pronounced "Plonk," ● Fig. 9.2). He introduced a radical new idea that explained the observed thermal radiation intensity distribution. In doing so, Planck took the first step toward a new theory in physics called **quantum physics.** Classically, an electron oscillator may vibrate at any frequency or have any energy up to some maximum value. But Planck's hypothesis stated that the energy was quantized, or that oscillators could have only discrete or certain amounts of energy. Moreover, the energy (E) of an oscillator

Figure 9.2 Max Planck (1858–1947).
While a professor of physics at the University of Berlin in 1900, Planck proposed that the energy of thermal oscillators existed only in discrete amounts, or quanta, thus introducing the idea of quantum physics. The important small constant *h* that appears over and over again in quantum physics is called Planck's constant. Planck was awarded the Nobel Prize in physics in 1918 for his contributions to quantum theory.

depended on its frequency (f) according to the relationship

$$E = hf \quad (9.1)$$

where h is a constant, called *Planck's constant*, with a value of 6.63×10^{-34} J-s (a very small number). With this concept, Planck's theory predicted the observed radiation curve as in Fig. 9.1.

Thus, Planck introduced the idea of a **quantum,** or a discrete amount of energy. This concept was in radical contrast to the classical idea that an electron oscillator, like a mass oscillating on a spring, could have continuously different energies.

In 1905 Albert Einstein used Planck's hypothesis to describe light in terms of particles or quanta (plural of *quantum*) rather than waves. He did so to explain what is called the **photoelectric effect,** which had been observed in the latter part of the nineteenth century. Electrons are observed to be emitted when certain metallic materials are exposed to light (hence the name *photoelectric*). This direct conversion of light (radiant energy) into electrical energy now forms the basis of photocells used in photographic light meters and solar energy applications (Fig. 9.3).

Some aspects of the photoelectric effect could not be explained by classical theory. For example, the amount of energy needed to free an electron from a photomaterial could be calculated. But according to classical theory, in which light is a wave with a continuous flow of energy, it would take an appreciable time for electromagnetic waves to supply the energy needed for an electron to be emitted. However, current flows from photocells almost immediately on being exposed to light. Also, it was observed that only light above a certain frequency would cause photoelectrons to be emitted. Classically, light of any frequency should be able to provide the needed energy.

Applying Planck's hypothesis, Einstein assumed that light and other electromagnetic radiation was quantized. He termed the quanta of light

Figure 9.3 Photoelectric effect in action.
Radiant (solar) energy is converted directly into electrical energy in the world's largest array of solar panels at Mount Lagune Air Force Station in California.

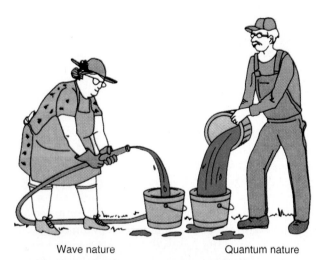

Figure 9.4 Wave and quantum analogy.
A wave supplies a continuous flow of energy, somewhat analogous to a stream of water. A stream of water would take an appreciable time to fill a bucket. A quantum supplies energy in a packet or bundle, analogous to a bucket of water. Dumping the water from a bucket fills the other one very quickly. This is analogous to a quantum supplying the energy in the photoelectric effect.

photons, which were "packets" or "particles" of energy. Planck's relationship ($E = hf$) was used, with the frequency f being taken as the frequency of the light. That is, a quantum of light contained a discrete amount of energy of $E = hf$.

The photon energy can also be expressed in terms of the wavelength of light. Recall for light $c = \lambda f$, so $f = c/\lambda$, and so $E = hf = hc/\lambda$. Thus, the shorter the wavelength (or higher the frequency) of light, the greater the energy of its photons. For example, photons of blue light have more energy than photons of longer wavelength red light.

By considering light to be composed of photons or quanta of energy, Einstein was able to successfully explain the photoelectric effect. (He received a Nobel Prize for this work, not for his more famous theory of relativity.) The classical time delay to get enough energy to free an electron is not a problem with quanta of energy. A photon with the proper amount of energy could provide the release energy instantaneously. An analogy of wave and quantum energies is shown in ● Fig. 9.4. Also, we can see why light above a certain frequency is required for photoemission. Since $E = hf$, light below a certain frequency would not have enough photon energy to free an electron.

The concept of light being composed of discrete packets or photons is confusing to most people. How can light be composed of "particles" when it shows wave phenomena such as diffraction, interference, and polarization? However, Einstein's quantum theory of the photoelectric effect was substantiated by experiment and satisfied the scientific method. Indeed, there seems to be a confused situation. Is light a wave or a particle?

The answer to this question is contained in the phrase the **dual nature of light,** which simply means that light apparently sometimes acts as a wave and sometimes acts as a particle—at least we must describe it so to explain different phenomena.

9.2 The Bohr Theory of the Hydrogen Atom

Learning Goals:

To explain Bohr's theory of atomic electronic structure.

To describe the structure of the hydrogen atom in terms of quantum numbers, states, and energy levels.

Quantum theory played an important role in the development of a simplified model of the atom. In the late 1800s, a great deal of experimental work was being done with gas-discharge tubes—for example, mercury vapor and neon tubes. When the light was analyzed with a spectrometer, discrete or line spectra were observed instead of the continuous spectra observed from incandescent sources such as filament lamps. That is, only spectral lines of certain frequencies or wavelengths were found (● Fig. 9.5). Scientists did not understand why only certain wavelengths were emitted by the atoms in various excited gases.

An explanation of the spectral lines observed for hydrogen was put forth in 1913 by the Danish physicist Niels Bohr (● Fig. 9.6). In 1911, the British physicist Lord Ernest Rutherford had shown that the protons of an atom were in a nucleus or central core in an atom. Being the simplest atom, hydrogen has only one proton in its nucleus and one associated electron.

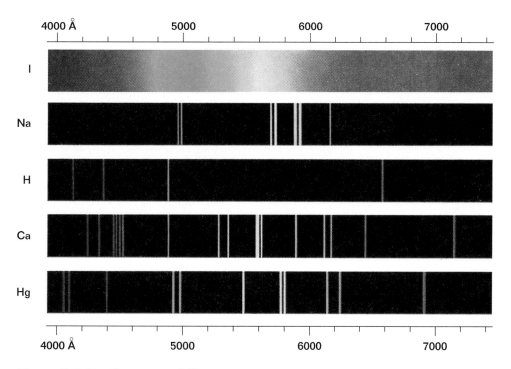

Figure 9.5 Continuous and discrete spectra.
A continuous spectrum of white light is shown at the top above the line spectra of light emitted from excited atoms in sodium (Na), hydrogen (H), calcium (Ca), and mercury (Hg) discharge tubes. Notice how each line spectrum is different (or characteristic) for a given element.

Figure 9.6 Niels Bohr (1885–1962).
Bohr was one of the foremost scientists of the twentieth century. His initial application of quantum theory to the hydrogen atom, for which he was awarded the Nobel Prize in 1922, led to the development of our present-day understanding of atomic structure and spectra. Bohr's subsequent work in nuclear theory played an important part in understanding nuclear fission. He was forced to flee his native Denmark during World War II because of his part in the anti-Nazi resistance movement. He came to the United States where he was involved in the atomic bomb project, as were many other foreign scientists.

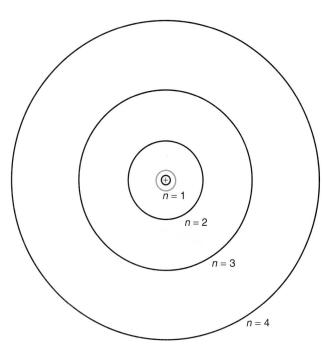

Figure 9.7 Bohr orbits.
The Bohr theory predicts only certain discrete orbits for the hydrogen electron. (Drawing not to scale.)

notice in the figure that the orbits are not evenly spaced.

There was still a classical problem with Bohr's theory. According to classical theory, an accelerating electron radiates electromagnetic energy. An electron in circular orbit has centripetal acceleration (Chapter 3), and hence should radiate energy. Such a loss of energy would cause the electron to spiral into the nucleus, similar to an Earth satellite in a decaying orbit because of atmospheric frictional losses.

But this doesn't happen in the atom, and Bohr made another nonclassical assumption in his theory. He postulated that the hydrogen electron does not radiate energy when in a bound, discrete orbit, but does so only when it makes a transition, or "quantum jump," from one discrete orbit to another.

The allowed orbits of the hydrogen atom are commonly expressed in terms of energy states or energy levels, with each state corresponding to a particular orbit (Fig. 9.8). Keep in mind that a particle (or satellite) in a circular orbit with a particular radius has a particular energy. We characterize the energy levels as being states in a potential well (Section 4.3). Like an object in a hole or real well, energy is required to lift it to a higher level, and if the top of the well is the zero reference

Bohr's theory assumed that the electron of the hydrogen atom revolved around the nuclear proton in a circular orbit, similar to planets orbiting the Sun (in nearly circular orbits) or satellites orbiting Earth. However, in the atom, the electrical force instead of the gravitational force supplies the necessary centripetal force. The radical part of the theory was that Bohr assumed the angular momentum of the electron was quantized. He apparently reasoned that a discrete line spectrum resulted from a quantum effect.

This assumption led to the prediction that the hydrogen electron could be in only discrete (specific) orbits with particular radii. (This is not the case for Earth satellites. For example, with proper maneuvering a satellite can have any orbital radius.)

The possible electron orbits were characterized by whole-number values, $n = 1, 2, 3, 4, \ldots$, where n is called the **principal quantum number** (Fig. 9.7). The lowest value ($n = 1$) has the smallest radius, and the radii increase with increasing principal quantum numbers. However,

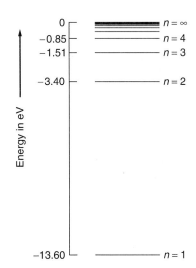

Figure 9.8 Energy levels.
The energy-level diagram for the hydrogen atom. Each orbit has a particular energy value, or level. The lowest level, $n = 1$, is called the ground state, and the higher levels are called excited states.

point, the energy levels in the well will have negative values.

In a hydrogen atom, the electron is normally at the bottom of the well, or in the **ground state** ($n = 1$), and must be given energy, or "excited," to raise it up in the well to a higher energy level or orbit. The states above the ground state ($n = 2, 3, 4, \ldots$) are called **excited states.**

The levels in the energy well somewhat resemble the rungs of a ladder. Just as a person going up and down a ladder must do so in discrete steps on the ladder rungs, so must a hydrogen electron be excited (or de-excited) by discrete amounts. However, as pointed out earlier, the energy "rungs" of the hydrogen atom are not evenly spaced—they get closer together as the n values increase. Also, if enough energy is supplied to excite the electron to the top of the well, the electron is no longer bound to the nucleus, and the atom is ionized.

A mathematical development of Bohr's theory is beyond the scope of this text; however, the important results are the predictions of the radii and the energies of the orbits. The radius of a particular orbit is given by

$$r_n = 0.53\, n^2 \text{ Å} \tag{9.2}$$

where n is the principal quantum number of an orbit, $n = 1, 2, 3, \ldots$, and r is the orbital radius, measured in angstrom units (Å, where 1 Å = 10^{-10} m).

For $n = 1$, we have $r_1 = 0.53$ Å; for $n = 2$, the orbit radius is $r_2 = (0.53)(2)^2 = (0.53)(4) = 2.12$ Å; and so on. A list of several allowed values of r is given in Table 9.1. Notice that the radii are not evenly spaced, as depicted in Fig. 9.7.

The total energy of an electron in one of the allowed orbits is given by

$$E_n = \frac{-13.6}{n^2} \text{ eV} \tag{9.3}$$

where n is the principal quantum number of the orbit, and

E is measured in electron volts (eV).

An *electron volt* is the amount of energy an electron acquires when it is accelerated through an electric potential of one volt. The eV is a common unit of energy in atomic and nuclear physics, and 1 eV is equal to 1.6×10^{-19} J.

The energy of a particular orbit or energy level is obtained by using the principal quantum number for that orbit. For example, for $n = 1$

$$E_1 = \frac{-13.6}{(1)^2} \text{ eV} = -13.6 \text{ eV}$$

and an electron in the ground state has an energy of -13.6 eV. In an excited state, for example, for $n = 3$

$$E_3 = \frac{-13.6}{(3)^2} \text{ eV} = \frac{-13.6}{9} \text{ eV} = -1.51 \text{ eV}$$

Other electron energy values are given in Table 9.1 for orbits near the nucleus. These values correspond to the energy levels in the diagram in Fig. 9.8. The minus signs, indicating negative energy values, show that the electron is in a potential well. With $n = \infty$, the electron is at the top of the well and $E = 0$.

But how did the theory stand up to experiment? Recall that Bohr was trying to explain discrete line spectra. According to the Bohr theory, an electron can make transitions only between the allowed orbits or energy levels. In these transitions, the total energy must be conserved. If the electron is initially in an excited state, it may lose energy by changing to a less excited (lower energy) state. In this case, the electron's energy loss is carried away by a photon—a quantum of light.

The total initial energy (E_{n_i}) must equal the total final energy, so we have

$$E_{n_i} = E_{n_f} + E_{\text{photon}}$$

or, the energy of the emitted photon is

$$E_{\text{photon}} = E_{n_i} - E_{n_f}$$

Table 9.1 Allowed Values of the Hydrogen Electron's Radius and Energy for Low Values of n

n	r_n	E_n
1	0.53 Å	-13.60 eV
2	2.12 Å	-3.40 eV
3	4.77 Å	-1.51 eV
4	8.48 Å	-0.85 eV
5	13.25 Å	-0.54 eV
$\vdots$	$\vdots$	$\vdots$
$\to \infty$	$\to \infty$	$\to 0$

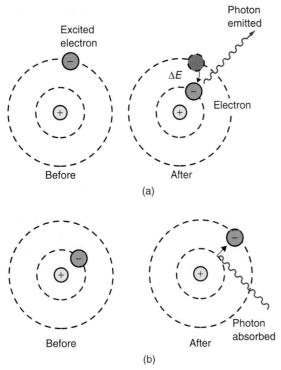

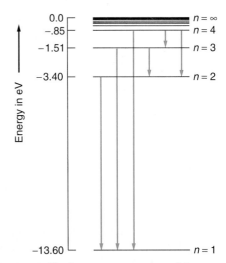

Figure 9.10 Hydrogen atom transitions. Some possible transitions for the hydrogen electron in returning to the ground state. The electron may make one or more "jumps" in the de-excitation process. The energy of an emitted photon is equal to the difference in the energy levels, so the lengths of the transition arrows give an indication of the relative energies of the photons.

Figure 9.9 Photon emission and absorption. (a) When an electron in an excited hydrogen atom makes a transition to a lower energy level, the atom loses energy by emitting a photon. (b) When a hydrogen atom absorbs a photon, the electron is excited into a higher energy level or orbit.

A schematic diagram of the process of photon emission is shown in Fig. 9.9a, and Fig. 9.9b illustrates the reverse process of photon absorption or excitation.

The transitions for photon emission are shown on an energy-level diagram in Fig. 9.10. The electron may "jump" down one or more energy levels in becoming de-excited. That is, the electron may go down the energy "ladder" using adjacent "rungs," or it may skip "rungs."

In this situation, discrete amounts of energy are given off, equal to the difference in the values of the energy levels. Recall that $E_{photon} = hf = hc/\lambda$, so we may write the preceding energy conservation equation as $hc/\lambda = E_{n_i} - E_{n_f}$, which gives us an expression for the wavelength of the photon:

$$\lambda = \frac{hc}{E_{n_i} - E_{n_f}}$$

Hence, the Bohr theory predicts that the hydrogen atom will emit light with discrete wavelengths corresponding to the discrete transitions. When the wavelengths for the transitions were computed using the energy values of the initial and final states, and compared to the wavelengths of the lines of the hydrogen spectrum, it was found that the theory agreed with experimental data (Fig. 9.11).

Transitions to a particular final state form a series of transitions. These series were named in honor of early spectroscopists who discovered or experimented with these wavelengths of light. For example, the series of lines in the visible spectrum, which corresponds to the transitions to a final state of $n = 2$, is called the *Balmer series*.

Thus, quantum theory and the quantum nature of light scored another success. As you might imagine, the energy-level arrangements for atoms other than hydrogen, those with more than one electron, are quite complex. Even so, the line spectra of various atoms are indicative of their energy-level spacings and provide characteristic "fingerprints" by which atoms and molecules can be spectroscopically identified.

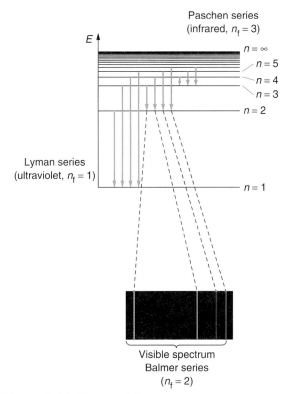

Figure 9.11 Spectral lines.
The transitions between discrete energy levels by the electron in the hydrogen atom give rise to discrete spectral lines. Transitions to a particular state form a series. For example, transitions to $n = 2$ are called the Balmer series; the corresponding spectral lines are in the visible spectrum.

9.3 Quantum-Physics Applications

Learning Goals:

To explain the principles of various applications, such as the microwave oven and the laser, by using quantum theory.

To describe how X-rays are produced.

Much of modern physics and chemistry is based on the study of energy levels of various atomic and molecular systems. When light is emitted, scientists study the emission spectrum to learn about the energy levels of the system (as in Fig. 9.11 for the hydrogen atom). Some chemists do research in the field of molecular spectroscopy—the study of energy levels of molecules and their associated spectra. Molecules can have quantized energy levels because of molecular vibrations, rotations, or excited atoms. Of course, different molecules have quite different spectra.

An *absorption spectrum* is obtained when light passes through a gas or liquid. Many of the wavelengths pass through, but photons that have the proper energy to excite the molecules to higher energy levels are absorbed. The absorption of a particular wavelength can be total or partial, depending on the absorptive material present and other factors. An absorption spectrum appears as dark lines (absorbed wavelengths) on a bright background.

The various gases of the atmosphere absorb light of particular wavelengths. The main absorbing gases are carbon dioxide (CO_2), water (H_2O), and ozone (O_3). The absorption properties of these gases are important atmospheric considerations, which will be discussed in Chapter 19.

The water molecule has some rotational energy levels spaced very closely together. The energy differences are such that microwaves (Section 6.2) are absorbed. Microwave photons have relatively low energies and are absorbed by water molecules. This principle forms the basis of the microwave oven, since all foods contain moisture. The water molecules (as some others) absorb microwave radiation, thereby heating and cooking the food. The interior metal sides of the oven reflect the radiation and remain cool.

Because it is the water content that is crucial in microwave heating, objects such as paper plates and ceramic or glass dishes do not get hot immediately in a microwave oven. However, they often become warm or hot after being in contact with hot food. Some people think that microwaves penetrate the food and heat it throughout. This is not the case. On the average, microwaves penetrate only a few centimeters before being completely absorbed. The interior of a large mass of food is heated by conduction as in a regular oven. For this reason, microwave-oven users are advised to let foods sit for a time after microwaving. Otherwise, the outside of the food may be quite hot, while the center is still cool.

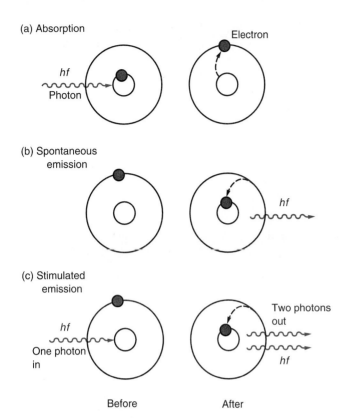

Figure 9.12 Spontaneous and stimulated emissions.
When an atom absorbs a photon and becomes excited (a), it may spontaneously return to its ground state in a short time with the emission of a photon (b). If, however, the excited atom is struck by a photon with the same energy as the absorbed photon, the atom is stimulated to emit a photon (c).

As a safety precaution, microwave ovens are constructed so they cannot be operated with the door open. Human tissue contains water and can be cooked just as easily as food can.

Another device based on energy levels is the laser. The development of the laser was a great success for the modern approach to science. Scientific discoveries have often been made accidentally, and even though some of these discoveries were put into practical use, no one fully understood how or why they worked; the discovery of X-rays is a good example. Similarly, early investigators often applied a trial-and-error approach until they discovered something that would work. Edison's invention of the incandescent lamp is a good example of this type of trial-and-error discovery. However, the laser was first developed "on paper" and then built with the expectation that it would work as predicted.

The word **laser** is an acronym for *l*ight *a*mplification by *s*timulated *e*mission of *r*adiation. (This amplification was first developed for microwave frequencies, and the first device was called a *maser*, where *m* is for *microwave*.) The amplification of light provides a very intense beam. Ordinarily, when an atom is excited, it "decays" or returns to the ground state in a very short time with emission of a photon. This process is called *spontaneous emission*. In this process, there is one photon in and one photon out (Fig. 9.12).

However, an excited atom can be stimulated to emit a photon. In a **stimulated emission** process—the key process of a laser—an excited atom is struck by a photon of the same energy of the allowed transition, and two photons are given off (one in, two out—amplification). Of course, in this process we do not get something for nothing. Energy is needed to initially excite the atom.

There are different types of lasers, but let's focus on the common helium-neon gas laser (Fig 9.13). The gas mixture is about 85% helium and 15% neon. The helium atoms are placed in an excited state by a radio frequency (r-f) voltage applied across the laser tube. Normally, the excited atoms would spontaneously emit photons almost immediately. However, some excited states are said to be *metastable*, meaning that the electron stays in such a state for a longer period of time (but still

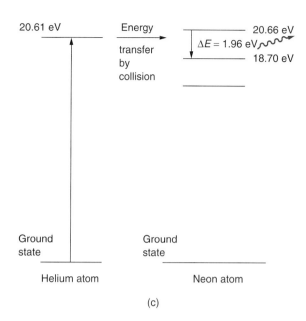

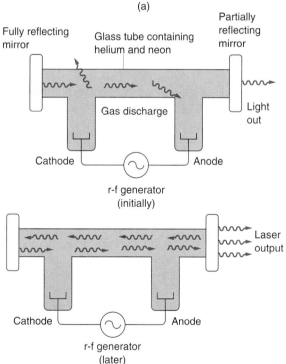

Figure 9.13 Laser action.
(a) The red light of a He-Ne laser beam passing through a transparent cell. (b) Schematic diagram of He-Ne laser tube. Stimulated emission and reflections from the end mirrors set up an intense beam along the axis of the tube. (c) The energy-level diagram for those levels involved in the lasing action of the helium-neon laser. See text for description.

only for a fraction of a second). The particular energy level to which the helium atom is excited is a metastable state.

As illustrated in the energy-level diagram in Fig. 9.13, the neon atom also has an energy level at about the same energy. With the metastable state condition, there is time for a helium atom to collide with an unexcited neon atom and transfer energy to it before emitting a photon spontaneously. The excited neon atom decays after a short time to another level, and the photon emitted in the transition has a wavelength that is in the visible red region of the spectrum.

The major amplification of light emitted by the neon atoms is accomplished by reflections from mirrors placed at each end of the laser tube. Reflecting back and forth in the tubes, the photons cause stimulated emissions, and an intense beam of light builds up parallel to the tube axis. Part of the beam emerges through one of the end mirrors, which is partially silvered. Because the light is amplified and very intense, a laser beam should never be directed toward a person's eyes (or reflected into the eyes). The direct viewing of a laser beam can cause serious eye damage.

The emitted light beam has some relatively unique properties. Light from common sources, such as an incandescent lamp, is said to be *incoherent*. In such light sources, the excitation occurs

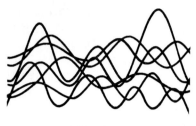

(a) Incoherent

(b) Coherent

Figure 9.14 Incoherent and coherent light.
(a) In incoherent light, waves have no particular relationship to each other and are random or "chaotic." (b) Waves of coherent light, on the other hand, have the same phase, wavelength, and direction.

randomly, and atoms emit randomly at different frequencies and the waves have no particular relationship to each other. As a result, incoherent light is "chaotic" (Fig. 9.14a). For example, if you threw a handful of gravel into a pond, the resulting waves would be incoherent.

The light waves from a laser, on the other hand, have the same frequency, and direction, and have a constant relationship to each other. Such light is said to be *coherent* (Fig. 9.14b). Because a laser beam is so directional, there is little spreading of the beam. This feature has permitted us to reflect a pulsed laser beam back to Earth from a mirror placed on the Moon by astronauts. Such experiments are used to accurately measure the distance between Earth and the Moon, so that small fluctuations in the Moon's orbit can be studied.

Lasers and laser light are used in an increasing number of applications; for instance, long-distance communications use laser beams in space and in optical fibers for telephone conversations. Lasers are used in the medical field as diagnostic and surgical tools. The intense heat that is produced by laser light on a small area can drill very tiny holes in metals and can weld machine parts. Laser "scis-

sors" are used to cut cloth in the garment industry. There are applications in surveying, weapons systems, chemical processing, photography, and holography (the process of making three-dimensional images). Laser printers are used in computer printouts.

You may own a laser yourself, a compact disc (CD) player. A laser "needle" is used to read the information (sound) stored on the disc in small dot patterns. The dots produce reflection patterns that are read by photocells and are converted to electronic signals. The laser in this case is a small, solid-state semiconductor laser.

Another common laser application is found in supermarkets and department stores (Fig. 9.15). You have no doubt seen the reddish glow of laser light in a supermarket checkout line as part of an optical scanner for reading the product codes on items. Each item has a particular code that identifies it to a computer that contains the programmed price of the item.

X-rays, another example of quantum phenomena, are used widely in medical and industrial fields. **X-rays** were discovered accidently in 1895 by German physicist Wilhelm Roentgen while he was working with gas-discharge tubes. He noticed

Figure 9.15 Optical scanning.
A laser scanner being used in a supermarket. The light reflected from the lighted areas of the code bar gives an identifying reflection pattern that is detected by the scanner. This information is sent to a computer that supplies the price of the item.

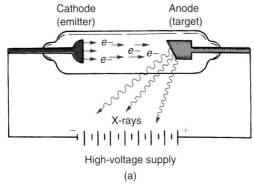

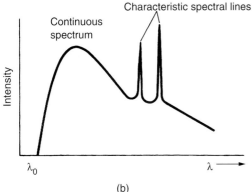

Figure 9.16 X-rays.
(a) X-rays are produced in a tube in which electrons from the cathode are accelerated toward the anode. On interacting with the atoms of the anode material, the electrons are slowed down and energy is emitted in the form of X-rays. (b) An X-ray spectrum. See text for description.

that a piece of fluorescent paper was glowing, apparently from having been exposed to some unknown radiation being emitted from the discharge tube. He called it X-radiation.

In the modern X-ray tube, electrons are accelerated through a large electrical voltage toward a metal target (Fig. 9.16a). When the electrons strike the target, they interact with the electrons in the atoms of the target material, and the electrical repulsion decelerates the incident electrons. The result is an emission of high-frequency X-ray quanta. X-rays are called *Bremsstrahlung* ("braking rays") in German.

An X-ray spectrum is illustrated in Fig. 9.16b. Notice that there is a cutoff wavelength λ_o below which no X-rays are emitted for a given tube voltage. Also, for very large tube voltages, there are intense "spikes" or spectral lines. These spikes are characteristic of the target material and are called *characteristic X-rays*. Both features of the X-ray spectrum can be explained by quantum considerations.

The low cutoff wavelength corresponds to the quantum of the highest frequency or energy ($E = hf = hc/\lambda$). Such quanta result when an incident electron is stopped completely and gives up all of its energy. There can be no quanta of greater frequency or lower (shorter) wavelength, because the maximum energy has been given up. The cutoff wavelength may be made smaller by increasing the tube voltage so as to give the electrons more energy.

In the atoms of a target material, there are many electrons. The electrons in the lower energy levels near the nuclear protons are shielded from the incident electrons by the atomic electrons in the outer orbits, and so there is very little interaction with the inner electrons.

At large tube voltages, however, the incident electrons may have sufficient energy to occasionally eject an electron from an inner orbit. This ejection leaves a vacancy that can be filled by an electron from a nearby outer orbit, and this leaves a vacancy in that orbit. The transitions in filling these vacancies in the inner orbits give rise to the characteristic spectral lines, similar to lines produced by transitions in the hydrogen atom.

Several more quantum effects are discussed in the chapter Highlight.

9.4 Matter Waves and Quantum Mechanics

Learning Goals:

To explain de Broglie's hypothesis concerning matter waves.

To analyze the meaning of the wave function ψ in the Schrödinger equation.

To explain Heisenberg's uncertainty principle and how it relates to measurement.

Matter Waves

As part of the dual nature of light, what was thought to be a wave sometimes acts as a "particle." Can the reverse be true? That is, can particles

HIGHLIGHT

Fluorescence and Phosphorescence

Most people know that the phenomena of fluorescence and phosphorescence have something to do with things glowing, but do you know what they refer to or the difference between them? Now that you know about energy levels, transitions, and the photon nature of light, you can understand what these terms mean.

Fluorescence is easily associated with common fluorescent lights—those long, white tubes that probably light your classroom. Fluorescent lights are much more efficient than incandescent light bulbs, which radiate primarily in the nonvisible, infrared region. Actually, the primary radiation emitted in a fluorescent tube is in the nonvisible, ultraviolet region. So where does the ultraviolet radiation come from?

This is where fluorescence comes in. In a **fluorescence** process, electrons that have been excited by absorbing quanta of energy return to the original state by two or more steps or transitions, like a ball bouncing down a flight of stairs. Since each downward transition is smaller than the initial upward transition, the emitted photons must have a lower frequency than the exciting photon. (Why?)

The ultraviolet (uv) region is at a higher frequency than the visible spectrum; and in some fluorescent substances, downward transitions lie in the visible region—hence we see visible radiation. In fluorescent lights, electrically excited mercury atoms in the tube

Figure 1 Fluorescence.
When illuminated with uv light, the crayons glow with visible colors because of fluorescent compounds.

emit ultraviolet radiation, which is absorbed by the fluorescent white material that coats the inside of the tube. This material absorbs the uv radiation and reradiates at frequencies in the visible region, giving visible light for reading and so on.

Other uses of fluorescence include the "black lights" used at discos and light displays. These devices use the emissions of radiations from tubes that emit in the near-violet (visible region) and ultraviolet (nonvisible region). Fluorescent materials in paints and dyes cause walls and clothes to glow and stand out in the visible region (Fig 1).

The conversion of the ultraviolet radiation to visible light is not complete, and the remnant of uv radiation is used to commercial advantage. In the grocery store, generally illuminated with economical fluorescent lights, some products, such as laundry detergents, appear to have extra-bright colored boxes. Now you know why: Manufacturers hoping to influence purchase selection use fluorescent inks in the box coloring and printing so as to make the packaging appear brighter, or "stand out" under the fluorescent lights.

Fluorescence occurs a great deal in nature. Several living organisms, such as butterflies, manufacture fluorescent pigments that emit visible radiations. Also, fluorescence provides a method of mineral identification (Chapter 20). Some minerals have fluorescent properties and some do not.

How about **phosphorescence**? Such materials are used in luminous toys, signs, and so on—things that "glow in the dark." When exposed to light, the atoms of a phosphorescent material are excited to higher energy levels, and some are in metastable states. Some of these states have a decay lifetime of several seconds, minutes, and even hours—a whole variety of metastable states.

As a result, photon-emitting transitions occur over relatively long periods of time, and an object phosphoresces, or glows, long after the exciting source has been turned off or removed. The source is often a lamp, which provides the needed photons.

Figure 9.17 Louis de Broglie (1892–1987). Prince Louis de Broglie, a French nobleman, studied medieval history at the Sorbonne. He enlisted in the French army and was a radioman in World War I. The experience created an interest in physics, and he presented his concept of matter waves in 1925 in a doctoral thesis. It was met with much skepticism. However, when it was shown that electrons can produce diffraction patterns, de Broglie's theory was taken more seriously. He was awarded the Nobel Prize for physics in 1929.

have a wave nature? This question was considered by the French physicist Louis de Broglie (dee-BROLY, Fig. 9.17), who in 1925 postulated that matter, as well as light, has properties of both waves and particles.

According to the hypothesis of de Broglie, any moving particle has a wave associated with it whose wavelength is given by

$$\lambda = \frac{h}{p} \quad (9.4)$$

where λ is the wavelength
p is the momentum of the moving particle
and h is Planck's constant.

The waves associated with moving particles are called **matter waves** or **de Broglie waves.**

Notice that the wavelength (λ) of a matter wave is inversely proportional to the momentum of the particle or object ($p = mv$)—the greater the momentum, the smaller the wavelength. However, Planck's constant is such a small number (on the order of 10^{-34}) that wavelengths of the matter waves are quite short. The longest wavelengths are for very small particles with little mass. For example, calculations show that a fast-moving electron has a de Broglie wavelength on the order of 10^{-11} m (0.1 Å or 0.01 nm), whereas a 1000-kg car traveling at 90 km/h (about 55 mi/h) has a wavelength on the order of 10^{-38} m. Because a wave is generally detected by the interaction with an object of about the same size as the wavelength, we can see why matter waves of common moving objects are not evident.

De Broglie's hypothesis was met with skepticism at first, but it was experimentally verified in 1927 by G. Davisson and L. H. Germer in the United States. They showed that a beam of electrons exhibited a diffraction pattern. Because diffraction is a wave phenomenon, the beam of electrons must have had wavelike properties.

For appreciable diffraction, a wave must pass through a slit with a width approximately the size of the wavelength. Visible light has wavelengths on the order of 400 nm to 700 nm, and slits with widths of these sizes can be made quite easily. However, as stated previously, a fast-moving electron has a wavelength on the order of 0.01 nm. Slits of this width cannot be made.

Fortunately nature has provided us with suitably small slits in the form of crystal lattices. The atoms in these crystals are arranged in rows (or some other orderly arrangement), and the rows of atoms form natural "slits." Davisson and Germer bombarded nickel crystals with electrons and obtained a diffraction pattern on a photographic plate. Diffraction patterns made by X-rays (electromagnetic radiation of very short wavelength) and an electron beam incident on a thin aluminum foil are shown in Fig. 9.18. The similarity in the diffraction patterns from the electromagnetic waves and electron particles is evident. Electron diffraction demonstrates the "dual nature of matter."

An electron microscope uses the concept of matter waves. A beam of electrons, rather than a

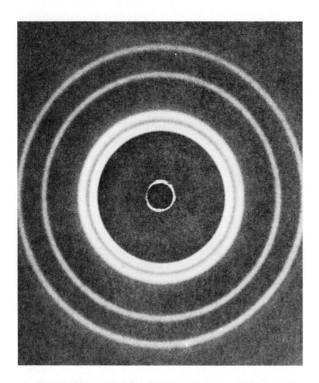

beam of light, is used to "view" an object. In one technique, the beam of electrons bounces off a surface. The beam is scanned across a specimen by means of deflecting coils, much as is done in a television tube. Surface irregularities cause directional variations in the intensity of the reflected electron beam, which gives a contrasted image.

The amount of fuzziness of an image is directly proportional to the wavelength that is used. A typical beam wavelength in an electron microscope is 1.0 nm. This length is quite short compared with the wavelength of visible light at about 500 nm. Hence, finer detail or greater resolution, as well as greater magnification, can be achieved with an electron microscope. Some electron photomicrographs are shown in ● Fig. 9.19.

Quantum Mechanics

De Broglie's hypothesis showed that waves were associated with moving particles, which somehow governs or describes the particle behavior. However, Eq. 9.4 gives only the wavelength of the matter wave and not its form. Waveforms can be represented mathematically, and such equations allow scientists to investigate a wave's behavior.

Another big step toward understanding the nature of atoms and nuclei was taken in 1926 when Erwin Schrödinger, an Austrian physicist (● Fig. 9.20), presented a widely applicable mathematical equation that gave new meaning to de Broglie's matter waves. This equation, known as the **Schrödinger equation,** is written in its simple form as

$$(E_k + E_p)\psi = E\psi$$

where E_k, E_p, and E are the kinetic energy, potential energy, and total energy, respectively. The symbol ψ (Greek letter psi) is called the **wave function** and represents the wave associated with a particle. The Schrödinger equation is basically a formulation of the conservation of energy.

The detailed form of the Schrödinger equation is quite complex, and the equation is difficult to solve. One relatively simple case for which the equation was first solved was that of the hydrogen atom. The possible energy levels were found to be the same as those Bohr had obtained in 1913. But there are additional results from the solution of the

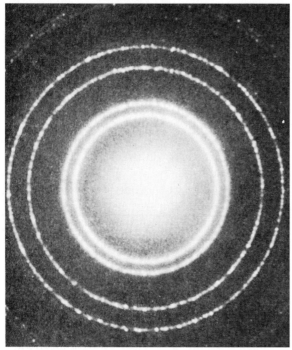

Figure 9.18 Diffraction patterns.
Two photographs showing diffraction patterns produced by X-rays and electrons. The X-ray pattern (top) can be explained by using a wave description of X-rays. The electron pattern (bottom) can be explained by using de Broglie or matter waves.

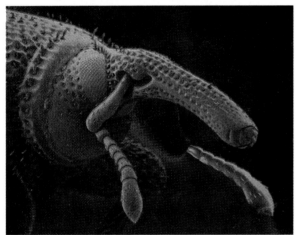

Figure 9.19 Electron micrographs. Scanning electron micrographs of a diamond stylus in the grooves of a phonograph record and of a maize weevil showing one of its compound eyes.

As a result of the dual nature of waves and particles, a new kind of physics based on the synthesis of wave and quantum ideas was born in the 1920s and 1930s. This was called **quantum mechanics,** and it replaced the classical-mechanics view that everything moved according to exact laws of nature with the concept of probability. For example, the quantum-mechanics analysis of the hydrogen atom using the Schrödinger equation predicts that the electron would most probably be in the discrete orbits predicted by the Bohr theory. This idea gives rise to the idea of an "electron cloud" around the nucleus, with the cloud density reflecting the probability that the electron would be in that region (Fig. 9.21b).

Figure 9.20 Erwin Schrödinger (1887–1961). Born in Austria, Schrödinger served in World War I and later became a professor of physics in Germany. In 1925, he published a paper in which moving particles were described mathematically as waves, making him one of the founders of quantum mechanics. Schrödinger left Germany in 1933 when the Nazis came to power, and was later a professor of physics at the University of Dublin.

Schrödinger equation that involve the wave function ψ.

At first, scientists were not sure how ψ should be interpreted. They finally concluded that ψ^2 (the wave function squared) represented the probability that the hydrogen electron would be at a certain distance from the nucleus. In Bohr's theory, the electron could be in circular orbits with discrete radii given by $r_n = 0.53$ Å. A plot of $r\psi^2$ versus r for the hydrogen electron is shown in Fig. 9.21a. It shows that the most probable radius for the hydrogen electron is one with $r = 0.53$ Å, which is the ground state for the hydrogen atom. The electron could be found at other radii, but with lower probability (that is, it is less likely).

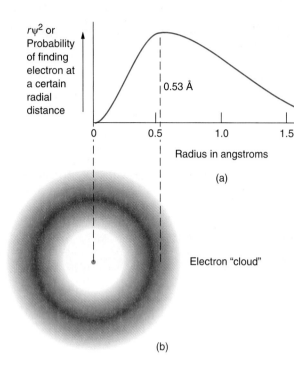

Figure 9.21 ψ^2 **Probability.**
(a) The square of the wavefunction (ψ^2) gives the probability of finding the electron at a particular radial distance. From a plot of $r\psi^2$ versus r, as shown here, the radius of the hydrogen atom with the greatest probability of containing the electron is at 0.53 Å, which corresponds to the first Bohr radius. (b) The probability of finding the electron at other radii gives rise to the concept of an electron "cloud," or probability distribution of the electron being at various radial distances.

There is another important aspect of quantum mechanics. According to classical mechanics, there is no limit to the accuracy of a measurement. The accuracy can be continually improved by refinement of the measurement instrument and/or procedure, to the point at which there may be no uncertainty in the measurement. This philosophy resulted in a deterministic view of nature. For example, if you know or measure the position and velocity of a particle exactly at a particular time, you can then determine where it will be in the future and where it was in the past (assuming no future or past unknown forces).

However, quantum theory predicts otherwise and sets limits on measurement accuracy. This idea was developed by the German physicist Werner Heisenberg (Fig. 9.22) and is called **Heisenberg's uncertainty principle,** which can be stated:

> It is impossible to simultaneously know a particle's exact position and velocity.

This concept is often illustrated with a simple example. Suppose you want to measure the position and momentum, mv (and therefore velocity), of an electron, as illustrated in   Fig. 9.23. If you are to see the electron and determine its location, at least one photon must bounce off the electron and come to your eye. In the collision process, some of the photon's energy and momentum are transferred to the electron. (This situation is analogous to a classical collision of particles or billiard balls. A collision involves the transfer of momentum and energy.)

After the collision, the electron recoils. Hence, in the process of trying to locate the position very accurately, which means that the uncertainty in position Δx is very small, a rather large uncertainty is caused in knowing the electron's velocity, and hence momentum, because the electron has recoiled at some angle to its original path. (Recall that $p = mv$, so $\Delta p = m\Delta v$.)

In the case of light, the position of an electron can be measured, at best, to an accuracy of about the wavelength λ of the light; that is, $\Delta x \approx \lambda$. The photon "particle" has a momentum of $p = h/\lambda$.

9.5 Multielectron Atoms and the Periodic Table

Learning Goals:

To describe the energy levels for multielectron atoms.

To explain the Pauli exclusion principle and its relationship in determining electron configurations of atoms in the periodic table.

Figure 9.22 Werner Heisenberg (1901–1976). At the age of 23, Heisenberg wrote a paper explaining quantum mechanics. His theory was quite different from that of Schrödinger, yet it produced the same results. Schrödinger later wrote a paper showing that his and Heisenberg's theories were mathematically the same. Heisenberg's name is chiefly associated with the uncertainty principle, which was conceived as a result of his quantum theory. During World War II, Heisenberg was in charge of the German nuclear energy program.

Bohr no doubt chose to analyze the hydrogen atom for an obvious reason—it's the simplest atom. It is increasingly difficult to analyze atoms with two or more electrons (multielectron atoms) and to determine their electronic energy levels. The difficulty arises because, in multielectron atoms, there are more electrical forces than in the hydrogen atom. There are forces among electrons; and in large atoms, electrons in outer orbits are "shielded" from the attractive force of the nucleus by electrons in inner orbits.

When the Schrödinger equation is solved for the hydrogen atom, several quantum numbers result. One is n, the *principal quantum number,*

(From the de Broglie hypothesis, $\lambda = h/mv = h/p$.) Similarly, the momentum of the electron will be uncertain by at least an amount $\Delta p \approx h/\lambda$.

The total uncertainty is the product of the individual uncertainties, and

$$(\Delta p)(\Delta x) = (m\Delta v)(\Delta x) \approx \left(\frac{h}{\lambda}\right)\lambda = h$$

or $\quad m(\Delta v)(\Delta x) \approx h$

Thus, Heisenberg's uncertainty principle states that the product of the minimum uncertainties are on the order of the value of Planck's constant ($\approx 10^{-34}$). This gives us the minimum uncertainties—that is, the best we can ever hope to measure simultaneously. The more accurately we measure the position (make Δx smaller), the greater the uncertainty in the velocity and momentum, $m\Delta v \approx h/\Delta x$, and vice versa. If we could measure the exact location of a particle ($\Delta x \to 0$), then we would have no idea about its velocity ($\Delta v \to \infty$).

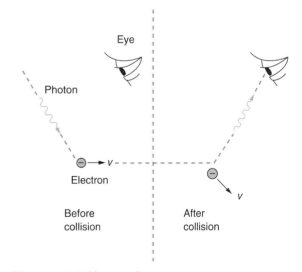

Figure 9.23 Uncertainty. Imagine trying to accurately determine the location of an electron with a single photon, which must strike the electron to become a detector. The electron recoils after the collision, and this introduces a great degree of uncertainty in knowing the electron's velocity or momentum.

which we discussed in the Bohr theory. However, three other quantum numbers arise, each with limitations on the values they can have. They are called the *orbital quantum number* (ℓ), the *magnetic quantum number* (m), and the *spin quantum number* (m_s). It is found that these quantum numbers apply to atoms other than hydrogen, and give us a basic insight into atomic physics. For the hydrogen atom, the principal quantum number alone determines the energy of the electron. However, for other atoms, those with two or more electrons, the orbital quantum number is also associated with the energy of each electron. The other two quantum numbers have nothing to do with the energy of atomic electrons.

A multielectron atom can be thought of in terms of the Bohr model with electrons orbiting the nucleus. As in the hydrogen atom, the lower the value of the principal quantum number n, the closer the orbit to the nucleus. However, in multielectron atoms, there are divisions or "orbitals" within a given orbit, corresponding to the different values of the orbital quantum number for that particular n. In terms of energy levels, this means that a principal quantum number n level contains orbital energy levels or subenergy levels. All such orbitals or energy levels with the same n value are said to be in an **electron shell**, such as the $n = 1$ shell, the $n = 2$ shell, and so on.

In a given shell, an orbital quantum number energy level is called an **electron subshell**. The orbital quantum number is designated by the letter ℓ, and can have values of 0, 1, 2, 3, . . . , $(n - 1)$. The last term means that the greatest value of ℓ is one less than the value of the principal quantum number n for a given energy level. For example, for the shell $n = 4$, there are four subshells corresponding to the numbers 0, 1, 2, and 3. Because of the use of historical spectroscopic notation, the values of ℓ are commonly designated by the letters s, p, d, f, g, . . . , corresponding to the values $\ell = 0, 1, 2, 3, 4, \ldots$, respectively. (The first four letters stood for the spectroscopic series called *s*harp, *p*rincipal, *d*iffuse, and *f*undamental.) After f and g, the letters continue alphabetically for higher values of ℓ. However, our discussion will be limited to the first few values.

CONFIDENCE QUESTION 9.1

What are the possible ℓ values for $n = 2$?

Because only n and ℓ are associated with the values of energy of the electrons in atoms other than hydrogen (n only), each atomic energy level is labeled by using these two quantum numbers. The common notation is to write n as a number, followed by the letter that stands for the value of ℓ. For example, 1s means the energy level for $n = 1$ and $\ell = 0$; 3d designates the energy level for $n = 3$ and $\ell = 2$; and 4p means the level with $n = 4$ and $\ell = 1$. This notation leads us to an energy-level diagram for multielectron atoms as shown in ● Fig. 9.24. Here we show the energy subshells for each shell. Note that they are not evenly spaced. Also, the 4s level is below the 3d level. This variation results from the electron "shielding" described earlier.

Now, suppose you were going to build up atoms by adding the proper number of electrons in the shells and subshells. That is, 1 electron for a hydrogen atom, 2 electrons for a helium atom, 3 electrons for a lithium atom, and so on. (For calcium you'd have to put in 20 electrons.) The question is, how would you fill up the energy levels? You'd probably start at the bottom. But how many electrons should go in each level?

We could label the electrons by the quantum numbers, a 1s electron or a 2p electron, indicating

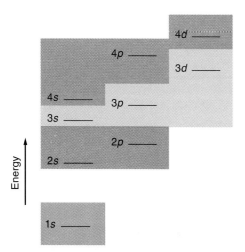

Figure 9.24 Multielectron atom energy levels.
A typical diagram for a multielectron atom (not to scale). In a multielectron atom, the energy levels depend on both the n and ℓ quantum numbers.

which energy level they were in; but recall that there are two other quantum numbers (labeled m and m_s). These provide two additional numbers to label electrons in a given energy level. That is, we have *sets* of quantum numbers, and for a particular subshell energy level, there can be only a certain number of sets. The number of sets increases as the you go up in the s, p, d, f levels.

In 1928, Wolfgang Pauli, a German physicist, made use of these sets of quantum numbers to provide an insight into how electrons are distributed in the energy levels of multielectron atoms. The **Pauli exclusion principle** states:

> **No two electrons can have the same set of quantum numbers.**

Hence, just as (x, y, z) designates a point in Cartesian coordinates, each different set of four quantum numbers (n, ℓ, m, m_s) corresponds to a different energy state that can be occupied by only one electron. Then, by filling up the lower energy levels with one electron for a particular set of quantum numbers, we can "build up" the various electron configurations for the atoms of various elements. This situation is analogous to having a set of mail boxes, each of which is designated by a particular set of numbers, and only one letter can be put in each box.

But, we still need to know how many sets of quantum numbers or states there are in each energy level. The various restrictions on the quantum numbers puts this at $2(2\ell + 1)$. Thus, for an s ($\ell = 0$) level, we can have $2[2(0) + 1] = 2$ states; for a p ($\ell = 1$) level, $2[2(1) + 1] = 6$, and so on; such that the number of states for the s, p, d, f levels are 2, 6, 10, and 14, respectively.

So, you start at the bottom and fill up energy levels with electrons according to these maximum numbers. That is, when a level is full, you go to the next higher one for the next electron. Figure 9.25 illustrates the electron energy-level diagram for three atoms: lithium (Li) with 3 electrons, neon (Ne) with 10 electrons, and sodium (Na) with 11 electrons.

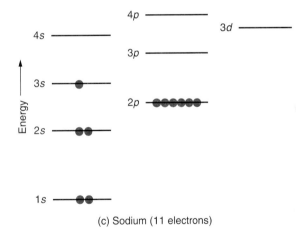

Figure 9.25 Electrons and energy levels.
The energy-level diagrams with electron occupancy for the ground states of lithium, neon, and sodium. The Pauli exclusion principle determines how many electrons can be in a particular energy level. (Energy-level spacings are not to scale.)

PERIODS	s		d										p					s
1	1 **H** $1s^1$																	2 **He** $1s^2$
2	3 **Li** $2s^1$	4 **Be** $2s^2$											5 **B** $2s^22p^1$	6 **C** $2s^22p^2$	7 **N** $2s^22p^3$	8 **O** $2s^22p^4$	9 **F** $2s^22p^5$	10 **Ne** $2s^22p^6$
3	11 **Na** $3s^1$	12 **Mg** $3s^2$											13 **Al** $3s^23p^1$	14 **Si** $3s^23p^2$	15 **P** $3s^23p^3$	16 **S** $3s^23p^4$	17 **Cl** $3s^23p^5$	18 **Ar** $3s^23p^6$
4	19 **K** $4s^1$	20 **Ca** $4s^2$	21 **Sc** $4s^23d^1$	22 **Ti** $4s^23d^2$	23 **V** $4s^23d^3$	24 **Cr** $4s^13d^5$	25 **Mn** $4s^23d^5$	26 **Fe** $4s^23d^6$	27 **Co** $4s^23d^7$	28 **Ni** $4s^23d^8$	29 **Cu** $4s^13d^{10}$	30 **Zn** $4s^23d^{10}$	31 **Ga** $4s^24p^1$	32 **Ge** $4s^24p^2$	33 **As** $4s^24p^3$	34 **Se** $4s^24p^4$	35 **Br** $4s^24p^5$	36 **Kr** $4s^24p^6$
5	37 **Rb** $5s^1$	38 **Sr** $5s^2$	39 **Y** $5s^24d^1$	40 **Zr** $5s^24d^2$	41 **Nb** $5s^14d^4$	42 **Mo** $5s^14d^5$	43 **Tc** $5s^14d^6$	44 **Ru** $5s^14d^7$	45 **Rh** $5s^14d^8$	46 **Pd** $4d^{10}$	47 **Ag** $5s^14d^{10}$	48 **Cd** $5s^24d^{10}$	49 **In** $5s^25p^1$	50 **Sn** $5s^25p^2$	51 **Sb** $5s^25p^3$	52 **Te** $5s^25p^4$	53 **I** $5s^25p^5$	54 **Xe** $5s^25p^6$
6	55 **Cs** $6s^1$	56 **Ba** $6s^2$	57 **La*** $6s^25d^1$	72 **Hf** $6s^25d^2$	73 **Ta** $6s^25d^3$	74 **W** $6s^25d^4$	75 **Re** $6s^25d^5$	76 **Os** $6s^25d^6$	77 **Ir** $6s^25d^7$	78 **Pt** $6s^15d^9$	79 **Au** $6s^15d^{10}$	80 **Hg** $6s^25d^{10}$	81 **Tl** $6s^26p^1$	82 **Pb** $6s^26p^2$	83 **Bi** $6s^26p^3$	84 **Po** $6s^26p^4$	85 **At** $6s^26p^5$	86 **Rn** $6s^26p^6$
7	87 **Fr** $7s^1$	88 **Ra** $7s^2$	89 **Ac**** $7s^26d^1$	104 **Unq** $7s^26d^2$	105 **Unp** $7s^26d^3$	106 **Unh** $7s^26d^4$	107 **Uns** $7s^26d^5$	108 **Uno** $7s^26d^6$	109 **Une** $7s^26d^7$	110 **Uun** $7s^16d^9$	111 **Uuu** $7s^16d^{10}$							

	f													
*Lanthanides	58 **Ce** $6s^24f^15d^1$	59 **Pr** $6s^24f^35d^0$	60 **Nd** $6s^24f^45d^0$	61 **Pm** $6s^24f^55d^0$	62 **Sm** $6s^24f^65d^0$	63 **Eu** $6s^24f^75d^0$	64 **Gd** $6s^24f^75d^1$	65 **Tb** $6s^24f^95d^0$	66 **Dy** $6s^24f^{10}5d^0$	67 **Ho** $6s^24f^{11}5d^0$	68 **Er** $6s^24f^{12}5d^0$	69 **Tm** $6s^24f^{13}5d^0$	70 **Yb** $6s^24f^{14}5d^0$	71 **Lu** $6s^24f^{14}5d^1$
**Actinides	90 **Th** $7s^25f^06d^2$	91 **Pa** $7s^25f^26d^1$	92 **U** $7s^25f^36d^1$	93 **Np** $7s^25f^46d^1$	94 **Pu** $7s^25f^66d^0$	95 **Am** $7s^25f^76d^0$	96 **Cm** $7s^25f^76d^1$	97 **Bk** $7s^25f^96d^0$	98 **Cf** $7s^25f^{10}6d^0$	99 **Es** $7s^25f^{11}6d^0$	100 **Fm** $7s^25f^{12}6d^0$	101 **Md** $7s^25f^{13}6d^0$	102 **No** $7s^25f^{14}6d^0$	103 **Lr** $7s^25f^{14}6d^1$

Figure 9.26 Electron periods.
A table of electron periods with the electron configuration (or latter part) for each atom.

Notice in Fig. 9.25 that the spaces between the energy levels are not equal. For instance, there is a large gap between the 1s and 2s levels. In general, there are large gaps between the s levels and the levels below them (except the 1s level). The gaps are smaller between the other levels, such as between the 4s–3d levels and the 3d–4p levels. The energy levels (such as 4s, 3d, and 4p) between the large gaps have approximately the same spacing and energy.

A set of energy levels that has about the same energy is called an **electron period.** For example, the energy gaps between periods are indicated by the vertical lines drawn between energy levels in the following electron configuration:

$$1s^2|2s^22p^6|3s^23p^6|4s^23d^{10}4p^6|\text{ etc.}$$

where the superscripts indicate the maximum number of electrons in each subshell level. When writing out the electron configuration of an atom in this notation, one begins with the lowest energy level and adds electrons. The superscript of the last subshell indicates the number of electrons in the outermost subshell. By adding the superscripts, the number of electrons in the electron configuration of a particular atom is obtained.

These periods form the basic of the chemical periodic table, which will be studied in Chapter 11.

The table was developed empirically in the 1800s, since the structure of the atom was not known at that time. The elements were arranged in horizontal rows in order of increasing atomic mass. When the chemical properties of the next element were similar to a previous element, that element was placed below the other. As a result, all the elements in a vertical column had similar properties. This periodic recurrence of similar chemical properties is related to the electron periods.

If the electron configurations are placed on the periodic table for the atom of each element, we find a very interesting occurrence (see Fig. 9.26). Notice that, in general, the outer subshell configurations of the atoms in a vertical column are the same or similar. Look at Li and Na in the far-left column. The total electron configurations for these elements were written previously. They both have a single electron in the outermost s subshell. In fact, all of the atoms in this column do.

Hence, we might conclude, and correctly so, that the similarity of the chemical properties of these elements is due to a similar outer electron configuration. Keep this idea in mind when you begin your study of chemistry in Chapter 11. But first, we'll look at the atomic nucleus, in Chapter 10.

Important Terms

These important terms are for review. After reading and studying the chapter, you should be able to define and/or explain each of them.

ultraviolet catastrophe
quantum physics
quantum
photoelectric effect
photon
dual nature of light
principal quantum number (n)
ground state
excited state
laser
stimulated emission
X-rays
fluorescence
phosphorescence

matter (de Broglie) waves
Schrödinger equation
wave function (ψ)
quantum mechanics
Heisenberg's uncertainty principle
electron shell
electron subshell
Pauli exclusion principle
electron period

Important Equations

Photon energy: $E = hf$

$(h = 6.63 \times 10^{-34}$ J-s$)$

Hydrogen electron radii:

$r_n = 0.53 \, n^2$ Å $n = 1, 2, 3, \ldots$

Hydrogen electron energy:

$E_n = \dfrac{-13.6}{n^2}$ eV

de Broglie wavelength:

$\lambda = \dfrac{h}{p} = \dfrac{h}{mv}$

Heisenberg's uncertainty principle: $m(\Delta v)(\Delta x) \approx h$

Number of possible states in a given ℓ subshell:

$2(2\ell + 1)$

Questions

9.1 The Dual Nature of Light

1. Planck's hypothesis was concerned with (a) the ultraviolet catastrophe, (b) the photoelectric effect, (c) the hydrogen atom, (d) uncertainty.
2. Light of which of the following colors has the greatest photon energy: (a) red, (b) orange, (c) yellow, (d) blue?
3. Why was Planck's hypothesis so radical at the time?
4. Is light a wave or a particle?
5. What proof can you give that light is a wave? a particle?

9.2 The Bohr Theory of the Hydrogen Atom

6. The Bohr theory was developed to explain (a) energy levels, (b) the photoelectric effect, (c) line spectra, (d) quantum numbers.
7. In which one of the following states does a hydrogen electron have the greatest energy: (a) $n = 1$, (b) $n = 3$, (c) $n = 5$, (d) $n = 7.5$?
8. Some hydrogen atoms have one proton and an electrically neutral particle, a neutron, in the nucleus. How would the neutron affect the Bohr theory?
9. According to the Bohr theory, what is the approximate diameter of a hydrogen atom in the ground state?
10. How did Bohr address the problem that, according to the classical approach, an orbiting electron should emit radiation?
11. How does the Bohr theory explain the discrete line spectrum of hydrogen?

9.3 Quantum-Physics Applications

12. The amplification of light by a laser depends on (a) the photoelectric effect, (b) microwave absorption, (c) spontaneous emission, (d) stimulated emission.
13. X-rays are produced by (a) photons, (b) electron interaction, (c) stimulated emission, (d) the dual nature of light.
14. Why does a microwave oven heat a potato but not a ceramic plate?
15. Why are microwave ovens constructed so they will not operate when the door is open?
16. What does the word *laser* stand for or mean?
17. What is unique about light from a laser source, and why should you never look directly into a laser beam?
18. Why are X-rays called "braking rays," and why does increasing the tube voltage shift the cutoff wavelength of an X-ray spectrum to a shorter wavelength?
19. What is the difference between fluorescence and phosphorescence?

9.4 Matter Waves and Quantum Mechanics

20. The square of the wave function of Schrödinger's equation is associated with (a) stimulated emission, (b) metastable states, (c) wavelength, (d) probability.

21. Limitations are put on measurements by (a) Schrödinger's equation, (b) de Broglie's hypothesis, (c) Heisenberg's principle, (d) Pauli's exclusion principle.
22. What is a matter wave?
23. How was a beam of electrons shown to have wavelike properties?
24. What is quantum mechanics?
25. How does the Heisenberg principle change the classical deterministic view of the universe?

9.5 Multielectron Atoms and the Periodic Table

26. The letter d designates which orbital quantum number: (a) 0, (b) 1, (c) 2, (d) 3?
27. How many possible sets of quantum numbers are there for a d subshell: (a) 2, (b) 6, (c) 8, (d) 10?
28. What did the letters s, p, d, and f originally stand for?
29. Which quantum number(s) determines the electron energy in (a) a hydrogen atom and (b) a multielectron atom?
30. Explain the Pauli exclusion principle.
31. Why are the possible sets of quantum numbers so important?
32. Distinguish between electron configuration, electron shell, electron subshell, and electron period.
33. Is it possible to have a $3f$ energy level? Explain.
34. What is the basis of the periodic table of elements, and what do the elements in a vertical column have in common?

Food for Thought

1. Is the graph in Fig. 9.21 complete? Explain.
2. Color television tubes are shielded to protect people from X-rays. How can a TV set produce X-rays?

Exercises

9.2 The Bohr Theory of the Hydrogen Atom

1. What is the radius of the electron orbit in the hydrogen atom for each of the following principal quantum numbers: (a) $n = 2$ (b) $n = 6$ (c) $n = 10$?

   Answer: (a) 2.12 Å

2. What is the energy of a hydrogen electron for each of the orbits designated by the following principal quantum numbers: (a) $n = 2$ (b) $n = 6$ (c) $n = 10$?

   Answer: (a) −3.40 eV

3. What is the ionization energy, in eV, for a hydrogen atom if the electron is in an orbit with a principal quantum number of (a) $n = 1$, (b) $n = 4$, and (c) $n = 10$? (The ionization energy is the amount of energy that would have to be given to the atom to ionize it, or remove the electron from the proton.)

4. What are the energies of the photons, in eV, emitted in the following transitions of an electron in a hydrogen atom?
   (a) $n = 4$ to $n = 2$ (c) $n = 3$ to $n = 1$
   (b) $n = 6$ to $n = 2$ (d) $n = 4$ to $n = 3$

   Answer: (b) 3.02 eV

9.5 Multielectron Atoms and the Periodic Table

5. What are the possible number of ℓ values for (a) $n = 3$ and (b) $n = 5$?

   Answer: (a) 3

6. What is the number of possible electron states for each of the electron subshells (a) $\ell = 3$ and (b) $\ell = 5$?

   Answer: (a) 14

7. How many electrons can occupy the $4d$ subshell?

   Answer: 10

8. Consider the following electron configuration:

$$1s^2 2s^2 2p^6 3s^2 3p^6 4s^2 3d^9 4p^1$$

   (a) To which atom does this configuration correspond?
   (b) Is the atom in the ground state?

   Answer: (a) Zn, (b) no

9. How many electrons would be in the outermost shell of the atom described in Exercise 8 if it were in the ground state?

10. Draw the ground-state energy-level diagrams (as in Fig. 9.25, for example) for each of the following atoms: (a) Al, (b) Cl, and (c) K.

11. Draw schematic diagrams of the electrons orbiting the nucleus in subshells for each of the following atoms: (a) Al, (b) Cl, and (c) K

Answers to Multiple-Choice Questions

1. a   6. c   12. d   20. d   26. c
2. d   7. c   13. b   21. c   27. d

Solutions to Confidence Question

9.1 Highest ℓ value is $(n - 1) = (2 - 1) = 1$, so $\ell = 0, 1$, where 0 and 1 represent s and p subshells, respectively.

Chapter 10

Nuclear Physics

10.1 The Atomic Nucleus
10.2 Radioactivity

Highlight: The Discovery of Radioactivity

10.3 Half-life and Radiometric Dating
10.4 Nuclear Reactions
10.5 Nuclear Fission

Highlight: The Building of the Bomb

10.6 Nuclear Fusion
10.7 Biological Effects of Radiation

Reading a daily newspaper will indicate that the atomic nucleus and its properties have an important impact on our society. The nucleus is involved with dating archeological objects and rock formations, treating and diagnosing cancer and other diseases, chemical analysis, radiation damage and nuclear bombs, the generation of electricity by nuclear energy, and even with the operation of the common household smoke detector. This chapter discusses these topics; and because of the significance of the history of the discovery of radioactivity and the building of the atomic bomb, we include a chapter Highlight on each event.

10.1 The Atomic Nucleus

Learning Goals:

To describe the structure and composition of atoms.

To calculate atomic masses of elements.

All matter encountered in day-to-day living is made up of atoms. An atom is composed of negatively charged particles, called **electrons,** that surround a positively charged nucleus. The **nucleus** is the central core of an atom. It consists of **protons,** which are positively charged, and **neutrons,** which have no electrical charge. Protons and neutrons are collectively called **nucleons.** The two have almost the same mass, and are about 2000 times more massive than an electron. Table 10.1 summarizes the basic properties of electrons, protons, and neutrons, all of which were discovered in England—the electron by J. J. Thomson in 1897, the proton by Ernest Rutherford in 1919, and the neutron by James Chadwick in 1932.

That the atom consists of a nucleus surrounded by orbiting electrons was discovered in 1911, also by Rutherford. He was curious about what would happen when energetic alpha particles (helium nuclei) were allowed to bombard a very thin sheet of gold. The experiment was conducted in a vacuum in an apparatus similar to that shown in Fig. 10.1a. The behavior of the alpha particles could be determined by a movable screen coated with zinc sulfide. When an alpha particle hit the screen, a flash of light was emitted that could be observed with a low-power microscope.

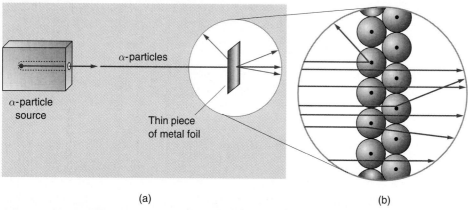

Figure 10.1 Rutherford's alpha scattering experiment.
See text for discussion.

(A similar phenomenon causes TV screens to glow when hit by moving electrons.) Rutherford found that by far the majority of the alpha particles went through the gold foil as if it were not even there. But a relatively few of the positively charged alpha particles were deflected off course, and about 1 out of 20,000 actually bounced back. Rutherford could explain this behavior only by assuming that the gold atoms had their positive charge concentrated in a small core of the atom, which he named the nucleus (Fig. 10. 1b).

An atom's nucleus has a diameter of about 10^{-14} m. In contrast, the atom's outer electrons have orbits with diameters of about 10^{-10} m (see Section 9.2). Thus the diameter of an atom is approximately 10,000 times the diameter of its nucleus. Most of an atom's volume consists of empty space. The electrical repulsion between an atom's electrons and those of adjacent atoms keeps matter from collapsing. Electron orbits determine the size (volume) of atoms, but the nucleons contribute over 99.9% of the mass, as a study of Table 10.1 shows. If nuclei could be packed together into a sphere the size of a ping-pong ball, the ball would have a mass of about 2.5 *billion* metric tons. Such a large density can occur in neutron stars (see Section 18.3), but no material on Earth is anywhere near so dense.

We designate the particles in an atom by certain numbers. The **atomic number,** symbolized by the letter Z, is the number of protons in the nucleus of atoms of that element. In fact, an **element** is defined as a substance in which all the atoms have the same number of protons (same atomic number). For an atom to be electrically neutral (have a total charge of zero), the number of electrons and protons must be the same. Therefore, the atomic number also represents the number of electrons in an atom. Electrons may be gained or lost by an atom, and the resulting particle (called an *ion*) will be electrically charged. However, because the number of protons has not changed, the particle is an ion of that same element. For example, if a *sodium* atom (Na) loses an electron, it becomes a *sodium ion* (Na^+), not an atom or ion of some other element.

Table 10.1 Major Constituents of an Atom

Particle (symbol)	Charge (C)	Electronic Charge	Mass (kg)	Mass (u)	Location
Electron (e)	-1.60×10^{-19}	-1	9.109×10^{-31}	0.00055	Outside nucleus
Proton* (p)	$+1.60 \times 10^{-19}$	$+1$	1.673×10^{-27}	1.00728	Nucleus
Neutron* (n)	0	0	1.673×10^{-27}	1.00867	Nucleus

*Modern theory views protons and neutrons as being composed of combinations of fundamental particles called *quarks*, which have fractional electronic charges of $+\frac{2}{3}$ and $-\frac{1}{3}$ (see Section 8.1).

The **neutron number** (N) is, of course, the number of neutrons in a nucleus. The **mass number** (A) is the number of protons plus neutrons in the nucleus; that is, the total number of nucleons. Atoms of the same element can be different because of different numbers of neutrons in their nuclei. Forms of atoms that have the same number of protons (same Z, same element) but differ in their number of neutrons (different N, different A) are known as the **isotopes** of that element. Even though only 109 elements are known, the total of their isotopes is about 2000.*

As shown below, the general designation for a specific isotope places the mass number (A) to the upper left of the chemical symbol (the X, for generality).

$$^{A}_{Z}X_{N}$$

mass number A, atomic number Z, chemical symbol X, atomic number N

The atomic number (Z) goes at the lower left. Sometimes, the neutron number (N) is added at the lower right, but it is usually omitted because the number of neutrons is readily determined by subtracting the atomic number Z from the mass number A.

$$N = A - Z \qquad (10.1)$$

So, for example, it is common to refer to the uranium isotope $^{238}_{92}U$. Because it is a simple matter to obtain an element's atomic number from a periodic table, a specific isotope of an element can also be represented by just the chemical symbol and mass number (for example, ^{238}U), or by the name of the element followed by a hyphen and the mass number (for example, uranium-238). The chemical symbols for all the elements are given in the periodic table on the inside front cover of the textbook, and an alphabetical listing is found in Appendix IX.

EXAMPLE 10.1 Determining the Composition of an Atom

Determine the number of protons, electrons, and neutrons in the fluorine atom $^{19}_{9}F$.

Solution

The atomic number Z is 9, so the number of protons is 9, as is the number of electrons. The mass number A is 19, so the number of neutrons $N = A - Z = 19 - 9 = 10$.

The answer is 9 protons, 9 electrons, and 10 neutrons.

CONFIDENCE QUESTION 10.1

Determine the number of protons, electrons, and neutrons in the uranium atom $^{238}_{92}U$.

The isotopes of an element have the same chemical properties because they have the same electron configuration, but they differ somewhat in physical properties because they have different masses. Figure 10.2 shows the atomic composition of the three isotopes of hydrogen. They even have their own names: $^{1}_{1}H$ is *protium* (or just *hydrogen*), $^{2}_{1}H$ is *deuterium* (D), and $^{3}_{1}H$ is *tritium* (T). The atomic nucleus in each case is referred to as a *proton*, *deuteron*, and *triton*, respectively. That is, a proton is the nucleus of a protium atom, and so on.

In a given sample of hydrogen, about one atom in 6000 is deuterium and about one atom in 10,000,000 is tritium. Protium and deuterium are stable atoms, whereas tritium is unstable (radioactive). *Heavy water* (D_2O) consists of molecules made up of two atoms of deuterium and one atom of oxygen.

The Atomic Mass

Generally, each element occurs naturally as a combination of its isotopes, and the mass of each isotope (the isotopic mass) can be determined using an instrument called a *mass spectrometer* (Fig. 10.3). The *weighted average* mass (in *unified atomic mass units*, u) of an atom of the element in a naturally occurring sample is called the **atomic mass of the element**, and is given under its symbol in the periodic table.

All atomic masses are based on the ^{12}C atom, which is assigned a relative atomic mass of *exactly* 12 u. Naturally occurring carbon has an atomic

---

*Information on the isotopes of the elements can be found in the Table of the Isotopes in the *Handbook of Chemistry and Physics* published yearly by the CRC Press.

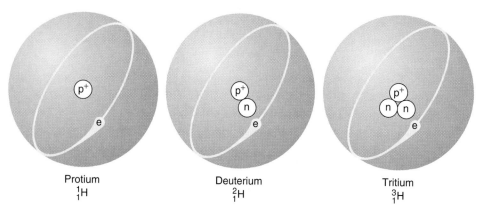

Figure 10.2 The three isotopes of hydrogen.
Each atom has one proton and one electron, but they differ in the number of neutrons in the nucleus. (*Note:* Not drawn to scale; the nucleus is shown much too large relative to the size shown for the whole atom.)

mass slightly greater than 12.0000 u because it contains not only ^{12}C but also a little ^{13}C and a trace of ^{14}C. An isotope's mass number closely approximates its *isotopic mass* (its actual mass in u), as you can see from Example 10.2.

EXAMPLE 10.2 Calculating an Element's Atomic Mass

Naturally occurring chlorine is a mixture consisting of 75.77% chlorine-35 (isotopic mass = 34.97 u) and 24.23% chlorine-37 (isotopic mass = 36.97 u). Calculate the atomic mass of the element chlorine.

Solution

Calculate the contribution each chlorine isotope makes to the atomic mass by multiplying the *fractional abundance* of each (the percentage abundance divided by 100) by its isotopic mass, then add the two answers to get the atomic mass of chlorine (35.46 u).

0.7577×34.97 u $= 26.50$ u (^{35}Cl)
0.2423×36.97 u $= \underline{8.96}$ u (^{37}Cl)
for a total of 35.46 u

CONFIDENCE QUESTION 10.2
Find the atomic mass of a hypothetical element *X* if it consists of 60.00% of ^{20}X (isotopic mass = 20.00 u) and 40.00% of ^{22}X (isotopic mass = 22.00 u).

The Strong Nuclear Force

In previous chapters we studied two fundamental forces of nature—electromagnetic and gravitational. The electromagnetic force between a proton and an electron in an atom is about 10^{39} times greater than the corresponding gravitational force. The electromagnetic force is the only important force on the electrons in an atom and is responsible for the structure of atoms, molecules, and, hence, matter in general.

In a nucleus the protons (and neutrons) are packed closely together. According to Coulomb's law (see Section 8.1), like charges repel each other, so the repulsive electric forces in a nucleus are huge. The nucleus should fly apart.

Obviously, the nucleus generally does remain together, and thus a third fundamental force must exist. This **strong nuclear force** (or just *nuclear force*) acts between nucleons; that is, between two protons, between two neutrons, and between a proton and a neutron. It holds the nucleus together. The exact equation describing the nucleon-nucleon interaction is unknown. However, for the very short nuclear distances of less than about 10^{-14} m, the interaction is very strongly attractive; in fact, it is the strongest fundamental force known. Yet at distances greater than about 10^{-14} m, the force is zero!

A typical large nucleus is illustrated in Fig. 10.4. A proton on the surface of the nucleus is attracted only by the six or seven nearest nucleons.

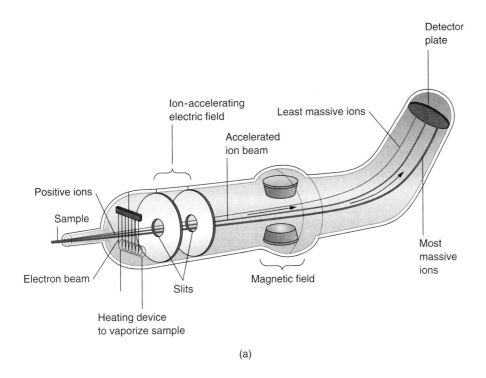

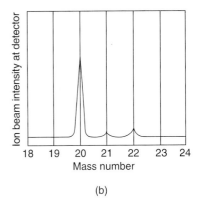

Figure 10.3 Schematic drawing of a mass spectrometer.
(a) A mass spectrometer separates and measures the masses of isotopes of an element by ionizing the atoms and sending them through a magnetic field. Because the ions have the same charge but different mass, they interact differently with the field and form separate beams. (b) A mass spectrogram shows the three isotopes of Ne and their relative abundances.

Because the strong nuclear force is a short-range force, only the nearby nucleons contribute to the attractive force.

On the other hand, the repulsive electrical force is long range and acts between any two protons, no matter how far apart they are in the nucleus. As more and more protons are added to the nucleus, the electric repulsive forces increase, yet the nuclear attractive forces remain constant because they are determined by nearest neighbors only. When the nucleus has more than 83 protons, the electric forces of repulsion overcome the nuclear attractive forces, and the nucleus is subject to spontaneous disintegration, or *decay*.

10.2 Radioactivity

Learning Goals:

To complete equations for radioactive decay.

To recognize which nuclides are radioactive.

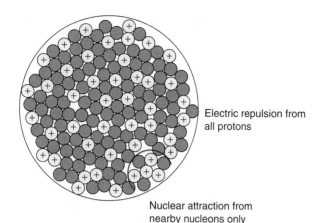

Figure 10.4 A multinucleon nucleus.
The protons on the surface of the nucleus, such as those shown in the red semicircle, are attracted by the strong nuclear force of only the six or seven closest nucleons; but they are electrically repelled by all other protons. When the number of protons exceeds 83, the electrical repulsion overcomes the nucleon attraction, and the nucleus is unstable.

A given species of nucleus, such as ^{238}U or ^{14}C, is referred to as a **nuclide**. Nuclides whose nuclei undergo spontaneous decay are called **radionuclides** (or, often, *radioisotopes*). The spontaneous process of nuclei undergoing a change by emitting particles or rays is called *radioactive decay*, or **radioactivity**.

Radioactivity was discovered in 1896 in France by Henri Becquerel, who found that uranium compounds spontaneously emit very penetrating radiation. In 1898 Marie and Pierre Curie announced the discovery of two new radioactive elements, radium and polonium. (See the chapter's first Highlight.)

Radioactive nuclei can disintegrate in three common ways: alpha decay, beta decay, and gamma decay (● Fig. 10.5). (Fission, another important decay process, will be discussed in Section 10.5.) In all decay processes energy is given off, usually in the form of energetic particles and heat. Equations for radioactive decay take the form

$$A \rightarrow B + b$$

In radioactive decay the original nucleus (*A*) is sometimes called the *parent* and the resulting nucleus (*B*) is referred to as the *daughter*. The *b* in the equation represents the emitted particle or ray.

Alpha decay is the disintegration of a nucleus into an *alpha particle* ($^4_2$He nucleus) and a nucleus of another element. Alpha decay is common for elements with atomic number greater than 83. An example is

$$^{232}_{90}\text{Th} \rightarrow ^{228}_{88}\text{Ra} + ^4_2\text{He}$$

Note in the above equation that the sum of the mass numbers on each side of the arrow is the same; that is, 232 = 4 + 228. Also, the sum of the atomic numbers on each side is the same; that is, 90 = 2 + 88. This principle holds for all nuclear decays and involves the conservation of nucleons and the conservation of charge, respectively.

> In a nuclear decay the sum of the mass numbers will be the same on each side of the arrow, as will be the sums of the atomic numbers.

EXAMPLE 10.3 Finding the Products of Alpha Decay

$^{238}_{92}$U undergoes alpha decay. Write the equation for the process.

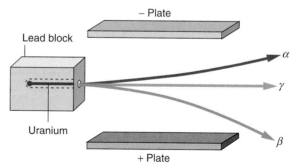

Figure 10.5 The three components of radiation from heavy radionuclides.
An electric field separates the rays from a sample of a heavy radionuclide, such as uranium, into alpha (α) particles (positively charged helium nuclei), beta (β) particles (negatively charged electrons), and neutral gamma (γ) rays (high-energy electromagnetic radiation).

Solution

STEP 1

Because this is a decay, write the original symbol followed by an arrow.

$$^{238}_{92}U \rightarrow$$

STEP 2

Because alpha decay involves the emission of $^{4}_{2}He$, write that symbol after the arrow, preceded by a plus sign and room for the symbol for the daughter.

$$^{238}_{92}U \rightarrow \underline{\quad} + ^{4}_{2}He$$

STEP 3

Determine the mass number, atomic number, and chemical symbol for the daughter. The sum of the mass numbers on the left is 238. The sum on the right must also be 238, and so far only the 4 for the alpha particle shows. Therefore, the daughter must have a mass number of $238 - 4 = 234$. By similar reasoning, the atomic number of the daughter must be $92 - 2 = 90$. From the periodic table, we find that the element with atomic number 90 is Th (thorium). The equation for the decay is

$$^{238}_{92}U \rightarrow ^{234}_{90}Th + ^{4}_{2}He$$

CONFIDENCE QUESTION 10.3

Write the equation for the alpha decay of $^{226}_{88}Ra$.

Beta decay is the disintegration of a nucleus into a *beta particle* ($^{0}_{-1}e$, an electron) and a nucleus of another element. An example is

$$^{14}_{6}C \rightarrow ^{14}_{7}N + ^{0}_{-1}e$$

Note that a beta particle, or electron, is assigned a mass number of 0 (because it contains no nucleons) and an atomic number of -1 (because its electric charge is opposite that of a proton's $+1$ charge). The mass numbers and atomic numbers on both sides of the arrow are equal in our example of beta decay, because $14 = 0 + 14$, and $6 = -1 + 7$. In beta decay, a neutron ($^{1}_{0}n$) is transformed into a proton and an electron at the nuclear surface. The proton remains in the nucleus, and the electron is emitted.

Gamma decay occurs when a nucleus emits a *gamma ray* (γ) and becomes a less energetic form of the same nucleus. A gamma ray, being a photon of high energy electromagnetic radiation, has no mass number or atomic number. An example is

$$^{204*}_{82}Pb \rightarrow ^{204}_{82}Pb + \gamma$$

The asterisk (*) following the mass number means that the nucleus is in an excited state, analogous to an atom being in an excited state with an electron in a higher energy level (see Section 9.2). Gamma decay generally occurs any time a nucleus is formed in an excited state; for example, as a product of alpha or beta decay. When the nucleus de-excites, one or more gamma rays are emitted and the nucleus is left in a state of lower excitation and, ultimately, in a "ground (or stable) state." Note the absence of the (*) in the symbol for the daughter.

A nucleus with atomic number greater than 83 is always radioactive and commonly undergoes a series of alpha, beta, and gamma decays until a stable nucleus is produced. For example, the series of decays beginning with uranium-238 and ending with stable $^{206}_{82}Pb$ is illustrated in ● Fig. 10.6. (Other similar decay series end with either $^{207}_{82}Pb$, $^{208}_{82}Pb$, or $^{209}_{83}Bi$.) Notice how the alpha (α) and beta (β) transitions are indicated in Fig. 10.6. The gamma decays that accompany the alpha and beta decays in the series are not apparent on the diagram, because the neutron and proton numbers do not change in gamma decay.

Recognizing Radionuclides

Which nuclides are radioactive and which stable? When we plot the number of protons (Z) versus the number of neutrons (N) for each stable nuclide, the points form a narrow band called the *belt of stability* (● Fig. 10.7). For comparison, the colored straight line in the figure represents equal numbers of protons and neutrons.

HIGHLIGHT

The Discovery of Radioactivity

In 1896 Henri Becquerel (beh-KREL) at the Museum of Natural History in Paris heard of Wilhelm Roentgen's (RUNT-gin's) recent discovery of X-rays and wondered if any of the fluorescent materials he was investigating might emit X-rays. He wrapped a photographic plate in black paper and put it in sunlight with a crystal of a fluorescent mineral on it. He knew that the ultraviolet light in sunlight could not penetrate the black paper, but that X-rays could. The film was indeed fogged, and Becquerel decided that the crystal was emitting X-rays as it fluoresced due to the sunlight hitting it. His experiments were interrupted by a series of cloudy days. In a drawer he left a fresh, wrapped plate with an unexposed crystal resting on it. Becoming impatient, he decided to develop the plate and found, to his surprise, that it was heavily fogged. Whatever radiation the crystal was giving off had nothing to do with fluorescence.

Becquerel (Fig. 1) traced the radiation to the uranium in the mineral, and found that some of the radiation consisted of electrons. Ernest Rutherford (Fig. 2), in England, had named that part of the radiation *beta rays*. Later, Rutherford showed that the two other parts, which he called *alpha rays* and *gamma rays*, were the nuclei of helium atoms and very high energy electromagnetic radiation, respectively (see Fig. 10.5). Because the alpha and beta "rays" were found to consist of speeding particles, the terms were changed to *alpha particles* and *beta particles*. Rutherford also introduced the concept of half-life and discovered both the atomic nucleus and the proton. He won the Nobel Prize in chemistry in 1908.

In 1898 Marie Curie (born Marie Sklodowska in 1867 in Warsaw, Poland) named Becquerel's uranium radiation phenomenon *radioactivity* (Fig. 3). Marie's story is interesting and inspirational. Her father was a physics teacher, and her mother was principal of a girls' school. Marie was able to obtain only a high school education because Poland at that time was part of the Russian Empire and repressed. Marie worked as a governess, and by 1891 had saved enough money to emigrate to Paris and enter the Sorbonne, a famous university. She lived frugally during this period (once fainting from hunger in the classroom), but graduated at the top of her class. In 1895 she married Pierre Curie, a physicist at the Sorbonne who was well known for his work on crystals and magnetism.

In 1897, for her doctoral dissertation, Marie began a search for naturally radioactive elements. She found only uranium and thorium, but noticed that some of the uranium ores she studied were much more radioactive than could be accounted for by their uranium and thorium content. Evidently, the ores contained traces of unknown elements that were intensely radioactive. At this point Pierre abandoned his own research and became his wife's assistant. In July of 1898 they isolated from the uranium ore a minute amount of a new element hundreds of times more radioactive than uranium. They called it *polonium* after Marie's native country. However, by no means did polonium account for all the intense radioactivity of the ore. In December 1898 they detected an even more radioactive element, which they named *radium*.

To get enough radium to investigate thoroughly, the Curies used their life savings to buy tons of uranium ore. They obtained permission to work in an old wooden shed with a leaky roof, no floor, and inadequate heat. For four years they processed the tons of ore into smaller and smaller samples of more and more intensely radioactive material. All this time, they had to take care of their baby, Irene. Eventually, eight tons of ore gave them about 10 mg of radium and a smaller amount of polonium. Despite the obvious chance of wealth, the Curies refused to patent their process. They were too ill in 1903 to journey to Stockholm to receive the Nobel Prize, which they shared with Becquerel.

It is evident from the abundance of white space all over Figure 10.7 that most theoretical combinations of protons and neutrons are not stable. Also, the increasing divergence of the belt from the $N = Z$ line shows that as more protons are packed into the nucleus, the neutron-to-proton ratio must increase for stability to result.

An inventory of the number of protons and number of neutrons in stable nuclides reveals an interesting pattern (Table 10.2). Most of the stable nuclides have both an even number of protons (p) and an even number of neutrons (n) in their nuclei.

Figure 1 Henri Becquerel.

Figure 2 Ernest Rutherford.

Figure 3 Marie Curie.

In 1906 Pierre was run over by a horse-drawn wagon as he stepped from a carriage into a Paris street. Marie, now known as Madame Curie, took over his professorship at the Sorbonne, becoming the first woman to teach there. In 1911 she received an unprecedented second Nobel Prize, this time in chemistry for her work on the properties of radium. Despite her fame, during World War I she and her daughter Irene organized medical units with X-ray equipment to locate shrapnel and broken bones in wounded soldiers.

In 1921 the women of the United States donated $100,000 (a fortune at that time) to finance the mining and purification of one gram of radium. Marie traveled to the United States to receive the radium from President Warren Harding.

Madame Curie's last years were spent in the supervision of the Paris Institute of Radium. She died in 1934 of leukemia (cancer of the blood), which probably was caused by overexposure to radioactivity. Her death came one year before Irene and her husband Frederick Joliot were awarded the Nobel Prize in chemistry for producing the first artificial radionuclide, phosphorus-30. Irene, like her mother, died of leukemia. During those days, the hazards of exposure to radioactive substances were not yet realized, and Madame Curie's early notebooks remain, literally, too "hot" to handle.

We refer to these as *even-even* nuclides. Practically all other stable nuclides are either *even-odd* or *odd-even*. Nature seems to dislike *odd-odd* nuclides, due apparently to the existence of energy levels in the nucleus that favor the pairing of two protons or of two neutrons. (Note that the descriptions such as *odd-odd* refer to the number of protons and neutrons, respectively, not to the atomic number and mass number. For example, an *odd* atomic number coupled with an *even* mass number means an *odd* number of protons but also an *odd* number of neutrons.)

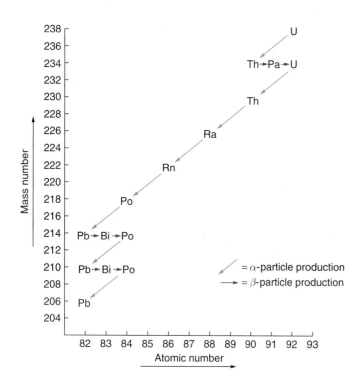

Figure 10.6 The decay of uranium-238 to lead-206.
Each radioactive nucleus in the series undergoes either alpha decay (blue arrows) or beta decay (red arrows). Finally, stable lead-206 is formed as the end product. (Gamma decays, which change only the energy of nuclei, are not shown.)

A nuclide will be radioactive if it meets any of the following criteria.*

1. Its atomic number is greater than 83.
2. It has fewer n than p in the nucleus (except for $^{1}_{1}$H and $^{3}_{2}$He).
3. It is an *odd-odd* nuclide (except for $^{2}_{1}$H, $^{6}_{3}$Li, $^{10}_{5}$B, and $^{14}_{7}$N).
4. It is an *odd-even* nuclide with mass number not within 1.5 units of the element's atomic mass.

EXAMPLE 10.4 Recognizing Radionuclides

Identify the radionuclide in each of the following pairs, and state your reasoning: (a) $^{208}_{82}$Pb and $^{222}_{86}$Rn, (b) $^{19}_{10}$Ne and $^{20}_{10}$Ne, (c) $^{63}_{29}$Cu and $^{64}_{29}$Cu, and (d) $^{108}_{47}$Ag and $^{110}_{47}$Ag.

---

*A fifth criterion, which we will not use because it is difficult to apply, is that the mass number of an *even-even* nuclide must still be relatively close to the element's atomic mass or the nuclide will be radioactive. The problem is that the meaning of "relatively close" differs depending upon the size of the atomic mass.

Solution

(a) $^{222}_{86}$Rn (Z above 83)
(b) $^{19}_{10}$Ne (fewer n than p)
(c) $^{64}_{29}$Cu (*odd-odd*)
(d) $^{110}_{47}$Ag (*odd-even*, but mass number is not within 1.5 units of Ag's atomic mass of 107.9 u in the periodic table)

CONFIDENCE QUESTION 10.4

Predict which two of the following nuclides are radioactive:

$^{232}_{90}$Th $^{24}_{12}$Mg $^{40}_{19}$K $^{31}_{15}$P

Table 10.2 The Pairing Effect in Stabilizing Nuclei

Proton Number (Z)	Neutron Number (N)	Number of Stable Nuclides
Even	Even	160
Even	Odd	52
Odd	Even	52
Odd	Odd	4

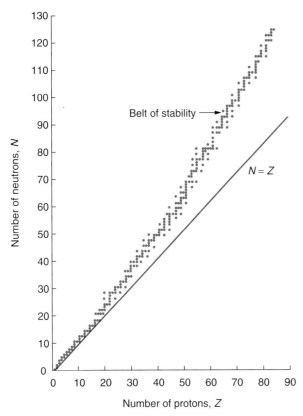

Figure 10.7 A plot of number of neutrons (N) versus number of protons (Z) for the stable nuclides.
Each dot represents a stable nuclide. They trace out a *belt of stability* that begins on a line where *N* and *Z* are equal and gradually diverges from the line as *Z* gets greater. Because all nuclides with Z > 83 are radioactive, the belt ends at that number of protons.

10.3 Half-life and Radiometric Dating

Learning Goals:

To apply the concept of half-life.

To explain radiometric dating.

Some samples of radionuclides take a long time to decay; others decay very rapidly (Table 10.3). The rate of decay of a given radionuclide is described by the term **half-life,** the time it takes for half of the nuclei of a sample to decay. That is, after one half-life has gone by, one-half of the original

Table 10.3 The Half-Lives of Some Radionuclides

Radionuclide	Half-Life
Beryllium-8	6.7×10^{-15} s
Oxygen-19	26.9 s
Technetium-104	18.3 min
Radon-222	3.82 d
Strontium-90	29 y
Carbon-14	5730 y
Uranium-238	4.46×10^9 y
Indium-115	4.4×10^{14} y

amount of radionuclide remains undecayed; after two half-lives, $\frac{1}{2} \times \frac{1}{2} = \frac{1}{4}$ of the original amount is undecayed; and so on (Fig. 10.8).

To determine the half-life of a radionuclide, we monitor the *activity*—the rate of emission of the decay particles, often measured in counts per minute (cpm)—with an instrument such as a *Geiger counter* (Fig. 10.9).* If half of the original nuclei of a sample decay in one half-life, then the activity decreases to one-half of its original amount during that time.

Let's take a look at some calculations involving half-life. The *original amount* of radionuclide sample is designated N_o, and the *final amount* of radionuclide sample is N. N_o and N can be given in several units (such as grams or number of atoms), but we will use either activity (cpm) or rational fractions, where N_o is defined to be $\frac{1}{1}$, and N is stated as $\frac{1}{2}$, $\frac{1}{4}$, and so on, of N_o. The *elapsed time* is the time between the measurement of N_o and N. If you are told a radionuclide's half-life is, say, 12 y, to keep the units straight, put it into a calculation as 12 y/half-life (12 years per half-life).

We will consider only problems in which the number of half-lives is a small whole number. Also, we will limit our "unknowns" to the final amount (N), the elapsed time, and the number of half-lives.

EXAMPLE 10.5 Finding the Final Amount

What fraction of a sample of a radionuclide (half-life = 8 d) will remain after 24 d?

---

*Activity can also be given in becquerels (1 Bq = one disintegration per second) or curies (1 Ci = 3.7×10^{10} disintegrations per second).

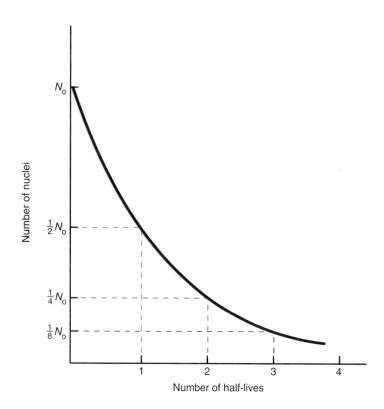

Figure 10.8 The decay curve for any radionuclide.
Starting with a number of nuclei, N_o, of a radionuclide, after one half-life has elapsed only one-half of the nuclei will remain undecayed. After two half-lives have gone by, only one-quarter of the nuclei will remain, and so on.

Solution

STEP 1

Find the number of half-lives that have passed in 24 d.

$$\frac{24 \text{ d}}{8 \text{ d/half-life}} = 3 \text{ half-lives}$$

STEP 2

Starting with the defined initial amount (N_o) of $\frac{1}{1}$, halve it three times (because three half-lives have passed).

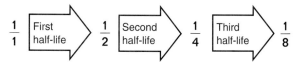

Thus, the answer is that $\frac{1}{8}$ of N_o remains.

CONFIDENCE QUESTION 10.5

For a radionuclide with half-life = 11 min, how many half-lives will go by in 22 min? What fraction of the original amount will be present after that time?

Radiometric Dating

The rate of decay of a given radionuclide is constant—unaffected by heat, pressure, electrical or magnetic fields, or the type of molecule of which it is a part. Therefore, this rate can serve as a clock for dating ancient rocks and the remains of long-dead plants and animals. The general name for such procedures is **radiometric dating.**

In the dating of rocks, the ratio of the amount of a daughter nuclide to the amount of its parent is measured. To date rocks with ages of millions or billions of years, radionuclides with very long half-lives must be used, or else not enough of the parent would remain to measure accurately.

For example, ^{238}U (half-life = 4.46×10^9 y) goes through a series of decays to form, ultimately, ^{206}Pb. Lead that has its origin in radioactive decay is known as *radiogenic lead;* nonradiogenic lead is termed *primordial lead.* If a rock contains uranium, its concentrations of ^{238}U and ^{206}Pb, together with the half-life of ^{238}U, can be used to calculate the rock's age. If a rock contains a 1-to-1 ratio of radiogenic ^{206}Pb to ^{238}U, the ^{238}U has been decaying in the rock for one half-life, or 4.46×10^9 y. The

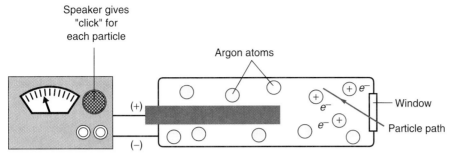

Figure 10.9 A schematic representation of a Geiger counter.
A high-energy particle from a radioactive source enters the window and ionizes argon atoms along its path. The ions and electrons formed produce a pulse of current, which is amplified and counted.

higher the ratio of radiogenic ^{206}Pb to ^{238}U, the older the rock (Fig. 10.10).

It is possible to tell how much ^{206}Pb in a rock is radiogenic by determining how much ^{204}Pb is present, because all ^{204}Pb is primordial and the ratio in which primordial ^{206}Pb and ^{204}Pb occurs is known. If a rock contains ^{238}U, then ^{235}U will also be present, and this radionuclide decays with a half-life of 7.04×10^8 y to ^{207}Pb. Thus the ratio of ^{207}Pb to ^{235}U can be used as a check on the age obtained by the ^{206}Pb/^{238}U ratio. Such checks substantiate the validity of the method. Dating of some types of rocks is achieved by use of other radionuclides, such as potassium-40 (half-life = 1.25×10^9 y), which decays to argon-40.

The oldest rocks on Earth have been dated at 3.8 billion y, which provides a minimum possible age of Earth—the time since the solid crust first formed. For meteorites, which are assumed to have solidified at the same time as other solid objects in the solar system including Earth, ages have been determined to be 4.4 billion to 4.6 billion y. The oldest Moon rocks returned to Earth by the Apollo missions were found to be 4.5 billion y old. These measurements and other evidence indicate that the age of Earth is about 4.6 billion y.

EXAMPLE 10.6 Dating a rock by use of ^{235}U

Uranium-235 has a half-life = 7.04×10^8 y and decays to ^{207}Pb. If analysis of a rock shows that enough radiogenic ^{207}Pb is present to indicate that only one-eighth of the original ^{235}U is undecayed, how old is the rock? (This is an example of finding the elapsed time.)

Solution

STEP 1

Start with $\frac{1}{1}$ and see how many times you must halve to get to $\frac{1}{8}$.

$$\frac{1}{1} \to \frac{1}{2} \to \frac{1}{4} \to \frac{1}{8}$$

If only one-eighth of the ^{235}U remains, the rock has existed for three half-lives. (Count the arrows in the sequence above.)

STEP 2

To find the rock's age, multiply the number of half-lives by the half-life.

(3 half-lives)(7.04×10^8 y/half-life) = 2.11×10^9 y

So the rock is about 2.1 billion years old.

CONFIDENCE QUESTION 10.6

Potassium-40 has a half-life of 1.25×10^9 y and decays to argon-40. If analysis of a rock shows that only one-fourth of the original ^{40}K is undecayed, about how old is the rock?

To find the age of organic (once-living) remains, such as charcoal, parchment, or bones, scientists use **carbon dating**, developed in 1950 by Willard F. Libby (Nobel Prize in chemistry, 1960). This procedure measures the activity of ^{14}C in the

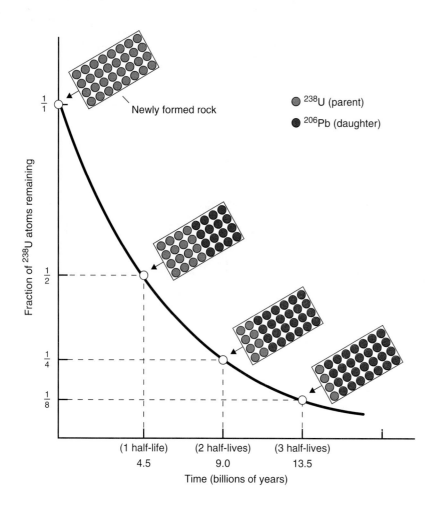

Figure 10.10 The decay curve for ^{238}U–^{206}Pb dating of a rock.
As time goes by, the original ^{238}U atoms (blue dots) in a newly formed rock slowly decay. After one half-life (4.5 billion years), half have changed into ^{206}Pb (red dots). As more time elapses, the ratio of radiogenic ^{206}Pb to ^{238}U increases at a predictable rate; so the ratio actually found in a rock can be used to tell its age.

ancient sample and compares it to the activity in present-day organic matter.

Throughout history, ^{14}C, a beta emitter with a half-life of 5730 y, has been produced by the action of neutrons on atmospheric nitrogen.

$$^{1}_{0}n + ^{14}_{7}N \rightarrow ^{14}_{6}C + ^{1}_{1}H$$

(The neutrons are formed by *primary cosmic rays,* streams of protons and other charged particles that pour into our atmosphere from the Sun and outer space. The primary cosmic rays collide with molecules in the atmosphere to produce *secondary cosmic rays,* some of which are neutrons.)

The newly formed ^{14}C reacts with oxygen in the air to form radioactive carbon dioxide, $^{14}CO_2$, which, along with ordinary $^{12}CO_2$, is used by plants in photosynthesis. About one out of every trillion (10^{12}) carbon atoms in plants is ^{14}C. Animals that eat the plants incorporate the ^{14}C in their cells, as do animals that eat the animals that ate the plants (● Fig. 10.11).

Thus all living matter has about the same level of radioactivity due to ^{14}C—an activity of about 15.3 counts per minute per gram of total carbon (15.3 cpm/g_C). Once an organism dies, it ceases to take in ^{14}C, but the ^{14}C in its remains continues to decay. Thus the longer the organism has been dead, the lower the activity of each gram of carbon in its remains.

EXAMPLE 10.7 Using Carbon Dating

Charcoal from an ancient campfire has a ^{14}C activity that is one-fourth that of new wood. About how old is the charcoal? The half-life of ^{14}C is 5730 y.

Solution

The ^{14}C has been decaying for a time period equal to two half-lives, shown by the number of arrows below.

$$\frac{1}{1} \rightarrow \frac{1}{2} \rightarrow \frac{1}{4}$$

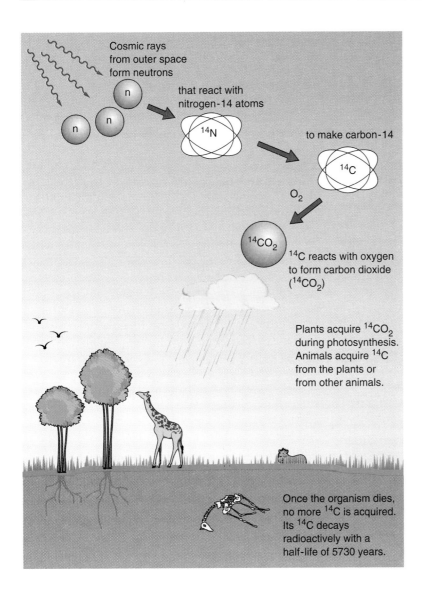

Figure 10.11 Carbon dating. An illustration of how carbon-14 forms in the atmosphere and enters the biosphere.

To find the charcoal's age, multiply the number of half-lives by the half-life.

(2 half-lives)(5730 y/half-life) = 11,460 y

So the charcoal is about 11,000 years old.

CONFIDENCE QUESTION 10.7

Suppose that one gram of carbon obtained by burning cloth wrappings of an Egyptian mummy is found to have one-half the carbon-14 activity of one gram of present-day carbon. About how old are the wrappings? The half-life of ^{14}C is 5730 y.

Organic remains up to about 50,000 years in age (nine half-lives) can be dated by this method. Beyond that age, less than 0.2% of the original ^{14}C is still undecayed, and the activity becomes too small to measure accurately.

Carbon dating assumes that the amount of ^{14}C in the atmosphere (and hence in the biosphere) has been the same throughout the past 50,000 y. However, because of changes in solar activity and Earth's magnetic field, it apparently has varied by as much as ±5%. California's bristlecone pines, which can live for up to five thousand years, allow us to correct for the slight changes occurring in the abundance of ^{14}C with time. By studying the ^{14}C

activity of samples taken from the annual growth rings in both dead and living trees, researchers have developed a calibration curve for ^{14}C dates as far back as about 5000 B.C. Therefore, carbon dating is most reliable for specimens no more than 7000 years old.

Scientists' confidence in the reliability of carbon dating was increased when the ages of specimens discovered in ancient Egyptian tombs calculated by carbon dating agreed with the chronological ages established by Egyptologists. A very recent method of calibration for as far back as 30,000 y dates ancient coral by both ^{14}C and ^{238}U methods and adjusts the ^{14}C date to the more reliable ^{238}U date.

A newer method of carbon dating relies on a specially designed mass spectrometer that separates and counts both the ^{14}C atoms and the ^{12}C atoms in a sample. By comparing the ratio of the isotopes in the specimen to the ratio in living matter, the time since its death can be calculated. This method utilizes far smaller samples, gives more precise dates, and can be used to date specimens as old as 70,000 y.

For centuries controversy has existed about whether the famous Shroud of Turin, a piece of linen cloth bearing the image of a man's head, is the burial cloth of Jesus of Nazareth (Fig. 10.12). Doubts about the cloth's authenticity were expressed only a few years after the shroud first surfaced in Lirey, France, about A.D. 1357. In 1988 three laboratories in Europe, England, and the United States, working on less than 50-mg samples of the shroud, independently showed by carbon-14 dating using the mass spectrometer that the flax from which the shroud was woven dates from between A.D. 1260 and 1390, a time far later than that of Jesus. The dating results are compatible with the 1978 findings of another researcher, who analyzed flecks of residue from the shroud and concluded that it was a very thin watercolor painting done in a style common to the fourteenth century.

Although the Shroud of Turin seems to be a forgery, carbon dating has helped establish the authenticity of another discovery with religious significance. In 1947, a shepherd boy found a number of Hebrew manuscripts of books of the Old Testament in a cave near the Dead Sea in the Middle East. Carbon dating established that these Dead Sea Scrolls, as they came to be called, were written at different times between about 200 B.C. and A.D. 100.

10.4 Nuclear Reactions

> **Learning Goals:**
>
> To complete equations for nuclear reactions.
>
> To state some uses of radionuclides.

Radioactive nuclei, through the emission of alpha and beta particles, spontaneously change, or *transmute*, into nuclei of other elements. Scientists wondered whether the reverse process was possible; that is, could a particle be added to a nucleus to change it into that of another element? The answer was yes.

Ernest Rutherford produced the first *nuclear reaction* in 1919 by bombarding nitrogen (^{14}N) gas with alpha particles from a radioactive source. Particles coming from the gas were identified as protons. Rutherford reasoned that an alpha parti-

Figure 10.12 The shroud of Turin.
This photographic negative of the shroud brings out the faded image more clearly.

10.4 Nuclear Reactions

Table 10.4 Common Particles in Nuclear Reactions

Name	Notation	Name	Notation
Proton	$^{1}_{1}\text{H}$	Neutron	$^{1}_{0}\text{n}$
Deuteron	$^{2}_{1}\text{H}$	Gamma ray	γ
Triton	$^{3}_{1}\text{H}$	Alpha particle (α)	$^{4}_{2}\text{He}$
Beta particle (β^{-}) (electron)	$^{0}_{-1}\text{e}$	Positron (β^{+})	$^{0}_{+1}\text{e}$

cle colliding with a nitrogen nucleus can occasionally knock out a proton. The result is an *artificial transmutation* of a nitrogen isotope into an oxygen isotope. The equation for the reaction is

$$^{4}_{2}\text{He} + ^{14}_{7}\text{N} \rightarrow ^{17}_{8}\text{O} + ^{1}_{1}\text{H}$$

As this equation indicates, the conservation of mass number and the conservation of atomic number hold in nuclear reactions, just as they do in nuclear decay.

The general form of a nuclear reaction is

$$a + A \rightarrow B + b$$

where a is the particle bombarding nucleus A to form nucleus B and particle b. Table 10.4 summarizes the notation for the most common particles encountered in nuclear reactions.

EXAMPLE 10.8 Completing an Equation for a Nuclear Reaction

Complete the equation for the proton bombardment of lithium-7.

$$^{1}_{1}\text{H} + ^{7}_{3}\text{Li} \rightarrow \underline{} + ^{1}_{0}\text{n}$$

Solution

The sum of the mass numbers on the left is 8. So far, only a mass number of 1 shows on the right, so the missing particle has a mass number of 8 − 1, or 7.

The sum of the atomic numbers on the left is 4. So far, the total showing on the right is 0. Thus the missing particle must have an atomic number of 4 − 0, or 4. The atom with mass number 7 and atomic number 4 is an isotope of Be (beryllium). The completed equation is

$$^{1}_{1}\text{H} + ^{7}_{3}\text{Li} \rightarrow ^{7}_{4}\text{Be} + ^{1}_{0}\text{n}$$

CONFIDENCE QUESTION 10.8

Complete the equation for the deuteron bombardment of aluminum-27.

$$^{2}_{1}\text{H} + ^{27}_{13}\text{Al} \rightarrow \underline{} + ^{4}_{2}\text{He}$$

The reaction in Rutherford's experiment was discovered almost by accident, because it took place so infrequently. One proton is produced for about every one million alpha particles that shoot through the gas. But think of the implications of its discovery: One element had been changed into another! This was the age-old dream of the alchemists, although their main concern was to change common metals, such as mercury and lead, into gold.

Such artificial transmutations are now common. Large machines called *particle accelerators* use electric fields to accelerate charged particles to very high energies (Fig. 10.13). The energetic particles are used to bombard nuclei and initiate nuclear reactions. Different reactions require different particles and different bombarding energies. One nuclear reaction that occurs when a proton strikes mercury-200 is

$$^{1}_{1}\text{H} + ^{200}_{80}\text{Hg} \rightarrow ^{197}_{79}\text{Au} + ^{4}_{2}\text{He}$$

Thus gold (Au) has indeed been made from another element. Unfortunately, making gold by this process costs more than it is worth.

Neutrons produced in nuclear reactions can be used to induce other nuclear reactions. Because they have no electric charge, they do not experience repulsive electrical interactions with nuclear protons, as would alpha-particle or proton projectiles. As a result, they are especially effective at penetrating the nucleus and inducing a reaction. For example,

$$^{1}_{0}\text{n} + ^{45}_{21}\text{Sc} \rightarrow ^{42}_{19}\text{K} + ^{4}_{2}\text{He}$$

Figure 10.13 The Fermi National Accelerator Laboratory at Batavia, Illinois.
The large circle is the main accelerator with a 6.4-km (4.0 mi) circumference. Magnets bend and focus a proton beam that travels around the circle 50,000 times a second. With each round-trip more energy is added by a radio-frequency system, producing highly energetic particles for use in nuclear research.

The *transuranium elements,* those with atomic numbers greater than 92, are all synthetic (as are Tc, Pm, At, and Fr). Elements 93 (neptunium) to 101 (mendelevium) can be made by bombarding a lighter nucleus with alpha particles or neutrons. For example,

$$^1_0 n + ^{238}_{92} U \rightarrow ^{239}_{93} Np + ^0_{-1} e$$

Beyond mendelevium, heavier bombarding particles are required. For example, element 103, lawrencium, is made by bombarding californium-252 with boron-10 nuclei.

$$^{10}_5 B + ^{252}_{98} Cf \rightarrow ^{257}_{103} Lr + 5\, ^1_0 n$$

Americium-241, a synthetic transuranium radionuclide (half-life = 432 y) is used in the most common type of home smoke detector. As the americium-241 decays, the alpha particles emitted ionize the air inside part of the detector. The ions carry a small current and allow the 9-V battery to power a closed circuit. If smoke enters the detector, the ions become attached to the smoke particles and slow down, causing the current to decrease and an alarm to sound.

Radionuclides have many uses in medicine, chemistry, industry, agriculture, and biology. As an example, a radioactive isotope of iodine, ^{123}I, is used in a diagnostic measurement connected with the thyroid gland. The patient is administered a prescribed amount of the ^{123}I and, like regular iodine in the diet, it is absorbed by the thyroid gland. This allows doctors to monitor the iodine intake of the thyroid because the radioactive iodine can be traced as it is released into the bloodstream in the form of protein-bound iodine.

Nuclear radiation can also be used to treat diseased cells, which generally can be damaged or destroyed by radiation more easily than healthy cells. By focusing an intense beam of radiation from cobalt-60 on a cancerous tumor, its cells can be destroyed and its growth impaired or stopped. Radiation can also be used in many types of medical diagnoses; as, for example, in PET scans (● Fig. 10.14).

In chemistry and biology, radioactive "tracers" such as ^{14}C and ^{3}H (tritium) are used to tag an atom in a certain part of a molecule so that it can be followed through a series of reactions. In this way, the reaction pathways of hormones, drugs, and other substances can be determined.

Neutron activation analysis is one of the most sensitive analytical methods in chemistry. A beam of neutrons irradiates the sample, and each constituent element forms a specific radionuclide that can be identified by the characteristic energies of the gamma rays it emits. One advantage is that the sample is not destroyed by this analytical technique. An interesting use is in the analysis of hu-

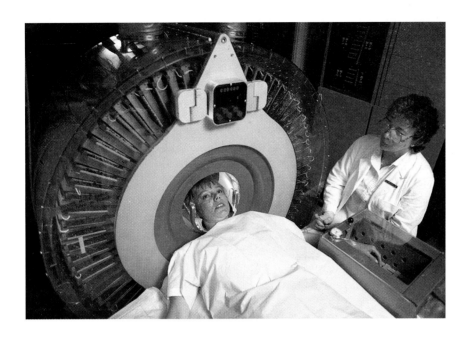

Figure 10.14 A PET machine being used for a brain scan.
Unlike conventional X-rays, PET (*positron emission tomography*) scans are not distorted by bone. Such scans are able to define very slight distinctions among tissues. (A positron, $^{0}_{+1}e$, is the antiparticle of the electron.)

man hair. By determining the amounts and position of elements in a strand of hair found, say, at a crime scene, hair can be matched with hair taken from a suspect. Also, as little as 10^{-9} g of arsenic can be detected in hair, so arsenic poisoners now run a high risk of detection.

In agriculture, less than lethal doses of radioactivity were used to cause sterility in male Mediterranean fruit flies in California and Florida, where they were destroying crops. When released, the sterilized males mated with females, but the eggs would not hatch, thus drastically reducing the number of fruit flies.

In industry, the penetrating ability of radiation has been used to gauge and adjust the thickness of metal sheets, plastic films, and cigarettes. Plutonium-238 powers a tiny battery used in heart pacemakers. In the next two sections we discuss the controlled and uncontrolled release of nuclear energy. Other uses of radioactivity, some very ingenious, are too numerous to mention.

10.5 Nuclear Fission

Learning Goals:

To explain how nuclear fission occurs.

To discuss how nuclear fission reactors operate.

Fission is the process in which a large nucleus is "split" into two intermediate-size nuclei, with the emission of neutrons and the conversion of mass into energy. As an example, consider the fission decay of ^{236}U. If ^{235}U is bombarded with low-energy neutrons, ^{236}U is formed momentarily.

$$^{1}_{0}n + ^{235}_{92}U \rightarrow ^{236}_{92}U$$

The ^{236}U immediately fissions into two smaller nuclei, emits several neutrons, and releases energy.

Figure 10.15a illustrates the following typical fission of ^{236}U.

$$^{236}_{92}U \rightarrow ^{140}_{54}Xe + ^{94}_{38}Sr + 2\,^{1}_{0}n$$

This is just one of many possible fission decays of ^{236}U. Another is

$$^{236}_{92}U \rightarrow ^{132}_{50}Sn + ^{101}_{42}Mo + 3\,^{1}_{0}n$$

EXAMPLE 10.9 Completing an Equation for Fission

Complete the following equation for fission:

$$^{236}_{92}U \rightarrow ^{88}_{36}Kr + ^{144}_{56}Ba + \underline{\quad}$$

Solution

The atomic numbers are balanced (92 = 36 + 56), so the other particle must have an atomic number of 0.

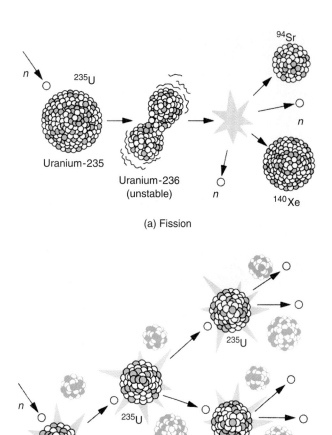

Figure 10.15 Fission and chain reaction. (a) In a fission reaction, such as that shown for uranium-235, a neutron is absorbed and the nucleus splits into two lighter nuclei with the emission of energy and two or more neutrons. (b) If the emitted neutrons cause increasing numbers of fission reactions, an expanding *chain reaction* occurs.

The mass number on the left is 236, and the sum of the mass numbers on the right is $88 + 144 = 232$. Hence if the mass numbers are to balance, there must be four additional units of mass on the right side of the equation. Since no particle with atomic number of 0 and mass number of 4 exists, the missing "particle" is actually four neutrons. The reaction is then

$$^{236}_{92}\text{U} \rightarrow ^{88}_{36}\text{Kr} + ^{144}_{56}\text{Ba} + 4\,^{1}_{0}\text{n}$$

The atomic numbers and mass numbers are both now balanced.

CONFIDENCE QUESTION 10.9

Complete the following equation for fission.

$$^{236}_{92}U \rightarrow ^{90}_{38}Sr + \underline{} + 2\,^{1}_{0}n$$

The fast-fissioning ^{236}U is an intermediate nucleus, and it is often left out of the equation for the neutron-induced fission of ^{235}U. That is, the equation for the reaction in Example 10.9 is usually written

$$^{1}_{0}n + ^{235}_{92}U \rightarrow ^{88}_{36}Kr + ^{144}_{56}Ba + 4\,^{1}_{0}n$$

Nuclear fission reactions have three important features:

1. The fission products are always radioactive, and many have half-lives of thousands of years. (This feature gives rise to nuclear waste disposal problems.)
2. Relatively large amounts of energy are produced (see Section 10.6).
3. Neutrons are released.

In an *expanding* **chain reaction,** one initial reaction triggers a growing number of subsequent reactions. In the case of fission, one neutron hits a nucleus of ^{235}U and forms ^{236}U, which can fission and emit two (or more) neutrons. These two neutrons can then hit two other ^{235}U nuclei, causing them to fission and release energy and four neutrons. These four neutrons can cause four fissions, releasing energy and eight neutrons, and so on (Fig. 10.15b). Each time a nucleus fissions, energy is released; and as the chain expands, the energy output increases.

For a *self-sustaining chain reaction* each fission event needs to cause only one more fission event. This leads to a steady release of energy, not a growing release.

Of course, the process of energy production by fission is not as simple as just described. For a self-sustaining chain reaction to proceed, a sufficient amount of fissionable material (^{235}U) must be present. Otherwise, too many neutrons would escape from the sample before reaction with a nucleus. The chain would be broken, so to speak. The minimum amount of fissionable material necessary to sustain a chain reaction is called the **critical mass.** The critical mass for pure ^{235}U is about 4 kg, approximately the size of a baseball. With a *subcritical mass*, no chain reaction occurs. With a *super-critical mass*, the chain reaction grows and, under optimal conditions, an explosion can occur.

Natural uranium is composed of 99.3% ^{238}U and only 0.7% of the fissionable ^{235}U isotope. So that more fissionable ^{235}U nuclei are present in a sample, the ^{235}U is concentrated, or "enriched." Enriched uranium used in U.S. nuclear reactors for the production of electricity is about 3% ^{235}U. Weapons-grade material is enriched to 90% or more, and this percentage provides many fissionable nuclei for a large and sudden release of energy. (The leftover ^{238}U is known as *depleted uranium* and can be used in armor piercing shells.)

In a fission, (or "atomic") bomb, a supercritical mass of highly enriched fissionable material must be formed and held together for a short time to get an explosive release of energy. Subcritical segments of the fissionable material in a fission bomb are kept separated before detonation so that a critical mass does not exist for the chain reaction. A chemical explosive is used to bring the segments together in an interlocking, supercritical configuration that holds them long enough for a large fraction of the material to undergo fission. The result is an explosive release of energy. The chapter's second Highlight discusses the building of the first atomic bombs.

Nuclear Reactors

A bomb is an example of *uncontrolled* fission. A nuclear reactor is an example of *controlled* fission, in which we control the growth of the chain reaction and the release of energy. The first commercial reactor for generating electricity went into operation in 1957 at Shippingsport, Pennsylvania. The basic design of a fission nuclear reactor is shown in Fig. 10.16.

Enriched uranium oxide fuel pellets are placed in metal tubes to form *fuel rods*, which are placed in the reactor core, where fission takes place. Also in the core are *control rods* made of neutron-absorbing materials such as boron (B) and cadmium (Cd). The control rods are adjusted (inserted or withdrawn) so that only a certain number of neutrons is absorbed, ensuring that the chain reaction releases energy at the rate desired. For a steady rate of energy release, one neutron from each fission event should initiate only one additional fission event. If more energy is needed, the rods are

HIGHLIGHT

The Building of the Bomb

In 1934 in Rome, Enrico Fermi and Emilio Segre bombarded uranium with neutrons in an attempt to produce transuranium elements. They succeeded in making element 93, neptunium, but were unable to identify other radioactive materials that were produced.

In 1938 in Berlin, Otto Hahn and Fritz Strassman repeated the experiment, and were surprised to find the element barium among the reaction products. Hahn described his findings in a letter to Lise Meitner, a former colleague who was an Austrian Jew but was living in Sweden. Hahn had helped Meitner escape when the Nazis annexed Austria in 1938. (Those of you who have seen the movie *The Sound of Music* are familiar with the annexation.)

Meitner surmised that Hahn had discovered a nuclear process in which the uranium atom was splitting. She informed her nephew, Otto Frisch, of her hypothesis, which she termed *nuclear fission*. Frisch was working with Niels Bohr in Denmark and was visiting Meitner over the Christmas season. Meitner and Frisch calculated that energy was released in relatively large amounts when uranium atoms split. Frisch passed the information on to Niels Bohr (Fig. 9.6), who was about to leave for a scientific conference at Princeton University.

When Bohr arrived at Princeton, he communicated the news of fission to Fermi, who had fled fascist Italy because his wife, Laura, was Jewish. (Fermi had persuaded Italian dictator Mussolini to let him take his family to Stockholm to see him receive the 1938 Nobel Prize in physics. After the ceremony, he and his family hastened to the United States.) Bohr returned to Denmark but fled to the United States when Nazi armies overran Denmark in 1940. He almost died when he passed into a coma due to lack of oxygen while flying in a small plane to England.

In 1939 at Columbia University, Walter Zinn (an American) and Leo Szilard (a Hungarian refugee) found that each fission produced more than one neutron, and thus a chain reaction could conceivably occur. Realizing that a chain reaction had the potential of tremendous explosive power, in 1939 Szilard and Fermi composed a letter to President Roosevelt, informing him of their fears that Germany might be working on such a bomb. Afraid that Roosevelt would pay no attention to their letter, they took it to Albert Einstein, who also had fled Nazi Germany and settled at Princeton University. Surely, they thought, Roosevelt would listen to the most famous scientist in the world. They were right. Einstein signed the letter, and Roosevelt took it seriously.

Late in 1941, the *Manhattan Project* began—in top secrecy. On December 2, 1942, a group under Fermi's direction achieved the first self-sustaining fission reaction in a squash court underneath the abandoned football stands at the University of Chicago (Fig. 1).

Figure 1
The first self-sustaining nuclear fission reaction occurred in Chicago on December 2, 1942, as depicted in this painting. Fermi is the partially bald gentleman at the center close to the rail.

The major hurdle in the building of the bomb was the production of the fissionable material needed. It was the ^{235}U and not the more abundant ^{238}U that was undergoing fission. The natural uranium had to be enriched from its normal 0.7% ^{235}U to about 90% ^{235}U. This was accomplished using gaseous diffusion of UF$_6$ at a sprawling, secret installation called Oak Ridge, which had been built in the hills of eastern Tennessee. (Separation by gaseous diffusion makes use of the ability of ^{235}UF$_6$ molecules to move a little faster through porous barriers than can the slightly heavier ^{238}UF$_6$ molecules.)

Plutonium-239, also fissionable and thus suitable for bomb-building, became available when it was found to be formed by the beta decay of the neptunium produced by bombardment of ^{238}U with neutrons. A series of large reactors was built at Hanford, Washington, to produce ^{239}Pu.

The first atomic bomb used ^{239}Pu as the fissionable material and was developed and tested in New Mexico in July of 1945. J. Robert Oppenheimer was the scientist in charge of the Los Alamos facility, and General Leslie Groves was then in charge of the entire Manhattan Project. The heat from the explosion vaporized the 30-m steel tower on which it was placed and melted the sand around the site. The light released was the brightest ever seen.

Some scientists were so awed by the blast that they argued against the bomb's use. However, fear of millions of casualties on both sides in an imminent invasion of Japan persuaded President Truman to order the dropping of two bombs. "Little Boy," a ^{235}U bomb (Fig. 2), was dropped on Hiroshima on August 6, 1945, from a B-29 bomber named the *Enola Gay*. The energy released was equivalent to that of 20,000 *tons* of TNT (thus it is called a 20-kiloton bomb). The casualties numbered 100,000. Three days later, at Nagasaki, a second bomb ("Fat Man," a ^{239}Pu bomb) was dropped. Five days after that, Japan surrendered.

As powerful as fission bombs are, it was only a few years before the construction of even more powerful nuclear bombs, based this time on *nuclear fusion*. The first "hydrogen bomb" was exploded on Eniwetok, a South Pacific atoll, by the United States in 1952. Its triggering mechanism was a fission bomb (Fig. 10.17). Although the power of fission bombs is rated in kilotons of TNT, the power of fusion bombs is given in megatons.

Many benefits exist in the applications of radioactivity in medicine and industry and for nuclear power. More lives have been saved by nuclear medicine than have been taken by nuclear warfare. Yet the fear persists that the awesome destructive power of nuclear energy could once again be used, especially as more and more countries acquire the technology to build bombs.

Figure 2
An atomic bomb (3-m long) of the "Little Boy" type.

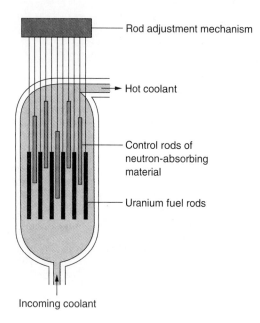

Figure 10.16 Nuclear reactor.
(a) A schematic diagram of the core of a nuclear reactor. The position of the control rods determines the level of energy production by regulating the number of neutrons available to cause additional fission. The fuel rods contain uranium oxide, enriched to about 3% $^{235}UO_2$. (b) A schematic diagram of the elements of an electrical generating plant. Note the core in place at the left. The heat from the fission of the ^{235}U atoms in the fuel rods is used to form steam, which drives a turbine and generates electricity.

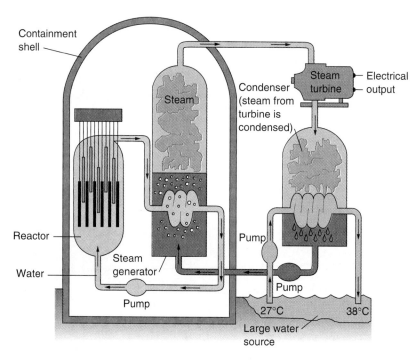

withdrawn farther. When fully inserted into the core, the control rods absorb enough neutrons to stop the chain reaction, and the reactor shuts down.

A reactor's core is basically a heat source, and the heat energy is removed by a coolant flowing through the core. In U.S. reactors the coolant is most commonly water. The coolant flowing through the hot fuel assemblies transfers heat that is used to produce steam to drive a turbogenerator that produces electricity.

In addition, the coolant acts as a *moderator*. The ^{235}U nuclei react best with "slow" neutrons. The neutrons emitted from the fission reactions are relatively "fast," with energies that are not best suited for ^{235}U fission. The fast neutrons are slowed down, or moderated, by energy losses through collisions. The neutrons transfer energy to the water molecules in collision processes. After only a few collisions, a neutron is slowed down to the point at which it can efficiently induce fission in the ^{235}U nuclei.

With a continuous-fission chain reaction, the possibility of a nuclear accident is always present, as occurred in 1979 at the Three Mile Island (TMI) nuclear plant in Pennsylvania and in 1986 with the reactor at Chernobyl in the Ukraine. The word *meltdown* is commonly used when discussing these accidents. If heat energy is not removed continuously from the core of a fission reactor, the fuel rods may fuse. In that event, the reaction could not be controlled with the control rods, and energy could no longer be removed by coolant flowing between the rods. The fissioning mass would become extremely hot, melt down through the floor of the containment vessel, and enter the environment. Even under the worst conditions, however, a reactor cannot explode like a nuclear bomb, because the fissionable material present is far from sufficient purity.

A partial meltdown occurred at TMI with a slight fusing of the fuel rods, but little radioactive material escaped into the environment. What did escape was primarily radioactive gases. Chernobyl, however, did experience a meltdown, explosions, and fire. This particular type of reactor used graphite (carbon) blocks for a moderator. Gas explosions caused the carbon to catch fire; radioactive material escaped with the smoke, and radioactive fallout spread over parts of Europe. Several hundred deaths occurred in the immediate region, and more will occur because of the long-term effects of radioactivity.

In addition to ^{235}U, the other fissionable nuclide of importance is ^{239}Pu (half-life = 2.4×10^4 y). This plutonium isotope is produced by bombardment of ^{238}U with fast neutrons. This means that ^{239}Pu is produced as nuclear reactors operate, because all of the neutrons are not moderated or slowed down. Because it is fissionable, ^{239}Pu extends the time before the refueling of a reactor.

In a *breeder reactor* this process is promoted, and fissionable ^{239}Pu is produced from ^{238}U, which is otherwise useless for energy production. It is said that if the depleted uranium present at Oak Ridge, Tennessee, were converted to ^{239}Pu, the fission of the ^{239}Pu could equal the energy output of all the oil in Saudi Arabia. The ^{239}Pu can be chemically separated from the fission by-products and used as the fuel in an ordinary nuclear reactor or in weapons. Its use in weapons is a major concern, because it can be obtained from regular reactors.

It takes about 20 breeder reactors running for one year to produce enough plutonium to fuel an additional reactor for a year. Breeders run at higher temperatures than conventional reactors, and are more difficult to control. The withholding of federal funds killed the Clinch River Breeder Project in Tennessee in 1983, but breeder research continues at Argonne National Laboratory in Illinois. Breeder reactors are currently being used in France and Germany.

10.6 Nuclear Fusion

Learning Goals:

To explain how nuclear fusion occurs.

To calculate mass and energy changes in nuclear reactions.

Fusion is the process in which smaller nuclei are merged to form larger ones, with the release of energy. Fusion is the source of energy of the sun and other stars. In the sun, the fusion process produces a helium nucleus from four protons (hydrogen nuclei). Also produced are two *positrons* ($_{+1}^{0}e$, the antiparticle of the electron). The process takes place in several steps, with the net result being

$$4\,_{1}^{1}\text{H} \rightarrow \,_{2}^{4}\text{He} + 2\,_{+1}^{0}e + \text{energy}$$

In the sun, about 600 million tons of hydrogen are converted to 596 million tons of helium *every second*. The other 4 million tons of matter are converted to energy. Fortunately, the sun has enough hydrogen to produce energy at its present rate for several billion more years.

Two other examples of fusion reactions are

$$_{1}^{2}\text{H} + \,_{1}^{2}\text{H} \rightarrow \,_{1}^{3}\text{H} + \,_{1}^{1}\text{H}$$

and

$$_{1}^{2}\text{H} + \,_{1}^{3}\text{H} \rightarrow \,_{2}^{4}\text{He} + \,_{0}^{1}n$$

In the first reaction, two deuterons fuse to form a triton and a proton. This is termed a D-D (deuteron-deuteron) reaction. In the second example (a D-T reaction), a deuteron and a triton form an alpha particle and a neutron.

Fusion involves no critical mass or size, because there is no chain reaction to maintain. However, the repulsive force between two positively charged nuclei opposes fusing. This force is smallest for hydrogen fusion because the nuclei contain only one proton. To overcome the repulsive forces, the kinetic energies of the particles must be increased by raising the temperature. Attaining a temperature on the order of millions of degrees is necessary to achieve fusion. At such high temperatures the hydrogen atoms are stripped of their electrons, and a **plasma** (a gas of electrons and protons or other nuclei) results. To achieve fusion, not only is a high temperature necessary but also the plasma must be confined at a high enough density for the protons (or other nuclei) to collide frequently.

Large amounts of fusion energy have been released in an uncontrolled manner in a hydrogen bomb (H-bomb), where a fission bomb is used to supply the energy needed to initiate the fusion reaction (Fig. 10.17). Unfortunately, controlled fusion for commercial use remains elusive.

Controlled fusion might be accomplished by steadily adding fuel in small amounts to a fusion reactor. The D-T reaction requires the lowest temperature of any fusion reaction. For this reason it is likely to be the first fusion reaction developed as an energy source. For practical purposes, the temperature will need to be about 100 million °C.

Major problems arise in reaching such temperatures and confining the high-temperature plasma. If the plasma touched the reactor walls, it would cool rapidly. However, the walls would not melt, because even though the plasma is nearly 100 million °C above the melting point of any material, the total quantity of heat that could be transferred from the plasma is very small because its concentration is extremely low.

One approach to the confinement problem is *magnetic confinement*.* Because a plasma is a gas of charged particles, it can be controlled and manipulated with electric and magnetic fields. A nuclear fusion reactor of the type called a *tokamak* uses a doughnut-shaped magnetic field to hold the plasma away from any material. Electric fields produce currents that raise the temperature of the plasma. One of the leading facilities for fusion research using magnetic confinement is the Princeton Plasma Physics Laboratory in New Jersey (Fig. 10.18).

Plasma temperatures, densities, and confinement times have been problems with magnetic confinement. However, in December 1993 there was a breakthrough. Scientists at Princeton fused deuterium and tritium nuclei in a magnetically confined plasma to produce energy at a record level of 6.4 million watts. Unfortunately, 24 million watts of energy had to be used to initiate the reaction. Many technical problems remain with fusion, and commercial energy production is not expected until well into the twenty-first century. Even so, fusion is a promising energy source because of its advantages over fission:

1. The low cost and abundance of deuterium. It is inexpensive to extract deuterium from ocean water, which is estimated to contain enough to supply the world's energy needs for more than a million years. Uranium is relatively scarce and expensive, and hazardous to mine.

---

*Another approach is *inertial confinement*, a technique in which simultaneous high-energy laser pulses from all sides cause a fuel pellet containing deuterium and tritium to implode, resulting in compression and high temperatures. If the pellet stays intact for a sufficient time, fusion might be initiated.

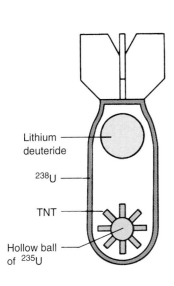

Figure 10.17 H-bomb.
The diagram at left shows the basic elements of a hydrogen bomb. To detonate an H-bomb, the TNT is exploded, forcing the ^{235}U together to get a supercritical mass and a fission explosion (a small atomic bomb, so to speak). The fusionable material is deuterium in the lithium deuteride (LiD). Heated to a plasma, D-D fusion reactions occur. Neutrons from the fission explosion react with the lithium to give tritium (^{3}H), and D-T fusion reactions also take place. The bomb is surrounded with ^{238}U, which tops off the explosion with a fission reaction. The result is shown in the photo.

2. Fewer nuclear waste disposal problems. Some fusion by-products are radioactive but have relatively short half-lives compared with those of fission wastes, which must be stored for thousands of years.
3. The impossibility of a runaway accident (unlike fission reactors).

Some of the disadvantages of fusion versus fission are:

1. Controlled fusion has not been proved practical for energy production, whereas controlled fission is currently being used extensively.
2. Fusion plants are projected to be more costly to build and operate than fission reactors.

Nuclear Reactions and Energy

In 1905 Albert Einstein published his special theory of relativity, which deals with the changes in mass, length, and time as an object's speed approaches the speed of light, c. The theory also predicted that mass (m) and energy (E) are not separate quantities but are related by the equation

$$E = mc^2 \qquad (10.2)$$

The predictions proved correct; scientists have indeed changed mass to energy, and, on a very small scale, converted energy to mass.

For example, a mass of 1.0 g (0.0010 kg) has an equivalent energy of

$$E = mc^2 = (0.0010 \text{ kg})(3.00 \times 10^8 \text{ m/s})^2$$
$$= 9.0 \times 10^{13} \text{ J}$$

This 90 *trillion* joules is the same amount of energy that is released by the explosion of about 20,000 *tons* of TNT! Such calculations convinced scientists that nuclear reactions in which just a small amount of mass was "lost" were a potential source of vast amounts of energy.

The units of mass and energy commonly used in nuclear physics are different from those used in

Figure 10.18 Magnetic confinement.
A magnetic confinement apparatus for fusion research at Princeton Plasma Physics Laboratory in New Jersey. The *tokamak* at the center has a magnetic confinement ring for the plasma. Beam injectors are seen around the periphery.

preceding chapters. Mass is usually given in atomic mass units, u, and energy is given in million electron volts, MeV. With these units, Einstein's equation reveals that 1 u of mass has the energy equivalent of 931 MeV, so we can say that there are 931 MeV/u.

To determine the change in mass and hence the energy released or absorbed in any nuclear process, just add up the masses of all reactant particles and subtract from that sum the total mass of all product particles. If a decrease in mass has resulted, the reaction is *exoergic* (releases energy) by that number of u times 931 MeV/u. If an increase in mass has taken place, the reaction is *endoergic* (absorbs energy) by that number of u times 931 MeV/u.

EXAMPLE 10.10 Calculating Energy in Nuclear Reactions

Calculate the mass lost and the corresponding energy released during this typical fission reaction.*

---

*Because we are dealing with differences in mass, either the masses of the atoms or the masses of just their nuclei can be used. The number of electrons is the same on each side of the equation and thus does not affect the mass difference. We will use the masses of the atoms, because they are more easily found in handbooks.

$$^{236}_{92}U \rightarrow {}^{88}_{36}Kr + {}^{144}_{56}Ba + 4\, {}^{1}_{0}n$$
(236.04556 u) (87.91445 u) (143.92284 u) (4 × 1.00867 u)

Solution

The total mass on the left of the arrow is 236.04556 u. Adding the masses of the particles on the right gives 235.87197 u. The difference (0.17359 u) has been converted to

$$(0.17359 \text{ u})(931 \text{ MeV/u}) = 162 \text{ MeV}$$

So during the reaction, 0.17359 u of mass is converted to 162 MeV of energy.

CONFIDENCE QUESTION 10.10

Calculate the mass lost and the corresponding energy released during a D-T fusion reaction.

$$^{2}_{1}H + {}^{3}_{1}H \rightarrow {}^{4}_{2}He + {}^{1}_{0}n$$
(2.0140 u) (3.0161 u) (4.0026 u) (1.0087 u)

Your answer to Confidence Question 10.10 shows that 17.5 MeV is released in the fusion reaction, which is less than the 162 MeV released in the fission reaction of Example 10.10. However, we started with only 5.03 u in the fusion reaction, but 236.05 u in the fission reaction. From a percentage standpoint, your calculations indicate that

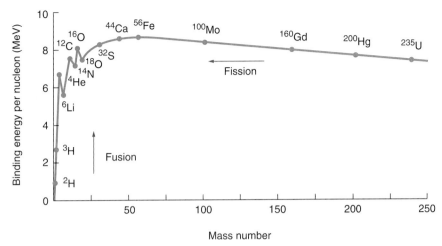

Figure 10.19 The relative stability of nuclei.
See text for discussion. (The relative stability is based on the *binding energy per nucleon*, which is the energy evolved in the formation of the nucleus divided by the number of nucleons.)

$0.0188/5.03 \times 100\% = 0.374\%$ of the initial mass is converted to energy in the fusion reaction, whereas only $0.1736/236.05 \times 100\% = 0.07354\%$ of the initial mass is converted to energy in the fission reaction. The bottom line is that *kilogram for kilogram* we can get more energy from fusion than from fission. (But recall that it takes a lot of energy to operate a fusion reactor.) Comparing fusion to energy production by ordinary chemical reactions, the fusion of one gram of deuterium releases the same amount of energy as the burning of nearly 20 tons of coal.

Figure 10.19 has been called the most significant graph in science because it shows that energy can be released both in nuclear fission *and* nuclear fusion. However, not every theoretical fission or fusion reaction releases energy. Note that fission of heavy nuclei at the far right of the curve to intermediate-size nuclei in the middle leads upward on the curve. Also, fusion of small nuclei on the left to larger nuclei farther to the right also leads upward on the curve. Any reaction that leads upward on the curve in Fig. 10.19 releases energy.

Any nuclear reaction in which the products are lower on the curve than are the reactants can proceed only with the net increase in mass and the corresponding net absorption of energy. One type of nucleus cannot give a net release of nuclear energy either by fission or fusion. Of course, it is the one at the top of the curve, ^{56}Fe (iron-56). You can't go higher than the top, so no net energy will be released either by splitting ^{56}Fe into smaller nuclei or by fusing several ^{56}Fe into a larger nucleus.

10.7 Biological Effects of Radiation

Learning Goal:

To discuss how radiation affects organisms.

Radiation that is energetic enough to knock electrons out of atoms or molecules and form ions is classified as *ionizing radiation*. Alpha particles, beta particles, neutrons, gamma rays, and X-rays all fall into this category. Such radiation can damage or even kill living cells, and is particularly harmful when it affects protein and DNA molecules involved in cell reproduction.

The effects of radiation on living organisms can be categorized as follows.

1. **Somatic effects** are short-term and long-term effects on the recipient of the radiation.
2. **Genetic effects** are defects in the recipient's subsequent offspring.

When discussing the somatic effects of radiation, the SI unit used for equivalent absorbed dose is the sievert (Sv).* This unit takes into consideration the relative ionizing power of each type of radiation and its ability to affect humans. For the dose levels encountered in natural background radiation and diagnostic procedures such as X-rays, the millisievert (mSv) is an appropriate unit. A mammography results in an exposure of about 0.75 mSv, and a dental X-ray about 0.3 mSv.

The average U.S. citizen is exposed to natural and man-made background radiation of about 3 mSv per year. Sources of natural radiation include cosmic rays from outer space. Frequent flying in jetliners or living in high-altitude cities such as Denver, Colorado, provide more exposure to this part of the background radiation. Cosmic rays also form radionuclides such as carbon-14 and potassium-40. Carbon and potassium are essential elements in living organisms, and so these radionuclides become a part of all living organisms and thus another source of natural background radiation.

Other sources of natural background radiation include radionuclides in the rocks and minerals in our environment. One of the decay products of ^{238}U is radon-222. Radon gas and its radioactive daughters can be breathed into the lungs where additional decays emit radiation. About 10,000 of the 130,000 annual lung cancer deaths in the United States are thought to be caused by indoor radon pollution. Exposure to radon varies greatly with location, because some soils and bedrocks are relatively rich in uranium, whereas others contain very little.

Artificial sources of radiation include X-rays and radionuclides used in medical procedures, fallout from nuclear testing, TVs, tobacco smoke, nuclear wastes, and emissions from power plants. Ironically, because fossil fuels contain traces of uranium and thorium and also of their daughters, more radioactivity is released into the atmosphere from power plants burning coal and oil than from nuclear power plants.

Shielding can sometimes be used to decrease radiation exposure. Alpha and beta particles are electrically charged and can be stopped by materi-

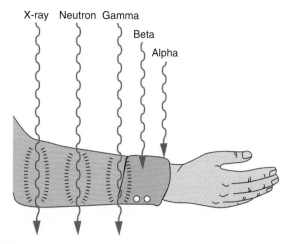

Figure 10.20 Penetration of radiation.
Alpha particles cannot penetrate clothing or skin. Beta particles can barely penetrate them. However, gamma rays, X-rays, and neutrons can penetrate an arm easily.

als such as paper and wood because of interactions with charged particles within these materials. Gamma rays, X-rays, and neutrons are not charged particles and are more difficult to stop (Fig. 10.20). Shielding by thick lead or concrete is required for protection from these kinds of radiation.

Table 10.5 lists the typical short-term somatic effects for an individual exposed to a single dose of radiation to the whole body. One-time exposure to radiation of up to 250 mSv gives no noticeable short-term somatic effects, but the cumulative effects of such exposures are not fully understood. Long-term somatic effects are for the most part an increased likelihood of developing cancer, particularly cancer of the blood or bone. Many early workers with radionuclides developed cancer from small doses of radiation repeated over long periods of time. Some developed cancer as long as 40 years after initial exposure.

A disturbing aspect of the situation is that there seems to be no lower limit, or *threshold*, below which the genetic effects of radiation are negligible. So, any increase in radiation level is taken seriously. The federal government maintains strict safety standards for occupational exposure to radiation, but just what levels of exposure are "safe" remains controversial.

---

*An older and still widely used unit is the rem (*r*oentgen *e*quivalent, *m*an), where 1 rem = 0.01 Sv.

10.6 Nuclear Fusion

40. What is a very hot gas of nuclei and electrons called?
    (a) tokamak   (b) laser
    (c) plasma   (d) maelstrom
41. Name the nuclear process that allows stars to emit such enormous amounts of energy.
42. In discussions of nuclear fusion, for what do the letters D and T stand?
43. What is a plasma? Briefly describe the main approach to forming and confining plasmas for controlled fusion.
44. Briefly, what are the advantages and disadvantages of energy production by fusion versus fission?
45. The equation $E = mc^2$ was developed by what scientist? Identify what each of the letters stands for.
46. What term is applied to a nuclear reaction if the products have more mass than the reactants? Less mass than the reactants?
47. Show by means of a sketch how nuclear energy can be released by both fission and fusion.

10.7 Biological Effects of Radiation

48. Which of the following units is most closely associated with the biological effects of radiation?
    (a) the curie   (b) the sievert
    (c) the becquerel   (d) the cpm
49. What number of mSv received all at once virtually assures death to a human?
50. Distinguish between the somatic effects and the genetic effects of radiation exposure.
51. Name two natural sources of radiation exposure and two artificial sources.

Food for Thought

1. Some short half-life radionuclides are found in nature. Shouldn't these have decayed a long time ago? Explain.
2. The term *China syndrome* is associated with nuclear reactors. What might that mean?
3. For fusion to start in stars, the temperature must be very high. What supplies the energy to raise the temperature?

Exercises

10.1 The Atomic Nucleus

1. For each of the following atoms, tell the number of nucleons, the number of protons, the number of electrons, the number of neutrons, and the chemical identity (symbol of element).
   (a) $^{7}_{3}X$ (b) $^{239}_{93}X$ (c) $^{31}_{15}X$
   (d) $^{34}_{16}X$ (e) $^{90}_{38}X$ (f) $^{235}_{92}X$
   (g) $^{11}_{5}X$ (h) $^{240}_{94}X$
   Answers: (a) 7, 3, 3, 4, Li (b) 239, 93, 93, 146, Np (c) 31, 15, 15, 16, P (d) 34, 16, 16, 18, S

2. On Earth, bromine occurs as 50.69% of ^{79}Br (isotopic mass = 78.918 u) and 49.31% of ^{81}Br (isotopic mass = 80.916 u). Calculate the atomic mass of bromine.
   Answer: 79.90 u

3. On Earth, potassium occurs as 93.26% of ^{39}K (isotopic mass = 38.964 u), 0.012% of ^{40}K (isotopic mass, 39.964 u), and 6.73% of ^{41}K (isotopic mass = 40.962 u). Calculate the atomic mass of potassium.

10.2 Radioactivity

4. Complete the following equations for nuclear decay, and state whether the process is alpha decay, beta decay, or gamma decay.
   (a) $^{47*}_{21}Sc \rightarrow ^{47}_{21}Sc +$ _____
   (b) $^{232}_{90}Th \rightarrow$ _____ $+ ^{4}_{2}He$
   (c) $^{47}_{21}Sc \rightarrow ^{47}_{22}Ti +$ _____
   (d) $^{237}_{93}Np \rightarrow$ _____ $+ ^{0}_{-1}e$
   (e) $^{210}_{84}Po \rightarrow ^{206}_{82}Pb +$ _____
   (f) $^{210*}_{84}Po \rightarrow ^{210}_{84}Po +$ _____
   Answers: (a) γ, gamma (b) $^{228}_{88}$Ra, alpha (c) $^{0}_{-1}$e, beta

5. Write the equation for
   (a) alpha decay of $^{226}_{88}$Ra.
   (b) beta decay of $^{60}_{27}$Co.

   Answer: (a) $^{226}_{88}\text{Ra} \rightarrow {}^{222}_{86}\text{Rn} + {}^{4}_{2}\text{He}$
   (b) $^{60}_{27}\text{Co} \rightarrow {}^{60}_{28}\text{Ni} + {}^{0}_{-1}e$

6. Thorium-229 ($^{229}_{90}$Th) undergoes alpha decay.
   (a) Write the equation.
   (b) The daughter formed in (a) undergoes beta decay. Write the equation.

7. Pick the radionuclide in each set. Explain your choice.
   (a) $^{249}_{98}$Cf $^{12}_{6}$C
   (b) $^{79}_{35}$Br $^{76}_{33}$As
   (c) $^{15}_{8}$O $^{17}_{8}$O
   (d) $^{31}_{15}$P $^{33}_{15}$P

   Answers: (a) $^{249}_{98}$Cf (Z > 83)
   (b) $^{76}_{33}$As (odd-odd)
   (c) $^{15}_{8}$O (fewer n than p)
   (d) $^{33}_{15}$P (odd-even, and A > 1.5 u from atomic mass)

8. Pick the radionuclide in each set. Explain your choice.
   (a) $^{107}_{47}$Ag $^{111}_{47}$Ag
   (b) $^{17}_{9}$F $^{32}_{16}$S
   (c) $^{209}_{83}$Bi $^{226}_{88}$Ra
   (d) $^{24}_{11}$Na $^{14}_{7}$N

10.3 Half-life and Radiometric Dating

9. Technetium-99 (half-life = 6.0 h) is used in medical imaging. How many half-lives would go by in 36 h?

   Answer: six half-lives

10. How many half-lives would have to elapse for a sample of a radionuclide to decrease from an activity of 160 cpm to 5 cpm?

11. A thyroid cancer patient is given a dosage of ^{131}I (half-life = 8.1 d). What fraction of the dosage of ^{131}I will still be in his thyroid after 24.3 days?

    Answer: $^{1}/_{8}$

12. A clinical technician finds the activity of a sodium-24 sample is 480 cpm. What will be the activity of the sample 75 h later if the half-life of sodium-24 is 15 h?

13. Tritium (half-life = 12.3 y) is used to verify the age of expensive brandies. If an old brandy contains only $^{1}/_{16}$ of the tritium present in new brandy, how long ago was it produced?

    Answer: 49 y

14. Living material gives 15.3 cpm/g_C. A skeleton of a prehistoric woman is found to yield 3.83 cpm/g_C from the decay of carbon-14 (half-life = 5730 y). What is the uncorrected carbon date of the skeleton?

15. Use the graph to find the half-life of radionuclide A and radionuclide B. All you need is a sound understanding of the definition of half-life.

    Answer: For A, half-life = 22 d, because half of 40 g is 20 g, and 20 g is reached after 22 d.

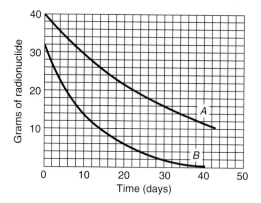

10.4 Nuclear Reactions

16. Complete the following nuclear reaction equations.
    (a) $^{4}_{2}\text{He} + {}^{14}_{7}\text{N} \rightarrow {}^{17}_{8}\text{O} + \underline{}$
    (b) $^{4}_{2}\text{He} + {}^{27}_{13}\text{Al} \rightarrow {}^{30}_{15}\text{P} + \underline{}$
    (c) $\underline{} + {}^{66}_{29}\text{Cu} \rightarrow {}^{67}_{30}\text{Zn} + {}^{1}_{0}\text{n}$
    (d) $^{1}_{0}\text{n} + {}^{235}_{92}\text{U} \rightarrow$
    $^{138}_{54}\text{Xe} + \underline{} + 5\,{}^{1}_{0}\text{n}$
    (e) $^{16}_{8}\text{O} + {}^{20}_{10}\text{Ne} \rightarrow \underline{} + {}^{12}_{6}\text{C}$
    (f) $^{1}_{0}\text{n} + {}^{28}_{14}\text{Si} \rightarrow \underline{} + {}^{1}_{1}\text{H}$
    (g) $\underline{} + {}^{230}_{90}\text{Th} \rightarrow {}^{223}_{87}\text{Fr} + 2\,{}^{4}_{2}\text{He}$
    (h) $^{4}_{2}\text{He} + {}^{65}_{30}\text{Zn} \rightarrow \underline{} + 2\,{}^{1}_{0}\text{n}$

    Answers: (a) $^{1}_{1}$H (b) $^{1}_{0}$n (c) $^{2}_{1}$H (d) $^{93}_{38}$Sr

10.5 Nuclear Fission

17. Complete the following equations for fission.
    (a) $^{240}_{94}\text{Pu} \rightarrow {}^{97}_{38}\text{Sr} + {}^{140}_{56}\text{Ba} + \underline{}$
    (b) $^{252}_{98}\text{Cf} \rightarrow \underline{} + {}^{142}_{55}\text{Cs} + 4\,{}^{1}_{0}\text{n}$

    Answer: (a) $3\,{}^{1}_{0}$n

10.6 Nuclear Fusion

18. One of the fusion reactions that takes place as a star ages is called the triple alpha process.

$$3\,{}^{4}_{2}\text{He} \rightarrow {}^{12}_{6}\text{C}$$

Calculate the mass loss (in u) and the energy produced in MeV each time the reaction takes place. (Isotopic mass of ^{4}He = 4.00260 u; isotopic mass of ^{12}C = 12.00000 u.)

Answer: 0.00780 u, 7.26 MeV

19. Calculate the change in mass (in u) and the change in energy (in MeV) in the reaction shown. Is the reaction exoergic or endoergic?

$${}^{1}_{0}\text{n} + {}^{16}_{8}\text{O} \rightarrow {}^{13}_{6}\text{C} + {}^{4}_{2}\text{He}$$

(1.00867 u)   (15.99491 u)   (13.00335 u)   (4.00260 u)

Answers to Multiple-Choice Questions

1. c   2. d   14. d   15. b   16. c
22. d   23. b   27. c   31. a   40. c
48. b

Solutions to Confidence Questions

10.1 92 protons, 92 electrons, 146 neutrons (238 – 92)

10.2 (0.6000 × 20.00 u) + (0.4000 × 22.00 u) = 12.00 u + 8.800 u = 20.80 u

10.3 ${}^{226}_{88}\text{Ra} \rightarrow {}^{222}_{86}\text{Rn} + {}^{4}_{2}\text{He}$

10.4 ${}^{232}_{90}\text{Th}$ (above Z = 83) and ${}^{40}_{19}\text{K}$ (*odd-odd* nuclide)

10.5 $\dfrac{22 \text{ min}}{11 \text{ min/half-life}}$ = 2 half-lives; $\frac{1}{4}$ of N_o remains.

10.6 If $\frac{1}{4}$ of N_o remains, then two half-lives have passed ($\frac{1}{1} \rightarrow \frac{1}{2} \rightarrow \frac{1}{4}$).
(2 half-lives)(1.25 × 10^9 min/half-life) = 2.50 × 10^9 y

10.7 A specimen whose carbon-14 activity is one-half of that of living matter must have died one half-life ago, or about 5730 y.

10.8 ${}^{2}_{1}\text{H} + {}^{27}_{13}\text{Al} \rightarrow {}^{25}_{12}\text{Mg} + {}^{4}_{2}\text{He}$

10.9 ${}^{236}_{92}\text{U} \rightarrow {}^{90}_{38}\text{Sr} + {}^{144}_{54}\text{Xe} + 2\,{}^{1}_{0}\text{n}$

10.10 (2.0140 u + 3.0161 u) − (4.0026 u + 1.0087 u) = 5.0301 u − 5.0113 u = 0.0188 u of mass lost (0.0188 u)(931 MeV/u) = 17.5 MeV of energy released

Chapter 11

The Chemical Elements

11.1 Classification of Matter
11.2 Names and Symbols of Elements
11.3 Occurrence of the Elements
11.4 The Periodic Table
Highlight: Mendeleev and the Periodic Table
11.5 Naming Compounds
11.6 Groups of Elements

The objects in our environment are composed of *matter*, which may be defined as anything that has mass. The division of physical science called **chemistry** deals with the composition and structure of matter, and the reactions by which substances are changed into other substances. The modern understanding of chemical reactions, based on the electron configuration of atoms, is relatively recent. However, chemistry had its beginnings early in history (see the Chapter 12 Highlight).

Chemistry is separated into five major divisions. *Physical chemistry,* the most fundamental of the five divisions, applies the theories of physics (especially thermodynamics) to the study of chemical reactions. *Analytical chemistry* identifies what substances are present in a material and determines how much of each substance is present. *Organic chemistry* is the study of compounds that contain carbon (see Chapter 14). The study of all other chemical compounds is called *inorganic chemistry.* *Biochemistry,* where chemistry and biology meet, deals with the chemical reactions that occur in living organisms. As you might expect, chemistry's five major divisions overlap, and there are other, smaller divisions such as polymer chemistry, nuclear chemistry, and so forth.

In this chapter we will examine how matter is classified by chemists, discuss elements, develop an understanding of the periodic table, and learn how compounds are named.

11.1 Classification of Matter

Learning Goals:

To explain how chemists classify matter.

To identify three types of solutions.

In Chapter 5 we saw that matter can be classified by its physical phase: solid, liquid, gas, or plasma. However, chemists find the scheme summarized in Fig. 11.1 particularly useful. It first divides matter into pure substances and mixtures.

A **pure substance** is a type of matter in which all samples have fixed composition and identical properties. Pure substances are either elements or compounds. An **element** is a substance in which all the atoms have the same number of protons, or same atomic number (Section 10.1). Therefore, all

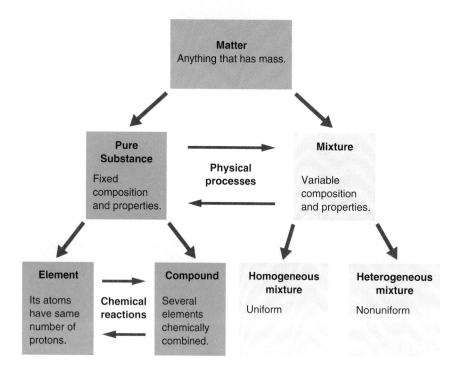

Figure 11.1 A chemical classification of matter. See text for description.

samples of a given element have a fixed composition and identical properties. A **compound** is a substance composed of two or more elements chemically combined in a definite, fixed proportion by mass. All samples of a given compound have identical properties that are usually very different from the properties of the elements of which it is composed. Table 11.1 compares the properties of the compound zinc sulfide (ZnS) with those of zinc (Zn) and sulfur (S), the two elements into which it decomposes in a fixed ratio by mass of 2.04 parts of zinc to 1 part of sulfur.

A compound can be broken into its component elements only by chemical processes, such as the passage of electricity through melted zinc sulfide. The formation of a compound from elements is also a chemical process.

A type of matter composed of varying proportions of two or more substances that are just physically mixed, *not* chemically combined, is called a **mixture.** Different samples of a type of mixture can have variable composition and properties. For example, a mixture of zinc and sulfur could consist of any mass ratio of zinc to sulfur; it would not be restricted to the 2.04 to 1 mass ratio found in the compound zinc sulfide. Mixtures are formed and broken down by physical processes such as dissolving and evaporation. For example, the mixture

Table 11.1 The Properties of a Compound Compared with Those of Its Component Elements

Property	Zinc sulfide	Zinc	Sulfur
Appearance	White powder	Silvery metal	Yellow powder or crystals
Density (g/cm^3)	3.98	7.14	2.07
Melting point (°C)	1700	281	113
Conducts electricity as a solid	No	Yes	No
Conducts electricity as a liquid	Yes	Yes	No
Soluble in carbon disulfide	No	No	Yes

of zinc and sulfur could be separated simply by adding carbon disulfide (CS_2) to dissolve the sulfur, filtering off the zinc, and then evaporating the carbon disulfide to leave a deposit of sulfur crystals.

A *heterogeneous mixture* is one that looks nonuniform. That is, it can be seen that at least two components are present. Examples are a pizza, a bottle of oil-and-vinegar salad dressing, and a pile of zinc and sulfur.

A mixture that looks uniform throughout is called a *homogeneous mixture,* or a **solution.** That is, it looks like it might be just one substance.* Most solutions we encounter are *aqueous solutions;* that is, the solution consists of one or more substances dissolved in water (such as coffee or salt water). However, from this definition a solution need not be a liquid; it may also be a solid or a gas. A metal *alloy* such as brass (a mix of copper and zinc) is an example of a solid solution. Air is an example of a gaseous solution (see Table 19.1). Each appears uniform, but different samples of coffee, salt water, brass, and air can have different compositions. In a solution containing two substances, the substance present in the larger amount is called the *solvent,* and the other substance is called the *solute*.

Aqueous Solutions

Let's take a moment to discuss aqueous solutions (abbreviated *aq*). Water is the solvent. When a solute dissolves in water and is thoroughly stirred, the distribution of its particles (molecules or ions) is the same throughout the solution. If the solution can dissolve more solute at the same temperature, it is an **unsaturated solution.** As more solute is added, the solution becomes more and more concentrated. Finally, when the maximum amount of solute is dissolved in the solvent, we have a **saturated solution.** Usually, some undissolved solute remains on the bottom of the container, and a dynamic (active) equilibrium is reached when solute dissolves and crystallizes at the same rate (Fig. 11.2).

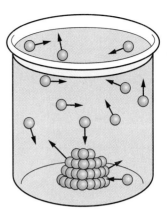

Figure 11.2 A saturated solution.
In a saturated solution containing excess solute, an equilibrium exists between the solute dissolving in the solvent and the solute crystallizing from the solvent.

The *solubility* of a particular solute is the amount of solute that will dissolve in a specified volume or mass of solvent (at a given temperature) to produce a saturated solution. Solubility depends on the temperature of the solution. If the temperature is raised, the solubilities of practically all solids increase.

When aqueous solutions are prepared at high temperatures and then cooled, solubility drops and may reach the saturation point, where excess solute begins crystallizing from the solution. However, if no crystals of the solid (or other nucleation sites, such as dust particles) are already present in the saturated solution, crystallization may not take place if the solution is carefully cooled. The solution will then contain a larger amount of solute than the solubility of the solute dictates. A solution that contains more than the normal maximum amount of dissolved solute at its temperature is a **supersaturated solution.** Such solutions are unstable, and the introduction of a "seed" crystal will cause the excess solute to crystallize (Fig. 11.3). This is the basic principle behind seeding clouds to cause rain. If the air is supersaturated with water vapor, the introduction of certain types of crystals into the clouds greatly increases the probability that the water vapor will form raindrops (see Chapter 19).

If a gas is being dissolved in water, the solubility increases in direct proportion to the pressure. This principle is used in the manufacture of soft drinks. Carbon dioxide (CO_2) is forced into the

---

*Technically, to be a true solution the components must be mixed on the atomic or molecular level, so that even a magnifying glass or microscope would not reveal that the sample was two substances mixed.

Figure 11.3 Supersaturated solution.
When a seed crystal is added to a supersaturated solution of sodium acetate, $NaC_2H_3O_2$, the excess solid quickly crystallizes, forming a saturated solution.

beverage under high pressure. Then the beverage is bottled and tightly capped to maintain pressure on the CO_2. Once the bottle is opened, the pressure inside the bottle is reduced to normal atmospheric pressure, and the CO_2 starts escaping from the liquid, as evidenced by rising bubbles. If the bottle is allowed to remain open for some time, most of the CO_2 escapes, and the drink tastes flat. The solubility of gases in most liquids decreases with increasing temperature. If an unopened soft drink is allowed to warm, the solubility of CO_2 decreases. When the bottle is opened, CO_2 may escape so fast that the beverage shoots out of the bottle.

11.2 Names and Symbols of Elements

Learning Goals:

To trace the development of the concept of element and tell how elements are named and symbolized.

To match the names and symbols of common elements.

The Greek philosophers who lived from about 600 to 200 B.C. were apparently the first people to speculate about what basic substance or substances make up matter. In the fourth century B.C., the Greek philosopher Aristotle developed the idea that all matter on Earth is composed of four "elements": earth, air, fire, and water. He was wrong on all four.

However, this idea of four elements was dominant for almost 2000 years. Then, in 1661, Robert Boyle (Fig. 11.4), an Irish-born chemist, developed a definition of element that took the concept from the realm of speculation and made it subject to laboratory testing. (This was the same era in which Isaac Newton was developing the laws that made physics a modern science.) In his book *The Sceptical Chemist*, Boyle proposed that the designation *element* be applied only to those substances that could not be separated into components by any method.

Boyle championed the need for experimentation in science. It was he who initiated the practice of carefully and completely describing experiments so that anyone might repeat and confirm them. This procedure became universal in science. Without it, progress probably would have continued at a snail's pace.

discovered. The Curies discovered radium and polonium in 1898. The first artificial element, technetium, was produced in 1937 by nuclear bombardment of the element molybdenum with deuterons. About 20 more synthetic elements have followed, giving a present total of 111 known elements. More are expected to be created by nuclear bombardment using particle accelerators (see Chapter 10).

With only a few exceptions, the approximately 90 naturally occurring elements in our environment, either singly or in chemical combination, are the components of all matter. The physical and chemical properties of the various elements affect us constantly. Everything we wear, eat, breathe, and use is made of elements. Even our bodies are composed of elements.

To designate the different elements, we use a symbol notation that was first conceived in the early 1800s by the Swedish chemist Jöns Jakob Berzelius (bur-ZEE-lee-us). He used one or two letters of the Latin name for each element. Thus sodium is designated Na for *natrium,* silver is Ag for *argentum,* and so forth (Table 11.2).

Since Berzelius' time, most elements have been symbolized by the first one or two letters of the English name. Examples are C for carbon, O for oxygen, and Ca for calcium. The first letter of a chemical symbol is always capitalized and the second is lowercase. Because of a conflict over the naming of element 104 (both a Russian and an American team claimed discovery and each

Figure 11.4 Robert Boyle (1627–1691).
He wrote *The Sceptical Chemist* in 1661 and discovered Boyle's law of gases: The volume and pressure of a gas are inversely proportional.

In the earliest of civilizations, 12 substances that later proved to be elements were isolated: gold, silver, lead, copper, tin, iron, carbon, sulfur, antimony, arsenic, bismuth, and mercury. Phosphorus is the first element whose date of discovery is known. It was isolated from urine by Hennig Brand, a German, in 1669, eight years after Boyle's definition of element. By 1751, platinum, cobalt, and nickel had been discovered. The rest of the 1700s saw the discovery of the gaseous elements hydrogen, oxygen, nitrogen, and chlorine, as well as eight more metals and the nonmetal tellurium.

About 1808, Humphry Davy, an English chemist, used electricity from a battery (a recent invention at the time) to break down compounds, thereby discovering six elements, the record for one person. By 1895 a total of 73 elements were known. From 1895 to 1898, the noble gases helium, neon, krypton, and xenon (ZEE-non) were

Table 11.2 Chemical Symbols from Latin Names

Modern Name	Symbol	Latin Name
Antimony	Sb	*Stibium*
Copper	Cu	*Cuprum*
Gold	Au	*Aurum*
Iron	Fe	*Ferrum*
Lead	Pb	*Plumbum*
Mercury	Hg	*Hydrargyrum*
Potassium	K	*Kalium*
Silver	Ag	*Argentum*
Sodium	Na	*Natrium*
Tin	Sn	*Stannum*
Tungsten	W	*Wolfram*

wanted a different name), in 1978 the International Union of Pure and Applied Chemistry (the IUPAC) recommended the name unnilquadium (abbreviated Unq). It has recommended similar names for elements 105 through 109, based on Latin words for the atomic numbers (*nil* for zero, *un* for one, *bi* for two, *tri* for three, *quad* for four, and so on.).

Appendix IX lists the first 103 elements in alphabetical order, and gives their symbols, atomic numbers, and atomic masses. Henceforth in the chemistry chapters, we will assume that you know the names and symbols of the 43 elements highlighted in Appendix IX. (Making and using flashcards is an efficient and effective way to learn them.) Inside the front cover, you will find a periodic table showing the positions of the 111 known elements, as well as their atomic numbers and atomic masses.

11.3 Occurrence of the Elements

Learning Goals:

To explain what elements are most common in our environment.

To define the terms *molecule* and *allotrope*.

About 74% of the mass of Earth's crust is composed of only two elements—oxygen (47%) and silicon (27%). Aluminum and iron, two important metals, are quite abundant also (● Fig. 11.5). Earth's core is thought to be about 85% iron and 15% nickel. Earth's atmosphere close to the surface consists of 78% nitrogen, 21% oxygen, and almost 1% argon, together with traces of other elements and compounds.

Living matter consists primarily of oxygen (65%), carbon (18%), hydrogen (10%), and nitrogen (3%). All the other elements combined make up the remaining 4%.

Analysis of electromagnetic radiation from stars, galaxies, and interstellar clouds of gas and dust (called *nebulae*) indicates that the simplest element of all, hydrogen, accounts for about 75% of the mass of elements in the universe, with about 24% being the next simplest element, helium. All the other elements account for only 1%. Chapter

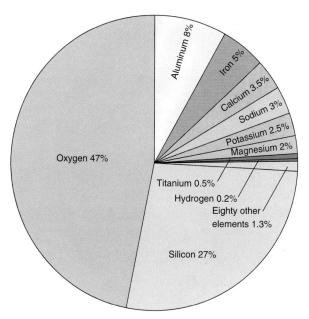

Figure 11.5 Relative abundance (by mass) of elements in Earth's crust.

18 discusses how elements are formed in stars, a process called *nucleosynthesis*.

Molecules

The atoms in a metal such as iron are packed together as shown in ● Fig. 11.6a, with each atom bonding equally to all its nearest neighbors. In other words, the individual unit of a metal is the atom. Thus for iron and other metallic elements, just writing their symbols adequately represents their compositions.

If you could see the particles in a noble gas like xenon, they would be flying around as individual atoms (Fig. 11.6b), so we represent them by the symbol, also. It is now possible to image and even manipulate individual xenon atoms with an instrument known as a scanning tunneling microscope.

However, if you could observe the individual particles in a sample of hydrogen gas, you would see that they are *diatomic*, consisting of particles made up of two atoms of hydrogen combined chemically (Fig. 11.6c). A **molecule** is an electrically neutral particle composed of two or more atoms chemically combined (● Fig. 11.7). If the atoms are all of the same element, it is a molecule

11.3 Occurrence of the Elements

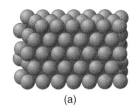

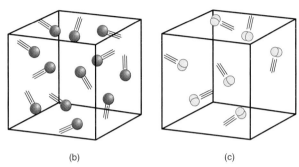

Figure 11.6 Atoms in iron, xenon, and hydrogen.
(a) In iron and other metallic elements, the individual atoms pack together in a repeating pattern, so they are represented by just the symbol (such as Fe). (b) Xenon and the other noble gases exist as single atoms, so they also are represented by just the symbol (such as Xe). (c) Hydrogen gas is composed of diatomic molecules, so the element is written as H_2 in chemical equations—not just H.

Table 11.3 Elements Existing as Diatomic Molecules

Element	Formula
Hydrogen	H_2
Nitrogen	N_2
Oxygen	O_2
Fluorine	F_2
Chlorine	Cl_2
Bromine	Br_2
Iodine	I_2

of an element (for example, H_2 or N_2). If the atoms are of different elements, it is a molecule of a compound (for example, H_2O or NH_3).

Table 11.3 lists the seven common elements that occur as diatomic molecules. Their atoms are too reactive to exist as independent individuals. Note that six of the seven are close together and form a "7" shape at the right of the periodic table, making them easy to recall. When writing chemical equations, as we will do in Chapter 13, the formulas of these seven elements are written in the diatomic form (such as $H_2 + Cl_2 \rightarrow 2\ HCl$).

Allotropes

Two or more forms of the same element that have different bonding structures in the same physical phase are called **allotropes.** Two of the three allotropes of carbon are shown in Fig. 11.8. Diamonds are composed of pure carbon. Figure 11.9a shows that each carbon atom in diamond bonds to four of its neighbors. Picture the carbon atom in the middle of a geometric structure called a regular tetrahedron, a figure having four faces that are identical equilateral triangles. The four bonds of the carbon atom point toward the corners of the tetrahedron and form an angle of 109.5° to

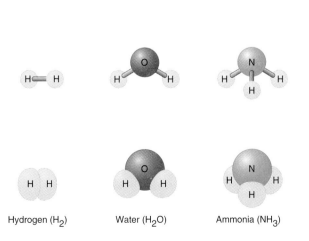

Figure 11.7 Representations of molecules.
Ball-and-stick (top) and space-filling (bottom) models of the element hydrogen and three compounds—water, ammonia, and methane.

Figure 11.8 Graphite and diamond.
Graphite is one component of pencil "lead." The diamond is synthetic. Carbon black (soot), charcoal, and coke are amorphous (seemingly noncrystalline) forms of graphite.

one another. This leads to a three-dimensional network that makes diamond the hardest substance known.

Graphite, the black slippery solid that is a major component of pencil "lead," is also pure carbon. Graphite conducts an electric current, whereas diamond does not. Figure 11.9b shows that the carbon atoms in graphite are bonded in a network of flat hexagons, giving an entirely different structure from that of diamond. The weak forces between each plane of hexagons allow them to slide easily over one another, thus giving graphite its characteristic slipperiness. In fact, graphite is used as a lubricant. Each carbon atom is bonded to three other carbon atoms that lie in the same plane, and thus each forms part of three hexagons. This leaves each carbon atom with one loosely held outer electron, which enables graphite to conduct an electric current.

By applying intense heat and pressure, it is possible to form diamonds from graphite in the laboratory. Conversely, when heated to 1000°C in the absence of air, diamond changes to graphite.

The *fullerenes*, a whole class of ball-like substances (such as C_{32}, C_{60}, C_{70}, and C_{240}), make up the third allotropic form of carbon. The most stable fullerene was first prepared in the laboratory in 1985 by vaporizing graphite with a laser. It is a 60-carbon-atom, hollow, soccerball-shaped molecule named *buckminsterfullerene*, or "buckyball" (Fig. 11.9c). It is named for the American engineer and philosopher Buckminster Fuller who invented the geodesic dome, the architectural principle of which underlies the structure of buckyballs. Scientists succeeded in isolating C_{60} in bulk in 1990, and found that it forms naturally in sooty flames like that of a candle. In various compound forms, C_{60} acts as an insulator, a conductor, a semiconductor, and a superconductor. Technological applications are highly probable.

Oxygen has two allotropes. It usually occurs as a gas of diatomic O_2 molecules, but can also exist as gaseous, very reactive, triatomic O_3 molecules named *ozone*. (Earth's ozone layer is discussed in Section 19.1.) Several other elements also exist in allotropic forms.

11.4 The Periodic Table

Learning Goals:

To trace the origin and describe the composition of the periodic table.

To use the periodic table to correlate and predict properties of elements.

By 1869 a total of 65 elements had been discovered. However, a system for classifying the elements had not been established. As discussed in the chapter Highlight, it remained for the Russian chemist Dmitri Mendeleev (men-duh-LAY-eff) to formulate a satisfactory classification scheme—the periodic table.

The *periodic table* puts the elements, in order of increasing atomic number, into seven horizontal rows called **periods.** As a result, the elements' properties show regular trends, and similar properties occur *periodically;* that is, at definite intervals. Later in this section, we will see some examples of these regular trends. The modern statement of the **periodic law** is:

The properties of elements are periodic functions of their atomic numbers.

The vertical columns in the periodic table are called **groups.** At present there is some confusion about the designation of the groups. In 1986 the IUPAC decreed that the groups be labeled 1 through 18 from left to right, as shown in the periodic table inside the front cover. However, for many years in the United States, the groups have been divided into A and B subgroups, as is also shown in your periodic table. Most chemistry textbooks in the United States still discuss the groups by using the A and B designation, so we will, also.

The structure of the periodic table reflects the elements' electron configurations on the subshell level, where s, p, d, and f subshells can contain a maximum of 2, 6, 10, and 14 electrons, respectively (see Section 9.5). As can be seen in Fig. 11.10, the block composed of Groups 1A and 2A has two columns, and the block composed of Groups 3A through 8A has six columns. The B groups comprise a block having 10 columns, and the lanthanides and actinides each have 14 members.

1. **Representative elements** are those in Groups 1A through 8A, shown in green in Fig. 11.10 and the periodic table inside the front cover. As Fig. 11.10 also shows, four of the representative element groups have special names: *alkali*

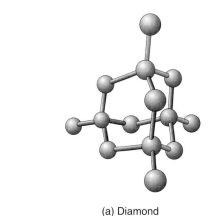

(a) Diamond

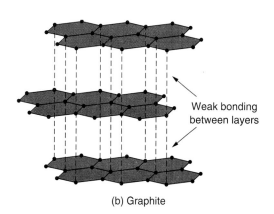

(b) Graphite

(c) Buckyball

Figure 11.9 Models of three allotropes of carbon.
(a) Diamond consists of a network of carbon atoms in which each atom bonds to four others in a tetrahedral fashion. (b) Graphite consists of a network of carbon atoms in which each atom bonds to three others and forms sheets of flat, interlocking hexagons. The sheets are held to each other only by weak electrical forces. (c) Buckminsterfullerene (buckyball), C_{60}, is a soccerball-like arrangement of interlocking hexagons and pentagons formed by carbon atoms.

HIGHLIGHT

Mendeleev and the Periodic Table

The story of the periodic table begins with Johann Dobereiner, a German chemist who in 1829 reported a relationship between the properties of certain elements and their atomic masses. Dobereiner arranged a few elements with similar properties into groups of three, which he called triads. But the arrangement failed when he applied the model to other known elements.

Thirty-five years later, in 1864, English chemist John Newlands observed that when the known elements were arranged according to increasing atomic mass, sometimes every eighth element had similar properties. Newlands called the relationship *the law of octaves*, comparing the elements to the notes in an octave of music. Although his model was successful with some groups of elements, it failed when he attempted to fit in all known elements.

The first detailed and useful periodic table also placed the elements in order of increasing atomic mass. It was published in 1869 by a Russian chemist named Dmitri Mendeleev (Fig. 1). The modern periodic table resembles the horizontal and vertical columns in his original table. (Julius Meyer, a German chemist and physician, published evidence for the periodic classification of elements about this same time, but his work did not contain the details and predictions described by Mendeleev.)

Figure 1.
Dimitri Mendeleev (1834–1907), developer of the first useful periodic table of the chemical elements.

Mendeleev's table was useful because of the importance he placed on the appearance of similar physical and chemical properties at regular intervals (that is, periodically) and because of his prediction of the properties of unknown elements. He even left vacancies in his table for them. When these elements, such as gallium, scandium, and germanium (which Mendeleev called eka-silicon, meaning "similar to silicon"), were discovered later, they had the properties and filled the vacancies Mendeleev had predicted. In Table 1, compare the properties Mendeleev predicted for eka-silicon with the properties of the element germanium (Ge) discovered in 1886, while Mendeleev was still living.

Mendeleev was unable to explain why similar chemical properties occurred at regular intervals with increasing atomic mass. He realized that in a few positions in his table the increasing atomic mass and the properties of the elements did not coincide. For example, on the basis of atomic mass, iodine should come *before* tellurium; but iodine's properties indicate that it should *follow* tellurium. Mendeleev thought that perhaps in these cases the atomic masses were incorrect. However, we now know that the periodic law is a function of the atomic number (and the corresponding electron configuration) of an element, not its atomic mass. Fortunately for Mendeleev, the atomic numbers generally increase as the atomic masses increase.

Mendeleev's periodic table, in modern form, is used by the entire scientific community. Element 101, mendelevium, is named for him.

Table 1 Predicted and Observed Properties of Germanium

Property	Mendeleev's Predictions for Eka-silicon (X)	Observed Properties of Germanium (Ge)
Atomic mass	72	72.6
Color	Dirty gray	Gray-white
Density	5.5 g/cm^3	5.35 g/cm^3
Oxide formula and density	XO$_2$; 4.7 g/cm^3	GeO$_2$; 4.23 g/cm^3
Chloride formula and density	XCl$_4$; 1.9 g/mL	GeCl$_4$; 1.84 g/mL
Boiling point of chloride	under 100°C	84°C

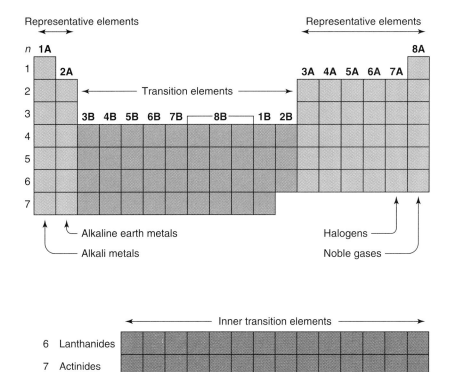

Figure 11.10 Names of specific portions of the periodic table.
The representative elements are shown in green, the transition elements in blue, and the inner transition elements in purple. Note the location of the alkali metals (Group 1A, except H), the alkaline earth metals (Group 2A), the halogens (Group 7A), the noble gases (Group 8A), and the lanthanides and actinides.

metals (Group 1A), *alkaline earth metals* (Group 2A), *halogens* (Group 7A), and *noble gases* (Group 8A).

2. **Transition elements,** shown in blue, are the groups designated by a numeral and the letter B. Such familiar elements as iron, copper, and gold are transition elements.

3. **Inner transition elements,** shown in purple, are generally placed in two rows at the bottom of the periodic table. Each row has its own name. The elements cerium (Ce) through lutetium (Lu) are called the *lanthanide series* or just *lanthanides*, whereas thorium (Th) through lawrencium (Lr) make up the *actinide series,* or *actinides.* (Except for uranium, few of the inner transition elements are well-known. However, the phosphors in color TV and computer screens are primarily compounds of lanthanide series elements.)

For a given element, the number of shells containing electrons is equal to the element's period number, which is shown on the left in Fig. 11.10.

Also, the total number of electrons (and protons) in an atom of a given element is equal to the element's atomic number (Z), which is given above the element's symbol in the periodic table. Thus sodium ($Z = 11$) is in the third period ($n = 3$) and has 11 electrons dispersed in three shells. Iodine ($Z = 53$) is in the fifth period ($n = 5$) and has 53 electrons dispersed in five shells. Figure 11.11 shows diagrams of the distribution of electrons into shells for the elements of Periods 1, 2, and 3.

The outer shell of an atom is known as the **valence shell,** and the electrons in it are called the **valence electrons.** The valence electrons are extremely important because they are the ones that form chemical bonds. Elements with the same number of valence electrons have similar chemical properties. We will limit our discussion of chemical bonding (Chapter 12) to the representative (A group) elements. The number of valence electrons of every element in a given A group is the same as the group number. The only exception is He, which has just two electrons in the outer shell although it is in Group 8A.

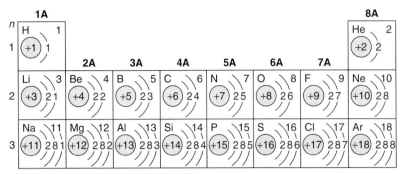

Figure 11.11 Shell distribution of electrons for Periods 1, 2, and 3.
Each atom's nucleus is shown in light blue. The number of shells containing electrons is the same as the period number. Each shell of electrons is shown as a partial circle, with a numeral showing the number of electrons in the shell. The number of valence electrons, those in the outer shell, is the same for all the atoms in the group and is equal to the group number (Exception: He).

Metals and Nonmetals

Let's now look at another method for classifying elements, namely, into *metals* and *nonmetals*. This early but still useful classification originally was done on the basis of certain distinctive properties (Table 11.4). Our modern definition is that a **metal** is an element whose atoms tend to lose valence electrons during chemical reactions. A **nonmetal** is an element whose atoms tend to gain (or share) valence electrons.

As shown in Fig. 11.12, the metallic character of the elements increases as you go down a group and from right to left across a period. As you would expect, the nonmetallic character of the elements increases in the reverse fashion, upward in a group and from left to right across a period. Cesium is the most metallic element, and fluorine is the most nonmetallic. (Francium, which technically might be the most metallic element, is a synthetic element and unavailable in amounts greater than a few atoms.)

Most elements are metals. The actual dividing line between metals and nonmetals cuts through the periodic table like a staircase (Fig. 11.12). The elements boron, silicon, germanium, arsenic, antimony, and tellurium are called *semimetals*, or *metalloids*. They are located next to the staircase line, and display properties of both metals and nonmetals. Several of them are of crucial importance as semiconductors in the electronics industry.

The elements can also be classified according to whether they are solids, liquids, or gases at room temperature and atmospheric pressure. Only two, bromine and mercury, occur as liquids. Eleven occur as gases: hydrogen, nitrogen, oxygen, fluorine, chlorine, and the six noble gases. All the rest are solids, with the vast majority being metallic solids. Figure 11.13 shows several elements.

Table 11.4 Some General Properties of Metals and Nonmetals

Metals	Nonmetals
Good conductors of heat and electricity.	Poor conductors of heat and electricity.
Malleable—can be beaten into thin sheets.	Brittle—if a solid.
Ductile—can be stretched into wire.	Nonductile.
Possess metallic luster.	Do not possess metallic luster.
Solids at room temperature (Exception: Hg).	Solids, liquids, or gases at room temperature.
Usually have 1 to 3 valence electrons.	Usually have 4 to 8 valence electrons.
Lose electrons, forming positive ions.	Gain electrons to form negative ions or share electrons.

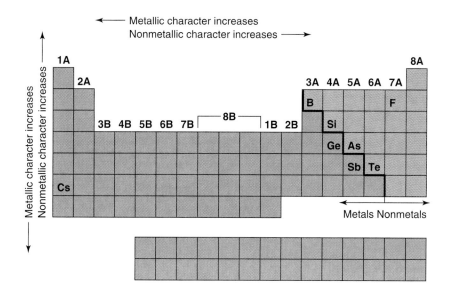

Figure 11.12 Metals and nonmetals in the periodic table.
The elements shown in blue, at the left of the staircase line, are metals. The nonmetals are shown in pink. Fluorine is the most reactive nonmetal, and cesium is the most reactive metal. Next to the staircase line are symbols identifying the six elements that are actually *semimetals* (which have both metallic and nonmetallic properties).

EXAMPLE 11.1 Identifying Some Properties of an Element

Figure 11.14 shows the interesting element gallium, Ga. Find Ga in the periodic table and answer the following:
(a) What are its atomic number (Z) and atomic mass?
(b) How many total electrons are in a Ga atom? How many protons?
(c) Is it a representative, transition, or inner transition element? Metal or nonmetal?
(d) What period is it in, and what is its group number?
(e) How many shells of a Ga atom contain electrons? How many valence electrons in it?

Solution

(a) $Z = 31$, and atomic mass is 69.7 u.
(b) 31 electrons and 31 protons (same as Z).
(c) Representative (its group number has "A" in it); metal (at *left* of staircase line).
(d) Period 4 (it is in the fourth horizontal row down).
(e) Four shells contain electrons (same as the period number). Three valence electrons (same as the group number, for representative elements).

CONFIDENCE QUESTION 11.1

Find the element phosphorus, P, in the periodic table and answer the following:
(a) What are its atomic number and atomic mass?
(b) How many total electrons are in a P atom? How many protons?
(c) Is it a representative, transition, or inner transition element? Metal or nonmetal?
(d) What period is it in, and what is its group number?
(e) How many shells of a P atom contain electrons? How many valence electrons in it?

Other Periodic Characteristics

We have mentioned several periodic characteristics, such as the number of valence electrons and metallic/nonmetallic character. Let's now examine the periodic nature of atomic size and ionization energy. Figure 11.15 illustrates the relative size of the atoms of the representative elements. Diameters of atoms range from about 0.74 Å (0.074 nm) for hydrogen to about 4.7 Å (0.47 nm) for cesium, Cs. (Recall that 1 Å = 10^{-10} m = 0.1 nm.)

Figure 11.15 shows that, generally speaking, the atomic size decreases from left to right in a period. Notice that each Group 1A atom is large with

Figure 11.13 Some metals and nonmetals.
Mercury (front center) is the only element that is a liquid metal. Copper (back center) is one of only two metals with a color other than silver or gray. (Can you name the other one?) Bromine (in the flask) is the only liquid nonmetal. Sulfur (back left) is a brittle, yellow, crystalline nonmetal. Iodine (front left) is a brittle, blue-black, crystalline nonmetal.

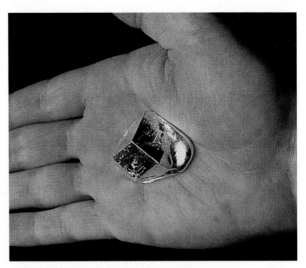

Figure 11.14 The element gallium, Ga.
Gallium's melting point is 30°C, which is above normal room temperature but below the temperature of the skin. Thus a solid cube of gallium will melt in the hand.

respect to atoms of the other elements of that period. This is because its one outer electron is loosely bound to the nucleus. As the charge on the nucleus increases (more protons) without adding an additional shell of electrons, the outer electrons are more tightly bound, thus decreasing the atomic size from left to right across the period. In a given *group*, the atomic size increases from top to bottom. This is logical because each successive element of the group has an additional shell containing electrons.

When an atom gains or loses electrons, it acquires a net electric charge and becomes an *ion*. The amount of energy it takes to remove an electron from an atom is called its **ionization energy**. The periodic nature of the ionization energy is apparent in ● Fig. 11.16. Notice that the Group 1A elements have the lowest ionization energies. Their one valence electron is in an outer shell, so it does not take much energy to remove the electron. Elements located to the right of the Group 1A element in a particular period have additional protons and electrons, but the electrons are added to the same shell. The added protons bind the electrons more and more strongly until the shell is completely filled. The Group 8A element in the period has the highest ionization energy. In general, the ionization energy increases from left to right in a period. The ionization energy decreases from top to bottom in a group because the electron to be removed is shielded from the attractive force of the nucleus by each additional shell of electrons.

11.5 Naming Compounds

Learning Goal:
To name simple, inorganic chemical compounds.

It proves difficult to go further in our discussion of chemistry without understanding the basics of compound *nomenclature* (from the Latin for "name" and "to call"). As you will see, the distinction we have just made between metals and nonmetals is critical in naming compounds.

Elements combine chemically to form compounds. A compound is represented by a *chemical formula*, which is written by putting the elements' symbols adjacent to each other, normally with the more metallic element first. A *subscript* (subscripted numeral) follows each symbol to desig-

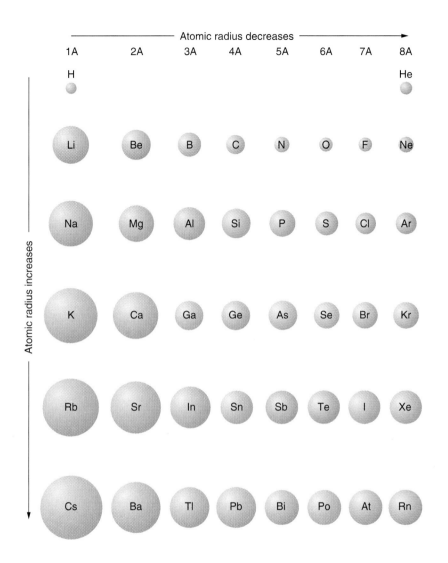

Figure 11.15 The relative sizes of atoms of the representative elements.
In general, atomic size decreases from left to right in a period and increases from top to bottom in a group. The diameter of the smallest atom, H, is about 0.74 Å (0.074 nm), whereas that of the largest atom shown, Cs, is about 4.7 Å (0.47 nm).

nate the number of atoms of the element in the formula. If an element has only one atom in the formula, the subscript "1" is not actually written, just understood. For instance, water is written H_2O, which means that a single molecule of water consists of two atoms of hydrogen and one atom of oxygen.

We often use chemical formulas, such as H_2O, NaCl, and Al_2O_3, to identify specific chemical compounds, but we still need names that unambiguously identify them. At this point, we come to polyatomic ions.

An **ion** is an atom, or chemical combination of atoms, having a net electric charge. An ion that is formed from a single atom (by gain or loss of electrons) is called a *monatomic ion*. An electrically charged combination of atoms is termed a **polyatomic ion.** The names and formulas of the common polyatomic ions you should know are given in Table 11.5.

Compounds with Special Names

Some compounds have such well-established special names that no systematic nomenclature can compete. Their names are just learned individually. Common examples are

H_2O	water
NH_3	ammonia
CH_4	methane
HCl(*aq*)	hydrochloric acid

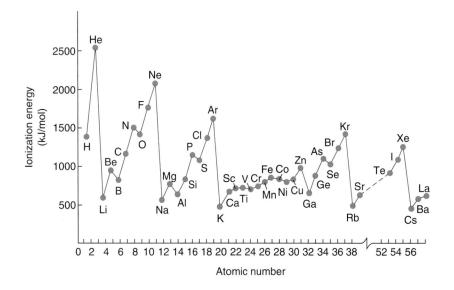

Figure 11.16 Ionization energy versus atomic number.
The ionization energy increases in a generally regular fashion across a period, being at its lowest for the Group 1A element and at its highest for the noble gas. It shows a decrease from top to bottom in a group.

Naming a Compound of a Metal and a Nonmetal

For the present, we will deal with naming compounds of metals that form only one ion, which are mainly those of Groups 1A (ionic charge 1+) and 2A (ionic charge 2+), plus Al, Zn, and Ag (ionic charge of 3+, 2+, and 1+, respectively). (Section 12.7 will discuss the nomenclature modification necessary for metals that form more than one ion.) To name a binary (two-element) compound of a metal combined with a nonmetal, first give the name of the metal, then give the name of the nonmetal with its ending changed to *-ide* (see Table 11.6). Examples are:

NaCl	sodium chloride
K$_2$S	potassium sulfide
Al$_2$O$_3$	aluminum oxide
Ca$_3$N$_2$	calcium nitride
LiH	lithium hydride

Naming Compounds of Two Nonmetals

In a compound of two nonmetals, the less nonmetallic element (the one farther left or farther down in the periodic table) is usually written first in the formula and named first. The second element is named using its *-ide* ending. Generally, two nonmetallic elements form several binary compounds, which are distinguished by using the Greek prefixes *mono-* (one), *di-* (two), *tri-* (three), *tetra-* (four), *penta-* (five), *hexa-* (six), *hepta-*

Table 11.5 Some Common Polyatomic Ions

Name	Formula	Name	Formula
Acetate	C$_2$H$_3$O$_2^-$	Carbonate	CO$_3^{2-}$
Hydrogen carbonate	HCO$_3^-$	Sulfate	SO$_4^{2-}$
Hydroxide	OH$^-$	Phosphate	PO$_4^{3-}$
Nitrate	NO$_3^-$	Ammonium	NH$_4^+$
Permanganate	MnO$_4^-$	Hydronium	H$_3$O$^+$

Note: Positive polyatomic ions end in *-ium*. Many negative polyatomic ions end in *-ate*. An older name for the hydrogen carbonate ion is bicarbonate.

Table 11.6 The *-ide* Nomenclature for Common Nonmetals

Element Name	*-ide* Name
Bromine	Bromide
Chlorine	Chloride
Fluorine	Fluoride
Hydrogen	Hydride
Iodine	Iodide
Nitrogen	Nitride
Oxygen	Oxide
Phosphorus	Phosphide
Sulfur	Sulfide

(seven), and *octa-* (eight). The prefixes designate the number of atoms of the element that occur in the molecule. The prefix *mono-* is always omitted from the name of the first element in the compound, and is usually omitted from the second. Examples are:

N_2O	dinitrogen oxide
NO_2	nitrogen dioxide
N_2O_4	dinitrogen tetroxide*
CO	carbon monoxide
CO_2	carbon dioxide
SO_3	sulfur trioxide
HCl	hydrogen chloride

Naming Compounds Containing Polyatomic Ions

For a compound of a metal combined with a polyatomic ion, simply name the metal and then the polyatomic ion. Examples are:

$ZnSO_4$	zinc sulfate
$NaC_2H_3O_2$	sodium acetate
$Ra(NO_3)_2$	radium nitrate
K_3PO_4	potassium phosphate

When hydrogen is combined with a polyatomic ion, the compound generally is named as an acid. The common examples are:

---

*You might have expected this to be dinitrogen *tetraoxide*, but when a prefix ending in *a* is attached to *oxide*, the *a* is dropped. This convention was adopted just to make pronunciation easier.

HNO_3	nitric acid
$HC_2H_3O_2$	acetic acid
H_2SO_4	sulfuric acid
H_2CO_3	carbonic acid
H_3PO_4	phosphoric acid

The only common positive polyatomic ion present in compounds is the ammonium ion, NH_4^+. (Hydronium ions, H_3O^+, are found only in water solutions.) If the ammonium ion is combined with a nonmetal, change the ending of the nonmetal to *-ide*. If it is combined with a negative polyatomic ion, simply name each ion. Examples are:

$(NH_4)_3P$	ammonium phosphide
$(NH_4)_3PO_4$	ammonium phosphate
NH_4OH	ammonium hydroxide

Figures 11.17 and 11.18 summarize the rules for the cases just discussed. Note the first question to ask yourself: "Is it a binary compound, or does it contain a polyatomic ion?" The naming of the complex compounds found in organic chemistry follows its own set of rules, as we will see in Chapter 14. And by the way, don't be upset if you run into a few compounds that seem to be named in violation of the rules given above. Space limitations prohibit complete coverage of the rules, and even scientists are not always consistent!

EXAMPLE 11.2 Naming Compounds

Name each of these six compounds in the preferred manner: (a) Na_2S, (b) SiO_2, (c) $ZnCO_3$, (d) NH_3, (e) NH_4OH, (f) H_2SO_4.

Solution

(a) Na_2S Sodium sulfide (a binary compound of a metal with a nonmetal, so the metal is named first, then the *-ide* name of the nonmetal is given).

(b) SiO_2 Silicon dioxide (a binary compound of two nonmetals, so the Greek prefix system is preferred).

(c) $ZnCO_3$ Zinc carbonate (a compound of a metal with the carbonate polyatomic ion, so name the metal and then name the polyatomic ion).

(d) NH_3 Ammonia. This common compound is always called by its special name.

276 CHAPTER 11 The Chemical Elements

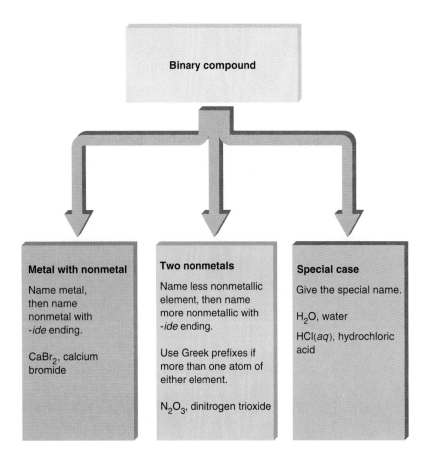

Figure 11.17 Naming binary compounds.
See text for discussion.

(e) NH_4OH Ammonium hydroxide (a compound that contains an ammonium ion with another polyatomic ion, so just name each one).

(f) H_2SO_4 Sulfuric acid (a compound of hydrogen with a polyatomic ion, so it is named as an acid).

CONFIDENCE QUESTION 11.2

Name the compounds N_2O_5 and $CaCl_2$ in the preferred manner.

11.6 Groups of Elements

Learning Goals:

To discuss the properties of four groups of elements.

To identify and tell the uses of some compounds.

Recall that a row of the periodic table is called a period, and a column of the table is termed a group. As we saw in Fig. 11.11, all the elements in a group have the same number of valence electrons. Therefore, if one element in a group reacts with a given substance, the other elements in the group usually react in an analogous manner with that substance. The formulas of the compounds produced will also be similar. In this section we discuss four of these groups of elements: the noble gases (8A), the alkali metals (1A), the halogens (7A), and the alkaline earth metals (2A).

The Noble Gases

The elements of Group 8A are known as the **noble gases.** They are monatomic; that is, they exist as single atoms in nature. The noble gases have the astounding chemical property of almost never forming compounds with other elements, or even

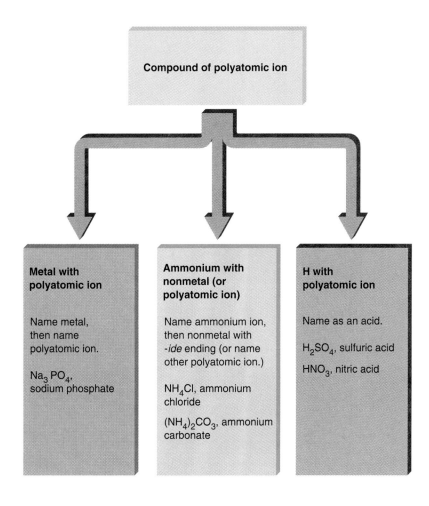

Figure 11.18 Naming compounds of polyatomic ions. See text for discussion.

bonding to themselves. We can conclude that electron configurations with eight electrons in the outer shell (or two if the element, like helium, uses only the first shell) are quite stable. This conclusion is of crucial importance when considering the bonding characteristics of atoms of other elements, as we will see in Chapter 12.

Helium is obtained from natural gas, and radon is a radioactive by-product of radium's decay, but the other noble gases are found only in the air. Recall that argon makes up almost 1% of air. A common use of the noble gases is in "neon" signs, which contain minute amounts of various noble gases or other gases in a sealed glass vacuum tube. When an electric current is passed through the tube, the gases glow. The particular color of the glow depends on the gas present (● Fig. 11.19).

The most striking physical property of the noble gases is their low melting and boiling points. For example, helium has the lowest melting point (m.p.) and boiling point (b.p.) known. It melts at −272 °C and boils at −269 °C, just one degree and four degrees, respectively, above absolute zero. For this reason liquid helium is used often in low-temperature research.

The Alkali Metals

The elements in Group 1A, except for hydrogen, are called the **alkali metals.** (Hydrogen, a gaseous *nonmetal*, is discussed at the end of this section.) Each alkali metal atom has only one valence electron. The atom tends to lose this outer electron quite easily, so the alkali metals react readily with other elements and are said to be *active* metals.

Sodium and potassium are abundant in Earth's crust, but lithium, rubidium, and cesium are rare. The alkali metals are all soft (● Fig.

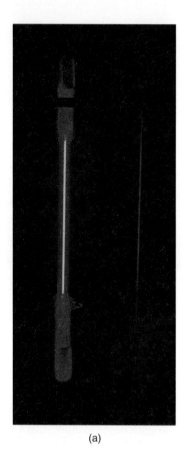
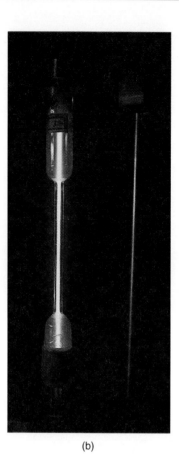

(a)   (b)

Figure 11.19 "Neon" lights.
When the atoms of a small amount of gas are subjected to an electric current, the gas glows, or fluoresces. The color of a particular light depends on the identity of the gas whose atoms are being excited. (a) Neon glows a beautiful orange-red; (b) argon emits a blue light.

11.20) and are so reactive with oxygen and moisture that they are stored under oil. The most common compound containing an alkali metal is table salt (sodium chloride, NaCl). Other alkali metal compounds are potash (potassium carbonate, K_2CO_3) and washing soda (sodium carbonate, Na_2CO_3). Two other common compounds of sodium are lye (sodium hydroxide, NaOH) and baking soda (sodium hydrogen carbonate, or sodium bicarbonate, $NaHCO_3$).

From seeing these formulas of sodium compounds, and by knowing that all the elements in a group produce similar compounds, we can predict the formulas of compounds of the other alkali metals. Thus we expect potassium chloride to have the formula KCl, rather than, say, K_2Cl or KCl_2. Similarly, lithium carbonate should be, and is, Li_2CO_3.

Some of the physical properties of the alkali metals are given in Table 11.7. Note the regular change of each property from top to bottom. This periodic trend can be used to estimate the melting and boiling points (at present unknown) of the rare, radioactive element francium. We expect all the properties of the alkali metals to follow a trend, so we would predict from Table 11.7 that francium's melting and boiling points should be about 25°C and 660°C, respectively.

The Halogens

The Group 7A elements are called the **halogens**. Halogen atoms have seven electrons in their valence shell and have a strong tendency to gain one more electron. They are active nonmetals and are present in nature only in the form of their compounds. They all consist of diatomic molecules (F_2, Cl_2, Br_2, and I_2).

Fluorine, a pale-yellow, poisonous gas, is the most reactive of all the elements. It corrodes even platinum, a metal that withstands most other chemicals. A stream of fluorine gas causes wood, rubber, and even water to burst into flame. Fluorine was responsible for the deaths of several very able chemists before it was finally isolated.

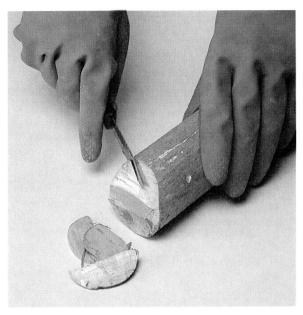

Figure 11.20 Cutting sodium.
The metallic element is so soft that it can be sliced with a knife. It is so reactive that an explosion or chemical burn can occur, so it must be handled with caution.

Chlorine, a pale-green, poisonous gas, was used as a chemical weapon in World War I. It is used as a purifying agent in swimming pools and in public water supplies in a ratio of about 1 part chlorine per 1 million parts water (1 ppm).

Bromine is a reddish-brown, foul-smelling, poisonous liquid used as a disinfectant, whereas iodine is a blue-black, brittle solid (see Fig. 11.13). The "iodine" found in a medicine cabinet is not the pure solid element; it is a *tincture*—iodine dissolved in alcohol. Most table salt is now iodized; that is, it contains 0.02% sodium iodide (NaI), which is added to supplement the human diet because an iodine deficiency causes thyroid problems. Astatine (At) is radioactive and so rare that it is seldom included in a discussion of the halogens.

Some formulas and names for other halogen compounds are:

$AlCl_3$ aluminum chloride

NH_4F ammonium fluoride

$CaBr_2$ calcium bromide

By analogy with these formulas, we can predict the correct formulas of many other halogen compounds; for example, AlF_3, NH_4Cl, and CaI_2.

The Alkaline Earth Metals

The elements in Group 2A are called the **alkaline earth metals.** Their atoms contain two valence electrons. They are active metals, but are not as chemically reactive as the alkali metals. They have higher melting points and are generally harder and stronger than their Group 1A neighbors.

Beryllium occurs in the mineral beryl, some varieties of which make beautiful gemstones such as aquamarines and emeralds (Fig. 11.21). Magnesium is used in safety flares because of its ability to burn with an intensely bright flame (see Fig. 13.5). Both magnesium and beryllium are used to make lightweight metal alloys. The common over-the-counter medicine called milk of magnesia is magnesium hydroxide, $Mg(OH)_2$.

Calcium is used by vertebrates in the formation of bones and teeth, both of which are mainly calcium phosphate. Calcium carbonate ($CaCO_3$) is a mineral present in nature in many forms. The shells of marine creatures, such as clams, are made of calcium carbonate, as are coral reefs. Limestone, a sedimentary rock often formed from seashells and thus from calcium carbonate, is used

Table 11.7 Physical Properties of the Alkali Metals

Element	Symbol	Atomic Number	Atomic Mass (u)	Melting Point (°C)	Boiling Point (°C)	Atomic Radius (Å)
Lithium	Li	3	6.94	181	1342	1.23
Sodium	Na	11	23.0	98	883	1.57
Potassium	K	19	39.1	63	760	2.03
Rubidium	Rb	37	85.5	39	686	2.16
Cesium	Cs	55	132.9	28	669	2.35
Francium	Fr	87	(223.0)	—	—	—

Figure 11.21 Emerald, the green gem variety of beryl.
Beryllium, an alkaline earth metal, comes primarily from the mineral named beryl, $Be_3Al_2Si_6O_{18}$.

Hydrogen

Although hydrogen is commonly listed in Group 1A, it is not considered an alkali metal. Hydrogen is a nonmetal that usually reacts like an alkali metal, forming HCl, H_2S, and so forth (similar to NaCl, Na_2S, etc.). Yet sometimes hydrogen reacts like a halogen, forming NaH (sodium hydride), CaH_2, and so forth (compare with NaCl and $CaCl_2$). It is the least dense of the elements and so was used in early dirigibles (Fig. 11.23). At room temperature hydrogen is a colorless, odorless gas and consists of diatomic molecules (H_2). It does not liquefy at normal atmospheric pressure until cooled to $-253°C$ (20 K).

Although hydrogen burns in air, it is actually a safer fuel than gasoline. It will probably be the main fuel used in transportation in the future. Since it burns to form only water, it is apparent how its use would help the pollution problem.

to build roads and to neutralize acid soils. Tums tablets are mainly calcium carbonate. Also composed of calcium carbonate are the impressive formations of stalactites and stalagmites in underground caverns.

Strontium compounds produce the red colors in fireworks, and barium compounds produce the green. When radioactive strontium-90 (from fallout) is ingested, it goes into the bone marrow because of its similarity to calcium. If enough strontium-90 is ingested, it can destroy the bone marrow or cause cancer. Fortunately, radioactive fallout is minimal now. A barium compound encountered all too frequently is barium sulfate ($BaSO_4$), the white material in the thick solutions a patient ingests prior to having X-rays taken of his or her gastrointestinal tract (Fig. 11.22).

Radium is an intensely radioactive element that glows in the dark. Its radioactivity is a deterrent to practical uses, although a few decades ago watch dials were painted with radium chloride ($RaCl_2$) so they could be read in darkness.

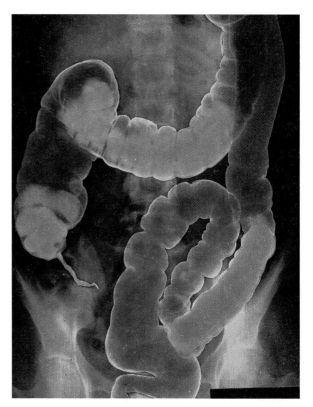

Figure 11.22 A common use of barium sulfate.
The contrast in this radiograph of a patient's large intestine is produced by the introduction of $BaSO_4$ which absorbs X-rays much better than do body tissue and bone.

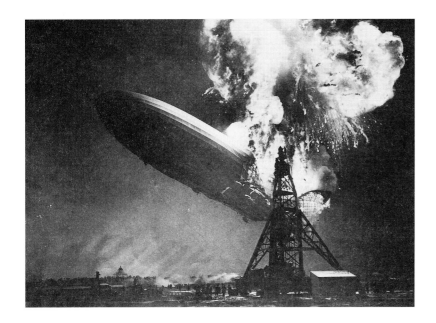

Figure 11.23 The *Hindenburg* disaster, Lakehurst, New Jersey, May 6, 1937.
The 800-foot-long airship burst into flame upon landing after a trip from Germany. The disaster killed thirty-five people and is responsible for the *"Hindenburg* syndrome"—reluctance to use hydrogen as a fuel. For size comparison, the Goodyear blimps are about 200 feet long. Airships nowadays use nonflammable helium to give them buoyancy.

IMPORTANT TERMS

These important terms are for review. After reading and studying the chapter, you should be able to define and/or explain each of them.

chemistry	allotropes	metal
pure substance	periods	nonmetal
element	periodic law	ionization energy
compound	groups	ion
mixture	representative elements	polyatomic ion
solution	transition elements	noble gases
unsaturated solution	inner transition elements	alkali metals
saturated solution	valence shell	halogens
supersaturated solution	valence electrons	alkaline earth metals
molecule		

QUESTIONS

1. Name the five major divisions of chemistry and describe the focus of each.

11.1 Classification of Matter

2. Which of the following is another name for any homogeneous mixture?
   (a) compound (b) colloid
   (c) solution (d) alloy

3. What characteristics distinguish pure substances from mixtures? Elements from compounds?
4. What type of process is involved in going from mixtures to pure substances, and vice versa? From elements to compounds, and vice versa?
5. Give one example each of a solid, liquid, and gaseous solution.

6. When sugar is dissolved in iced tea, what is the solvent and what is the solute? When would you become aware that the solution was saturated?
7. Explain what is meant by a supersaturated solution, and tell how one might be formed.
8. What is the basic principle behind seeding clouds to cause rain?
9. Why does an opened soft drink go flat after a period of time?

11.2 Names and Symbols of Elements

10. Which of the following scientists in 1661 defined *element* in a manner that made it subject to laboratory testing?
    (a) Boyle (b) Davy
    (c) Berzelius (d) Mendeleev
11. Which of the following is the symbol for arsenic?
    (a) Ar (b) AS (c) As (d) aR
12. Who were the first people to speculate about the basic composition of matter? Approximately when in history was this?
13. What view did the Greek philosopher Aristotle hold concerning the composition of matter?
14. For what does Humphry Davy hold a record, and what invention did he use?
15. How many elements are known at present? About how many occur naturally in our environment? Name the first artificial element.
16. Who first conceived the symbol notation used today for elements? Why are the symbols for some elements less easy to learn?

11.3 Occurrence of the Elements

17. Which one of these elements normally exists as a gas of diatomic molecules?
    (a) iodine (b) argon
    (c) sulfur (d) chlorine
18. Graphite is a(n) _____ of carbon.
    (a) isotope (b) allotrope
    (c) isomer (d) compound
19. Name the elements that predominate in (a) Earth's crust (two), (b) Earth's core (two), (c) living matter (four), (d) the air (three), and (e) the universe as a whole (two). In each case, list the most predominant element first.

11.4 The Periodic Table (and Highlight)

20. Consider the element calcium. Which statement is *false*?
    (a) It is in Period 4.
    (b) A Ca atom has electrons in 4 shells.
    (c) It is in Group 2A.
    (d) A Ca atom has 20 valence electrons.
21. Which one of the following elements has the greatest ionization energy?
    (a) Li (b) F (c) Rb (d) I
22. Name the scientist who receives the major credit for the development of the periodic table. In what year was his work published?
23. At first, the periodic table placed the elements in order of increasing atomic mass. What property is now used instead of atomic mass, and why?
24. State the periodic law and briefly explain what it means.
25. What formal term is applied (a) to the horizontal rows of the periodic table, and (b) to the vertical columns?
26. Why are chemists so interested in the number of valence electrons in atoms?
27. Inspection of the periodic table shows four major groupings of columns: 2 columns on the left, 6 on the right, 10 in the middle, and 14 below. What causes these groupings?
28. Name the two series of elements into which the inner transition elements are divided.
29. How do metals differ from nonmetals in regard to (a) number of valence electrons, (b) conductivity of heat and electricity, and (c) phase?
30. How does metallic character change (a) from left to right across a period and (b) down a group? Name the most metallic element and the most nonmetallic element.
31. About six elements are intermediate in metallic and nonmetallic properties. What are these called and where, in general, are they located in the periodic table?
32. List the 2 elements that are liquids and the 11 that are gases at room temperature and atmospheric pressure.
33. How does the size of atoms change (a) from left to right across a period and (b) down a group?

11.5 Naming Compounds

34. Which of the following is the preferred name for Na_2SO_4?
    (a) sodium sulfide (b) disodium sulfide
    (c) disodium sulfate (d) sodium sulfate
35. Distinguish between (a) an atom and a molecule, (b) an atom and an ion, and (c) a molecule and a polyatomic ion.

36. Identify two groups of metals, and three other common metals, that form only one ion in their compounds. What is the ionic charge for each?
37. When naming binary compounds, why must you be able to tell whether an element is a metal or a nonmetal?

11.6 Groups of Elements

38. Which of the following elements will be most like F in its chemical properties?
    (a) Cl   (b) Ne   (c) O   (d) H
39. Briefly, why do elements in a given group have similar chemical properties?
40. Name one unusual chemical property and one unusual physical property of the noble gases.
41. Name the four common halogens and tell the normal phase of each. Which one is the most reactive of all elements?
42. Explain why sodium iodide is added to table salt.
43. Name the major compound (a) in bones and (b) in the shells of shellfish.
44. What is the *Hindenburg* syndrome?

Food for Thought

1. Why do some periodic tables list hydrogen at the top of both Groups 1A and 7A?
2. A search is on for the element with atomic number 114. Which element should it most resemble?
3. Homogenized milk is composed of microscopic globules of fat suspended in a watery medium. Is homogenized milk a true solution (homogeneous mixture)? Explain.

Exercises

11.1 Classification of Matter

1. Classify each of the following materials as an element, compound, heterogeneous mixture, or homogeneous mixture: (a) pencil lead, (b) water, (c) diamond, (d) soil, (e) a fried egg, (f) ozone, (g) air, (h) brass, and (i) cane sugar ($C_{12}H_{22}O_{11}$).
   Answer: (a) homogeneous mixture (b) compound (c) element (d) heterogeneous mixture

11.2 Names and Symbols of Elements

2. Give the symbol for each element: (a) sulfur, (b) sodium, (c) aluminum, (d) iron, (e) radon, and (f) barium.
3. Give the name for each element: (a) N, (b) K, (c) Zn, (d) Be, (e) Au, and (f) Ar.

11.4 The Periodic Table

4. Use the periodic table to find the atomic mass, atomic number, number of protons, and number of electrons for an atom of (a) lithium, (b) gold, (c) neon, and (d) lead.
   Answer: (a) 6.94 u, 3, 3, 3; (b) 197.0 u, 79, 79, 79

5. Refer to the periodic table and give the period and group number of each of the following elements: (a) magnesium, (b) zinc, (c) tin, and (d) helium.
   Answer: (a) 3, 2A (b) 4, 2B

6. Classify each of these elements as representative, transition, or inner transition. State whether each is a metal or a nonmetal, and give its normal phase: (a) krypton, (b) iron, (c) iodine, (d) strontium, (e) uranium, (f) hydrogen, (g) mercury, (h) sodium, and (i) lutetium (symbol: Lu).
   Answer: (a) representative, nonmetal, gas (b) transition, metal, solid

7. From the list of elements in Exercise 6, identify the alkali metal, the alkaline earth metal, the halogen, the noble gas, the actinide, and the lanthanide.
   Answer: sodium is the alkali metal

8. Tell the total number of electrons, the number of valence electrons, and the number of shells containing electrons for (a) a silicon atom, (b) a sulfur atom, (c) a calcium atom, and (d) an arsenic atom.
   Answer: (a) 14, 4, 3 (b) 16, 6, 3

9. Arrange in order of increasing *nonmetallic* character (a) the Period 4 elements Ge, Ca, Mn, Se, and Cu, and (b) the Group 6A elements Se, Po, O, and S.
   Answer: (a) Ca, Mn, Cu, Ge, Se (b) Po, Se, S, O

10. Arrange in order of increasing *metallic* character (a) the Period 3 elements P, Cl, Na, Ar, and Al, and (b) the Group 2A elements Sr, Ba, Be, and Mg.

11. Arrange in order of increasing ionization energy (a) the Period 5 elements Sn, Sr, Xe, Rb, and I, and (b) the Group 8A elements Ar, He, Kr, and Ne.
    Answer: (a) Rb, Sr, Sn, I, Xe (b) Kr, Ar, Ne, He

12. Arrange in order of increasing atomic size (a) the Period 2 elements F, C, Ne, and Be, and (b) the Group 4A elements Si, Sn, Pb, and Ge.

11.5 Naming Compounds

13. Give the preferred names for (a) $CaBr_2$, (b) PBr_5, (c) $ZnSO_4$, (d) KOH, (e) $AgMnO_4$, (f) IF_7, (g) $(NH_4)_3PO_4$, (h) Na_3P, (i) CCl_4, and (j) N_2S_5.
    Answer: (a) calcium bromide
    (b) phosphorus pentabromide
    (c) zinc sulfate (d) potassium hydroxide
    (e) silver permanganate (f) iodine heptafluoride
    (g) ammonium phosphate (h) sodium phosphide
    (i) carbon tetrachloride (j) dinitrogen pentasulfide

14. Give the preferred names for (a) N_2O, (b) $RaCl_2$, (c) $Al_2(CO_3)_3$, (d) NH_4OH, (e) Li_2S, (f) SO_3, (g) Ba_3N_2, (h) $Ba(NO_3)_2$, (i) SiF_4, and (j) S_2Cl_2.

15. Name each of the following compounds. (*Hint:* The names end in *acid*.) (a) H_2SO_4, (b) HNO_3, (c) HCl(*aq*), (d) H_3PO_4, (e) $HC_2H_3O_2$, (f) H_2CO_3
    Answer: (a) sulfuric acid
    (b) nitric acid (c) hydrochloric acid

11.6 Groups of Elements

16. Sodium forms the compounds Na_2S, Na_3N, and $NaHCO_3$. What are the formulas for lithium sulfide, lithium nitride, and lithium hydrogen carbonate?
    Answer: Li_2S, Li_3N, $LiHCO_3$

17. Magnesium forms the compounds $Mg(NO_3)_2$, $MgCl_2$, and $Mg_3(PO_4)_2$. What are the formulas for barium nitrate, barium chloride, and barium phosphate?

Answers to Multiple-Choice Questions

2. c   10. a   11. c   17. d   18. b
20. d   21. b   34. d   38. a

Solutions to Confidence Questions

11.1 (a) $Z = 15$; atomic mass is 31.0 u
   (b) 15 electrons; 15 protons
   (c) representative; nonmetal
   (d) Period 3; Group 5A
   (e) 3 shells; 5 valence electrons

11.2 N_2O_5 is a binary compound of two nonmetals, and since each has more than one atom showing in the formula, Greek prefixes are needed for both. The answer is *dinitrogen pentoxide*. (Note the dropping of the *a* from the prefix *penta*-). $CaCl_2$ is a binary compound of a metal and a nonmetal. Name the metal, then use the *-ide* ending for the nonmetal. The answer is *calcium chloride*.

Chapter 12
Chemical Bonding

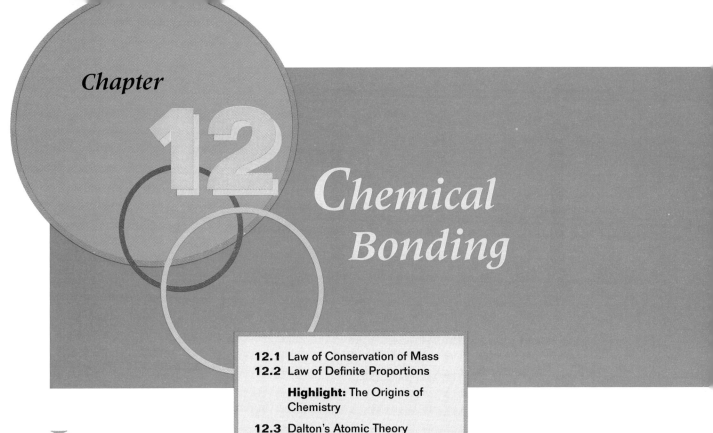

- 12.1 Law of Conservation of Mass
- 12.2 Law of Definite Proportions

Highlight: The Origins of Chemistry

- 12.3 Dalton's Atomic Theory
- 12.4 Ionic Bonding
- 12.5 Covalent Bonding
- 12.6 Metallic Bonding and Hydrogen Bonding
- 12.7 The Stock System of Nomenclature

In this chapter we focus on chemical bonding and its role in compound formation. Even a cursory knowledge of chemistry shows that virtually everything in nature depends on chemical bonds. For example, the proteins, carbohydrates, fats, and nucleic acids that make up living matter are complex molecules held together by chemical bonds. Simpler molecules such as the ozone, carbon dioxide, and water in the air absorb radiant energy, thus keeping Earth at a liveable temperature. Various molecules and ions bond together to form the compounds that comprise the minerals and rocks of Earth. Potassium ions in our heart cells help maintain the proper contractions, and in the extracellular fluids they help control nerve transmissions to muscles. Metallic bonding holds together the atoms in the metals that play such an important role in our civilization; and were it not for the hydrogen bonding that causes one water molecule to attract four others, water would not be a liquid and life would not exist on Earth.

Chemists often use powerful computers and the techniques of quantum mechanics when investigating chemical bonding, which results from the electromagnetic forces among the various electrons and nuclei of the atoms involved. Although much remains to be understood, they are often able to synthesize new compounds having a particular desired property.

We begin our study of chemical bonding with a discussion of two basic laws that describe mass relationships in compounds and helped lead John Dalton to the atomic theory.

12.1 Law of Conservation of Mass

Learning Goal:
To state and use the law of conservation of mass.

If the total mass involved in a chemical reaction is precisely measured before and after the reaction takes place, the most sensitive balances cannot

286 CHAPTER 12 Chemical Bonding

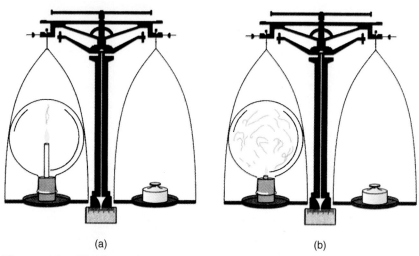

Figure 12.1 The law of conservation of mass.
When a candle is burned in an airtight container of oxygen, there is no detectable change in mass, as illustrated by the balance's pointer being in the same place (a) before the reaction and (b) after the reaction.

detect any change (● Fig. 12.1). This generalization is known as the **law of conservation of mass:**

> **No detectable change in the total mass occurs during a chemical reaction.**

As discussed in the chapter Highlight, this law was discovered in 1774 by the father of chemistry, Antoine Lavoisier (lah-vwah-zee-AY). Now, two centuries later, Lavoisier's law of conservation of mass is still valid and useful.

> **EXAMPLE 12.1 Using the Law of Conservation of Mass**

In a chemical reaction involving carbon and oxygen, laboratory measurements showed that the complete burning in oxygen of 4.09 g of carbon gave 15.00 g of carbon dioxide as the only product.

$$\text{Carbon} + \text{oxygen} \rightarrow \text{carbon dioxide}$$

$$4.09 \text{ g} + \quad ? \quad \rightarrow \quad 15.00 \text{ g}$$

How many grams of oxygen gas must have reacted?

Solution

The total mass of substances before and after reaction must be equal. Because carbon dioxide was the only product and weighed 15.00 g, the total mass of carbon and oxygen that combined to produce it must have been 15.00 g. Since carbon's mass was 4.09 g, the oxygen that reacted with it must have contributed 15.00 g − 4.09 g, or 10.91 g.

CONFIDENCE QUESTION 12.1
If 111.1 g of calcium chloride are formed when 40.1 g of calcium are reacted with chlorine, what mass of chlorine combined with the calcium?

12.2 Law of Definite Proportions

> **Learning Goals:**
> To calculate the formula masses of compounds.
> To state and use the law of definite proportions.

Formula Mass

Recall from Section 10.2 that the atomic masses of elements are based on a scale that assigns the ^{12}C atom the value of exactly 12 u. The atomic masses for most elements have been determined to several decimal places, but for convenience we will round off these values to the nearest 0.1 u, as shown in the periodic table on the inside front cover. For example, we will consider the atomic masses of hydrogen, oxygen, chlorine, and calcium to be 1.0 u, 16.0 u, 35.5 u, and 40.1 u, respectively. This is the atomic mass (abbreviated AM) of the element; that is, the average mass assigned to one atom of the naturally occurring element.

The **formula mass** (abbreviated FM) of a compound or element is the sum of the atomic masses given in the formula of the substance. The formula mass of $CaCO_3$ is thus 40.1 u + 12.0 u + (3 × 16.0 u) = 100.1 u. As other examples, the formula mass of O_2 is 16.0 u + 16.0 u = 32.0 u, and that of methane (swamp gas, CH_4) is 12.0 + (4 × 1.0 u) = 16.0 u. If the formula of an element under consideration is given by just the element's *symbol* (for example Fe or Xe), then the formula mass is just the atomic mass.*

EXAMPLE 12.2 Calculating the Formula Mass of a Compound

Calculate the formula mass of lead chromate, $PbCrO_4$, the bright yellow compound used in paint for the yellow lines on streets.

Solution

Look up the atomic masses of Pb, Cr, and O in the periodic table. The formula shows one atom of Pb (207.2 u), one atom of Cr (52.0 u), and four atoms of O (16.0 u), so FM = 207.2 u + 52.0 u + (4 × 16.0 u) = 323.2 u.

---

*Some terminology in chemistry has changed in the past few years. For example, *atomic weight* is being replaced by *atomic mass*, and *molecular weight* and *formula weight* are being replaced by *formula mass*. You may still hear or read these older terms because changes like these take a while to catch on.

CONFIDENCE QUESTION 12.2

Calculate the formula mass of hydrogen sulfide, H_2S, the gas that gives rotten eggs their offensive odor.

Law of Definite Proportions

In 1799, the French chemist Joseph Proust ("proost") discovered the **law of definite proportions:**

> **Different samples of a pure compound always contain the same elements in the same proportion by mass.**

For example,

9 g H_2O is composed of 8 g oxygen and 1 g hydrogen.

18 g H_2O is composed of 16 g oxygen and 2 g hydrogen.

36 g H_2O is composed of 32 g oxygen and 4 g hydrogen.

In each case the ratio by mass of oxygen to hydrogen is 8 to 1.

Because the elements in a compound are there in a particular proportion or ratio by mass, they are also present in a certain percentage by mass. The general equation for calculating the percentage of any component in a total (such as the percentage of sophomores in a class of students) is

$$\% \text{ component} = \frac{\text{amount component}}{\text{total amount}} \times 100\%$$

Of course, if one of the other two quantities is the unknown, the equation can be rearranged and solved for that unknown.

The percentage by mass of an element X in a compound can be calculated from the compound's formula:

$$\% X \text{ by mass} = \frac{(\text{atoms of } X \text{ in formula}) \times (\text{AM}_X)}{\text{FM}_{cpd}} \times 100\% \quad (12.1)$$

where AM_X is the atomic mass of element X, and FM_{cpd} is the formula mass of the compound.

HIGHLIGHT

The Origins of Chemistry

When people tamed fire they, in a sense, began using chemistry. Moving forward in history, Egyptian hieroglyphs from 3400 B.C. show wine presses. Making wine, of course, required a chemical fermentation process. By about 2000 B.C. the Egyptians and Mesopotamians possessed the knowledge required to produce and work the metals gold, silver, lead, iron, copper, and tin, as well as bronze, an alloy of copper and tin. Other examples of ancient chemistry in Egypt, China, and elsewhere are the preparation of dyes, glass, pottery, and embalming fluids. However, these developments were achieved by trial and error without benefit of guidance from any valid theory of matter.

The Greeks were the first to attempt an explanation of how chemical changes occur. Chapter 11 mentioned the efforts of Aristotle and other Greeks to understand elements. The Greeks also speculated on whether matter is continuous (infinitely divisible into smaller pieces) or discontinuous (ultimately, an indivisible particle would be reached). Aristotle promoted the continuous theory, but Democritus (about 480–370 B.C.) thought matter was composed of ultimately indivisible (Greek, *atomos*) particles. The main shortcoming of the ancient Greeks was their reliance on logic to such a degree that they deemed it totally unnecessary to conduct systematic experimentation.

As the Greek culture faded, Roman civilization ascended. Although the Romans were skilled in political, military, and, to some extent, technological affairs, they did little to advance scientific knowledge. In Europe, the Roman civilization was succeeded by the Dark Ages, when learning was at a low ebb. Arabic cultures carried the torch of learning, developing algebra, the concept of zero, and alchemy.

Alchemy was a pseudoscience whose main goals were to change common metals into gold and to find an "elixir of life" that could restore an aging human body to its youthful vigor. Alchemy was a mixture of mysticism and experimentation, often done in secrecy with few records being kept. It flourished from approximately A.D. 500 to 1600. The Arabs learned of it in the seventh century when they overran Egypt, and they carried it with them during their other conquests. By the fourteenth century, alchemy was practiced throughout the civilized world. Although many alchemists were charlatans, the finest practitioners were dedicated experimentalists. They never reached their goals, but they contributed to the real science of chemistry by discovering several important compounds and elements, and by developing new laboratory equipment and techniques.

Modern science differs from the approach of the ancient Greeks by its reliance not only on logic but also on the systematic gathering of facts by observation and experimentation. Modern chemistry is distinguished from alchemy by having reasonable objectives and avoiding mysticism, superstition, and secrecy.

Figure 1
Antoine Lavoisier (1743–1794), known as the father of chemistry, and his wife Marie-Anne.

The foundations of modern chemistry were laid in the sixteenth and seventeenth centuries. Georg Bauer (1494–1555), a German, developed a systematic method for extracting metals from ores. Paracelsus (1493–1541), a Swiss physician, insisted that the main goal of alchemy should be to prepare medicines with which to treat diseases. In Chapter 11 we noted Robert Boyle's contribution to the concept of element in 1661. In addition, it was Boyle who first performed truly quantitative physical experiments, finding the inverse relationship between the pressure and volume of a gas.

In the seventeenth and eighteenth centuries the phenomenon of combustion was studied intensely. Georg Stahl (1660–1734), a German, explained combustion by theorizing that a substance called *phlogiston* was emitted into the air when a material burned. If the material was in a closed container, it presumably stopped burning because the air around it became saturated with phlogiston. Oxygen gas, discovered in 1774 by the Englishman Joseph Priestley, vigorously supported combustion and was, in fact, first called "dephlogisticated air." The phlogiston theory was discarded thanks to the quantitative experiments of the wealthy French nobleman, Antoine Lavoisier, in that same year.

Lavoisier performed the first quantitative chemical experiments to explain combustion and to settle the question of whether mass was gained, lost, or not changed during a chemical reaction. He commissioned the construction of the most accurate balances ever built (up to that time), and this instrument revolutionized chemistry almost as much as the telescope did astronomy. Unlike some others who had tried to answer the same question, Lavoisier understood the nature of gases and took care to do his experiments in closed containers so that no substances could enter or leave.

After many experiments, he was able to reason inductively from his specific findings on individual reactions to the general case: *No detectable change in mass occurs during a chemical reaction* (the law of conservation of mass). During the course of these investigations, Lavoisier established conclusively that when things burn, they are not losing phlogiston to the air but gaining oxygen from it. Sometimes the burned materials gained mass by forming solid oxides and sometimes they lost mass by forming gaseous oxides. (The former would have required phlogiston to have a negative mass.) Several chemists before Lavoisier had devoted themselves to measurement, but it was the successes of Lavoisier that caused its importance to be recognized by chemists generally. He made chemistry a modern science, just as Galileo and Newton had done for physics more than a century earlier.

In addition to introducing quantitative methods into chemistry, discovering the role of oxygen in combustion, and finding the law of conservation of mass, Lavoisier established the principles for naming chemicals, and in 1789 wrote the first modern chemistry textbook. Because of his many important contributions, Lavoisier is justly referred to as the father of chemistry.

Lavoisier invested in a private tax-collection firm and married the daughter of one of its executives (Fig. 1). Because of this, and because he had made an enemy of a would-be scientist named Marat, who became a leader of the French Revolution, Lavoisier was guillotined in 1794 during the last months of the Revolution. When he objected to the arresting officer that he was a scientist, not a "tax-farmer," the officer replied that "the republic has no need of scientists." Isaac Asimov reports that within two years of Lavoisier's death, the regretful French were unveiling busts of him.*

Hard on the heels of Lavoisier's work came Joseph Proust's discovery of the law of definite proportions in 1799. In 1803 John Dalton resurrected the idea of the atom, proposed by Democritus 2200 years before. Unlike Democritus, Dalton had sound experimental support for the existence of atoms. With the aid of the atomic theory and the commitment to quantitative experimentation, chemistry began to advance in great strides.

---

*Isaac Asimov, *Asimov's Biographical Encyclopedia of Science & Technology*, second revised edition, Doubleday & Company, Inc., New York, 1982. This highly recommended book provides an entertaining and educational look at the lives and achievements of 1510 great scientists from ancient times to the present.

Figure 12.2 Dry ice.
Solid carbon dioxide has a temperature of about −78°C. It is so cold that it must never touch the bare skin. Like mothballs, it sublimes (passes directly from the solid phase to the gaseous phase). The white fog is water vapor condensing from the air in contact with the cold CO_2.

EXAMPLE 12.3 Finding the Percentage by Mass from a Compound's Formula

"Dry ice" is solid carbon dioxide (■ Fig. 12.2). Find the percentage by mass of carbon and oxygen in CO_2.

Solution

The periodic table shows the atomic masses of C and O as 12.0 u and 16.0 u, respectively. Therefore, the formula mass of CO_2 is 12.0 u + 16.0 u + 16.0 u = 44.0 u. Using Eq. 12.1:

$$\% \text{ by mass of C} = \frac{1 \times AM_C}{FM_{CO_2}} \times 100\%$$

$$= \frac{1 \times 12.0 \text{ u}}{44.0 \text{ u}} \times 100\% = 27.3\%$$

Because oxygen is the only other component, the percentage of O and the percentage of C must add up to 100.0%. Thus the percentage of O by mass must be 100.0% − 27.3%, or 72.7% O. (*Note:* When finding these percentages on a calculator, do *not* hit the % key after punching in 100.)

CONFIDENCE QUESTION 12.3

Find the percentage by mass of aluminum and oxygen in aluminum oxide, Al_2O_3, the major compound in rubies.

When a compound is broken down, its elements are found in a definite proportion by mass. Conversely, when the same compound is made from its elements, the elements will combine in that same proportion by mass. Often the elements that are combined to form a compound are not mixed in the correct proportion. In this case one of the elements, called the **limiting reactant**, will be used completely. The other, the **excess reactant**, will be only partially used.

In ■ Fig. 12.3a, 10.00 g of Cu wire reacts completely with 5.06 g S to form 15.06 g of CuS. None of either reactant is left over. In Fig. 12.3b, 10.00 g of Cu reacts with 7.06 g S, but this is not the proper ratio. In accordance with the law of definite proportions, only 5.06 g of S can combine with 10.00 g of Cu. Therefore, 15.06 g of CuS is again formed, and the other 2.00 g of sulfur is in excess. (Note that the law of conservation of mass is satisfied, since the total mass before and after is 17.06 g.) In Fig. 12.3c, it is the copper that is the excess reactant, and both the law of conservation of mass and the law of definite proportions again hold.

12.3 Dalton's Atomic Theory

Learning Goal:

To state the postulates of Dalton's atomic theory.

In 1803, John Dalton (■ Fig. 12.4) proposed the following postulates to explain the laws of conservation of mass and definite proportions.

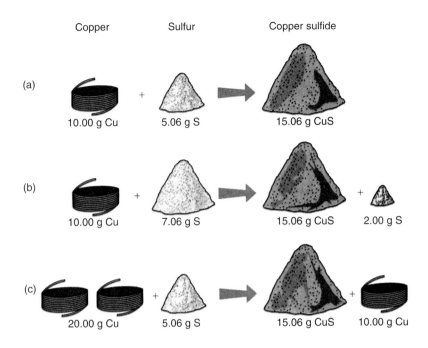

Figure 12.3 The law of definite proportions.
When Cu and S react to form a specific compound, the law of definite proportions states that they always must react in the same ratio. If the ratio in which they are mixed is different, part of one reactant will be left over.

1. Each element is composed of small indivisible particles called atoms, which are identical for that element but are different (particularly in their masses and chemical properties) from atoms of other elements.
2. Chemical combination is the bonding of a definite number of atoms of each of the combining elements to make one molecule of the formed compound.
3. No atoms are gained, lost, or changed in identity during a chemical reaction.

A little thought shows us that if a chemical reaction is just a rearrangement of atoms, then the law of conservation of mass is explained.

These postulates also explain the law of definite proportions. If different samples of a particular compound are made up of various numbers of the same basic molecule, it is easily seen that the mass ratio of the elements will be the same in every sample of that compound. For example, each molecule of H_2O is composed of one atom of oxygen (AM 16.0 u) and two atoms of hydrogen (AM 1.0 u, total mass of hydrogen = 2 × 1.0 u = 2.0 u). Thus each molecule of H_2O is composed of 16.0 parts oxygen by mass and 2.0 parts hydrogen by mass, or a ratio of 8.0 to 1.0. Because any pure sample of a

Figure 12.4 John Dalton (1766–1844).
Born the son of a poor weaver in the village of Eaglesfield, England, Dalton was a child prodigy who opened his own school at the age of 12 and became a professor at the University of Manchester in his mid-twenties. Dalton was a Quaker and continued to live a very simple life even after he became famous for his atomic theory.

molecular compound is simply a very large collection of identical molecules, the proportion by mass of any element in a molecular compound will be the proportion by mass that it has in an individual molecule of that compound.

Since his postulates explained these two laws, it is no wonder that Dalton believed so strongly that atoms exist and behave as he stated. In addition, Dalton saw a way to test his postulates—a necessary step in the scientific method. If atoms existed and formed molecules of two compounds in which the number of atoms of one element were the same (as for carbon in CO and CO_2) and the number of atoms of the other element were different (as for oxygen in CO and CO_2), then the *mass ratio* of the second element in the two compounds would have to be a small whole number ratio (in this case, 1 to 2), thus reflecting the small whole number *atom ratio* of the second element (in this case 1 oxygen atom to 2 oxygen atoms) in the two molecules.

Experiments by other scientists verified Dalton's predicted law (called the *law of multiple proportions*) and showed that his ideas about atoms were substantially correct. As with all scientific theories, modification has occurred as new evidence became available. Still, Dalton's atomic theory is the cornerstone of chemistry. Whenever an explanation of a chemical occurrence is sought, our thoughts immediately turn to the concept of atoms.

12.4 Ionic Bonding

Learning Goals:

To describe ionic bonding.

To write formulas of ionic compounds.

To describe the characteristics of ionic compounds.

We saw in Chapter 11 that elements in the same group have the same number of valence (outer) electrons and form compounds with similar formulas; for example, the chlorides of Group 1A: LiCl, NaCl, KCl, RbCl, and CsCl. Because of this behavior, we conclude that these valence electrons are the ones involved in compound formation.

Recall that the Group 8A elements, the noble gases, are unique in that, except for several compounds of Xe and of Kr, they do not bond chemically with atoms of other elements. Also, all noble gases are monatomic; their atoms do not bond to one another to form molecules. We conclude that their electron configurations (8 electrons in the outer shell, except 2 for He) are uniquely stable.

The formation of the vast majority of compounds is explained by combining these two conclusions into the **octet rule:**

> **In forming compounds, atoms tend to gain, lose, or share valence electrons to achieve electron configurations of the noble gases. That is, they tend to get eight electrons (an octet) in the outer shell (except for H, which tends to get two electrons in the outer shell, like the configuration of He).**

Individual atoms can achieve a noble gas electron configuration in two ways: by transferring electrons or by sharing electrons. Bonding by transfer of electrons is discussed in this section, and bonding by sharing of electrons is treated in Section 12.5.

In the transfer of electrons, one or more atoms lose their valence electrons, and another one or more atoms gain these same electrons to achieve noble gas electron configurations. Compounds formed by this electron transfer process are called **ionic compounds.** This name is used because the loss or gain of electrons destroys the electrical neutrality of the atom and leads to the net positive or negative electric charge that is characteristic of an ion.

In Chapter 11 we saw that atoms of metals generally have low ionization energies and thus tend to lose electrons and form positive ions. On the other hand, atoms of nonmetals tend to gain electrons to form negative ions. A nonmetal atom that needs only one or two electrons to fill its outer shell can easily acquire the electrons from atoms that have low ionization energy—that is, atoms with only one or two electrons in the outer shell, such as those of Groups 1A and 2A. Atoms of some elements gain or lose as many as three electrons.

To illustrate, let us consider how common table salt, NaCl, is formed. When chlorine, a green

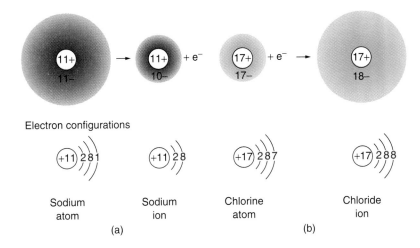

Figure 12.5 The formation of (a) a sodium ion and (b) a chloride ion.
After the transfer of one electron from the Na atom to the Cl atom, both ions have the electron configurations of a noble gas (eight electrons in the outer shell).

gas composed of Cl_2 molecules, is united with the metal sodium, an energetic chemical reaction forms the ionic compound sodium chloride. Figure 12.5a illustrates the loss of the valence electron from a neutral sodium atom to form a sodium ion with a 1+ charge. The neutral atom has the same number of protons (11) as electrons (11). Thus the net electric charge on the atom is zero. After the loss of an electron, the number of protons (11) is one more than the number of electrons (10), leaving a net charge on the sodium ion of 11 − 10, or 1+. Similarly, Fig. 12.5b illustrates the gain of an electron by a neutral chlorine atom (17 p, 17 e) to form a chloride ion (17 p, 18 e) with a charge of 17 − 18, or 1−. The net electric charge on an ion is the number of protons minus the number of electrons.

As the Na^+ and Cl^- ions are formed, they are bonded by the attractive electric force between the positive and negative ions. Figure 12.6a shows that actual crystals of sodium chloride have a cubic structure. Figure 12.6b shows a model of sodium chloride that illustrates how the sodium ions and chloride ions arrange themselves in an orderly manner to form a cubic crystal. (A *crystal* is a solid whose external symmetry reflects an orderly, geometric, internal arrangement of atoms, molecules, or ions.) Note that there are no neutral units of fixed size; it is impossible to associate any one Na^+ with a particular Cl^-. Therefore, it is considered inappropriate to refer to a "molecule" of sodium chloride or of any other ionic compound. Instead, we generally refer to one sodium ion and one chloride ion as being a *formula unit* of sodium chloride, the smallest combination of ions that gives the formula of the compound (see Fig. 12.6b). Similarly, the formula $CaCl_2$ means that the compound calcium chloride has one calcium ion (Ca^{2+}) for every two chloride ions (2 Cl^-). The three ions are a formula unit of calcium chloride.

The basic ideas of compound formation that we have been discussing are due mainly to the work of Gilbert Newton Lewis in 1916 (Fig. 12.7). Lewis electron-dot symbols are helpful in discussing chemical bonding. In a **Lewis symbol,** the nucleus and inner electrons of an atom are represented by the element's symbol, and the valence electrons are shown as dots (Table 12.1). The dots are arranged in four groups of one or two dots around the symbol. It makes no difference on what side of the symbol various electron-dots are put, except that they are left unpaired to the extent possible.

Lewis structures use Lewis symbols to show valence electrons in molecules and ions of compounds. In a Lewis structure, a shared electron pair is indicated by two dots halfway between the atoms or, more often, by a dash connecting the atoms. Unshared pairs of valence electrons (called *lone pairs*, those not used in bonding) are shown as belonging to the individual atom or ion. Lewis structures are two-dimensional representations, and so do not show the actual three-dimensional nature of molecules.

The formation of sodium chloride can be illustrated using Lewis symbols and structures. The

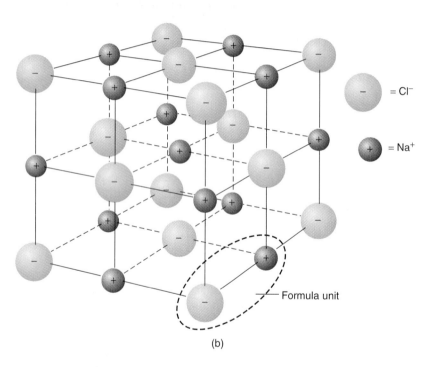

Figure 12.6 Sodium chloride (NaCl).
(a) Crystals of the mineral *halite*, natural sodium chloride. (b) Schematic diagram of the sodium chloride crystal. The oval of dashes indicates what is meant by a formula unit of NaCl.

sodium and chlorine atoms of Fig. 12.5 are represented in the Lewis symbols as

$$\text{Na} \cdot \quad \cdot \ddot{\underset{..}{\text{Cl}}}:$$

The sodium and chloride ions of Fig. 12.5 are represented in the Lewis structure of salt as

$$\text{Na}^+ \quad :\ddot{\underset{..}{\text{Cl}}}:^-$$

where the + and − indicate the same charge as that on a proton or an electron, respectively. If the charge of the ion is greater than 1+ or 1−, it is represented by a numeral followed by the appropriate sign, as shown in the Lewis structure of magnesium oxide (one magnesium ion and one oxide ion).

$$\text{Mg}^{2+} \quad :\ddot{\underset{..}{\text{O}}}:^{2-}$$

The charge on most ions of the representative elements can be determined easily. First, metals form positive ions, or **cations** (CAT-eye-ons), whereas nonmetals form negative ions, or **anions** (AN-eye-ons). Second, the positive charge on the metal's ion will be equal to the atom's number of valence electrons (its group number), whereas the

Figure 12.7 Gilbert Newton Lewis (1875–1946). Lewis, one of America's greatest chemists, revolutionized chemistry with his 1916 paper proposing that atoms can be bonded by sharing pairs of electrons. He developed the electron-dot symbols and structures that bear his name, and made outstanding contributions to acid-base theory and chemical thermodynamics.

negative charge on the nonmetal's ion will be the atom's number of valence electrons (its group number) minus 8. In accordance with the octet rule, this gives the metal or nonmetal ion eight electrons in the outer shell; that is, an electron configuration isoelectronic with a noble gas. (Atoms or ions that have the same electron configuration are termed *isoelectronic*.)

For example, aluminum, a metal in Group 3A, has three valence electrons. When the atom loses three electrons, an aluminum ion with a charge of 3+ is formed. The Al^{3+} is isoelectronic with the noble gas neon (Ne). Sulfur, a nonmetal in Group 6A, has six valence electrons. When forming an ion, a sulfur atom gains two additional electrons. The sulfide ion (S^{2-}) thus has a negative charge of 6 −8, or 2−, and is isoelectronic with the noble gas argon (Ar).

Table 12.2 gives the generalized Lewis symbol for elements of each representative element group and the characteristic ionic charge for each group. (For a specific element, the X would be replaced with the element's symbol.) The metals of Groups 1A, 2A, and 3A tend to lose their valence electrons and form positive ions, whereas the nonmetals of Groups 5A, 6A, and 7A form negative ions as they gain enough additional valence electrons to get eight in the outer shell. Group 8A elements already have eight valence electrons, and thus form no ions. Table 12.2 indicates that the situation in Group 4A is complicated, because it is hard to either lose or gain four electrons.*

Table 12.1 Lewis Symbols for the First Three Periods of Representative Elements

1A	2A	3A	4A	5A	6A	7A	8A
H·							He:
Li·	·Be·	·B·	·C·	·N·	:Ö·	:F̈·	:N̈e:
Na·	·Mg·	·Al·	·Si·	·P̈·	:S̈·	:C̈l·	:Är:

*For simplicity, Sn and Pb are often treated as if their atoms not only form ions with a 2+ charge but also lose four electrons and form ions with a 4+ charge. However, the bonding in the latter case is actually covalent and not ionic.

Table 12.2 Characteristic Ionic Charge for the Representative Elements

1A	2A	3A	4A	5A	6A	7A	8A
$X\cdot$	$\cdot X \cdot$	$\cdot \dot{X} \cdot$	$\cdot \dot{\underset{\cdot}{X}} \cdot$	$\cdot \ddot{\underset{\cdot}{X}} \cdot$	$:\ddot{\underset{\cdot}{X}}\cdot$	$:\ddot{\underset{\cdot\cdot}{X}}\cdot$	$:\ddot{\underset{\cdot\cdot}{X}}:$
1+	2+	3+ (Also 1+ for Tl; B forms no ions)	2+ and 4+ (C and Si form no ions)	3− (3+ for Sb and Bi)	2−	1−	0

In the formation of simple ionic compounds, one element loses its valence electrons and the other element gains them, resulting in ions. The ions are then held together in the crystal lattice by the electrical attractions among them—the **ionic bonds.** The formation of NaCl can be represented as

$$Na\cdot + \cdot\ddot{\underset{\cdot\cdot}{Cl}}: \rightarrow Na^+ \;\; :\ddot{\underset{\cdot\cdot}{Cl}}:^-$$

In this example the sodium atom lost one electron and the chlorine atom gained one. Thus we see that for every sodium ion there is one chloride ion. In every ionic compound, the total charge in the formula adds up to zero and the compound exhibits electrical neutrality. Thus, in the case of NaCl, the ratio of Na^+ to Cl^- must be 1 to 1 so that the compound will exhibit net electrical neutrality.

Another example of an ionic compound is calcium oxide, CaO. Its formation is represented as

$$\cdot Ca \cdot + \cdot \ddot{\underset{\cdot\cdot}{O}}: \rightarrow Ca^{2+} \;\; :\ddot{\underset{\cdot\cdot}{O}}:^{2-}$$

In the formation of this compound, the calcium atom has lost two electrons and the oxygen atom has gained two. Thus to get a net electrically neutral compound, we again need a 1 to 1 ratio of ions of each element.

Now consider what happens when calcium and chlorine react. A calcium atom has two electrons to lose, but each atom of chlorine can gain only one. We must have two atoms of chlorine to accept both of the electrons of the calcium atom and give a net electrically neutral compound. We have

$$\cdot Ca\cdot + \cdot\ddot{\underset{\cdot\cdot}{Cl}}: + \cdot\ddot{\underset{\cdot\cdot}{Cl}}: \rightarrow Ca^{2+} \;\; :\ddot{\underset{\cdot\cdot}{Cl}}:^- \;\; :\ddot{\underset{\cdot\cdot}{Cl}}:^-$$

Thus the formula for calcium chloride is $CaCl_2$. All of the ions in the compound have noble gas configurations, and the total charge is zero.

We now can see how formulas of ionic compounds arise. The numbers of atoms of the various elements involved in the compound are determined by the requirements that the total electrical charge be zero and that all the atoms have noble gas electron configurations.

For ionic compounds to be electrically neutral, each formula unit must have an equal number of positive and negative charges. Thus, to write correct formulas for ionic compounds, the anions and cations must be shown in the smallest whole number ratio that will equal zero charge. The principle, which works for both monatomic and polyatomic ions, can be mastered by studying Table 12.3, and it is critical for understanding later topics that you do so.

The only other information required to become adept at writing the formulas of ionic compounds is knowledge of the charges on common polyatomic ions (refer back to Table 11.5) and on ions of the representative elements (follow the pattern in Table 12.2).

> **EXAMPLE 12.4 Writing Formulas for Ionic Compounds**
>
> Write the formulas for (a) sodium sulfide and (b) calcium phosphate, the major component of bones.
>
> **Solution**
>
> (a) Na is in Group 1A and so has an ionic charge of 1+. Sulfur is in Group 6A and thus its ions have a charge of 2−. Neutrality can be achieved with two Na^+ and one S^{2-}, so the correct formula is Na_2S.
>
> (b) Calcium is in Group 2A and so has an ionic charge of 1+. The phosphate ion is PO_4^{3-} (Table 11.5).

Table 12.3 Formulas of Ionic Compounds

General Cation Symbol	General Anion Symbol	Cation to Anion Ratio for Neutrality	General Compound Formula	Specific Example of Compound
M^+	X^-	1 to 1	MX	NaF
M^{2+}	X^{2-}	1 to 1	MX	MgO
M^{3+}	X^{3-}	1 to 1	MX	AlN
M^+	X^{2-}	2 to 1	M_2X	Na_2O
M^{2+}	X^-	1 to 2	MX_2	MgF_2
M^+	X^{3-}	3 to 1	M_3X	Na_3N
M^{3+}	X^-	1 to 3	MX_3	AlF_3
M^{2+}	X^{3-}	3 to 2	M_3X_2	Ca_3N_2
M^{3+}	X^{2-}	2 to 3	M_2X_3	Al_2O_3

Neutrality can be achieved with three Ca^{2+} and two PO_4^{3-}, so the correct formula is $Ca_3(PO_4)_2$.

CONFIDENCE QUESTION 12.4

Write the formulas for the ionic compounds formed by combining Al^{3+} ions with (a) Cl^-, (b) O^{2-}, and (c) P^{3-}.

Very strong forces of attraction exist among oppositely charged ions, so ionic compounds are always crystalline solids with high melting and boiling points. Another important property of ionic compounds is their behavior when an electric current is passed through them.

If a lightbulb is connected to a battery by two wires, the bulb glows. Electrons flow from the negative terminal of the battery, through the lightbulb, and back to the positive terminal. If one wire is cut, the electrons cannot flow and the lightbulb will not glow. If the ends of the cut wire are inserted into a solid ionic compound, the bulb does not light, because the ions are held in place and cannot move. However, if the cut wires are inserted into a *melted* ionic compound, such as molten (liquid) NaCl, the bulb lights (Fig. 12.8).

Ionic compounds in the liquid phase conduct an electric current because the ions are now free to move and carry charge from one wire to the other. This is the crucial test of whether or not a compound is ionic: When melted, does it conduct an electric current? If so, it is ionic. Figure 12.9 illustrates how the conduction takes place. The reason that negative ions are called anions is because they flow toward the positive wire, which is termed the *anode*. Similarly, cations are positive ions and flow toward the negative wire, the *cathode*.

Many ionic compounds dissolve in water, especially if their ions have only single charges and thus are not attracted to one another quite as much as if the charges were greater. The mechanism by which this dissolving process occurs is discussed in Section 12.6. Because in a water solution the ions are free to move, such solutions also conduct

Figure 12.8 Ionic compounds conduct electricity when melted (but not when solid).
When the compound melts, the ions are free to move, the circuit is completed, and the bulb lights.

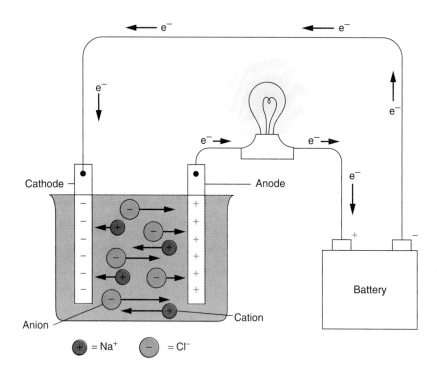

Figure 12.9 How melted NaCl conducts electricity. The negative ions (anions) move toward the anode, and the positive ions (cations) move toward the cathode. Chemical reactions of the ions at the electrodes allow electrons to flow in the wire, as indicated by the lighted bulb.

an electric current. (Table 12.5 on page 302 summarizes the properties of ionic compounds and compares them to those of covalent compounds.)

12.5 Covalent Bonding

Learning Goals:

To write molecular formulas and Lewis structures for covalent compounds.

To describe the characteristics of covalent compounds.

In the formation of ionic compounds electrons are transferred, producing ions. In the formation of **covalent compounds,** pairs of electrons are shared, producing molecules. When a pair of electrons is shared by two atoms, we say that a **covalent bond** exists between the atoms. If the covalent bond is between identical atoms, a molecule of an element is formed. As examples, let's examine the H_2 and Cl_2 molecules.

Hydrogen gas, H_2, is the simplest example of a molecule with a single covalent bond. When two hydrogen atoms are brought together, the result is attraction between opposite electrons and protons, and repulsion between the two electrons and between the two protons. Once the nuclei are 0.74 Å apart, attraction is at a maximum and repulsion at a minimum (Fig. 12.10). The two electrons no longer orbit individual nuclei. They are shared equally by both nuclei, and it is this sharing that holds the atoms together.

In Lewis symbols, the separated hydrogen atoms are written as

$$H\cdot \quad \cdot H$$

The Lewis structure of the H_2 molecule is written as

$$H:H$$

The two dots between the hydrogen atoms indicate that these electrons are being shared. The single covalent bond (one shared pair of electrons) can also be represented by a dash.

$$H-H$$

Not all atoms will share electrons. For instance, a helium atom, with two electrons already filling its valence shell, does not form a stable molecule with another helium atom or with any other

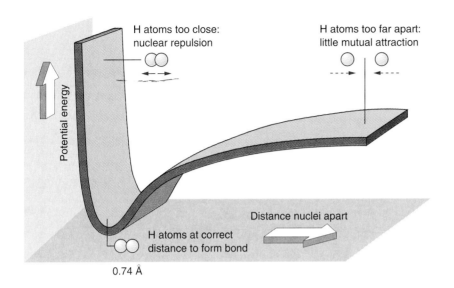

Figure 12.10 Bonding in the H₂ molecule.
A stable chemical bond is formed between two hydrogen atoms when they are an optimum distance apart (0.74 Å). If they are any farther apart, there is little attraction; if they are any closer together, the nuclei repel too strongly.

atom. Two hydrogen atoms attract each other, but two helium atoms repel. The molecule H_2 is stable but He_2 is not. Sometimes it takes complicated quantum mechanical calculations to determine which molecules are stable and which are not. However, the basic octet rule serves as an excellent guide. Stable covalent molecules are formed when the atoms share electrons in such a way as to give all atoms a noble gas configuration.

Now consider the bonding in Cl_2. In Lewis symbols, two chlorine atoms are shown as

$$:\ddot{C}l\cdot \quad \cdot\ddot{C}l:$$

Each chlorine atom requires one electron to complete its octet. If each shares its unpaired electron with the other, a Cl_2 molecule is formed.

$$:\ddot{C}l\!:\!\ddot{C}l: \quad \text{or} \quad :\ddot{C}l\!-\!\ddot{C}l:$$

In the Cl_2 molecule, each chlorine atom has six electrons (three lone pairs) plus two shared electrons, giving each an octet of electrons. (The shared electrons are counted for each atom.)

Covalent bonds between atoms of different elements form molecules of compounds. Consider the gas HCl, hydrogen chloride. The H and Cl atoms have the Lewis symbols

$$H\cdot \quad \cdot\ddot{C}l:$$

If each shares its one unpaired electron with the other, they both have noble gas configurations, and the Lewis structure is

$$H\!:\!\ddot{C}l: \quad \text{or} \quad H\!-\!\ddot{C}l:$$

Up to this point, we have seen that the noble gases tend to form no covalent bonds, that H forms one bond, and that Cl and the other elements of Group 7A form one bond. Let's take a look at the covalent bonding tendencies of oxygen, nitrogen, and carbon—representatives of Groups 6A, 5A, and 4A, respectively.

One of the most common molecules containing oxygen is water, H_2O. The Lewis symbols of the individual atoms are

$$H\cdot \quad \cdot\ddot{O}:$$
$$\dot{H}$$

When they combine, two single bonds are formed to the oxygen atom as each hydrogen atom shares its electron with the oxygen atom, which shares two of its electrons.

$$H\!:\!\ddot{O}: \quad \text{or} \quad H\!-\!\ddot{O}:$$
$$\ddot{H} \qquad\qquad\qquad\; |$$
$$\qquad\qquad\qquad\qquad H$$

Oxygen and the other Group 6A elements tend to form two covalent bonds.

One of the most common compounds containing a nitrogen atom is ammonia, NH_3. The nitrogen atom has five valence electrons ($\cdot\dot{N}\cdot$) and shares one each of its three unpaired electrons with three different H atoms (H·).

$$H:\overset{..}{\underset{H}{N}}:H \quad \text{or} \quad H—\overset{H}{\underset{|}{\overset{|}{N}}}—H$$

Nitrogen and the other Group 5A elements tend to form three covalent bonds.

Before leaving ammonia, let's mention that it is an important industrial compound and can be used as a fertilizer (Fig. 12.11). Ammonia is the gas that comes from smelling salts and from some household cleaners. During World War I, the development of the Haber process for making ammonia directly from nitrogen and hydrogen gas allowed Germany to produce explosives and extend the war for a full two years.

A common and simple compound of carbon is methane, CH_4. By now, you can predict that a carbon atom with its four valence electrons (·C·) will share one each with four different hydrogen atoms (H·).

$$H:\overset{H}{\underset{H}{\overset{..}{C}}}:H \quad \text{or} \quad H—\overset{H}{\underset{H}{\overset{|}{\underset{|}{C}}}}—H$$

Carbon and the other Group 4A elements tend to form four covalent bonds.

Covalent bonding is encountered mainly, but not exclusively, in compounds of nonmetals with one another. Writing Lewis structures for covalent compounds requires an understanding of the number of covalent bonds normally formed by common nonmetals. Table 12.4 summarizes our discussion of the number of covalent bonds to be expected from the elements of Groups 4A through 8A. (Exceptions are uncommon in Periods 1 and 2, but occur with more frequency starting with Period 3.)

When an element has two, three, or four valence electrons, its atoms will sometimes share more than one of them with another atom. Thus double bonds and triple bonds are possible (but, for reasons involving the geometry of bonding orbitals, not quadruple bonds). Consider the bonding in carbon dioxide, CO_2. The Lewis symbols for two oxygen atoms and one carbon atom are

$$:\overset{..}{O}· \quad ·\overset{.}{C}· \quad ·\overset{..}{O}:$$

To get a stable molecule, eight electrons are needed around each atom. If only two electrons are shared

Figure 12.11 Fertilizing a field with ammonia, NH_3.
Ammonia gas can be liquefied easily for storage or transportation. Here, ammonia from a tank is injected into the soil, where it dissolves in the moisture present.

Table 12.4 Number of Covalent Bonds Formed by Common Nonmetals

4A	5A	6A	7A	8A
·Ẋ·	·Ẍ·	:Ẍ·	:Ẍ:	:Ẍ:
4 bonds	3 bonds	2 bonds	1 bond	0 bonds
—C�—	—N̈—	:Ö—	:F̈—	:N̈e:

between each oxygen atom and the carbon atom, we get

$$:\ddot{O}:\dot{C}:\ddot{O}:$$

This structure is unstable because the carbon atom has only six electrons around it and each oxygen atom has only seven. To get eight electrons around each atom, we must have four electrons shared between the carbon atom and each oxygen atom. The correct Lewis structure of the CO_2 molecule is

$$:\ddot{O}::C::\ddot{O}: \quad \text{or} \quad :\ddot{O}=C=\ddot{O}:$$

A sharing of two pairs of electrons between two atoms produces a *double bond*, represented by a double dash. A sharing of three pairs of electrons between two atoms produces a *triple bond*, represented by a triple dash. Nitrogen gas, N_2, is an example of a molecule with a triple bond. Two nitrogen atoms may be represented as

$$:\ddot{N}· \quad ·\ddot{N}:$$

To satisfy the octet rule, each nitrogen atom must share three electrons. The Lewis structure for the N_2 molecule is

$$:N:::N: \quad \text{or} \quad :N\equiv N:$$

Sometimes, a molecule must have a combination of single, double, or even triple bonds in order to satisfy the octet rule. For example, in the Lewis structure of iodoethyne (HC_2I) the carbon atoms are connected by a triple bond, whereas the other elements are held to the carbon atoms by single bonds.

$$H:C:::C:\ddot{I}: \quad \text{or} \quad H-C\equiv C-\ddot{I}:$$

EXAMPLE 12.5 Drawing Lewis Structures for Simple Covalent Compounds

Draw the Lewis structure for these two covalent compounds. (a) Chloroform, $CHCl_3$, was used as an anesthetic but has now been replaced by less toxic compounds. (b) Hydrogen cyanide, HCN, smells like bitter almonds and is the lethal gas used for gas chamber executions.

Solution

In each case, proceed in this general fashion. Write the Lewis symbol for each atom in the formula. Realize that atoms forming only one bond can never connect two other atoms, they can only stick on the outside, so to speak. (This is like saying that a one-armed person can't grab *two* people and "connect" them.) Try connecting the atoms using single bonds, remembering the number of covalent bonds each atom normally forms. If each atom gets a noble gas configuration, your job is done. If not, see if a double or triple bond will work.

(a) $CHCl_3$. C (Group 4A) forms four bonds, H forms one bond, and Cl (Group 7A) forms one bond. Only C can be the central atom, thus

$$H· \quad ·\dot{C}· \quad ·\ddot{Cl}: \quad ·\ddot{Cl}: \quad ·\ddot{Cl}:$$

gives

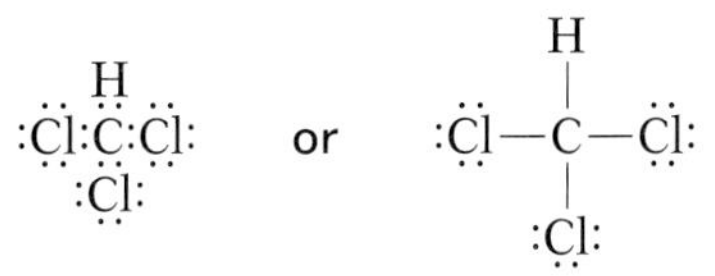

$$:\ddot{Cl}:\ddot{C}:\ddot{Cl}: \quad \text{or} \quad :\ddot{Cl}-\overset{H}{\underset{:\ddot{Cl}:}{C}}-\ddot{Cl}:$$
$$\quad\;\;:\ddot{Cl}:$$

(b) HCN. H forms one bond, C (Group 4A) forms four, and N (Group 5A) forms three. The H has to be on the outside (why?), so the middle atom must be either C or N. Only with C in the middle and by using a triple bond between C and N can the octet rule be satisfied, thus

$$H· \quad ·\dot{C}· \quad ·\ddot{N}·$$

gives $\quad H:C:::N: \quad \text{or} \quad H-C\equiv N:$

CHAPTER 12 Chemical Bonding

Table 12.5 Comparison of Properties of Ionic and Covalent Compounds

Ionic Compounds	Covalent Compounds
Crystalline solids (made of ions)	Gases, liquids, or solids (made of molecules)
High melting and boiling points	Low melting and boiling points
Conduct electricity when melted	Poor electrical conductors in all phases
Many soluble in water but not in nonpolar liquids	Many soluble in nonpolar liquids but not in water

CONFIDENCE QUESTION 12.5

Dilute solutions of hydrogen peroxide, H_2O_2, are used to bleach and disinfect. Concentrated solutions have been used as rocket propellants. Draw the Lewis structure for H_2O_2.

Because of the nature of the bonding involved, covalent compounds have quite different properties from those of ionic compounds. Unlike ionic compounds, covalent compounds are composed of individual molecules with a specific molecular formula. For example, carbon tetrachloride consists of individual CCl_4 molecules, each composed of one carbon atom and four chlorine atoms. Although each covalent bond is strong *within* a molecule, the various molecules in a sample of the compound only weakly attract *one another*. Therefore, the melting points and boiling points of covalent compounds are generally low compared with those of ionic compounds. For instance, CCl_4 melts at −23°C and is a liquid at room temperature, whereas the ionic compound NaCl melts at 800°C. Many covalent compounds occur as liquids or gases at room temperature. Covalent compounds do not conduct electricity well, no matter what phase they are in. The general properties of ionic and covalent compounds are summarized in Table 12.5.

Some compounds contain both ionic and covalent bonds. Sodium hydroxide (lye), NaOH, is an example. We saw in Chapter 11 that it is not uncommon for several atoms to form a polyatomic ion such as OH^-. The atoms *within* the polyatomic ion are covalently bonded, but the whole aggregation behaves like an ion in forming compounds. Because strong covalent bonds are present between atoms within polyatomic ions, it is difficult to break them up. Therefore, in chemical reactions they frequently act as a single unit. The Lewis structure of sodium hydroxide is

$$Na^+[:\ddot{O}:H]^-$$

A covalent bond exists between the O and H in the hydroxide ion, OH^-, but the hydroxide ions and the sodium ions are bound together in a crystal lattice by ionic bonds, and thus NaOH is an ionic compound.

Follow these rules to predict whether a particular compound is ionic or covalent.

1. Compounds formed of only nonmetals are covalent (except ammonium compounds).
2. Compounds of metals and nonmetals are generally ionic (many exceptions exist).
3. Compounds of metals with polyatomic ions are ionic.
4. Compounds that are gases, liquids, or low-melting-point solids are covalent.
5. Compounds that conduct an electric current when melted are ionic.

EXAMPLE 12.6 Predicting Bonding Type

Predict which compounds are ionic and which are covalent: (a) KF, (b) SiH_4, (c) $Ca(NO_3)_2$, (d) compound *X*, a gas at room temperature, (e) compound *Y*, which melts at 900°C and then conducts an electric current.

Solution

(a) KF is probably ionic (metal and nonmetal).
(b) SiH_4 is covalent (only nonmetals).
(c) $Ca(NO_3)_2$ is ionic (metal and polyatomic ion).
(d) *X* is covalent (as are all substances that are gases or liquids at room temperature).
(e) *Y* is ionic (the compound has a high melting point, and the melt conducts electricity).

CONFIDENCE QUESTION 12.6

Is PCl_3 ionic or covalent in bonding? What about MgF_2?

Polar Covalent Bonding

In covalent bonding, the electrons involved in the bond between two atoms are shared. However, unless the atoms are of the same element, the bonding electrons will spend more time around the more nonmetallic element. That is, the sharing is unequal. Such a bond is called a **polar covalent bond,** a term that indicates that it has a slightly positive end and a slightly negative end.

Electronegativity (abbreviated EN) is a measure of the ability of an atom to attract electrons in the presence of another atom. It shows a definite periodic trend. Electronegativity increases from left to right within a period and decreases down a group, just as does nonmetallic character. Figure 12.12 shows the numerical values calculated for the electronegativities of the representative elements.

Consider the covalent bond in HCl. The chlorine atom (EN = 3.0) is more electronegative than the hydrogen atom (EN = 2.1). Although the two bonding electrons are shared between the two atoms, they tend to spend more time at the chlorine end than at the hydrogen end of the molecule. We get a polar bond in which the polarity could be represented by an arrow:

$$H \xrightarrow{+} \ddot{\underset{..}{Cl}}:$$

The head of the arrow points to the more electronegative atom and denotes the negative end of the bond. The "feathers" of the arrow make a plus-sign (at left above) that indicates the positive end of the bond.

EXAMPLE 12.7 Showing the Polarity of Bonds

Use arrows to show the polarity of the covalent bonds in (a) water, H_2O, and (b) carbon tetrachloride, CCl_4.

Increasing electronegativity →

↓ Decreasing electronegativity

			H 2.1			
Li 1.0	Be 1.5	B 2.0	C 2.5	N 3.0	O 3.5	F 4.0
Na 0.9	Mg 1.2	Al 1.5	Si 1.8	P 2.1	S 2.5	Cl 3.0
K 0.8	Ca 1.0	Ga 1.6	Ge 1.8	As 2.0	Se 2.4	Br 2.8
Rb 0.8	Sr 1.0	In 1.7	Sn 1.8	Sb 1.9	Te 2.1	I 2.5
Cs 0.7	Ba 0.9	Tl 1.8	Pb 1.9	Bi 1.9	Po 2.0	At 2.2
Fr 0.7	Ra 0.9					

Figure 12.12 Electronegativity values.
In general, electronegativity increases across a period and decreases down a group.

Solution

(a) Oxygen (EN = 3.5) is more electronegative than hydrogen (EN = 2.1), so the arrows denoting the polarity of the bonds would point as shown here.

$$H \xrightarrow{+} \ddot{O}: \\ \updownarrow \\ H$$

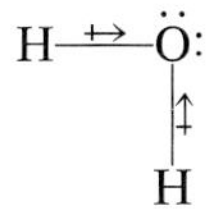

(b) Chlorine (EN = 3.0) is more electronegative than carbon (EN = 2.5), so the arrows denoting the polarity of the bonds would point as shown here.

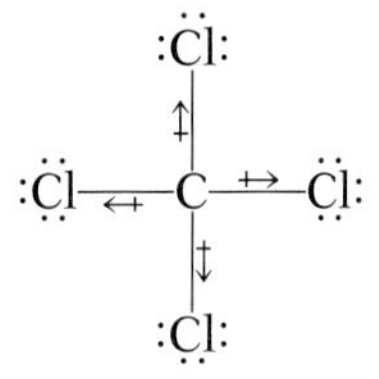

$$:\ddot{Cl}: \\ \updownarrow \\ :\ddot{Cl} \leftrightarrow C \xrightarrow{+} \ddot{Cl}: \\ \updownarrow \\ :\ddot{Cl}:$$

CONFIDENCE QUESTION 12.7

Use arrows to show the polarity of the covalent bonds in ammonia, NH_3.

If the bonds in a molecule are nonpolar, the molecule can only be *nonpolar* (no positive and negative ends). If a molecule has a positive end and a negative end, we say it has a *dipole*, or is a **polar molecule.** The slightly negative end of the polar molecule is often denoted by a δ– (delta minus) and the positive end by a δ+ (delta plus). Consider the HCl molecule with its one polar bond. With only one polar bond present, it should be obvious that the chlorine end of this molecule must be slightly negative and the hydrogen end slightly positive.

$$\delta + \quad H \!\rightarrow\! \ddot{C}\ddot{l}\!: \quad \delta -$$

If the molecule has more than one polar bond, a nonpolar molecule will result if the geometry of the molecule causes the polarities of the bonds to cancel. If they do not cancel, a polar molecule will result. Consider the water molecule. If it were linear, it would be nonpolar. The polarity of the two bonds would cancel, because the center of the positive charges of the bonds would be exactly the same place as the center of the negative charges—right in the middle of the molecule.

$$H \!\rightarrow\! \ddot{O} \!\leftarrow\! H$$

But if the water molecule is angular, the bond polarities reinforce instead of canceling. The center of positive charge is midway between the two hydrogen atoms. The center of negative charge is at the oxygen atom. The water molecule is indeed angular (the bond angle is 105°), so water molecules are polar.

$$\underset{H \quad\quad H}{\overset{\ddot{O}}{\diagdown\!\!\diagup}}$$

EXAMPLE 12.8 Predicting the Polarity of Molecules

The carbon tetrachloride molecule is tetrahedral, as shown in the accompanying sketch. All the bond angles are 109.5°, and the bonds are polar. Is the molecule polar?

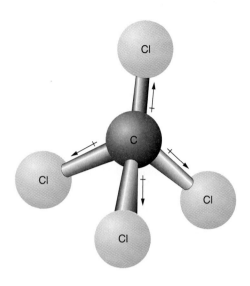

Solution

The polar bonds cancel. The center of positive charge is at the C, and the center of negative charge is also there. Thus the molecule is nonpolar.

CONFIDENCE QUESTION 12.8

Which one of the two 1,2-dichloroethenes illustrated is polar? (Both are "flat" molecules.)

$$\underset{(a)}{\overset{:\ddot{C}l:\quad\quad :\ddot{C}l:}{\underset{H\quad\quad\quad H}{\diagdown\!\diagup\;C\!=\!C\;\diagdown\!\diagup}}} \quad\quad \underset{(b)}{\overset{:\ddot{C}l:\quad\quad H}{\underset{H\quad\quad\quad :\ddot{C}l:}{\diagdown\!\diagup\;C\!=\!C\;\diagdown\!\diagup}}}$$

The polarity of the molecules of a compound can be tested by bringing an electrically charged rod close to the compound and seeing if attraction occurs. Because water molecules are polar, a thin stream of water is attracted to a charged rod, but a similar stream of nonpolar carbon tetrachloride is not attracted (Fig. 12.13).

12.6 Metallic Bonding and Hydrogen Bonding

Learning Goal:

To describe metallic bonding and hydrogen bonding.

(a)

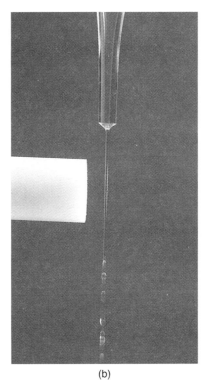
(b)

Figure 12.13 Polar and nonpolar liquids.
(a) A stream of polar water molecules is deflected toward an electrically charged rod, but (b) a stream of nonpolar CCl_4 molecules is undeflected.

Metallic Bonding

Metals are malleable, ductile, and conduct heat and electricity very well in all directions (Section 11.4). Most are durable and have high melting points. These properties indicate that the atoms in metals are difficult to separate (the bonding is strong) but not very difficult to move in various directions (the bonding has no preferred direction), and that there must be some charged particles that can move freely.

The simplest model that explains **metallic bonding** pictures the metal atoms as spherical positively charged ions packed closely together in a regular crystal lattice. Surrounding the ions is a "sea" of mobile valence electrons (Fig. 12.14). These electrons are responsible for the ease of conduction of heat and electricity, and the positive ions can move in any direction in response to hammering or pulling with no significant change in the bonding. As you would expect, this simple model has to become more complex to explain more detailed aspects of metallic bonding.

Pure gold is said to be 24 karat (kt). This means that 24 of 24 parts, or 100%, of a sample is gold. In 18-kt gold, 18 parts are gold and the other 6 are silver or copper. Unlike compounds, alloys contain no fixed, definite proportion of elements. If we use the positive-ions-in-a-sea-of-electrons model, it is easy to envision how atoms of several metals could join in virtually any ratio to form metallic crystals.

Steel is an interesting and useful alloy consisting mainly of iron, Fe. Pure iron is fairly soft, ductile, and malleable due to the absence of strong directional bonding forces. However, when some relatively small atoms of carbon are added to iron, the carbon atoms fit into holes between the iron ions in the crystal lattice. The carbon atoms form strong directional bonds with the iron ions, making steel much harder and stronger than pure iron. Small amounts of additional elements, such as Mn, Si, and Cr, are added to some steels to give them the characteristics desired for special purposes.

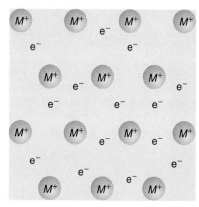

Figure 12.14 Metallic bonding.
In metal, the valence electrons (e−) are delocalized and move throughout the lattice of positive ions (M^+). This "sea" of electrons bonds the metal ions together.

Hydrogen Bonding

Why does water dissolve table salt but not oil? The polar nature of water molecules causes them to interact with an ionic substance such as salt. The positive ends of the water molecules attract the negative ions, and the negative ends attract the positive ions (Fig. 12.15). If the attraction of the water molecules can overcome the attraction among the ions, the salt dissolves. As Fig. 12.15 shows, the negative ions move into solution surrounded by several water molecules with their positive ends pointed toward the ion. Just the opposite is true for the positive ions. Such attractions are called *ion-dipole interactions*.

As you would expect, the molecules of two polar substances have a *dipole-dipole interaction* and tend to dissolve in one another. In general, however, polar substances and nonpolar substances do not dissolve in one another, because the polar molecules tend to gather together and exclude the nonpolar molecules. Similarly, it would not be surprising to find that two nonpolar substances mix well, such as oil and gasoline. The nonpolar molecules in oil have no more affinity for one another than they do for the nonpolar molecules in gasoline. These are examples of the well-known principle, *like dissolves like*.

A **hydrogen bond** is a special kind of dipole-dipole interaction that can occur whenever hydrogen atoms are covalently bonded to small, highly electronegative atoms (O, F, or N). Because the bond is so polar and a hydrogen atom is so small, the partial positive charge on it is highly concentrated. Thus it has an electrical attraction for close O, F, or N atoms in the same or neighboring molecules.

For small molecules, such as H_2O, HF, and NH_3, hydrogen bonding is a weak force of attraction between a hydrogen atom in *one* molecule and an O, F, or N atom in *another* molecule. Such hy-

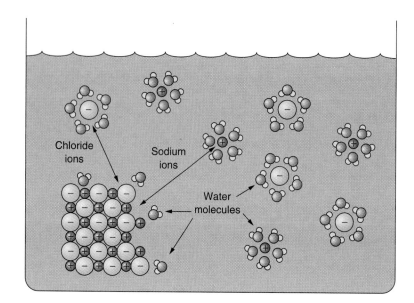

Figure 12.15 Sodium chloride dissolving in water.
The negative ends of the polar water molecules attract and surround the positive sodium ions (purple spheres). The positive ends of the water molecules attract and surround the negative chloride ions (green spheres).

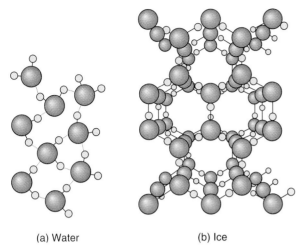

Figure 12.16 Hydrogen bonding in water and ice.
(a) Weak forces of attraction between the hydrogen atom of one water molecule and the oxygen atom of another molecule cause water to be a liquid. At room temperature, about 80% of the molecules are hydrogen-bonded at any one time. (b) At or below 0°C, the extent of hydrogen bonding is virtually 100%, and the molecules assume a highly ordered, hexagonal, open structure.

drogen bonds are *inter*molecular forces, not *intra*molecular forces.*

Hydrogen bonds are strong enough (about 5% to 10% of the strength of a covalent bond) to have a pronounced effect on the properties of the substance. Consider water, the most common example of a compound whose molecules are hydrogen bonded (Fig. 12.16a). Most substances with a formula mass as low as water's 18 u are gases. Without hydrogen bonding, water at room temperature would be a gas, not a liquid, because there would be so little attraction among the water molecules. There is no hydrogen bonding in hydrogen sulfide, H_2S, and it is indeed a gas at room temperature (boiling point of $-61°C$, compared to water's boiling point of 100°C).

---

*For large molecules such as DNA and proteins, hydrogen bonds exist between some of the H atoms and O or N atoms in other parts of the same molecule that have twisted around into close proximity. The famous "unzipping" of DNA molecules is actually a breaking of intramolecular hydrogen bonds.

Almost all solids sink to the bottom in a sample of their liquid; why does ice float in water? The two hydrogen atoms of a water molecule and the two lone pairs of electrons on the oxygen atom point toward the corners of a somewhat irregular tetrahedron. This gives the hydrogen bonds in water a three-dimensional directional character. As the temperature of a sample of water drops to the freezing point, the hydrogen bonds align more and more in a hexagonal manner. (The hexagonal nature of the hydrogen bonds in water is manifested in the hexagonal shapes associated with snowflakes.) This alignment gives ice an open structure (Fig. 12.16b), increases its volume, and makes it less dense than liquid water. Thus ice floats on water (Fig. 12.17). If ice did not float, the consequences for aquatic life would be severe, because ponds would freeze from the bottom up and not form an insulating ice cover first.

12.7 The Stock System of Nomenclature

Learning Goal:
To name compounds of metals that form more than one ion.

Figure 12.17 Hydrogen bonding and density.
Hydrogen bonding and the shape of the water molecule cause ice to have an open structure, making it less dense than liquid water. Thus ice floats, as shown on the left. For virtually every other substance, such as the benzene (C_6H_6) shown on the right, the solid is more dense than the liquid and thus does not float.

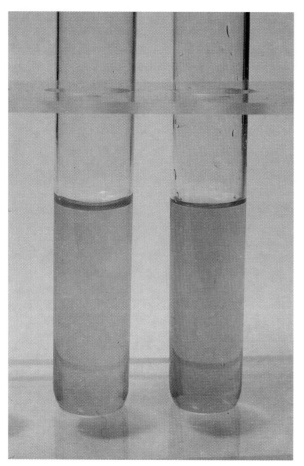

Figure 12.18 The ions of iron.
Solutions containing Fe^{2+} are usually pale green; those of Fe^{3+} are generally yellow-brown.

Figure 12.19 Skywriting.
TiO_2 is formed when $TiCl_4$ reacts with moisture.

Section 11.5 discussed the basic rules for naming compounds, and stated that the rules must be expanded for metals that form more than one ion and hence more than one compound with a given nonmetal or polyatomic ion. To distinguish the compounds, use the **Stock system:** a Roman numeral giving the number of the metal's ionic charge is placed in parentheses directly after the metal's name; for example

    $FeCl_2$ iron(II) chloride
    $FeCl_3$ iron(III) chloride

Because the chloride ions have a 1− charge and the compound must be electrically neutral, the iron ion in $FeCl_2$ must have a charge of 2+. Similar reasoning indicates that the iron ion in $FeCl_3$ must have a charge of 3+. The two compounds, $FeCl_2$ and $FeCl_3$, have entirely different properties (Fig. 12.18).

EXAMPLE 12.9 Naming Compounds of Metals that Form Several Ions

The compound that forms skywriting and smoke screens has the formula TiO_2 (Fig. 12.19). Titanium is a metal that forms several ions, so it would be preferable to name TiO_2 using the Stock system. What is its Stock system name?

Solution

Oxygen has a 2− ionic charge. Thus the two O atoms contribute a total of 4−. The one Ti atom must contribute 4+. Therefore, its ionic charge is 4+, and the compound's Stock system name is titanium(IV) oxide.

CONFIDENCE QUESTION 12.9

Under an older system of nomenclature, SnF_2 and SnF_4 are named stannous fluoride and stannic fluoride, respectively. What would be their Stock system names?

IMPORTANT TERMS

These important terms are for review. After reading and studying the chapter, you should be able to define and/or explain each of them.

law of conservation of mass
formula mass
law of definite proportions
alchemy
limiting reactant
excess reactant
octet rule

ionic compounds
Lewis symbol
Lewis structures
cations
anions
ionic bonds
covalent compounds

covalent bond
polar covalent bond
electronegativity
polar molecule
metallic bonding
hydrogen bond
Stock system

IMPORTANT EQUATION

Percentage by Mass of Element *X* from Compound's Formula:

$$\% X \text{ by mass} = \frac{(\text{atoms of } X \text{ in formula}) \times \text{AM}_X}{\text{FM}_{\text{cpd}}} \times 100\%$$

QUESTIONS

1. Which one of the three forces (gravitational, electromagnetic, strong nuclear) is responsible for chemical bonding?

12.1 Law of Conservation of Mass (and Highlight)

2. Which of the following discovered the law of conservation of mass in 1774?
   a. Dalton   b. Lavoisier
   c. Lewis   d. Proust
3. State the law of conservation of mass and give an example.
4. Give four reasons why Lavoisier is generally designated the "father of chemistry."
5. What was the major shortcoming of the ancient Greek approach to answering scientific questions?
6. What were the two main goals of alchemy? What was the new goal that Paracelsus emphasized for alchemy?
7. During approximately what historical time span did alchemy flourish? Briefly, how does modern chemistry differ philosophically from alchemy?
8. Lavoisier named a new element "oxygen" and tried to take credit for its discovery. Who actually discovered the element, and what did that scientist call it?

12.2 Law of Definite Proportions

9. A sample of compound *AB* decomposes to 48 g of *A* and 12 g of *B*. If another sample of the same compound *AB* decomposes to 16 g of *A* predict the number of g of *B*.
   a. 12   b. 8   c. 4   d. 3
10. State the law of definite proportions and give an example. Who discovered this law?
11. What is the chemical formula for dry ice? What are its two most striking physical properties?
12. When a sample of *A* reacts with one of *B*, a new substance *AB* is formed and no *A* but a little of *B* is left. Which is the limiting reactant?

12.3 Dalton's Atomic Theory

13. Dalton proposed that during a chemical reaction
    (a) no atoms are formed.
    (b) no atoms are destroyed.
    (c) no atoms are changed in identity.
    (d) all of the preceding answers.
14. State the three postulates of Dalton's atomic theory.

12.4 Ionic Bonding

15. What is the normal charge on an ion of sulfur?
    (a) 6+   (b) 6−   (c) 2+   (d) 2−
16. An ionic compound formed between a Group 2A element M and a Group 7A element X would have the general formula
    (a) M_2X (b) M_7X_2 (c) MX_2 (d). M_2X_7
17. State the octet rule. What common element is an exception to the rule, and why?
18. Which electrons of an atom generally take part in the formation of compounds?
19. How is the number of valence electrons in an atom related to its tendency to gain or lose electrons during compound formation?
20. Ionic compounds are formed when individual atoms achieve a noble gas configuration by what general process?
21. Of what particles would a formula unit of sodium oxide (Na_2O) consist?
22. During discussions of chemical bonding, the name of what scientist will likely be brought up? What was his nationality?
23. What does the element's symbol in a Lewis symbol stand for? The number of dots used is the same as the atom's number of _____ (two words).
24. What is the difference between a Lewis symbol and a Lewis structure?
25. What is the basic difference in ions formed by metals and ions formed by nonmetals? What are the meanings of the terms *cation* and *anion*?
26. What is meant when we say F^-, Ne, and Na^+ are *isoelectronic*?
27. State the two principles used in writing the formulas of ionic compounds.
28. What is the crucial experimental test of whether or not a compound is ionic?

12.5 Covalent Bonding

29. How many covalent bonds are normally formed by a nitrogen atom?
    (a) 3   (b) 4   (c) 2   (d) 5
30. Which of the following is definitely a covalent compound?
    (a) NF_3 (b) $TiCl_2$ (c) Na_2O (d) $CaSO_4$
31. Briefly, how are covalent compounds formed, and what principle is used to predict if a formula represents a stable molecule?
32. What is the basic difference distinguishing single, double, and triple bonds?
33. How does electronegativity vary, in general, from left to right in a period and down a group? What element has the highest electronegativity?
34. A covalent bond in which the electron pair is unequally shared is called by what name?
35. Could a molecule composed of two atoms joined by a polar covalent bond ever be nonpolar? Explain.
36. Ammonia, NH_3, is pyramidal, as shown in Fig. 11.7. Use arrows to show the polarity of the bonds. Will the molecule be polar or nonpolar? Explain.
37. Compare ionic and covalent compounds with regard to (a) phase and (b) electrical conductivity.
38. A certain compound has a boiling point of −10°C. Is it ionic or covalent?

12.6 Metallic Bonding and Hydrogen Bonding

39. "A lattice of positive ions in a sea of electrons" is a description of which of the following types of bonding?
    (a) ionic   (b) metallic
    (c) covalent   (d) hydrogen
40. State the short general principle of solubility and explain what it means.
41. What is a *hydrogen bond*? About how strong are hydrogen bonds relative to covalent bonds?

12.7 The Stock System of Nomenclature

42. Which of the following is the formula for iron(III) bromide?
    (a) FeBr   (b) Fe_3Br
    (c) Fe_2Br_3 (d) $FeBr_3$
43. When is the use of the Stock system of nomenclature preferred?

Food for Thought

1. Titan, the major moon of Saturn, is thought to have an atmospheric composition and temperature such that seas of liquid methane (CH_4) exist. An artist's conception of the surface's appearance shows a beautiful view of a blue methane sea with "methanebergs" floating around. Where did the artist go wrong?

2. Boron trifluoride, BF_3, is an exception to the octet rule. There is a single bond from each fluorine atom to the boron atom, which has a share in only six electrons. The three bonds are very polar, yet the molecule itself is nonpolar. What must be the geometry of the molecule and the angle between the bonds?

Exercises

12.1 Law of Conservation of Mass

1. Volcanoes emit large quantities of hydrogen sulfide gas (H_2S), which reacts with the oxygen in the air to form water and sulfur dioxide (SO_2). Every 68 tons of H_2S reacts with 96 tons of oxygen and forms 36 tons of water. How many tons of SO_2 are formed?

    Answer: 128 tons

2. Silver utensils are not attacked by oxygen unless H_2S or sulfur-containing foods such as eggs or mustard are present. The reaction that causes the tarnishing of silver in the presence of oxygen and H_2S leads to the formation of silver sulfide (the tarnish) and water. If 432 g of silver react with 68 g of H_2S and 32 g of oxygen, 36 g of water are formed. How many grams of silver sulfide are formed?

3. Calculate the formula mass (to the nearest 0.1 u) of the following compounds:
    (a) carbon dioxide, CO_2
    (b) methane, CH_4
    (c) magnesium phosphate, $Mg_3(PO_4)_2$
    (d) sodium chloride, NaCl
    (e) benzene, C_6H_6
    (f) calcium acetate, $Ca(C_2H_3O_2)_2$

    Answer: (a) 44.0 u (b) 16.0 u (c) 262.9 u

12.2 Law of Definite Proportions

4. Find the percentage by mass of Cl in (a) $MgCl_2$ if it is 25.5% Mg by mass, and (b) $AlCl_3$ if it is 20.2% Al by mass.

    Answer: (a) 74.5%

5. Determine the percentage by mass of each element in the following compounds.
    (a) table salt, NaCl
    (b) iron(III) oxide, Fe_2O_3
    (c) baking soda, $NaHCO_3$
    (d) sugar, $C_{12}H_{22}O_{11}$

    Answer: (a) 39.3% Na, 60.7% Cl (b) 69.9% Fe, 30.1% O

6. In a lab experiment, 6.1 g of Mg reacts with sulfur to form 14.1 g of magnesium sulfide. (a) How many grams of sulfur reacted, and how do you know? (b) How much magnesium sulfide would be formed if 6.1 g of Mg were reacted with 10.0 g of S, and how do you know?

    Answer: (a) 8.0 g, law of conservation of mass (b) Still 14.1 g. The law of definite proportions indicates the proper ratio is 6.1 g Mg to 8.0 g S, so 10.0 g − 8.0 g = 2.0 g of S is in excess.

7. In a lab experiment, 7.75 g of phosphorus reacts with bromine to form 67.68 g of phosphorus tribromide. (a) How many grams of bromine reacted, and how do you know? (b) How much phosphorus tribromide would be formed if 10.00 g of phosphorus reacted with 59.93 g of bromine, and how do you know?

12.4 Ionic Bonding

8. Referring only to a periodic table, give the ionic charge expected for each of these representative elements: (a) S, (b) K, (c) Br, (d) N, (e) Mg, (f) Ne, (g) C, (h) Al, (i) F, (j) O, (k) Ga, (l) He, (m) P, (n) Ca, (o) Rb, and (p) Si.

    Answer: (a) 2 − (b) 1 + (c) 1−
    (d) 3 − (e) 2 + (f) 0 (g) 0 (h) 3 +

9. Write the Lewis symbols and structures showing how Na₂O forms from sodium and oxygen atoms.

10. Referring to a periodic table and recalling the polyatomic ions in Table 11.5, predict the formula for the ionic compounds:
    (a) cesium iodide
    (b) barium fluoride
    (c) aluminum nitrate
    (d) lithium sulfide
    (e) beryllium oxide
    (f) ammonium sulfate
    (g) potassium permanganate
    (h) ammonium phosphate
    (i) potassium nitride
    (j) strontium acetate
    (k) gallium (Ga) hydroxide
    (l) aluminum phosphide
    (m) sodium carbonate
    (n) magnesium bromide

    Answer: (a) CsI (b) BaF₂ (c) Al(NO₃)₃ (d) Li₂S (e) BeO (f) (NH₄)₂SO₄ (g) KMnO₄

12.5 Covalent Bonding

11. Referring only to a periodic table, give the number of covalent bonds expected for each of these representative elements: (a) S, (b) Ne, (c) Br, (d) N, (e) C, (f) H, (g) Ar, (h) Te, (i) F, (j) Si, and (k) P.

    Answer: (a) 2 (b) 0 (c) 1 (d) 3 (e) 4

12. Draw the Lewis structures for (a) the rocket fuel hydrazine, N₂H₄, and (b) formaldehyde, H₂CO, a compound whose odor is known to most biology students. In each case show a structure with all dots, and then one with both dots and dashes.

    Answer: (a)

    H:N̈:N̈:H   and   H—N̈—N̈—H
     H H | |
                       H   H

13. Use your knowledge of the appropriate number of covalent bonds to predict the formula for a simple compound formed between each of the following elements.
    (a) hydrogen and bromine
    (b) nitrogen and chlorine
    (c) hydrogen and sulfur
    (d) carbon and chlorine
    (e) carbon and sulfur
    (f) carbon and hydrogen
    (g) bromine and chlorine
    (h) sulfur and chlorine
    (i) nitrogen and hydrogen
    (j) silicon and oxygen

    Answer: (a) HBr (b) NCl₃ (c) H₂S (d) CCl₄ (e) CS₂

14. Predict which of these compounds are ionic and which are covalent. State your reasoning.
    (a) N₂H₄   (b) NaF
    (c) Ca(NO₃)₂   (d) CBr₄
    (e) C₁₂H₂₂O₁₁   (f) (NH₄)₃PO₄
    (g) CaO   (h) PBr₃
    (i) C₆H₄Cl₂   (j) K₃N
    (k) H₂S   (l) NaC₂H₃O₂

    Answer: (a) covalent, two nonmetals (b) probably ionic, metal and nonmetal (c) ionic, metal and polyatomic ion (d) covalent, two nonmetals (e) covalent, all nonmetals (f) ionic, two polyatomic ions

15. Use arrows to show the polarity of each bond in the following molecules. (*Hint*: Use the general periodic trends in electronegativity. Refer to electronegativity values in Fig. 12.12 in cases of uncertainty.)
    (a) TeCl₂
    (b) NH₃
    (c) GeBr₄
    (d) BrCl
    (e) NI₃
    (f) SiH₄

    Answer: (a) Cl is higher and farther right than Te in the periodic table, so it is the more electronegative and the arrows would point to it. (b) H is higher, but N is farther right. Figure 12.12 shows H with EN = 2.1 and N with EN = 3.0, so N is the more electronegative, and the arrows would point to it. (c) Br is farther right than Ge and at the same height in the chart, so Br is more electronegative, and the arrows would point to it.

12.7 The Stock System of Nomenclature

16. Name these compounds using the Stock system.
    (a) Ni(OH)$_2$
    (b) PbCl$_4$
    (c) Sn(NO$_3$)$_4$
    (d) AuI$_3$
    (e) Fe$_2$S$_3$
    (f) TiBr$_4$
    (g) HgF$_2$
    (h) Sn(C$_2$H$_3$O$_2$)$_2$

    Answer: (a) nickel(II) hydroxide (b) lead(IV) chloride (c) tin(IV) nitrate (d) gold(III) iodide

17. Write the formula for each of the following:
    (a) iron(II) oxide
    (b) mercury(II) sulfate
    (c) chromium(III) oxide
    (d) gold(I) permanganate
    (e) manganese(IV) sulfide
    (f) copper(II) chloride

    Answer: (a) FeO (b) HgSO$_4$ (c) Cr$_2$O$_3$

Answers to Multiple-Choice Questions

2. b 9. c 13. d 15. d 16. c 29. a 30. a 39. b 42. d

Solutions to Confidence Questions

12.1 The answer is 71.0 g of chlorine. Because the compound is composed of only calcium and chlorine, the difference in mass between 111.1 g of compound and 40.1 g of calcium must be the mass of the chlorine.

12.2 The answer is 34.1 u. FM = (2 × 1.0 u for H) + 32.1 u for S = 34.1 u.

12.3 The answer is 52.9% Al and 47.1% O.

$$\%Al = \frac{2 \times 27.0 \text{ u}}{102.0 \text{ u}} \times 100\% = 52.9\%;$$

$$\%O = 100.0\% - 52.9\% = 47.1\%$$

12.4 Al is in Group 3A, so will form Al^{3+}. Cl (Group 7A) will form Cl$^-$. To get a neutral compound requires 3 Cl$^-$ for each Al^{3+}, so the formula is AlCl$_3$. O (Group 6A) will form O^{2-}. To get neutrality takes two Al^{3+} and three O^{2-}, so the formula is Al$_2$O$_3$. P (Group 5A) will form P^{3-}. To get a neutral compound requires one Al^{3+} and one P^{3-}, so the formula is AlP.

12.5 O (Group 6A) forms two bonds and H forms one bond. Only the O atoms can connect to two atoms, thus the structure of H$_2$O$_2$ must be

H:Ö: or   H—Ö:
 :Ö:H |
                  :Ö—H

12.6 PCl$_3$ is *covalent* (two nonmetals). MgF$_2$ is probably ionic (metal and nonmetal).

12.7 Nitrogen (EN = 3.0) is more electronegative than hydrogen (EN = 2.1), so the arrows denoting the polarity of the bonds would point as shown here.

$$H \longrightarrow \ddot{N} \longleftarrow H$$
$$|$$
$$H$$

12.8 Compound (a) is the polar one. The Cl side of the molecule will be negative relative to the H side. In Compound (b), both the center of negative charge and center of positive charge will be at the center of the molecule.

12.9 The Stock system names are tin(II) fluoride and tin(IV) fluoride. Sn is the symbol for the metal tin, and F is the symbol for the nonmetal fluorine. The name of the nonmetal changes to its -*ide* form, so both compounds are tin fluorides. Since the ionic charge of F is always 1−, tin's ionic charge must be 2+ in the first compound and 4+ in the second.

Chapter 13

Chemical Reactions

13.1 Chemical and Physical Properties and Changes
13.2 Chemical Equilibrium
13.3 Balancing Equations
13.4 Energy and Rate of Reaction
13.5 Acids and Bases
13.6 Single-Replacement Reactions
13.7 Electrochemistry

Highlight: The Mole and Avogadro's Number

Now that elements, compounds, and chemical bonding have been discussed, the groundwork is laid for an examination of chemical reactions. Our environment is composed of atoms that combine with one another to form molecules, which in turn react with other atoms or molecules to produce the many substances we need and use. In producing new products, energy is released or absorbed.

For example, green plants absorb carbon dioxide from the air and, with energy from the Sun and chlorophyll as a catalyst, react with water from the soil to form glucose (a carbohydrate) and oxygen. This complex chemical reaction is called *photosynthesis,* and the Sun's energy is stored in chemical bonds as *chemical energy.* Animal life on our planet inhales the oxygen and ingests the plants' carbohydrates to obtain energy. Chemical changes in the animals' cells then return water and carbon dioxide to the environment. This interaction between plants and animals and their environment constitutes an ecosystem.

In this chapter, we will discuss general types of chemical reactions and the major principles that underlie chemical change. The chapter Highlight discusses the mole and Avogadro's number.

13.1 Chemical and Physical Properties and Changes

Learning Goals:

To distinguish between chemical and physical properties.

To differentiate between chemical and physical changes.

The characteristics of a substance are known as its properties. *Physical properties* are those that do not describe the chemical reactivity of the substance. Among them are density, hardness, phase, color, melting point (m.p.), boiling point (b.p.), electrical conductivity, thermal conductivity, and specific heat. These properties can be measured without causing new substances to form.

Figure 13.1 The chemical reaction by which Joseph Priestley discovered oxygen in 1774. Heating mercury(II) oxide, the orange powder, decomposes it to silvery liquid mercury and colorless oxygen gas.

Chemical properties are those that describe the transformation of one substance into another (the chemical reactivity). As examples, a chemical property of wood is that it burns when heated in the presence of air, a chemical property of water is that it can be decomposed by electricity into hydrogen and oxygen, and a chemical property of zinc is that it reacts with hydrochloric acid to form hydrogen and zinc chloride.

Changes that alter the chemical composition of a substance are called *chemical changes* or, more often, **chemical reactions.** That is, new substances are formed by the changes. The burning of wood, the rusting of iron, the explosion of gunpowder, the souring of milk, and the decomposition of mercury(II) oxide to form mercury and oxygen gas are examples of chemical changes (● Fig. 13.1).

Changes that do not alter the chemical composition are classified as *physical changes.* Examples are the freezing of water (it is still H_2O), dissolving of table salt in water (it is still sodium ions and chloride ions), heating of an iron bar (still Fe), evaporation of rubbing alcohol (still C_3H_8O), and crushing of sugar crystals (still $C_{12}H_{22}O_{11}$).

13.2 Chemical Equilibrium

Learning Goals:

To define the terms *equilibrium* and *reversible reaction.*

To use Le Châtelier's principle in a qualitative fashion.

When two or more substances are brought together, they may react chemically to form one or more new substances. Consider the reaction shown by the generalized chemical equation:

$$A + B \rightarrow C + D$$
Reactants   Products

The arrow indicates the direction of the reaction and has the meaning of "react to form" or "yield," so the equation is read "Substances A and B react to form substances C and D." The original substances A and B are called the **reactants,** and the new substances C and D are called the **products.** In any chemical reaction, three things take place:

1. The reactants disappear or are diminished.
2. New substances appear as products that have different chemical and physical properties from the original reactants.
3. Energy (heat, light, electricity, sound) is either released or absorbed, although sometimes the energy change is too small to be detected easily.

Now let's turn our original reaction around, so that when C and D are combined, A and B are formed as the products. We get

$$C + D \rightarrow A + B$$

If we start with A and B, we get C and D; if we start with C and D, we get A and B. This is called a **reversible reaction,** and can be written

$$A + B \rightleftarrows C + D$$

The *double arrow* means that the reaction may proceed in either direction.

In such a reversible reaction, all four substances (A, B, C, and D) are present in the reaction vessel. On a molecular scale, molecules of A and B are interacting to form molecules of C and D (we called this the *forward reaction*). At the same time, molecules of *C* and *D* are interacting to form molecules of *A* and *B* (this is termed the *reverse reaction*). When the two competing reactions are occurring at the same rate, we say that the system is in equilibrium. **Equilibrium** is a dynamic process in which the reactants are interacting to form the products at the same rate that the products are interacting to form the reactants. Although no net change occurs in the number of molecules of each reactant and product present, individual molecules are changing back and forth continuously because of interactions with one another. The amounts of substances present at equilibrium depend upon the temperature, pressure, and the concentrations of the reactants and products.

A simple example of a reversible reaction involves brown nitrogen dioxide gas (NO_2) and colorless dinitrogen tetroxide gas (N_2O_4).

$$2\ NO_2(g) \rightleftarrows N_2O_4(g) + \text{heat}$$

(A *g* in parentheses after a formula means the substance is a gas, an *s* means solid, an *l* means liquid, and an *aq* means aqueous solution.) If either substance is put into a closed container, it is not long before equilibrium is established and molecules of both substances are present and changing back and forth at the same rate.

In 1888 French chemist Henri Le Châtelier (luh-shah-tuh-LYAY) discovered a powerful law that forecasts the direction taken by a chemical reaction in response to a particular change of conditions. It guides chemists in producing desired products with a minimum of waste. **Le Châtelier's principle** states:

> **Whenever the conditions of a system at equilibrium (temperature, pressure, concentration) are changed, the system will shift in the direction that counteracts the change.**

For example, in the reversible reaction just mentioned,

$$2\ NO_2(g) \rightleftarrows N_2O_4(g) + \text{heat}$$
(brown)   (colorless)

the forward reaction liberates heat to the surroundings, and the reverse reaction absorbs heat from the surroundings. If the system is at equilibrium and then heated, the equilibrium shifts to the left to absorb heat; that is, the reverse reaction is favored and more NO_2 will be formed at the expense of some N_2O_4 (Fig. 13.2). Also, if the system is put under pressure, the equilibrium will shift to reduce the pressure by decreasing the number of gas molecules present, because pressure is directly proportional to the number of gas molecules. The forward reaction produces one gas molecule (N_2O_4) from two (2 NO_2), so an increase in pressure will shift this system toward the right (Fig. 13.3).

(a)

(b)

Figure 13.2 The effect of temperature on equilibrium between brown NO_2 and colorless N_2O_4.
(a) At a higher temperature, the equilibrium shifts toward more NO_2 because energy is absorbed by a shift in that direction. (b) At a lower temperature, the equilibrium shifts in favor of more N_2O_4 because energy is liberated by a shift in that direction. (Notice the lessening of the characteristic brown color of NO_2.)

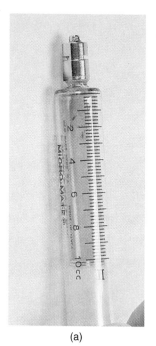
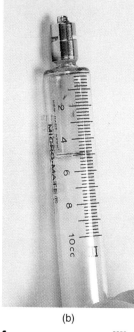

(a)   (b)

Figure 13.3 The effect of pressure on equilibrium between brown NO_2 and colorless N_2O_4. (a) Brown NO_2 gas and colorless N_2O_4 gas are in equilibrium. (b) A few seconds after the pressure has been increased (by decreasing the volume), the equilibrium has shifted to produce fewer molecules in the syringe and thus to lower the pressure. That is, because two gaseous NO_2 molecules form only one gaseous N_2O_4 molecule, the equilibrium shifts toward more of the colorless N_2O_4 and less of the brown NO_2. The lighter color in (b) than in (a) confirms the direction of equilibrium shift.

EXAMPLE 13.1 Using Le Châtelier's Principle

Consider the equation for the formation of sulfur trioxide from sulfur dioxide and oxygen.

$$2\ SO_2(g) + O_2(g) \rightleftarrows 2\ SO_3(g) + \text{heat}$$

If the system were at equilibrium and the vessel were cooled, would the equilibrium shift to the right, to the left, or not shift at all?

Solution

The system would shift in such a way as to counteract being cooled. That is, it would liberate heat by shifting to the right, forming more sulfur trioxide.

CONFIDENCE QUESTION 13.1

If the system in Example 13.1 were at equilibrium and the volume of the vessel was decreased (thereby increasing the pressure), would the equilibrium shift to the right, to the left, or not shift at all? (*Hint:* Note that all the substances are gases.)

Every reaction is to some extent reversible. That is, there will always be present some of every kind of atom or molecule that can possibly occur in a reaction. For example, consider the reaction

$$2\ HI \rightarrow H_2 + I_2$$

The equation says that hydrogen iodide decomposes into hydrogen and iodine. However, any sample containing both hydrogen and iodine will also contain some hydrogen iodide, even though it may be a very small amount. To show explicitly that this is the case, we could write the reaction with a smaller arrow going from right to left:

$$2\ HI \rightleftarrows H_2 + I_2$$

Writing the reaction this way emphasizes its reversibility, however slight. In the discussions that follow we will generally omit the smaller arrow from right to left. It will be understood that at equilibrium some molecules of all the various reactants and products are always present in the reaction.

13.3 Balancing Equations

Learning Goals:

To explain why chemical equations must be balanced.

To balance chemical equations.

A chemical reaction is simply a rearrangement of atoms in which some of the original chemical bonds are broken and new bonds are formed to give different chemical structures (Fig. 13.4). Only an atom's valence electrons are directly involved in a chemical reaction. The nucleus, and hence the atom's identity as a particular element, is unchanged.

A chemical equation can be written for each chemical reaction. The correct chemical formulas for the reactants and products must be used and

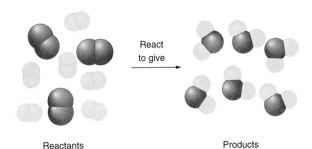

Reactants → Products

Figure 13.4 A chemical reaction, such as hydrogen and oxygen forming water, is a rearrangement of atoms.
Bonds are broken in the reactants, and new bonds are formed to give the products. No atoms can be lost, gained, or changed in identity.

cannot be changed. For example, the decomposition of hydrogen iodide is written $HI \rightarrow H_2 + I_2$. However, until the equation is *balanced*, it does not express the actual *ratio* in which the substances react and form. Most chemical reactions can be balanced by trial and error, using three simple principles.

1. The same number of atoms of each element must be represented on each side of the reaction arrow, because no atoms can be gained, lost, or changed in identity during a chemical reaction.

The equation $HI \rightarrow H_2 + I_2$ is unbalanced because two atoms of both H and I are represented on the right side, but only one of each is shown on the left side.

2. You may manipulate only the **coefficients**—the numbers in front of the formulas, which designate the relative amounts of the substances—and not the *subscripts*, which denote the correct formulas of the substances.

Thus, you *cannot* balance the equation above by changing the formula of HI to H_2I_2. However, you can place a coefficient of 2 before the formula of HI. The 2 HI represents two molecules of hydrogen iodide, each made up of one hydrogen atom and one iodine atom. This gives $2\,HI \rightarrow H_2 + I_2$ and balances the equation. (Just as with subscripts, a coefficient of 1 is not written, just understood.)

3. The final set of coefficients should be whole numbers (not fractions) and should be the smallest whole numbers that will do the job.

For example, $2\,HI \rightarrow H_2 + I_2$ is appropriate, but not $HI \rightarrow \frac{1}{2} H_2 + \frac{1}{2} I_2$ or $4\,HI \rightarrow 2\,H_2 + 2\,I_2$.

The following tips will help.

1. You must be able to count atoms. Consider $4\,Al_2(SO_4)_3$. The subscript 2 multiplies the Al; the subscript 4 multiplies the O; the subscript 3 multiplies everything in parentheses; the coefficient 4 multiplies the whole formula. Therefore, a total of 8 Al atoms, 12 S atoms, and 48 O atoms are on hand. (If you were counting sulfate ions, there would be 12.)

2. Start with an element that is present in only one place on each side of the arrow. For example, when balancing $C + SO_2 \rightarrow CS_2 + CO$, start with S or O, not C.

3. Find the lowest common denominator for each element that is present in only one place on each side, and insert coefficients in such a way as to get the same number of atoms of that element on each side.

   For example, in $C + SO_2 \rightarrow CS_2 + CO$, two atoms of sulfur show on the product side and only one shows on the reactant side. The *lowest common denominator* (smallest whole number into which each goes a whole number of times) is two. So, put a coefficient 2 in front of the SO_2, giving $C + 2\,SO_2 \rightarrow CS_2 + CO$. Next, take care of the oxygen: Four O show on the reactant side (in $2\,SO_2$) and only one on the product side (in CO). Putting a 4 before the CO gives $C + 2\,SO_2 \rightarrow CS_2 + 4\,CO$. Finally, balance the carbons. There are now five C on the product side and one on the reactant side, so place a 5 before the C on the reactant side to get

$$5\,C + 2\,SO_2 \rightarrow CS_2 + 4\,CO$$

A quick recheck shows that all is in balance (two S, four O, five C) and that it is not possible to get a set of smaller coefficients by dividing them by, say, 2 or 3.

4. When *polyatomic ions* remain intact during the reaction, balance them as a unit.

   In Al + H_2SO_4 → $Al_2(SO_4)_3$ + H_2, for example, you would balance Al atoms, H atoms, and SO_4^{2-} (sulfate ions). (What coefficient would you put before the H_2SO_4?)

5. If you come to a point where everything would be balanced except that a *fractional* coefficient has to be used in one place, multiply all the coefficients by whatever number is in the denominator of the fraction.

   For example, in C_2H_2 + O_2 → CO_2 + H_2O, putting a 2 in front of the CO_2 and leaving an understood 1 in front of both H_2O and C_2H_2 would require a $\frac{5}{2}$ in front of the O_2 (the oxygen is left for last because it is present in two places on the product side). This would give $C_2H_2 + \frac{5}{2} O_2 \rightarrow 2 CO_2 + H_2O$. But we generally don't want fractional coefficients, so multiply the whole equation by 2 (the number in the denominator), thus getting

   $$2\ C_2H_2 + 5\ O_2 \rightarrow 4\ CO_2 + 2\ H_2O$$

For practice in this fundamental chemical skill, let's go through the process of balancing two more equations in Example 13.2. Each example reaction also illustrates a basic type of reaction that we want you to understand.

1. A **combination reaction** occurs when at least two reactants combine to form just one product: $A + B \rightarrow AB$.
2. A **decomposition reaction** occurs when only one reactant is present and breaks into two (or more) products: $AB \rightarrow A + B$.

EXAMPLE 13.2 Balancing Equations

(a) An example of a combination reaction is the reaction of hydrogen and oxygen to form water. Balance the equation. (Recall that elemental hydrogen and oxygen exist as diatomic molecules.)

$$H_2 + O_2 \rightarrow H_2O$$

(b) A common method of producing oxygen in the laboratory is the decomposition reaction of potassium chlorate ($KClO_3$) by heating. Balance the equation.

$$KClO_3 \rightarrow KCl + O_2$$

Solution

(a) The hydrogens are balanced (two on the left and two on the right), but there are two oxygens on the left and only one on the right. Since the lowest common denominator of 2 and 1 is 2, to balance the oxygen we place a coefficient 2 in front of the H_2O.

$$H_2 + O_2 \rightarrow 2\ H_2O$$

This step balances the oxygen, but now the hydrogen is not balanced (four on the right and two on the left). To rebalance the hydrogen, we place a coefficient 2 in front of the H_2. The balanced equation is

$$2\ H_2 + O_2 \rightarrow 2\ H_2O$$

(b) Each element is present in only one place on each side, so it does not matter which one we start with. We see that the potassium and chlorine atoms are balanced already. Only the oxygen is not. With three atoms of O on the left and two on the right, the lowest common denominator is six. Therefore, we put a 2 in front of the $KClO_3$ and a 3 in front of the O_2, giving

$$2\ KClO_3 \rightarrow KCl + 3\ O_2$$

The oxygen is balanced at six atoms on each side, but the potassium and chlorine are now unbalanced. However, we can balance them by putting a 2 in front of the KCl. The balanced equation is

$$2\ KClO_3 \rightarrow 2\ KCl + 3\ O_2$$

CONFIDENCE QUESTION 13.2

Magnesium and oxygen gas undergo a reaction to form white magnesium oxide. An intensely bright light is emitted, as shown in ● Fig. 13.5. Balance the equation: $Mg + O_2 \rightarrow MgO$. Is this a combination reaction, or a decomposition reaction?

13.4 Energy and Rate of Reaction

Learning Goals:

To describe the role of energy in chemical reactions.

To describe the factors that affect the rate of a reaction.

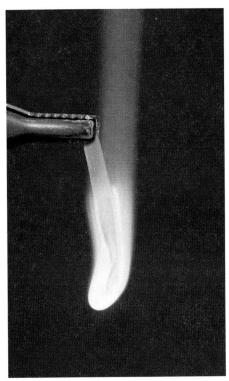

Figure 13.5 A combination reaction.
Magnesium metal and oxygen in the air combine to form magnesium oxide (the "smoke"). Magnesium is used in fireworks because of the bright light produced by the reaction.

All chemical reactions involve a change in energy. The energy is either released or absorbed in the form of heat, light, electrical energy, or sound. If energy is released to the surroundings in a chemical reaction, it is an **exothermic reaction** (Fig. 13.6). If energy is absorbed from the surroundings, it is an **endothermic reaction.** An example of a common exothermic reaction is the burning of natural gas, which is composed primarily of methane, CH_4.

$$CH_4 + 2\,O_2 \rightarrow CO_2 + 2\,H_2O + \text{energy}$$

An example of an endothermic reaction is the production of ozone, the triatomic molecule of oxygen.

$$3\,O_2 + \text{energy} \rightarrow 2\,O_3$$

This reaction occurs in the upper atmosphere, where the energy is provided by the ultraviolet radiation from the Sun. It also occurs near electric discharges, and ozone's pungent odor can be detected when we are near electrical sparking. Ozone is a worrisome pollutant at Earth's surface, but its presence in the upper atmosphere provides us with vital protection from ultraviolet radiation (see Chapter 19).

The energy associated with a chemical reaction is related to the bonding energies between the atoms that form the molecules. During a chemical reaction, some chemical bonds are broken and others are formed. Energy must be absorbed to break bonds, and energy is released when bonds are formed. If more energy is released than absorbed, the reaction is exothermic; if more is absorbed than is released, the reaction is endothermic. The oxygen-ozone reaction is illustrated in Fig. 13.7. Energy is released when the bonds are formed in the ozone molecules, but the amount is less than that absorbed in breaking the oxygen molecules' bonds. Thus there is a net absorption of energy from the surroundings, and the reaction is endothermic.

When methane burns in air, energy is released. The chemical energy of the bonds in the products is less than the chemical energy of the bonds in the reactants. That is, more energy is given off when the new bonds form than is absorbed in breaking the old bonds. This is illustrated in Fig. 13.8.

We are all aware of the necessity of striking a match before it ignites. In this example we must contribute some energy—through friction—to initiate the chemical reaction of a match burning. Another example is the burning of methane gas in, say, a gas stove. A flame or spark is necessary to ignite the methane, because the C—H and O—O bonds must be broken initially. Once the gas is ignited, the net energy released breaks the bonds of still more CH_4 and O_2 molecules, and the reaction proceeds continuously, giving off energy in the form of heat and light.

The energy necessary to start a chemical reaction is called the **activation energy.** It is a measure of the minimum kinetic energy colliding molecules must possess in order to react chemically. However, once the activation energy is supplied, an exothermic reaction can release more energy than was supplied. Think of a boulder resting next to a low barrier wall at the edge of a cliff. Without the initial input of energy to raise the boulder to the

 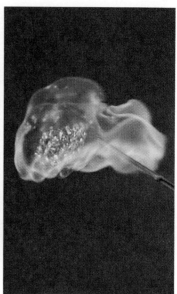

Figure 13.6 The reaction of a mixture of hydrogen and oxygen to form water.
Hydrogen gas (left) is creating soap bubbles, which are ignited with a candle (right). The reaction is exothermic (liberates energy).

top of the barrier, it cannot fall over the cliff and release a much larger amount of energy than was initially supplied.

A similar situation exists for an exothermic chemical reaction. Once the activation energy is supplied, the formation of new chemical bonds can release more energy than was absorbed in breaking the original bonds. This is usually perceived on a large scale as the surroundings becoming warmer. In endothermic reactions, more energy is absorbed to break chemical bonds than is generated when new bonds are formed. In this case there is a net intake of energy of reaction (E_R) from the surroundings, which is usually perceived on a large scale as a cooling of the surroundings. The activation energy (E_{act}) is not fully recovered. Thus the "humps" (called *energy barriers*) in Figures 13.7 and 13.8 indicate the activation energy that must be supplied for the reaction to proceed.

An explosion happens when an exothermic chemical reaction takes place so rapidly that a large change in the volume of the gases occurs, with an almost instantaneous liberation of the energy of the reaction. A **combustion reaction,** in which a substance reacts with oxygen to burst into flame and form an oxide, proceeds more slowly than an explosion and yet is still quite rapid. Common examples of combustion are the burning of natural gas, coal, paper, and wood. All carbon-hydrogen compounds (hydrocarbons) and carbon-hydrogen-oxygen compounds produce energy and give carbon dioxide and water when burned completely.

EXAMPLE 13.3 Complete Combustion of Hydrocarbons

One of the components of gasoline is the hydrocarbon named heptane, C_7H_{16}. Write the balanced equation for its complete combustion.

Solution

Write the formula for heptane plus that of oxygen gas, O_2, followed by a reaction arrow.

$$C_7H_{16} + O_2 \rightarrow$$

The products of complete hydrocarbon combustion are always CO_2 and H_2O, so write their formulas on the product side.

$$C_7H_{16} + O_2 \rightarrow CO_2 + H_2O$$

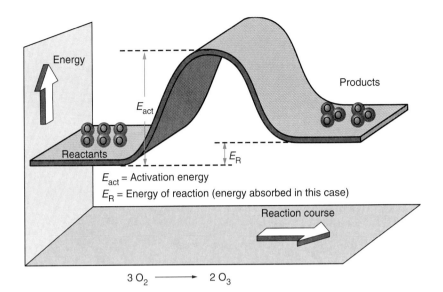

Figure 13.7 Endothermic reaction.
In the oxygen-ozone reaction, there is a net absorption of energy (E_R in this figure) because the bonds in two ozone molecules have a higher total energy than the bonds in three oxygen molecules.

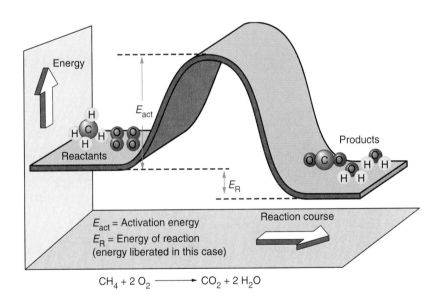

Figure 13.8 Exothermic reaction.
This reaction results in a net release of energy (E_R) because the bonds in the products (CO_2 and H_2O) have less total energy than the bonds in the reactants (CH_4 and O_2).

Now balance the equation, starting with either C or H, and leaving O until last. The answer is

$$C_7H_{16} + 11\ O_2 \rightarrow 7\ CO_2 + 8\ H_2O$$

CONFIDENCE QUESTION 13.3

Write and balance the equation for the complete combustion of the hydrocarbon named propane, C_3H_8, a common fuel gas.

When gasoline in the form of heptane (C_7H_{16}) is burned completely, the combustion reaction is

$$C_7H_{16} + 11\ O_2 \rightarrow 7\ CO_2 + 8\ H_2O$$

Given insufficient time or oxygen for complete combustion, sooty black carbon (C) and the poisonous gas carbon monoxide (CO) will be products, also. The automobile engine has insufficient oxygen to burn the hydrocarbon completely, so the reaction is more commonly

$$C_7H_{16} + 9\ O_2 \rightarrow 4\ CO_2 + 2\ CO + C + 8\ H_2O$$

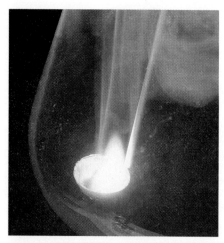 

Figure 13.9 Concentration reaction rate.
The effect of concentration on reaction rate is apparent from the greater intensity of light coming from phosphorus burning in 100% oxygen (left) as compared to burning in air's 21% oxygen (right).

The black color of some exhaust gases indicates the presence of large amounts of carbon and shows that oxidation is incomplete. Because of the significant amounts of CO formed, running an automobile engine in a closed garage can be fatal. When tobacco burns, CO is one of the health-damaging gases released.

Rate of Reaction

The rate of a reaction depends on temperature, concentration, surface area, and the possible presence of a catalyst (a substance that can speed up a reaction). To react, molecules (or atoms, or ions) must collide in the proper orientation and with enough kinetic energy to break bonds—the activation energy. The kinetic energy involved in any given collision may or may not be great enough for reaction. Molecules in reactions that have low activation energies react readily, because a greater number of the collisions are *effective*—have at least the minimum energy and collide at the right places in the molecules to break bonds and form new substances. Recall from Chapter 5 that if heat is added to a substance and raises its temperature, the average speed and kinetic energy of the molecules will be increased. For chemical reactions, this added heat will result in more collisions and harder collisions, and the reaction rate will increase dramatically.

In fact, for many reactions, a 10°C increase in temperature leads to an approximate doubling of the reaction rate. For example, the reaction $2\ NO_2(g) + F_2(g) \rightarrow 2\ NO_2F(g)$ proceeds 1.8 times as fast at 37°C as it does at 27°C, which is close to a doubling of the rate. (Snakes, lizards, and other cold-blooded creatures warm themselves in the Sun in order to speed up their metabolism and become more active.)

Generally, the greater the concentration of the reactants, the greater the rate of the reaction. The greater concentration means that the molecules are packed more closely. Thus more collisions will occur each second, so naturally the reaction rate should increase (Fig. 13.9). Astronauts Gus Grissom, Roger Chaffee, and Ed White lost their lives in 1967 when fire broke out in the Project Apollo spacecraft in which they were training on the ground. The environment of the capsule was 100% oxygen, and thus everything burned furiously. Since then, an environment with much smaller oxygen concentration has been used in space capsules.

Surface area can play a surprisingly important role in the rate of a reaction. You would be startled to see a lump of coal or a pile of grain explode when a match was held to it. But get finely divided coal dust or grain dust in the air, and its enormous surface area will give a combustion reaction that

Figure 13.10 A grain-dust explosion destroys several silos.
If finely divided grain dust is suspended in air, perhaps in a grain elevator, the enormous surface area of the dust particles can cause such a fast reaction that an explosion occurs.

takes place with explosive speed (Fig. 13.10). Knowledge of basic scientific principles can sometimes mean the difference between life and death.

In some reactions the activation energy required can be lowered by adding a **catalyst,** a substance that increases the rate of reaction but is not itself consumed in the reaction.* Some catalysts act by providing a surface on which the reactants are concentrated. The majority of catalysts work by providing a new reaction pathway with lower activation energy; in effect, they lower the energy barrier (Fig. 13.11).

By definition, catalysts are not consumed in the reaction, but they are involved in it. That is, they unite with a reactant to form an intermediate substance that takes part in the chemical process and then decomposes to release the catalyst in its original form. For example, in the manufacture of sulfuric acid, sulfur dioxide must react with oxygen to form sulfur trioxide:

$$2\ SO_2 + O_2 \rightarrow 2\ SO_3 \quad \text{(slow)}$$

Yet, this reaction is very slow unless nitrogen oxide (NO) is added as a catalyst. The reaction then proceeds in two fast steps. The NO combines with oxygen to form NO_2. The NO_2 then reacts with SO_2 to form SO_3, releasing the NO catalyst, which can be used again. Adding together the two fast reactions shows that the net result is the same as the one slow reaction:

$$2\ NO + O_2 \rightarrow 2\ NO_2 \quad \text{(fast)}$$

$$2\ SO_2 + 2\ NO_2 \rightarrow 2\ SO_3 + 2\ NO \quad \text{(fast)}$$

$$2\ SO_2 + O_2 \rightarrow 2\ SO_3 \text{ net reaction (fast)}$$

Another example of the use of a catalyst occurs in the decomposition of potassium chlorate ($KClO_3$). When heated to 400°C, the white crystals break down, producing potassium chloride and releasing oxygen.

$$2\ KClO_3 + \text{heat} \rightarrow 2\ KCl + 3\ O_2$$

The process, however, is slow and heat requirements are high. If a small amount of manganese(IV) oxide (MnO_2) is mixed with the $KClO_3$, the reaction takes place rapidly at 250°C. The manganese(IV) oxide is not consumed in the reaction but acts only as a catalyst. The presence of a catalyst is indicated by placing its formula over the reaction arrow.

$$2\ KClO_3 + \text{heat} \xrightarrow{MnO_2} 2\ KCl + 3\ O_2$$

A common example of catalysis is the use of catalytic converters in cars. Beads of a platinum (Pt) or palladium (Pd) catalyst are packed into a

---

*A catalyst never slows down a reaction. A substance that reduces the rate of a chemical reaction is called an *inhibitor.* It generally acts by tying up a catalyst for the reaction or interacting with a reactant to reduce its concentration.

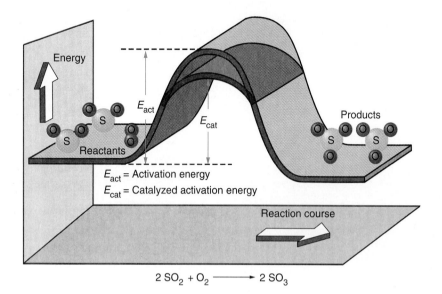

Figure 13.11 How a catalyst works.
A catalyst generally operates by providing a new reaction pathway with a lower activation energy requirement ($E_{cat} < E_{act}$). Thus more collisions possess enough energy to break bonds, and the reaction rate increases.

chamber through which the exhaust gases must pass before they leave the tailpipe. During the passage through the converter, CO and NO are changed to CO_2 and N_2, which are normal components of the atmosphere. This results in a great decrease in air pollution.

Catalysts are used extensively in manufacturing, and they also play an important part in biochemical processes. The human body has many thousands of biological catalysts called *enzymes* that act to control various physiological reactions. The names of enzyme catalysts usually end in *-ase*. During digestion, lactose (milk sugar) is broken down in a reaction catalyzed by the enzyme lactase. Many infants and adults, particularly those of African and Asian descent, have a deficiency of lactase and thus are unable to digest milk. Glowworms and fireflies use enzymes to catalyze *chemiluminescence* reactions in which light (but little or no heat) is emitted from the excited molecular products of a reaction (● Fig. 13.12). The bombardier beetle also uses enzymes (● Fig. 13.13).

13.5 Acids and Bases

Learning Goals:

To describe the properties of acids and bases.

To write equations for three types of double-replacement reactions.

The classification of substances as acids or bases originated early in the history of chemistry. An acid, when dissolved in water, has the following properties:

1. Conducts electricity.
2. Changes the color of litmus dye from blue to red.
3. Tastes sour (but never taste an acid or anything else in a lab).
4. Reacts with a base to neutralize its properties.
5. Reacts with active metals to liberate hydrogen gas.

A base, when dissolved in water, has the following properties:

1. Conducts electricity.
2. Changes the color of litmus dye from red to blue.
3. Reacts with an acid to neutralize its properties.

One of the first theories formulated to explain acids and bases was put forth in 1887 by Svante Arrhenius (ar-RAY-nee-us), a Swedish chemist (● Fig. 13.14). He proposed that the characteristic properties of aqueous solutions of acids and bases are due to the hydrogen ion (H^+) and the hydroxide ion (OH^-), respectively.

According to the *Arrhenius acid-base concept*, when a substance such as colorless, gaseous hy-

Figure 13.12 Chemiluminescence.
Fireflies employ a light-emitting reaction between the compounds luciferin and adenosine triphosphate (ATP). The enzyme catalyst is named luciferase.

Figure 13.13 Bombardier beetle.
An enzyme catalyzes the highly exothermic reaction that gives the hot spray of chemicals with which the beetle protects itself.

drogen chloride (HCl) is added to water, most HCl molecules ionize into H^+ ions and Cl^- ions, which exist in the solution in equilibrium with the HCl molecules that are not ionized.

$$HCl \rightleftharpoons H^+ + Cl^-$$

The acidic properties of HCl are due to the H^+ ions. Actually, when hydrogen chloride is placed in water, hydrogen ions are transferred from the HCl to the water molecules, as shown by the following equation:

$$HCl + H_2O \rightleftharpoons H_3O^+ + Cl^-$$

This reaction proceeds almost entirely from left to right. The H_3O^+, called the **hydronium ion,** and Cl^- are formed in this reaction (Fig. 13.15). An Arrhenius **acid** is a substance that gives hydrogen ions, H^+, (or hydronium ions, H_3O^+) in water.

Acids are classified as strong or weak. A *strong acid* is one that ionizes almost completely in solution (Fig. 13.16a). For example, HCl, nitric acid (HNO_3), and sulfuric acid (H_2SO_4) are common strong acids. In water, they ionize almost completely, as the equilibrium arrow showed in the case of HCl. A *weak acid* is one that does not ionize to any great extent; that is, only a small number of its protons unite with H_2O to form H_3O^+ ions (Fig. 13.16b). Acetic acid, $HC_2H_3O_2$, is the most common weak acid. In aqueous solution we have

$$HC_2H_3O_2 + H_2O \rightleftharpoons H_3O^+ + C_2H_3O_2^-$$

Note how the equilibrium arrow indicates that ionization is not substantial for acetic acid in water.

When pure sodium hydroxide (NaOH), a white solid, is added to water, it dissolves, releasing Na^+ and OH^- into the solution. The basic properties of

Figure 13.14 Svante Arrhenius (1859–1927).
Arrhenius postulated the existence of ions in 1884, even before the discovery of the electron, and wrote his doctoral thesis explaining electrolytes on the basis of ions. His professors disagreed and only reluctantly granted his degree. After his model was confirmed by additional evidence, he won the Nobel Prize in 1903. He also developed the idea of activation energy and the acid-base theory that bears his name.

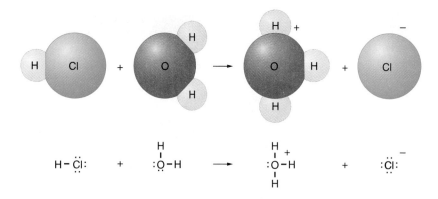

Figure 13.15 The reaction of hydrogen chloride with water.
When gaseous hydrogen chloride (HCl) molecules are added to water (H_2O), hydronium ions (H_3O^+) and chloride ions (Cl^-) are formed.

NaOH solutions (commonly known as lye) are due to the hydroxide ions, OH^-. An Arrhenius **base** is a substance that produces hydroxide ions, OH^-, in water. However, a substance need not *initially* contain hydroxide ions to have the properties of a base. For example, ammonia (NH_3) contains no OH^-, but in water solutions it is a weak base. Its molecules react to a slight extent with water to form OH^-.

$$NH_3 + H_2O \rightleftarrows NH_4^+ + OH^-$$

Water ionizes, but only slightly, as shown in the following equation.

$$H_2O + H_2O \rightleftarrows H_3O^+ + OH^-$$

Thus, all aqueous solutions contain both H_3O^+ and OH^-. For pure water the concentrations of H_3O^+ and OH^- are equal, and the liquid is *neutral*. An *acidic solution* contains a higher concentration of H_3O^+ than OH^-. If the concentration of OH^- is higher than that of H_3O^+, it is a *basic solution* (Fig. 13.17).

It is common practice to designate the relative acidity or basicity of a solution by citing its **pH**, which is a measure (on a logarithmic scale) of the concentration of hydrogen ion (or hydronium ion) in the solution. A pH of 7 indicates a neutral solution. Values from 6 down to −1 indicate increasing acidity, with each drop of 1 in value meaning a *tenfold* increase in acidity. Similarly, pH values from 8 up to 15 indicate increasing basicity, with each increase of 1 in value meaning a tenfold increase in basicity. Figure 13.18 illustrates this concept and shows the pH values of some common solutions.

Most body fluids have a normal pH range, and a continued deviation from the normal usually indicates some disorder in a body function. Thus the

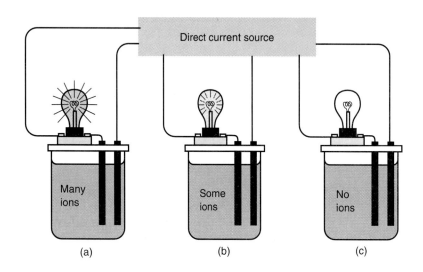

Figure 13.16 Strong and weak acids.
(a) A strong acid such as HCl ionizes almost completely in water, so the bulb glows brightly. (b) A weak acid such as acetic acid only ionizes slightly, so the bulb glows dimly. (c) A nonconductor such as sugar remains as molecules in water solution, so the bulb does not glow at all.

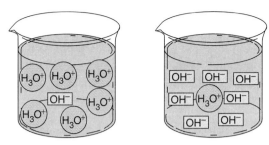

Figure 13.17 Acidic and basic solutions.
An acidic solution (left) has a higher concentration of H_3O^+ than OH^-. In a basic solution (right) the situation is reversed.

pH value can be used as a means of diagnosis. For example, if the pH of blood is not between 7.35 and 7.45, illness or death can result.

The pH of a solution is usually measured by a pH meter that can find the value to several decimal places (● Fig. 13.19). However, approximate values can be found by the use of an acid–base indicator, which is a chemical that changes colors over a narrow pH range. A common indicator is *litmus*, an organic compound derived from lichens, which is red below pH 5 and blue above pH 8. The red and blue litmus papers used in labs are just short strips of paper impregnated with litmus. Red litmus turns blue in base, blue litmus turns red in acid, and neither changes color significantly if the solution is about neutral.

Acids, particularly sulfuric acid, are important in industry. Sulfuric acid is used in refining petroleum, "pickling" steel, and manufacturing fertilizers and numerous other products. A dilute solution of hydrochloric acid is present in the human stomach to help digest food. Many weak acids are present in our foods; for example, citric acid ($H_3C_6H_5O_7$) in citrus fruits, carbonic acid (H_2CO_3) and phosphoric acid (H_3PO_4) in soft drinks, and acetic acid ($HC_2H_3O_2$) in vinegar.

An important property of an acid is the disappearance of its characteristic properties when brought into contact with a base, and vice versa. This is called an **acid-base reaction** (● Fig. 13.20). The H^+ of the acid unites with the OH^- of the base to form water, and the cation of the base combines with the anion of the acid to form a salt. Thus we can generalize: An acid and a hydroxide base react to give water and a salt. Of course, you are aware that NaCl is called "salt," but actually

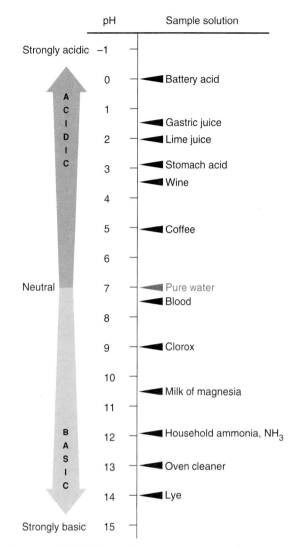

Figure 13.18 Acids, bases, and pH values.

there are many salts. A **salt** is an ionic compound composed of any cation except H^+ and any anion except OH^-. Examples are potassium chloride (KCl) and calcium phosphate, $Ca_3(PO_4)_2$. The salt may remain dissolved in water (KCl does), or it may form solid particles (precipitate) if it is insoluble like $Ca_3(PO_4)_2$.

EXAMPLE 13.4 Acid-Base Reactions

If you have an "acid stomach" (excess HCl), you might take a milk of magnesia tablet, $Mg(OH)_2$, to neutralize it. Write the balanced chemical equation for the reaction.

Figure 13.19 A pH meter.
The digital display gives the pH of the solution in the beaker. Is it acidic, neutral, or basic?

Solution

You know that HCl is an acid and that $Mg(OH)_2$ is a base (note the hydroxide ion). Therefore, the products of this acid–base reaction must be water and a salt. Write the formulas for both reactants, put in the reaction arrow, and then write the formula for water.

$$HCl + Mg(OH)_2 \rightarrow H_2O + \text{(a salt)}$$

A crucial step is writing the correct formula for the salt (see Section 12.4). Write the cation from the base first in the salt formula, then follow it with the anion of the acid: $Mg^{2+}Cl^-$, in this case. You can see that a 1 to 1 ratio of ions does not give electrical neutrality; two Cl^- are needed for each Mg^{2+}. So the correct formula for the salt is $MgCl_2$. Add it to the product side, and then balance the equation. (*Note:* If you have trouble balancing the equation for an acid–base reaction, you probably have an incorrect formula for the salt.) The answer is:

$$2\ HCl + Mg(OH)_2 \rightarrow 2\ H_2O + MgCl_2$$

CONFIDENCE QUESTION 13.4

Aluminum hydroxide is another popular antacid ingredient. Complete and balance the equation for the reaction of this base with stomach acid:

$$HCl + Al(OH)_3 \rightarrow$$

A chemical common to many households is sodium hydrogen carbonate, $NaHCO_3$, commonly known as baking soda. Acids act on the hydrogen carbonate ion of baking soda to give off carbon dioxide, a gas (Fig. 13.21).

$$NaHCO_3 + H^+ \text{ (from an acid)} \rightarrow$$
$$Na^+ + H_2O + CO_2(g)$$

This reaction is involved in the leavening process in baking. Baking powders contain baking soda plus an acidic substance such as $KHC_4H_4O_6$ (cream of tartar). When this combination is dry, no reaction occurs; but when water is added, CO_2 is given off. This is called an **acid-carbonate reaction:** An acid and a carbonate (or hydrogen carbonate) react to give carbon dioxide, water, and a salt. Figure 13.22 shows how this type of reaction can be used to neutralize acid spills.

EXAMPLE 13.5 Acid-Carbonate Reactions

Another way to relieve an over-acid stomach is to take an antacid tablet that contains calcium carbon-

Figure 13.20 Acid-base reaction.
Slaked lime, $Ca(OH)_2$, is added to a lake to neutralize acid rain.

Double-Replacement Reactions

Acid-base and acid-carbonate reactions are types of **double-replacement reactions**—those in which the positive and negative components of the two compounds "change partners"; that is,

$$AB + CD \rightarrow AD + CB$$

Note, for example, in the acid-base reaction of HCl and NaOH, the positive part of the acid (H$^+$) attaches to the negative part of the base (OH$^-$) to form HOH (another way of writing H$_2$O). And the positive part of the base (Na$^+$) attaches to the negative part of the acid (Cl$^-$).

$$HCl + NaOH \rightarrow HOH + NaCl$$

As another example, in the acid-carbonate reaction of H$_2$SO$_4$ and K$_2$CO$_3$, the positive part of the acid (H$^+$) attaches to the negative part of the carbonate (CO$_3{}^{2-}$) to form H$_2$CO$_3$ (which immediately decomposes to CO$_2$ and H$_2$O). The positive part of the carbonate (K$^+$) attaches to the negative part of the acid (SO$_4{}^{2-}$).

$$H_2SO_4 + K_2CO_3 \rightarrow CO_2 + H_2O + K_2SO_4$$

Figure 13.21 An acid-carbonate reaction. HCl(aq) reacts with NaHCO$_3$ to release CO$_2$ gas. H$_2$O and NaCl are the other products.

ate, CaCO$_3$. Write the balanced equation for the reaction between stomach acid (HCl) and such an antacid (CaCO$_3$).

Solution

This is an acid-carbonate reaction, so the products are carbon dioxide, water, and a salt. Start by writing the formulas for both reactants, the reaction arrow, and the formulas for carbon dioxide and water.

$$HCl + CaCO_3 \rightarrow CO_2 + H_2O + \text{(a salt)}$$

As in the previous example, figure out the correct formula for the salt. The Ca^{2+} and Cl$^-$ will form CaCl$_2$. Add the formula of the salt to the equation, and then balance it.

$$2\,HCl + CaCO_3 \rightarrow CO_2 + H_2O + CaCl_2$$

CONFIDENCE QUESTION 13.5

Complete and balance the equation for the reaction of nitric acid with sodium carbonate, the reaction shown in Fig. 13.22.

$$HNO_3 + Na_2CO_3 \rightarrow$$

Figure 13.22 Another acid-carbonate reaction. Nitric acid (20,000 gal) was spilled from a railroad tank car in Denver in 1983. Firefighters used an airport snow blower to throw sodium carbonate (Na$_2$CO$_3$) onto the HNO$_3$ and neutralize it.

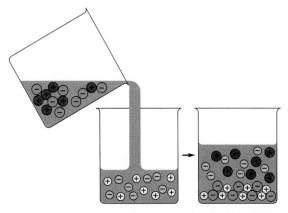

Figure 13.23 A double-replacement precipitation reaction.
The positive ions of one soluble substance combine with the negative ions of another soluble substance to form a precipitate when the solutions are mixed. The other positive and negative ions generally remain dissolved.

Now switch ion partners and put them together so that the positive ion is first and in such a ratio that each compound will be net electrically neutral. To show that the lead(II) iodide is insoluble, place (*s*), for solid, after its formula. To show that potassium nitrate is soluble, put (*aq*) after its formula. Finally, balance the equation. The answer is

$$2\ KI(aq) + Pb(NO_3)_2(aq) \rightarrow PbI_2(s) + 2\ KNO_3(aq)$$

CONFIDENCE QUESTION 13.6
Complete and balance the equation for the double-replacement reaction between aqueous solutions of sodium sulfate and barium chloride:

$$Na_2SO_4(aq) + BaCl_2(aq) \rightarrow$$

When aqueous solutions of two salts are mixed, the "changing of partners" often results in the formation of a **precipitate** (*s* or *ppt*), an insoluble solid that appears when two liquids (usually aqueous solutions) are mixed. Figure 13.23 illustrates how a double-replacement precipitation reaction might look if we could see the ions. The formation of a precipitate of insoluble lead(II) iodide and soluble potassium nitrate from solutions of potassium iodide and lead(II) nitrate is shown in Fig. 13.24 and discussed in Example 13.6. A useful, but not complete, generalization about water solubility and insolubility is that nitrates, acetates, ammonium, and alkali metal compounds are soluble in water.

EXAMPLE 13.6 Double-Replacement Precipitation Reactions

Write the equation for the double-replacement reaction shown in Fig. 13.24.

Solution

First write the correct formulas for the reactants using your knowledge of nomenclature and ionic charges.

$$KI(aq) + Pb(NO_3)_2(aq) \rightarrow$$

13.6 Single-Replacement Reactions

Learning Goals:
To define the terms oxidation and reduction.
To write equations for single-replacement reactions.

We have seen that combustion reactions involve a combination with oxygen. When oxygen combines with another substance (or when an atom or ion loses electrons), we call the process **oxidation**. For instance, the equation for the oxidation of carbon to form carbon dioxide is

$$C + O_2 \rightarrow CO_2$$

When oxygen is removed from a compound (or when an atom or ion gains electrons), the process is called **reduction**. An important reduction process in the steel-making industry is the heating of hematite ore (Fe_2O_3) with coke (C) in a blast furnace:

$$2\ Fe_2O_3 + 3\ C \rightarrow 4\ Fe + 3\ CO_2$$

Oxygen is removed from the iron oxide, so we say that the iron oxide has been reduced. At the same time, oxygen has reacted with the carbon to form

13.6 Single-Replacement Reactions

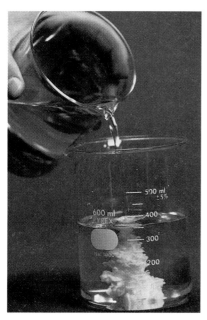

Figure 13.24 The yellow precipitate is lead(II) iodide.
When a clear, colorless solution of potassium iodide is poured into a clear, colorless solution of lead(II) nitrate, yellow lead(II) iodide precipitates.

Oxidation and reduction occur in all **single-replacement reactions,** in which one element replaces another that is in a compound.

$$A + BC \rightarrow B + AC$$

We will deal only with single-replacement reactions in which element A is a metal. In such cases, A will lose its valence electrons and thus be oxidized. B will gain these electrons and be reduced.

The relative *activity* of any metal is its tendency to lose electrons to ions of another metal or to hydrogen ions. It is determined by placing the metal in a solution containing the ions of another metal and observing whether the test metal replaces the one in solution. If it does, it is more active, because it has given electrons to the ions of the metal in solution. For example, Fig. 13.25 shows that when zinc metal is placed in a solution containing Cu^{2+} ions, the zinc loses electrons to the Cu^{2+} ions, producing copper metal and Zn^{2+} ions. On the other hand, copper metal placed in a solution of Zn^{2+} leads to no reaction. Thus zinc is more active than copper. If similar experiments are carried out for all metals (and hydrogen), an **activity series** can be obtained (Fig. 13.26).

carbon dioxide; that is, the carbon has been oxidized. We call such a chemical change an *oxidation-reduction reaction,* or for short, a **redox reaction.**

Redox reactions do not always involve the gain or loss of oxygen. So that the term will apply to a larger number of reactions, redox reactions are also identified in terms of electrons lost (oxidation) or gained (reduction) by atoms or ions. For example, when sodium (Na) combines with chlorine (Cl_2) to form sodium chloride, the sodium atoms lose electrons and are thus said to be oxidized to Na^+. The chlorine atoms gain electrons and are reduced to Cl^-. Thus we have

$$2\ Na + Cl_2 \rightarrow 2\ NaCl$$

Because all the electrons lost by atoms or ions must be gained by other atoms or ions, it follows that oxidation and reduction occur at the same time and at the same rate. Most combination and decomposition reactions (Section 13.3) are also redox reactions, but double-replacement reactions are not.

Figure 13.25 Zinc replaces copper ions.
When a strip of Zn (left) is placed in aqueous, blue $CuSO_4$ (center), it does not take long for the single replacement redox reaction to form metallic copper and colorless $ZnSO_4$ solution (right).

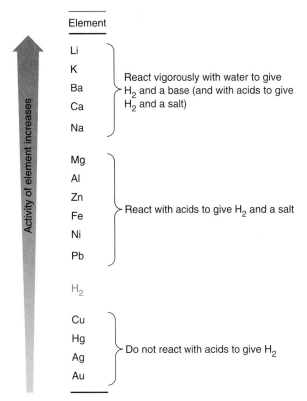

Figure 13.26 A simplified activity series.
In aqueous solutions, activity of the elements increases as you go up the table. For example, Fe is more active than Cu, and Zn is more active than Fe.

EXAMPLE 13.7 Single-Replacement Reactions

Refer to the activity series (Fig. 13.26) and predict if placing copper metal in a solution of silver nitrate will lead to a reaction. If you predict it will, complete and balance the equation.

Solution

Copper is above silver in the activity series, and thus a reaction will occur as shown in Fig. 13.27. The copper atoms will be oxidized to Cu^{2+} ions, and the Ag^+ ions will be reduced to Ag atoms. The nitrate ions, NO_3^-, stay intact and dissolved in solution, but we can write them as if they were joined to the Cu^{2+} ions to give $Cu(NO_3)_2$. The unbalanced equation for the reaction is

$$Cu + AgNO_3(aq) \rightarrow Ag + Cu(NO_3)_2(aq)$$

Figure 13.27 A single-replacement reaction between a copper wire and a solution of silver nitrate.
One product, metallic silver, is evident on the surface of the wire in the solution. Copper(II) nitrate is the other product and remains dissolved in solution, making its presence known by the characteristic blue color of Cu^{2+}.

Because the nitrate ions stay intact during the reaction, they can be balanced as a unit. Two nitrates are showing on the product side, so place a 2 before $AgNO_3$ on the left. Complete the balancing by adding a 2 before the Ag on the right.

$$Cu + 2\,AgNO_3(aq) \rightarrow 2\,Ag + Cu(NO_3)_2(aq)$$

CONFIDENCE QUESTION 13.7

Suppose you place a strip of aluminum metal in a solution of copper(II) sulfate, $CuSO_4(aq)$, and also put a strip of copper metal in a solution of aluminum sulfate, $Al_2(SO_4)_3(aq)$. Refer to Fig. 13.26 and predict in which case a single-replacement reaction will take place, then complete and balance the equation for the reaction.

Figure 13.26 states that active metals, such as those of Group 1A and those toward the bottom of Group 2A, react vigorously with water and produce

Figure 13.28 Alkali metals and water.
Sodium metal reacts so vigorously with water that the hydrogen gas produced bursts into flame. A solution of sodium hydroxide (a base) is also formed in this single-replacement reaction.

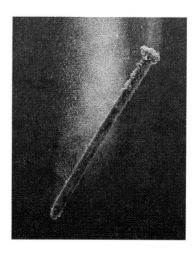

Figure 13.29 Metals and acids.
A single-replacement reaction occurs when an iron nail reacts with an aqueous solution of sulfuric acid to form H_2 gas and aqueous $FeSO_4$.

hydrogen gas and the metal hydroxide (Fig. 13.28). (The metal hydroxide is a base, or *alkali*, and that is why these two groups have "alkali" and "alkaline" in their names.) For example,

$$2\,K + 2\,HOH \rightarrow H_2 + 2\,KOH(aq)$$

(The water has been written "HOH" to emphasize the single-replacement nature of the reaction.)

Figure 13.26 also states that the metals above hydrogen in the activity series will undergo a single-replacement reaction with acids to give hydrogen gas and a salt of the metal (Fig. 13.29). For example,

$$Fe + H_2SO_4(aq) \rightarrow H_2 + FeSO_4(aq)$$

This is an appropriate place to summarize the reaction types we have covered in this chapter. Please examine Table 13.1 carefully.

Table 13.1 A Summary of Reaction Types

Reaction Type	Example
Combination	$2\,Mg + O_2 \longrightarrow 2\,MgO$
Decomposition	$2\,HgO \longrightarrow 2\,Hg + O_2$
Hydrocarbon combustion (complete)	$C_2H_4 + 3\,O_2 \longrightarrow 2\,CO_2 + 2\,H_2O$
Single-replacement	
(a) two metals	$Zn + CuSO_4 \longrightarrow Cu + ZnSO_4$
(b) alkali metal and water	$2\,Na + 2\,H_2O \longrightarrow H_2 + 2\,NaOH$
(c) metal and acid	$Fe + 2\,HCl \longrightarrow H_2 + FeCl_2$
Double-replacement	
(a) precipitation	$BaCl_2 + Na_2SO_4 \longrightarrow BaSO_4(s) + 2\,NaCl$
(b) acid-base	$2\,HCl + Ca(OH)_2 \longrightarrow 2\,H_2O + CaCl_2$
(c) acid-carbonate	$H_2SO_4 + Na_2CO_3 \longrightarrow H_2O + CO_2 + Na_2SO_4$ (The H_2CO_3 formed decomposes.)

HIGHLIGHT

The Mole and Avogadro's Number

The SI defines the **mole** (abbreviated mol) as the quantity of a substance that contains as many elementary units as there are atoms in exactly 12 g of carbon-12. This turns out to be 6.02×10^{23} carbon-12 atoms, and this huge number is referred to as **Avogadro's number,** symbolized N_A.

The concept of mole is like that of dozen. Just as when you hear *dozen* you think 12 units, when you hear *mole* you should think 6.02×10^{23} units. Also, a mole of any substance has a mass equal to the same number of grams as the formula mass (see Section 12.2) of the substance. For example, a mole of carbon atoms (FM 12.0 u) is 6.02×10^{23} atoms and 12.0 g; a mole of water (FM 18.0 u) is 6.02×10^{23} water molecules and 18.0 g; a mole of sodium chloride (FM 58.5 u) is 6.02×10^{23} pairs of sodium ions and chloride ions and 58.5 g. (See Fig. 1.)

The number 6.02×10^{23} is called Avogadro's number in honor of Italian physicist Amedeo Avogadro (Fig. 2), and not because he discovered it. Avogadro was the first person to use the term *molecule*. In 1811 he used the concept of molecules of elements to explain the newly discovered law of combining gas volumes, a law that threatened to torpedo Dalton's atomic theory. (Dalton at first proposed that all elements consisted of independent atoms, and thus thought the formula for hydrogen gas was H, and not H_2 as Avogadro proved later.)

How can we comprehend such a large number as 6.02×10^{23}? Suppose we want to count this number of carbon-12 atoms (1 mole or 12 g). If each of the almost 6 billion persons on Earth counted, uninterrupted, one atom each second, it would take us about 3 million years to count all 6.02×10^{23}.

How do we *know* how many particles are in one mole? The unit for measuring electric charge is the coulomb (C), and it takes 96,485 C to reduce 1

Figure 1 A mole of six substances. Each sample consists of 6.02×10^{23} units.
From left to right: 58.5 g NaCl, 18.0 g H_2O, 74.1g 1-butanol (C_4H_9OH), 12.0 g carbon, 342 g cane sugar ($C_{12}H_{22}O_{11}$), and 180 g aspirin ($C_9H_8O_4$).

mole of singly charged ions to atoms. For example, by electrolysis (chemical change accomplished by an electric current) 96,485 C will produce 1 mole of sodium (23.0 g) from molten sodium chloride (NaCl). It takes one electron to reduce each sodium ion to a sodium atom:

$$Na^+ + e^- \rightarrow Na$$

Thus 96,485 C must be the total charge on one mole of electrons. The electron has a negative charge of 1.6022×10^{-19} C. Therefore, Avogadro's number (N_A) must be

$$N_A = \frac{96485 \text{ C/mol}}{1.6022 \times 10^{-19} \text{C/electron}} = 6.0220 \times 10^{23} \text{ electrons/mol}$$

This is the number of electrons in a mole of electrons, or the number of units in a mole of anything.

Figure 2 Amadeo Avogadro (1776–1856).

13.7 Electrochemistry

Learning Goal:
To describe the relationship between electricity and chemical reactions.

Chemistry is involved with electric charges. Nowhere is that more apparent than in **electrochemistry,** the study of chemical reactions that involve the consumption or production of electric current. If the direct current from a battery or other source is *consumed* while causing a chemical reaction, the process is termed **electrolysis** and takes place in an *electrolytic cell*. Conversely, if the chemical reaction *produces* an electric current, it occurs in a *voltaic cell*. All electrochemical reactions are redox reactions, because electron transfer always occurs. The two electrodes of an electrolytic or voltaic cell are called the *cathode*, where reduction occurs, and the *anode*, where oxidation occurs.

In Figure 12.9 we saw that melted ionic compounds, such as molten NaCl, conduct an electric current. When the Na^+ cations reach the cathode, they are reduced to Na atoms by gaining electrons. When the Cl^- anions reach the anode, they are oxidized to Cl_2 gas by losing electrons. This is an example of electrolysis.

Electrolysis can also take place in water solution. Substances that dissolve in water to give solutions that conduct an electric current are called *electrolytes*. All ionic compounds that are water soluble are electrolytes. Any covalent substance, such as HCl or NH_3, that reacts with water to form ions is also classified as an electrolyte.

Sending an electric current through a water solution of an electrolyte will either reduce (at the cathode) any metallic cation present or, if that cation is more active than hydrogen (above H in the activity series), hydrogen ions in the water probably will be reduced to hydrogen gas. Similarly, at the anode, either the anion of the electrolyte will be oxidized or the OH^- ions (always present in any water solution) will be oxidized to O_2 gas. The concentration of the electrolyte also plays a part in determining exactly what products are formed. Figure 13.30 illustrates the electrolysis of water. Hydrogen gas is produced at the cath-

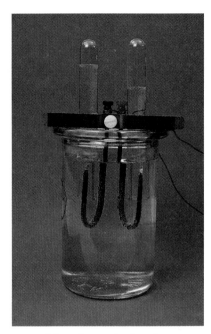

Figure 13.30 The electrolysis of water.
Oxygen (left tube) and hydrogen (right tube) are formed in a 1 to 2 volume ratio as an electric current decomposes water: $2 H_2O(l) \rightarrow O_2(g) + 2 H_2(g)$.

ode and oxygen gas at the anode. Note the 1 to 2 volume ratio of O_2 to H_2, as predicted by the balanced equation for the reaction ($2 H_2O \rightarrow O_2 + 2 H_2$).

In a voltaic cell, a spontaneous chemical reaction produces an electric current. Recall how zinc metal replaced copper from a copper sulfate solution (Fig. 13.25). By proper arrangement of the reactants, the transfer of electrons can be made to take place indirectly and thus supply an electric current (Fig. 13.31). A strip of Zn is immersed in $ZnSO_4(aq)$, and a strip of Cu is placed in $CuSO_4(aq)$. The two strips are connected by wires to a voltmeter, and the two cells are connected by a *salt bridge* that allows the passage of ions, so that ions of one charge do not accumulate in either beaker and stop the reaction. The 1.10-V difference in potential on the meter indicates that electricity is passing through the external wire and could be used to light a bulb, ring a buzzer, or operate some other small gadget.

Let's examine the operation of lead storage batteries, which for many years have been used in cars. A 12-V *lead storage battery* consists of six sets

Figure 13.31 A voltaic cell.
See text for description.

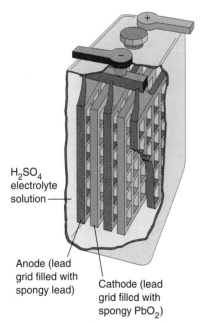

Figure 13.32 One of six cells of a 12-V lead storage battery.

of lead grids in a dilute sulfuric acid solution (Fig. 13.32). One grid of each set contains spongy lead, and the other contains lead(IV) oxide, PbO_2. These grids are mounted in a solution of H_2SO_4, which acts as an electrolyte by ionizing in water to yield H^+ and SO_4^{2-} ions. At each lead grid the reaction during discharge is

$$Pb + SO_4^{2-} \rightarrow PbSO_4 + 2\ e^- \quad \text{(anode)}$$

At each PbO_2 grid the reaction during discharge is

$$PbO_2 + 4\ H^+ + SO_4^{2-} + 2\ e^- \rightarrow \\ PbSO_4 + 2\ H_2O \quad \text{(cathode)}$$

This leads to a net reaction of

$$Pb + PbO_2 + 2\ H_2SO_4 \rightarrow 2\ PbSO_4 + 2\ H_2O$$

If an external electrical potential is applied to a discharged storage battery, it can be recharged. The reactions that occur during charging are the reverse of those that occur during discharge. During discharge, the sulfuric acid is consumed. Conversely, it is regenerated by the charging process. Because a high H_2SO_4 content indicates a charged state, the condition of a lead storage battery can be determined by measuring the density of the sulfuric acid solution. The higher the density, the more the battery is charged. When the auto engine is running, the battery is being charged (provided the alternator is working properly). The battery is discharging when the car is being started.

Important Terms

These important terms are for review. After reading and studying the chapter, you should be able to define and/or explain each of them.

chemical properties	endothermic reaction	double-replacement reactions
chemical reactions	activation energy	precipitate
reactants	combustion reaction	oxidation
products	catalyst	reduction
reversible reaction	hydronium ion	redox reaction
equilibrium	acid	single-replacement reactions
Le Châtelier's principle	base	activity series
coefficients	pH	electrochemistry
combination reaction	acid-base reaction	electrolysis
decomposition reaction	salt	mole
exothermic reaction	acid-carbonate reaction	Avogadro's number

Questions

1. Name the reactants, products, and catalyst for photosynthesis. What is the source of the necessary energy?

13.1 Chemical and Physical Properties and Changes

2. The density of lead is 11.3 g/cm³. This statement is an example of a _____ of lead.
   (a) physical property
   (b) chemical property
   (c) physical change
   (d) chemical change

3. When iron rusts in the presence of oxygen and water, a _____ is occurring.
   (a) physical property
   (b) chemical property
   (c) physical change
   (d) chemical change

4. Hydrogen is a (1) colorless, (2) odorless (3) gas that has a (4) very low density and (5) reacts with oxygen to form water and (6) with oils to form fats. Which of these six properties are physical and which chemical?

13.2 Chemical Equilibrium

5. An increase in pressure on the system $PCl_5(g) \rightleftarrows PCl_3(g) + Cl_2(g)$ at equilibrium would cause (a) a shift to the right, (b) a shift to the left, or (c) no shift.

6. What three things occur in every chemical reaction?

7. When acetic acid and ethyl alcohol are heated together, they change in part to water and ethyl acetate. But when water and ethyl acetate are heated together, they change in part to acetic acid and ethyl alcohol. What is a reaction of this type called?

8. State Le Châtelier's principle. For pressure changes to affect a reaction at equilibrium, a shift must change the number of molecules in what phase?

13.3 Balancing Equations

9. When the equation $MnO_2 + CO \rightarrow Mn_2O_3 + CO_2$ is balanced, the sum of all written and "understood" coefficients is
   (a) 9.   (b) 4.   (c) 5.
   (d) none of the previous answers.
10. In a chemical reaction, what happens to chemical bonds and the identities of atoms?
11. When balancing chemical reactions, some numbers can be manipulated and some cannot. Distinguish these types of numbers by name, and give an example of each.
12. The following reaction occurs when a butane cigarette lighter is operated: $C_4H_{10} + O_2 \rightarrow CO_2 + H_2O$. When balancing the equation, you should *not* start with which element?
13. How many lead, nitrogen, and oxygen atoms are indicated by $2\ Pb(NO_3)_2$? How many nitrate ions?
14. What is inappropriate about each of the "balanced" equations shown?
   (a) $C_4H_{10} + \frac{13}{2} O_2 \rightarrow 4\ CO_2 + 5\ H_2O$
   (b) $4\ H_2O \rightarrow 2\ O_2 + 4\ H_2$
   (c) $Na + H_2O \rightarrow NaOH + H$
15. Use the letters *A*, *B*, and *C* to describe the general appearance of (a) combination reactions and (b) decomposition reactions.

13.4 Energy and Rate of Reaction

16. Which of the following is a substance that increases the rate of a reaction, but is not consumed?
   (a) reactant   (b) catalyst
   (c) allotrope   (d) redox agent
17. Distinguish between exothermic and endothermic reactions.
18. What is absorbed during bond breaking, but liberated during bond formation?
19. A collision between two molecules that have the potential to react may or may not result in a reaction. Explain.
20. List the four major factors that influence the rate of a chemical reaction.
21. Explain why chemical reactions proceed faster as (a) the temperature is increased, and (b) the concentrations of the reactants are increased.
22. Name the two products of the complete combustion of hydrocarbons or carbon-hydrogen-oxygen compounds. What is the other reactant, and how do the products change when this reactant is in short supply?
23. Heating a mixture made of lumps of sulfur and zinc does not lead to reaction nearly as fast as heating a mixture made of powdered sulfur and powdered zinc. Explain.
24. What is the role of a catalyst in a chemical reaction? Describe, in brief, how it accomplishes its role.
25. What do sucrase and cholinesterase have in common?
26. Tell what is indicated by each of these six symbols sometimes seen in chemical reactions: (*aq*), (*s*), (*l*), (*g*), $\rightarrow$, $\rightleftarrows$, and by a chemical formula written *above* a reaction arrow.

13.5 Acids and Bases

27. What would be the pH of a solution ten times as acidic as one of pH 4?
   a. 3   b. 14   c. 5   d. −6
28. List five general properties of acids and three of bases.
29. In the Arrhenius theory, how are acids and bases defined? Distinguish among hydrogen ions, hydronium ions, and hydroxide ions.
30. Who first postulated the existence of ions? In what year? What was his nationality?
31. Aqueous solutions of four acids (nitric, acetic, hydrochloric, sulfuric) are tested for electrical conductivity. Which solution gives results different from the other three, and why?
32. What is the pH of a neutral aqueous solution? How many times as acidic is a solution of pH 3 than one of pH 6?
33. What color will litmus be in a solution of pH 10? Of pH 4?
34. What is the name of the electronic instrument used to measure pH? Is the solution shown in Fig. 13.19 acidic, neutral, or basic?
35. What is the chemical identity of (a) stomach acid, (b) milk of magnesia, (c) lye, (d) baking soda, and (e) vinegar?
36. The reaction of an acid with a hydroxide base always gives what two products? When writing an equation for such a reaction, what is the most common mistake made?

37. The reaction of an acid with a carbonate or hydrogen carbonate always gives what three products?
38. Use the letters *A*, *B*, *C*, and *D* to illustrate the general format of a double-replacement reaction.
39. Describe what is seen in a precipitation reaction. What is happening on the atomic level to cause what is observed?

13.6 Single-Replacement Reactions

40. *Oxidation* can be defined as which of the following?
    (a) a gain of electrons (b) a loss of oxygen
    (c) a loss of electrons (d) both (a) and (b)
41. The reaction $3\ Zn + 2\ Au(NO_3)_3(aq) \rightarrow 2\ Au + 2\ Zn(NO_3)_2(aq)$ will occur if Zn is (a) above, (b) to the right of, (c) below, or (d) to the left of Au in the activity series.
42. Describe oxidation and reduction from the standpoint of gain and loss of (a) oxygen and (b) electrons.
43. Use the letters *A*, *B*, and *C* to illustrate the general format of a single-replacement reaction.
44. The metals of Groups 1A react with water to give what two products? What general type of reaction does this exemplify?
45. Metals above hydrogen in the activity series react with acids to give what two products? What general type of reaction does this exemplify?

13.7 Electrochemistry (and Highlight)

46. One mole of hydrogen peroxide, H_2O_2, would consist of how many grams?
    (a) 6.02×10^{23} (b) 17.0
    (c) $34.0 \times 6.02 \times 10^{23}$ (d) 34.0
47. One mole of hydrogen peroxide, H_2O_2, would consist of how many molecules?
    (a) 6.02×10^{23} (b) 1
    (c) $34.0 \times 6.02 \times 10^{23}$ (d) 34.0
48. What is electrochemistry? Basically, what happens in an electrolytic cell as contrasted with a voltaic cell?

Food for Thought

1. The human body converts sugar into carbon dioxide and water at body temperature, 98.6°F or 37.0°C. Why are much higher temperatures required for the same conversion in the laboratory?
2. Why does an aqueous solution of table salt conduct electricity, whereas an aqueous solution of table sugar does not?
3. Why are automobile batteries so heavy, and why must you be careful not to come in contact with any of the liquids in them?
4. Why are the precious metals gold and silver found as elements in nature, whereas the metals sodium and magnesium are found in nature only in compounds?

Exercises

13.1 Chemical and Physical Properties and Changes

1. Identify each of the following as a physical or chemical change:
   (a) melting ice
   (b) burning a match
   (c) fermenting wine
   (d) dissolving sugar in water
   (e) rusting steel
   (f) magnetizing a sewing needle

   Answer: (a) physical (b) chemical (c) chemical

13.2 Chemical Equilibrium

2. Use Le Châtelier's principle to predict whether the equilibrium will shift to the right, shift to the left, or not shift for each change applied to the following system in equilibrium.
   $N_2(g) + 3\,H_2(g) \rightleftarrows 2\,NH_3(g) + \text{heat}$
   (a) More N_2 is injected into the reaction vessel.
   (b) More NH_3 is injected into the vessel.
   (c) The vessel is heated.
   (d) The pressure is increased.
   (e) Some H_2 is removed from the vessel.

   Answer: (a) right (b) left (c) left (d) right (e) left

3. Use Le Châtelier's principle to predict whether the equilibrium will shift to the right, shift to the left, or not shift for each change applied to the following system in equilibrium.
   $N_2(g) + O_2(g) + \text{heat} \rightleftarrows 2\,NO(g)$
   (a) More O_2 is injected into the reaction vessel.
   (b) Some NO is removed from the vessel.
   (c) The vessel is cooled.
   (d) The pressure is increased.
   (e) Some N_2 is removed from the vessel.

13.3 Balancing Equations

4. Balance the following chemical equations. Check each answer against the set of coefficients given.
   (a) $CuCl_2 + H_2 \rightarrow Cu + HCl$
   (b) $SiO_2 + HF \rightarrow SiF_4 + H_2O$
   (c) $Fe + H_2O \rightarrow Fe_3O_4 + H_2$
   (d) $Al + H_2SO_4 \rightarrow Al_2(SO_4)_3 + H_2$
   (e) $Al + O_2 \rightarrow Al_2O_3$
   (f) $CaC_2 + H_2O \rightarrow Ca(OH)_2 + C_2H_2$
   (g) $KNO_3 \rightarrow KNO_2 + O_2$
   (h) $C_6H_6 + O_2 \rightarrow CO_2 + H_2O$

   Answer: (a) 1, 1, 1, 2 (b) 1, 4, 1, 2 (c) 3, 4, 1, 4 (d) 2, 3, 1, 3 (e) 4, 3, 2 (f) 1, 2, 1, 1 (g) 2, 2, 1 (h) 2, 15, 12, 6

5. Balance the following chemical equations.
   (a) $SO_2 + O_2 \rightarrow SO_3$
   (b) $NH_3 + O_2 \rightarrow N_2 + H_2O$
   (c) $C_4H_{10} + O_2 \rightarrow CO_2 + H_2O$
   (d) $NH_4Cl + CaO \rightarrow NH_3 + CaCl_2 + H_2O$
   (e) $CO_2 + NaOH \rightarrow NaHCO_3$
   (f) $Pb(NO_3)_2 \rightarrow PbO + NO_2 + O_2$
   (g) $MnO_2 + HCl \rightarrow H_2O + MnCl_2 + Cl_2$
   (h) $Al + Fe_3O_4 \rightarrow Al_2O_3 + Fe$
   (i) $Ba(C_2H_3O_2)_2 + H_2SO_4 \rightarrow BaSO_4 + HC_2H_3O_2$

6. Identify the following reactions from Exercise 4 as combination, decomposition, or hydrocarbon combustion: 4(e), 4(g), and 4(h).

   Answer: 4(e) combination, 4(g) decomposition, 4(h) combustion

7. Identify the following reactions from Exercise 5 as combination, decomposition, or hydrocarbon combustion: 5(a), 5(c), 5(e), and 5(f).

8. (a) Nitrogen and hydrogen react to give ammonia in a combination reaction. Write and balance the equation.
   (b) Electrolysis can decompose KCl into its elements. Write and balance the equation.

   Answer: (a) $N_2 + 3\,H_2 \rightarrow 2\,NH_3$, (b) $2\,KCl \rightarrow 2\,K + Cl_2$

9. (a) Aluminum and bromine react to give aluminum bromide in a combination reaction. Write and balance the equation.
   (b) Heating decomposes $MgCO_3$ to carbon dioxide and magnesium oxide. Write and balance the equation.

13.4 Energy and Rate of Reaction

10. A reaction that takes 12 s to occur at 20°C would be expected to take about how many seconds at (a) 30°C and (b) 40°C? State the general rule you used, and give the underlying reason for the change in reaction rate.

    Answer: (a) 6 s

11. Write and balance the reaction for the complete combustion of (a) pentane, C_5H_{12}, and (b) ethane, C_2H_6.

    Answer: (a) $C_5H_{12} + 8\,O_2 \rightarrow 5\,CO_2 + 6\,H_2O$

13.5 Acids and Bases

12. Complete and balance the following acid-base and acid-carbonate reactions.
    (a) $HNO_3 + KOH \rightarrow$
    (b) $HC_2H_3O_2 + K_2CO_3 \rightarrow$
    (c) $H_3PO_4 + NaOH \rightarrow$
    (d) $CaCO_3 + H_2SO_4 \rightarrow$
    (e) $HCl + Ba(OH)_2 \rightarrow$
    (f) $HCl + Al_2(CO_3)_3 \rightarrow$
    (g) $H_3PO_4 + LiHCO_3 \rightarrow$
    (h) $Al(OH)_3 + H_2SO_4 \rightarrow$

    Answer: (a) $HNO_3 + KOH \rightarrow H_2O + KNO_3$
    (b) $2 HC_2H_3O_2 + K_2CO_3 \rightarrow CO_2 + H_2O + 2 KC_2H_3O_2$
    (c) $H_3PO_4 + 3 NaOH \rightarrow 3 H_2O + Na_3PO_4$
    (d) $CaCO_3 + H_2SO_4 \rightarrow CO_2 + H_2O + CaSO_4$

13. Complete and balance the following double-replacement precipitation reactions. (*Hint:* Knowing that all nitrates, acetates, ammonium, and alkali metal compounds are water soluble will help you identify the precipitate.)
    (a) $AgNO_3(aq) + KCl(aq) \rightarrow$
    (b) $Ba(C_2H_3O_2)_2(aq) + K_2CO_3(aq) \rightarrow$
    (c) $Na_2CrO_4(aq) + Pb(NO_3)_2(aq) \rightarrow$
    (d) $K_3PO_4(aq) + CuSO_4(aq) \rightarrow$

    Answer: (a) $AgNO_3(aq) + KCl(aq) \rightarrow AgCl(s) + KNO_3(aq)$
    (b) $Ba(C_2H_3O_2)_2(aq) + K_2CO_3(aq) \rightarrow BaCO_3(s) + 2 KC_2H_3O_2(aq)$

13.6 Single-Replacement Reactions

14. Refer to the activity series (Fig. 13.26) and predict in each case whether the single-replacement reaction shown will, or will not, actually occur.
    (a) $Na + KCl(aq) \rightarrow K + NaCl(aq)$
    (b) $Ni + CuBr_2(aq) \rightarrow Cu + NiBr_2(aq)$
    (c) $2 Al + 6 HCl(aq) \rightarrow 3 H_2 + 2 AlCl_3(aq)$
    (d) $Zn + Fe(NO_3)_2(aq) \rightarrow Fe + Zn(NO_3)_2(aq)$
    (e) $Pb + FeCl_2(aq) \rightarrow Fe + PbCl_2(aq)$
    (f) $2 Ag + 2 HNO_3(aq) \rightarrow H_2 + 2 AgNO_3(aq)$

    Answer: (a) will not (b) will (c) will

15. Refer to the activity series (Fig. 13.26). Complete and balance the equation for the following single-replacement reactions.
    (a) $Na + H_2O \rightarrow$
    (b) $Ni + Hg(NO_3)_2(aq) \rightarrow$
    (c) $Zn + H_2SO_4(aq) \rightarrow$
    (d) $Mg + HCl(aq) \rightarrow$
    (e) $Al + FeSO_4(aq) \rightarrow$
    (f) $Ca + H_2O \rightarrow$

    Answer: (a) $2 Na + 2 H_2O \rightarrow H_2 + 2 NaOH$
    (b) $Ni + Hg(NO_3)_2(aq) \rightarrow Hg + Ni(NO_3)_2(aq)$
    (c) $Zn + H_2SO_4(aq) \rightarrow H_2 + ZnSO_4(aq)$

Answers to Multiple-Choice Questions

2. a   3. d   5. b   9. c   16. b
27. a   40. c   41. a   46. d   47. a

Solutions to Confidence Questions

13.1 The equilibrium would shift *to the right*, because that would use up three gas molecules (two of SO_2 and one of O_2) while forming only two gas molecules (of SO_3). The reduction of number of gaseous molecules would reduce the pressure, counteracting the pressure increase imposed on the system.

13.2 $2\,Mg + O_2 \rightarrow 2\,MgO$; a combination reaction

13.3 $C_3H_8 + 5\,O_2 \rightarrow 3\,CO_2 + 4\,H_2O$

13.4 $6\,HCl + 2\,Al(OH)_3 \rightarrow 6\,H_2O + 2\,AlCl_3$

13.5 $2\,HNO_3 + Na_2CO_3 \rightarrow CO_2 + H_2O + 2\,NaNO_3$

13.6 $Na_2SO_4(aq) + BaCl_2(aq) \rightarrow BaSO_4(s) + 2\,NaCl(aq)$

13.7 Examination of the activity series shows that Al is higher than Cu, so a reaction takes place only in the first beaker. It may be written:

$2\,Al + 3\,CuSO_4(aq) \rightarrow Al_2(SO_4)_3(aq) + 3\,Cu$

Chapter 14

Organic Chemistry

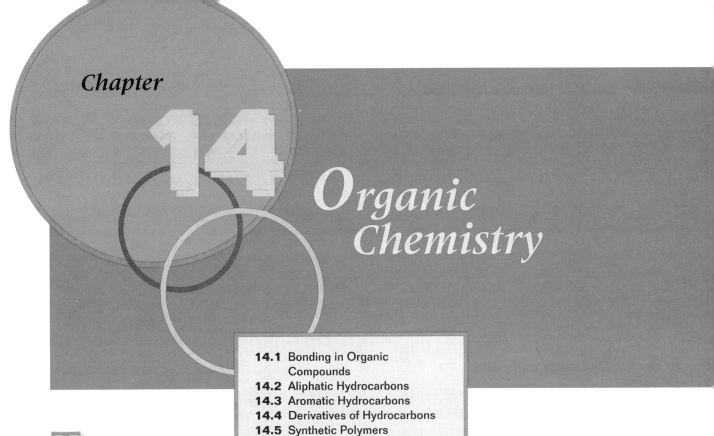

- **14.1** Bonding in Organic Compounds
- **14.2** Aliphatic Hydrocarbons
- **14.3** Aromatic Hydrocarbons
- **14.4** Derivatives of Hydrocarbons
- **14.5** Synthetic Polymers

Highlight: Drugs

The majority of chemicals are organic compounds. Scientists originally defined these as compounds of plant or animal origin—hence the name. However, it proved possible to make many of them in the laboratory from minerals (inorganic compounds). Therefore, the definition of *organic compounds* has changed to just *compounds that contain carbon*, and their study is known as **organic chemistry.** The study of the chemical compounds and reactions that occur in living cells is now called *biochemistry*. It is a close relative of organic chemistry, because most compounds present in living cells contain carbon.

Millions of organic compounds have been identified, and others are being added to the list continually as they are isolated from natural sources or synthesized in the laboratory. Many of them are found in things we use daily—food, fuels, drugs, detergents, perfumes, synthetic fibers, and so on. Carbon is the basic constituent of the complex molecules of carbohydrates, fats, proteins, and nucleic acids.

Organic chemistry is a gigantic field. In this chapter we introduce enough of the fundamental concepts, compounds, and reactions to impart a basic appreciation and understanding of this fascinating area.

14.1 Bonding in Organic Compounds

Learning Goal:

To distinguish the number of bonds formed by atoms of elements commonly found in organic compounds.

In addition to carbon, the most common elements in organic compounds are hydrogen, oxygen, nitrogen, sulfur, and the halogens. Because these are all nonmetals, it is no wonder that organic compounds are covalent in bonding. Examination of the Lewis symbols and application of the octet rule show that these elements should bond as summarized in Table 14.1 on the following page. The information in Table 14.1 should be learned at once.

Any structural formula that follows the bonding rules probably represents a known or possible compound. Any structure drawn that breaks one of these rules is unlikely to represent a real compound.

Table 14.1 Numbers and Types of Bonds for Common Elements in Organic Compounds

Element	Total Number of Bonds	Distribution of Total Number of Bonds and Examples		
C	4	4 singles —C— (with vertical bonds)	2 singles, 1 double —C=	1 single, 1 triple* —C≡
N	3	3 singles —N—	1 single, 1 double —N=	1 triple N≡
O (or S)	2	2 singles O—	1 double O=	
H or halogens	1	1 single H—, Cl—, etc.		

Note: A quadruple bond to a carbon atom is impossible because of geometric considerations.

EXAMPLE 14.1 Spotting Bogus Structural Formulas

Three structural formulas are shown. Which two probably do not represent a real compound? Explain why.

(a) Structure with H—O, C=C—N, with H's attached

(b) H—C(H)(F)—C≡C—O—C(H)(H)—H

(c) H—C(H)(S—H)—C(H)(H)—Br—C(=O)—N—H

Solution

Structure (a) cannot be correct. Although the nitrogen has three bonds, the hydrogens one each, and the carbons four each, the oxygen has three bonds when it should have only two.

In structure (b), each hydrogen and halogen has one bond, each carbon has four, and the oxygen has two. This should be a valid structure and represent a real compound.

Structure (c) cannot be correct. Although each C has four bonds, each H one, and the S and O two each, the N has only two bonds rather than three, and the Br has two bonds but should have only one.

CONFIDENCE QUESTION 14.1

Is this a bogus or a correct structural formula? Explain.

H—C(H)(H)—C(=O)—C(F)(H)—Br

14.2 Aliphatic Hydrocarbons

Learning Goals:

To describe the four major classes of aliphatic hydrocarbons.

To explain and use the concept of structural isomers.

The simplest organic compounds are the **hydrocarbons**, which contain only carbon and hydrogen. For purposes of classification, all other organic compounds are considered to be *derivatives* of hydrocarbons. Thus, the first class of compounds discussed in organic chemistry is the hydrocarbons.

An **aromatic hydrocarbon** is one that contains one or more benzene rings, which are symbolized by

Hydrocarbons having no benzene rings are **aliphatic hydrocarbons**, which are divided into four subclasses: alkanes, cycloalkanes, alkenes, and alkynes (Fig. 14.1). We will examine each of these subclasses, discuss aromatic hydrocarbons, and then go on to derivatives of hydrocarbons.

Alkanes

The **alkanes** are hydrocarbons that contain only single bonds. They are said to be *saturated* hydrocarbons because their hydrogen content is at a maximum. Alkanes have a composition that satisfies the general formula

$$C_nH_{2n+2}$$

where n = the number of carbon atoms
   $2n + 2$ = the number of hydrogen atoms

The formula states that the number of hydrogen atoms present in a particular alkane is twice the number of carbon atoms, plus two. Note how it applies to the alkanes listed in Table 14.2.

Methane (CH_4) is the first member ($n = 1$) of the alkane series. Ethane (C_2H_6) is the second member, propane (C_3H_8) the third, and butane (C_4H_{10}) the fourth. After butane, the number of carbon atoms is indicated by Greek prefixes such as *penta-*, *hexa-*, and so on. The names of alkanes always end in -ane. Methane through butane are gases, pentane through about $C_{17}H_{36}$ are liquids, and the rest are solids. The alkanes generally are colorless and, because they are nonpolar, do not dissolve in water.

Table 14.2 The First Eight Members of the Alkane Series

Name	Molecular Formula	Condensed Structural Formula
Methane	CH_4	CH_4
Ethane	C_2H_6	CH_3CH_3
Propane	C_3H_8	$CH_3CH_2CH_3$
Butane	C_4H_{10}	$CH_3(CH_2)_2CH_3$
Pentane	C_5H_{12}	$CH_3(CH_2)_3CH_3$
Hexane	C_6H_{14}	$CH_3(CH_2)_4CH_3$
Heptane	C_7H_{16}	$CH_3(CH_2)_5CH_3$
Octane	C_8H_{18}	$CH_3(CH_2)_6CH_3$

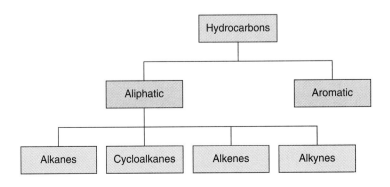

Figure 14.1 Classification of hydrocarbons.
See text for description.

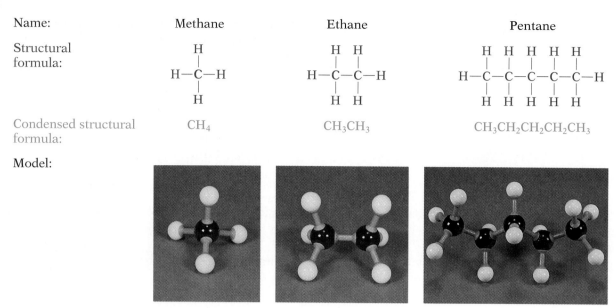

Figure 14.2 Structural formulas, condensed structural formulas, and ball-and-stick models of three alkanes.

A hydrocarbon's structure is easy to visualize when we write its **structural formula** (a graphic representation of the way the atoms are connected to one another) instead of its *molecular formula* (which tells only the type and number of atoms). For example, the structural formulas for methane, ethane, and pentane are shown in the second row in Fig. 14.2. Each dash represents a covalent bond (two shared electrons). To save time and space, *condensed* structural formulas often are used, as in the third row of Fig. 14.2 (and the third column of Table 14.2).

The fourth row of Fig. 14.2 shows ball-and-stick models of methane, ethane, and pentane. Note carefully that the four single bonds of each carbon point to the corners of a regular tetrahedron, which is a geometric figure with four identical equilateral triangles as faces. Carbon's four single bonds form angles of 109.5°, and not 90° as may appear from two-dimensional structural formulas. Figure 14.3, which shows a ball-and-stick model and a space-filling model of methane, emphasizes the tetrahedral geometry of four single bonds to a carbon atom.

The alkanes make up many well-known products. Methane is the principal component of natural gas, which is used in many homes for heating and cooking. Propane and butane are also used for that purpose. Petroleum is made up chiefly of alkanes, but it contains other classes of hydrocarbons, also. The crude oil must be refined; that is, separated into fractions by distillation. Each fraction is still a very complex mixture of hydrocarbons. Gasoline contains many of the alkanes from pentane to decane ($n = 5$ to $n = 10$). At oil refineries, additional gasoline is made by catalytic "cracking" of larger alkanes into smaller ones.

Kerosene contains the alkanes with $n = 10$ to 16. The alkanes with higher values of n make up other products such as diesel fuel, fuel oil, petroleum jelly, paraffin wax, and lubricating oil. The largest alkanes make up asphalt. The alkanes are also used as starting materials for many other products, such as paints, plastics, drugs, detergents, insecticides, and cosmetics.

The alkanes are highly combustible. Like all hydrocarbons, when ignited they react with the oxygen in air, forming carbon dioxide and water

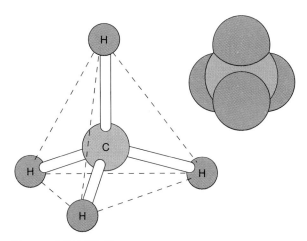

Figure 14.3 Methane.
The tetrahedral geometry of four single bonds to a carbon atom is emphasized in the ball-and-stick and space-filling models for methane.

and releasing heat. If the combustion is not complete, carbon monoxide and black, sooty carbon are formed. Otherwise alkanes are not very reactive, because any reaction would involve breaking the strong C—H and C—C single bonds.

The way chains are built up in three dimensions is illustrated by butane :

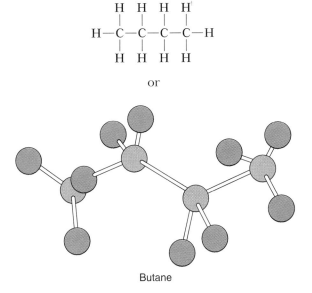

Butane

Each end carbon atom has three hydrogen atoms attached, and two hydrogens are bonded to each of the middle two carbon atoms.

However, another arrangement of the atoms of the butane molecule is possible. Isobutane (2-methylpropane) also has the molecular formula C_4H_{10} (4 carbon atoms and 10 hydrogen atoms), but its structural formula is

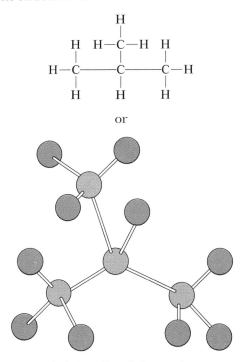

Isobutane (2-methylpropane)

Isobutane's structural formula is different from butane's because three of the carbon atoms have three hydrogens attached, whereas the fourth carbon atom has only one hydrogen bonded to it. That the structures are different can be seen from the two-dimensional representations of the molecules and from the three-dimensional, ball-and-stick models. Isobutane has a *branched-chain* structure, whereas butane has a *continuous-chain* (or *straight-chain*) structure. Although they have the same molecular formula, these compounds have different physical and chemical properties (for example, butane boils at −0.5°C and has a density of 0.58 g/mL at 20°C, whereas isobutane boils at −11.6°C and has a density of 0.55 g/mL at 20°C).

Isobutane and butane are **structural isomers,** compounds that have the same *molecular* formula but differ in *structural* formula. Structural isomers exist whenever two or more structural formulas can be built from the same number and type of atoms without violating the octet rule.

The phenomenon of isomerism is somewhat akin to using the same amounts of wood and brick to build houses that are entirely different. Because of the ability of carbon atoms to bond to many other carbon atoms and atoms of other elements in so many different ways, the number of possible organic compounds is infinite. Just a simple example will point this out. As noted, C_4H_{10} has only 2 isomers. However, C_5H_{12} has 3, C_6H_{14} has 5, $C_{10}H_{22}$ has 75, and it is calculated that $C_{18}H_{38}$ can be arranged in 60,523 different ways!

Because of the number and complexity of organic compounds, a consistent method of nomenclature had to be developed so that communication would be effective. The IUPAC system for nomenclature of organic compounds actually begins with the rules for the alkanes. Let's examine the basic rules and see how they allow the writing of a structure from the name, and vice versa. If these rules seem confusing the first time you read them, don't worry. Some examples will make them clear.

1. Find the longest continuous chain of carbon atoms (the "backbone") and name the compound as a derivative of that alkane.
2. Add, as prefixes to the name of the chain, the positions and names of the branches (substituents) that replace hydrogen atoms on the chain. When more than one type of substituent is present, either on the same carbon atom or on different carbon atoms, the substituents are listed alphabetically. If more than one of the same type of substituent is present, use the prefixes *di-*, *tri-*, *tetra-*, *penta-*, and so forth to indicate how many.
3. Number the carbon atoms on the backbone chain by counting from the end of the chain nearest the substituents. The position of attachment of each substituent is identified by the number of the carbon atom in the chain. Each substituent must have a number. Commas are used to separate numbers from other numbers, and hyphens are used to separate numbers from names.

A substituent that contains one less hydrogen atom than the corresponding alkane is called an **alkyl group**, and is given the general symbol R.

The name of the alkyl group is obtained by dropping the *-ane* suffix and adding *-yl*. For example, *methane* becomes *methyl*, and *ethane* becomes *ethyl*. The open bonds in the methyl and ethyl groups indicate that these groups are bonded to another atom; they do not have an independent existence. If the substituent is a halogen atom, it is named *fluoro-*, *chloro-*, *bromo-*, or *iodo-*. An alkane in which one or more of the hydrogen atoms has been replaced by a halogen atom is called an *alkyl halide*.

EXAMPLE 14.2 Drawing a Structure from a Name

Draw the structural formula for the alkyl halide named 1,2-dichloro-2-methylbutane.

Solution

The end of the name is *butane*, so draw a continuous chain of four carbon atoms joined by three single bonds, and add enough bonds to each C atom so that all have four.

Place a chlorine atom on carbon 1 and another on carbon 2 (you may number from either end). Now, attach a methyl group to carbon 2. Hydrogen atoms are added to the remaining bonds, giving

CONFIDENCE QUESTION 14.2

The octane rating for gasoline assigns a value of 100 to the combustion of the "octane" whose IUPAC name is 2,2,4-trimethylpentane. Draw the structure of this important hydrocarbon.

Cycloalkanes

The **cycloalkanes,** members of a second series of saturated hydrocarbons, have the general molecular formula C_nH_{2n} and possess *rings* of carbon atoms, with each carbon atom bonded to a total of four carbon or hydrogen atoms. The smallest possible ring occurs with cyclopropane, C_3H_6; then come cyclobutane, cyclopentane, and so forth. Note the inclusion of the prefix *cyclo* when naming cycloalkanes. The shorthand structural formulas for cyclopropane and cyclopentane show them just as a triangle and pentagon, respectively.

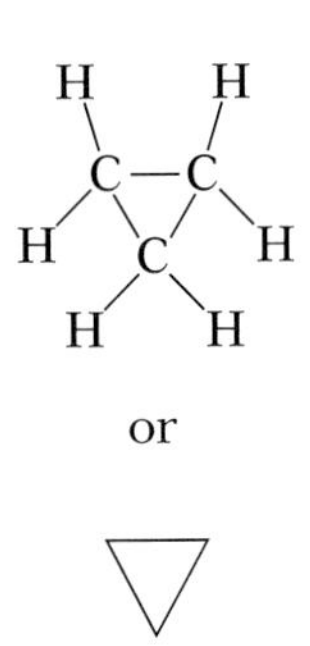

Cyclopropane 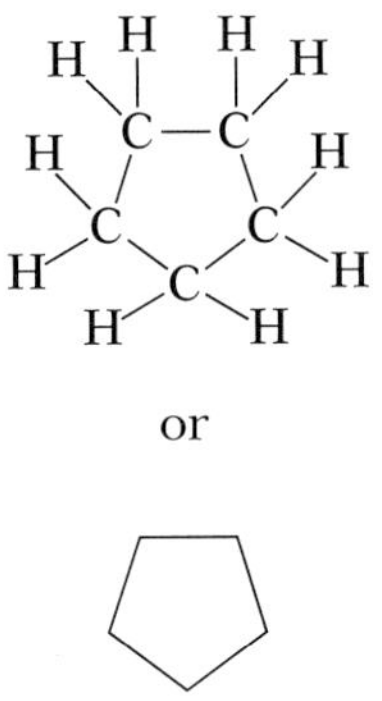   Cyclopentane

Quickly you will come to recognize automatically that a carbon atom is at each corner of the shorthand figure, and that enough hydrogen atoms are assumed to be attached to each carbon to give a total of four single bonds.

Alkenes

Hydrocarbons that have a double bond between two carbon atoms are called **alkenes.** It is as if a hydrogen atom has been removed from each of two adjacent carbon atoms in an alkane, allowing the two C atoms to form an additional bond between them. So the general formula for the alkene series is C_nH_{2n} (the same as for cycloalkanes). The series begins with ethene, C_2H_4 (also known by the common name ethylene), shown by the structural formula

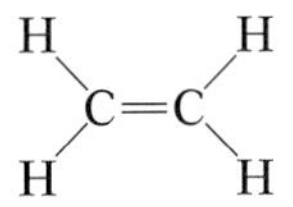

Ethene (ethylene)

Some of the simpler alkenes are listed in Table 14.3. Note that the *-ane* suffix for alkane names is changed to *-ene* for alkenes. A number preceding the name indicates the carbon atom on which the double bond starts. The carbons in the chain are numbered starting at the end that gives the double bond the lower number. For example, 1-butene and 2-butene have the following structural formulas:

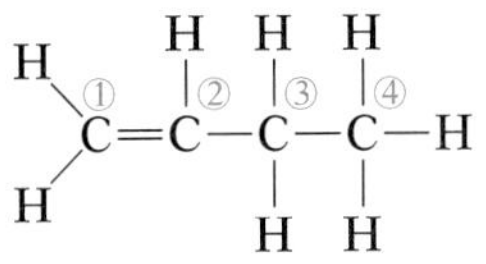

1-Butene 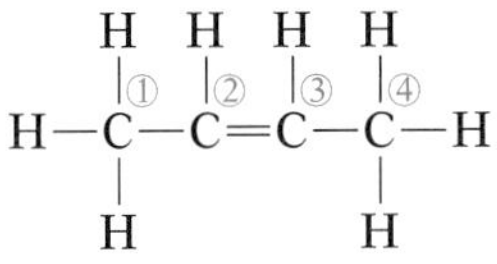   2-Butene

Alkenes are termed *unsaturated* hydrocarbons because under proper conditions additional hydrogen can be added to them. Alkenes are very reactive, and two characteristic reactions are the *addition* of hydrogen (using a platinum catalyst) and halogens to the double-bonded carbons.

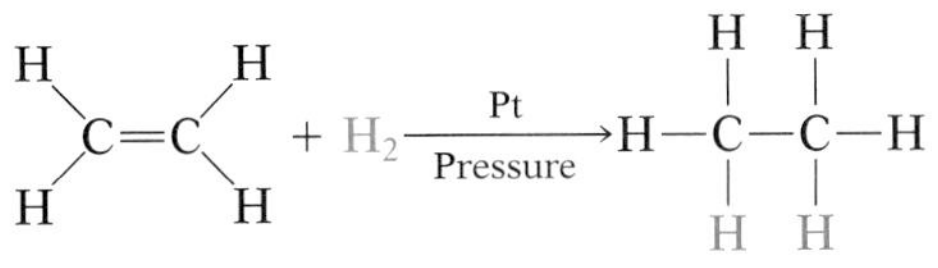

Ethene   Ethane
Addition of hydrogen

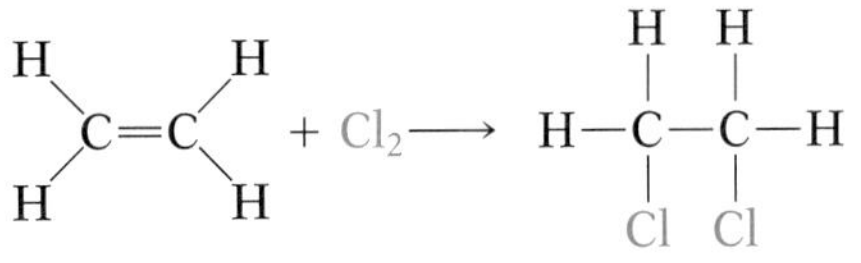

Ethene   1,2-Dichloroethane
Addition of halogens

Alkene molecules can also bond to one another in long chains and sheets by using the double bonds, as we will see in Section 14.5.

Table 14.3 Some Members of the Alkene Series

Name	Molecular Formula	Condensed Structural Formula
Ethene (ethylene)	C_2H_4	$CH_2{=}CH_2$
Propene	C_3H_6	$CH_3CH{=}CH_2$
1-Butene	C_4H_8	$CH_3CH_2CH{=}CH_2$
2-Butene	C_4H_8	$CH_3CH{=}CHCH_3$
1-Pentene	C_5H_{10}	$CH_3(CH_2)_2CH{=}CH_2$

Alkynes

Hydrocarbons that have a triple bond between two carbon atoms are called **alkynes.** It is as if two hydrogen atoms have been removed from each of two adjacent carbon atoms in an alkane, allowing the two C atoms to form two additional bonds between them. So the general formula for the alkyne series is C_nH_{2n-2}. The simplest alkyne is ethyne (common name acetylene, ● Fig. 14.4):

$$H-C{\equiv}C-H$$

Ethyne (acetylene)

Figure 14.4 An iron worker cutting steel with an oxyacetylene torch.
The combustion of acetylene (ethyne) produces intense heat.

Some members of the alkyne series are listed in Table 14.4. Note that the nomenclature for alkynes follows rules similar to those of the alkenes. (Cycloalkenes and cycloalkynes do exist, but we will not discuss them.)

Alkynes are unsaturated and, like alkenes, add hydrogen and halogens. Alkynes can add *two* molecules of hydrogen or halogen across the triple bond. For example,

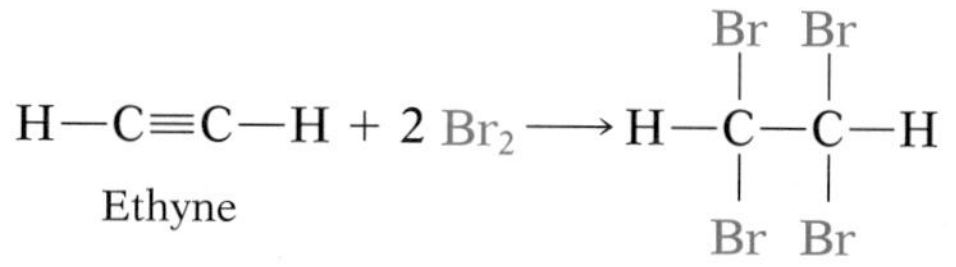

$$H-C{\equiv}C-H + 2\ Br_2 \longrightarrow H-\underset{Br}{\underset{|}{\overset{Br}{\overset{|}{C}}}}-\underset{Br}{\underset{|}{\overset{Br}{\overset{|}{C}}}}-H$$

Ethyne   1,1,2,2-Tetrabromoethane

14.3 Aromatic Hydrocarbons

Learning Goals:

To identify the structures of benzene and its relatives.

To state some uses and properties of aromatic hydrocarbons.

Hydrocarbons that possess one or more benzene rings are called *aromatic hydrocarbons* (many have pungent aromas). The most important one is benzene (C_6H_6) itself, a clear, colorless liquid with a distinct odor. It is a **carcinogen** (a cancer-causing agent), and has the structural formula

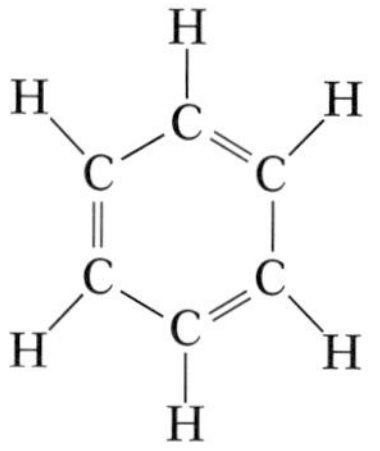

Benzene (Lewis structure)    Benzene (old symbol)    Benzene (modern symbol)

The Lewis structure and old symbol indicate that the ring has alternating single and double bonds between the carbon atoms. However, the properties of the benzene molecule and advanced bonding theory show that six electrons are shared by *all* the atoms in the ring, forming an electron

Table 14.4 Some Members of the Alkyne Series

Name	Molecular Formula	Condensed Structural Formula
Ethyne (acetylene)	C_2H_2	$HC{\equiv}CH$
Propyne	C_3H_4	$CH_3C{\equiv}CH$
1-Butyne	C_4H_6	$CH_3CH_2C{\equiv}CH$
2-Butyne	C_4H_6	$CH_3C{\equiv}CCH_3$
1-Pentyne	C_5H_8	$CH_3(CH_2)_2C{\equiv}CH$

Methylbenzene (toluene)

Naphthalene

Phenanthrene

cloud that extends above and below the plane of the ring (Fig. 14.5). Additional explanation would take more space than is justified. Suffice it to say that this sharing of six "delocalized" electrons by all the ring atoms lends a special stability to benzene and its relatives.

Some other aromatic hydrocarbons are methylbenzene (or toluene, used in model airplane glue), naphthalene (used in mothballs), and phenanthrene (used in the synthesis of dyes, explosives, and drugs). Polycyclic hydrocarbons such as phenanthrene are so stable that they have even been identified in interstellar space, dust shells around stars, and meteorites.

Benzene is obtained from coal tar, a by-product of soft coal. When other atoms or groups of atoms are substituted for one or more of the hydrogen atoms in the benzene ring, a vast number of different compounds can be produced. These compounds include such things as perfumes, explosives (Fig. 14.6), drugs, solvents, insecticides, lacquers, and a host of others. An example is the explosive called TNT.

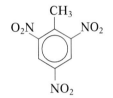

TNT (2,4,6-trinitrotoluene)

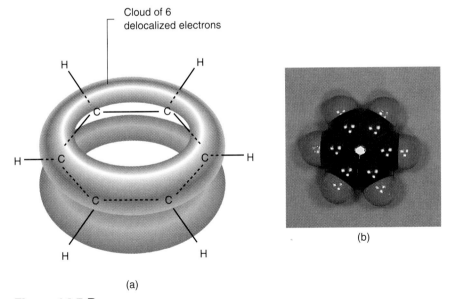

Figure 14.5 Benzene.
(a) A representation of benzene shows that it is a flat molecule with six delocalized electrons forming an electron cloud above and below the plane of the ring. (b) A space-filling model of benzene.

Figure 14.6 Explosives demolish a building.

14.4 Derivatives of Hydrocarbons

Learning Goals:

To identify the structures of some derivatives of hydrocarbons.

To complete equations for the formation of esters and amides.

Alkyl Halides

The general formula for an **alkyl halide** is R—X, where X is a halogen atom and R is the hydrocarbon portion of the molecule. The alkyl halides called CFCs are often in the news. The **CFCs** are **c**hloro**fluoro**carbons, such as dichlorodifluoromethane (freon-12), and are used in air conditioners, refrigerators, heat pumps, and so forth. They are unreactive gases at Earth's surface.

Dichlorodifluoromethane (a CFC)

The major problem with the use of CFCs is that they travel upward into the stratosphere where Earth's protective ozone (O_3) layer is formed by the action of sunlight on ordinary oxygen (O_2). The ozone layer absorbs much of the solar ultraviolet rays (UV) that can harm plant and animal life on Earth's surface.

$$O_3(g) + UV \rightarrow O(g) + O_2(g)$$

When a CFC molecule reaches the stratosphere, UV radiation can break off a chlorine atom. The chlorine atom can react with an O_3 molecule to form chlorine monoxide (ClO).

$$Cl(g) + O_3(g) \rightarrow ClO(g) + O_2(g)$$

The ClO can react with an O atom to regenerate the Cl atom.

$$ClO(g) + O(g) \rightarrow Cl(g) + O_2(g)$$

The Cl atom can then destroy another O_3 molecule, and so on and on. A single Cl atom can survive in the stratosphere for years and destroy thousands of O_3 molecules before it finally reacts with some other chemical species and is removed from the cycle. Fortunately, substitutes for CFCs are being synthesized, and CFCs are being phased out rapidly. (See Section 19.5.)

Chloroform is another familiar alkyl halide. It was used in the past as a surgical anesthetic but has been found to be a carcinogen. A similar compound, carbon tetrachloride, was used in the past in fire extinguishers and fabric cleaners. However, it causes liver damage so its use has been greatly reduced. An interesting use of some liquid fluorocarbons (compounds containing only C and F) is shown in Fig. 14.7.

Chloroform (trichloromethane)

Carbon tetrachloride (tetrachloromethane)

$$-\overset{|}{\underset{|}{C}}-\overset{|}{\underset{|}{C}}-\overset{|}{\underset{|}{C}}-$$

Count the open bonds. They total eight, just right for the attachment of seven H atoms and one F. (If there were too many open bonds, an alkene, alkyne, or ring structure would be tried.)

Ignore the H atoms for the time being and see how many different ways you can put on the F. You will find two different ways. (*Note:* You must remember the tetrahedral geometry of the four bonds to C. There is *no difference* in the bonds you are drawing "out," "up," or "down." Each bond is really at 109.5° from another, not 90° or 180°. Also, in this case, there would be no difference in putting the F on one end of the chain than on the other; that is, "3-fluoropropane" is really 1-fluoropropane.)

$$-\overset{①|}{\underset{|F}{C}}-\overset{②|}{\underset{|}{C}}-\overset{③|}{\underset{|}{C}}- \quad -\overset{①|}{\underset{|}{C}}-\overset{②|}{\underset{|F}{C}}-\overset{③|}{\underset{|}{C}}-$$

Add the seven hydrogens to each structural formula, name each compound, and the question is answered.

$$\text{H}-\overset{\overset{H}{|}}{\underset{\underset{F}{|}}{C}}-\overset{\overset{H}{|}}{\underset{\underset{H}{|}}{C}}-\overset{\overset{H}{|}}{\underset{\underset{H}{|}}{C}}-\text{H} \qquad \text{H}-\overset{\overset{H}{|}}{\underset{\underset{H}{|}}{C}}-\overset{\overset{H}{|}}{\underset{\underset{F}{|}}{C}}-\overset{\overset{H}{|}}{\underset{\underset{H}{|}}{C}}-\text{H}$$

1-Fluoropropane   2-Fluoropropane

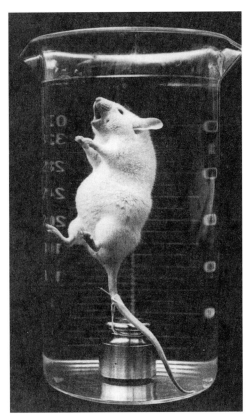

Figure 14.7 Liquid fluorocarbons.
Oxygen is so soluble in this liquid fluorocarbon that the submerged mouse can breathe by absorbing oxygen from it. When the mouse is taken out, the fluorocarbon in its lungs vaporizes and it starts breathing air once more. Perhaps you saw the movie "The Abyss," in which undersea miners were depicted as using this technique.

EXAMPLE 14.3 Drawing Structural Isomers

To establish a better understanding of structural isomers, let's draw and name the structural formulas for the two isomers of C_3H_7F.

Solution

Carbon atoms form four bonds, but H and F can form only one each. Thus H and F can never connect two other atoms; they can only stick to the backbone. Therefore, first draw the three-carbon backbone using single bonds, and add enough bonds so that each carbon has four.

CONFIDENCE QUESTION 14.3

Two isomers of $C_2H_4Cl_2$ exist. Draw the structure and give the IUPAC name for each.

Alcohols

Alcohols are organic compounds containing a **hydroxyl group**, —OH, attached to an alkyl group. The general formula for an alcohol is R—OH, and their IUPAC names end in *-ol*. Thousands of alcohols exist, some with only one hydroxyl group and others with two or more. The simplest alcohol is methanol, which is also called methyl alcohol or wood alcohol. It is poisonous, but it has its uses (Fig. 14.8).

Figure 14.8 Methanol is used for fuel in some types of racing cars.

Methanol (methyl alcohol): H—C(H)(H)—O—H or CH₃OH

Ethanol (CH$_3$CH$_2$OH) is also called ethyl alcohol or grain alcohol. It is a colorless liquid that mixes with water in all proportions and is the least toxic and most important of all alcohols. It is used in alcoholic beverages and in the production of many substances, including perfumes, dyes, and varnishes. *Gasohol* is a mixture of ethanol and gasoline.

Ethanol (ethyl alcohol): CH$_3$CH$_2$OH

Ethylene glycol, a compound widely used as an antifreeze, is an example of an alcohol with two hydroxyl groups. It causes kidney failure if ingested.

Ethylene glycol (1,2-ethanediol)

Carbohydrates, an important class of compounds in living matter, contain multiple hydroxyl groups in their molecular structures, and their names end in *-ose*. One of the primary uses of carbohydrates in cells is as an energy source. The most important carbohydrates are the sugars, the starches, and cellulose. Two important simple sugars are the isomers glucose ($C_6H_{12}O_6$) and fructose ($C_6H_{12}O_6$). When a glucose molecule is bonded to a fructose molecule by removal of the three colored atoms, a molecule of sucrose (common cane sugar) is formed.

Glucose

Fructose

Fructose (fruit sugar) is the sweetest of all sugars and is present in fruits and honey. Glucose, also known as dextrose, is found in sweet fruits, such as grapes and figs, and in flowers and honey. Carbohydrates must be digested into glucose for circulation in the blood. It is normally present in the blood to the extent of 0.1%, but occurs in much greater amounts in persons suffering from diabetes. Hospitalized patients are sometimes fed intravenously with glucose solutions because glucose requires no digestion. Glucose is formed in plants by the action of sunlight and chlorophyll on carbon dioxide from the air and water from the soil. The energy from the sunlight is stored as the potential energy of the chemical bonds.

Starch consists of long chains of glucose units and has the general formula $(C_6H_{10}O_5)_n$, where n may take on values up to 3000. It is a noncrystalline substance formed by plants in seeds, tubers, and fruits. In the digestion process, the starches are converted to glucose. Glycogen (animal starch), a smaller and more highly branched polymer of glucose, is stored in the liver and muscles of animals as a reserve food supply that is easily converted to energy.

Cellulose, another polymer of glucose, has the same general formula, $(C_6H_{10}O_5)_n$, as starch. However, its structure is slightly different, so it has different properties. Cellulose is the main component of the cell walls of plants and received its name for this reason. It is the most abundant organic substance found in our environment. Cellulose cannot be digested by humans, because our digestive system does not contain the necessary enzymes to break the linkages in the molecular chain. The bacteria in the digestive tracts of termites and herbivores (such as cows or deer) do have the necessary enzymes to break the linkages between the glucose units and thus obtain nutrition from cellulose. Cellulose is contained in many commercial products, such as rayon, cellophane, explosives, and paper.

Amines

Amines (ah-MEANS) are organic compounds that contain nitrogen and are basic (alkaline). The general formula for an **amine** is R—NH_2, but one or two more alkyl groups could be attached to the nitrogen atom of the **amino group,** —NH_2, in place of one or more hydrogen atoms. (Recall that a nitrogen atom forms three bonds.) Examples are methylamine, dimethylamine, and trimethylamine. A condensed formula is shown for dimethylamine, and an even more condensed formula represents trimethylamine.

Methylamine Dimethylamine Trimethylamine

The simple amines have strong odors. The odor of raw fish comes from the amines it contains. Decaying flesh forms putresine (1,4-diaminobutane) and cadaverine (1,5-diaminopentane), two amines with especially foul odors.

Putresine (1,4-diaminobutane)

Amines occur widely in nature as drugs, amino acids, and vitamins. They have many applications as medicinals and as starting materials for synthetic fibers. Aniline (aminobenzene) is the starting material for a whole class of synthetic dyes, as well as many other useful compounds.

Aniline (aminobenzene)

Amphetamines are synthetic amines, such as Benzedrine, that are powerful stimulants of the central nervous system. They raise the level of glucose in the blood, thereby fighting fatigue and reducing appetite. Although these drugs have legitimate medical usages, they are addictive and can lead to insomnia, excessive weight loss, and paranoia.

Benzedrine

Carboxylic Acids

The **carboxylic acids** (CAR-box-ILL-ic) contain a **carboxyl group** and have the general formula RCOOH, in which the bonding is as shown.

Carboxyl group   Carboxylic acid (general formula)

The simplest carboxylic acid, formic acid (methanoic acid), is the cause of the painful discomfort from insect bites or bee stings. Vinegar is a 5% solution of acetic acid (ethanoic acid) in water. The structural formulas for formic and acetic acids are

$$\underset{\substack{\text{Formic acid}\\\text{(methanoic acid)}}}{\text{H}-\overset{\overset{\text{O}}{\|}}{\text{C}}-\text{O}-\text{H}} \quad \underset{\substack{\text{Acetic acid}\\\text{(ethanoic acid)}}}{\text{CH}_3-\overset{\overset{\text{O}}{\|}}{\text{C}}-\text{O}-\text{H}}$$

Esters

An **ester** is a compound that has the general formula

$$\text{R}-\overset{\overset{\text{O}}{\|}}{\text{C}}-\text{O}-\text{R}'$$

where R and R′ (read "R prime") are any alkyl groups. Both R and R′ may be identical, but they are usually different.

Unlike amines, most esters possess pleasant odors. The fragrances of many flowers and the pleasing taste of ripe fruits are due to one or more esters. For example, bananas contain the ester named amyl acetate.

$$\underset{\text{Amyl acetate}}{\text{CH}_3-\overset{\overset{\text{O}}{\|}}{\text{C}}-\text{O}-\text{CH}_2\text{CH}_2\overset{\overset{\text{CH}_3}{|}}{\text{C}}\text{HCH}_3}$$

Beeswax and other waxes are esters. Wintergreen mints and Pepto-Bismol get their odor from the ester named methyl salicylate, commonly called *oil of wintergreen*.

Methyl salicylate (oil of wintergreen)

A carboxylic acid and an alcohol react to give an ester and water. This is referred to as *ester formation*. The reaction mixture must be heated, and sulfuric acid is used as a catalyst. In the general case

$$\underset{\text{Carboxylic acid}}{\text{R}-\overset{\overset{\text{O}}{\|}}{\text{C}}-\text{O}-\text{H}} + \underset{\text{Alcohol}}{\text{H}-\text{O}-\text{R}'} \xrightarrow{\text{H}_2\text{SO}_4}$$

$$\underset{\text{Water}}{\text{H}_2\text{O}} + \underset{\text{Ester}}{\text{R}-\overset{\overset{\text{O}}{\|}}{\text{C}}-\text{O}-\text{R}'}$$

Note carefully that the net result of the reaction (proved by tracing oxygen isotopes) is that the —OH from the carboxyl group unites with the H from the hydroxyl group to form water. The remaining two fragments then bond together by means of the bond (dash) that was attached to the H of the hydroxyl group of the alcohol.

EXAMPLE 14.4 Writing the Equation for an Ester Formation

Complete the equation for the sulfuric acid catalyzed reaction between acetic acid and ethanol.

$$\underset{\text{Acetic acid}}{\text{CH}_3-\overset{\overset{\text{O}}{\|}}{\text{C}}-\text{O}-\text{H}} + \underset{\text{Ethanol}}{\text{H}-\text{O}-\text{CH}_2\text{CH}_3} \xrightarrow{\text{H}_2\text{SO}_4}$$

Solution

"Lasso" the —OH from the acid and the H from the hydroxyl group of the alcohol to form water. Attach the acid fragment to the alcohol fragment by means of the bond (dash) that was attached to the H of the hydroxyl group of the alcohol. This gives the ester called ethyl acetate. You would recognize its odor as that of fingernail polish remover.

$$\underset{\text{Acetic acid}}{\text{CH}_3-\overset{\overset{\text{O}}{\|}}{\text{C}}-\text{O}-\text{H}} + \underset{\text{Ethanol}}{\text{H}-\text{O}-\text{CH}_2\text{CH}_3} \xrightarrow{\text{H}_2\text{SO}_4}$$

$$\underset{\text{Water}}{\text{H}_2\text{O}} + \underset{\text{Ethyl acetate}}{\text{CH}_3-\overset{\overset{\text{O}}{\|}}{\text{C}}-\text{O}-\text{CH}_2\text{CH}_3}$$

CONFIDENCE QUESTION 14.4

Complete the equation for this ester formation reaction.

$$\text{C}_6\text{H}_5\text{-CH}_2\overset{\overset{\text{O}}{\|}}{\text{C}}\text{-O-H} + \text{H-O-CH}_3 \xrightarrow{\text{H}_2\text{SO}_4}$$

Fats are esters composed of the trialcohol named glycerol, $\text{C}_3\text{H}_5(\text{OH})_3$, and long-chain carboxylic acids known as *fatty acids*. A typical fatty acid is stearic acid, $\text{C}_{17}\text{H}_{35}\text{COOH}$, a component of beef fat. Stearic acid's structure is that of a long-chain hydrocarbon containing 17 carbon atoms attached to a carboxyl group. It is written here in condensed form to save space. When stearic acid is combined with glycerol, the triester named glyceryl stearate (a fat) is obtained. The reaction is

$$3\left[\text{C}_{17}\text{H}_{35}\overset{\overset{\text{O}}{\|}}{\text{-C-OH}}\right] + \begin{array}{c}\text{H-O-CH}_2\\ \text{H-O-CH}\\ \text{H-O-CH}_2\end{array} \longrightarrow$$

Stearic acid   Glycerol

$$3\text{ H}_2\text{O} + \begin{array}{c}\text{C}_{17}\text{H}_{35}\overset{\overset{\text{O}}{\|}}{\text{-C-O-CH}_2}\\ \text{C}_{17}\text{H}_{35}\overset{\overset{\text{O}}{\|}}{\text{-C-O-CH}_2}\\ \text{C}_{17}\text{H}_{35}\overset{\overset{\text{O}}{\|}}{\text{-C-O-CH}_2}\end{array}$$

Glyceryl stearate (a fat)

Fats that come from animals are generally solids at room temperature, but those that come from plants and fish are usually liquid. These liquid fats are usually referred to as *oils*. Their molecules are composed of hydrocarbon chains with double bonds between some of the carbon atoms; that is, they are unsaturated. These oils can be changed to solid (saturated) fats by a process called *hydrogenation*. In this process hydrogen is added to the carbon atoms that have the double bonds, and the hydrocarbon chains become saturated, or nearly so. Thus liquid fats (oils) are esters of glycerol and unsaturated acids, and solid fats are esters of glycerol and saturated acids. When cottonseed oil (a liquid) is hydrogenated, margarine (a solid) is obtained. The reaction is

$$\begin{array}{c}\text{CH}_3(\text{CH}_2)_7\text{CH=CH}(\text{CH}_2)_7\text{COOCH}_2\\ \text{CH}_3(\text{CH}_2)_7\text{CH=CH}(\text{CH}_2)_7\text{COOCH} + 3\text{ H}_2 \rightarrow\\ \text{CH}_3(\text{CH}_2)_7\text{CH=CH}(\text{CH}_2)_7\text{COOCH}_2\end{array}$$

Cottonseed oil

$$\begin{array}{c}\text{CH}_3(\text{CH}_2)_{16}\text{COOCH}_2\\ \text{CH}_3(\text{CH}_2)_{16}\text{COOCH}\\ \text{CH}_3(\text{CH}_2)_{16}\text{COOCH}_2\end{array}$$

Margarine

Fats and oils are used in the diets of humans and other organisms. In the digestive process, the fats are broken down into glycerol and acids, which are absorbed into the bloodstream and oxidized to produce energy that may be used immediately or stored for future use. It is well established that a diet heavy in saturated fats is unhealthy because the fats lead to a buildup of cholesterol, a waxy substance that can clog the arteries. Fats are also used by the body as insulation to prevent loss of heat, and are important components of cell membranes.

When fats are treated with sodium hydroxide (lye), the ester linkages break to give glycerol and sodium salts of fatty acids. The sodium salts of fatty acids are **soap**. A typical soap is sodium stearate, whose condensed structural formula is

$$\text{CH}_3(\text{CH}_2)_{16}\overset{\overset{\text{O}}{\|}}{\text{C}}\text{-O}^-\text{Na}^+$$

Sodium stearate

If you wish to dissolve stains made by a nonpolar compound such as grease, you must use either a nonpolar solvent or a soap or detergent. One end of the soap or detergent molecule is highly polar and dissolves in the water, whereas the other part of the molecule is a long, nonpolar, hydrocarbon chain that dissolves in the grease (Fig. 14.9). The grease is then emulsified and swept away by rinsing.

Soaps have the disadvantage of forming precipitates when used in acidic solutions or with

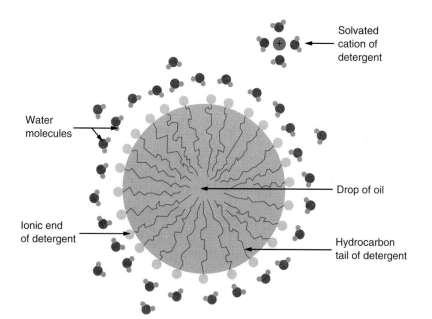

Figure 14.9 Like dissolves like.
The ionic ends of detergent molecules dissolve in the polar water, and the long, nonpolar chains of the detergent molecules dissolve in the grease. The emulsified grease droplets can then be rinsed away.

hard water, which contains ions of calcium, magnesium, and iron. The modern **synthetic detergents,** which are soap substitutes, contain a long hydrocarbon chain that is nonpolar (such as $C_{12}H_{25}$—) and a polar group such as sodium sulfate (—OSO_3^- Na^+). The detergent sodium lauryl sulfate has the condensed structural formula

$$CH_3(CH_2)_{11}-O-SO_3^-Na^+$$

Sodium lauryl sulfate

Synthetic detergents do not form precipitates with the calcium, magnesium, and iron ions. Therefore, they are effective cleansing agents in hard water.

Amides

Amides (AM-eyeds), another class of nitrogen-containing organic compounds, have the general formula

$$R-\overset{\overset{O}{\|}}{C}-\overset{\overset{H}{|}}{N}-R'$$

The reaction for *amide formation* is very similar to that of ester formation. A carboxylic acid and an amine react to give water and an amide. The general reaction is

$$R-\overset{\overset{O}{\|}}{C}-O-H + H-\overset{\overset{H}{|}}{N}-R' \rightarrow$$

Carboxylic acid   Amine

$$H_2O + R-\overset{\overset{O}{\|}}{C}-\overset{\overset{H}{|}}{N}-R'$$

Water   Amide

EXAMPLE 14.5 Writing the Equation for an Amide Formation

Complete this equation for the reaction between acetic acid and methylamine.

$$CH_3-\overset{\overset{O}{\|}}{C}-O-H + H-\overset{\overset{H}{|}}{N}-CH_3 \rightarrow$$

Acetic acid   Methylamine

Solution

"Lasso" the —OH from the acid and an H from the amino group to form water. Attach the acid fragment to the amine fragment by means of the bond (dash) left on the amine fragment by removal of the H. That gives the amide called methyl acetamide.

$$CH_3-\overset{O}{\underset{\|}{C}}-O-H + H-\overset{H}{\underset{|}{N}}-CH_3 \rightarrow$$

Acetic acid   Methylamine

$$H_2O + CH_3-\overset{O}{\underset{\|}{C}}-\overset{H}{\underset{|}{N}}-CH_3$$

Water   Methyl acetamide

CONFIDENCE QUESTION 14.5

Complete this equation for an amide formation reaction.

$$\text{C}_6\text{H}_5-CH_2\overset{O}{\underset{\|}{C}}-O-H + H-\overset{H}{\underset{|}{N}}-CH_3 \rightarrow$$

An **amino acid** is an organic compound that contains both an amino group and a carboxyl group. Over 20 natural amino acids exist, 8 of which are essential in the human diet. The simplest amino acids are glycine and alanine. When a molecule of glycine combines with a molecule of alanine, a molecule of water is eliminated and an amide linkage, —CONH—, is formed in the resulting molecule called a dipeptide:

$$HN-\overset{H}{\underset{H}{\overset{|}{C}}}-\overset{O}{\underset{\|}{C}}-OH + H-\overset{H}{\underset{|}{N}}-\overset{H}{\underset{CH_3}{\overset{|}{C}}}-\overset{O}{\underset{\|}{C}}-OH \rightarrow$$

Glycine   Alanine

$$H_2O + H-\overset{H}{\underset{H}{\overset{|}{N}}}-\overset{H}{\underset{|}{C}}-\overset{O}{\underset{\|}{C}}-\overset{H}{\underset{CH_3}{\overset{|}{N}}}-\overset{H}{\underset{|}{C}}-\overset{O}{\underset{\|}{C}}-OH$$

Water   A dipeptide

This process can be repeated by linking more amino acid molecules to each end of such a dipeptide, eventually forming a protein. **Proteins** are extremely long-chain polyamides formed by the enzyme-catalyzed condensation of amino acids under the direction of nucleic acids in the cell (or the biochemist in the lab). Their formula masses range from a few thousand (insulin, 6000 u) up to millions for the most complex (hemocyamine, 9 million u). Proteins function in living organisms as both structural components (such as muscle fiber) and as enzymes (biological catalysts).

14.5 Synthetic Polymers

Learning Goals:

To explain the importance of polymers.

To explain how addition and condensation polymers are formed.

Chemists have long tried to duplicate the compounds of nature. As the science of chemistry progressed and formulas and basic components became known, chemists were able to synthesize some of these natural compounds by reactions involving the appropriate elements or compounds. During this early trial-and-error period, there were, no doubt, as many accidental (serendipitous) as deliberate discoveries.

From the attempts to synthesize nature's compounds, synthetics were discovered. A **synthetic** is a material whose molecule has no duplicate in nature. The first synthetic was a polymer prepared by Leo Baekeland in 1907 and commercially known as Bakelite, a common electrical insulator.

This discovery set off serious efforts to prepare synthetic materials. Chemists became aware that substituting different atoms or groups in a molecule would change its properties. For example, substituting a chlorine atom for a hydrogen atom in ethane produces chloroethane, which has very different properties from ethane. By knowing the general properties of the substituted groups, a chemist often can tailor a chemical molecule to satisfy a given requirement.

As a result of this scientific approach, multitudes of synthetic compounds have been constructed. Probably the best known of these is the group of synthetic polymers that can be molded and hardened—the *plastics*. Plastics have become an integral part of our modern life, being used in

Figure 14.10 The formation of polyethylene. Polyethylene and similar polymers are formed from large continuous tubes blown from the hot, liquid polymer.

end can attach to another monomer, and so on and on. The polymerization of ethylene (ethene) to polyethylene is shown below, where the subscripted n means that the unit shown in the brackets is repeated thousands of times. (The atoms that eventually terminate the polymer are not usually shown.)

$$\underset{\text{Ethylene (ethene)}}{\overset{H}{\underset{H}{>}}C=C\overset{H}{\underset{H}{<}}} \xrightarrow{\text{Catalyst}} \underset{\text{Polyethylene}}{\left[\begin{array}{c} H\ H \\ |\ \ | \\ -C-C- \\ |\ \ | \\ H\ H \end{array}\right]_n}$$

Polyethylene is the simplest synthetic polymer. Because of its chemical inertness, it is used for chemical storage containers and many other packaging applications (Fig. 14.10).

Another common polymer that has found many uses, especially in coating cooking utensils, is Teflon, a hard, strong, chemically resistant, fluorocarbon resin with a high melting point and low surface friction. Its monomer (tetrafluoroethene) and the polymer structure are

$$\underset{\text{Tetrafluoroethene}}{\overset{F}{\underset{F}{>}}C=C\overset{F}{\underset{F}{<}}} \xrightarrow{\text{Catalyst}} \underset{\text{Teflon}}{\left[\begin{array}{c} F\ F \\ |\ \ | \\ -C-C- \\ |\ \ | \\ F\ F \end{array}\right]_n}$$

Natural rubber consists of a polymer of isoprene. One rubber molecule contains about 2000 isoprene units!

$$\underset{\text{Isoprene}}{CH_2=\overset{\overset{CH_3}{|}}{C}-CH=CH_2} \rightarrow$$

clothing, shoes, buildings, autos, sports, art, electrical appliances, toothbrushes, toys, and myriad other things.

Some plastics have been used as heat shields for space vehicle reentry, where the temperatures are in excess of 8000°C. These materials transmit virtually no heat, because slow, layer-by-layer decomposition of their molecules uses excess heat as the latent heat of vaporization. Let's now examine how, in general, polymers are formed, and discuss a few of the more common ones.

Molecules containing large numbers of atoms and having exceedingly high formula masses often are made up of repeating units of smaller molecules that have bonded together to form long, chainlike structures. The fundamental repeating unit is termed the **monomer**, and the long chain made up of the repeating units is called the **polymer**.

The two major types of polymers are addition polymers and condensation polymers. **Addition polymers** are those formed when molecules of an alkene monomer add to one another. Under proper reaction conditions, often including a catalyst, one bond of the monomer's double bond opens up. This allows the monomer to attach itself by single bonds to two other monomer molecules, then each

$$\underset{\text{Polyisoprene (rubber)}}{\left[-CH_2-\overset{\overset{CH_3}{|}}{C}=CH-CH_2-\right]_n}$$

When the supply of natural rubber to the United States was cut off early in World War II, a synthetic rubber called *neoprene* was developed. (Both the polymer and the monomer are called neoprene.) Neoprene is even better than natural

Figure 14.11 Nylon, a polyamide, sythesized in 1935 by Dr. Wallace Carothers at DuPont.
Here a strand of nylon is being drawn from the interface of the two reactants, where reaction is occurring.

Figure 14.12 A scanning electron micrograph of Velcro.
The "hooks" of one nylon surface entangle with the loops of the other.

rubber, which decomposes rather quickly. Note the similarity of the structures of isoprene and neoprene.

$$CH_2=\underset{\underset{Cl}{|}}{C}-CH=CH_2$$

Neoprene (the monomer)

Condensation polymers are constructed from molecules that have two or more reactive groups. One molecule attaches to another by an ester or amide linkage. Water is the other product, hence the name condensation polymer. Of course, if a monoacid reacts with a monoalcohol or monoamine, the reaction stops with the condensation of the two molecules and there is no chance to form a long-chain polymer. However, if a diacid reacts with a dialcohol or a diamine, the reaction can go on and on. An example is the formation of the polyester named Dacron from thousands of molecules of the monomers.

HOOC—⟨◯⟩—COOH + HO(CH$_2$)$_2$OH →

Terephthalic acid   Ethylene glycol

$$H_2O + \left[\underset{\|}{\overset{O}{C}}-⟨◯⟩-\underset{\|}{\overset{O}{C}}O-(CH_2)_2-O \right]_n$$

Water   Dacron (a polyester)

Another example is the formation of the polyamide named nylon (Fig. 14.11). Both Dacron and nylon are widely used as synthetic fibers.

HOOC(CH$_2$)$_4$COOH + H$_2$N(CH$_2$)$_6$NH$_2$ →

Adipic acid   Hexamethylenediamine

$$H_2O + \left[\underset{\|}{\overset{O}{C}}(CH_2)_4\underset{\|}{\overset{O}{C}}-\underset{|}{\overset{H}{N}}(CH_2)_6\underset{|}{\overset{H}{N}} \right]_n$$

Water   Nylon (a polyamide)

Velcro is made of two nylon strips, one strip having thick loops that are slit open to form "hooks" and the other having thin, closed loops that entangle the slit fibers when the sides are pressed together (Fig. 14.12). The inventor of Velcro took his idea from noticing how cockleburrs clung to his clothing when he walked through a field.

HIGHLIGHT

Drugs

A drug is a compound that can produce a physiological change in human beings or animals. Many are extracted from living sources such as bacteria and plants, but many others are synthetic.

The development of new drugs is relatively slow compared with that of other materials, such as plastics. That is because caution must be exercised in administering new drugs, and many experiments must first be run on laboratory animals in an effort to find any side effects that may be harmful to humans.

Occasionally, these side effects are latent and emerge only after widespread use, as in the case of thalidomide, a sleep-inducing compound that causes deformities in human fetuses. In 1962, Dr. Frances O. Kelsey received the President's Award for Distinguished Federal Civilian Service from President Kennedy for her role in keeping thalidomide off the market in the United States. Often no conclusive evidence exists that a compound is harmful to humans, but preventive action is taken anyway, as in the case of the banning of cyclamate sweeteners when research showed that large amounts caused cancer in rats.

Aspirin, or acetylsalicylic acid, is used primarily as an *analgesic*, a drug that relieves pain without dulling consciousness. Aspirin also is an *antipyretic*, a fever reducer. It is estimated that enough acetylsalicylic acid is produced in the United States to supply every person in the country with 400 aspirin tablets each year.

Figure 1 The hemlock Socrates drank contained the deadly alkaloid, coniine.
In 1787 the French artist Jacques David (dah-VEED) painted *The Death of Socrates.* The same artist painted the picture of Lavoisier and his wife shown in the Chapter 12 Highlight.

An *antibiotic* is a compound that interferes with the growth or survival of one or more microorganisms. The first antibiotics, or "wonder drugs," were developed about 50 years ago. Their discovery was one of the major milestones in medicine because, prior to that time, death was quite common from bacterial infections. Probably the best known antibiotics are the penicillins, which are derived from molds of a certain type. Many other wonder drugs have been developed; for example, the *mycins*, which have proved to be more effective than penicillin in many cases. A present concern is that many strains of bacteria have developed resistance to antibiotics, and so bacterial infections may once again become a serious medical problem.

Another group of drugs, called *alkaloids*, are nitrogen-containing, basic (alkaline) compounds found in plants. They include the well-known substances caffeine, nicotine, morphine, and cocaine. The hemlock Socrates drank contained the deadly alkaloid named *coniine* (Fig. 1).

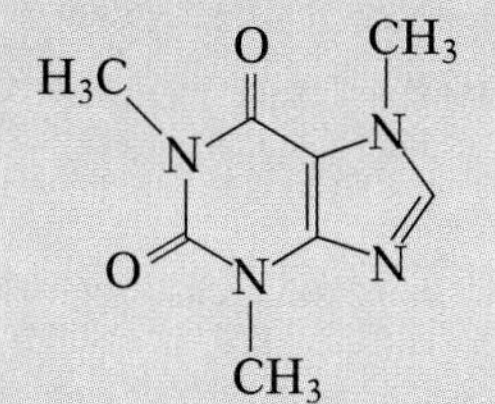

Caffeine

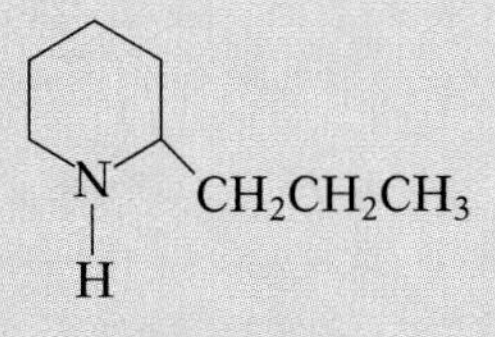

Coniine

Tobacco contains approximately 0.5% to 5% nicotine in its leaves. Because it is a dangerous poison, nicotine has little beneficial use as a drug, but it is an effective insecticide. Morphine is an addictive narcotic present in crude opium, which is extracted from the Oriental poppy. It is used to relieve pain and induce sleep.

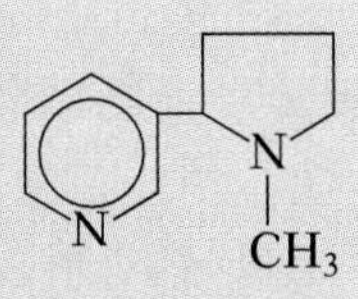

Nicotine

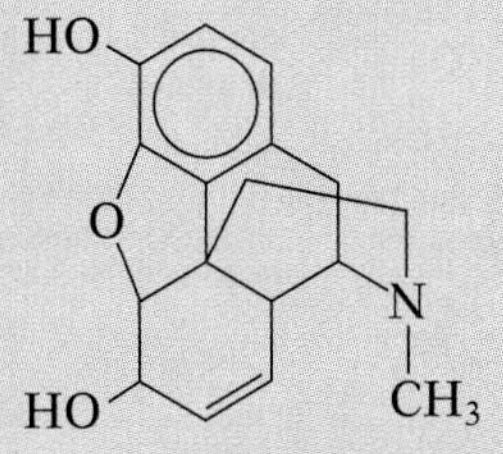

Morphine

Concern has increased over the misuse of drugs, especially heroin and cocaine. Once a person is addicted to a drug, the craving is insatiable. Many addicts commit crimes to support their habit, thus creating a serious social problem. Heroin, produced synthetically from morphine by an ester formation reaction with acetic acid, is a narcotic that depresses the central nervous system. Cocaine is a stimulant, but its effects are short-lived and depression follows. Habitual use causes chronic depression, hallucinations, and psychoses.

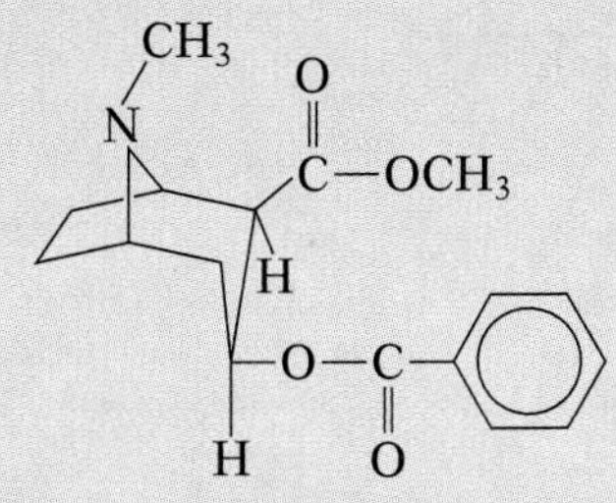

Cocaine

Of a different social impact are the drugs contained in birth control pills. These drugs are synthetics and have proved to be more effective than similar natural compounds. The first one to be synthesized was norethynodrel.

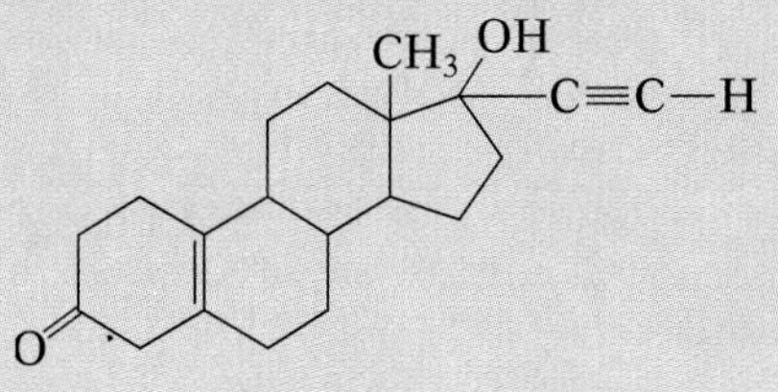

Norethynodrel

Scientists had known for some time that the hormone progesterone would prevent ovulation in the human female about 85% of the time, but it is hardly an effective contraceptive. By synthesizing compounds similar in structure to progesterone, researchers found norethynodrel, which turned out to be virtually 100% effective at one-third the dosage of progesterone when given with a small amount of another hormone, estrogen.

Important Terms

These important terms are for review. After reading and studying the chapter, you should be able to define and/or explain each of them.

organic chemistry
hydrocarbons
aromatic hydrocarbon
aliphatic hydrocarbon
alkanes
structural formula
structural isomers
alkyl group
cycloalkanes
alkenes
alkynes
carcinogen
alkyl halide
CFCs
alcohols
hydroxyl group
carbohydrates
amine
amino group
carboxylic acids
carboxyl group
ester
fats
soap
synthetic detergents
amides
amino acid
proteins
synthetic
monomer
polymer
addition polymers
condensation polymers
drug

Questions

1. Distinguish between organic chemistry and biochemistry.

14.1 Bonding in Organic Compounds

2. How many covalent bonds does a carbon atom form?
   a. 1   b. 2   c. 3   d. 4
3. Tell the number of covalent bonds formed by an atom of each of these common elements in organic compounds: C, H, O, S, N, halogens.

14.2 Aliphatic Hydrocarbons

4. C_6H_{14} is the molecular formula for which of the following?
   a. hexane   b. hexene
   c. hexyne   d. benzene
5. Which of the following is an ethyl group?
   a. CH_3—   b. CH_3CH_2—
   c. CH_2CH_3—   d. none of the previous answers
6. What structural feature distinguishes an aromatic hydrocarbon from an aliphatic hydrocarbon?
7. Give the general molecular formulas for alkanes, cycloalkanes, alkenes, and alkynes. Write the structural formula for one example of each of these classes.
8. Name and give the molecular formulas for the first eight members of the alkane series. Which one is the principal component of natural gas?
9. Describe the geometry and bond angles when a carbon atom forms four single bonds.
10. What term is applied to two or more compounds having the same molecular formula but different structural formulas?
11. When drawing a structure from the name of an alkane or alkane derivative, what part of the name should you look at first?
12. Use both full and condensed structural formulas to show the difference in methane and a methyl group, and ethane and an ethyl group.
13. In a structural formula, what does R stand for?
14. Name the compound represented by a square, and give its molecular formula and full structural formula.
15. Both ethene and ethyne are often called by their more common names. Tell the common names, and draw the structural formula for each compound.
16. Distinguish between saturated and unsaturated hydrocarbons.
17. Describe two types of addition reactions that alkenes and alkynes undergo. Why can't alkanes undergo addition reactions?

14.3 Aromatic Hydrocarbons

18. Which of the following is the most common aromatic compound?
    (a) Benzedrine   (b) ethylene
    (c) butane   (d) benzene
19. Is benzene a solid, liquid, or gas? What is its source? Give its molecular formula and the preferred representation of its structural formula.

14.4 Derivatives of Hydrocarbons

20. Which is the general formula for an alcohol?
    (a) RX (b) ROR
    (c) ROH (d) RCOOH
21. A carboxylic acid and an amine react to give water and which of the following?
    (a) ether (b) ester
    (c) alkyl group (d) amide
22. A *carcinogen*
    (a) fights bacteria. (b) depletes ozone.
    (c) ripens fruit. (d) causes cancer.
23. Give the general formula for an alkyl halide.
24. What does CFC stand for? What is the primary use of CFCs? Why is their use a problem?
25. Give the general formula for an alcohol. Name the characteristic group it contains. What suffix is used in the IUPAC name of alcohols?
26. Give the general formula for an amine. Name the characteristic group it contains. What property makes the simple amines unpopular?
27. Give the general formula for a carboxylic acid. Name the characteristic group it contains. Name and give the structural formula for the carboxylic acid found in vinegar.
28. Give the general formula for an ester. Why are esters popular? Name the ester found in wintergreen mints.
29. What two classes of organic compounds react to form esters? What is the other product? Use general formulas to show how the reaction takes place.
30. What two classes of organic compounds react to form amides? What is the other product? Use general formulas to show how the reaction takes place.
31. Name the type of biochemical compound that (a) contains an abundance of hydroxyl groups, (b) is a triester of glycerol, (c) has multiple amide linkages.
32. What two simpler sugars combine to form sucrose? Which is the monomer of both starch and cellulose? Why can herbivores digest cellulose but humans cannot?
33. Name the monomers of proteins. Name and write the structural formula for the simplest one of these monomers.
34. What are the basic differences and similarities in structure and physical properties of fats and oils? How is an oil converted to a fat?
35. What is the relationship between a fat and a soap, and between a soap and a synthetic detergent?

14.5 Synthetic Polymers (and Highlight)

36. Which of the following is an addition polymer?
    (a) nylon (b) Teflon
    (c) Dacron (d) Velcro
37. Describe from a structural standpoint the basic difference in the method of formation of addition and condensation polymers.
38. Name a well-known synthetic fiber that is a polyester and one that is a polyamide. Name two addition polymers.
39. What is an analgesic? Name the most widely used analgesic.
40. What three characteristics classify a compound as an alkaloid? Name two common alkaloids.

Food for Thought

1. Although any life elsewhere in the universe is probably based on carbon, science fiction writers have speculated that it could be based on another element. What element do you think they chose?
2. Why are fabrics made of Acrilan no longer used in airplanes? (*Hint:* See Exercise 17.)
3. What class of compounds discussed in this chapter do you think the first *vitamins* belonged to?

EXERCISES

14.1 Bonding in Organic Compounds

1. At least one bonding rule is broken in each of the following structural formulas. Identify exactly why each formula is bogus.

(a) S—H—C—C—Cl—N(H)(H) (with H's on carbons)

(b) benzene ring with two CH₃ groups on the same carbon

(c) H—O=C—H

(d) H—C—C—H with N(H)(H)(H) attached

(e) H—C—Cl—C—H (with H's)

Answer: (a) S needs two bonds, an H has one bond too many, the Cl has one bond too many. (b) Each of the ring carbons of benzene already has three bonds, so only one methyl group can be attached to any one ring carbon. (There is an "understood" H on each of the other carbons of the ring.)

14.2 Aliphatic Hydrocarbons

2. Classify each of the following hydrocarbon structural formulas as an alkane, cycloalkane, alkene, alkyne, or aromatic.

(a) H₃C\C=C/CH₃ with H's

(b) cyclopentane with two CH₃ groups

(c) CH₃—C≡C—CH₃

(d) CH₃CH₂CH₃

(e) benzene ring with CH₂CH₃

(f) CH₃CH₂CH₂CH₂CH₃

(g) benzene ring with three CH₃ groups (H₃C, CH₃, CH₃)

(h) cyclopropane—CH₂CH₃

(i) H\C=C/CH₂CH₃ with H's

(j) H—C≡C—CH₂CH₃

Answer: (a) alkene (b) cycloalkane (c) alkyne (d) alkane (e) aromatic

3. Use each of the following names to classify each hydrocarbon as an alkane, cycloalkane, alkene, alkyne, or aromatic.

(a) 2-methylbutane
(b) 3-methyl-1-pentyne
(c) 1,1-dimethylcyclobutane
(d) 3-octene
(e) 1-ethyl-2,3-dimethylbenzene
(f) 2-methyl-2-hexene
(g) 1,3-dimethylcyclohexane
(h) 3,3,4-triethylhexane
(i) ethylbenzene
(j) 2-heptyne

Answer: (a) alkene (b) alkyne (c) cycloalkane (d) alkene (e) aromatic

4. Name each of the compounds with the following structural formulas. Which represent the same compound?

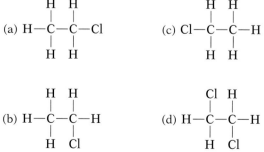

Answer: All are chloroethane except (d), which is 1,2-dichloroethane.

5. Name each of the compounds with the following structural formulas. Which represent the same compound?

(a) H—C(H)(H)—C(H)(H)—C(H)(H)—Cl

(b) H—C(H)(H)—C(H)(Cl)—C(H)(H)—H

(c) CH$_3$CH$_2$CH$_2$Cl

(d) H—C(Cl)(H)—C(H)(H)—C(H)(H)—H

6. Name and give the structural formulas of the two structural isomers that could result from the substitution of one chlorine atom for one hydrogen atom of (a) the butane molecule, and (b) the isobutane (2-methylpropane) molecule.

Answer: (a) CH$_3$CH$_2$CH$_2$CH$_2$Cl (1-chlorobutane) and CH$_3$CH$_2$CH(Cl)CH$_3$ (2-chlorobutane)

7. Given the IUPAC name, draw the structural formula for each alkyl halide.
   (a) 1,1-dibromo-2-fluorobutane
   (b) 1-chloro-2-ethylcyclopentane
   (c) 1,3,5-trichlorobenzene
   (d) 1-bromo-3-chlorobenzene
   (e) 1,1-dichlorocyclobutane
   (f) 2,2-diiodo-3-methylhexane

Answer:
(a) H—C(Br)(Br)—C(F)(H)—C(H)(H)—C(H)(H)—H

(b) cyclopentane with Cl and CH$_2$CH$_3$ substituents

(c) benzene ring with three Cl substituents at 1,3,5 positions

8. Two structural isomers of continuous-chain butenes exist: 1-butene and 2-butene. (a) How many structural isomers of continuous-chain pentenes exist? Name each and draw its structural formula. (b) How many structural isomers of continuous-chain hexenes exist? Name each and draw its structural formula.

Answer:

(a) 1-pentene H$_2$C=CH—CH$_2$CH$_2$CH$_3$

and 2-pentene CH$_3$—CH=CH—CH$_2$CH$_3$

(3-Pentene would be the same as 2-pentene, and 4-pentene the same as 1-pentene.)

9. Complete the equations and name the products.

(a) $\overset{H}{\underset{H}{>}}C=C\overset{CH_3}{\underset{H}{<}} + H_2 \xrightarrow[\text{Pressure}]{\text{Pt}}$

(b) CH$_3$—C≡C—CH$_3$ + 2 Cl$_2$ →

(c) $\overset{H}{\underset{H}{>}}C=C\overset{CH_3}{\underset{H}{<}} + Br_2 \rightarrow$

(d) CH$_3$—C≡C—CH$_3$ + 2 H$_2$ $\xrightarrow[\text{Pressure}]{\text{Pt}}$

Answer:

(a) H—C(H)(H)—C(H)(H)—C(H)(H)—H; propane

(b) CH$_3$—C(Cl)(Cl)—C(Cl)(Cl)—CH$_3$; 2,2,3,3-tetrachlorobutane

14.3 Aromatic Hydrocarbons

10. Three isomeric dimethybenzenes exist. Draw the structural formula and give the full name for each.

Answer:

1,2-dimethyl-benzene   1,3-dimethyl-benzene   1,4-dimethyl-benzene

11. Draw the structural formulas and give the names for all possible triethylbenzenes.

14.4 Derivatives of Hydrocarbons

12. Identify each structural formula as belonging to an alkyl halide, alcohol, amine, carboxylic acid, ester, or amide.

    (a) $CH_3CH_2NH_2$

    (b) $CH_3CH_2\overset{\overset{O}{\|}}{C}-\overset{\overset{H}{|}}{N}-CH_3$

    (c) 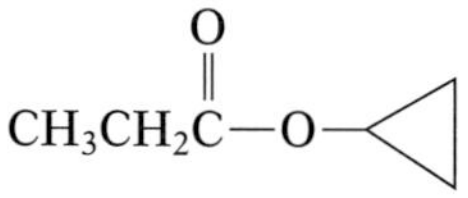 $CH_3CH_2\overset{\overset{O}{\|}}{C}-O-\triangleleft$

    (d) $CH_3CH_2\overset{\overset{O}{\|}}{C}-O-H$

    (e) $CH_3\underset{\underset{Cl}{|}}{C}HCH_3$

    (f) $CH_3\underset{\underset{OH}{|}}{C}HCH_2CH_3$

    (g) 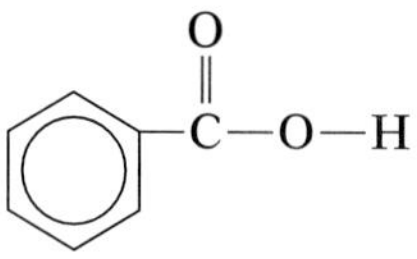 phenyl–$\overset{\overset{O}{\|}}{C}-O-H$

    (h) cyclopentyl–$\overset{\overset{O}{\|}}{C}-O-CH_2CH_3$

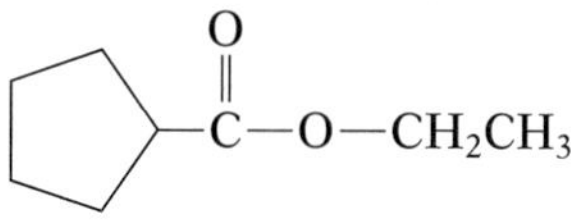

    (i) $CH_3CH_2CH_2OH$

    (j) CF_3CF_3

    (k) 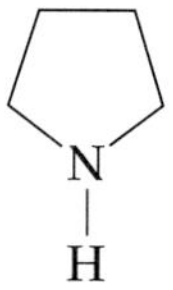 pyrrolidine (N–H)

    (l) 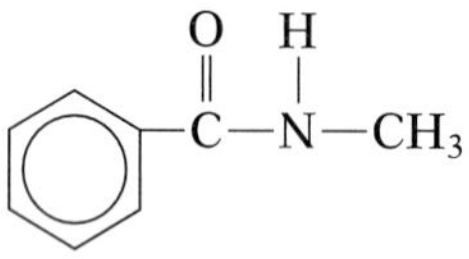 phenyl–$\overset{\overset{O}{\|}}{C}-\overset{\overset{H}{|}}{N}-CH_3$

    Answer: (a) amine (b) amide (c) ester
    (d) carboxylic acid
    (e) alkyl halide (f) alcohol

13. Draw the structural isomers for each of the following: (a) C_2H_6O (two), (b) C_3H_9N (four), (c) dimethylcyclobutane (three), (d) C_5H_{12} (three), and (e) C_4H_8 (five).

    Answer: (a) CH_3CH_2-OH and CH_3-O-CH_3
    (b) $CH_3CH_2CH_2NH_2$, $CH_3CH(NH_2)CH_3$, $CH_3CH_2NHCH_3$, $(CH_3)_3N$

14. Complete the equations.

    (a) $CH_3CH_2\overset{\overset{O}{\|}}{C}-O-H + H-OCH_2CH_3 \xrightarrow{H_2SO_4}$

    (b) 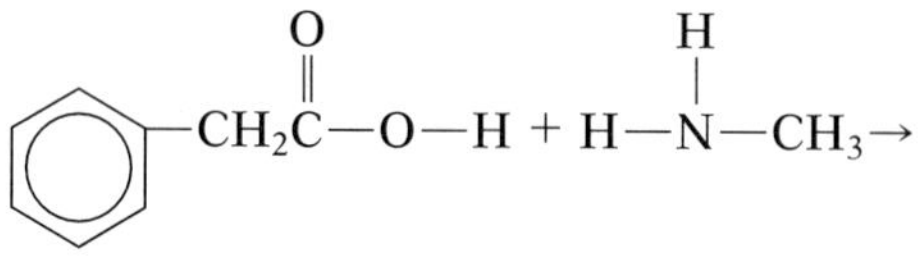 phenyl–$CH_2\overset{\overset{O}{\|}}{C}-O-H + H-\overset{\overset{H}{|}}{N}-CH_3 \rightarrow$

    (c) $CH_3\underset{\underset{CH_3}{|}}{C}H\overset{\overset{O}{\|}}{C}-OH + H-OCH_2$–phenyl $\xrightarrow{H_2SO_4}$

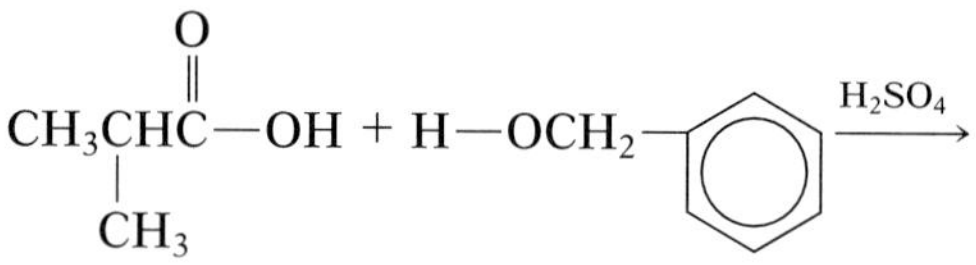

    (d) 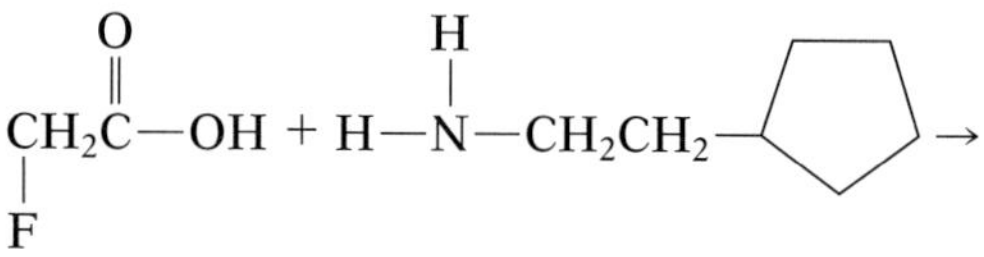 $\underset{\underset{F}{|}}{C}H_2\overset{\overset{O}{\|}}{C}-OH + H-\overset{\overset{H}{|}}{N}-CH_2CH_2$–cyclopentyl $\rightarrow$

    Answer: (a) $CH_3CH_2\overset{\overset{O}{\|}}{C}-O-CH_2CH_3 + H_2O$

    (b) 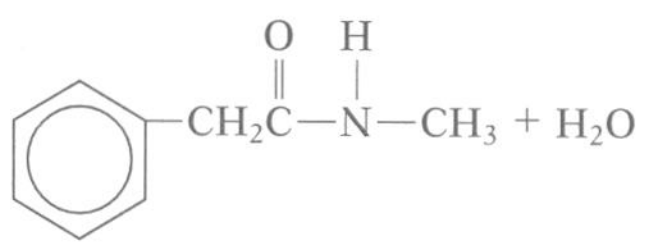 phenyl–$CH_2\overset{\overset{O}{\|}}{C}-\overset{\overset{H}{|}}{N}-CH_3 + H_2O$

15. Identify each of the following structural formulas as a sugar, a saturated fat, an amino acid, or a soap.

(a) $\text{CH}_2\text{CHC}-\text{O}-\text{H}$ with SH and NH$_2$ groups, and a C=O
$$\text{CH}_2\text{CH}(\text{SH})\text{CH}(\text{NH}_2)\text{C}(=O)-O-H$$

(b) [cyclic sugar structure with HO, CH$_2$OH, H, OH groups]

(c) $\text{CH}_3(\text{CH}_2)_{12}\overset{O}{\underset{\|}{C}}-O^-\,\text{Na}^+$

(d)
$$\text{CH}_3(\text{CH}_2)_{12}\overset{O}{\underset{\|}{C}}-O-\text{CH}_2$$
$$\text{CH}_3(\text{CH}_2)_{12}\overset{O}{\underset{\|}{C}}-O-\text{CH}$$
$$\text{CH}_3(\text{CH}_2)_{12}\overset{O}{\underset{\|}{C}}-O-\text{CH}_2$$

14.5 Synthetic Polymers

16. Polystyrene, or styrofoam, is an addition polymer made from the monomer named styrene. Show by an equation how styrene polymerizes to polystyrene.

[Structure of styrene: C$_6$H$_5$–CH=CH$_2$]

Answer:
$$\text{C}_6\text{H}_5\text{–CH=CH}_2 \rightarrow \left[-\text{CH}_2-\text{CH}(\text{C}_6\text{H}_5)-\right]_n$$

17. Acrilan is an addition polymer made from the monomer named *cyanoethene*. Show by means of an equation how cyanoethene polymerizes to Acrilan.

[Structure: CH$_2$=CH–CN]

18. Which two of these compounds would you choose for the monomers, if you wished to form a condensation polymer of the polyester type? Of the polyamide type? In each case, draw the structural formula for the polymer.

(a) $\text{C}_6\text{H}_5-\overset{O}{\underset{\|}{C}}-O-H$

(b) $\text{H}-O-\overset{O}{\underset{\|}{C}}\text{CH}_2\text{CH}_2\overset{O}{\underset{\|}{C}}-O-H$

(c) $\text{H}-O-\overset{O}{\underset{\|}{C}}\text{CH}_2\overset{O}{\underset{\|}{C}}-O-\text{CH}_3$

(d) [cyclopentane]–CH$_2$NH$_2$

(e) $\text{CH}_3\text{CHCHCH}_3$ with OH OH

(f) $\text{CH}_2\text{CH}_2\text{CH}_2\text{CH}_2$ with NH$_2$ NH$_2$

Answer: For the polyester, use (b) and (e), because a *di*acid and a *di*alcohol are needed. The structural formula of the polyester would be

$$\left[-\overset{O}{\underset{\|}{C}}\text{CH}_2\text{CH}_2\overset{O}{\underset{\|}{C}}-O-\overset{\text{CH}_3}{\underset{}{C}}-\overset{\text{CH}_3}{\underset{}{C}}-O-\right]_n$$

Answers to Multiple-Choice Questions

2. d	4. a	5. b	18. d	20. c
21. d	22. d	36. b		

Solutions to Confidence Questions

14.1 This structural formula is bogus, because the oxygen atom has only one bond, whereas it should have two. Also, one carbon atom has only three bonds, whereas it should have four.

14.2 The compound 2,2,4-trimethylpentane has five carbons in a chain connected by single bonds. Two methyl groups are attached to carbon 2, and one methyl group is on carbon 4. Hydrogen atoms are at the end of all remaining bonds necessary to give each carbon atom four bonds. The final structure is

$$\begin{array}{c} H CH_3 H CH_3 H \\ | | | | | \\ H-C-C-C-C-C-H \\ | | | | | \\ H CH_3 H H H \end{array}$$

14.3 When the two carbon atoms are connected by single bonds, enough bonds remain to connect six singly-bonded atoms. Four hydrogen atoms and two chlorine atoms fill that requirement, so the only question is how can the two chlorines be attached to give the two isomers? The result is 1,1-dichloroethane and 1,2-dichloroethane, as shown.

$$\begin{array}{cc} \begin{array}{c} H H \\ | | \\ Cl-C-C-H \\ | | \\ Cl H \end{array} & \begin{array}{c} H H \\ | | \\ H-C-C-H \\ | | \\ Cl Cl \end{array} \end{array}$$

14.4 "Lasso" the —OH from the acid and the H from the alcohol to form water, then connect the remaining fragments by the bond from the alcohol's oxygen atom to obtain the products shown.

$$H_2O + \langle\bigcirc\rangle\text{—}CH_2\overset{\overset{\displaystyle O}{\|}}{C}\text{—}O\text{—}CH_3$$

14.5 "Lasso" the —OH from the acid and the H from the amine to form water, then connect the remaining fragments by the bond from the amine's nitrogen atom to obtain the products shown.

$$H_2O + \langle\bigcirc\rangle\text{—}CH_2\overset{\overset{\displaystyle O}{\|}}{C}\text{—}\overset{\overset{\displaystyle H}{|}}{N}\text{—}CH_3$$

Chapter 15

The Solar System

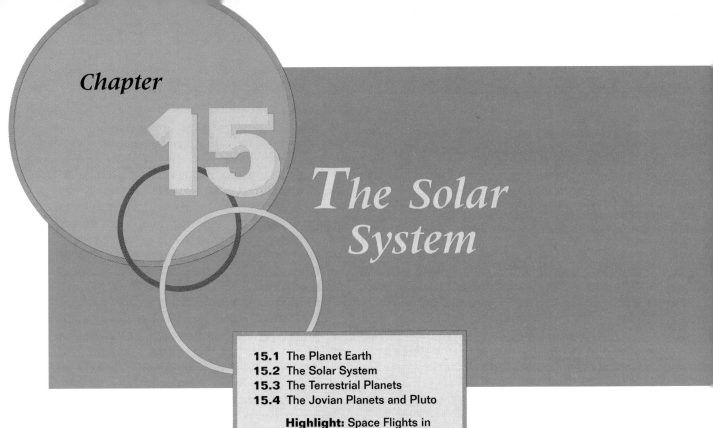

- 15.1 The Planet Earth
- 15.2 The Solar System
- 15.3 The Terrestrial Planets
- 15.4 The Jovian Planets and Pluto

Highlight: Space Flights in the Solar System

- 15.5 Other Solar System Objects
- 15.6 The Origin of the Solar System
- 15.7 Other Planetary Systems

As we go about our daily lives, we often lose our sense of curiosity and wonder. But on a clear dark night when we gaze toward the heavens, we are struck by the grandeur of the starry sky. We marvel at our seeming insignificance, and we begin to ask ourselves questions. How many bright starry objects are there, and how far away are they? What makes them shine? Why do they appear to move from east to west? What is the composition and structure of these shining objects?

The science concerned with such questions is astronomy, the "science of stars." But astronomy deals with more than just stars. It is concerned with planets and their moons, stars, comets, asteroids, star clusters, galaxies, quasars, black holes, interstellar material, plus the origin and evolution of all celestial objects. We define **astronomy** as the scientific study of the universe beyond Earth's atmosphere. The **universe** is defined as everything that is—all energy, matter, and space.

The Babylonians and Greeks were two early civilizations that watched the skies and originated many of the words used today in describing our universe, but their limited observations were made with the unaided eye. With the invention of the telescope and the development of photography and charge-coupled devices (CCDs), more accurate measurements of celestial objects have become possible.

In 1957 the first artificial satellite was launched by the Soviet Union. Shortly thereafter, the United States expanded its space program. The National Aeronautics and Space Administration (NASA) was formed, and astronauts and many artificial satellites were launched into space. Most of these satellites orbited Earth, but some have been sent to the Moon, Mars, Venus, Mercury, Jupiter, Saturn, Uranus, and Neptune. As these satellites transmitted pictures and data back to Earth, our knowledge of the solar system exploded. In fact, the well-read college student of today knows more about our solar system than the most distinguished scientist knew in 1960.

As we continue to discover and learn more about the universe, we gain a greater understanding of our lives and planet Earth. And with every discovery our sense of curiosity increases, because with each discovery new channels of thought are opened.

15.1 The Planet Earth

Learning Goals:

To identify Earth's physical properties.

To define and explain two of Earth's major motions.

Earth is an oblate spheroid, flattened at the poles and bulging at the equator. Its shape is due primarily to its rotation on its axis. Although the difference in the diameter at the poles and at the equator is about 43 km (27 mi), it is very small compared with the total diameter of Earth, which is about 12,900 km (8000 mi). The ratio of 43 to 12,900 is approximately 1/300, which is a rather small fraction. If Earth were represented by a basketball, which is approximately 0.25 m in diameter, the eye would not detect a difference of 0.25/300 m in the diameter. Earth is a more nearly perfect sphere than the average basketball.

Full Earth, as viewed from the Moon, appears four times as large in diameter and more than 60 times as bright as the full Moon at a distance of 384,000 km (240,000 mi). Earth appears much brighter because the clouds and water areas are much better reflecting surfaces than the dull, dark surface of the Moon.

The fraction of incident sunlight reflected by an object is called its **albedo.** Earth's albedo is 0.33, and the Moon's albedo is only 0.07. This number indicates that the Moon's surface reflects 7% of the incoming sunlight falling on its surface. The planet Venus, the third brightest object in the sky (only the Sun and Moon are brighter) has an albedo of 0.76.

Earth is one of nine planets that revolve around the Sun. The Sun (a star) plus these nine planets and their 62 or more satellites, plus thousands of asteroids, countless comets, innumerable meteoroids, and interplanetary dust make up the **solar system.**

Although we are unable to sense directly the motion of our home planet, Earth is undergoing several motions simultaneously. Two that have major influences on our daily lives will be explained in this section: (1) the daily rotation of Earth on its axis and (2) the annual revolution of Earth around the Sun. A third motion, precession, is discussed in Section 16.5.

Earth revolves eastward around the Sun and sweeps out a plane called the *orbital plane* or *ecliptic plane.* This motion of Earth produces an apparent annual westward motion of the Sun on the celestial sphere, the apparent sphere of the sky. The apparent annual path of the Sun on the celestial sphere is called the **ecliptic.** The word is derived from eclipse, because eclipses of the Sun and Moon occur when the Moon is on or near the ecliptic or great circle forming the apparent annual path of the Sun.

Earth is rotating eastward around a central internal axis that is tilted 23.5° from a line that is perpendicular to its orbital plane. Later discussion will show why the 23.5° tilt of the axis and the revolving of Earth around the Sun are the reasons for the four seasons we experience annually.

When studying astronomy, one must know the difference between rotating and revolving. A mass is said to be in **rotation** when it spins on an internal axis. An example is a spinning toy top or a ferris wheel at an amusement park. Revolving, or **revolution,** is the movement of one mass around another. Earth revolves around the Sun, and the Moon revolves around Earth.

The fact that Earth rotates on its axis was not generally accepted until the nineteenth century. A few scientists had considered the possibility, but no definite proof was available to support their beliefs; therefore, their ideas were not accepted.

In 1851 an experiment demonstrating the rotation of Earth was performed in Paris by Jean Foucault (foo-KOH) (1819–1868), a French physicist, using a 61 m (200 ft) pendulum. Today any pendulum used to demonstrate the rotation of Earth is called a **Foucault pendulum.** Even more noticeable results can be seen if the experiment is performed at Earth's North or South Pole.

Picture a large one-room building with a ceiling over 61 m high located at the North Pole (Fig. 15.1). Fastened to the ceiling precisely above the North Pole is a swivel support having very little friction, from which a 61-m, fine steel wire is attached. Connected to the lower end of the wire is a massive iron ball with a short, sharp steel needle attached permanently to its underside. On the floor, under the pendulum, is a layer of fine sand that is slightly furrowed by the needle as the pendulum swings back and forth.

15.1 The Planet Earth 375

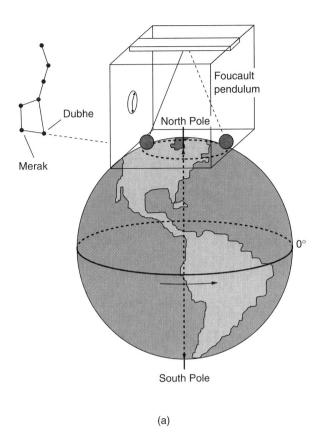

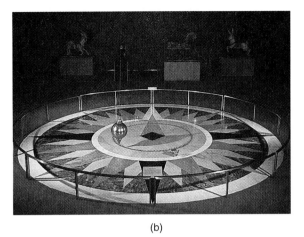

(b)

Figure 15.1 Foucault pendulum.
(a) The drawing illustrates a Foucault pendulum positioned in a room at the North Pole of Earth. To an observer in the room, the pendulum will appear to change its plane of swing by 360° every 24 h. The Big Dipper is shown with stars Dubhe and Merak labeled. See the text for an explanation of the drawing. (b) The photograph shows a Foucault pendulum at the Smithsonian Institution in Washington, D.C. As the pendulum swings back and forth, its plane of swing appears to change as noted by the consecutive knocking over of the red markers positioned in a circle.

Someone starts the pendulum swinging by displacing it to one side with a strong fine thread, with one end attached to the side of the ball and the other end attached to one wall of the building where a 24-h wall clock is mounted. To prevent any sideways motion, the iron ball is allowed to become motionless before it is released by burning the thread. Extreme care is taken to prevent any lateral external forces from being applied to the upper support point of the 61-m wire. As the pendulum swings freely back and forth, the needle point traces its path in the layer of sand.

After a few minutes the plane of the swinging pendulum appears to be rotating clockwise, as shown by the markings in the sand. At the end of one hour, the plane has rotated 15° clockwise from its original position. When six hours have elapsed, the plane of the pendulum appears to have rotated 90° clockwise and is parallel to the wall that holds the 24-h clock. With the passing of each hour, the plane appears to rotate another 15° clockwise. At the end of 24 hours it has made an apparent rotation of 360°.

A person who believes in a motionless Earth would argue that the pendulum actually rotated 360°, because one rotation of the swinging pendulum has been observed by anyone stationed in the large room. A different view can be obtained if we make the walls of the building out of a transparent material such as glass and perform the experiment sometime during the winter months for the Northern Hemisphere. The North Pole has 24 hours of darkness during these months, and the stars are always visible with clear skies. When starting the pendulum this time, we take care to place the iron ball in direct line with the stars Dubhe and Merak, the pointers in the cup of the Big Dipper. As the minutes pass, we observe, as before, the apparent rotation of the plane of the swinging pendulum in a clockwise direction in reference to the large room and clock on the wall.

We also observe, through the transparent walls of the room, that the pendulum still swings in the same direct line with the stars Dubhe and Merak; that is, the pendulum, Dubhe, and Merak are in the same plane. The pendulum has not rotated in reference to fixed stars. No forces have been acting on the pendulum to change its plane of swing. Only the force of gravity has been acting vertically downward to keep it swinging. Therefore, the pendulum does not rotate, but the building and Earth rotate eastward, or turn counterclockwise, as viewed from above the North Pole, once during the 24-hour period. Earth's rotation has a major influence on weather, deflecting winds from their paths, and causing cyclones and other cyclic storms (Chapter 19).

The Foucault pendulum is an experimental proof of Earth's rotation on its axis. What experimental observation would prove that Earth revolves around the Sun?

As Earth orbits the Sun once a year, the apparent positions of nearby stars change with respect to more distant stars. This effect is called parallax. In general, **parallax** is the apparent motion, or shift, that occurs between two fixed objects when the observer changes position. To see parallax for yourself, hold your finger at a fixed position in front of you. Close one eye, move your head from side to side, and notice the apparent motion between your finger and some distant object. Note also that the apparent motion becomes less as you move your finger farther away. Figure 15.2 is an illustration of the parallax of a nearby star as measured from Earth in relation to stars that are more distant.

When speaking or writing about distances of planets in the solar system, we sometimes use a unit of length called the **astronomical unit** (AU), which is the average distance between Earth and the Sun. One AU is 1.5×10^8 km (9.3×10^7 mi). The concept of the astronomical unit is illustrated in Fig. 15.2.

The motion of Earth as it revolves around the Sun leads to an apparent shift in the positions of the nearby stars with respect to the stars that are more distant. Because the stars are at very great distances from Earth, the parallax angle is very small.

The parallax of the distant stars illustrated in Fig. 15.2 cannot be seen with the unaided eye. It was first observed with a telescope in 1838 by Friedrich W. Bessel (1784–1846), a German astronomer and mathematician. The observation of parallax was indisputable proof the Earth really does go around the Sun. Today, the measurement

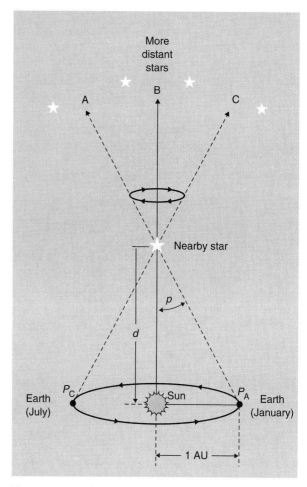

Figure 15.2 Stellar parallax.
The parallax of a star is the apparent displacement of a star that is located fairly close to Earth with respect to more distant stars. When the observer is at P_A, the star appears in the direction A. As Earth revolves counterclockwise, the star appears to be displaced and appears in the direction indicated for different positions of Earth. Positions P_A and P_C are six months apart. The angle of parallax p is also shown.

of the parallax angle is the best method we have of determining the distances to nearby stars.

A second proof of Earth's orbital motion around the Sun is the telescopic observation of a systematic change in the position of all stars annually. The observed effect (called the aberration of starlight) is due to the finite speed of light and the motion of Earth around the Sun. The *aberration of starlight* is defined as the apparent displacement in the direction of light coming from a star because of the orbital motion of Earth.

If you have driven a car during a snowstorm or rainstorm, you have observed the aberration of snowflakes or raindrops. Assume rain is falling vertically and your car is at rest. You will observe raindrops falling vertically. Start the car and drive north at a slow speed. Now you will observe the raindrops coming toward the windshield of the car at a slight angle from the vertical. If you increase the speed of the car, the raindrops will seem to be coming toward you at a greater angle from the vertical. Stop the car, and you will see the raindrops falling vertically again. If you travel east, south, west, or any direction, the effect will be similar. It is the motion of the car that produces the apparent change in the direction of the raindrops.

The great distance to stars is measured in a unit called the *parsec*. The name comes from the first three letters of the word parallax plus the first three letters of the measuring unit, the second, which is used to measure angle.

A circle contains 360°. A degree is divided into 60 equal divisions, each of which is called a minute; and the minute is further divided into 60 equal divisions, each of which is called a second. Thus one second is an angular measurement equal to 1/3600 of a degree. One **parsec** (pc) is defined as the distance to a star when the star exhibits a parallax of 1 s. The parsec is explained in greater detail in Chapter 18.

The Greek mathematician and astronomer Eratosthenes calculated the circumference of Earth about 250 B.C. Eratosthenes was living in Alexandria, Egypt, which was located 5000 stadia (the *stadium* was a Greek measure of length) almost due north of Syene (now Aswan), Egypt. He had received reports that deep wells at Syene were lighted all the way to the bottom on the first day of summer, which meant the Sun was directly overhead (on the zenith) there. Eratosthenes discovered that, on the same day of the year at Alexandria, a vertical stick cast a shadow that positioned the Sun 7.2° south of his zenith. Thus, Eratosthenes knew that Alexandria and Syene were separated by 7.2° of latitude or 1/50 of a circle. Thus, he was able to calculate the circumference of Earth to be 250,000 stadia (5000 × 50).

After repeated measurements Eratosthenes increased the circumference to 252,000 stadia. If we assume the stadium used (more than one were in use at the time) was 1/6 km, the circumference calculated by Eratosthenes is 42,000 km. Figure 15.3 illustrates the principles involved in Eratosthenes' calculations.

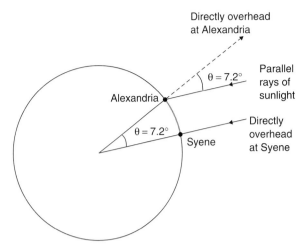

Figure 15.3 The method Eratosthenes used to determine the size of Earth.
The 7.2° zenith angle measured at Alexandria by Eratosthenes was equal to the angle θ, since the rays of sunlight come in parallel to one another. The 7.2° angle is $\frac{1}{50}$ of a circle.

The 1/6 km conversion factor used to convert stadia to kilometers is based on an assumption. The exact length of the Greek unit is unknown. Thus, we are in doubt about Eratosthenes' unit of measurement and calculated value for Earth's circumference. Today we realize the importance of an international system of standard units of measurement.

CONFIDENCE QUESTION 15.1

(a) Specify the albedo, diameter (km), and shape of planet Earth. (b) Define and explain rotation and revolution of Earth, and specify the direction of each.

15.2 The Solar System

Learning Goals:

To describe the composition and structure of the solar system.

To state and explain Kepler's laws of planetary motion.

Figure 15.4 Nicolaus Copernicus.
Copernicus, a Polish astronomer, developed mathematical proof of the heliocentric theory.

The solar system is a complex system of moving masses held together by gravitational forces. At the center of this complex system is a star called the Sun. Revolving around the Sun are 9 rotating planets, their 62 satellites (moons), thousands of asteroids, numerous comets and meteoroids, plus gases and very small masses that are referred to as interplanetary dust particles.

The revolving motion of planet Earth was a concept not readily accepted by most people. In early times most people were convinced that Earth was motionless and that the Sun, Moon, planets, and stars revolved around Earth, which was considered the center of the universe. This concept or theory of the solar system was called the Earth-centered model, or **geocentric theory.**

Nicolaus Copernicus (1473–1543), a Polish astronomer (Fig. 15.4), developed the Sun-centered or **heliocentric theory** of the solar system. Although he did not prove that Earth revolves around the Sun, he did provide mathematical proofs that could be used to predict future positions of the planets.

After the death of Copernicus, the study of astronomy was continued and developed by several scientists, three of whom made their contributions in the last half of the sixteenth century. Notable among these men was Tycho Brahe (1546–1601), a Danish astronomer who built an observatory on the island of Hven near Copenhagen and spent most of his life observing and studying the stars and planets (Fig. 15.5). Brahe is considered the greatest practical astronomer since the Greeks. His measurement of the planets and stars, all made with the unaided eye (the telescope had not been invented), proved to be more accurate than any previously made. Brahe's data, published in 1603, were edited by his colleague Johannes Kepler (1571–1630), a German mathematician and astronomer who had joined Brahe during the last year of his life. After Brahe's death, his lifetime of observations was at Kepler's disposal, and it proved very useful in providing the data necessary

Figure 15.5 Tycho Brahe.
Brahe, a Danish astronomer, is known for his very accurate observations, made with the unaided eye, of the positions of stars and planets. The instrument shown, a large quadrant, was used to make these measurements. Light passes through a small window (upper left in photo) and onto the quadrant.

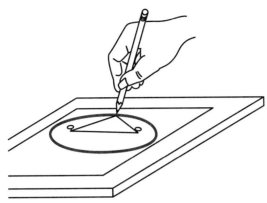

Figure 15.6 Ellipse.
An ellipse can be drawn by using two thumbtacks, a loop of string, a pencil, and a sheet of paper.

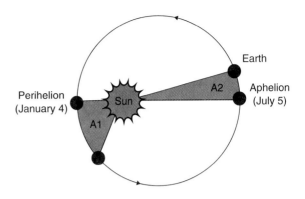

Figure 15.7 Kepler's law of equal areas.
An imaginary line joining a planet to the Sun sweeps out equal areas in equal periods of time. Area A1 equals area A2. Earth has a greater orbital speed in January than in July.

for the formulation of the laws we know today as Kepler's laws of planetary motion.

Kepler was very interested in the irregular motion of the planet Mars. He spent considerable time and energy before he came to the conclusion that the uniform circular orbit proposed by Copernicus was not a true representation of the observed facts. Perhaps because he was a mathematician, he proposed a simple geometric figure that would give the correct solution to his calculations while removing all the epicycles used by Copernicus. Kepler's first law, known as the law of **elliptical paths,** states:

> **All planets move in elliptical paths around the Sun, with the Sun at one focus of the ellipse.**

Note that there is nothing at the other focus of the ellipse.

An ellipse is a figure that is symmetrical about two unequal diameters (Fig. 15.6). An ellipse can be drawn by using two thumbtacks, a closed piece of string, paper, and a pencil. The points where the two tacks are positioned are called the foci of the ellipse.

Kepler's first law gives the shape of the orbit but fails to predict when the planet will be at any position in the orbit. Kepler, aware of this, set about to find a solution from the mountain of data he had at his disposal. After a tremendous amount of work, which was carried out with no indication that a solution was possible, he discovered what is now known as Kepler's second law of planetary motion, or the **law of equal areas,** which states:

> **An imaginary straight line joining a planet to the Sun sweeps out equal areas in equal periods of time.**

From Fig. 15.7 we can see that the speed of the revolving body will be greatest when the planet is closest to the Sun, and its speed the slowest when the planet is farthest away. The law provided a way for determining the speed, which allows the position of the planet to be predicted at some future time.

After the publication of his first two laws in 1609, Kepler began a search for a relationship between the motions of the different planets and an explanation to account for these motions. Ten years later he published *De Harmonica Mundi (Harmony of the World)*, in which he stated the **harmonic law:**

> **The ratio of the square of the period and the cube of the semimajor axis (one-half the longer axis of an ellipse) is the same for all planets.**

This law can be written as

$$\frac{T^2}{R^3} = k$$

where T = the period,

R = the semimajor axis,

k = a constant that has the same value for all planets.

EXAMPLE 15.1 Calculating the Period of Revolution of a Planet

Use Kepler's harmonic law and calculate the period of revolution, in years, for the planet Mars.

Solution

STEP 1

Determine the value of the constant k in Kepler's harmonic law using Earth data. (T = 1 year, R = 1 astronomical unit)*

$$\frac{T^2}{R^3} = k = \frac{(1 \text{ year})^2}{(1 \text{ AU})^3} = \frac{1 \text{ y}^2}{(\text{AU})^3}$$

Since $k = \frac{1 \text{ y}^2}{(\text{AU})^3}$, $T^2 = \frac{1 \text{ y}^2}{(\text{AU})^3} R^3$.

STEP 2

Use data from Table 15.1 on page 382 concerning Mars and substitute in Kepler's law.

$$T^2 = kR^3 = k(1.524 \text{ AU})^3 = 3.54 \text{ y}^2$$
$$T = 1.88 \text{ y}$$

CONFIDENCE QUESTION 15.2

(a) Calculate the period of revolution, in years, for the planet Venus. (b) How does your calculated value compare with the value given in Table 15.1?

Galileo Galilei (1564–1642), Italian astronomer, mathematician, and physicist who is usually called Galileo, was one of the greatest scientists of all time (● Fig. 15.8). The most important of his many contributions to science were in

---

*One astronomical unit, abbreviated AU, is equal to the mean distance from Earth to the Sun.

Figure 15.8 Galileo Galilei.
Galileo Galilei, the great Italian scholar, was the first to use the newly invented telescope to observe the planets and stars.

the field of mechanics. He originated the basic ideas for the formulation of Newton's first two laws of motion, and he founded the modern experimental approach to scientific knowledge. The motion of objects, especially the planets, was of prime interest to Galileo. His concepts of motion and the forces that produce motion opened up an entirely new approach to astronomy. In this field he is noted for his contribution to the heliocentric theory of the solar system.

In 1609 Galileo became the first person to observe the Moon and planets through a telescope. With the telescope he discovered four of Jupiter's 16 moons, thus proving that Earth was not the only center of motion in the universe. Equally important was his discovery that the planet Venus went through a change in phase similar to that of the Moon, as called for by the heliocentric theory, but contrary to the geocentric theory, which called for a new or crescent phase of Venus at all times.

Sir Isaac Newton (1642–1727), the English physicist regarded by many as the greatest scientist the world has known, formulated the principles of gravitational attraction between bodies and established physical laws determining the magnitude and direction of the forces that cause the planets to move in elliptical orbits in accordance with Kepler's laws. Newton invented calculus and used it to help explain Kepler's first law. He also used the law of conservation of angular momentum to explain Kepler's second law.

To explain Kepler's third law, Newton showed that the constant in Kepler's equation was

$$\frac{T^2}{R^3} = \frac{4\pi^2}{Gm_{Sun}} = k$$

where T = period of a planet,

R = mean distance between the planet and the Sun,

G = gravitational constant,

m_{Sun} = mass of the Sun.

Newton's explanations of Kepler's laws unified the heliocentric theory of the solar system and brought an end to the confusion of the past. He gave us an ordered system of the Sun and planets satisfactory for the present time.

Today, our solar system is known to consist of one star (the Sun, which contains 99.87% of the material of the system), nine planets (including Earth), about five dozen satellites (our Moon is an example), thousands of asteroids (Ceres is the largest, with a diameter of 940 km), billions of comets, and countless meteoroids. The distribution of the remaining 0.13% of the solar system's mass is shown in Table 15.1. Note that more than half the remaining mass is concentrated in Jupiter.

Planets that have orbits smaller than Earth's are classified as "inferior," and those with orbits greater than Earth's as "superior." Another method is to classify Mercury, Venus, Earth, and Mars as the inner or **terrestrial planets** because they resemble Earth. Jupiter, Saturn, Uranus, and Neptune are classified as outer or **Jovian planets** because they resemble Jupiter. (The Roman god Jupiter was also called Jove.) Pluto does not resemble Earth or Jupiter, and some astronomers have suggested that it be classified as an asteroid.

The relative distances of the planets from the Sun are shown in Fig. 15.9. The orbits are all elliptical, but nearly circular, except for that of Pluto. Pluto's orbit actually goes inside Neptune's orbit. The actual position of Pluto is shown for 1990. We see that for several years (until 1999) Neptune will be farther from the Sun than Pluto. Study Fig. 15.9 for a minute and notice how far from the Sun Jupiter is compared with Mars. Note that the distance from Saturn to Neptune is greater than that from the Sun to Saturn. The distinction between the four inner planets and the five outer planets can also be seen in this illustration.

When viewed from above the solar system (that is, looking down on the North Pole of Earth), the planets all revolve counterclockwise around the Sun. This motion is west-to-east (eastward) revolving or **prograde motion.** The planets also rotate with a counterclockwise or prograde motion when viewed from above the North Pole, with the exception of Venus, Uranus, and Pluto. They have **retrograde motion**—that is, the motion is east-to-west (westward) or clockwise as viewed from above the North Pole of Earth.

The relative sizes of the planets are shown in Fig. 15.10. Note the huge size of the outer planets compared with the size of the inner planets. Compare the size of Pluto with the size of the Jovian planets.

The inclinations of the orbits of the planets relative to Earth's orbit are shown in Fig. 15.11. Note that the solar system is contained within a disk shape rather than a spherical shape. Also, note the large angle between Pluto's orbital plane and the ecliptic.

A simple way to remember the order of some of the planets is to realize that the first letters of the words *S*aturn, *U*ranus and *N*eptune spell *SUN*.

A simple method to remember the approximate distance of the planets from the Sun is given by Bode's law. Johann Bode (1747–1826), a German astronomer and editor of a German astronomical journal, was a strong supporter, but not the discoverer, of the method. Bode's law, in reality, is not a law. That is, it does not represent a physical property of the solar system. The so-called law is a simple method to be followed in calculating the approximate radial distance of some planets from the Sun. See Table 15.1 for a comparison of the results of Bode's law with actual distances. The

Table 15.1 The Solar System

| Name | Semimajor Axis | | | Diameter | | Mass with Respect to Earth | Density (g/cm³) (Water = 1) |
	Million km	Astron. Units	Titius-Bode Law	km	Earth = 1		
Sun				1.39×10^6		332,000.	1.4
Mercury	57.9	0.387	0.4	4878	0.38	0.055	5.4
Venus	108.2	0.723	0.7	12,102	0.95	0.82	5.2
Earth	149.6	1.00	1.0	12,756	1.00	1.00	5.5
Mars	227.9	1.524	1.6	6790	0.53	0.11	3.9
Asteroids	414	2.767	2.8				
Jupiter	778	5.203	5.2	142,980	11.2	318.	1.3
Saturn	1427	9.539	10.0	120,536	9.41	94.3	0.7
Uranus	2871	19.18	19.6	51,118	3.98	14.54	1.2
Neptune	4497	30.06	38.8	49,500	3.81	17.2	1.7
Pluto	5913	39.53	77.2	2300	0.27	0.0025	

*Retrograde motion is clockwise as observed from above the North Pole.

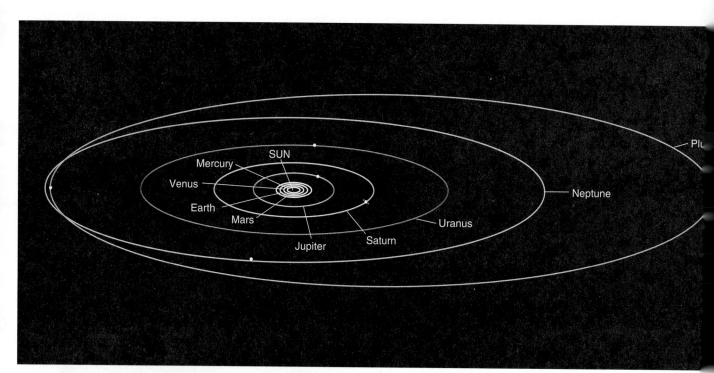

Figure 15.9 The orbits of the planets, drawn to scale.
The orbits of the planets are all counterclockwise when viewed from above the North Pole of Earth. The terrestrial planets are near the Sun, with orbits close together. The Jovian planets are far from the Sun, with orbits widely separated. Note that Pluto is presently inside the orbit of Neptune.

Period of Revolution	Period of Rotation	Inclination of Axis with the Vertical	Inclination of Orbit with the Ecliptic	Surface Gravity (Earth = 1)	Magnetic Field	Satellites
88 days	25 days 59 days	less than 28°	7°	0.38	Yes Weak	None
225 days	243 days (retrograde)*	3°	3.4°	0.91	No	None
365.242 days	24 h	23.5°	0°	1.00	Yes	1
687 days	24.6 h	24°	2°	0.38	Very weak	2
5 years typical			10° average		No	
12 years	10 h	3.1°	1.3°	2.53	Yes	16 or more
29.5 years	10.7 h	26.7°	2.5°	1.07	Yes	19 or more
84 years	17.24 h (retrograde)*	82°	0.8°	0.92	Yes	15 or more
165.6 years	16 h	29°	1.8°	1.18	Yes	8 or more
248 years	6.4 days (retrograde)*	43°	17°	0.09	?	1

method was first published by Johann Daniel Titius (1729–1796), a German physicist and mathematician, in 1766. Many astronomers refer to the method more correctly as the **Titius-Bode law**. The method was very useful in the discovery of the asteroids (starlike objects) located between Mars and Jupiter. Values for the distances in astronomical units from the Sun to the different planets, as obtained by the Titius-Bode law, are given in Table 15.1. The law gives the distance from the Sun to the planets when the figures 0, 3, 6, 12, 24, and so on (doubling the number each time, except for the zero) are added to 4, and the sum divided by 10. Although the law has no physical interpretation, it does provide an easy method for remembering the distance from the Sun to most of the planets.

The period of time required for a planet to travel one complete orbital path is referred to as either the sidereal or synodic period. The **sidereal period** is defined as the time interval between two successive *conjunctions* of the planet with a star (planet and star are together on the same meridian) as observed from the Sun. The **synodic period** is the time interval between two successive *conjunctions* of the planet with the Sun (planet and Sun on same meridian) as observed from Earth.

The relationship between the sidereal and synodic periods of a planet is illustrated in Fig. 15.12. At position P_1 Mercury revolves eastward (counterclockwise) around the Sun through 360° back to position P_1. This motion and time period is represented by the solid-color circle in Fig. 15.12. This revolution is the sidereal period for Mercury—the time (88 Earth days) that Mercury requires to make one revolution around the Sun. During this time (88 days) Earth revolves approximately 87° eastward to position P_2. Mercury continues revolving eastward from position P_1 to position P_3. During this same period Earth revolves from position P_2 to position P_3. This movement and time period are represented by the broken color lines in Fig. 15.12. At position P_3 an observer on Earth now again sees Mercury on the meridian with the Sun. The total time for Mercury to revolve from P_1 back to P_1 then to P_3 is the synodic period, equal to 116 Earth days. This is the time it takes the planet Mercury to make one orbit around the Sun as observed from Earth, or the time from the conjunction at position P_1 to the next conjunction at position P_3. The true period of revolution is the sidereal period.

Opposition is the term used to describe the position of a planet when the planet has a celestial longitude of 180° from that of the Sun; that is, the planet is on the opposite side of Earth from the Sun.

Figure 15.10 The solar system.
The nine planets and the Sun are drawn to scale, and their colors are similar to their surface colors.

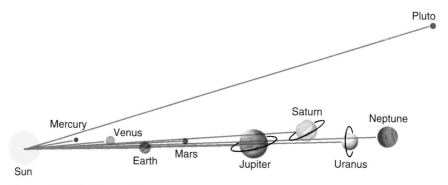

Figure 15.11 The solar system.
This diagram shows the inclination of the planets' orbits with the orbital plane of Earth.

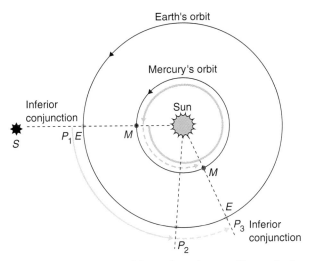

Figure 15.12 The sidereal and synodic periods of the inferior planet Mercury.
Mercury and Earth travel the distances shown in color in equal time periods. See the text for an explanation of the diagram.

Figure 15.13 The terrestrial planets.
The four terrestrial planets, showing their approximate color and relative size. Earth's Moon is shown for comparison.

CONFIDENCE QUESTION 15.3

(a) What is the closest possible approach of any two terrestrial planets? Name the planets and the distance of closest approach in astronomical units. (b) Why is the k in Kepler's third law equal to $\frac{1\,y^2}{(AU)^3}$ when the distance from Earth to the Sun is measured in astronomical units and Earth's revolving time is measured in years?

15.3 The Terrestrial Planets

> **Learning Goal:**
> To list and compare the physical characteristics of the terrestrial planets.

As you learned earlier, Mercury, Venus, Earth, and Mars are called the terrestrial planets because their physical and chemical characteristics resemble Earth in certain respects. Figure 15.13 shows these four planets in comparison to planet Earth's Moon. All four are relatively small in size and mass. They are composed of rocky material (silicates) and metals (their cores are mostly iron and nickel). All are relatively dense (average density about 5.0 g/cm³), have solid surfaces, and weak magnetic fields. Their orbits are comparatively close together, and they are relatively close to the Sun. None has a ring system, and only Earth and Mars have moons. Although the terrestrial planets have some similarities, they are also very different. Planet Earth alone is blessed with an abundance of surface water and an atmosphere with 21 percent oxygen. The other terrestrials have no surface water and no free oxygen in their atmospheres. Earth also has a rotation period (24 h) and a tilt of its axis (23.5°) that provide excellent distribution of the solar energy radiating on its surface.

Mercury

Mercury is the closest planet to the Sun and has the shortest period of revolution (88 days). The early Greeks named the planet after Mercury, the speedy messenger of the gods, and it is the fastest moving of the planets because of its position closest to the Sun. Refer to Section 3.5 concerning conservation of angular momentum.

Mercury, at its greatest eastern or western elongation, can be seen only just after sunset or just before sunrise. The *elongation* (the greatest angular distance between Mercury and the Sun as viewed from Earth) is only 28°. When Mercury is near eastern elongation, it will appear above the western horizon just after sunset. At western elongation Mercury will be on the eastern horizon shortly before sunrise.

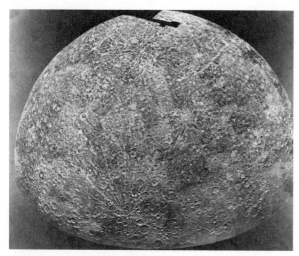

Figure 15.14 Mercury.
This photomosaic illustrates that the planet's surface is heavily cratered, indicating little geologic activity since the solar system was formed. The close-up photograph was taken by the *Mariner 10* spacecraft.

ments are believed to be atoms ejected from Mercury's surface by the solar wind.

The most perplexing property of Mercury is its weak magnetic field, which was first detected in 1974 and confirmed in 1975. Mercury's magnetic field is about 1% as strong as Earth's. We believe Earth's magnetic field is caused by Earth's rapid rotation, but with Mercury's slow rotation, no magnetic field was expected. The origin of Mercury's weak magnetic field remains a mystery.

Venus

Venus is our closest planetary neighbor, approaching Earth at a distance of 26 million miles at inferior conjunction. It is the third brightest object in the sky, exceeded only by the Sun and Moon. Because of its brightness, it was named in honor of the Roman goddess of beauty.

The relative position of Venus with respect to the Sun and Earth, and the important positions with respect to the Sun as observed from Earth, are shown in Fig. 15.15. When Venus is at superior conjunction, it is in full phase (full illumination of the side facing Earth) for an observer located on Earth, but it is not visible to the observer at this time because of the brightness of the Sun. As Venus moves eastward, it appears to the Earth observer as the "evening star" until it reaches inferior conjunction. The greatest eastern elongation (greatest angular distance from the Sun) occurs 220 days after superior conjunction, but maximum brightness does not occur until Venus has about 39° elongation from the Sun. This occurs about 36 days before and after inferior conjunction.

The appearance of Mercury is similar to that of the Moon, as can be seen from Fig. 15.14. However, Mercury has a very high density, almost as high as the density of Earth. This high density indicates that it probably has an inner core of iron, as does Earth.

Mercury's rotation period is exactly two-thirds as great as its period of revolution. Thus it rotates exactly three times while circling the Sun twice. This period probably results from tidal gravitational effects from the Sun. As the planet rotates, the side facing the Sun has temperatures of approximately 700 K (427°C), while the dark side is at about 100 K (−173°C).

Because of Mercury's small size and high daylight surface temperature, the planet should not possess an atmosphere. But the *Mariner 10* mission produced ultraviolet spectral data indicating Mercury had an extremely thin, temporarily held atmosphere, which was mainly hydrogen and helium captured from the solar wind.[*] Earth-based equipment has detected an extremely thin atmosphere of sodium and potassium. These two ele-

Venus and Earth resemble one another in many ways. They have similar properties, such as average density, mass, size, and surface gravity. But the similarities end there. Venus is covered with a dense atmosphere whose composition is 96% carbon dioxide, some nitrogen (less than 4%), and traces of argon, oxygen, and water vapor. At the surface of Venus the atmospheric pressure is a tremendous 90 atm and the temperature 750 K, or about 480°C. The high temperature is due mainly to the large amount of carbon dioxide in the atmosphere, which produces a "greenhouse effect" (see Section 19.2), so life as we know it cannot exist. Both temperature and pressure decrease with increase in altitude. The temperature at the top of the atmosphere is 220 K.

---

[*]The solar wind is composed of charged particles (mainly protons) moving away from the Sun with speeds of about 400 km/s. See Section 18.1 for more data on the solar wind.

15.3 The Terrestrial Planets   **387**

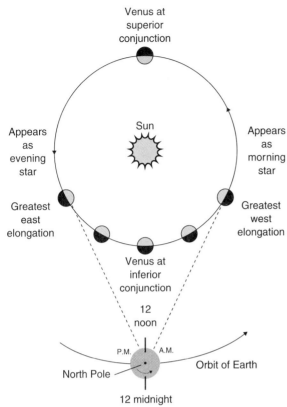

Figure 15.15 Venus.
The positions Venus must have in order to appear as either our evening or morning star.

planets. Why Venus is rotating in retrograde motion and very slowly is a mystery. One suggested answer is the possibility that Venus was struck by a large object during the formation of the solar system, an impact that stopped the planet's rotation and produced a slow rotation in the opposite direction.

Although thick clouds conceal the surface of Venus from our eyes and light-sensitive instruments, surface features have been and are being obtained with radar (*radio detecting and ranging*) imaging. The *Magellan* spacecraft, shown in Fig. 15.16, is the latest Venus probe. The spacecraft was launched from the space shuttle *Atlantis* in May 1989. It traveled a 948-million-mile journey to the planet, and entered a nearly polar orbit around Venus on August 10, 1990.

The *Magellan* radar images, which are two-dimensional black-and-white images, reveal Venus' surface as hot black rock with relatively few large craters. About 1000 craters larger than a few miles

The surface of Venus can never be seen by an observer on Earth because of dense thick clouds that cover the planet. The clouds are composed mainly of sulfuric acid (H_2SO_4) droplets, along with some water droplets. The droplets do not fall out as rain because of the extremely high atmospheric pressure. The clouds occur in four layers, beginning at about 31 km (19 mi) above the surface and extending upward another 37 km (23 mi). The top layer of clouds contains large amounts of yellowish sulfur dust, giving Venus its yellowish or yellow-orange color when viewed from Earth (Fig. 15.13). Orbital spacecraft observations of the outer cloud layer revealed that Venus' atmosphere makes one rotation every four Earth days in retrograde direction. This rotation is extremely fast compared with the −243 days for rotation of the solid planet. (The minus sign indicates retrograde motion.) Refer to Table 15.1 and note the slow rotation of Venus with respect to the other eight

Figure 15.16 The *Magellan* spacecraft being released from the shuttle *Atlantis*.
The radar-mapping *Magellan* entered orbit around Venus on August 10, 1990. After mapping 98% of Venus' surface, NASA ground controllers destroyed the space probe in October 1994 by sending it to a searing death in Venus' atmosphere.

Figure 15.17a Radar image of Venus.
Data collected by the *Magellan* spacecraft were used for mapping this surface view of the northern hemisphere of Venus. The north pole is at the center of the image. Zero° longitude is at the bottom and 90°E is at the three o'clock position. The bright spot just south of the north pole is the Maxwell highlands. The bright areas around the lower-right side are the Aphrodite highlands.

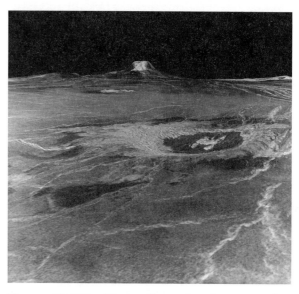

Figure 15.17b Three-dimensional perspective view of Venus' surface.
A portion of the Western Elistla Regio is shown in this image. The impact crater Cunitz, which is 48.5 km (30 mi) in diameter is shown. Gula Mons, a 3 km (1.86 mi) high volcano located at approximately 22° N, 359° E, is shown at the upper left. See Fig. 15.17a for longitude reference.

in diameter were detected. The largest of these (named Mead) has a diameter of 274 km (170 mi). Astronomers think that craters smaller than about four miles in diameter do not appear in the image because small incoming objects have not struck the surface. The smaller objects have been consumed in Venus' thick atmosphere. Other surface features identified were fractures and fault lines with high walls and cliffs, mountain chains that extend hundreds of miles in length, and volcanic plains that cover more than 80% of the planet's surface. No active volcanoes appeared in *Magellan* radar images, but most surface rocks appear to be volcanic in origin, indicating that volcanism was the last geological process to take place on the planet. No surface feature in the images appears older than about a billion years, and most features are thought to be approximately 400 million years old.

Radar images of Venus' surface from the radar-mapping *Magellan* can be transformed into three-dimensional views with color added to enhance details. Except for some areas where data are missing, Fig. 15.17a shows radar images of Venus' northern hemisphere. The north pole is at the center of the radar image. Maxwell Montes, the highest mountain on Venus, can be seen slightly south of the north pole. Its summit rises 11 km (6.8 mi) above the average level of the planet's plains. The top of the highlands is believed to be composed of iron oxide or iron sulfide, which are good reflectors of radar wavelengths. Thus, they appear much brighter than the surrounding regions. Another high region, Aphrodite* Terra, can be seen extending along the right edge of the radar image. This region is just north of the equator. Figure 15.17b shows radar images of the crater Cunitz (named

---

*The name of the goddess of love in Greek mythology (corresponding to Venus in Roman mythology). Surface features on Venus are named mainly for famous women or female mythological characters.

for mathematician Maria Cunitz), which is approximately 48 km (30 mi) in diameter. At the top left is volcanic mountain Gula Mons, named after an Assyrian goddess.

Mars

Viewed from Earth, Mars has a reddish color and was named after the bloody Roman god of war. Mars is about 1.5 times as far from the Sun as Earth. It is tilted on its axis at an angle of 24°, which is very close to Earth's 23.5° angle of tilt. Mars rotates once every 24.5 h, which is very close to a single Earth day. It takes 687 days (about 23 Earth months) for Mars to go around the Sun.

Mars has two small satellites, or moons, named Phobos ("fear") and Deimos ("panic") after the mythical horses that pulled the chariot of the god Mars. The moons are very small. Phobos is about 32 km (20 mi) across and Deimos is about half that size. They are irregularly shaped and extensively cratered. Both revolve eastward around Mars. Phobos circles Mars in only 7 h and 39 min, and Deimos revolves once in 30 h and 18 min. Like our Moon, they keep one side always facing the planet. The periods of rotation and revolution are equal.

The mass of Mars is about one-tenth as large as that of Earth. Its density is also much less (3.9 g/cm^3). This low density indicates that, unlike Earth, Mars probably does not have a large iron core in the center.

Very little is known about the internal structure of the planet. Since its density is fairly high, its internal composition is believed to be of rocky and metallic materials. Mars does have a very weak magnetic field, less than 0.004 times as strong as Earth's.

The surface temperatures of Mars range from a low at the south polar cap of 130 K (−143°C) to a high in the equatorial regions of 290 K (17°C).

The surface of Mars has many craters similar to those of the Moon. The largest of the so-called basins, about 1600 km (1000 mi) across, is located in the southern hemisphere from 50° to 90° east longitude. Its smoothness is attributed to dust that has settled out from dust storms.

The two most outstanding features of the surface of Mars are the polar caps and the 12 or more dead volcanoes (Fig. 15.18). In winter the polar

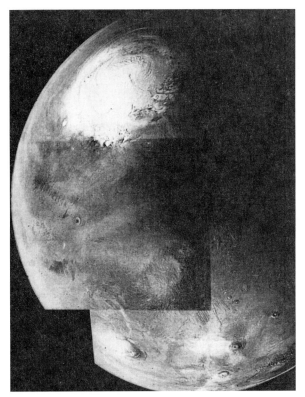

Figure 15.18 Photomosaic of Mars.
This mosaic of three photos of Mars shows the northern hemisphere from the polar cap to a few degrees south of the equator. The huge Martian volcanoes are visible at the bottom of the photo.

caps are composed of frozen carbon dioxide (CO_2) and water ice. In summer the frozen CO_2 changes to vapor, leaving behind a residual polar cap of water ice, which is probably hundreds of meters thick. Olympus Mons, "Mount Olympus," shown in Fig. 15.19, is the largest known volcano in the solar system. It rises 24 km (15 mi) above the plain and has a base with a diameter greater than 600 km (372 mi). The volcano is crowned with a 70 km wide (43 mi) crater. The largest volcano on Earth is Mauna Loa on the island of Hawaii. Mauna Loa's base rests on the ocean floor 5 km (3.1 mi) below the surface of the Pacific Ocean and extends upward another 4.2 km (2.6 mi) above the level of the ocean. Thus Mauna Loa is about one-third the height of Olympus Mons.

Another major feature of the surface is the large canyon called *Valles Marineris*, shown in Fig. 15.20. The canyon is about 4000 km (2500 mi)

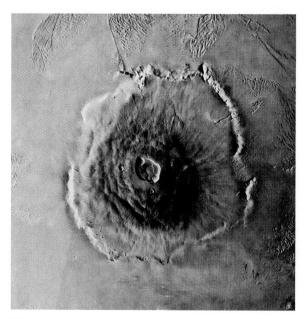

Figure 15.19 Olympus Mons, "Mount Olympus."
This huge Martian volcano (the largest in the solar system) is about 24 km (15 mi) high and 600 km (372 mi) wide at its base. The caldera is 70 km (43 mi) across at the summit.

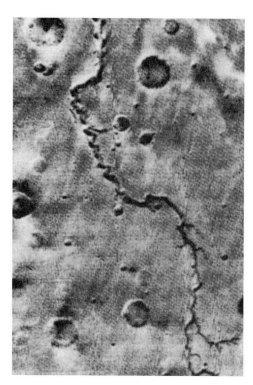

Figure 15.21 An ancient channel on Mars.
Though not unique on the Martian surface, this meandering "river" is the most convincing piece of evidence that a fluid once flowed on Mars, draining a large area and eroding a deep channel. The feature is some 575 km (357 mi) long and 5 to 6 km (3 to 3.7 mi) wide.

Figure 15.20 Valles Marineris.
This enhanced color mosaic shows the great canyon Valles Marineris. The canyon is 4000 km (2500 mi) in length. Geologists believe that it is a fracture in the planet's crust caused by internal forces.

long and 6 km (3.7 mi) deep. Its length is equivalent to the width of the United States. The canyon is almost four times as deep as the Grand Canyon. This tremendous gash in Mars' surface is thought to be a gigantic fracture formed by stress within the planet.

Interesting features found on the planet's surface are shown in ● Fig. 15.21 and ● Fig. 15.22. They are believed to be formed by a moving liquid, and water is the chief suspect. Today the atmospheric pressure on Mars is too low for water to exist, although traces of water have been detected in the atmosphere.

The atmosphere of Mars is much more tenuous than that of Earth. It is composed of 95% carbon dioxide, 2.7% nitrogen, 1.6% argon, and traces of oxygen, carbon monoxide, krypton, and xenon. The atmosphere pressure is slightly less than 1% of that on Earth.

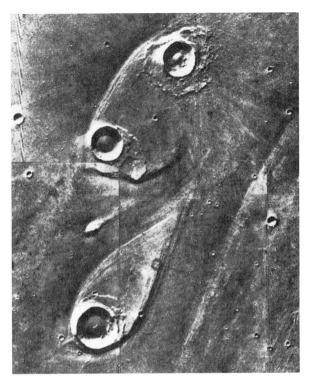

Figure 15.22 A flood plain on Mars.
These teardrop-shaped features indicate that a fluid flowed on the surface of Mars. The patterns indicate that the craters were there before the fluid flowed.

Figure 15.23 The rocky, reddish surface of Mars.
This *Viking 2* photo shows a ground-level view of the Martian soil on the Utopia plains. The fine grain soil and rock appear reddish due to the presence of iron ore. The sky appears pink due to dust particles in the thin atmosphere. The horizon appears tilted because the landing spacecraft is tilted about 8°.

Our best photographs of the surface of Mars have come from the *Viking* landers. Figure 15.23 shows the first close-up photograph ever taken of the surface of Mars, courtesy of *Viking 2*. The *Viking* missions carried out numerous experiments searching for evidence of life on Mars. None was found, but the possibility has not been ruled out completely.

CONFIDENCE QUESTION 15.4

(a) Name three physical properties that the terrestrial planets possess that classify them as very similar. (b) Name two physical properties that planet Earth possesses that make it very different from the other terrestrial planets.

15.4 The Jovian Planets and Pluto

Learning Goals:

To list and compare the physical properties of the Jovian planets.

To compare the major differences between the terrestrial and the Jovian planets.

The four Jovian planets are large compared to the terrestrial planets, are gaseous, and have no solid surface. Figure 15.24 shows the four Jovian planets. They are composed mainly of hydrogen and helium, and all have a very low density (average density 1.2 g/cm^3). The planets possess strong magnetic fields, have many moons and rings, are far from the Sun, with orbits far apart. All four planets are believed to have rocky cores with a layer of ice above the rocky core. Upper layers of molecular and metallic hydrogen apply high pressure and create high temperature to the ice layers and rock core, producing ice and rock that are much different from the rock and ice on Earth. The rock is believed to be composed mainly of iron, oxygen, and silicon, whereas the ice is believed to be composed of carbon, nitrogen, and oxygen integrated with hydrogen. Thus, the Jovian planets are very different from the terrestrial planets. Table 15.2 lists the significant differences between the terrestrial and Jovian planets.

Figure 15.24 The Jovian planets.
This montage shows the four Jovian planets. Cloud patterns are easily seen and their colors are approximately true. The planets are not shown to scale.

Pluto does not resemble Earth or Jupiter, and some astronomers have suggested that Pluto be classified as an asteroid.

When the planets first began to coalesce around 5 billion years ago, the predominant elements were the two least massive—hydrogen and helium. The heat from the Sun allowed these two elements to escape from the inner planets. That is, the speeds of the molecules of these elements were sufficient to allow them to escape the planets' gravitational forces. Thus the inner planets were left with mostly rocky cores, giving these planets a high density. The four large outer planets were much colder, and they retained their hydrogen and helium, which now surround their rocky cores. Thus the four large outer planets consist primarily of hydrogen and helium in various forms, and this composition gives them much lower densities.

Because of their larger masses and greater gravitational forces, the large outer planets also have many more moons than the smaller inner planets. In fact, we now know that Jupiter, Saturn, Uranus, and Neptune have large satellite systems with at least 16, 19, 15, and 8 moons, respectively. All of the Jovian planets have rings. No rings exist around the terrestrial planets.

Jupiter

Jupiter, named after the supreme Roman god of heaven because of its brightness and giant size, is the largest planet of the solar system, in both volume and mass. The motion about its axis is faster than that of any other planet, since it takes only 10 h to make one rotation. Jupiter possesses more than half of the total angular momentum of the solar system.

Jupiter's diameter is 11 times as large as that of Earth, and it has 318 times as much mass. However, its density is only 1.3 g/cm^3. Jupiter consists of a rocky core, a layer of hydrogen in liquid metallic form (because it is at high pressure and temperature), and an outer layer of molecular hydrogen. Above the molecular hydrogen is a thin layer of clouds composed of hydrogen, helium, methane, ammonia, and several other substances. The interior structure of Jupiter is shown in Fig. 15.25. The mean surface temperature at the top of the clouds is about 125 K ($-148°C$).

Jupiter has the interesting property of actually giving off twice as much heat as it receives from the Sun. Since this heat is probably radiated from gravitational energy released during the planet's

Table 15.2 Significant Differences Between Terrestrial and Jovian Planets

Terrestrial Planets	Jovian Planets
Small diameter, approximately 5000 to 13000 km	Large diameter, approximately 50,000 to 143,000 km
Rocky	Gaseous, mainly hydrogen and helium
Solid surface	No solid surface
Relatively high density (3.9–5.5 g/cm^3)	Relatively low density (0.7–1.7 g/cm^3)
Relatively close to Sun	Great distance from Sun
Relatively high temperature environment	Cold temperature environment
Close together Greatest separation less than 1.2 AU	Widely separated Greatest separation 25 AU
Weak magnetic field, if any	Strong magnetic field
Only three moons	Many moons (61 known)
No rings	All have rings
Slow rotation	Fast rotation

formation, Jupiter is, in a sense, a star-like body as well as a planet. If Jupiter had about 80 times more mass, nuclear reactions similar to those in the center of the Sun could have started in its interior.

In Fig. 15.24 the cloud features are easily seen. The clouds of Jupiter show a structure of many patterns—bands, ovals, light and dark areas in white, yellow, orange, red, and brown colors. Convection currents exist, and we see the tops of updrafts (the lighter areas) and downdrafts (the darker areas). The Great Red Spot, which appears yellow in this photograph, stands out. The spot has an erratic movement and changes color and shape. In addition, it sometimes completely disappears. The most recent theory of the Great Red Spot states that it is a huge counterclockwise storm similar to a hurricane on Earth but lasting for hundreds of years.

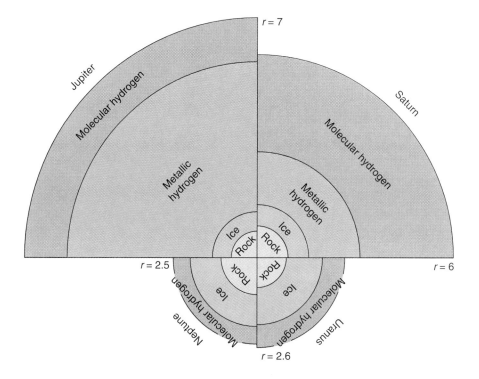

Figure 15.25 Internal structure of the Jovian planets.
The greater part of Jupiter's interior is liquid metallic hydrogen. The interior of Saturn is mostly hydrogen. Internal pressure is great enough within Jupiter and Saturn to liquefy hydrogen, which acts as a liquid metal. Uranus and Neptune, much smaller, consist largely of ice composed of the elements carbon, nitrogen, and oxygen in combination with hydrogen. The approximate radius (r) of each planet is given in units of 10^4 km.

Figure 15.26 Jupiter's faint ring.
The ring shown in this photograph has been drawn in to show its size and location relative to the planet. The ring lies in Jupiter's equatorial plane about 50,000 km above the planet's clouds. The ring was discovered by the *Voyager 1* spacecraft.

The uppermost cloud layer observed either from *Voyager*, Hubble, or Earth's surface, is composed of frozen ammonia crystals. The temperature at this layer is about 140 K (−133°C). The clouds of ammonia crystals should appear white instead of yellow, orange, red, and brown mixtures. Other chemicals must also be present, but presently they have not been identified.

Jupiter possesses a tremendous magnetic field that, at the top of the atmosphere, is 10 times as strong as Earth's field. Jupiter's magnetic poles are reversed from those on Earth, and are 10° from its geographic poles, which are defined by its rotation axis. Its rotation axis is inclined by only a few degrees, so Jupiter does not experience seasonal effects as Earth and Mars do.

Jupiter has many moons, 16 or more depending on where the distinction is made between a large rock and a small moon. In addition, Jupiter has a very faint planetary ring of dust particles some 50,000 km above the planet's clouds that is bright enough to be seen from Earth. Jupiter's ring is shown in Fig. 15.26.

Jupiter's 16 moons (4 large and 12 small) can be arranged into four groups of four moons each. The four closest to the planet are small (average diameter 88 km) and have prograde orbits. Next comes the four largest moons of Jupiter, which were first discovered by Galileo in 1609. They are sometimes called the Galilean moons of Jupiter. In order of increasing distances from Jupiter they are Io, Europa, Ganymede (the largest moon in the solar system), and Callisto. They were photographed close up by the *Voyager 1* and *Voyager 2* spacecraft in 1979. They all have diameters between about 3200 and 5300 km, and prograde orbits that range from 422,000 km for Io to 1,880,000 km for Callisto. Photos of the moons are shown in Fig. 15.27. Jupiter's eight outermost moons are very small (average diameter 63 km) and move in eccentric orbits. They are arranged in two groups of four each. The first group, moving in prograde orbits with periods of about 250 days, are approximately 11.5 million km from Jupiter. The second group, moving retrograde, have orbital periods of about 700 days and average about 23 million km from the planet. All eight are believed to be objects captured by Jupiter's strong gravity, either as single objects or as two separate objects that later broke apart.

One of the most spectacular findings of the *Voyager* missions was that Io has many active volcanoes on it. The volcanoes occur because the

Figure 15.27 The Galilean moons of Jupiter.
The moons were photographed by the *Voyager 1* spacecraft. They are shown to scale as they would appear from a distance of one million kilometers. Clockwise from the upper right, they are Europa, Callisto, Ganymede, and Io.

Figure 15.28 Saturn and its rings.
This photograph was taken by the *Voyager 2* spacecraft as it approached the planet. The color, sort of a butterscotch, is approximately correct for the human eye. Note three of the planet's moons at the bottom of the photo and a fourth casting a shadow on Saturn's clouds.

gravitational attraction of other nearby moons, notably Europa, causes Io's orbit to vary so that it is closer to, then farther from, Jupiter. The resulting changes in Jupiter's gravitational force cause stresses in the interior rock of Io, and a great deal of frictional heat is generated, resulting in volcanoes.

Europa is very bright, mainly because its surface is covered with a mantle of ice approximately 242 km (150 mi) thick. Ganymede and Callisto, the two largest satellites, have cratered surfaces similar to that of Earth's moon.

Saturn

The most distinctive feature of Saturn is its system of three prominent rings, which can be seen in Fig. 15.28. These rings have been viewed by Earth-based observers for many years and are the most spectacular celestial sight that can be seen with a small telescope. The rings, inclined by 27° to Saturn's orbital plane, are identified by the letters A, B, and C. The outer ring, shown in Fig. 15.28, is the A ring, and it is separated from the bright B ring by the broad dark region known as the Cassini division. The Cassini division is named in honor of G. D. Cassini, an Italian astronomer, who discovered the dark region in 1675. The gap is about 4800 km (3000 mi) wide. The B ring, which contains the largest number of particles per unit volume, is the brightest ring.

The inner C ring has a very low particle density and appears with less intensity than do rings A and B, because only a small amount of light is reflected from the smaller number of particles. The small dark region near the outer edge of the A ring is known as the Encke division. Although the dark regions appear to be without particles, the divisions or gaps do contain a few tiny particles, which were discovered by the *Voyager 1* spacecraft. The rings, which are less than 50 m (164 ft) thick, are believed to be composed of particles of ice and ice-coated rocks ranging in size from a few micrometers to approximately 10 m (33 ft) in diameter.

Within a certain distance of any planet's center, called the *tidal stability limit* or the **Roche limit,** the planet's tidal forces* (due only to gravity) acting on a large, solid, revolving object will tear the object apart, because the tidal forces are greater than the binding forces (due only to gravity) holding the object together. The Roche limit is directly proportional to the radius of the planet and is a function of the density of the planet and the orbiting object. The Roche limit for Saturn is about 2.5. Thus, a large solid object within 2.5 times the planet's radius will be broken into fragments. Saturn's outer A ring is at a distance of 2.3 times the radius of Saturn.

We can only speculate that the ring fragments of Saturn are the remnants of a moon that was torn apart, or perhaps they are particles that could not assemble into a satellite because of tidal forces. The small particles of ice and ice-coated rocks that make up the rings of Saturn are held together by cohesive, electromagnetic, and self-gravitational forces that are greater than the planet's tidal forces.

The *Voyager 1* and *Voyager 2* flights showed the structure of the rings to be very complicated systems of many individual ringlets (Fig. 15.29). This highly enhanced color view was assembled from visible light and ultraviolet frames. The composite photograph, taken from *Voyager 2* at a distance of 5.5 million mi, shows the possible variations in the chemical composition from one part of

---

*Tidal force is a differential gravitational force that tends to deform or stretch a body. See Section 17.6 for a specific explanation of tidal force.

Figure 15.29 Saturn's rings in false color.
Possible variations in chemical composition from one part of Saturn's ring system to another are visible in this *Voyager 2* photograph as subtle color variations that can be recorded with special computer-processing techniques. This highly enhanced color view is assembled from colorless, orange, and ultraviolet frames obtained at a distance of 5.5 million mi. The C ring and Cassini division appear blue in the photo.

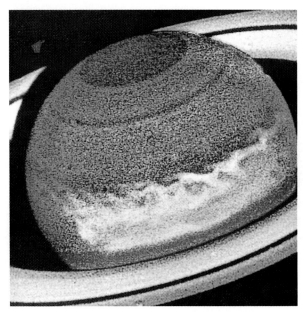

Figure 15.30 Saturn's Great White Spot.
The Hubble Space Telescope's Wide-Field Planetary Camera used blue and infrared light to record this view of Saturn. This picture combines two colors to show the lower parts of the clouds in blue and the region with high clouds in red.

Saturn's ring system to another. Special computer-processing techniques exaggerate subtle color variations in the photograph.

The structure of Saturn itself is somewhat similar to that of Jupiter. That is, it has a small solid core surrounded by a layer of metallic hydrogen and an outer layer of liquid hydrogen and helium. Saturn's density is only 0.7 g/cm³, so it would float in water. Its temperature near the top of the clouds is approximately 120 K (−153°C). Like Jupiter, Saturn radiates more heat than it gets from the Sun.

Saturn's mass is 95 times that of Earth, and its diameter is 9 times larger. It rotates about once every 11 h. It has a magnetic field that is 1000 times stronger than that of Earth but only 0.05 times as strong as Jupiter's.

Saturn's atmosphere is similar in chemical composition to that of Jupiter. The atmospheric surface has a banded structure, but it appears dull and not as bright as Jupiter's atmosphere because Saturn's surface temperature is lower than that of Jupiter's.

Figure 15.30 shows an image of Saturn taken by Hubble on November 9, 1990, with the Wide-Field Planetary Camera. The photo shows with clarity the banded structure and the Great White Spot, a rare and unusual cloud formation occurring over the equatorial region. A combination of blue and infrared light on the Planetary Camera was used to record this view. The photo combines the two colors to show the lower parts of the cloud in blue and the region with high clouds in red. The clouds are believed to be composed mainly of ammonia ice crystals.

The flow pattern of Saturn's atmosphere is complex; the planet's winds move with high speeds, some as high as 1600 km/h (1000 mi/h). These winds are generated by a combination of convection currents and rapid planetary motion.

Outside the main visible rings lie 17 or more moons. Many of them are quite small, with diameters between about 30 and 100 km (roughly 20 and 60 mi). Most of these small moons were discovered in 1980 by *Voyager 1*. The moons Mimas, Enceladus, Tethys, Dione, Rhea, Hyperion, and Iapetus are similar in two respects. They all have diameters between 385 and 1530 km (240 and 950 mi). Because their densities are between 1 and 2 g/cm³, they are all believed to be composed of rock and ice.

Except for the distant moons, Phoebe and (possibly) Hyperion, all the moons of Saturn rotate once each revolution, similar to our own Moon. Thus they always keep the same face toward Saturn. Except for the two distant moons, Phoebe and Iapetus, all the moons of Saturn move in nearly circular orbits, revolve in the same direction, and lie along the planet's equatorial plane. Phoebe revolves in the opposite direction to that of the other 17 moons, and its orbit is inclined by 150°. Thus Phoebe is probably a moon that was somehow captured by the Saturnian system in the distant past.

The most interesting moon of Saturn is Titan. Titan is the largest moon of Saturn, with a diameter of 5150 km (3193 mi) and a density of 1.9 g/cm^3. It is the only satellite known to have a dense, hazy atmosphere, which is probably due to its low surface temperature of 94 K. The main constituent of the atmosphere is nitrogen (about 90 percent), argon (less than 10 percent), methane (CH_4, less than two percent), and traces of hydrocarbons.

Uranus

Uranus was discovered in 1781 by William Herschel (1738–1822), an English astronomer. The name, chosen in keeping with the tradition of naming planets for the gods of mythology, was first suggested by Johann Bode. Uranus was the father of the Titans and the grandfather of Jupiter.

Figure 15.31 shows two pictures of Uranus taken by *Voyager 2* when the spacecraft was 9.1 km

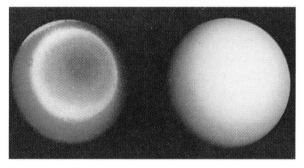

Figure 15.31 Uranus.
These two pictures of Uranus were taken by the narrow-angle camera of *Voyager 2* when the spacecraft was 9.1 ×10^6 km (5.7 × 10^6 mi) from the planet. The photo on the right shows Uranus as the human eye would see it from the vantage point of the spacecraft. The photo on the left employs enhanced color. See the text for details.

(5.7 million mi) from the planet. The right photo has been processed to show Uranus as human eyes would see it from the spacecraft. The blue-green color results from the absorption of red light by methane gas in Uranus' atmosphere. The darker shading at the lower left of the disk corresponds to the day-night boundary. The boundary line dividing day and night on the surface of a planet or moon is called the **terminator.** The left photo uses false colors and contrast enhancement to show distinctive details in the polar region. The false-color picture reveals a dark polar hood surrounded by a series of progressively lighter concentric bands. One possible explanation is the brownish haze or smog concentrated over the polar region.

The internal structures of Uranus and Neptune are similar, but differ from those of Jupiter and Saturn. Uranus and Neptune are much smaller and less massive than Jupiter and Saturn. See Table 15.1 for a comparison of physical properties. Also, their rocky cores are relatively large compared with their total size. See Figure 15.25.

From a study of the physical properties, the internal structure of Uranus is calculated to be in three layers. The inner rocky core, which is about 13,000 km (8100 mi) in diameter, contains about 25% of the planet's mass and is probably composed of iron and silicon. The rocky core is surrounded by a liquid mantle approximately 8000 km (5000 mi) deep, composed of water, ammonia, and methane ice. The mantle makes up 65% of the planet's mass. The outer layer, the atmosphere, is about 11,000 km (6800 mi) thick and is composed mostly of molecular hydrogen (84 percent), helium (14 percent), and a small percentage of methane.

Uranus has a ring system that is very thin. The nine major rings are shown in Fig. 15.32, which is a montage of Uranus, with its outer veil removed, and the *Voyager 2* spacecraft. The planet does have clouds at lower elevations with flow patterns that move in the same direction as the planet's rotation. A new, very faint ring discovered by *Voyager 2*, called 1985U1R, is not shown in the montage.

The rings of Uranus are composed mainly of boulder-size particles 1 m or larger in diameter, with very few dust-size particles present. Because of the lack of dust particles, the rings do not have good reflective properties like the rings of Saturn, which are filled with tiny particles 1 cm and

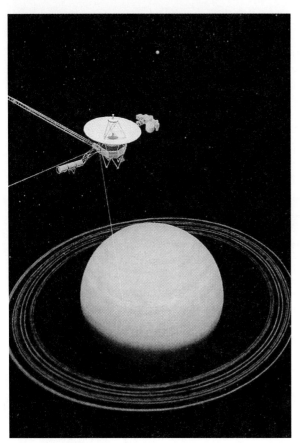

Figure 15.32 Montage of Uranus, its nine rings, and *Voyager 2*.
A very faint ring discovered by *Voyager 2* is not shown in the montage. The spacecraft's cameras also discovered 10 new satellites circling Uranus.

smaller. *Voyager 2* also recorded some very narrow sections of rings.

Voyager 2 and *Voyager 1* are identical spacecraft. *Voyager 2*, shown in Fig. 15.32, was launched in 1977 on a grand tour of the Jovian planets using gravity-assisted encounters with all four planets: Jupiter in 1979, Saturn in 1980, Uranus in 1986, and Neptune in 1989. The two spacecraft have taken numerous photographs and collected thousands of bits of data with their sensors. Information given in this textbook comes mainly from the *Voyager* spacecraft. Both *Voyagers* are now headed out of the solar system.

Uranus has several other interesting features besides its rings. Its rotation axis is inclined 82° with respect to its orbital motion, which positions the Sun nearly overhead at the north and south poles as Uranus revolves around the Sun. The planet rotates retrograde with a period of 17.24 h. It has a magnetic field that is about 50 times as strong as Earth's. An unusual feature of Uranus' magnetic field is its orientation with respect to the rotational axis of the planet. The magnetic field is tipped 55° to the rotational axis. The clouds of Uranus are deep within its atmosphere and are nearly invisible. Thus the planet appears as a bland, almost featureless, blue-green disk. Its atmospheric temperature, which is fairly uniform over the whole planet, is about 59 K (−223°C). For some reason, the planet radiates more energy than it receives from the Sun.

Uranus has five major satellites. They are, in order of distance from the planet, Miranda, the smallest and closest, having a diameter of 481 km (298 mi) and orbiting at a distance of 129,000 km (80,000 mi); Ariel; Umbriel; Titania, the largest, with a diameter of 1587 k (986 mi); and Oberon, the most distant, orbiting at 584,000 km (362,000 mi). The satellites have densities of about 1.6 g/cm^3, which indicate a composition of a mix of ice (water) and rock.

The cameras of *Voyager 2* recorded 10 new satellites circling Uranus inside the orbit of Miranda. The first was discovered in 1985 and named 1985U1. This satellite has a diameter of about 165 km (102 mi). The other nine satellites, discovered in 1986, are named 1986U1 through 1986U9. They are very small, with diameters less than 100 km (62 mi).

The surface features of the satellites show that, with the exception of Umbriel, the moons have been tectonically active in the past. Figure 15.33 shows Miranda's surface, with large curvilinear regions of grooves and ridges plus regions that appear chevron-shaped. The satellite's surface is pockmarked with craters. The large crater shown in the lower-right region of Fig. 15.33 is about 24 km (15 mi) in diameter. Photos obtained from *Voyager 2* show the surface has very deep valleys and high cliffs ranging in height from 0.5 to 5 km (0.3 to 3 mi). The geologic forms on Miranda are some of the most bizarre in the solar system.

The surfaces of Ariel and Titania are pitted with craters having numerous valleys and fault scarps cutting across the highly pitted terrain. Uranus' most distant moon, Oberon, has several large impact craters in the planet's icy surface.

Figure 15.33 Miranda, Uranus' fifth largest satellite.
Uranus' innermost large moon, Miranda, is roughly 300 miles in diameter and exhibits a variety of geological forms—some of the most bizarre forms in the solar system. Chevron-shaped regions and folded ridges in circular racetrack patterns are visible on the satellite's surface. There are large scarps, or cliffs, ranging up to 3 mi in height; they are clearly visible in the lower right part of the photo. Next to them is a deep canyon approximately 30 mi wide.

Neptune

Neptune was discovered in 1846 by John G. Galle (1812–1910), a German astronomer at the Berlin Observatory. Partial credit is also shared by Englishman John Couch Adams and Frenchman U. J. J. Leverrier, two mathematicians. Using Newton's law of gravitation, Adams and Leverrier made calculations that produced information on where to look for a supposed planet that was disturbing the motion of Uranus. The name Neptune was proposed by D. F. Arago, a French physicist who had suggested that Leverrier begin the critical calculations.

The planet cannot be observed with the unaided eye, and appears to have a greenish hue when viewed through a telescope. The physical makeup of Neptune is similar to that of Uranus. Methane and hydrogen have been detected spectrographically, so Neptune definitely has a gaseous atmosphere.

On August 25, 1989, *Voyager 2* arrived at Neptune. Cameras on board took thousands of photographs, then sent back to Earth a photographic record of Neptune's clouds, storms, the Great Dark Spot (see Fig. 15.34) similar to Jupiter's Great Red Spot, large wind systems, eight satellites, five rings, and a thin layer of dust.

Neptune can be regarded as a twin to Uranus. Not only are they similar in size and in composition of their atmospheres, but it is believed that their internal structures are also similar. Each planet has a rocky core surrounded by a liquid mantle of water, methane, and ammonia. The mantle is surrounded by a layer of gas composed mainly of hydrogen and helium. Data from *Voyager 2* revealed Neptune's magnetic field to be comparable to that of its twin planet. Neptune's magnetic field is tipped at 55° relative to its axis of rotation. Uranus' magnetic field is tipped at 60°. The magnetic field of each planet is offset from the center of the planet.

Figure 15.34 shows the Great Dark Spot that is located near 21° south. This spot covers an area 12,000 km (7500 mi) in longitude and 8000 km (5000 mi) in latitude. The size varies with time in both directions. A second dark spot is located at

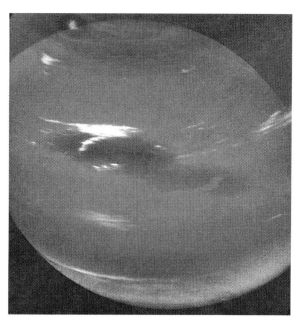

Figure 15.34 Voyager's image of Neptune's atmosphere.
This image of Neptune was taken when the *Voyager 2* spacecraft was 6.1 million km (3.8 million mi) from the planet. The dominant storm system in the atmosphere is the Great Dark Spot.

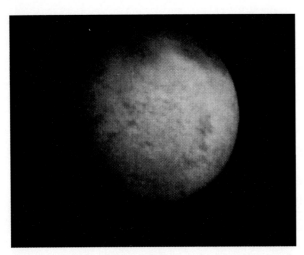

Figure 15.35 Triton, Neptune's largest satellite.
Triton's polar cap of frozen nitrogen tends to be pinkish in color. The satellite is primarily a white object with a pinkish cast in some areas. The land patterns are strange and complex, and of unknown origin.

51° south and has at its center a bright white spot. This white spot is a cloud that is believed to have formed from a rising convection current containing methane. The cloud is several miles higher than the dark spot.

Because of its thick methane-rich hydrogen cloud cover, Neptune's surface features were not photographed. However, excellent photographs were obtained of its largest moon, Triton, which is slightly smaller than Earth's moon and orbits Neptune in retrograde. Surface details are shown in Fig. 15.35. Note the white polar cap of frozen nitrogen and the strange, complex land forms. The surface temperature is about 37 K, which is below the freezing point of nitrogen gas. Triton has a thin gaseous atmosphere composed of nitrogen with a small amount of methane.

Another of Neptune's satellites, Nereid, is very small. Its diameter is estimated to be less than 645 km (400 mi). Nereid has the most eccentric orbit (0.749) of any satellite in the solar system. The highly elliptical orbit takes the tiny moon from 1.4 to over 9.7 million km from Neptune.

Orbiting Neptune is a system of five rings and a sheet of dust in the equatorial region. The rings are not optically visible from Earth. They were believed to exist from their occultation of starlight, but the observations were questionable. In 1989 *Voyager 2* took photographs of the rings and confirmed their reality. They were labeled 1989N1R through 1989N5R. The three brightest rings (N1R, N2R, N3R) were named Adams, Leverrier, and Galle, respectively, after the three nineteenth-century astronomers who took part in Neptune's discovery.

Pluto

Pluto, named for the Greek god of the underworld, is the most distant planet from the Sun. It was discovered by C. W. Tombaugh in 1930 at the Lowell Observatory in Arizona, after a thorough search near the position predicted by theoretical calculations. The planet had been predicted because discrepancies appeared in the orbital motions of Uranus and Neptune. General information concerning Pluto is given in Table 15.1. Because of Pluto's small size and great distance from the Sun, very little is known about its surface features. Spectroscopic investigations indicate the planet is covered with methane ice. The surface temperature ranges from about 50 K near aphelion to 60 K near perihelion. Pluto is presently in the warmest part of its orbit. See Fig. 15.9.

In June 1978 a satellite of Pluto was discovered by James W. Christy of the U.S. Naval Observatory. Named Charon (KEHR-on), this moon is about half the size of Pluto, making it the largest satellite in relation to its parent planet. Simultaneously, Pluto was found to be much smaller than previously believed. Its diameter is now thought to be about 2300 km, making it the smallest planet. The image shown in Fig. 15.36 was taken by Hubble and for the first time shows Charon separate from Pluto.

The planet is so far away that its temperature is difficult to measure. Also, the diameter, density, and period of rotation given in Table 15.1 are estimated values and may change when more precise data are obtained.

Some astronomers believe that Pluto was once a moon of Neptune, for three reasons:

1. Pluto is much smaller than the four other outer planets (see Fig. 15.10).
2. Pluto does not lie along the planetary disk (see Fig. 15.11).
3. Pluto's orbit is highly elliptical. It actually goes inside the orbit of Neptune (see Fig. 15.9).

15.5 Other Solar System Objects

Figure 15.36 Pluto and its satellite Charon.
Pluto is the first solar system object to be observed by the Hubble Space Telescope. This photo shows, for the first time, Pluto separate from its satellite. The circular halo around the planet is caused by a defect (spherical aberration) in the telescope's primary mirror.

Pluto went inside Neptune's orbit in late 1978 and will exit in the year 1999. Thus for more than 20 years Neptune will be the most distant planet from the Sun.

There is one reason to doubt that Pluto was once a moon of Neptune. For this to have been true, Pluto must have been very close to Neptune, and the present orbits suggest that it was not close in the past. Detailed studies show that in their present orbits Pluto and Neptune can never come closer than 18 AU. Some astronomers believe that Pluto and Charon resemble a double asteroid, and that Pluto should no longer be classified as a major planet.

Pluto is the only planet that has not been visited by a space probe. A fly-by mission to Pluto is being developed at the Jet Propulsion Laboratory in California with hopes of launching a 100- to 150-kg spacecraft in 1998. The trip to the planet will take six to eight years depending on the rocket used for launching and the total weight of the spacecraft. As stated above, Pluto is presently near perihelion (the warmest part of its orbit) and now is the best time for obtaining data about the planet's surface and thin atmosphere.

Much of the information about the solar system has been obtained by robot probes to the planets. The spacecrafts carry sophisticated equipment that collects gravitational, electromagnetic, temperature, and other data that is transmitted by radio waves back to Earth. Over the past several years, astronauts have been sent to the Moon, robot probes have analyzed the surface of Mars, fly-by missions have radar-mapped the surface of Venus; and fly-by missions have photographed and collected other data on Mercury, Jupiter, Saturn, Uranus, and Neptune. See the Highlight on space flights in the solar system.

Planet X

Is there a Planet X? No one knows. We have neither persuasive theoretical justification to believe that a planet beyond Pluto does not exist, nor positive evidence that a planet beyond Pluto *does* exist. Discrepancies in the orbital motion of Neptune and Uranus, thought to be due to the gravitational influence of some unknown mass have been repudiated as erroneous. Modern measurements in their orbital motion show no discrepancies. Presently, it is generally accepted that the discrepancies do not, and did not, exist.

CONFIDENCE QUESTION 15.5

(a) Compare the physical properties of the terrestrial planets with the Jovian planets. Give at least five characteristics that show a difference between the two groups. Refer to Table 15.2. (b) What is the approximate duration of one day at the north pole of Uranus?

15.5 Other Solar System Objects

Learning Goal:

To describe the minor objects—asteroids, meteoroids, comets, and interplanetary dust—and list their physical characteristics.

The Sun is the predominant mass of the solar system and holds the system together with its strong gravitational field whose influence may extend

HIGHLIGHT

Space Flights in the Solar System

Planet Earth orbiting the Sun has both kinetic and potential energy. Kinetic energy is due to its orbital motion, and potential energy is due to its position in the Sun's gravitational field. The greater the distance a planet is from the Sun, the greater the total energy. In reference to Earth, Mercury and Venus, having smaller orbits, have smaller total energies. Planets with orbits greater than Earth's have greater total energies.

Viewing the solar system from the North Celestial Pole, Earth orbits the Sun in a counterclockwise or eastward direction with an average orbital speed of 29 km/s (65,000 mi/h). If we want to send a space probe to Mercury or Venus, the probe must lose energy. Since our launching platform is moving with a speed of 29 km/s, we want to launch the probe in the direction opposite Earth's orbital motion. This will lower its speed with respect to the Sun and decrease its energy. Similarly, if we want to send a space probe to Mars or any planet with an orbit greater than Earth's, the probe must gain energy. Thus, we launch the probe in the direction of Earth's orbital motion to increase its speed and energy.

To send a space probe to Mercury, we must give it a speed, relative to Earth, of 7.3 km/s (16,000 mi/h) in the direction opposite Earth's orbital motion. This speed, plus other parameters, will place the probe in an elliptical orbit that coincides with Mercury's orbit.

Before we can send a space probe off to Mercury, it has to be launched from Earth's surface with enough speed to escape Earth's gravity. The minimum vertical launch speed required for a space probe to escape Earth's gravity is 11.2 km/s (25,000 mi/h). Since this is greater than 7.3 km/s, the speed of the space probe must be reduced so that it can be sent on the proper course to Mercury after escaping Earth's gravity. The speed of the probe must also be adjusted to compensate for the gravitational force of Mercury.

Another parameter is launch time. Since Earth and Mercury are moving eastward around the Sun at different speeds, they must be in proper relative positions at the time of launch, or the space probe will miss its target. A term often used to describe this period is "launch window." This term refers to the interval of time when the space probe and the target planet are located in the proper positions for a successful mission.

Space probe flights have been made to all planets in the solar system except Pluto. All of these flights have taken place in the ecliptic plane to take advantage of Earth's orbital motion to assist in projecting the probes on their journeys.

Scientists also want to examine all areas of the Sun. To accomplish this, the space probe must circle the Sun from pole to pole. This means that the probe has to be propelled out of the ecliptic plane. Launching a probe from Earth to enter a polar orbit around the Sun requires a velocity component of 29 km/s opposite Earth's orbital motion plus a perpendicular velocity to send the probe out of the ecliptic plane. Presently no launch vehicle can produce the thrust necessary to achieve these velocities. An alternative is to use the planet Jupiter to assist the probe to obtain the necessary velocities.

On October 6, 1991, the *Ulysses* space probe was released from the space shuttle *Discovery* and propelled into space on its way to the Sun by way of Jupiter. *Ulysses* arrived at Jupiter in February 1992 and looped around Jupiter in the opposite direction to the planet's orbital motion, thereby reducing its energy. At the same time, Jupiter's gravity flung *Ulysses* down out of the ecliptic plane toward the south polar region of the Sun, where it arrived in June 1994.

The probe's instruments, including solar wind plasma and ion detectors, magnetometers, energetic particle detectors, radio and plasma wave instruments, solar X-ray and gamma-ray burst detectors, and cosmic dust sensors, are examining every latitude of the Sun by circling it, pole to pole, from a distance of approximately 2 AU.

The space probe has detected shock waves millions of kilometers in diameter, which are generated by material ejected from the Sun's southern latitudes. Solar winds with speeds nearly double those ejected from the equatorial region have been detected flowing through large holes in the Sun's southern corona, and measurements indicate the Sun's magnetic field is not as strong as it is at the poles.

---

outward to 100,000 AU and beyond. The Sun supplies energy to all members of the system, controlling their temperatures as well as their orbital motions.

Thus far this chapter has dealt with the planets, their satellites, and ring systems. This section will consider asteroids, meteoroids, comets, and interplanetary dust. The solar wind, a stream of charged particles expelled from the Sun and flowing outward through the solar system, will be discussed in Chapter 18.

Asteroids

The Titius-Bode law calls for the next planet beyond Mars to be 2.8 AU from the Sun, but no large planetary body is found at this distance. In 1801, the first of the many planetary bodies was discovered by Giuseppi Piazzi, an Italian astronomer. This small body is named Ceres after the protecting goddess of Sicily. Ceres is slightly more than 940 km (583 mi) in diameter, has an orbital period of 4.6 y, and is the largest of more than 2000 named and numbered objects that orbit the Sun between Mars and Jupiter. These objects are called **asteroids,** or minor planets. The three largest, in order of size, are Ceres, Pallas, and Vesta. They have diameters between 500 to 1000 km. Ceres and Pallas have very low albedos. Only Vesta, which has a relatively high albedo, can be seen with the unaided eye and then only when it is simultaneously at opposition and perihelion.

The first close-up image of an asteroid was taken on October 29, 1991, by the Jupiter-bound *Galileo* spacecraft. Figure 15.37 is a portrait of the irregularly shaped asteroid (named 951 Gaspra), which is about 11 km (7 mi) wide and 19 km (12 mi) long. Gaspra is classified as a stony S-type asteroid. Its surface is covered to a depth of about 0.9 m (3 ft) with a loose rocky gray material called *regolith*.

The diameters of the known asteroids range from that of Ceres (940 km) down to a few kilometers, but most asteroids are probably less than a few kilometers in diameter. There are perhaps billions the size of boulders, marbles, and grains of sand. Only the largest asteroids are spherical in shape. Eros, which can approach Earth to within 1 AU, is roughly rod shaped. Many others are irregular in shape.

Like the planets, asteroids revolve counterclockwise around the Sun, as seen from the North Celestial Pole (Earth's North Pole extended into space), with an average inclination to the ecliptic plane of 10°. About one hundred thousand asteroids exist that can be detected with Earth-based telescopes. The total mass of all the asteroids orbiting between Mars and Jupiter is much less than the mass of Earth's Moon. Although most asteroids move in an orbit between Mars and Jupiter, some have orbits that range beyond Saturn or inside the orbit of Mercury.

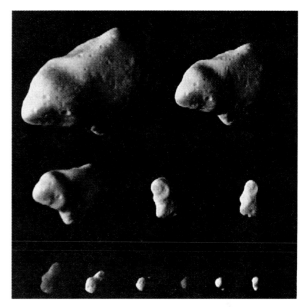

Figure 15.37   Multiple images of asteroid Gaspra.
These are approximately true color views of Gaspra taken at various distances by the Jupiter-bound *Galileo* spacecraft on October 29, 1991. These are the first close-up views of an asteroid. Gaspra is a stony S-type asteroid whose surface is covered with rocks moderately less gray than those on Earth's Moon. The asteroid is approximately 19 km (12 mi) long and 11 km (7 mi) wide and irregular in shape.

Asteroids are believed to be early solar system material that never collected into a single planet. One piece of evidence supporting this view is that there seem to be several different kinds of asteroids. Those at the inner edge of the belt seem to be stony, whereas the ones farther out are darker, indicating more carbon content. A third group may be composed mostly of iron and nickel.

In 1977 astronomers discovered a small (less than 320 km in diameter), dark object between Saturn and Uranus that they classified as an asteroid and named Chiron (KI-ron). Chiron has a very elliptical orbit with a period of 50.7 y. The orbit takes Chiron between 8.5 AU and 18.8 AU from the Sun. Chiron is presently heading closer to the Sun, and it will reach perihelion in 1996.

Since its discovery, Chiron has been varying in brightness and is increasing in overall brightness as it approaches the Sun. It is evidently giving off gas and dust particles as radiation from the Sun strikes its surface. These particles increase Chiron's reflectivity of sunlight. Because of the observed change in its appearance, Chiron is now classified as a comet rather than an asteroid. Comets are discussed later in this section.

Meteoroids

Meteoroids are interplanetary metallic and stony objects that range in size from a fraction of a millimeter to a few hundred meters. They circle the Sun in elliptical orbits and strike the Earth from all directions with very high speeds. Their high speed, which is increased by Earth's gravitational force, produces great frictional heating when the meteoroids enter Earth's atmosphere.

A meteoroid is called a **meteor,** or "shooting star," when it enters Earth's atmosphere and becomes a luminous trail of light because of the tremendous heat generated by friction with the air. Most meteors are vaporized in the atmosphere, but some larger ones may survive the flight through the atmosphere and strike Earth's surface. They are then known as **meteorites.**

When a large meteorite strikes Earth's surface, a large hole, called a crater, is created. ● Figure 15.38 is a photograph of a large meteorite crater near Winslow, Arizona, which scientists estimate to be more than 25,000 years old.

The meteoroids are members of the solar system that probably come from the remains of comets and fragments of shattered asteroids. The largest known meteorite has a mass of more than 55,000 kg and fell in southwest Africa. The largest known meteorite found in North America, with a mass of about 36,000 kg, was found near Cape York, Greenland, in 1895 and is on display at the Hayden Planetarium in New York City.

Meteorites vary in size and shape and are classified into three broad groups: stones, irons, and stony-irons. About 94% of all meteorites that fall on Earth are stones. Stones are composed of silicate minerals and other minerals and have the appearance of ordinary rocks found in Earth's crust. Thus, they are difficult to identify by the layperson. Irons are mostly iron with 5 to 20% nickel. Stony-irons, as the name implies, are mixtures of iron and stony materials. A stone meteorite is shown in Fig. 15.39.

Figure 15.38 The Barringer Meteorite Crater near Winslow, Arizona.
The crater is 1300 m (4264 ft) across, 180 m (590 ft) deep, and its rim is 45 m (148 ft) above the surrounding land.

Comets

Comets are named from the Latin word *komets,* which means "long-haired." They are also the solar system members that periodically appear in our sky for a few weeks or months, then disappear. A **comet** is a reasonably small object (with typical diameter 5 to 20 km) composed of dust and ice that revolves about the Sun in a highly elliptical orbit. As it comes near the Sun some of the surface vaporizes to form a gaseous head, or *coma,* and usually a long *tail.*

A comet consists of four parts: (1) the *nucleus,* typically a few kilometers in diameter and composed of rocky or metallic material and solid ices of water, ammonia, methane, and carbon dioxide; (2) the head, or *coma,* which surrounds the nucleus and can be as much as several hundred kilometers in diameter, and is formed from the nucleus as it approaches within about five astronomical units of the Sun; (3) long, voluminous, and magnificent *tails,* also formed from the coma by solar winds and radiation, composed of either ionized molecules or dust or a combination of both, and

Figure 15.39 Antarctica meteorite.
This 15-pound meteorite was formed from two different basalt rocks. The meteorite, which is approximately 1,300 million years old, was collected from the Elephant Moraine region of Antarctica near the United States McMurdo base.

which can be millions of kilometers in length; and (4) a spherical cloud of hydrogen, believed to be formed from the dissociation of water molecules in the nucleus, surrounding the coma. The sphere of hydrogen in some comets may have a radius exceeding that of the Sun (Fig. 15.40).

Comets are seen by reflected light from the Sun and fluorescence of some of the molecules comprising the comet. As comets approach and move around the Sun, the amount of material in the coma and tail increases (Fig. 15.41). This increase in size is evidently caused by the Sun, perhaps by (1) the solar wind and (2) radiation pressure generated by the radiant energy given off by the Sun.

Scientists believe that only a thin outer shell of the comet's nucleus is heated. As it moves toward its closest approach (perihelion) to the Sun, the increasing solar radiation causes the surface materials to melt and evaporate. The evaporating surface particles form the coma and the long tail, which is driven away from the Sun by the solar wind and radiation pressure. Each time the comet passes near the Sun on its long journey through space, it loses part of its mass. Eventually, the comet loses most of its mass and disappears as a comet, but the rocky or metallic material continues to move around the Sun.

Halley's comet, named after Edmond Halley (1656–1742), a British astronomer, is the brightest and best-known comet. Halley (HAL-ee) was the first to suggest and predict the periodic appearance of comets. He observed the comet that bears his name in 1682 and, using Newton's laws of motion, correctly predicted it would return in 76 years. Halley's comet has appeared every 76 years, including 1910 and 1986.

On July 16, 1994, comet Shoemaker-Levy (SL9) crashed into Jupiter's atmosphere. The comet was broken into 21 major fragments and thousands of minor pieces by Jupiter's strong gravitational field during a close encounter with Jupiter in 1992. Eight impact sites are visible in Fig. 15.42. After extensive study of all collected data, astronomers will be able to learn about the composition of the comet and Jupiter's atmosphere.

Scientists believe that comets originate and evolve from dirty, icy objects that were part of the primordial debris thrown outward into interstellar space when the solar system was formed. These dirty, icy objects are believed to be more dirt than ice and are presently called frozen mudballs. We observe these objects when they enter the vicinity

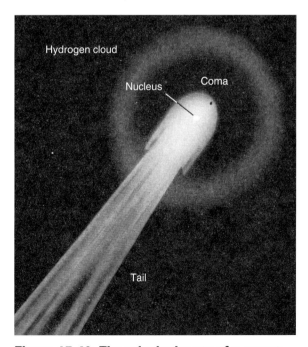

Figure 15.40 The principal parts of a comet.

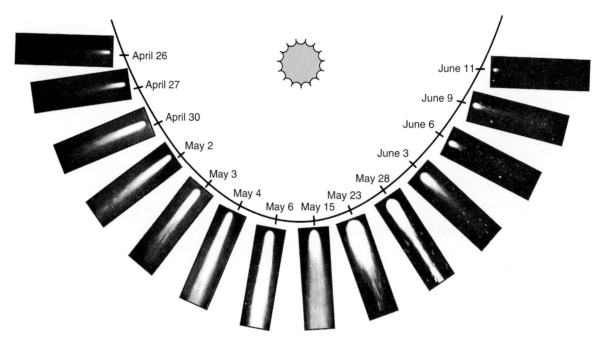

Figure 15.41 Halley's comet.
These fourteen views of Halley's comet were taken between April 26 and June 11, 1910. Note the change in size of the coma and tail.

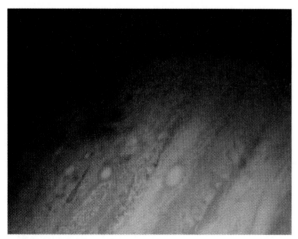

Figure 15.42 Comet Shoemaker-Levy fragments impact Jupiter's atmosphere.
This is Hubble's image of several of the impact sites of comet Shoemaker-Levy (SL9) which crashed into Jupiter's atmosphere on July 16, 1994. The comet was broken into 21 major fragments and this is a photo showing some of the impacts. One of the impact sites was as large as the planet Earth. This picture is a composite of separate images taken through different color filters to create "true color" representations of Jupiter's clouds.

of the Sun and develop a long plasma tail, which is composed of the gas that is expelled from the comet's head due to heat received from the Sun. The **Oort cloud** is defined as the source region for the observed comets and is named in honor of Jan Hendrik Oort (1900–1989), the Dutch astronomer who proposed its existence in 1950. The highly eccentric orbits of typical comets indicate their greatest distance from the Sun to be about 50,000 AU.

The Oort cloud is near the outer limits of the Sun's gravitational influence, and the objects within this region can be perturbed by passing stars. The passing stars' gravitational field starts the cometary material inward on its fall toward the Sun or outward beyond the influence of the Sun's gravitational field. Very few of these cold hard objects become captured comets that come close enough to the Sun to receive sufficient heat to vaporize the icy material. Most pass undetected beyond the orbit of Pluto. Fewer than a thousand comets have been seen and recorded by astronomers, and only one-tenth of these make repeated trips near the Sun.

What objects, if any, are located in the vast volume of space between Pluto's orbit (39 AU) and 50,000 AU? This vast volume of space is called the *Kuiper belt* in honor of Gerard P. Kuiper (1905–1972), a Dutch-born American astronomer who suggested that the space is a region containing solar debris, which serves as a reservoir of available cometary material. The debris cannot be detected presently by astronomers because of its great distance and small size. The objects in this volume of space are held more firmly by the Sun's gravity and not influenced by nearby stars.

Where is the outer boundary of the solar system? As indicated above, the Oort comet cloud is located in the vicinity of 50,000 AU. Does the Sun's gravitational force field have any influence beyond this limit? The answer is yes, according to astrophysicists who have made some computer simulations that indicate the outer limit for the influence of the Sun's gravitational field is between 80,000 and 100,000 AU. This limit is where the Sun's gravitational force field is balanced by that of our Milky Way galaxy. Beyond this boundary the Sun's gravitational force field cannot hold any object.

Interplanetary Dust

In addition to the planets and other large bodies discussed thus far, the large volume of space of the solar system is occupied by very small solid particles known as *micrometeoroids* or **interplanetary dust.** There are two celestial phenomena, which can be observed with the unaided eye or photographed, that show that the dust particles do exist.

Perhaps you have observed on a very clear dark night, just after sunset in the western sky, *zodiacal light*—a faint band of light along the zodiac (ecliptic). The band of light can also be seen just before sunrise. The faint glow is due to reflected sunlight from dust particles. The zodiacal light is shown in Fig. 15.43.

The other phenomenon is called the *gegenshein,* which means "counterglow." This faint glow is observed on the ecliptic exactly opposite the Sun. It is more difficult to observe than zodiacal light and appears as a diffuse oval spot with an average angular size of 6 by 9 degrees. The gegenshein is also due to reflected sunlight from dust particles.

Figure 15.43 The zodiacal light.
This photo shows the scattering of sunlight by dust particles in orbit around the Sun. The reflected sunlight can be seen best about an hour before sunrise or an hour after sunset. This photo is overexposed. The zodiacal light does not appear this bright.

CONFIDENCE QUESTION 15.6

(a) What are the observed physical properties of a comet located at 50,000 AU, and (b) near the Sun? (c) Explain the difference between a meteoroid, meteor, and meteorite.

15.6 The Origin of the Solar System

Learning Goal:

To explain the most valid and acceptable theory for the origin of the solar system.

Any theory proposed to explain the origin and development of the solar system must account for the system as it presently exists. The preceding sections have given a general description of the system in its present state, which, according to our best measurements, has lasted for about 5 billion years.

If an acceptable theory for the origin is to have validity, the following questions concerning major properties of the solar system must have valid answers.

1. What is the origin of the material used to create the system?

2. What were the forces that acted to form the system?
3. Why are the planets in isolated orbits that are almost circular and located nearly in the same plane?
4. Why is the revolution of the Sun, planets, and satellites of the planets in the same direction?
5. Why is the rotation of the Sun, all planets except three, and nearly all satellites of the planets in the same direction?
6. What determined the chemical and physical properties of the planets, and why are the terrestrial planets so different from the Jovian planets?
7. What is the origin of the asteroids, which have properties that are different from the terrestrial and Jovian planets?
8. How do the comets and meteoroids fit in with the theory?

At present, the formation of the solar system is believed by most astronomers of have begun with a large swirling volume of cold gases and dust, a rotating **solar nebula,** positioned in space among the stars of the Milky Way. Through the process of condensation and accretion the nebula evolved into the system we observe today. The process is known, by some astronomers, as the **condensation theory.** This premise is supported by the fact that today we observe many such nebulae throughout the universe. What initiated the process to form the nebula from interstellar matter is unknown. A good candidate is a shock wave from a supernova passing through the interstellar matter.

The interstellar dust played a major role in the condensation process by allowing condensation to take place before the gas had a chance to disperse. The collection of particles was slow at first but became increasingly faster as the central mass became larger. As the particles moved inward, the rotation of the mass had to increase to conserve angular momentum. Because of the rapid turning, the cloud began to flatten and spread out in the equatorial plane. See Fig. 15.44. Kepler's third law states that the central part must move faster than the outer parts. This motion set up shearing forces, which, coupled with variations in density, produced the formation of other masses that moved around the large central portion, the protosun, sweeping up more material and forming the protoplanets.

The protoearth was perhaps 1000 times more massive than planet Earth. Temperature played a major role in the condensation process to form protoplanets. The terrestrial protoplanets, receiving more heat than the Jovian protoplanets, were greatly affected by this heating and formed out of rocky or metallic material. The Jovian protoplanets, being at great distances and receiving little radiation, were formed in a cold environment out of low-density, icy material. They appear today in a similar protoplanet stage. Planetary satellites are believed to have formed from similar accretion of matter surrounding their protoplanets. See the Highlight in Section 17.2 for a discussion of the origin of Earth's Moon. Theory suggests that formation of the solar system took place over a 100-million-year period beginning about five billion years ago.

During the early stages of development, the space between the protosun and the protoplanets was filled with large amounts of gas and dust, shielding the protoplanets from the protosun, which was beginning to fuse hydrogen into helium and radiate energy. With the passing of time the space between the protosun and protoplanets became transparent due to the accumulation of material by the planets and the ejection of dust and gas by radiation from the Sun and by the solar wind.

The asteroids, which have irregular sizes and shapes, are believed to be the remnants of a planet that never formed. Meteoroids are thought to be small asteroids, and by studying their composition we have important clues to the origin of our solar system.

As indicated in Section 15.5, comets are believed to be planetesimals[*] that formed mostly among the Jovian planets and were gravitationally ejected into space beyond the orbit of Pluto. Other cometary material is also believed to have originated and to still exist in the vast volume of space between Pluto's orbit and the Oort comet cloud.

The above discussion for the origin of the solar system is a theoretical explanation that is constructed from observational evidence. Much of the knowledge concerning the formation process

---

[*]Planetesimals are objects that range from grain size up to hundreds of kilometers in diameter that were created during the planet-forming process in the solar nebula.

Figure 15.44 The formation of the solar system.
This drawing illustrates the formation of the solar system according to the condensation theory. The protostellar cloud of gas and dust (1), in gravitational collapse, develops into a flattened, rotating disk (2)—the primordial nebula. Additional contraction produced a disk (solar nebula), whose mass is condensed and accreted to give birth to the planets in the low-temperature regions while the Sun formed in the central part.

comes from complete chemical and physical examination of meteorites, which are the least changed fragments of the initial solar system. These fragments show detailed chemical and geologic properties of the solar system.

There are unanswered questions concerning solar system formation, and research is ongoing to find the answers. As mentioned above, astronomers do not know what initiates the formation of a star from interstellar matter. Does gravitational attraction act alone to bring about condensation and accretion, or are there other factors involved such as a shock wave moving through the nebular cloud? What is the role of electric and magnetic fields? These and other questions remain to be answered.

CONFIDENCE QUESTION 15.7

(a) What physical property played a major role in the formation of the protoplanets? (b) How did this physical property affect the composition of planet Earth?

15.7 Other Planetary Systems

Learning Goal:

To describe the latest information concerning other planetary systems.

Are there other planetary systems in the universe? We have little reason to believe that our solar system is one of a kind. Statistical analysis indicates that planetary systems like ours are probable. The search for planets of other stars is being conducted by many scientists, but as of this date they have yet to produce conclusive evidence for the existence of a single one.

As mentioned in the previous section, our solar system is believed to have originated from a rotating solar nebula that was disk-shaped during the early stage of formation. Figure 15.45 shows a star called Beta Pictoris that is about 50 light-years (ly) from Earth. This star, with a supposedly dust-laden disk, is the best clue we have that other planetary systems may exist. Much better evidence must be obtained before a definite answer is known concerning the existence of other planetary systems.

The latest reported candidate for an extrasolar planetary system comes from two U.S. astronomers. They reported in 1992 the discovery of two objects revolving about a type of star called a pulsar. One orbits the pulsar every 66.6 days and the other every 98.2 days. The method of detection was an observable Doppler shift in the spectrum of the pulsar. See the discussion below for the cause of the shift. The pulsar is named PSR 1257 +12. The numbers refer to its celestial location (DEC +12, RA 12 h 57 min). See Section 18.2.

Pulsars are identified as rapidly rotating stars with precise periods. PSR 1257 +12 rotates 161 times a second. The rotation rate is observable because the pulsar emits pulses of radio radiation as beams that sweep past Earth like the beam of a rotating searchlight. The beams show a regular variation in pulse arrival time. The variations are deciphered as indicating gravitational disruptions by two rotating objects about the pulsar. More research is being conducted to detect any gravitational effect between the two revolving objects. This will provide confirming evidence of the two presumed planets.

Detecting extrasolar planets is no easy task. One method used to detect a star with a companion planet is to observe a star's motion. A star with a large planet has a small wobble superimposed on its motion that is due to gravitational effects. The change in motion is very small and difficult to detect, especially if the observations are made by Earth-bound telescopes. NASA plans to launch a new infrared space telescope facility in the late 1990s that will greatly improve the observations.

Another technique used to search for a star-planet system is the observation of Doppler shifts in the spectrum of a star. Because of gravitational pull, a planet revolving around a star causes the star to move toward and away from the astronomer's telescope as the planets and star move around their center of mass. These periodic changes are detected as periodic shifts in the spectrum of the star.

A different approach to the discovery of other planetary systems is the search for alien signals from extraterrestrial intelligence. Scientists have been scanning the skies for years hoping to detect a radio signal from another solar system. The early searches were in the electromagnetic spectrum close to 21 cm, the wavelength at which interstellar

Figure 15.45 Beta Pictoris disk.
This optical photograph shows material around Beta Pictoris, a star about 50 ly from the solar system. Light from the star has been blocked out by a coronagraph, an instrument placed in front of the optical telescope. More detailed observation of the material may reveal a planetary system.

hydrogen emits radiation, and near 18 cm, the wavelength radiated by the hydroxyl (OH) radical. Together, these form the water molecule (HOH). When the background noise level of the Milky Way is plotted as a function of wavelength, there is a minimum at these two wavelengths. That is, this is the quietest part of the electromagnetic spectrum and is referred to as the *water hole*.

The search for extraterrestrial intelligence, called SETI, will become more sophisticated in the late 1990s, when NASA begins using equipment that can scan eight megachannels at the same time, over frequencies ranging from 1.2 to 10 GHz. The NASA project will scan the entire sky, aiming at 773 sunlike stars within 100 ly of the Sun. Perhaps the broad comprehensive search will detect evidence of extraterrestrials from other planetary systems.

The private scientific community is also involved in searching for radio signals from beyond our atmosphere. Project META began in 1985 but as of this date (1995) nothing has been found. META stands for Megachannel Extra Terrestrial Assay. There are actually over nine million channels. The META project is funded by The Planetary Society headed by president Carl Sagan. There is also a META II project that scans the southern skies with an 8.4-multichannel system analyzer located near Buenos Aires.

The next advance in the program is the operation of the BETA (Billion-channel Extra Terrestrial Assay). Actually there are only 240 million channels. The project is a joint venture of The Planetary Society, NASA, and the Bosack/Kruger Foundation. The electronic equipment, which is expected to begin operation in late 1995, will cover the entire water hole. This includes all wavelengths between hydrogen and the hydroxyl radical.

Our receiving technology is becoming very sophisticated. Presently antennae, detectors, and computers are merging into one system for collecting data over a broad spectrum of wavelengths, time, and space.

CONFIDENCE QUESTION 15.8

(a) What is the water hole? (b) What is the frequency range (in gigahertz) of the water hole?

Important Terms

These important terms are for review. After reading and studying the chapter, you should be able to define and/or explain each of them.

astronomy	heliocentric theory	Roche limit
universe	Kepler's law of elliptical paths	terminator
albedo	Kepler's law of equal areas	asteroids
solar system	Kepler's harmonic law	meteoroids
ecliptic	terrestrial planets	meteor
rotation	Jovian planets	meteorites
revolution	prograde motion	comet
Foucault pendulum	retrograde motion	Oort cloud
parallax	Titius-Bode law	interplanetary dust
astronomical unit	sidereal period	solar nebula
parsec	synodic period	condensation theory
geocentric theory		

Important Equations

Kepler's third law: $$\frac{T^2}{R^3} = k$$

Newton's version of Kepler's third law:

$$\frac{T^2}{R^3} = \frac{4\pi^2}{Gm_{Sun}}$$

Questions

15.1 The Planet Earth

1. Which one of the following is *not* a major component of the solar system?
   (a) comet
   (b) satellite
   (c) meteoroid
   (d) star
   (e) meteorite

2. Which one of the following does *not* describe one of Earth's motions?
   (a) rotation
   (b) precession
   (c) parallax
   (d) revolution

3. State and explain two major motions of planet Earth.

4. Before 1900, what proof was there that Earth (a) rotated and (b) revolved?

5. What is the ecliptic?

6. What does the parsec measure? State the definition of one *parsec*.

7. State why the parallax of a star cannot be seen with the unaided eye.

8. The observation of the aberration of starlight by astronomers provided proof of one of Earth's motions. Name the motion.

9. Define one *astronomical unit*.

15.2 The Solar System

10. The solar system
    (a) is a heliocentric system.
    (b) is held together by gravitational forces.
    (c) contains planets classified as terrestrial and Jovian.
    (d) was formed about 5 billion years ago.
    (e) all of the above
11. Which of the following is *not* a significant difference between the terrestrial and Jovian planets?
    (a) density
    (b) number of moons
    (c) size
    (d) rotation period
    (e) none of the above
12. An imaginary line joining a planet to the Sun sweeps out equal areas in equal periods of time. This is a statement of Kepler's _____ law.
    (a) first
    (b) second
    (c) third
    (d) none of the above
13. Name the major components that comprise the solar system.
14. What is the significant difference between the geocentric and the heliocentric models of the solar system?
15. What is the explanation of Kepler's second law? (*Hint:* Refer to a conservation law.)
16. Give five major differences between the terrestrial and the Jovian planets.
17. State the difference between prograde and retrograde motion.
18. Why is the Titius-Bode law not a "law of nature" but only a method or rule?
19. Distinguish between sidereal and synodic period.
20. Which planet is presently the farthest planet from the Sun? Explain.

15.3 The Terrestrial Planets

21. The terrestrial planets
    (a) are all relatively small and have relatively small mass.
    (b) are dense and rocky.
    (c) have physical and chemical properties similar to Earth.
    (d) all of the above
22. Which of the following statements concerning terrestrial planets is false?
    (a) All have permanent or interim atmospheres.
    (b) All have magnetic fields.
    (c) All rotate counterclockwise as viewed from above the North Pole.
    (d) They are relatively close to the Sun with orbits close together as compared to the outer planets.
23. Name the terrestrial planets. Why are they called the terrestrial planets?
24. Give the physical characteristics of Mercury's tenuous atmosphere.
25. Why does Venus have a high surface temperature?
26. What is the evidence that a fluid once flowed on Mars?
27. How do the atmospheric pressures on Venus, Earth, and Mars differ?

15.4 The Jovian Planets and Pluto

28. Which of the following statements concerning Jovian planets is false?
    (a) All are composed mainly of hydrogen and helium.
    (b) All have strong magnetic fields.
    (c) All rotate faster that the terrestrial planets.
    (d) All have a relatively high temperature environment.
    (e) All have rings.
29. The largest satellite in the solar system orbits the planet _____.
    (a) Jupiter
    (b) Saturn
    (c) Uranus
    (d) Neptune
30. Name the Jovian planets in order of increasing size.
31. Describe the internal structure of the Jovian planets.
32. Give another name for *tidal stability limit*.
33. What are tidal forces?
34. What is unusual about the satellite Io?
35. What is unique about the satellite Titan?
36. Give the characteristics of Pluto that classify it as (a) a planet and (b) an asteroid.
37. Titan is the third largest moon in the solar system. Jovian moons Ganymede and Callisto are larger. Why does Titan have a thick atmosphere when the two larger moons do not?
38. How do the interior structures of Uranus and Neptune differ from those of Jupiter and Saturn?
39. How do the magnetic fields of Uranus and Neptune differ from those of the other planets?

15.5 Other Solar System Objects

40. Asteroids
    (a) are believed to be initial solar system material that never collected into a single planet.
    (b) are located mainly in orbits around the Sun between Mars and Jupiter.
    (c) range in size from hundreds of kilometers down to the size of sand grains.
    (d) are odd and irregular in shape.
    (e) all of the above
41. Comets
    (a) are composed of dust and ice.
    (b) revolve around the Sun in highly elliptical orbits.
    (c) are observable only when relatively close to the Sun.
    (d) usually have a long tail when they are close to the Sun.
    (e) all of the above
42. How do asteroids differ from meteoroids?
43. Name and describe the largest asteroid.
44. How many asteroids have been named and numbered? Where are they generally located?
45. Describe the physical characteristics of a comet at (a) aphelion and (b) perihelion.
46. What is the origin of most observed comets?
47. What is the Oort comet cloud? Where is it located?
48. What is the difference between meteoroids, meteors, and meteorites?
49. What is the evidence that interplanetary dust exists?

15.6 The Origin of the Solar System

50. Which of the following is *not* a basic factor that must be taken into account in the formation of an acceptable theory for the origin of the solar system?
    (a) the forces needed to form the system
    (b) the origin of the material used to create the system
    (c) the present size and structure of the system
    (d) the presence of interstellar dust
    (e) the origin of elements
51. The outer boundary of the solar system is thought to be as much as _____ astronomical units from the Sun.
    (a) 10,000
    (b) 40,000
    (c) 75,000
    (d) 100,000
52. Describe the solar nebula.
53. What role do dust particles play in the condensation theory for the formation of the solar system?
54. Where do scientists obtain the best evidence concerning the condensation and accretion processes in the formation of the solar system?
55. What causes interstellar matter to form a rotating nebula?
56. How many years were required for the formation of the solar system?

15.7 Other Planetary Systems

57. Which of the following is *not* a method for detecting an extrasolar planetary system?
    (a) the observation of a star's motion
    (b) the observation of Doppler shifts in the spectrum of a star
    (c) the observation of a very large orbiting planet
    (d) the detection of alien electromagnetic signals
58. A frequency of one gigahertz (1 GHz) is equal to _____ hertz.
    (a) 10^6
    (b) 10^9
    (c) 10^{10}
    (d) 10^{12}
59. What evidence do astronomers have concerning the actual existence of another planetary system?
60. What method was used by astronomers to detect that pulsar 1257 +12 has planets revolving around it?
61. What is the BETA project?
62. What frequency range (in gigahertz) is covered by the BETA project?

Food for Thought

1. Name two major factors that function to maintain a planet's atmosphere. Give reasons for your answer.
2. How is density used to determine the physical makeup of a planet?
3. Name three sources from which scientists can obtain primitive material in order to gain knowledge about the origin of the solar system.
4. Comment on why the terrestrial planets are vastly different from the Jovian planets.

Exercises

1. How does the period of revolution of a planet vary with its distance from the Sun? (In equation $T^2 = kR^3$, if $T = 1$ y and $R = 1$ AU, then $k = \frac{1\,y^2}{(AU)^3}$.)

    Answer: T varies as the square root of R^3

2. Determine the period of revolution of Earth, if its distance from the Sun were 2.0 AU. Assume the mass of the Sun remains the same.

    Answer: 2.8 y

3. Determine the period of revolution of Earth, if its distance from the Sun were 4.0 AU. Assume mass of the Sun remains the same.
4. Determine the period of revolution of Earth, if the Sun's mass is 0.5 times its present value. Assume the distance to the Sun remains 1 AU.

    Answer: 1.4 y

5. Determine the period of revolution of Earth, if the Sun's mass is 4 times its present value. Assume the distance to the Sun remains 1 AU.
6. Use Kepler's third law to show that the closer a planet is to the Sun, the shorter is its period.
7. Use Kepler's third law to show that the closer a planet is to the Sun, the faster is its speed around the Sun.
8. Determine the distances from the Sun to the terrestrial planets and the asteroids, using the Titius-Bode law. Compare your answers with the actual distances.
9. Determine the distances from the Sun to the Jovian planets, using the Titius-Bode law. Compare your answers with the actual distances.
10. Using the Titius-Bode law, determine the distance from the Sun to Pluto, and compare it with the actual distance.
11. Refer to Table 15.1 and draw a diagram showing how far each inner planet goes around the Sun in 60 days.
12. Refer to Table 15.1 and draw a diagram showing how far each outer planet goes around the Sun in 10 years.
13. List the following distances in order of increasing length: (a) Sun to Earth, (b) Mars to Jupiter, (c) Jupiter to Saturn, (d) Saturn to Uranus.
14. List the planets in order of increasing distance from the Sun.
15. List the planets in order of decreasing size (largest first).
16. List the planets in order of decreasing density (densest first).
17. List the years of the next three appearances of Halley's comet.

Answers to Multiple-Choice Questions

1. e   2. c   10. e   11. e   12. b   21. d   22. c
28. d   29. a   40. e   41. e   50. e   51. d   57. c
58. b

Solutions to Confidence Questions

15.2 (a) Planet Earth has an albedo of 0.33, a diameter of 12,714 km from pole to pole, and the shape of an oblate spheroid.

(b) Rotation is the spinning of a body about an internal axis. Earth rotates in a counterclockwise direction as viewed from above Earth's North Pole. Revolution is the movement of one body around another. Earth revolves around the Sun in a counterclockwise direction.

15.2 (a) $T^2 = kR^3$

$T^2 = \dfrac{1 \text{ y}^2}{(\text{AU})^3} (0.723 \text{ AU})^3$

$T = 0.615$ y

(b) They are approximately the same.

15.3 (a) Earth and Venus (closest distance = 41.4 km)

(b) Kepler's third law can be written as $T^2 = kR^3$, where T is the period of revolution, and R the average distance from Earth to the Sun. If $T = 1$ y and $R = 1$ AU, then k will equal $\dfrac{1 \text{ y}^2}{(\text{AU})^3}$.

15.4 (a) The terrestrial planets are all relatively small in size. They are composed of rocky material. All have relatively high densities. They are relatively close to the Sun and have orbits that are relatively close together.

(b) Planet Earth has an abundance of water and 21% oxygen in its atmosphere. The other terrestrial planets have no water or oxygen.

15.5 (a) Sizes, densities, rotation periods, atmospheres, and the number of satellites.

(b) One-half the period of revolution, or 42 years.

15.6 (a) The cometary material at 50,000 AU is hypothesized to be icy dusty objects that are asymmetric in shape. The objects vary in size from 5 to 20 km, but some may be larger.

(b) Near the Sun the surface of the cometary objects vaporizes and forms a coma and a long ion tail directed away from the Sun.

(c) A meteoroid is a particle in space before it encounters Earth. A meteor is a meteoroid that enters Earth's atmosphere and appears luminous as a shooting star. A meteorite is the remains of a meteoroid that strikes Earth's surface.

15.7 (a) Temperature; (b) The terrestrial planets being relatively close to the Sun received large amounts of heat energy from the Sun and formed out of rocky and metallic material.

15.8 (a) Hydrogen atoms radiate naturally at 21 cm wavelength, and hydroxyl radicals radiate near 18 cm. Hydrogen (H) and hydroxyl (OH) form water (HOH). The water hole is the band of electromagnetic radiation between these two wavelengths. This is a region of the electromagnetic spectrum in which we believe aliens may be broadcasting signals.

(b) The 18 cm wavelength has a frequency of 1.67 GHz. The 21 cm wavelength has a frequency of 1.43 GHz.

Chapter 16

Place and Time

16.1 Cartesian Coordinates
16.2 Latitude and Longitude
16.3 Time

Highlight: The Concept of Time

16.4 The Seasons
16.5 Precession of Earth's Axis
16.6 The Calendar

In physical science we observe and examine events that take place in our environment. These events occur at different places and at different times. Some occur nearby and are observed immediately, whereas others occur at great distances and are not observed until a later time. For example, you see a flash of lightning, but you do not hear the sound of thunder until later. A star explodes at some great distance from Earth, and years later the event is observed as the radiation reaches Earth. Thus the events or happenings taking place in our environment are separated in space and time.

Albert Einstein was the first to point out that space and time are related and exist as a single entity. Einstein associated gravitational fields with space and developed the concept of four-dimensional space, which has given us an entirely different concept concerning the reality of our environment.

In this chapter we introduce the concepts used in two- and three-dimensional reference systems. These concepts are then expanded to explain the place (location) of objects on Earth's surface, and the place of planets, stars, galaxies, and other celestial objects beyond Earth's atmosphere.

Our five senses make it possible for us to know and understand objects and their place relative to one another in the physical world. The concept of time, on the other hand, is rather elusive. We relate to the daily cycle of the Sun and the yearly cycle of Earth revolving around the Sun, and we use these periodic changes to measure what we call time. Thus, we think of time in reference to changes we observe in our environment. This chapter introduces and explains the time concepts that are used in our daily living.

16.1 Cartesian Coordinates

Learning Goal:
To explain the Cartesian coordinate system.

The location of an object in our environment requires a reference system that has one or more dimensions. A one-dimensional system is depicted by the number line shown in Figure 16.1, which illustrates two fundamental features of every coordinate system. A straight line is drawn, which may extend to plus infinity in one direction and minus

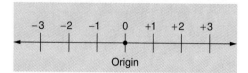

Figure 16.1 A one-dimensional reference system.

infinity in the opposite direction. For the line to represent a coordinate system, an origin must be indicated, and unit length along the line must be expressed. Temperature scales, left-right, above ground-below ground, time past-time future, and profit-loss are all examples of one-dimensional coordinate systems.

A two-dimensional system is shown in Fig. 16.2, in which two number lines are drawn perpendicular to each other and the origin assigned at the point of intersection. The two-dimensional system is called a **Cartesian coordinate system,** in honor of the French philosopher and mathematician René Descartes (1596–1650), the inventor of coordinate geometry. It is also referred to as a *rectangular coordinate system*. The horizontal line is normally designated the *x*-axis and the vertical line, the *y*-axis. Every position or point in the plane is assigned a pair of coordinates (*x* and *y*), which gives the distance from the two lines, or axes. The *x* number gives the distance from the *y*-axis, and the *y* number gives the distance from the *x*-axis. Many of the cities in the United States are laid out in the Cartesian coordinate system. Usually, one street runs east and west, corresponding to the *x*-axis, while another street runs north and south, which corresponds to the *y*-axis.

Our interest in the Cartesian coordinate system results from our desire to determine the location of any position on the surface of the spherical Earth and the location of any object on the celestial sphere.

A *sphere* is a three-dimensional surface where all points on the surface are equidistant from a fixed point called the *center*. A *great circle* is a circle described on a spherical surface by a plane passing through the center of the sphere. Two particular great circles are used as reference lines in the designation of position on Earth's surface: one the great circle around Earth, equally distant from the North and South Poles, called the *equator*, and the great circle, called a *meridian*, that passes through both the North and South Poles and also Greenwich, England.

CONFIDENCE QUESTION 16.1

The Cartesian coordinate system is known by another name. (a) What is the name? (b) Give the reason for the name.

16.2 Latitude and Longitude

Learning Goals:

To define and explain latitude and longitude.

To solve latitude and longitude exercises relative to Earth's surface.

The location of an object on the surface of Earth is accomplished by means of a coordinate system known as latitude and longitude. Because Earth is turning about an axis, we can use as reference points the *geographic poles*, which are defined as those points on the surface of Earth where the axis projects from the sphere. The *equator*, defined in respect to the poles, is an imaginary line circling

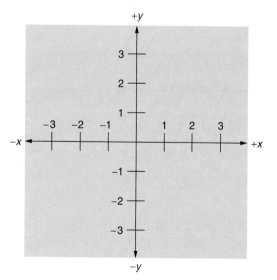

Figure 16.2 A two-dimensional reference system.

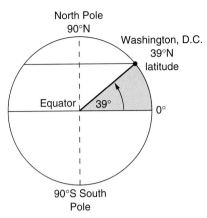

Figure 16.3 Diagram showing the latitude of Washington, D.C.

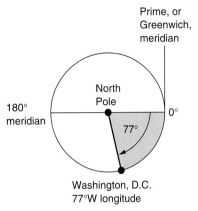

Figure 16.4 Diagram showing the longitude of Washington, D.C.

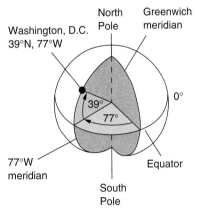

Figure 16.5 Diagram showing the latitude and longitude of Washington, D.C.

Earth at the surface, halfway between the north and south geographic poles. The equator is a great circle; that is, a circle on the surface of the Earth located in a plane that passes through the center of the Earth. Any such plane would divide Earth into two equal halves.

The **latitude** of a surface position is defined as the angular measurement in degrees north and south of the equator. The angle is measured from the center of Earth relative to the equator (Fig. 16.3). Lines of equal latitude are circles drawn around the surface of the sphere parallel to the equator. Any number of such circles can be drawn. The circles become smaller as the distance from the equator becomes greater. These circles are called **parallels,** and when we travel due east or west, we follow a parallel. Latitude has a minimum value of 0° at the equator and a maximum value of 90° north or 90° south at the poles.

Imaginary lines drawn along the surface of Earth running from the geographic North Pole, perpendicular to the equator, to the geographic South Pole are known as **meridians.** Meridians are half-circles, which are portions of a great circle, because the circle is located in the same plane as the center of Earth. An infinite number of lines can be drawn as meridians. **Longitude** is the angular measurement, in degrees, east or west of the reference meridian, which is called the prime, or Greenwich, meridian. Longitude has a minimum value of 0° at the prime meridian and a maximum value of 180° east and west (Fig. 16.4).

The latitude and longitude of one point (Washington, D.C., 39°N, 77°W) are shown in Figs. 16.3 and 16.4. The latitude and longitude shown in Figs. 16.3 and 16.4 are combined in Fig. 16.5 and shown in a cutaway view of Earth. The **Greenwich,** or **prime, meridian** was chosen as the zero meridian because a large optical telescope was located at Greenwich, England, and because England ruled the seas at the time the coordinate system of latitude and longitude was originated. The main purpose of the system at the time was to determine the location of ships at sea.

CONFIDENCE QUESTION 16.2

Is there a place on Earth's surface where 1° of latitude and 1° of longitude have approximately the same value?

EXAMPLE 16.1 Determining Your Location After Traveling Meridians and Parallels

Suppose you start at latitude 20°N, longitude 75°W and travel 600 nautical miles* due north, then 600 nautical miles due east, then 600 nautical miles south, then 600 nautical miles due west. Where will you arrive in respect to your starting point?

Solution

The key to solving this exercise is knowing that meridians are one-half of a great circle, and they come together at the poles (see Fig. 16.11). Also, parallels are circles. The equator is the largest, and the circles become smaller as the latitude increases.

Draw a diagram showing the meridians and the parallels then record the data given in the exercise on the diagram.

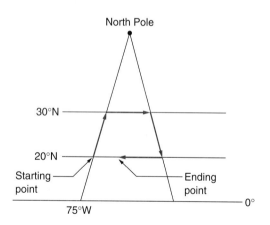

The completed diagram shows you the answer. When traveling north or south, you travel a great-circle route. The distance (600 nautical miles) traveled is the same when going north and south, and the angle (10°) passed through is the same. When traveling the parallels, your distance (600 nautical miles) is the same whether traveling east or west. But, the angle passed through traveling east is greater than traveling west because a smaller circle is traveled.

CONFIDENCE QUESTION 16.3

Where will you be with respect to your starting point, if you start at 90°W, 20°S and travel 400 nautical miles due north, then 400 nautical miles due east, then 400 nautical miles due south, then 400 nautical miles due west?

16.3 Time

Learning Goal:

To explain and solve exercises pertaining to local solar and standard time.

The continuous measurement of time requires the periodic movement of some object as a reference. On October 8, 1964, the 12th General Conference on Weights and Measures, meeting in Paris, adopted an atomic definition of the second as the international unit of time. The definition is based on a change in energy levels of the cesium-133 atom. The exact wording of the definition for the atomic second is as follows: "The standard to be employed is the transition between the two hyperfine levels $F = 3$, $M_F = 0$ of the fundamental state $2s_{1/2}$ of the atom of cesium-133 undisturbed by external fields, and the value 9,192,631,770 Hz is assigned." When an electron transition from one energy state to another occurs, the atom emits or absorbs radiation, whose frequency is proportional to the energy difference in the two states. The cesium-133 atom provides a highly accurate and stable reference frequency of 9,192,631,770 cycles per second, which can be referred to by electronic techniques. The National Institute of Standards and Technology, NIST-III, a cesium beam with a 3.66-m interaction region, is shown in Fig. 16.6.

For everyday purposes, we are interested in Earth as a time reference, because our daily lives are influenced by the day and its subdivisions of hours, minutes, and seconds. The day has been de-

---

*A nautical mile is a unit of distance equal to the length of one minute of arc on a great circle. Since 60 minutes of arc equals one degree, one degree equals 60 nautical miles. Also 60 nautical miles equals 69 statute or land miles, which equals 111 meters.

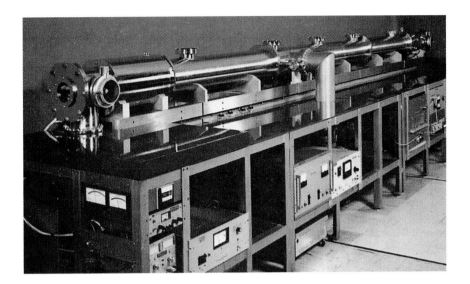

Figure 16.6 National Institute of Standards and Technology (formerly the National Bureau of Standards) cesium-beam frequency generator for establishing the time standard.

fined in two basic ways. In the first definition, the solar day has been defined as the elapsed time between two successive crossings of the same meridian by the Sun. This period is also known as the **apparent solar day,** since this is what appears to happen. Because Earth travels in an elliptical orbit, the orbital speed of Earth is not constant; therefore, the apparent days are not the same in duration. To remedy this situation the **mean solar day** is computed from all the apparent solar days during a one-year period. That is, mean solar time is the apparent solar time averaged uniformly.

In the second definition, the **sidereal day** has been defined as the elapsed time between two successive crossings of the same meridian by a star other than the Sun. Figure 16.7 illustrates the difference between solar and sidereal days.

Because Earth rotates 365.242 times during one revolution, the magnitude of the angle through which Earth revolves in one day is

$$\frac{360°}{365.242 \text{ days}} = 0.985°$$

or slightly less than 1° per day. Earth must rotate through an angle of this same magnitude for the completion of one rotation with respect to the Sun. Therefore, the solar day is longer than the sidereal day by approximately 4 min, because the Earth rotates 360°/24 h, or 15°/h, or 1°/4 min.

The earliest measurement of solar time was accomplished with a simple device known as a gnomon (from the Greek *gnomon,* meaning "a way of knowing"), which is a vertical rod erected on level ground that casts a shadow when the Sun is shining (Fig. 16.8). The vertical pointer of the sundial is called a gnomon. By the third century B.C. the water clock had been invented. In hot and dry regions of Earth, falling sand was used to count the hours. Candles were also used, because they burn at a fairly constant rate.

In the fourteenth century the first mechanical clock was invented, but it was not accurate. (See the chapter Highlight.) In the sixteenth century

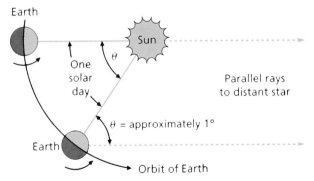

Figure 16.7 The difference between the solar day and the sidereal day.
One rotation of Earth on its axis with respect to the Sun is known as one solar day. One rotation of Earth on its axis with respect to any other star is known as one sidereal day. Note that Earth turns through an angle of 360° for one sidereal day and approximately 361° for one solar day.

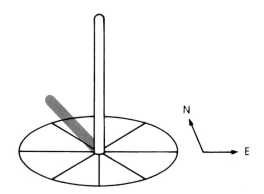

Figure 16.8 A gnomon.
This device is simply a vertical rod positioned so as to cast a shadow to indicate the time of day.

Galileo originated the plans for the construction of a pendulum clock. The story is written that while attending church, Galileo noticed a large hanging lamp swinging back and forth at regular intervals of time as measured by his pulse rate. This model provided him with the incentive to experiment with the pendulum and thus paved the way for the building of the first pendulum clock by the Dutchman Christian Huygens after Galileo's death.

Today, we have electric clocks, which are controlled by the frequency of the alternating current delivered by the power company. We also have wristwatches with quartz-crystal control, which are precise to one part in 10^9. Still greater precision is obtained by the atomic clocks mentioned earlier in this chapter.

The 24-h day, as we know it, begins at midnight and ends 24 h later at midnight. By definition, when the Sun is on the meridian, it is 12 noon local solar time at this meridian. The hours before noon are designated A.M. (*ante meridiem*, before midday) and those after noon, P.M. (*post meridiem*, after midday). The time of 12 o'clock should be stated as 12 o'clock noon or 12 o'clock midnight, with the dates. For example, we should write 12 o'clock midnight, December 3–4, to distinguish that time from, say, 12 o'clock noon, December 4.

Our modern civilization runs efficiently because of our ability to keep accurate time. Since the late nineteenth century, most of the countries of the world have adopted the system of *standard time zones*. This scheme theoretically divides Earth into 24 time zones, each containing 15° of longitude or 1 hour, since the planet rotates 15° per hour. The first zone begins at the prime meridian, which runs through Greenwich, England, and extends 7.5° each side of the prime meridian. The zones continue east and west from the Greenwich meridian, with the centers of the zones being multiples of 15°. The actual widths of the zones vary because of local conditions, but all places within a zone have the same time, which is the local time of the central meridian of that zone. For example, Washington, D.C., is located at 77°W longitude, which is within 7.5° of the 75°W meridian (● Fig. 16.9).

● Figure 16.10 shows the time and date on Earth for any Tuesday at 7 A.M. (PST), 8 A.M. (MST), 9 A.M. (CST), and 10 A.M. (EST). As Earth turns eastward, the Sun appears to move westward, taking 12 noon with it. Twelve midnight is 180° or 12 h eastward of the Sun; and as 12 noon moves westward, 12 midnight follows, bringing the new day.

When you travel westward into a different time zone, the time kept by your watch will be 1 h fast or ahead of the standard time of the westward zone; therefore, you must move the hour hand back 1 h if the watch is to have the correct time. This process will be necessary as you continue westward through additional time zones. A trip all the way around Earth in a westward direction will mean the loss of 24 h, or one complete day. When you travel eastward, the opposite is true; that is, your watch will be 1 h slow for each zone, and the hour hand is set ahead 1 h.

A better understanding of why a day is lost in traveling around Earth in a westward direction can be obtained if we take a make-believe trip. Suppose we leave Dulles International Airport in Washington, D.C., by jet plane at exactly 12 noon local mean solar time on Tuesday, December 5, and fly westward at a speed equal to the apparent westward speed of the Sun. Because Washington is located at 39°N latitude, the plane must travel the 39°N parallel westward at about 800 mi/h. As we leave the airport and fly westward, we observe the Sun out the left window of the plane about 32° above the southern horizon. One hour after leaving the airport, we notice the Sun can still be seen out the left window at the same altitude. Six hours later, with our watches indicating 7 P.M., the Sun still has the same apparent position as observed

HIGHLIGHT

The Concept of Time

In Chapter 1 time was defined as the continuous forward flowing of events. This definition implies that time relates to motion, has a forward direction, and cannot be quantized.

Things change from one position to another, and an interval of time is noted for the change to occur. This is something we observe in our physical world, and we call this changing of position *motion*. Everything appears to be in motion. The things we say are at rest are at rest only in reference to something else. A book on a table is at rest, but the table and book are on a moving Earth. Our thoughts concerning the concept of time are meaningless unless we include the concept of motion.

Do the events that take place in our physical world have a forward direction? That is, does time flow in a forward direction? If the answer is yes, how can we differentiate between forward and backward events? The answer can be found in the concept of entropy.

One of the most important laws of nature is the second law of thermodynamics, which can be and has been stated in many ways. One way, which gives the direction of time, is with the entropy concept. Entropy is a measure of disorder (see Chapter 5), and the second law of thermodynamics states that the total entropy of the universe increases in every natural process. Thus, as events occur (the flow of time), disorder becomes greater.

We do not observe and cannot measure extremely short (less than 10^{-13}s) intervals of time. Does this mean that things we cannot measure are meaningless? Not necessarily. For example, calculations concerning the Big Bang theory for the origin of the universe specify that during the extremely brief interval 1.35×10^{-43}s

Figure 1 Fourteenth-century astronomy clock.

(known as Planck time) following the Big Bang all forces were unified. Although we cannot measure this time interval, scientists use it in theories concerning the origin and expansion of the universe.

Is time absolute? The answer is a definite no. One of the outcomes from Albert Einstein's special theory of relativity was time dilation. This concept refers to the measurement of time by observers in different reference frames. For example, clocks run slower or time is stretched out in a moving reference frame as observed by a stationary observer. Also, biological processes proceed at a slower rate on a moving reference frame than on a stationary one. See Appendix VIII for an equation showing how to calculate relativistic time dilation.

Scientists performed an experiment that measured the time difference between clocks, four moving and one stationary. Four cesium-beam atomic clocks were flown on jet flights around the world twice, once eastward and once westward. The times recorded by the clocks on the jet flights were compared with the corresponding clock at the U.S. Naval Observatory, and time differences were recorded that were in good agreement with relativity theory.

Einstein's special theory of relativity also binds three-dimensional space with time, ranking time as a basic reference coordinate and labeling it as a fourth dimension. Thus, to give the complete location of an event, we must state where it is in three spatial dimensions plus the time it is there. Space and time are relative concepts. They are dependent on the relative velocity of the reference frame in which they are measured. Space and time are related in a fundamental way in what we call the four-dimensional continuum, or space-time.

Although the concept of time is difficult to comprehend, the existence of time, as we experience it, is related to change. We, as individuals, experience the universe in our own time-frame, which is different from everybody else's.

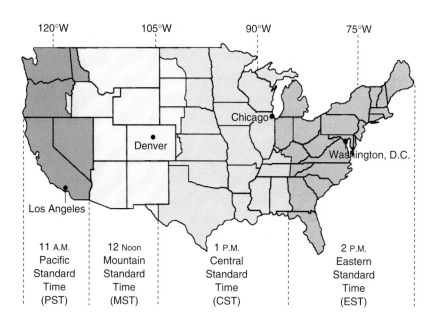

Figure 16.9 Time zones of the conterminous United States. Theoretical time zones are shown. Actual boundaries are irregular.

from the left window of the plane. Because the plane is flying at the same apparent speed as the Sun, the Sun will continue to be observed out the left window of the plane. Twenty-four hours later we arrive back in Washington with the Sun in the same apparent position. During the 24-h trip, our time remained at 12 noon local mean solar time. If we had been observing the time with a sundial, it would have remained at 12 noon. We are aware of the passing of the 24 h because we kept track of the times with our watches, but we did not see the Sun set or rise, and we did not pass through 12 midnight; therefore, the time to us is still 12 noon Tuesday, December 5. To friends meeting us at the airport, it is 12 noon Wednesday, December 6.

To remedy situations like this one, the **International Date Line** (IDL) was established at the 180° meridian. When one crosses the IDL traveling westward, the date is advanced into the next day; and when one crosses the IDL traveling eastward, one day is subtracted from the present date. A day begins and ends at the International Date Line (Fig. 16.10).

If the local solar time is known at one longitude, the local time at another longitude can be determined by remembering that there are 15° for

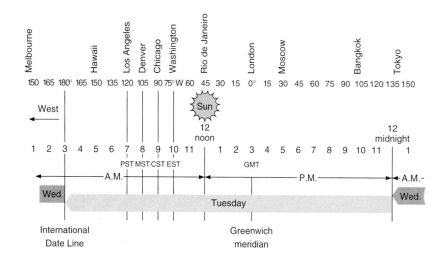

Figure 16.10 Diagram showing the times and dates on Earth for any Tuesday at 10 A.M. EST. As time passes, the Sun appears to move westward; thus, 12 noon moves westward and midnight follows (180° or 12 hours) behind, bringing the new day, Wednesday, with it.

each hour of time, or 4 minutes for each degree. Should the calculation extend through midnight or should the International Date Line be crossed, the date would change.

The following examples are explained using the views presented by the common highway map that gives a bird's-eye view of Earth's surface. On the highway map, the vertical lines represent meridians and north is toward the top of the map. East is to the right and west is to the left.

The following longitude exercises involve four quantities—the longitude and time at one meridian and the longitude and time at a second meridian. Of these quantities, three are given, and you are asked to find the fourth.

EXAMPLE 16.2 Calculating Local Solar Time and Date at a Given Longitude

What is the local solar time and the date at 118°W, when the local solar time at 77°W is 12 noon on December 10?

Solution

STEP 1

Draw two vertical lines, which represent meridians, a small distance apart and record at the lines the given data plus what you are to find.

118°W   77°W

$t = ?$ $t = 12$ noon

date = ?   date = December 10

STEP 2

Determine the difference in longitude of the two meridians.

$$118°W - 77°W = 41°$$

STEP 3

Determine the amount of time corresponding to 41°. Since 15° is equivalent to one hour and 1° is equivalent to 4 minutes, the time difference is

$$30° = 2 \text{ hours}$$

$$41° - 30° = 11° = 44 \text{ minutes}$$

STEP 4

Determine the time at 118°W. Since the Sun appears to travel westward due to Earth's eastward rotation, the Sun has not arrived at 118°W. Therefore, longitude 118°W will have A.M. time. The Sun will be over the 118°W meridian in two hours and 44 minutes. Therefore, we subtract this time from 12 noon.

$$12 \text{ noon} - 2 \text{ h and } 44 \text{ min} = 9:16 \text{ A.M.}$$

STEP 5

Determine the date at 118°W. Neither midnight nor the 180° meridian is between the two given longitudes. Thus, we do not cross either in traveling from 77°W to 118°W. The date is the same as the date at 77°W, December 10.

EXAMPLE 16.3 Calculating Standard Time and the Date at a Given Longitiude

What is the standard time and the date at Paris, France (2°E) when the standard time at Washington, D.C. (77°W) is 9 P.M. on October 13?

Solution

STEP 1

Draw two vertical lines, which represent meridians, a small distance apart and record at the lines the given data plus what you are to find.

77°W   2°E

$t = 9$ P.M. $t = ?$

date = October 13   date = ?

STEP 2

Change the given longitude values. Since the exercise has given the standard time at one meridian and is asking for the standard time at the second meridian, cross out on your step 1 diagram the longitudes given and write the value of the central meridian where the given longitudes obtain their standard time.

The value for the central meridian representing standard time must be a multiple of 15 (beginning at zero). Thus, the 2°E meridian obtains its standard time from the 0° meridian and the 77°W meridian obtains its standard time from the 75°W meridian.

STEP 3

Determine the angular difference between the two standard meridians.

$$75°W - 0° = 75°$$

STEP 4

Determine the amount of time corresponding to 75°.

75° divided by 15° per hour = 5 hours

STEP 5

Determine the time at 0° meridian that has the standard time for Paris. Nine P.M. at Washington means the Sun passed the 75°W meridian nine hours ago. Since 0° is five hours east of 75°W, the Sun passed 0° five hours earlier. Adding five hours to 9 P.M. gives the time at Paris to be 2 A.M.

STEP 6

Determine the date at Paris. Since we passed through midnight going east, we advance to the next day, October 14.

CONFIDENCE QUESTION 16.4

(a) The date is October 7 and the local solar time at Washington, D.C. (77°W) is 2:04 P.M. What is the corresponding time and date at Paris, France (2°E)?
(b) When the standard time and date at Washington, D.C. (77°W) is 4 A.M. October 6, what is the corresponding standard time and date in Honolulu, Hawaii (158°W)?

Another way to visualize longitude problems is shown in Fig. 16.11. The drawing represents Earth as viewed from above the North Pole. All 24 central meridians are drawn, but only every other one has the longitude designated. The standard time at all designated longitudes is given at the instant the Sun is overhead at 30°W. Note in this

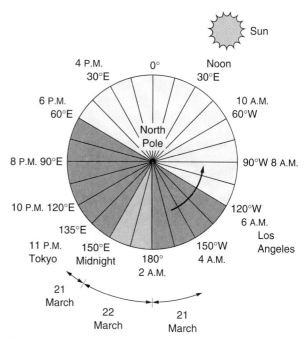

Figure 16.11 Finding the time in Tokyo.
An example of finding the time and date in Tokyo, knowing that the time and date in Los Angeles is 6 A.M., March 21. The North Pole is at the center of the circle.

drawing, as in Fig. 16.10, that as Earth rotates counterclockwise or eastward, the Sun appears to move westward bringing 12 noon. Because of this motion, 12 midnight also appears to move westward, bringing the new day, March 22.

An examination of Fig. 16.11 shows that when the standard time at Los Angeles is 6 A.M. on March 21, the standard time at Tokyo is 11 P.M. on March 21. If you should travel westward from Los Angeles to Tokyo with no change in the Sun's position, you will change the date of March 21 to March 22, then back to March 21. This double shift will occur because you will cross both the 180° meridian (the IDL) and the 12 midnight meridian.

During World War I, the clocks of many countries were set ahead one hour during the summer months to give more daylight hours in the evening, thus conserving fuel used for generating electricity for lighting. This practice has now become standard for all but three or four of the 50 states of the United States. During the summer months in this country, time known as **Daylight Saving Time** (DST) begins at 2 A.M. on the first Sunday of April and ends at 2 A.M. on the last Sunday of October.

The change to Daylight Saving Time helps conserve energy; it also reduces injuries and saves lives by preventing early-evening traffic accidents.

16.4 The Seasons

Learning Goals:

To explain Earth's four seasons.

To describe how daylight hours vary with an observer's latitude and the day of the year.

The spinning Earth is revolving around the Sun in an orbit that is elliptical yet nearly circular. When the Earth makes one complete orbit around the Sun, the elapsed time is known as one *year*. We are concerned with two different definitions of the year in this text. The **tropical year,** or the year of the seasons, is the time interval from one vernal equinox to the next vernal equinox. That is, the tropical year is the elapsed time between one northward crossing of the Sun above the equator and the next northward crossing of the Sun above the equator. In respect to the rotation period of Earth, the tropical year is 365.2422 mean solar days.

The **sidereal year** is the time interval for Earth to make one complete revolution around the Sun with respect to any particular star other than the Sun. The sidereal year has a period equal to 365.2563 mean solar days. This period is approximately 20 min longer than the tropical year. The reason for the difference will be explained later in Section 16.5.

The axis of the spinning Earth is not perpendicular to the plane swept out by the Earth as it revolves around the Sun; it is tilted 23.5° from the vertical, as illustrated in Fig. 16.12. This position of the axis with respect to the orbital plane produces a change in the Sun's overhead position throughout the year and causes our changing seasons. Figures 16.13 and 16.14 illustrate the apparent positions of the Sun over a period of one year.

In the summer in the Northern Hemisphere, the Sun's rays are most direct on the Northern Hemisphere. Thus it is hotter in the summer when the Sun's rays are most direct, and it is colder in the winter when the Sun's rays are the least direct. When it is summer in the Northern Hemisphere, it

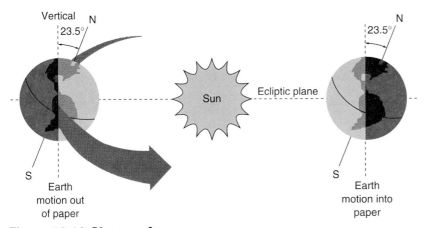

Figure 16.12 Change of seasons.
As Earth revolves around the Sun, Earth's axis remains tilted 23.5° from the vertical. This inclination of the axis, in conjunction with Earth's motion, causes a change in seasons on Earth. When Earth is at the position shown with its motion out of the page, the Northern Hemisphere has summer and the Southern Hemisphere has winter. When Earth is at the position shown with its motion into the paper, the Northern Hemisphere has winter and the Southern Hemisphere has summer.

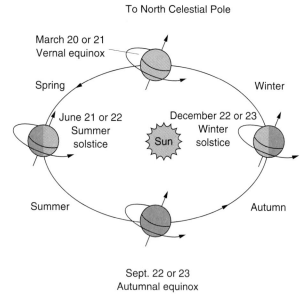

Figure 16.13 Earth's positions, relative to the Sun, and the four seasons.
As Earth revolves around the Sun, its north-south axis remains pointing in the same direction. On March 21 and September 21, the Sun is directly above the equator and everywhere on Earth has 12 hours of daylight and 12 hours of darkness. On June 21 the Sun's declination is 23.5°N, and the Northern Hemisphere has more daylight hours than dark hours; it is then summer in the Northern Hemisphere and winter in the Southern Hemisphere. On December 23 the Sun's declination is 23.5°S, and the Southern Hemisphere has more daylight hours than dark hours; it is then winter in the Northern Hemisphere and summer in the Southern Hemisphere. The Sun is drawn slightly off center to indicate that Earth is slightly closer to the Sun during winter in the Northern Hemisphere than during summer.

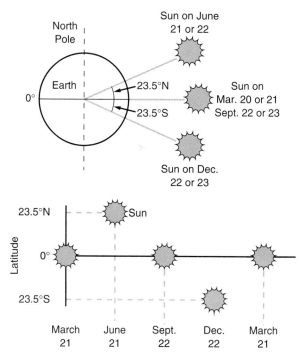

Figure 16.14 Diagrams of the Sun's position (degrees latitude) at different times of the year.
The upper drawing shows a greatly magnified Earth with respect to the Sun and gives the Sun's position on the dates indicated. The lower graph plots the Sun's position (degrees latitude) versus time (months).

is winter in the Southern Hemisphere, and vice versa.

The noon Sun's overhead position is never greater than 23.5° latitude, and the Sun always appears due south at 12 noon local solar time for an observer located in the continental United States. When the Sun is at 23.5° north or south, it is at its maximum latitude—the farthest point from the equator. This farthest point of the Sun from the equator is known as the *solstice* (meaning the Sun appears to stand still). The most northern point is called the **summer solstice,** and the most southern position is known as the **winter solstice.** This discussion applies only to the Northern Hemisphere. In the Southern Hemisphere dates for summer and winter solstices are reversed from those shown in Fig. 16.13.

As Earth revolves around the Sun during one year, the Sun's position overhead varies from 23.5° north to 23.5° south of the equator. The **vernal (spring) equinox** is the point on the ecliptic where the Sun crosses the celestial equator from south to north. This occurs on or about March 21. The **autumnal (fall) equinox** is the point on the ecliptic where the Sun crosses the celestial equator from north to south. This occurs on or about September 22. At the moment of both the vernal equinox and autumnal equinox, all latitudes except the geographic poles have approximately 12 hours of daylight and 12 hours of night. (Equinox means "equal night.") Observers at the poles see the Sun on their horizon throughout the day. When the Sun crosses the equator, sunrise will take place at 6 A.M. local solar time, and sunset will take place at 6 P.M. local

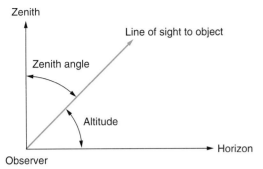

Figure 16.15 Zenith angle and altitude.
Because the zenith is perpendicular to the horizon, the zenith angle plus the altitude equals 90°.

solar time for all latitudes except the geographic poles where the Sun appears on the horizon during the 24-h day.

When the Sun is observed at 12 noon local solar time, it is on the observer's meridian and appears at its maximum altitude above the southern horizon on that day for all observers north of the Sun. The angle measured from the horizon to the line of sight to the Sun at noon is called its **altitude,** and the angle from the **zenith** (position directly overhead) to the line of sight to the Sun at noon is called its **zenith angle.** See Fig. 16.15. The zenith is 90° from the *horizon* (the dividing line where the Earth and sky appear to meet); therefore, the sum of the zenith angle and the altitude is 90°. The altitude of the Sun can easily be determined by measurement with a sextant. If the Sun's position is known, the observer's latitude can be determined. The relation between the above terms is illustrated in Fig. 16.16.

The maximum and minimum altitudes of the Sun for an observer in Washington, D.C. (39°N), can be determined by using data from Fig. 16.16. The solutions are shown in Figs. 16.17 and 16.18. The relationship between the two solutions is illustrated in Fig. 16.19. The altitude of the Sun for all other days of the year, as observed from Washington, would be between these two values. A similar solution will give the Sun's altitude from any latitude.

EXAMPLE 16.4 Use Cross-section Diagrams and Calculate the Observer's Latitude When the Sun's Position Is Known

What is the latitude of an observer who sees the Sun 39° due south of his zenith at the time of the vernal equinox?

Solution

STEP 1

Draw two horizontal lines, which represent parallels, a small distance apart. Since the observer is looking south to see the Sun, he or she must be north of the Sun. Designate the

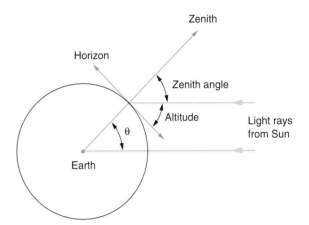

Figure 16.16 The relationship among zenith angle, altitude, horizon, and the altitude angle.
Because the incoming rays of light from the Sun are parallel, the angle θ and the zenith angle are equal in magnitude.

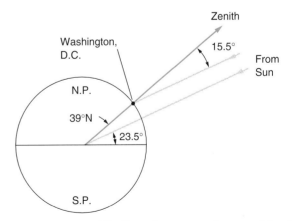

Figure 16.17 Finding the approximate altitude of the Sun as observed from Washington, D.C., on June 21.
The angle between the Sun and the observer is 39° − 23.5° = 15.5°, which makes the altitude of the Sun 90° − 15.5° = 74.5°.

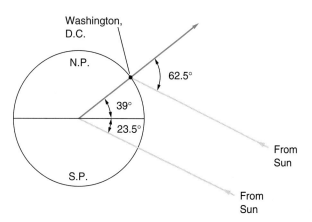

Figure 16.18 Finding the approximate altitude of the Sun as observed from Washington, D.C., on December 21.
The angle between the Sun and the observer is 39° + 23.5° = 62.5°, which makes the altitude of the Sun 90° − 62.5° = 27.5°.

upper line for the observer, and the lower line for the Sun. Record the given data plus what you are to find at the lines. The Sun is crossing the equator at the time of the vernal equinox.

STEP 2
Determine the angle between the two parallels. Since the incoming light rays from the Sun are coming in parallel to one another, the angle between the two drawn lines is equal to the zenith angle. (See Fig. 16.16.) Thus, the observer is at 39°N.

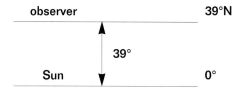

EXAMPLE 16.5 Calculating the Observer's Latitude When Polaris, the Pole Star, Is Visible

Calculate the latitude of an observer who sees Polaris due north of his or her zenith at a zenith angle of 50°.

Solution

STEP 1
Draw two horizontal lines, which represent parallels, a small distance apart. Since the observer is looking north, the lower line represents his or her latitude. Record, at the lines, the given data plus what you are to find. The star Polaris is above the Earth's North Pole.

Polaris	90°N
observer	latitude = ?

STEP 2
Determine the angle between the two parallels. Light rays coming from Polaris are parallel to one another; therefore, the angle between the two lines in your drawing is equal to the zenith angle (Fig. 16.16). Thus, the observer is 50° south of 90°N, and 90° − 50° = 40°N.

CONFIDENCE QUESTION 16.5
(a) What is the zenith angle of Polaris, the pole star, for each of the following: an observer at 90°N, an observer at 0°? (b) What is the altitude of Polaris, the pole star, for an observer at 40°N?

CONFIDENCE QUESTION 16.6
(a) What is the maximum altitude of the Sun for an observer at the North Pole? (b) On what month and day does the maximum altitude occur? (c) What is the maximum zenith angle of the Sun for an observer at the equator? (d) How many times during the year does the Sun appear at the maximum zenith angle for an observer at the equator? (e) State the months and dates of maximum zenith angle for an observer at the equator.

The seasons have a tremendous effect on the lives of everyone. Our yearly life cycles are ordered by the season's progressions. Many of our holidays were originally celebrated as commemorating a certain season of the year. The celebration that has

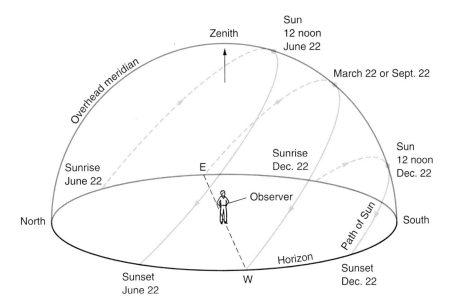

Figure 16.19 The Sun's apparent path. Diagram showing the apparent path of the Sun across the sky one June 21 and December 21, as observed from Washington, D.C. (39°N latitude).

evolved into our Easter holiday was originally a celebration of the coming of spring and a renewal of nature's life. Halloween originally commemorated the beginning of the winter season, while Thanksgiving commemorated the end of the harvest. The ancient festival of the winter solstice (on December 21 or December 22) and the beginning of the northward movement of the Sun has evolved into our Christmas holiday on December 25. The reasons for celebrating our various holidays have changed over the course of time, but the original dates were set by nature's annual timepiece—the movement of Earth around the Sun.

Our daily lives are greatly influenced by the number of daylight hours we enjoy. We appreciate the daylight hours for working, playing, and especially driving. Most of us avoid as much as possible driving during the dark hours. We hurry home from work to be out of traffic during darkness.

The number of daylight hours at any place on Earth depends on the latitude and the day of the year. The duration of daylight is a function of latitude because Earth's axis is tilted with reference to the incoming rays of the Sun, and daylight hours depend on the date because the orientation of Earth's axis relative to the Sun's rays changes continuously as Earth orbits the Sun. Since Earth is a great distance from the Sun, the light rays incident on Earth's surface are approximately parallel. Therefore one-half of Earth's surface will be illuminated (daylight) all the time and one-half will be in darkness all the time (see Fig. 16.12). Since Earth is rotating, most places on the planet's surface will receive sunlight each day. All places will experience daylight or darkness whose duration depends on the latitude and the day of the year.

Many persons think that the closer you are to the equator the more daylight hours you will experience each day. This is true during the fall and winter months for those persons living in the Northern Hemisphere. During the spring and summer months, New York has more daylight hours than Florida. The Sun rises earlier and sets later in New York than in Florida during these months. Note in Fig. 16.13 that on June 21 or 22 Earth's axis is tilted toward the Sun and on December 22 or 23 Earth's axis is pointed away from the Sun, whereas on March 20 or 21 and September 22 or 23 Earth is broadside to the Sun. Thus during the spring and summer months the Northern Hemisphere will experience more daylight hours than dark hours. Note in Fig. 16.13 that the equator has 12 hours of daylight and 12 hours of darkness every day of the year. As Earth rotates, the light and dark areas shown in the diagram are always equal for an observer at the equator. During the spring and summer months, the North Pole has 24 hours of daylight every day. Thus during these months in the northern latitudes there is a minimum number of daylight hours (12) at the equator and a maximum number (24) at the North Pole. Thus, a person traveling continuously northward

Table 16.1 Duration of Daylight Hours* at Certain Northern Latitudes.

	On June 21	On December 22
90°N	24 hours	0 hours
60°N	19 hours	6 hours
50°N	16 hours	8 hours
40°N	15 hours	9 hours
30°N	14 hours	10 hours
20°N	13 hours	11 hours
10°N	12.5 hours	11.5 hours
0°N	12 hours	12 hours

*The daylight hours are approximate values rounded off to whole numbers. Values for other latitudes can be roughly interpolated.

Figure 16.20 Precession of a top.

from the equator will experience, each day, more daylight hours.

Table 16.1 gives the number of daylight hours for some northern latitudes on June 21 and December 22. The daylight hours are rounded off for simplicity. Values for other latitudes can be roughly interpolated. To find the approximate time of sunrise and sunset, divide the hours of daylight by 2, then subtract the half-value from 12 noon to obtain sunrise. Add the half-value to 12 noon to obtain sunset. Remember there are approximately the same number of daylight hours before noon as in the afternoon. Also, there are approximately the same number of dark hours before midnight as after midnight. Table 16.1 can also be used for the Southern Hemisphere. Simply label the June column December and the December column June and change all N's to S's.

CONFIDENCE QUESTION 16.7

(a) Give the approximate latitude of your hometown.
(b) Determine the time of sunrise at your hometown using today's date.

16.5 Precession of Earth's Axis

Learning Goals:

To explain precession.

To describe the precession of Earth's axis.

Many of us are acquainted with the action of a toy top that has been placed in rapid motion and allowed to spin about its axis. After spinning a few seconds, the top begins to wobble or do what physicists call "precess" (Fig. 16.20). The top, a symmetrical object, will continue to spin about a vertical axis if the center of gravity remains above the point of support. When the center of gravity is not in a vertical line with the point of support, the axis slowly changes its direction. This slow rotation of the axis is called **precession.**

Because Earth is spinning rapidly, it bulges at the equator and cannot be considered a perfect sphere. The Moon and Sun apply a gravitational torque to Earth; this torque tends to bring Earth's equatorial plane into its orbital plane. Because of this torque, the axis of Earth slowly rotates clockwise or westward about the vertical or the North Ecliptic Pole (Fig. 16.21). The period of the precession is 25,800 years; that is, it takes 25,800 years for the axis to precess through 360°. As the axis changes its direction, the equinoxes move westward along the ecliptic. This movement is called the *precession of the equinoxes*.

Because the precession is clockwise and Earth is revolving counterclockwise around the Sun, the tropical year is approximately 20 min shorter than the sidereal year. As the axis precesses, Polaris will no longer be the north star. The star Vega in the constellation Lyra will be the north star some 12,000 years from now. The Southern Cross, a constellation of stars located within 27° of the present South Celestial Pole, will then be visible from Washington, D.C.

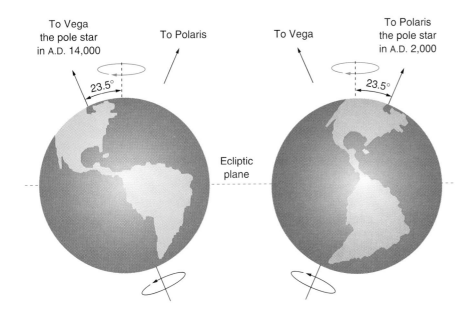

Figure 16.21 Precession of Earth's axis clockwise about a line perpendicular to Earth's orbit. Earth's axis is presently pointing toward the star Polaris, which we call the pole, or north, star. In approximately 12,000 years the axis will be pointing toward the star Vega in the constellation Lyra.

Our calendar is based on the seasons of the year. We want our summers to be warm and our winters to be cold. As Earth precesses, we define June 21 to be the date when the Northern Hemisphere tilts toward the Sun at a maximum angle. Thus, as Earth precesses, the stars seen at various seasons will change. This change is shown in Fig. 16.22. The stars we see on summer nights will slowly change within a period of 25,800 years. In the year A.D. 14,900 stars seen on summer nights will be our present winter night stars. Because of precession, the 12 constellations in the zodiac will slowly cycle through different months, with a one-month change occurring every 2150 years (25,800/12). The stars seen overhead on June 21 some 2150 years ago are seen overhead on July 21 today and will be seen overhead on August 21 in another 2150 years.

In the early 1970s a popular song called "The Age of Aquarius" took its title from the fact that the constellation Aquarius was moving from its January 21 to February 21 time slot of thousands of years ago toward the March 21 to April 21 time slot. Because older cultures started the year on March 21, the "age" or 2150-year period took on aspects of the March 21 to April 21 constellation. The song spoke of "the dawning of the age of Aquarius," which we see would slowly occur as Earth precessed.

CONFIDENCE QUESTION 16.8

Why is the tropical year shorter than the sidereal year?

16.6 The Calendar

Learning Goals:

To describe a brief history of the calendar.

To identify the origin of the names for the days of the week.

The continuous measurement of time requires the periodic movement of some object as a reference. Various lengths of time had a direct influence on the life of ancient men and women. Because of these influences, they had more than one reference for measuring the events of their lives. It is reasonable to believe that the first unit for the measurement of time was the day; because of the daily need for food, people spent the daylight hours hunting. During the dark hours they slept.

A longer period of reference was based on the periodic movement of the Moon. Some societies probably used the Moon as a basic division of time, because they were unable to count the days over a long period. The first appearance of the crescent

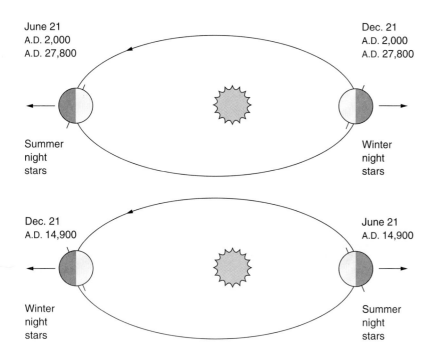

Figure 16.22 Precession. The precession of Earth's axis will cause future generations to see different stars in the summer than are seen today. After 12,900 years, our winter night stars will be their summer night stars, and vice versa. After another 12,900 years, the constellations will be similar to what we see now.

Moon and the time of the full Moon were times of worship for many primitive tribes.

Our month of today originated from the periodic movement of the Moon, which requires 29.5 solar days to orbit Earth. The plans for the first calendar seem to have originated before 3000 B.C. with the Sumerians, who ruled Mesopotamia. Their calendar was based upon the motion of the Moon, which divided the year into 12 lunar months of 30 days each. Because 30 × 12 = 360 days, and the year actually contains 365.25 days, corrections had to be made to keep the calendar adjusted to the seasons. How the Sumerians made the corrections remains a mystery, but their successors, the Babylonians, adjusted the length of the months and added an extra month when needed. This Babylonian calendar set the pattern for many of the calendars adopted by ancient civilizations.

The calendar we use today originated with the Romans. The early Roman calendar contained only 10 months, and the year began with the coming of spring. The months were named March, April, May, June, Quintilis, Sextilis, September, October, November, and December. The winter months of January and February did not exist. This was the period of waiting for spring to arrive. About 700 B.C. Numa, who reigned in Rome, added the month of January at the beginning of the year, before March; and February at the end of the year, after December. About 275 years later, in 425 B.C., the two months were changed to the present order.

The so-called Julian calendar was adopted in 45 B.C., during the reign of Julius Caesar. Augustus Caesar, who ruled the Roman Empire after Julius, renamed the month Quintilis as July in honor of Julius, and the month Sextilis as August in honor of himself. He also removed one day from February, which he added to August to make it as long as the other months.

The Julian calendar had 365 days in a year. In every year divisible by 4, an extra day was added to make up for the fact that it takes approximately 365.25 days for Earth to orbit the Sun. Thus, in 1997, 1998, and 1999 there are 365 days. In 1996 and 2000 there are 366 days. The Julian calendar was fairly accurate and was used for over 1600 years.

In 1582 Pope Gregory XIII (Fig. 16.23) realized that the Julian calendar was slightly inaccurate. The vernal equinox was not falling on March 21, and religious holidays were coming at the wrong time. A discrepancy was found, and the Pope decreed that 10 days would be skipped to correct it. The discrepancy arose because there are

16.6 The Calendar

Figure 16.23 Pope Gregory XIII.
In 1582 he issued a proclamation to drop 10 days from the calendar and to have leap years 97 times (instead of 100) every 400 years. The Gregorian calendar is now used worldwide.

365.2422 days in a year and not 365.25, as the Julian calendar used.

Pope Gregory set the calendar straight and devised a method to keep it correct. He decreed that every 400 years, 3 leap years would be skipped. The leap years to be skipped were the century years not evenly divisible by 400. The present corrections make the calendar accurate to one day in 3300 years. Our present-day calendar with these leap-year designations is called the *Gregorian calendar*.

Pope Gregory's reform of 1582 was not universally accepted. Most Protestant countries did not immediately go along. England and the American colonies finally changed their calendars in 1752, when there was an 11-day discrepancy. As a correction, September 2 was followed by September 14, and some people thought they were losing 11 days of their life. Other problems also arose when landlords asked for a full month's rent and banks sought to collect a full month's interest. In fact, riots were touched off in London over the change. When Russia and China accepted the Gregorian calendar in the early 1900s, it marked the first time that the whole world used the same calendar. There are other calendars still in use for religious purposes, but for civil events the Gregorian calendar is used around the world.

In our calendar we have seven days in each week. The origin of the seven-day week is not definitely known, and not all cultures had a seven-day week. One possible origin of the seven-day week is that is takes approximately seven days for the Moon to go through one quarter-revolution.

A more likely origin is the nighttime sky. As the ancients watched the sky night after night, there were exactly seven celestial bodies that moved relative to the fixed stars. These seven objects visible to the unaided eye are the Sun, Moon, and the five visible planets: Mars, Mercury, Jupiter, Venus, and Saturn.

Our present days of the week can still be connected by their names to the Sun, Moon, and five visible planets. Table 16.2 shows the heavenly object, English name, French name, and Saxon name for the seven days of the week. Note that either the English or French word is similar to the name commonly used for the heavenly object.

Our English days, Tuesday, Wednesday, Thursday, and Friday, come from the words Tiw, Woden, Thor, and Fria, who were Nordic gods. Woden was the principal Nordic god; Tiw and Thor were the gods of law and war: and Fria was the goddess of love. Sunday, Monday, and Saturday retain their connection to the Sun, Moon, and Saturn.

CONFIDENCE QUESTION 16.9

Why is it necessary to keep adjusting the calendar?

Table 16.2 The Days of the Week

Heavenly Object	English Name	French Name	Saxon Name
Sun	Sunday	dimanche	Sun's day
Moon	Monday	lundi	Moon's day
Mars	Tuesday	mardi	Tiw's day
Mercury	Wednesday	mercredi	Woden's day
Jupiter	Thursday	jeudi	Thor's day
Venus	Friday	vendredi	Fria's day
Saturn	Saturday	samedi	Saturn's day

Important Terms

These important terms are for review. After reading and studying the chapter, you should be able to define and/or explain each of them.

Cartesian coordinate system
latitude
parallels
meridians
longitude
Greenwich meridian, prime meridian
apparent solar day
mean solar day
sidereal day
International Date Line
Daylight Saving Time
tropical year
sidereal year
summer solstice
winter solstice
vernal equinox
autumnal equinox
altitude
zenith
zenith angle
precession

Questions

16.1 Cartesian Coordinates

1. A coordinate system must have
   (a) an indicated origin.
   (b) two dimensions.
   (c) a unit length expressed.
   (d) both (a) and (c).
2. A Cartesian coordinate system
   (a) is a two-dimensional system.
   (b) normally designates the horizontal line the *x*-axis.
   (c) normally designates the vertical line the *y*-axis.
   (d) all of the above
3. Name two fundamental features of every coordinate system.
4. Give three examples of a one-dimensional reference system.
5. (a) What two streets in your city or town divide the city into four quadrants?
   (b) Name the four quadrants.

16.2 Latitude and Longitude

6. Latitude
   (a) is a linear measurement.
   (b) can have greater numerical values than longitude.
   (c) is measured in an east-west direction.
   (d) can have negative values.
   (e) none of the above
7. Longitude
   (a) can have a maximum value of 180°.
   (b) is an angular measurement.
   (c) has east or west values.
   (d) is measured along parallels.
   (e) all of the above
8. What are the minimum and maximum values for latitude and longitude?
9. Can one travel continuously eastward and circle Earth? Why or why not?
10. Can one travel continuously southward and circle Earth? Why or why not?
11. How is 0° defined for latitude and longitude?
12. What is the name of a line of equal longitude?

16.3 Time

13. The 24-h apparent solar day
    (a) begins at midnight.
    (b) begins first at the 180° meridian.
    (c) is not exactly 82,400 seconds every solar day.
    (d) all of the above
14. The number of daylight hours at a specific place on Earth's surface is dependent on the
    (a) latitude.
    (b) date.
    (c) height above sea level.
    (d) all of the above
15. Why is there a time difference between the solar and the sidereal day?
16. How are the boundaries of standard time zones determined?
17. How many time zones are there in the 48 contiguous (adjoining) states?
18. What do A.M. and P.M. mean?
19. Is it correct to state the time as 12 A.M.? Explain your answer.

20. What are some advantages of Daylight Saving Time?
21. What is the direction of rotation (clockwise or counterclockwise) of the shadow cast by a sundial located at (a) 30°N and (b) 30°S?
22. Explain how a gnomon can be used to determine true north.
23. When does Daylight Saving Time begin and end?

16.4 The Seasons

24. The seasons
    (a) are a function of the inclination of Earth's axis.
    (b) are a function of Earth's revolving around the Sun.
    (c) would be more severe if the solstices were at 25°.
    (d) all of the above
25. For an observer at 40°N the Sun
    (a) is never directly overhead.
    (b) has an altitude of 50° on March 21.
    (c) has a zenith angle of 63.5° on December 22.
    (d) all of the above
26. State when each of the following occur in the Northern Hemisphere: autumnal equinox, spring equinox, winter solstice, summer solstice.
27. Explain the difference between one sidereal year and one tropical year.
28. The year is divided into four seasons. How are the divisions determined?
29. How would the seasons be modified if Earth's axis were tilted at 10° instead of 23.5°?
30. What is the altitude of Polaris (the north star) for an observer at the equator (0° latitude)? at the North Pole (90°N)?
31. What is the altitude of Polaris (the north star) for an observer at Washington, D.C. (39°N)?

16.5 Precession of Earth's Axis

32. Precession of Earth's axis
    (a) is counterclockwise as viewed from above the North Pole.
    (b) changes the angle between the axis and the vertical.
    (c) has no important effect on Earth's seasons.
    (d) none of the above
33. Precession of Earth's axis
    (a) is also called precession of the equinoxes.
    (b) is in a westward direction.
    (c) makes the tropical year approximately 20 minutes shorter than the sidereal year.
    (d) all of the above
34. Define and explain precession.
35. How long does it take Earth to precess one time?
36. What evidence is there to support precession of Earth's axis?

16.6 The Calendar

37. A natural unit of the calendar is the
    (a) day, which is based on the period of rotation of Earth.
    (b) month, which is based on the period of revolution of the Moon.
    (c) year, which is based on the period of revolution of Earth around the Sun.
    (d) all of the above
38. The Gregorian calendar
    (a) is our present calendar.
    (b) is accurate to one day in 6000 years.
    (c) skips leap year every century year.
    (d) all of the above
39. What is the origin of the month?
40. What is the origin of the seven-day week?
41. How often was there a leap year in the Julian calendar?
42. How often is there a leap year in the Gregorian calendar?
43. What are the origins of the dates for Halloween (October 31) and Christmas (December 25)?

Food for Thought

1. (a) What is the number of degrees between the Arctic Circle and the Tropic of Cancer? (b) If Earth's axis were tilted at 10°, what would be the number of degrees between the Arctic Circle and the Tropic of Cancer?
2. (a) Why does Atlanta, Georgia, have more daylight hours than Orlando, Florida, on April 21? (b) Why does Atlanta have fewer daylight hours than Orlando on October 21?
3. If Earth's axis were tilted at 15°, how would the number of daylight hours at 40°N on May 21 be changed?
4. (a) What is the direction of movement of the vernal equinox?
   (b) What effect does this have on the seasons? Why?

Exercises

16.2 Latitude and Longitude

1. What is the shortest distance in nautical miles between place A at 40°N, 75°W and place B at 28°N, 75°W?

   Answer: 720 nautical miles

2. What is the shortest distance in nautical miles between place C at 80°N, 90°E and place D at 70°N, 90°W?

3. How far away is the point at 90°N, 130°E from 90°N, 150°E?

4. Draw a diagram and explain why the points 60°N, 130°E and 60°N, 150°E are closer together than the points 30°N, 130°E and 30°N, 150°E.

5. What are the latitude and longitude of the point on the Earth that is opposite Washington, D.C. (39°N, 77°W)?

   Answer: 39°S, 103°E

6. What are the latitude and longitude of the point on the Earth that is opposite Atlanta, GA (34°N, 84°W)?

7. One nautical mile is a length unit of one minute of arc of a great circle. Sixty minutes of arc equal one degree.
   (a) Determine the shortest distance in nautical miles between Washington, D.C. (39°N), and the equator.
   (b) The nautical mile is 69/60 of a statute (land) mile. Determine the distance in part (a) in statute miles.

8. Suppose you start at Washington, D.C. (39°N, 77°W), and travel 300 nautical miles due north, then 300 nautical miles due west, then 300 nautical miles due south, then 300 nautical miles due east. Where will you arrive with respect to your starting point—at your starting point, or north, south, east, or west of your starting point?

   Answer: west

9. Suppose you start at Atlanta, GA (34°N, 84°W) and travel 300 nautical miles due west, then 300 nautical miles due north, then 300 nautical miles due east, then 300 nautical miles due south. Where will you arrive with respect to your starting point—at your starting point, or north, south, east, or west of your starting point?

16.3 Time

10. A professional basketball game is to be played in Portland, Oregon. It is televised live in New York beginning at 9 P.M. EST. What time must the game begin in Portland?

    Answer: 6 P.M. PST

11. If the polls close during a presidential election at 7 P.M. EST in New York, what is the time in California?

12. If an Olympic event begins at 10 A.M. on July 28 in Los Angeles (34°N, 118°W), what time and date will it be in Moscow (56°N, 38°E)?

    Answer: 9 P.M., July 28

13. When it is 9 A.M. on November 26 in Moscow (56°N, 38°E), what time and date is it in Tokyo (36°N, 140°E)?

    Answer: 3 A.M., Nov. 27

14. When it is 10 P.M. on February 22 in Los Angeles (34°N, 118°W), what time and date is it in Tokyo (36°N, 140°E)?

15. What is the altitude angle of the Sun on March 21 for someone at the North Pole? Answer: 0°

16. What is the altitude angle of the Sun for someone at 34°N latitude on (a) March 21 and (b) June 21?

    Answer: (a) 56°

17. What is the altitude angle of the Sun for someone at 34°N latitude on (a) September 22 and (b) December 22?

18. What is the latitude of someone in the United States who sees the Sun at an altitude angle of 71.5° on June 21?

    Answer: 42°N

19. What is the latitude of someone in the United States who sees the Sun at an altitude angle of 31.5° on December 22?

20. How many days are in each of the following years: 1995, 1996, 2000, 2001, 2004, 2100, 2200, 2300, 2400?

21. Determine the month and day when the Sun is at maximum altitude for an observer at Washington, D.C. (39°N). What is the altitude of the Sun at this time?

    Answer: on or about June 21; 74.5°

22. Determine the month and day when the Sun is at minimum altitude for an observer at Washington, D.C. (39°N). What is the altitude of the Sun at this time?
23. Is the difference between the maximum and minimum altitude of the Sun as determined in Exercises 18 and 19 equal to twice the angle Earth's axis is tilted from the vertical?
24. Determine the approximate number of daylight hours at 40°N on June 21.
    Answer: 15 hours
25. Estimate the approximate time of sunrise at 30°N on December 22.
    Answer: 7 A.M.
26. Estimate the approximate time of sunset at 25°N on June 21.

Answers to Multiple-Choice Questions

1. d	2. d	6. e	7. e	13. d
14. d	24. d	25. d	32. d	33. a
37. d	38. d			

Solutions to Confidence Questions

16.1 (a) Rectangular coordinate system (b) The x- and y-axes are perpendicular to one another, forming right angles (like the sides of a rectangle).
16.2 Yes, at the equator where the only parallel is a great circle. All meridians are one-half of a great circle.
16.3 west of the starting point
16.4 (a) 7:20 P.M., October 7, (b) 9:00 P.M., October 5
16.5 (a) 90°, 0°, (b) 40°
16.6 (a) 23.5°, (b) June 21, (c) 23.5°, (d) two, (e) June 21 ± 2 days, December 21 ± 2 days.
16.7 The answer is dependent on the latitude of the student's hometown.
16.8 The tropical year—the period of revolution of Earth (eastward) around the Sun with respect to the vernal equinox—is shorter than the sidereal year because the vernal equinox precesses westward.
16.9 The sidereal year and the solar year are not exactly the same number of days.

Chapter 17

The Moon

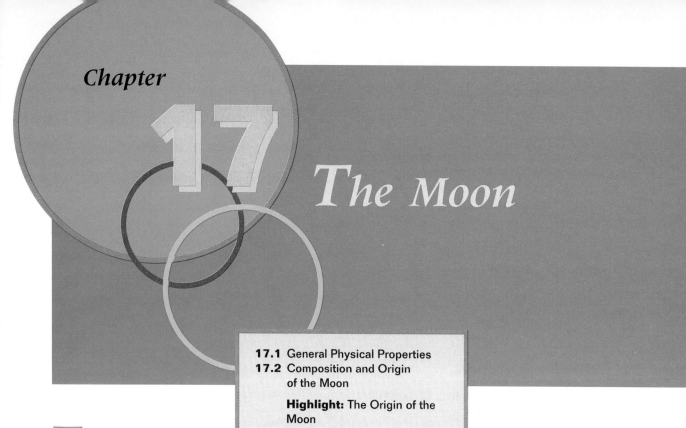

17.1 General Physical Properties
17.2 Composition and Origin of the Moon

Highlight: The Origin of the Moon

17.3 Lunar Motions
17.4 Phases of the Moon
17.5 Solar and Lunar Eclipses
17.6 Ocean Tides

The exact origin of the word *moon* seems to be unknown, but many writers believe it is related to the measurement of time. We do know that the length of our present month is based on the motion and phases of the Moon, that primitive people worshipped the Moon, and that many societies today base their religious ceremonies on the new and full phases of the Moon. We also know that the human reproductive cycle is synchronized to the lunar cycle, with the ovaries producing ova about every 28 days.

On July 20, 1969, humans first landed on the Moon, and Apollo 11 astronauts placed a retroreflector array on the Moon's surface. The retroreflector (an optical reflector designed to return the reflected ray of a laser beam parallel to the incident ray) is part of a lunar-ranging experiment that measures the distance to the Moon with a precision of ± 15 cm.

Measurements with this high degree of precision can be taken over long periods of time and will show the variation in the orbital distance of the Moon in great detail. Such measurements can be used to (1) determine the rate of continental drift on Earth (latitude and longitude of a place can be determined with great accuracy); (2) detect any change in the location of the North Pole; (3) determine the orbit of the Moon more exactly; and (4) determine whether the gravitational constant (G) is decreasing very slowly with time.

The Moon appears as the second-brightest object in the sky because it is very close to us. The Moon's average distance from Earth is about 384,000 km (239,000 mi). Because of its nearness and its influence on our lives, this chapter is devoted to the study of the Moon. Figure 17.1 shows an astronaut with the Lunar Rover collecting samples of the lunar surface.

17.1 General Physical Properties

Learning Goal:

To describe the general physical properties of the Moon.

The Moon at its brightest is a wondrous sight as it reflects the Sun's light back to our eyes. The Moon appears quite large to Earth observers. In fact, our

17.1 General Physical Properties

Figure 17.1 Geologist-Astronaut Harrison Schmitt on the Moon.
This surface view of the moon was taken during a survey trip by geologist and astronaut Harrison Schmitt during the Apollo 17 mission. Schmitt is shown standing by the roving vehicle, where he first spotted orange soil.

Moon is the largest of any inner planet's. Mercury and Venus have no moons, and those of Mars are quite small. Our Moon is the fifth largest in the solar system.

The Moon revolves around Earth in approximately 29.5 solar days, and it rotates at the same rate as it revolves. For this reason we see only one side of the Moon. An observer on the side that faces Earth would always be able to see Earth, but the Sun would appear to rise and set and rise again once every 29.5 days. Thus all sides of the Moon are heated by the Sun's rays.

The Moon is nearly spherical, with a diameter of 3476 km (2160 mi), a distance slightly greater than one-fourth Earth's diameter. The slow rotation of the Moon, coupled with the tidal bulge caused by Earth's gravitational pull on the solid material, produces an oblateness that is very small. The best measurements indicate a difference of less than a mile between the polar and equatorial diameters.

The mass of the Moon is $\frac{1}{81}$ that of Earth, and its average density is 3.3 g/cm^3. (Earth's average density is 5.5 g/cm^3.) The surface gravity of the Moon is only one-sixth that of Earth. Therefore, your weight on the surface of the Moon would be about one-sixth of your weight on Earth's surface (Fig. 1.4). The Moon's interior is thought to be made up of a small, perhaps solid, iron-rich core, a solid mantle rich in silicates, and a crust that is about 60 km (37 mi) thick on the near side and 150 km (93 mi) thick on the far side (Fig. 17.2).

The Moon does not possess a magnetic field; at least, none was detected by instruments carried by the Apollo astronauts. Surface rocks brought back by astronauts show some magnetism, indicating that the Moon had a slight magnetic field when the rocks solidified. The origin of this previous magnetic field is not known.

Except for its phases, the Moon's most predominant feature is the appearance of its surface, which is marked with craters, plains, rays, rilles, mountain ranges, and faults. These features vary in size, shape, and structure. The most outstanding

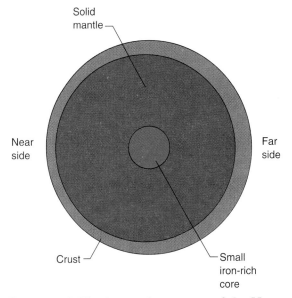

Figure 17.2 The internal structure of the Moon.
Geologists theorize that the Moon consists of a solid mantle rich in silicate materials, a crust composed mostly of low-density feldspar minerals that varies in depth from about 60 km on the side facing Earth to at least 150 km on the farside, and a small, perhaps solid, iron-rich core with a radius of 300 to 400 km.

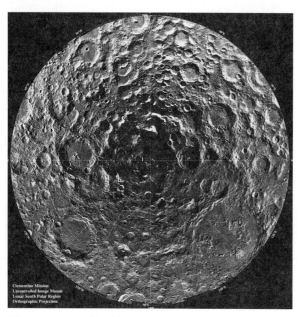

Figure 17.3 The Moon's south pole.
The images forming this photograph were taken by the *Clementine* spacecraft (the first ever taken of the Moon's south pole) and show the multiring South Pole-Aiken basin. Another multiring basin is shown at the four o'clock position. See the text for details.

are the craters that are clearly visible to an Earth observer with low-power binoculars or a telescope.

Craters and Basins

The word **crater** (Greek *krater*) means bowl-shaped, and the lunar craters are small and large diameter depressions believed to be caused by the impact and explosion of small and large objects that have come from space. The craters are rather shallow (their depths are small in comparison with their diameters), and their floors are located below the lunar surface (● Fig. 17.3).

The slope of a rim, with a width of about one-fifth the width of the crater from crest to crest, is greater on the inside than on the outside. Measuring the volume of the material in the rim and comparing it with the volume of the crater hole shows that they are approximately the same. This result supports the impact hypothesis.

The largest impact features are the multiring basins. Figure 17.3, which is compiled from images taken by the spacecraft *Clementine* during its 71 days in lunar orbit in February and April, 1994,

shows the South Pole-Aiken basin. The basin (unknown before *Clementine*'s voyage) is the largest known in the solar system. The basin has a diameter of 2500 km (1550 mi) and a depth that averages about 12 km (7.4 mi). Its location at the South Pole plus its great depth keeps the basin floor from sunlight and view. Another multiring basin is shown at the four o'clock position in Fig. 17.3. This is the second-youngest impact basin on the moon. The basin has a diameter of 320 km (198 mi) and is named Schrödinger after Erwin Schrödinger (1887–1961), German professor of physics and co-founder of quantum mechanics (Chapter 9). The youngest impact basin, Orientale, which is about 1000 km (1600 mi) across, is located at 20°S, 96°W. The basin was formed 3.8 billion years ago.

Plains

The lunar surface contains thousands of craters ranging in diameter from a few feet to the 240-km (150-mi) Clavius. The large flat areas called *maria* (an Italian word meaning "seas"), named by Galileo, are believed to be craters formed by the impact of huge objects from space and that later were filled with lava. These areas, which are now called *plains*, appear to be very dark because the Moon's surface is a poor reflector. The average albedo is about 0.07. The surface reflects only 7% of the light received from the Sun. The plains, which are similar to black asphalt, reflect very little light. Fourteen major plains on the side facing Earth cover 50% of the visible lunar surface. Most of the plains are located in the northern hemisphere and can be clearly seen with the unaided eye during the full phase of the Moon.

The surface of the Moon is a terrain of rolling, rounded knolls composed of a layer of loose debris or soil called *regolith*, which has a depth less than 10 m (33 ft) on the flat lunar plains. The lunar highlands, because they are older, have a thicker layer of regolith. The rock samples brought back by Apollo astronauts are similar to the volcanic rock found on Earth.

Rays

Some craters are surrounded by streaks, or **rays,** that extend outward over the surface. They are believed to be pulverized rock that was thrown out when the crater was formed. The rays appear much brighter than the crater, and we know that powdered rock reflects light better than regular-sized rock. The rays also become darker with age. Photographs show that in cases where rays from one crater overlap those of another, the ones on top appear to be brighter.

The ray system of a crater has an average diameter of about 12 times the diameter of the crater. Lunar photographs also show that the ray systems are marked with small craters called *secondary craters*, which are believed to have been formed by debris thrown out from the primary crater during the explosion caused by an impinging object from space.

Rilles

The lunar surface also has long narrow trenches, or valleys, called **rilles.** They vary from a few meters to about 5 km (3 mi) in width and extend hundreds of miles in length with little or no variation of width. Schröter's Valley, the largest sinuous rille on the Moon, is 160 km long, about 7 km wide (width varies), and up to 1300 m deep (Fig. 17.4). A sinuous rille is a winding valley, which resembles a channel cut by a flow of fluid such as lava. Some rilles are rather straight, whereas others follow a circular path. They have very steep walls and fairly flat bottoms.

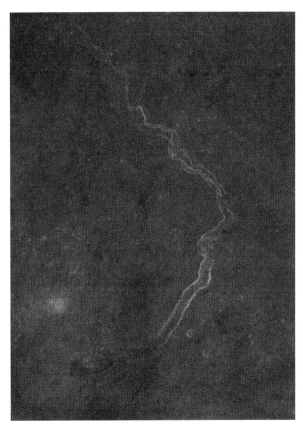

Figure 17.4 Schröter's Valley.
This valley is the largest sinuous rille on the Moon. The rille is located on the Aristarchus plateau. See the text for details. The photo was taken by the *Clementine* spacecraft.

Mountain Ranges

The *mountain ranges* on the lunar surface have peaks as high as 6100 m (20,000 ft), and all formations seem to be components of circular patterns bordering the great plains. This pattern indicates that they were not formed and shaped by the same processes as mountain ranges on Earth, which were formed by internal forces.

The craters and other surface features of the Moon were formed a long time ago when the solar system was filled with large amounts of matter. They are much the same now as they were when formed, because of the near-absence of erosion. Some changes have resulted from the impact of projectiles from space.

A **fault** is a break or fracture in the surface of the Moon along which movement has occurred. The motion along a fault can be vertical, horizontal, or parallel. Several faults are observed on the lunar surface. A very large cliff on the eastern side of Mare Nubium is the result of slippage of the Moon's crust along a fault. This cliff (called the Straight Wall) is about 113 km (70 mi) long and 244 m (800 ft) high, with its side inclined about 40° to the horizontal.

CONFIDENCE QUESTION 17.1

(a) Why do observers on Earth see only one side of the Moon? (b) Are all surface areas of the Moon heated by the Sun's radiation? Why or why not?

17.2 Composition and Origin of the Moon

Learning Goals:

To identify the composition of the Moon.

To present and explain the latest theory for the origin of the Moon.

Before the Apollo program to land an astronaut on the Moon was begun in the early 1960s, very little was known about the origin and history of the Moon. Exploration by the Luna and Apollo programs has changed all that.

The first landing on the Moon was July 20, 1969, when the landing craft of Apollo 11 settled in Mare Tranquillitatis at 0.67°N, 23.49°E. (The 0° longitude meridian designates the middle of the Moon facing Earth, and the 180° meridian designates the middle of the far side.) After that, five other Apollo lunar landing missions (Apollo 12, 14, 15, 16, and 17) were completed. The astronauts collected and brought back to Earth 379 kg of lunar material and erected on the Moon's surface 2104 kg of scientific instruments that will collect data for many years.

The rock samples, such as the one shown in Fig. 17.5, have enabled us to have a much better understanding of the Moon's origin and history.

Figure 17.5 Scientists examine a lunar sample. The samples are stored in an atmosphere of dry nitrogen, thus isolating them from oxygen and moisture (water) to prevent chemical reactions.

(See the chapter Highlight.) Samples taken from the plains and lowlands have yielded ages that are considerably younger than those from the highlands. Rocks from the highlands were formed between 4.4 and 3.9 billion years ago, whereas the rocks from the plains or the lowlands have ages between 3.8 and 3.1 billion years. No rocks older than 4.4 billion years or younger than 3.1 billion years have been found.

Almost all of the craters on the Moon are now known to have resulted from the bombardment of meteorites of various sizes. Because the Moon has no atmosphere or water on its surface, very little erosion takes place. Therefore, once formed, a crater remains for billions of years or until a meteorite hits to form a crater on top of it. In contrast, Earth has only a few remaining meteorite craters. For the most part, the craters left by meteorites striking Earth have been eroded away.

The Moon's plains and a few of its craters (about 1%) were produced by volcanic eruptions. The plains are composed of black volcanic lava that covered many craters. Most of the plains are on the near side of the Moon. The fact that fewer volcanic eruptions occurred on the far side is probably correlated with the fact that Moon's crust is thicker there (Fig. 17.2).

The oldest rocks on the Moon were formed about 4.4 billion years ago when the Moon's crust became cool enough to solidify. Between 4.4 and 3.9 billion years ago, the Moon was bombarded intensely by many meteorites. This was the period when most of its craters were formed. The Moon's surface was virtually pulverized, leaving little evidence of the original crust.

During the period 3.9 to 3.1 billion years ago, the Moon's interior had heated up enough from radioactive effects to cause volcanic eruptions that formed many plains. The lava had flowed from these eruptions and covered much of the Moon's lowlands. During this period, meteorite bombardment became less intense, because fewer and fewer rock fragments were left near the Earth-Moon system.

After 3.1 billion years ago, the Moon's mantle had become so thick that it could no longer be penetrated by molten rock. The Moon has been geologically quiet since that time. Meteorites have continued to bombard the surface and have formed a layer of dust several feet thick on the surface.

HIGHLIGHT

The Origin of the Moon

What is the origin of the Moon? Earth's Moon is an anomaly in the solar system because the Earth-Moon system is one of a kind. Earth is the only inner planet that has a large satellite, and the only planet in the solar system that has a satellite as large as one-fourth the diameter of the planet. Mercury and Venus do not have satellites, and Mars has two small ones that appear to be asteroids captured from the nearby asteroid belt. The outer planets are huge bodies with satellites that are small compared to the planets. Pluto and its relatively large satellite Charon are an oddity, and both are considered by some astronomers to be asteroids.

Several theories have been suggested for the origin of the Moon. Some of the facts that must be considered when formulating a theory are the following:

1. The lunar samples collected by the Apollo astronauts show that the Moon has a similar chemical composition to Earth's mantle (the interior region of Earth between the core and the crust). (See Section 21.1.)
2. The percentage abundances of the isotopes of oxygen (^{16}O, ^{17}O, ^{18}O) found in the lunar rock samples are similar to those of Earth, indicating that Earth and the Moon were formed at about the same distance from the Sun.
3. Water was not found in the lunar rock samples.
4. The lunar rocks show a small percentage of volatile elements (those that are driven off by extreme heating, such as sodium and potassium) and a high abundance of refractory elements (elements that are not easily vaporized).
5. Compared to Earth, the Moon samples show a lower abundance of iron.
6. The Moon has an average density of 3.3 g/cm^3. Earth's average density is 5.5 g/cm^3, the crust about 3.0 g/cm^3, and the core about 15 g/cm^3.
7. The oldest Earth and lunar rocks were formed at about the same time.

One explanation for the origin of the Moon is that Earth and the Moon were created about the same time, and the Moon coalesced from particles revolving around Earth. This theory fails to explain why Earth has a great abundance of iron and water, whereas the Moon contains very little iron and no water.

The preferred theory at present is the **great impact theory.** This theory proposes that a planet-sized object collided with Earth, and the impact ejected enough matter (most of it coming from Earth's mantle) into orbit to form the Moon. This would account for the similar densities of Earth's mantle and the Moon and for the low abundance of iron. Also, the impact of a large object generates tremendous heat, which would drive off water and other volatile substances.

The impact theory raises the possibility that other collisions or near-collisions may have taken place in the solar system. For example, a collision or near-collision could account for the large inclination of Uranus' axis and the planet's retrograde motion. Also, Venus' retrograde motion may have been caused by an outside disturbance. The planet Mercury, which has only an iron core and a thin crust, may have lost its mantle due to a collision.

The impact theory is gaining support within the scientific community, but lunar samples were collected from only nine surface areas by the Luna and Apollo programs. This is a very small sample of the Moon's surface. Perhaps when additional information is obtained from the Moon, an improved theory can be established.

CONFIDENCE QUESTION 17.2

(a) What is the present preferred theory for the origin of the Moon? (b) Why is this theory preferred over others?

17.3 Lunar Motions

Learning Goals:

To describe the Moon's orbit and motions.

To explain sidereal and synodic months.

The Moon revolves eastward around Earth in an elliptical orbit in a little over 29.5 solar days or almost 27.33 sidereal days. Its orbital plane does not coincide with that of Earth but is tilted at an angle of approximately 5° with respect to Earth's orbital plane (Fig. 17.6). The 5° tilt allows the Moon to be overhead at any latitude between 28.5°N and 28.5°S. The Moon rotates eastward as it revolves, making one rotation during one revolution. Figure 17.6 is an illustration of the eastward motion of the Moon and the inclination of the orbital plane to the orbit of Earth.

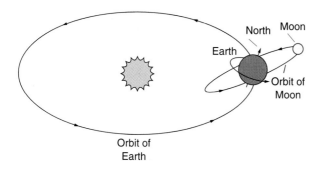

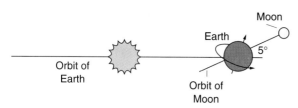

Figure 17.6 The relative motions of the Moon and Earth.
The top diagram is a view from above Earth's orbital plane; the lower diagram is a view from within Earth's orbital plane.

Because the Moon revolves in an elliptical orbit, its distance from Earth varies. At the closest point, called **perigee,** the Moon is 363,000 km (225,000 mi) from Earth's center. At the farthest point, called **apogee,** it is 405,000 km (251,000 mi) from Earth's center. The average distance from Earth's center to the Moon is about 384,000 km (239,000 mi), but 3.8×10^5 km (2.4×10^5 mi) will be used as the value for solving problems in this text.

As stated above, there are two different lunar months. The period of the Moon with respect to a star other than the Sun is approximately 27.33 days; this is called the sidereal period, or **sidereal month.** It is the actual time taken for the Moon to revolve 360°. The period of the Moon with respect to the Sun is approximately 29.5 days. This period is called the **synodic month,** or the month of the phases. The Moon revolves more than 360° during the synodic month (Fig. 17.7).

To an observer on Earth, the Moon appears to rise in the east and set in the west each day. This apparent motion of the Moon is due to Earth rotating eastward on its axis once each day. The times at which the Moon rises and sets are discussed in the following section.

CONFIDENCE QUESTION 17.3

(a) How often would an observer on the Moon see sunrise? (b) An observer sees the Sun while standing on the lunar terminator, the boundary line dividing day and night on the surface of a planet or the Moon. Where is the Sun in the observer's sky?

17.4 Phases of the Moon

Learning Goals:

To define and explain the phases of the Moon.

To solve exercises relating to different phases of the Moon.

The most outstanding feature presented by the Moon to an Earth observer is the periodic change in its appearance. One-half of the Moon's surface is always reflecting light from the Sun, but only once during the lunar month does the observer see all of the illuminated half. Throughout most of the Moon's period of revolution, only a portion of its illuminated side is presented to us.

The starting point for the periodic, or cyclical, motion of the Moon is arbitrarily taken at the new-phase position. The new phase of the Moon occurs when Earth, Sun, and Moon are in the same plane, with the Moon positioned between the Sun and Earth. They are not necessarily in a straight line. At this position the dark side of the Moon is toward Earth, and the Moon cannot be seen from this planet because the Sun is on the observer's meridian with the Moon. The new moon occurs at 12 noon local solar time.

The **new moon** occurs for an instant—the instant it is on the same meridian as the Sun. We often speak, however, of a phase of the Moon lasting for a full day of 24 h.

The Moon revolves eastward from the new-phase position, and for the next 7.375 solar days (one-fourth of 29.5 days) it is seen as a waxing crescent moon. The term **waxing moon** means that the illuminated portion of the Moon is appearing larger each passing night for an observer

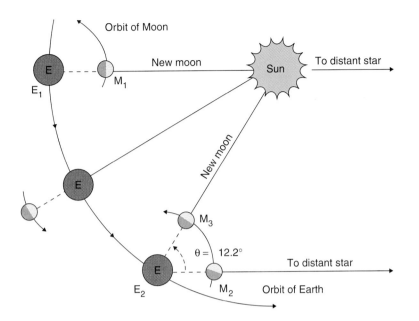

Figure 17.7 Diagram illustrating the difference between the sidereal and synodic months. With Earth at position E_1 and the Moon at position M_1, the Sun, Moon, and Earth are all in the same plane. At this time the Moon is in its new phase. As Earth revolves eastward to position E_2, the Moon has revolved through 360° in approximately 27.3 days to position M_2, and 1 sidereal month—one revolution with respect to a distant star—has elapsed. The Moon must revolve through 360° plus the angle θ before arriving at position M_3. At this time the Sun, Moon, and Earth will be in the same plane, the Moon will be in new phase, and 1 synodic month will have passed. The time for 1 synodic month is approximately 29.5 days.

on Earth; **waning moon** means that the illuminated portion is appearing smaller each passing night for an observer on Earth. A **crescent moon** is seen when less than one-quarter of the illuminated surface is facing an observer on Earth. A **gibbous moon** is seen when more than one-quarter of the illuminated surface is facing an observer on Earth. Figures 17.8 and 17.9 illustrate how the phases occur and how they appear to an observer to Earth.

The Moon is in the *waxing crescent phase* and appears as a crescent moon to an Earth-bound observer when it is less than 90° east of the Sun. The Moon is in **first-quarter phase** when it is 90° east of the Sun and appears as a quarter moon on the observer's meridian at 6 P.M. local solar time with the illuminated side toward the west. The first-quarter phase has a duration of only an instant, because the Moon can only be 90° east of the Sun for an instant (Fig. 17.8).

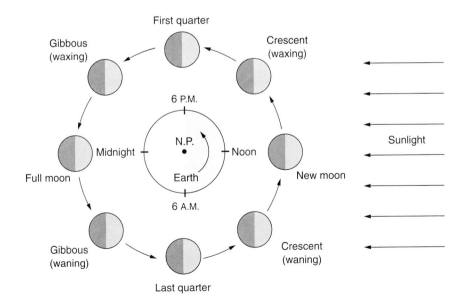

Figure 17.8 Phases of the Moon. Diagram shows the position of the Moon relative to Earth and the Sun during one lunar month as observed from a position in space above Earth's North Pole.

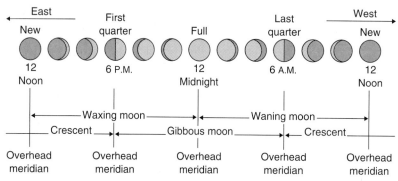

Figure 17.9 Phases of the Moon.
Diagram illustrating the phases of the Moon as observed from any latitude north of 28.5°N. The observer is looking south; therefore, east is on the left. The Sun's position can be determined by noting the local solar time the Moon is on the overhead meridian. The time period represented in the drawing is 29.5 days. Compare this drawing with Fig. 17.8.

From the first-quarter position, the Moon enters the *waxing gibbous phase* for 7.375 solar days. During this phase it appears larger than a quarter moon but less than a full moon. When the Moon is 180° east of the Sun, it will be in full phase and will appear as a **full moon** to the Earth-bound observer. The full moon appears on the observer's meridian at 12 midnight local solar time.

From the full-phase position, the Moon enters the *waning gibbous phase* and remains in that phase for 7.375 solar days. The appearance of the Moon during this phase is the same as in the waxing gibbous phase, except that the illuminated side is toward the east, and the Moon is seen in the sky at a different time. When the Moon is 270° east of the Sun, it will be in the last-quarter phase. The **last-quarter phase** appears on the observer's meridian at 6 A.M. local solar time, with the illuminated side of the Moon toward the east.

From the last-quarter position, the Moon enters the *waning crescent phase* and remains in this phase for 7.375 solar days. During this phase, the illuminated portion appears smaller than a quarter moon. It looks like the waxing crescent moon, except that the illuminated side is toward the east, and the Moon appears in the sky at a different time.

Figure 17.10 Phases of the Moon.
These eight photographs show the Moon at different times of the lunar month. They are arranged so that the views, from Earth, of the phases are the way they appear in close-up views by the unaided eye. Compare these photos with Fig. 17.9.

Waxing crescent
4 days

First quarter
7 days

Waxing gibbous
10 days

Full moon
14 days

Figure 17.9 illustrates the Moon's appearance and position above the southern horizon as observed from a latitude greater than 28.5° north. The Moon is shown in the first drawing on the left in the new-phase position. It is shown on the observer's meridian at 12 noon local solar time and shaded gray, illustrating that it cannot be seen at this time because it is on the same meridian as the Sun. The next two positions illustrate the waxing crescent phase. Note that the illuminated area (colored yellow) appears larger for an Earth observer as the Moon approaches the first-quarter phase, and that the illuminated side is toward the west where the Sun is located.

When the Moon is in first-quarter phase, it is on the observer's meridian at 6 P.M. local solar time. The Sun will be at or near the western horizon at this time. The Moon revolves eastward, entering the waxing gibbous phase, as shown by the next two positions. Note that the illuminated area is larger than a quarter moon and still increasing in size of face. The illuminated side is still toward the west.

The next position shows the Moon at full phase and on the observer's meridian at 12 midnight. If the date is at the time of the vernal (spring) or autumnal (fall) equinox, the Moon will rise on the eastern horizon at 6 P.M., when the Sun is setting in the west; and the Moon will set at 6 A.M., as the Sun is seen rising on the eastern horizon.

The Moon continues revolving eastward, entering the waning gibbous phase. Note that the size of the illuminated area is decreasing for an Earth observer and that the illuminated side is toward the east—just the opposite from the waxing gibbous moon. When the Moon is 90° west of the Sun (same as 270° east of the Sun), it will be on the observer's meridian at 6 A.M. local solar time, as shown in the next position. The Moon appears as a quarter moon, but the illuminated side is toward the east. The Sun will be rising at or near this time.

After the third-quarter phase, the Moon enters the waning crescent phase and the size of its face continues to decrease. Note that the illuminated side of the waning crescent phase is toward the east. The last position shows the Moon back to the new-phase position. Figure 17.10 shows eight photographs of the Moon as it appears in a close-up view by the unaided eye. Compare these photographs with Fig. 17.9

Figure 17.8 and Fig. 17.9 do not record the time a waxing or a waning moon is on the overhead meridian, but a close examination of the drawings indicates the time. If the Moon is observed on the overhead meridian at 3 P.M. local solar time, the Moon must be in the waxing crescent phase. Observe in both drawings that 3 P.M. is between 12 noon and 6 P.M., and coincides with the position of the waxing crescent moon in the drawing. If the Moon is observed on the overhead meridian at 9 P.M. local solar time, the Moon must be in the waxing gibbous phase.

Table 17.1 summarizes the times for the various phases of the Moon to rise, be overhead, and set. An example of what an observer in the United States sees when looking at the first-quarter phase is shown in Fig. 17.11. Figures similar to 17.11 for the other phases can be drawn by using the information in Table 17.1.

Waning gibbous
18 days

Last quarter
22 days

Waning crescent
26 days

Waning crescent
28 days

CHAPTER 17 The Moon

Table 17.1 Times for the Various Phases of the Moon to Rise, Be Overhead, and Set When the Sun is at the Vernal or Autumnal Equinox

Phase	Approximate Rising Time	Approximate Time Overhead	Approximate Setting Time
New moon	6 A.M.	Noon	6 P.M.
First-quarter moon	Noon	6 P.M.	Midnight
Full moon	6 P.M.	Midnight	6 A.M.
Last-quarter moon	Midnight	6 A.M.	Noon

Because the Moon revolves around Earth every 29.5 solar days, it gains 360° on the Sun in that time, or 12.2°/day. Thus the Moon is on the observer's meridian about 50 min later each day, because Earth must rotate through 360° plus 12.2° before the Moon appears on the overhead meridian (● Fig. 17.12). The average time of moonrise is thus delayed about 50 min each day. The actual time depends on the latitude of the observer, with greater variation noted in the higher altitudes. The variation depends on the angle between the Moon's path and the horizon.

The approximate altitude of the full moon can be found by recognizing that the full moon will be on the opposite side of Earth from the Sun (● Fig. 17.13). Thus when the Sun is low in the sky in the winter, the full moon will be high in the sky. In the summer the Sun is high in the sky and the full moon is low in the sky.

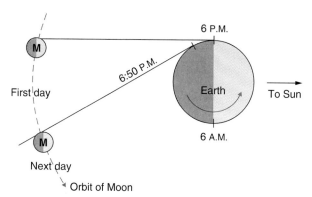

Figure 17.12 Moon's rising time.
The Moon rises about 50 min later each day because as Earth rotates, the Moon is revolving around Earth. For example, the full moon rises at about 6 P.M. on March 21 and at about 6:50 P.M. on March 22.

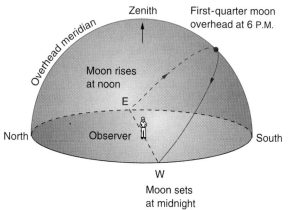

Figure 17.11 The rising and setting of the first-quarter moon.
Diagram illustrates the first-quarter moon rising, on the overhead meridian, and setting during the time of the vernal or autumnal equinox for an observer in the United States. The side of the Moon facing west is illuminated by sunlight.

What is the maximum altitude of any phase of the Moon as observed from the United States? The answer, of course, depends on the latitude of the observer. The closer the observer's latitude is to 28.5°N, the greater the maximum altitude of the Moon. If the Moon is overhead at 28.5°N and the observer's latitude is 28.5°N, then the zenith angle is zero and the Moon's altitude is 90°. The following example illustrates how to determine the maximum altitude of the Moon.

EXAMPLE 17.1 Calculating the Maximum Altitude of the Moon as Observed from a Given Latitude

Calculate the maximum altitude of the full moon as observed from Washington, D.C. (39°N, 77°W).

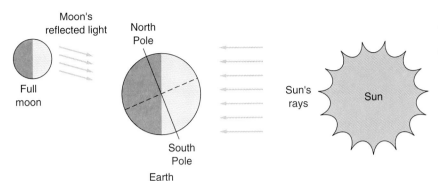

Figure 17.13 Relative positions of the full moon, Earth, and the Sun. This side view of Earth, the full moon, and the Sun during the winter months in the Northern Hemisphere shows the Sun's rays striking the Southern Hemisphere most directly. The Moon's reflected light falls most directly on the Northern Hemisphere.

Solution

Maximum altitude refers to the maximum angle above the horizon or the minimum zenith angle. To be at the minimum zenith angle, the full moon must be as close to 39°N (the observer's latitude) as possible. This is accomplished when the Sun is as far south as possible, since the full moon is 180° from the Sun's position. The Sun's most southern position is latitude 23.5°S on December 21 ± 2 days.

STEP 2

Calculate the zenith angle.

$$\text{zenith angle} = 39°N - 28.5°N = 10.5°$$

STEP 3

Calculate the altitude of the Moon.

$$\begin{aligned}\text{altitude of the moon} &= 90° - \text{zenith angle} \\ &= 90° - 10.5° = 79.5°\end{aligned}$$

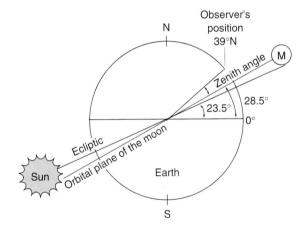

CONFIDENCE QUESTION 17.4

(a) Where is the starting point for naming the phases of the Moon? (b) What is the phase of the Moon that is observed on the overhead meridian at 3 A.M., and at 9 A.M.?

CONFIDENCE QUESTION 17.5

(a) Determine the maximum altitude of the first-quarter moon as observed from Washington, D.C. (39°N, 77°W). (b) During what month can this occur?

STEP 1

Draw a diagram that illustrates the data. The orbital plane of the Moon must be oriented so that the full Moon is at 28.5°N latitude. *Note:* Over one precession cycle of 18.6 y, the Moon's northern latitude is 23.5°N ± 5° or varies between 18.5°N and 28.5°N.

17.5 Solar and Lunar Eclipses

Learning Goal:

To describe and explain solar and lunar eclipses.

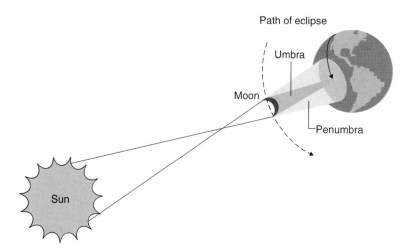

Figure 17.14 A total solar eclipse.
This diagram shows the positions of the Sun, Moon, and Earth during a total solar eclipse. The umbra and penumbra are, respectively, the dark and semidark shadows cast by the Moon on the surface of Earth.

The word **eclipse** means the darkening of the light of one celestial body by another. The Sun provides the light by which we see objects in our solar system. That is, nonluminous objects in our solar system are observed by reflected light from the Sun. Because the light from the Sun falls on objects in the solar system, objects cast shadows that extend away from the Sun. The size and shape of the shadow depend on the size and shape of the object and its distance from the Sun. Earth and the Moon, being spherical bodies, cast conical shadows, as viewed from space.

If we examine the shadow cast by Earth or the Moon, we discover two regions of different degrees of darkness. The darkest and smallest region is known as the **umbra** (Fig. 17.14). An observer located within this region is completely blocked from the Sun during a solar eclipse. The semidark region is called the **penumbra.** An observer positioned in this region can see a portion of the Sun during a solar eclipse.

A **solar eclipse** occurs when the Moon is at or near new phase and is in or near the ecliptic plane. When these two events occur together, the Sun, Moon, and Earth are nearly in a straight line. The Moon's shadow will then fall upon Earth, and the Sun's rays will be hidden from those observers in the shadow zone. A *total eclipse* occurs in the umbra region and a *partial eclipse* in the penumbra region (Figs. 17.14,   17.15, and   17.16).

The length of the Moon's shadow varies as the Moon's distance to the Sun varies. The average length of the Moon's umbra is 375,000 km (233,000 mi), which is slightly less than the mean distance between Earth and the Moon. Because the umbra is shorter in length than the mean distance from Earth to the Moon, an eclipse of the Sun can occur in which the umbra fails to reach Earth. An observer positioned on Earth's surface directly in line with the Moon and Sun sees the Moon's disk pro-

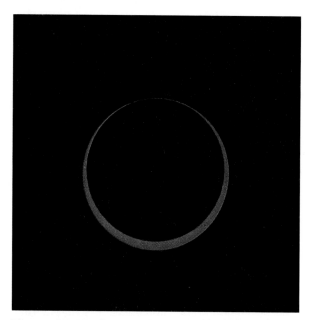

Figure 17.15 An annular eclipse of the Sun.
When the umbra of the Moon's shadow does not reach all the way to the surface of Earth, we observe an annular eclipse of the Sun. This photograph was taken at 1:44 P.M. EST on May 11, 1994, at Concord, New Hampshire.

Figure 17.16 Total solar eclipse showing solar corona.
The solar corona, which can be photographed during a total solar eclipse, is composed of hot gases that extend millions of miles into space. This photo, which was taken on March 7, 1970, at Nantucket, Massachusetts, shows only the brightest inner part of the solar corona.

jected against the Sun, and a bright ring, or annulus, appears outside the dark Moon. This condition is called an **annular eclipse** (Fig. 17.15).

Around the zone of the total (very dark) annular eclipse appears the larger semidark region of the penumbra. The penumbra region may be as large as 9600 km (6000 mi) in diameter at the surface of Earth. The maximum diameter of the umbra at Earth's surface is about 274 km (170 mi). This maximum value can exist only when the Sun is farthest from Earth, which is in early July, and the Moon is at perigee, or at its closest distance to Earth.

The motion of the Moon and Earth are such that the shadow of the Moon moves generally eastward during the time of the eclipse with a speed of 1700 to 2000 km/h depending on latitude. Thus the region of total eclipse does not remain long at any one place. The greatest possible value is about 7.5 min, and the average is about 3 or 4 min.

A **lunar eclipse** occurs when the Moon is at or near full phase and is in or near the ecliptic plane (● Fig. 17.17). The Sun, Earth, and Moon will be positioned in a nearly straight line, with Earth between the Sun and Moon. Thus the shadow formed by Earth conceals the face of the Moon. The average length of Earth's shadow is about 1.4 million km (860,000 mi), and the diameter of the shadow at the Moon's position is great enough to place the Moon in total eclipse for a time slightly greater than 1.5 h. A partial eclipse of the Moon can last as long as 3 h 40 min.

The orbital plane of the Moon is inclined to the ecliptic (the apparent annual path of the Sun) at an angle slightly greater than 5°. Therefore, the path of the Moon crosses the ecliptic at two points as it makes its monthly journey around Earth. The points where the Moon's path crosses the ecliptic are known as *nodes*. The point of crossing going northward is called the *ascending node,* and the point of crossing going southward is called the *descending node* (● Fig. 17.18). A solar or lunar eclipse can occur only at or near the nodal points, because Earth, Moon, and Sun must be in a nearly straight line. This positioning occurs only at or near the points where the Moon crosses the ecliptic.

The orbital plane of the Moon is precessing westward or clockwise if viewed from above the orbital plane. The precession of the Moon's orbit causes the nodal points to move westward along the ecliptic, making one complete cycle in 18.6 y.

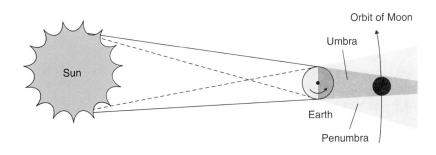

Figure 17.17 A lunar eclipse.
This diagram shows the positions of the Sun, Earth, and Moon during an eclipse of the Moon.

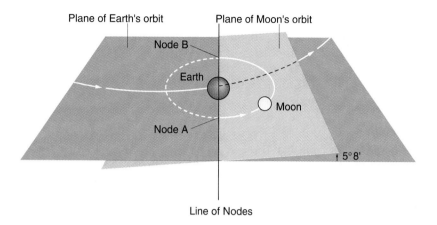

Figure 17.18 The 5°8' angle between the orbital planes of Earth and the Moon. This diagram illustrates the intersection of Earth's and the Moon's orbital planes. The angle between the planes is 5°8'. The Moon passes through the intersection twice during its monthly journey around Earth. The intersecting points are called *nodes*.

Predicting total solar eclipses is complicated, because the line of nodes gradually moves westward and changes the nodes' directions in space. Data concerning the next six total solar eclipses are given in Table 17.2 on page 456.

CONFIDENCE QUESTION 17.6

(a) A lunar eclipse is beginning to occur. Is the Moon entering Earth's shadow from the east or from the west? (b) During a lunar eclipse, which planets can the Moon never block from sight? (c) You are on the surface of the Moon during a total solar eclipse. What is your view of Earth at this time?

17.6 Ocean Tides

Learning Goals:

To define and describe tidal force.

To explain ocean tides.

Anyone who has been to the seashore for a day's visit is aware of the rising and falling of the surface level of the ocean. The alternate rise and fall of the ocean's surface level is called the *tides*.

People related tides to the passage of the Moon in the first century A.D., but all efforts to explain the phenomenon failed until the seventeenth century, when Newton applied his law of universal gravitation to the problem. He related the alternate rise and fall of the ocean's surface level to the motions of Earth, Moon, and Sun. A few of the many factors contributing to the height that the ocean rises and falls at a particular location are:

1. the force of attraction between the Moon and solid Earth.
2. the rotation of Earth on its axis.
3. the position of Earth, Moon, and Sun with respect to one another.
4. the varying distance between Earth and Moon.
5. the inclination of the Moon's orbit.
6. the varying distance between Earth and Sun.
7. the variation in the shape of coastlines and the relief of ocean basins.

There are generally two high and two low tides daily because of the Moon's gravitational attraction and the motion of the Moon and Earth.

An understanding of the two daily tides can be clarified by visualizing what shape Earth and its surface water would take if there were no external gravitational forces and Earth had no motion (●Fig. 17.19a). If Earth did not rotate, there would be no centripetal force. On a nonrotating Earth, with no external gravitational forces, no forces would be exerted on Earth or the surface water, and hence no tides would occur. How, then, do the gravitational force of the Moon and the motion of Earth produce the tides?

The answer can be found, and the reason for two daily tides at a given location on Earth's surface can be explained, if the Moon's mass is considered to be concentrated at a point (Fig. 17.19b). The Sun is also a factor in causing tides, but it is left out of this explanation to simplify the results.

The magnitude and direction of the Moon's gravitational forces acting on Earth at points A, B, C, D, and O are shown. The magnitude is indicated by the length of the arrow drawn to represent the gravitational force. Note that the forces at A, O, and B are all in the same direction toward the Moon, but the magnitudes are not the same. The force at A is greatest because it is closest to the Moon.

Remember that Newton's law of gravitational attraction says the force between two masses is inversely proportional to the square of the distance between the two masses. Thus the force at O is less than the force at A, and the force at B is less than the force at O. Also, forces at C and D have approximately the same magnitude but are slightly less than the force at O. The direction of the forces at C and D are toward the Moon, as shown. There is a net differential force between points A and B that stretches solid Earth. This differential gravitational force is called the **tidal force**.* Varying forces on different parts of Earth due to the attraction of the Moon give rise to other tidal forces. The net effect of all tidal forces causes Earth to round out (bulge) in the direction toward and opposite the Moon. The effect of the tidal forces is much more noticeable on the oceans, since water flows easily over Earth's solid surface. Figure 17.19c shows the tidal forces with their vertical and horizontal components.

Although the horizontal components are very small, the forces acting over a period of hours will produce movement of the water that results in tidal bulges in the oceans. It is important to note that the tidal bulges are due to the horizontal components of the tidal forces. The high tides are not a result of the Moon's gravitational forces lifting the ocean's water away from Earth. The horizontal components cause the water to flow over Earth's surface toward areas nearest the Moon and toward areas opposite the Moon. Thus, the oceans rise higher in these areas and will be correspondingly lower in areas where the water is flowing from.

When the Sun, Earth, and Moon are positioned in a nearly straight line, the gravitational force of the Moon and Sun combine to produce higher high tides and lower low tides than usual.

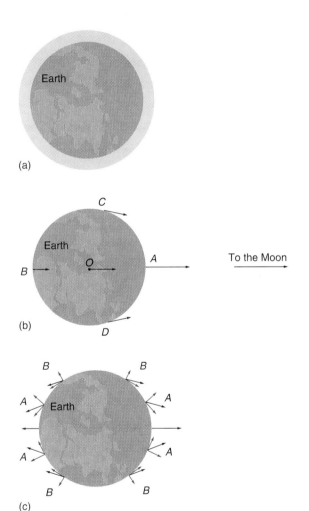

Figure 17.19 Tides on Earth.
(a) The shape of stationary Earth with all water surfaces unaffected by any external gravitational forces. (b) Gravitational forces of the Moon (arrows) acting on Earth, which is considered as having a mass concentrated at a point. (c) The resultant tidal forces (arrows) act on Earth to produce an oblate (elongated at the center) contour. This view of Earth is from above the North Pole. See text for details.

---

*The tidal force—a differential gravitational force—occurs when two separate bodies interact gravitationally, and the mass of body 1 (Earth) produces a difference in the gravitational force from position to position within body 2 (Moon). The force difference tends to stretch or change the shape of body 2, producing *body tides*. There are bulges that rise in a solid body. For example, since the masses that constitute Earth are at varying distances from the masses that make up the Moon, Earth's gravitational forces vary throughout the Moon, producing body tides in the Moon. Likewise the Moon's gravitational force varies throughout Earth producing body and ocean tides.

Table 17.2 Total Solar Eclipses from 1995 through 2002

Date	Total Duration (min)	Location Where Visible
October 24, 1995	2.4	South Asia
March 9, 1997	2.8	Siberia, Arctic
February 26, 1998	4.4	Central America
August 11, 1999	2.6	Central Europe, Central Asia
June 21, 2001	4.9	Southern Africa
December 4, 2002	2.1	South Africa, Australia

That is, the variations between high and low tides are greatest at this time. These tides of greatest variation are called **spring tides.** They occur at the new and full phases of the Moon and they have no relationship to the spring season. When the Moon is at first- or last-quarter phase, the Sun and Moon are 90° with respect to Earth. At these times the tidal forces of the Moon and Sun tend to cancel one another, and there is a minimum difference in the height of the surface of the ocean. In this case the tides are known as **neap tides.**

Note that two spring tides and two neap tides take place each lunar month, because the Moon passes through each of its phases once each month. A spring tide occurs at new moon, a neap tide at first quarter, another spring tide at full moon, and a second neap tide at last quarter (Fig. 17.20).

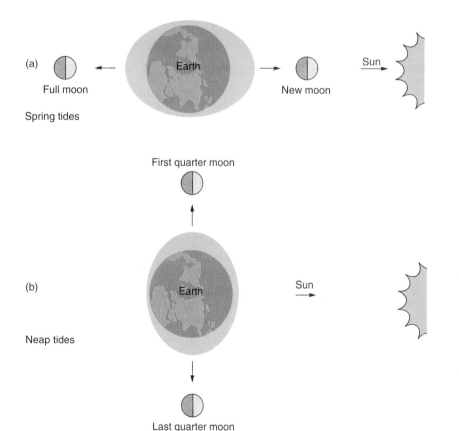

Figure 17.20 Spring and neap tides.
This diagram shows the relative position of Earth, the Moon, and the Sun at the times of spring and neap tides. (a) During spring tides, the Sun and Moon are aligned, which causes more extreme tidal effects. (b) During neap tides, the Sun and Moon are 90° to each other and reduce the tidal forces, so the tides are more moderate.

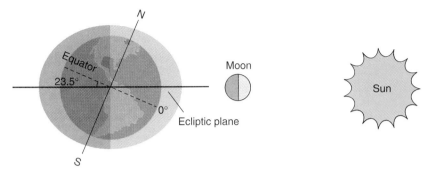

Figure 17.21 Spring tide at the time of summer solstice.
This diagram shows the position of Earth, the new moon, and the Sun at the time of summer solstice. The maximum height of the tidal bulge is at 23.5°N and 23.5°S, because of the Sun is overhead at 23.5°N. If a spring tide were to occur three months later, when the Sun is on the equator, the maximum height of the bulge would be in the equatorial region. (The tidal bulge is exaggerated in the diagram.)

The height of the tide also varies with latitude (Fig. 17.21). The tide is highest at the Moon's overhead position and on the other side of Earth opposite the position of the Moon. The time of high tide does not correspond to the time of the meridian crossing of the Moon. The bulge is always a little ahead (eastward) of the Moon because of Earth's rotation. Because Earth rotates faster than the Moon revolves, Earth carries the tidal bulge forward in the direction it is rotating, which is eastward.

The action of the tides produces a retarding motion of Earth's rotation, slowing it and lengthening the solar day about 0.001 s per century. Because the conservation of angular momentum applies, the decrease of Earth's angular momentum must appear as an increase in the Moon's angular momentum. A measurement of the Moon's orbit shows that the semimajor axis is increasing about 0.5 in./y. Thus, 1 billion years ago the solar day was 2.8 h shorter, and the Moon was 12900 km (8000 mi) closer to Earth.

CONFIDENCE QUESTION 17.7

What role do the horizontal components of the tidal forces play in causing ocean tides?

IMPORTANT TERMS

These important terms are for review. After reading and studying the chapter, you should be able to define and/or explain each of them.

craters
rays
rille
fault
great impact theory
perigee
apogee
sidereal month
synodic month
new moon
waxing moon
waning moon
crescent moon
gibbous moon
first-quarter phase
full moon
last-quarter phase
eclipse
umbra
penumbra
solar eclipse
annular eclipse
lunar eclipse
tidal force
spring tides
neap tides

Questions

17.1 General Physical Properties

1. Which of the following is *not* a general physical property of the Moon's surface?
   (a) mountain ranges
   (b) craters
   (c) plains
   (d) rays
   (e) none of the above
2. The internal structure of the Moon
   (a) consists of a solid mantle rich in silicates.
   (b) has a crust that varies in depth (60 km earthside, 150 km farside).
   (c) includes a small iron-rich core.
   (d) all of the above.
3. What is the numerical value of the Moon's synodic period of (a) revolution and (b) rotation?
4. What is the diameter of the Moon in miles and kilometers?
5. How does the surface gravity of the Moon compare with the surface gravity of Earth?
6. What is the average distance between the Moon and Earth in miles and kilometers?
7. Distinguish between the Moon's surface features rays and rilles.
8. What causes each of the following on the Moon: (a) craters, (b) plains, (c) rays, and (d) rilles?
9. Name the three major components of the Moon's internal structure.
10. An examination of the planets and their moons reveals something unique about the Earth-Moon system. What is it?

17.2 Composition and Origin of the Moon

11. Which of the following statements is false?
    (a) The average density of the Moon is less than Earth's.
    (b) Experimental evidence indicates the oldest Earth and lunar rocks were formed about the same time.
    (c) The Moon samples show a large abundance of iron.
    (d) The percentage abundance of oxygen isotopes found in lunar samples is similar to that of Earth.
12. Which of the following must be considered when formulating a theory concerning the origin of the Moon?
    (a) similar chemical composition of Earth's mantle
    (b) no water found in lunar samples
    (c) lunar rocks showing a small percentage of volatile elements
    (d) Earth and Moon rocks approximately the same age
    (e) all of the above
13. Give the age of the oldest rocks found on the Moon during the Apollo 11 mission.
14. What is the present preferred theory for the origin of the Moon?
15. Distinguish between volatile elements and refractory elements.
16. (a) What is the approximate age of the Moon?
    (b) How does this compare with the age of Earth?
17. How does the average density of the Moon compare with the average density of Earth's mantle?
18. Why can scientists learn more about the early history of our planet by studying rocks from the Moon than by studying rocks from Earth?
19. How do the chemical properties of the Moon's composition compare to Earth's mantle?
20. How does the abundance of iron and water on Earth compare with that found in rock samples taken from the Moon's surface?

17.3 Lunar Motions

21. Which of the following statements is false?
    (a) The Moon rotates and revolves eastward as observed from above the North Pole.
    (b) The orbital plane of the Moon is tilted about 5° to Earth's orbital plane.
    (c) The difference between the sidereal and synodic month is two days.
    (d) The Moon revolves in an elliptical orbit.
    (e) None of the above.
22. The rising of the Moon in the east and setting in the west is due to
    (a) the orbital motion of the Moon.
    (b) the rotational motion of the Moon.
    (c) Earth's rotation.
    (d) none of the above.

23. What is the direction of rotation and revolution of the Moon?
24. What is the period of rotation with respect to the period of revolution?
25. (a) What is the numerical value of the Moon's sidereal period?
    (b) How does the sidereal period compare with the synodic period?
    (c) Explain the difference.
26. What is the numerical value of the angle between the Moon's orbital plane and Earth's orbital plane?
27. Why does the Moon appear to rise in the east and set in the west every day?
28. How often would an observer on the Moon see (a) sunrise and (b) earthrise?

17.4 Phases of the Moon

29. During one month the Moon passes through _____ phases.
    (a) four
    (b) six
    (c) eight
    (d) none of the above
30. The angular difference between the maximum and minimum altitude of the full moon as observed from 35°N is _____ degrees.
    (a) 19
    (b) 26.5
    (c) 57
    (d) 64.5
31. How often do we have a full moon?
32. Determine when each of the following occurs: (a) the setting of the new moon, (b) the rising of the full moon, and (c) the setting of the last-quarter moon.
33. Why does the Moon rise about 50 min later each day?
34. Which phase of the Moon (a) is overhead at 6 P.M., (b) sets at 6 A.M., and (c) rises at midnight?
35. What is the difference between a waxing and a waning moon?
36. Why is the full moon higher in the sky in winter than in summer?

17.5 Solar and Lunar Eclipses

37. An eclipse
    (a) occurs due to the darkening of the light of one celestial body by another.
    (b) requires the interplay of a minimum of three bodies.
    (c) can be total, partial, or annular.
    (d) all of the above.
38. Which of the following is *not* a contributing factor in causing eclipses.
    (a) rotation of Earth about an axis
    (b) the inclination of the Moon's orbit
    (c) the varying distance between Earth and Moon
    (d) the varying distance between Earth and Sun
    (e) none of the above
39. State the positions of the Sun, Moon, and Earth during a (a) solar eclipse and (b) lunar eclipse.
40. Distinguish between umbra and penumbra.
41. What is the period of precession of the Moon's orbital plane? What is the direction of the precession?
42. Why do eclipses (solar and lunar) only take place at or near the nodal points of the Moon's and Earth's orbital planes?
43. What is an annular eclipse?
44. Why doesn't an eclipse occur every time there is a new or full moon?
45. Why do lunar eclipses last much longer than solar eclipses?

17.6 Ocean Tides

46. Tidal bulges in the ocean are
    (a) due to the Moon's gravitational force lifting the ocean water away from solid Earth.
    (b) due to the horizontal components of the tidal forces that cause water to flow over Earth's surface toward areas nearest the Moon and toward areas opposite the Moon.
    (c) mainly due to gravitational forces between the Sun and Earth.
    (d) none of the above.
47. Tidal forces
    (a) are differential gravitational forces.
    (b) cause ocean tides.
    (c) cause solid Earth to stretch.
    (d) are all of the above.
48. What is the origin of tidal forces?
49. What are some factors that contribute to the height of ocean tides?
50. During which phases of the Moon do spring and neap tides occur?
51. Why are there two high and two low tides each day?

Food for Thought

1. An observer on Earth sees the Moon in the waxing crescent phase. At the same time, what phase of Earth will an observer on the Moon see?
2. What must be the relationship with respect to each of the distances between Sun, Moon, and Earth in order to obtain the longest duration of solar eclipse? That is, should the Sun be at its greatest and the Moon at its greatest, or the Sun at its closest and the Moon at its closest, or the Sun at its greatest and the Moon at its closest, or the Sun at its closest and the Moon at its greatest?
3. Why is the surface of the Moon being struck by particles from space to a greater extent than is Earth's surface?
4. An observer on the Moon sees planet Earth. How often would the observer see earthrise?

Exercises

17.3 Lunar Motions

1. How many days are in 12 lunar months (synodic months)?
2. Are there more sidereal or synodic months in 1 y? State the number difference, and explain your answer.

17.4 Phases of the Moon

3. (a) Determine the maximum altitude of the full Moon as observed from Chicago, Ill. (42°N, 88°W). (b) What is the approximate date this can occur?

   Answer: (a) 76.5°, (b) December 21

4. (a) Determine the maximum altitude of the full Moon as observed from Nashville, Tenn. (36°N, 87°W). (b) What is the approximate date it can occur?

5. (a) Determine the maximum altitude of the first-quarter moon as observed from Atlanta, Ga. (34°N, 84°W). (b) What is the approximate date it can take place?

   Answer: (a) 84.5°, (b) March 21

6. (a) Determine the maximum altitude of the last-quarter moon as observed from Denver, Colo. (40°N, 105°W). (b) What is the approximate date it can take place?

7. Consider a person in the United States who sees the first-quarter phase.
   (a) Which side of the Moon is illuminated, east or west?
   (b) What phase does an observer in Australia see at the same time, and which side is illuminated?

   Answer: (a) west, (b) first-quarter phase; west (left) side

8. Consider a person in the United States who sees the last-quarter phase.
   (a) Which side of the Moon is illuminated?
   (b) What phase does an observer in Australia see at the same time, and which side is illuminated?

9. An observer at 28.5°S sees the Moon on his zenith. If the Moon appears on the southern horizon at this time for an observer in the Northern Hemisphere, what is the latitude of this observer?

10. An observer at 40°S sees the Moon in Exercise 9 due north of his or her position at this same time. What is the altitude of the Moon for this observer?

17.5 Solar and Lunar Eclipses

11. Draw a diagram illustrating a total solar eclipse. Include the orbital paths of Earth and Moon, and indicate the approximate time of day the eclipse is taking place.

12. Draw a diagram illustrating a total lunar eclipse. Include the orbital paths of Earth and Moon, and indicate the approximate time of day the eclipse is taking place.

17.6 Ocean Tides

13. A high tide is occurring at Washington, D.C. (39°N, 77°W).
    (a) What other longitude is also experiencing high tide?
    (b) What two longitudes are experiencing low tide?

    Answer: (a) 103°E, (b) 13°E and 167°W

14. A low tide is occurring at Los Angeles (34°N, 118°W).
    (a) What other longitude is also experiencing low tide?
    (b) What two longitudes are experiencing high tide?

Answers to Multiple-Choice Questions

1. e   11. c   21. e   29. c   37. e   46. b
2. d   12. e   22. c   30. c   38. a   47. d

Solutions to Confidence Questions

17.1 (a) The Moon's revolving period and rotation period are the same. Therefore, the same side of the Moon always faces Earth.
   (b) Yes. The Moon makes a complete rotation with respect to the Sun.

17.2 (a) The great impact theory
   (b) The theory is preferred over others because data from the Moon landings provided facts that support this theory rather than competing theories.

17.3 (a) Every 29.5 days (b) On the horizon

17.4 (a) New phase that begins at 12 noon local solar time.
   (b) Waning gibbous at 3 A.M., waning crescent at 9 A.M.

17.5 (a) The first quarter moon must be at its most northern latitude (28.5°N) for an observer at Washington, D.C. (39°N) to see the Moon at minimum zenith angle or maximum altitude.

   zenith angle = 39° − 28.5° = 10.5°
   maximum altitude = 90° − 10.5° = 79.5°

   (b) The first-quarter moon is the Moon 90° east of the Sun. Therefore, the Sun must be at the equator (0°) for the Moon to be at 28.5°N. This occurs on March 20 or 21.

17.6 (a) West; Earth rotates and revolves counterclockwise or eastward. The Moon also revolves eastward around Earth and enters Earth's shadow from the west. See Fig. 17.17.
   (b) The Moon can never occult Mercury or Venus because their orbits are smaller than Earth's.
   (c) See Fig. 17.14 for the answer. *Note:* Earth's atmosphere is a good reflector of sunlight.

17.7 The horizontal components of the tidal forces move the ocean water, which results in tidal bulges.

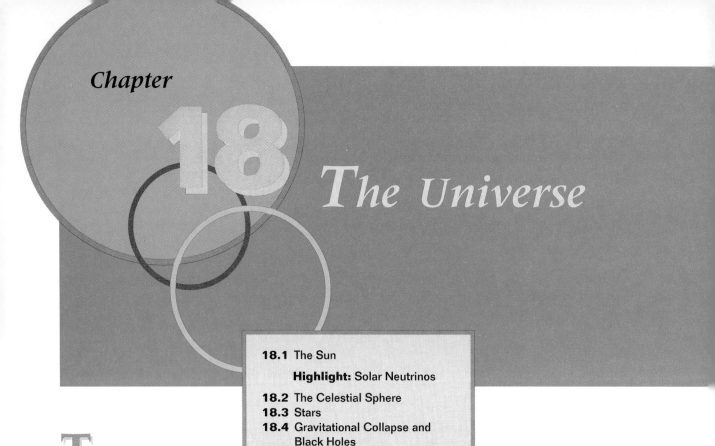

Chapter 18

The Universe

18.1 The Sun

Highlight: Solar Neutrinos

18.2 The Celestial Sphere
18.3 Stars
18.4 Gravitational Collapse and Black Holes
18.5 Galaxies
18.6 Quasars
18.7 Cosmology

Highlight: Other Cosmologies

The study of the stars is the oldest science. Thousands of years ago under the clear desert skies of the Near East, people watched the stars in awed wonder. The earliest scientists plotted the positions and brightness of the stars as Earth went through its calendar of seasons. The Sun and Moon were worshipped as gods, and the days of the week were named after the Sun, Moon, and visible planets, as discussed in Section 16.6. Yet it is only recently that we have understood the most basic nature of stars.

Sixty years ago not even Einstein knew what made the stars shine. Today, we know that all stars go through a cycle of stages. Stars are born, they radiate energy, and then they expand, contract, possibly explode, and eventually die out. Our knowledge of how all this happens has been made possible through our study of the atomic nucleus and by applying the laws of science from many diverse fields.

Our knowledge of the universe is also growing. We are learning more and more of its secrets, and we are beginning to understand how our present civilization fits into an overall scheme in the harmony of the universe. We now have an imperfect theory of evolution—a theory concerned not so much with the origin of humans but with the origin of elements of our solar system. We can now begin to understand how carbon, nitrogen, and oxygen atoms were produced in the stars. It is these atoms that were necessary to produce life itself. As these theories evolve, new mysteries appear, but the continued search for truth is invigorating.

The stars and galaxies of our universe give off different kinds of electromagnetic radiation. This radiation was discussed in Chapter 6 and includes radio waves, microwaves, infrared, visible, ultraviolet, X-rays, and gamma rays. Up until about 50 years ago we looked only at the visible light given off by the stars. With the advent of radio telescopes, quasars and pulsars were discovered in 1960 and 1968, respectively. Most of the other regions of the electromagnetic spectrum are absorbed by our atmosphere. Satellites and balloons going above our atmosphere have enabled us to study other forms of radiation emitted by stars, and new developments are occurring frequently.

What is the universe? What are its composition, structure, size, and age? What are its primary building components, and what is the composition of these building components? What are stars? What are galaxies? What is the history of the universe, and what is its future? These are some of the topics introduced and discussed in this chapter.

18.1 The Sun

Learning Goals:

To describe the Sun's structure and list its physical properties.

To give the fusion reactions for changing hydrogen to helium with the liberation of radiant energy.

The **Sun** is a star—a self-luminous sphere of gas held together by its own gravity and energized by nuclear reactions in its interior. Viewed from Earth, at its mean distance, the Sun's angular diameter is approximately 0.5°. This ordinary star of the Milky Way galaxy is the most important object in the solar system to us because it supplies heat, light, and other radiation for the processes of life on Earth. The Sun has a diameter of 1.4×10^6 km, a mass of 2.0×10^{30} kg, rotates on its axis every 25 Earth days, and moves through space with its family of planets at a speed of approximately 250 km/s (150 mi/s) around the galactic center, completing one orbit in about 200 million years. The Sun's equator is inclined about 7° from the orbital plane of Earth. The rotational period of the Sun given above is the period at its equator. The period of rotation is longer at higher latitudes.

A cross-sectional view of the Sun is shown in Fig. 18.1. The Sun's temperature is believed to be about 15 million kelvins at its center, and decreases radially outward to the visible surface of the Sun, which is called the **photosphere**. The temperature of the photosphere has been measured at about 6000 K.

The interior of the Sun is so hot that individual atoms do not exist, because high-speed collisions continually knock the electrons loose from the atomic nuclei. There is no way to measure the interior temperatures, but computer models display a core temperature of about 15 million kelvins, and a density of 150 g/cm³. This is more than 13 times the density of lead. The interior is composed of

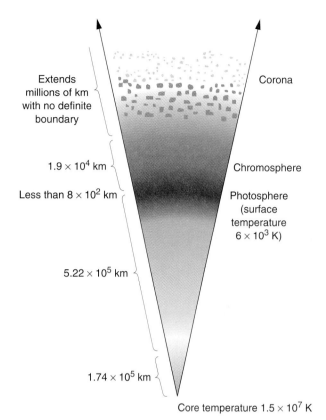

Figure 18.1 A radial cross-section of the Sun. The boundary between layers is not sharply defined.

high-speed nuclei and electrons moving about more or less independently, similar to a gas. A gas, you will recall, is composed of rapidly moving atoms or molecules. The Sun has very rapidly moving positively charged nuclei and negatively charged electrons. These high-speed charged nuclei and electrons form a fourth phase of matter called a **plasma**. The Sun is a plasma with an average density of 1.4 g/cm³. Note that this density is 1.4 times as great as water!

The Sun is composed (by number of atoms) of about 91% hydrogen, 8.7% helium, and 0.1% heavier elements, the most abundant being carbon, oxygen, nitrogen, and neon. We believe that the interior of the Sun has a similar composition, although we have no good experimental evidence for this belief. Note that when we discuss elements in the interiors of stars, we are really speaking about the nuclei of the elements, because the electrons are stripped off the nuclei and are speeding around independent of them.

The photosphere, viewed through a telescope with appropriate filters, has a granular appearance. The granules are hot spots (about 100 K higher than the surrounding surface) that are a few hundred miles in diameter and last only a few minutes. Extending more than 19,000 km (12,000 mi) above the photosphere lies the **chromosphere** (meaning *color* and *sphere*), which is composed mainly of hydrogen. The temperature of the chromosphere, which is possibly heated by energy from shock waves, averages about 5×10^4 K. The chromosphere can be seen as a thin, red crescent for only the few seconds during which the photosphere has been concealed from view during a solar eclipse. At the time of a total solar eclipse, the chromosphere and photosphere are hidden by the Moon, and the *corona* (outer solar atmosphere) can be seen as a white halo (Fig. 17.16). The corona receives energy possibly by shock waves, and its temperature exceeds 1×10^6 K. This extreme temperature is sufficient to give protons, electrons, and ions enough energy to escape the Sun's atmosphere. The particles are projected into space, giving rise to a radial flow, which is controlled by the Sun's magnetic field. This outward flow of charged particles is called the **solar wind,** and wind speeds exceed 400 km/s (893,000 mi/h) as the solar wind passes Earth's orbit on its way through the solar system. Measurements made by the *Voyager* spacecraft have confirmed that the solar wind and the accompanying magnetic field extend outward to at least 50 AU from the Sun's surface. This volume of space surrounding the Sun in which the winds and magnetic field exist is called the *heliosphere*.

A very distinct feature of the Sun's surface is the periodic occurrence of sunspots. **Sunspots** are patches (some thousands of miles in diameter) of cooler material on the surface of the Sun. Each has a central darker part, called the *umbra*, and a lighter border, called the *penumbra*. Figure 18.2 is a photograph of the whole solar disk and an enlargement of a very large sunspot. These large sunspots last for several weeks before disappearing from view.

The number of sunspots appearing on the Sun varies over a 22-year period. A period begins with the appearance of a few spots or groups near 30° latitude in both hemispheres on the Sun. The number of spots increases, with a maximum generally between 100 and 200 occurring about 4 y later near

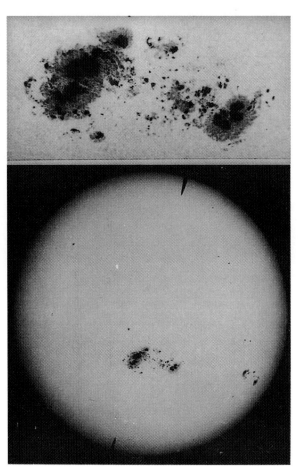

Figure 18.2 The very large sunspot group of March–April 1947.
The photo at the top is an enlargement of the large group just below the center of the Sun in the bottom photo.

an average latitude of 15°. As time passes, the number of spots decreases until, in about 4 more years, only a few are observed near 8° latitude.

About this same time a few spots begin to appear at 30°, and the number begins to increase again, indicating an 11-y cycle. But here is a notable difference. All sunspots have an associated magnetic field that is different in appearance from the previous ones. Studies indicate that if sunspots had a north magnetic pole during the initial increase and decrease, the next 11-y cycle will show a south magnetic pole associated with the sunspots.

The 22-y or 11-y sunspot cycle has been observed continuously since about 1715. Galileo saw sunspots through a telescope in 1610, and they were possibly observed even before that without telescopes. There were reports of their observation after Galileo, but from 1645 to 1715 hardly any sunspots were reported. During this 70-y period, very few northern lights were seen in northern Europe, where a "Little Ice Age" occurred. The evidence seems to indicate that the Sun's activity is not always as regular as it appears to be.

Another distinct feature of the Sun's surface is the appearance of *prominences* that seem to be connected with violent storms in the chromosphere. They are very evident to the astronomer during solar eclipses, at which time they appear as great eruptions at the edge of the Sun. They are red, have an associated magnetic field, and may appear as streamers, loops, spiral or twisted columns, fountains, curtains, or haystacks. They extend outward for thousands of miles from the surface, occasionally reaching a height of 1.6 million km (1.0 million mi). An extraordinarily large prominence is shown in Fig. 18.3.

The chief property of the Sun, of course, is the fact that it radiates energy. However, it was not until 1938 that scientists came to understand the source of the radiation. We now know that the Sun radiates energy because of nuclear fusion reactions inside its core (see Chapter 10).

The Sun's core is made up mostly of hydrogen nuclei or protons, or in nuclear notation, $_1^1H$. These protons are moving at very high speeds and they occasionally fuse together, as shown in Fig. 18.4. The products of this nuclear reaction are a deuteron (proton and neutron together, or $_1^2H$), a positive electron (or positron, designated $_{+1}^0e$), and a neutrino (designated by the symbol ν). A **neutrino** is an elementary particle that has no charge, has no mass (or very little), travels at or near the speed of light, and hardly ever interacts with other particles such as electrons or protons. (See the chapter Highlight.) The first reaction (shown below) is fairly rare, and for this reason the reaction is relatively slow. In fact, scientists believe that the Sun has been radiating energy for about 5 billion years.

Once the deuteron is formed, it quickly reacts with a proton to form a helium-3 nucleus ($_2^3He$) and gamma rays, designated by the symbol γ. Next, two helium-3 nuclei fuse to form the more common helium-4 nucleus and two protons.

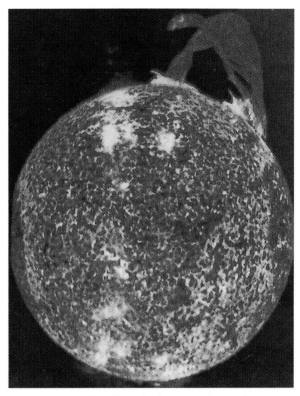

Figure 18.3 The Sun during a major solar eruption.
This ultraviolet photograph showing several flares and a large prominence was taken from the *Skylab* orbiting space station.

In each of these three fusion reactions, energy is liberated by the conversion of mass. These three reactions are collectively called the **proton-proton chain** and can be written as

$$_1^1H + _1^1H \rightarrow _1^2H + _{+1}^0e + \nu + \text{energy} \quad \text{(slow)}$$

$$_1^2H + _1^1H \rightarrow _2^3He + \gamma + \text{energy} \quad \text{(fast)}$$

$$_2^3He + _2^3He \rightarrow _2^4He + _1^1H + _1^1H + \text{energy} \quad \text{(fast)}$$

If we multiply the first two reactions by 2 and add both sides, we get the net reaction, which is

$$4\,_1^1H \rightarrow _2^4He + 2(_{+1}^0e + \gamma + \nu) + \text{energy}$$

In the net reaction four protons or hydrogen nuclei react to form a helium nucleus, two positive electrons (positrons), two high-energy gamma rays, two neutrinos, and a great deal of energy. The energy factor simply means that the particles on the right possess more kinetic and radiant energy

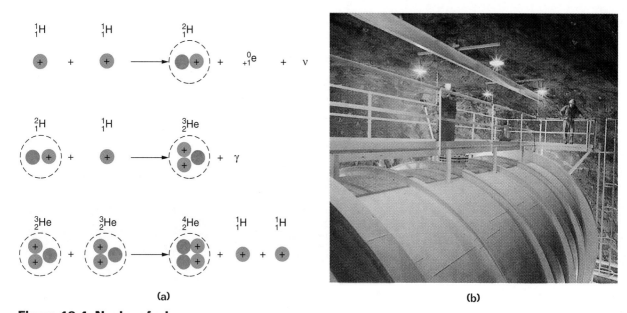

Figure 18.4 Nuclear fusion.
(a) The reactions that make up the proton-proton chain. See text for description. (b) The Brookhaven solar neutrino experiment apparatus. (See the chapter Highlight.)

than the particles on the left. In our reaction we have converted mass into energy in conformity with Einstein's equation $E = mc^2$.

Every second in the Sun's interior about 6.0×10^{11} kg of hydrogen are being converted into helium and energy. Even at this rate we expect the Sun to radiate energy from hydrogen fusion for about another five billion years.

CONFIDENCE QUESTION 18.1

Give a brief description of the Sun's physical properties.

18.2 The Celestial Sphere

Learning Goals:

To describe the celestial sphere and explain its functions.

To list some familiar constellations and asterisms.

A view of the stars on a clear night makes a deep impression. The stars appear as bright points of light on a huge dome overhead. As the time of night passes, the dome seems to turn westward as part of a great sphere, with the observer at the center. The apparent motion of the stars is due to the eastward rotation of Earth. The stars all appear to be mounted on a very large sphere with Earth at the center.

This huge, apparently moving, imaginary sphere has been named the **celestial sphere,** and the way it appears depends on the observer's position on Earth. An observer positioned at 90°N (the North Pole) would see Polaris, the north star, directly overhead. From this latitude all stars on the celestial sphere north of the celestial equator appear to move in concentric circles about the north star, never going below the horizon (Fig. 18.5); they never set. An observer located at 40°N latitude would observe the north star 40° above the northern horizon, and all stars within 40° of the north star would appear to move in concentric circles, never going below the horizon (Fig. 18.6). Detailed observations reveal the celestial sphere to rotate (apparently) about an axis that is an extension of Earth's polar axis, with the celestial equator lying in the same plane as Earth's equator.

The position of a star or other object beyond our solar system is determined with the assignment of three space coordinates. The first and sec-

HIGHLIGHT

Solar Neutrinos

We now have a good understanding of the basic concept of how a star radiates energy. A star radiates energy because of nuclear fusion reactions inside its core. But do we have any direct evidence of this reaction? To get direct evidence, scientists are conducting several solar neutrino experiments in an effort to detect neutrinos from the Sun's core.

One such experiment is being conducted by Raymond Davis, Jr., of the Brookhaven National Laboratory. The data collected thus far by Davis do not agree with theoretical calculations. He is finding fewer neutrinos than calculated by a factor of 3 to 1.

In his experiment, Davis uses 100,000 gal of fluid tetrachloroethylene (C_2Cl_4) in a tank located deep within a South Dakota gold mine to detect the neutrinos. When a neutrino interacts with a chlorine atom (only high-energy neutrinos are able to do so), the atom is transformed into an atom of argon that can be removed from the tank by bubbling helium gas through the fluid. (See Fig. 18.4b.)

Davis plans to increase the sensitivity of his equipment to detect neutrinos by using the element gallium in place of chlorine. With gallium, low-energy as well as high-energy neutrinos will be detected. Perhaps the new experimental results will agree more closely with calculated values.

The major difference between photons and neutrinos is that photons react much more readily with matter than do neutrinos. In fact, a photon produced at the center of the Sun will interact countless times and take possibly millions of years before it finally reaches the Sun's surface. A neutrino, on the other hand, will zip right through the Sun and Earth and most neutrino-detection devices without interacting. Of course, occasionally a neutrino will interact with a detector, and we can say that it has been detected. In solar neutrino experiments, only a few neutrinos are detected in a month's time.

So far, the results of the solar neutrino experiments have been perplexing. They indicate that fewer neutrinos are being detected than should be according to current theory. Either our understanding of neutrinos has to be improved or some unsolved problems remain with regard to our understanding of the structure and dynamics of the interior of a star.

---

ond are *declination* and *right ascension,* which are angular coordinates representing the direction of the star with respect to the Sun. The third coordinate is *distance,* which determines the star's linear distance from the Sun (Fig. 18.7).

Declination (DEC) is the angular measure in degrees north or south of the celestial equator. It has a minimum value of zero at the celestial equator and increases to a maximum of 90° north and 90° south. All angles measured north of the equator have (+) values, and all angles measured south of the equator have (−) values. For example, the star in Fig. 18.7 has a declination of +37°.

Right ascension (RA) is the angular measure in hours, with the hours divided into minutes and seconds. Right ascension begins with 0 h at the celestial prime meridian and continues eastward to a maximum value of 24 h, which coincides with the starting point. The **celestial prime meridian** is an imaginary half-circle running from the North Celestial Pole to the South Celestial Pole and crossing perpendicular to the celestial equator at the point of the vernal equinox.

The distance coordinate is usually measured in astronomical units, in parsecs, or in light-years. We have defined an **astronomical unit** (AU) as the mean distance of Earth from the Sun, which is 1.5×10^8 km (9.3×10^7 mi) as measured by radar. A **light-year** (ly) is the distance traveled by light in one year. One light-year equals approximately 9.5×10^{12} km (6×10^{12} mi) calculated by multiplying the speed of light by the number of seconds in one year. One **parsec** (pc) is defined as the distance to a star when the star exhibits a parallax of one second of arc (Fig. 18.8). One parsec equals 3.26 ly, or 206,265 AU:

$$1 \text{ pc} = 3.26 \text{ ly}$$
$$= 2.06 \times 10^5 \text{ AU}$$

The star in Fig. 18.8, observed from two positions, appears to move against the background of

Figure 18.5 Star trails.
This time-exposure photograph shows star trails at the South Celestial Pole.

more distant stars. This apparent motion is, as you know, called *parallax*. The angle p measures the parallax in seconds of arc. The definition of a parsec provides an easy method for determining the distance to a celestial object, because merely taking the reciprocal of the angle p, measured in seconds, gives the distance in parsecs. That is,

$$d = \frac{1}{p} \qquad (18.1)$$

where d = distance in parsecs,

p = parallax angle in seconds of arc.

EXAMPLE 18.1 Calculating the Distance (in parsecs) to a Star

What is the distance to the star Proxima Centauri if the annual parallax is 0.762 second of arc?

Solution

$$d = \frac{1}{p} = \frac{1}{0.762} = 1.31 \text{ pc}$$

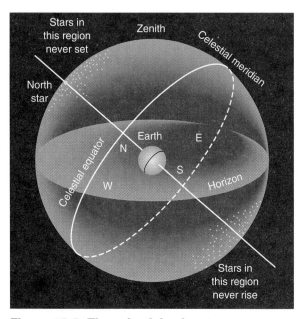

Figure 18.6 The celestial sphere.
This drawing illustrates the celestial sphere as seen by an observer on Earth at a latitude of 40°N.

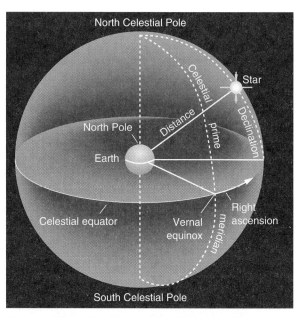

Figure 18.7 The three celestial coordinates.
This drawing illustrates the three celestial coordinates: declination (DEC), right ascension (RA), and distance.

Prominent groups of stars in the celestial sky appear to an Earth observer as distinct patterns. These groups, called **constellations,** have names that can be traced back to the early Babylonian and Greek civilizations. Although the constellations have no physical significance, today's astronomers find them useful in referring to certain areas of the sky. In 1927 astronomers set specified boundaries for the 88 constellations so as to encompass the complete celestial sphere.

We are all aware of the apparent daily motion of the Sun across the sky, and when the Moon is visible, we are aware of its apparent motion. The constellations also appear to move across the sky from east to west, if one observes the stars for an hour or two. Their daily motion is due to the eastward rotation of Earth on its axis. The constellations also have an annual motion resulting from Earth's motion around the Sun. We observe the constellations Pisces, Aquarius, and Capricornus in the autumn night sky. In the winter months Orion (the Hunter) is seen, along with Gemini, Taurus, and Aries. Sagittarius (the Archer) is a summer constellation. For other constellations see Appendix XI. Some other familiar constellations are Andromeda, Cassiopeia, Cygnus, Ursa Major, and Ursa Minor.

Some familiar star groups also are part of a constellation or part of different constellations. These groups are called *asterisms*. The Big Dipper, which is part of Ursa Major, is an example. See Fig. 18.9. In terms of brightness, six of the stars in the Big Dipper are second magnitude (Section 18.3) and the other is third magnitude. Another example is the Summer Triangle, which is formed by very bright (first-magnitude) stars Altair, Deneb, and Vega. These stars are in three different constellations: Altair is in Aquila (the Eagle), Deneb is in Cygnus (the Swan), and Vega is in Lyra (the Lyre).

The *zodiac* is a section (actually a volume) of the sky extending around the ecliptic 8° above and 8° below the ecliptic plane (Fig. 18.10). The zodiac is divided into 12 equal sections, each 30° wide and 16° high. Each section has its apex at the Sun and extends outward to infinity. The boundaries of the zodiac were specified such that the Sun, Moon, visible planets, and most of the asteroids travel within its limits. Occasionally, however, the planets Pluto and Venus are outside the boundaries of the zodiac.

CONFIDENCE QUESTION 18.2

At a given latitude, how many degrees can a star be from the celestial pole and be above the horizon at all times—that is, be at a position so that, to an observer, the star never sets?

18.3 Stars

Learning Goals:

To identify and explain the different classes of stars.

To explain the Hertzsprung-Russell (H-R) diagram.

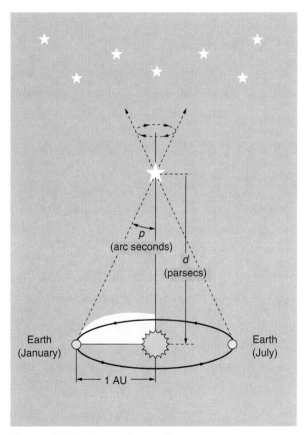

Figure 18.8 Annual parallax of a star.
The shaded angle represents the annual parallax of the star, measured in arc seconds. By definition, when angle p is equal to 1 arc second, the distance d to the star is equal to 1 pc. The basic relationship can be written $d = 1/p$.

Figure 18.9 Ursa Major and Ursa Minor.
The Big Dipper and the Little Dipper make up parts of the constellations Ursa Major and Ursa Minor, respectively. Therefore they are considered asterisms rather than constellations. These constellations can be seen throughout the year from the United States.

tude, which are stars barely visible to the unaided eye. About 6000 stars are visible to the unaided eye.

A modified version of Hipparchus' scale is used today. When a comparison was made between a first-magnitude star and a sixth-magnitude star, the brighter first-magnitude star gave off about 100 times as much radiant energy. From this observation a definition of the magnitude scale was made in which each magnitude difference is equal to the fifth root of 100. This definition can be written as

$$\text{magnitude difference} = \sqrt[5]{100} = 2.512$$

For example, a first-magnitude star is 2.512 times as bright as a second-magnitude star, and 2.512 × 2.512 times as bright as a third-magnitude one, and so on. Note that in giving the apparent magnitude, the greater the negative number, the brighter the star, and the greater the positive number, the dimmer the star. On this scale the brightest object in the sky, our Sun, has a magnitude of −26.7; the full moon a magnitude of −12.7; the planet Venus a magnitude of −4.2; and the brightest star, Sirius, which is 8.7 ly distant, has an apparent magnitude

Hipparchus of Nicaea, a Greek astronomer and mathematician, was antiquity's greatest known observer of the stars. He measured the celestial latitude and longitude of more than 800 stars and compiled the first star catalog, which was completed in 129 B.C. He assigned the stars, with respect to their brightness, to six magnitudes. The *apparent brightness*, or *magnitude*, of a star or other celestial object is its brightness as observed from Earth. The brightest stars were listed as stars of the first magnitude, those not quite as bright as second magnitude, the next less bright as third magnitude, and so on, down to the sixth magni-

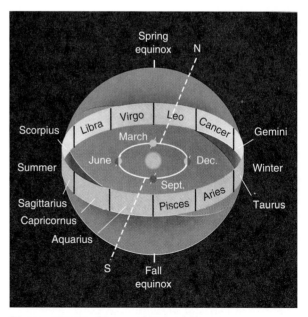

Figure 18.10 Signs of the zodiac.
The drawing illustrates the boundaries of the zodiacal constellations. Each of the 12 sections of the zodiac is 30° wide and 16° high, or 8° above and below the ecliptic plane.

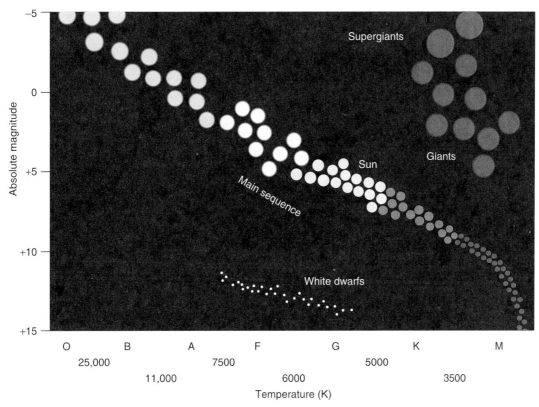

Figure 18.11 Hertzsprung-Russell (H-R) diagram.
This diagram shows the absolute magnitude, spectral class, temperature, and color of various classes of stars. Note that the temperature increases from right to left. See the text for details.

of −1.43. Sirius is the brightest star that can be seen from Earth, except for the Sun, that can be seen by an observer in the United States. Alpha Centauri A (magnitude −0.01) and its close companion Alpha Centauri B (magnitude 1.33) revolve around each other. To the unaided eye, they appear as a single star. Close to Alpha Centauri A and B is a faint red dwarf star called Proxima Centauri (magnitude 11), which is slightly closer to Earth. The three stars, which are 4.3 ly away, are the closest stars to Earth. They are located near 14 h right ascension and −61° declination.

The energy output of a star is measured by its absolute magnitude. *Absolute magnitude* is defined as the apparent magnitude a star would have if it were placed 10 pc (32.6 ly) from Earth. The absolute magnitude of the Sun is about 4×10^{26} J/s, which is measured by determining the amount of energy falling on Earth's surface and knowing the distance from Earth to the Sun. If the annual parallax (from which distance is calculated) and the apparent brightness of a star can be measured, the absolute magnitude can be calculated.

When the absolute magnitudes or brightnesses of stars are plotted against their surface temperatures or colors, we get an **H-R diagram,** named after Ejnar Hertzsprung, a Danish astronomer, and Henry Russell, an American astronomer. An H-R diagram for stars is shown in Fig. 18.11. Note that the temperature axis is reversed. That is, the temperature increases to the left instead of to the right.

Most stars on an H-R diagram get brighter as they get hotter. These stars form the **main sequence,** a narrow band going from upper left to lower right. Stars above the main sequence that are cool and yet very bright must be unusually large to be so bright. So they are called **red giants.**

Stars below the main sequence that are hot yet very dim must be very small, so they are called **white dwarfs.**

The spectra (Section 7.2) of stars vary considerably, and this variation is mostly a function of the temperature in the stars' outer layers. The pattern of the absorption lines in the spectrum can be used to determine a star's temperature. Stars are placed in seven different spectral classes that range from type O to type M and indicate a temperature range of 50,000 to 2000 K. The spectral classes are shown on the H-R diagram in Fig. 18.11.

Spectral type O represents stars whose temperatures are greater than 25,000 K. Spectral types B and A represent hot stars whose temperatures range from 25,000 to 7500 K. Types O, B, and A stars are blue. Spectral type F stars have a temperature range of approximately 7500 to 6000 K and appear blue-white. Spectral type G stars have a temperature range of 6000 to 5000 K and appear white to orange. The Sun is a type G star. Type K stars have a temperature range of 5000 to 3500 K and appear orange to red. Type M stars have temperatures less than 3500 K and appear red. The majority of known stars are small, cool, red, type M stars called *red dwarfs*.

Stars are composed of plasma having a chemical composition of mostly hydrogen, with some helium and a very small percentage of other elements. Their surface temperatures range from about 50,000 to 2000 K. They vary in mass and size. Most have a mass of between 0.1 and 5 solar masses. They range in size from white dwarfs, which are about 13,000 km (8000 mi) in diameter, to supergiants such as Antares, Betelgeuse, and Rigel, which are millions of kilometers in diameter.

Most stars are part of multiple systems. A *binary star* system consists of two stars orbiting each other. Systems with three or more stars also occur, but they are not nearly as common as binary stars.

Many stars in the sky are observed to vary in brightness over a period of time. The first one noted was Delta Cephei in the constellation Cepheus. Delta Cephei varies in brightness with a period of approximately 5.4 days. From its least bright magnitude of 4.3, it gradually becomes brighter for about 2 days and attains a magnitude of 3.6, after which it decreases in brightness until it reaches its minimum in another 5.4 days. The cycle begins again and goes through the same sequence. Presently over 500 known stars vary in magnitude with a fixed period of between 1 and 50 days, and they are known as **cepheid variables,** after Delta Cephei.

The importance of the cepheid variables lies in the fact that a definite relationship exists between the period and the average absolute brightness. Generally, the longer the period, the brighter the cepheid variable. By measuring the period, astronomers can calculate absolute brightness. Then distance can be computed from absolute brightness. Thus astronomers can calculate the distance to any galactic system in which a cepheid variable can be detected.

The period-luminosity relationship for cepheid-variable stars was discovered in 1912 by an American astronomer, Henrietta Swan Leavitt (1868–1921). This relationship provided Edwin P. Hubble (1889–1953), an American astronomer at the Mount Wilson Observatory in California, with the knowledge to show that some observed white patches of light (called *nebulae*, plural for *nebula*, the Latin word for cloud) were actually galaxies beyond the Milky Way. Reflection nebulae are clouds of interstellar gas, dust, or both that shine by reflected light from stars.

The distance to more remote galaxies is calculated by means of the red shift in the galaxy's spectrum. When a photograph is taken of an excited element in the gaseous phase by a spectrograph located in an Earth-based laboratory, a normal-line emission spectrum is obtained. When a similar spectrograph is taken of a galaxy containing this same element, the spectrum may show a displacement of the normal lines toward the red end of the spectrum (longer wavelengths) or toward the blue end of the spectrum (shorter wavelengths). The displacement is a lengthening of the wavelength due to the expansion of the universe. This shift in spectral lines toward longer wavelengths is referred to as the *red shift*, because the shift is toward the red end of the electromagnetic spectrum. When a galaxy is approaching Earth, the lines of the spectrum shift toward the blue end of the spectrum, indicating a shortening of wavelength. When the galaxy is receding from Earth, the lines shift to the red end of the spectrum, indicating a lengthening of the wavelength. See Fig. 18.12. More information concerning the red shift will be given in Section 18.7.

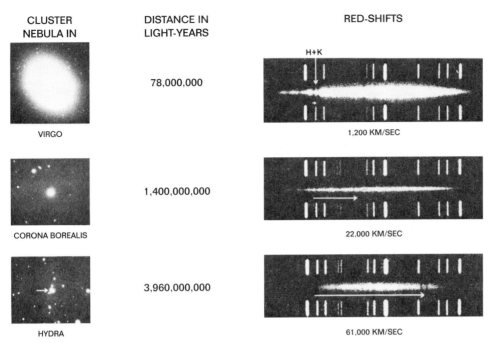

Figure 18.12 Three elliptical galaxies and their respective red shifts. On the left are three individual elliptical galaxies. From top to bottom they show the galaxies at increasing distance from the observer. On the right, the spectrum (the broad white band) of each galaxy is shown between an upper and lower comparison spectrum. The H and K lines of ionized calcium are the two dark vertical lines in the galaxy's spectrum. The arrows indicate the shift in the calcium H and K lines. The red shifts are expressed as velocities.

The displacement for two lines in the calcium spectrum, labeled H and K, for three different galaxies is shown in Fig. 18.12. A small shift of the lines means a low recessional velocity, and so on. The interesting point concerning the red shift is a correlation of velocity and the apparent brightness of the galaxy. The fainter the galaxy, the greater the velocity. Generally speaking, the farther away a galaxy is located, the less its apparent brightness. Thus, a velocity and magnitude correlation yields a velocity and distance correlation. This provides us with a means for measuring great distances that are far beyond the limits of cepheid-variable observation.

Many stars appear dim and insignificant but suddenly increase in brightness in a matter of hours by a factor of 100 to 1 million. A star undergoing such a drastic change in brightness is called a nova, or "new" star. A *nova* is the result of an explosion on the surface of a white dwarf star, caused by matter falling onto its surface from the atmosphere of a larger binary companion. A nova is not a new star, but a faint white dwarf that increases temporarily in brightness.

Occasionally, a star explodes and throws off large amounts of material that may be so great that the star is destroyed. Such a gigantic explosion is known as a **supernova**. Only three supernovae have been observed in our galaxy. One of the most celebrated is the Crab Nebula in the constellation Taurus. This nebula is expanding at the rate of approximately 113 million km per day. Because we know the average angular radius and the expansion rate, the original time of the explosion can be calculated. The result yields about 900 y, which agrees closely with Chinese and Japanese records that report the appearance of a bright new star in the constellation Taurus in A.D. 1054.

On February 24, 1987, a supernova was observed in the Large Magellanic Cloud (LMC). It is

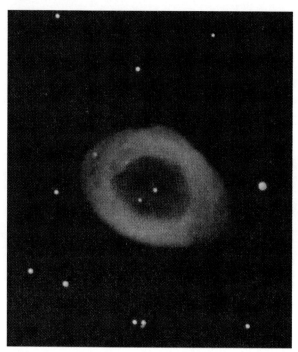

Figure 18.13 Planetary nebula.
This is the Ring nebula in the constellation Lyra, which is located about 5000 ly from Earth. The nebula is about 0.65 ly in diameter, has a shell structure, but appears as a ring because the layers of glowing matter are very thin and are visible only at the edges.

Figure 18.14 Great Nebula in Orion.
Considered the brightest of the diffuse nebulae, this one is located near the middle star in Orion's sword (Appendix XI) and is a gaseous and dusty nebula.

the first supernova observed in this galaxy, which is the closest galaxy (170,000 ly away) to our Milky Way. The supernova was designated 1987A, because it was the first one to be observed in 1987.

The discovery of supernova 1987A is important to astronomers because they are observing a stellar explosion from the beginning, and they will be able to observe and examine the debris coming from the explosion and the remains of the star after the debris has cleared. Thus it will provide a history of a stellar explosion and supply valuable information concerning many branches of astrophysics.

In addition to the novae and supernovae, there are the planetary nebulae, which possess a large, slowly expanding ringlike envelope, as shown in Fig. 18.13. These nebulae, as observed with a telescope, appear greenish. The Orion nebula shown in Fig. 18.14, one of the brightest in the night sky, is known as an emission nebula. This type of nebula is more irregular and turbulent than the planetary nebulae. An *emission nebula* is a glowing cloud of interstellar gas. The gas glows due to the presence of neighboring stars that ionize the gas. The Orion nebula is over 25 ly in diameter and contains hundreds of stars. It is in a dense core of material found within a molecular cloud such as the Orion nebula that the formation of a star begins.

Astronomers know that stars are born, radiate energy, expand, possibly explode, and then die. This is their general life cycle. However, the exact details depend on a star's initial composition (the percentage amounts of hydrogen, helium, and heavier elements) and on its mass. The greater the mass of a star, the faster it moves through its life cycle.

The general evolution of a star with a mass typical of our Sun is shown in Fig. 18.15. The birth of a star, according to accepted theory, begins with the condensation of interstellar material (mostly hydrogen) in nebulae, because of the gravitational attraction between the interstellar material, radiation pressure from nearby stars, and supernova shock waves. The size of the star formed depends on the total mass available, which in turn

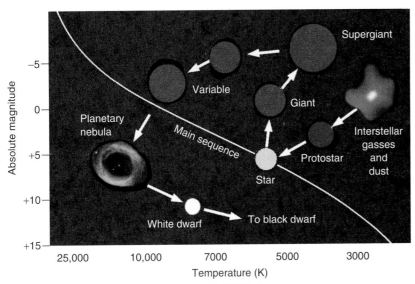

Figure 18.15 Hertzsprung-Russell (H-R) diagram.
This diagram illustrates the evolution of a low-mass star similar to the Sun.

determines the rate of contraction. As the interstellar mass condenses and loses gravitational potential energy, the temperature rises and the material gains thermal energy. As the star continues to decrease in size, the temperature continues to increase until a thermonuclear reaction begins and hydrogen is converted to helium, as discussed in Section 18.1. The star's position in the main sequence of the H-R diagram, its temperature, and the radiation given off are determined by the mass of the star. On the main sequence of the H-R diagram, the star continues to convert hydrogen into helium. This process is called *hydrogen burning,* but it is a nuclear "burning," or fusion, not a chemical fire. The hydrogen-burning stage lasts for billions of years—possibly 10 to 15 billion years for a star like our Sun.

As the hydrogen in the core is converted into helium, the core begins to contract and heat up. This heats the surrounding shell of hydrogen and causes the fusion of the hydrogen in the shell to proceed at a more rapid rate. This rapid release of energy causes the star to expand and enter the first red giant phase of its evolution. From the red giant phase the star expands further into the red supergiant phase after which the star becomes very unstable and the outer layers expand forming a beautiful expanding shell of matter called a *planetary nebula.** See Fig. 18.15 for the path taken during the expanding process. The expanding shell of matter eventually diffuses into interstellar space, and the star core becomes a white dwarf.

When a star becomes a white dwarf, it is very small. It has gravitationally collapsed as far as it can and still obey the laws of physics. The star is about the size of Earth and is so dense that a single teaspoonful of matter weighs five tons. Because it is very small, the star is not very luminous.

More massive stars develop into supergiants that may become supernovae. When a massive star's nuclear fuel becomes depleted, the interior collapses catastrophically to form a small-diameter neutron star. During the process the outer layers of the star bounce off the inner rigid core, then expand into space, destroying the star and giving rise to a supernova. See Fig. 18.17. The supernova phase does not occur for the vast majority of stars.

Our bodies and the universe in which we live are made of elements. The origin of the elements is

---

*The name is misleading. The expanding layers of the star have nothing to do with planets. Astronomers in the eighteenth century viewed these objects as planet-like objects, not as points of light, like stars.

Figure 18.16 The Very Large Array (VLA) Radio Telescope.
This Y-shaped array is constructed with three arms, each containing nine parabolic reflectors 80 ft in diameter. Each railroad arm extends 13 mi into the desert near Socorro, New Mexico. The VLA is computer-linked as a single antenna and produces very sharp images of distant celestial objects.

a major concern to astronomers. They believe that hydrogen and most of the helium were the original elements, and the other elements were created from hydrogen and helium by nuclear fusion in stars. The creation of the nuclei of elements within stars is called **nucleosynthesis.** The process takes place in the interior of stars and in supernovae. The physics of the process is extremely complicated and will not be dealt with here.

Large-mass stars (between 10 and 25 times the Sun's mass) have more gravitational attraction, and, after a supernova explosion, they collapse to a size of approximately 20 km in diameter. The electrons and protons in this superdense star combine to form neutrons, and this **neutron star** is composed of about 99% neutrons. A teaspoonful of a neutron star would weigh one billion tons. Because the angular momentum of the star must be conserved, the small size of the neutron star dictates that it must be spinning rapidly. Rapidly rotating neutron stars give out radio radiation in pulses and are called **pulsars.** The radio radiation is detected and measured on Earth by large radio telescopes (Fig. 18.16). They pulse with a constant period that may be between 0.03 and 4 s. One of the fastest-spinning pulsars discovered is located at the center of the Crab Nebula (Fig. 18.17). It is identified with the remains of the A.D. 1054 supernova. The period of this pulsar is slowly increasing, indicating that the rotating neutron star is gradually slowing down.

If the remaining core after a supernova explosion is greater than about three times the mass of the Sun, the star is believed to end up as a gravitationally collapsed object even smaller and more dense than a neutron star. Such a star is so dense that light cannot escape from its surface because of its intense gravitational field. Thus it would appear black and is called a *black hole*. (Section 18.4.)

After a star explodes in a supernova and the core goes into its end state, what becomes of the ejected material? It is thrown into space (Fig. 18.17) and eventually becomes seed material for future stars. In fact, because of its surface composition and age, we believe our own Sun is the product of several generations of stellar evolution. That is, one or more stars went through their life cycles and exploded to form supernovae. The ejected material later coalesced to create our Sun and solar system.

18.4 Gravitational Collapse and Black Holes

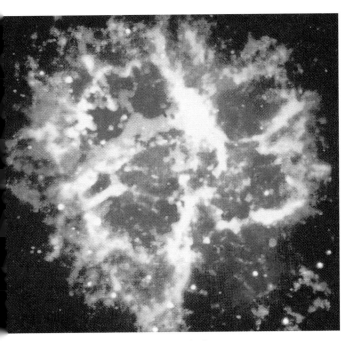

Figure 18.17 Crab Nebula.
The Crab Nebula is the remnant of a supernova explosion observed by Chinese astronomers in A.D. 1054. False colors have been added to the photograph to bring out details.

CONFIDENCE QUESTION 18.3

State in what order the following possible stages of a star occur: hydrogen burning, cepheid variable, white dwarf, gravitational accretion, and red giant.

18.4 Gravitational Collapse and Black Holes

Learning Goals:

To describe gravitational collapse.

To give evidence to support the belief that black holes exist.

The possibility of a rather remarkable phenomenon has been proposed to explain some strange astronomical findings. The phenomenon is called **gravitational collapse** and is the collapse of a very massive body because of its attraction for itself.

Recall that Newton's law of universal gravitation was given as

$$F = G \frac{m_1 m_2}{r^2} \quad (18.2)$$

where F = force of gravity between m_1 and m_2,

m_1 = mass of first object,

m_2 = mass of second object,

G = universal gravitational constant,

r = distance between m_1 and m_2.

From Eq. 18.2, it is evident that if the distance r is made smaller, then the force of gravitation increases.

In a large mass such as Earth or the Sun, all parts are always attracting each other. Normally, in large masses, electromagnetic forces tend to keep the various particles (such as atoms) apart. What would happen if the body were so massive that the gravitational forces of attraction were stronger than the electromagnetic forces of repulsion? All parts would be drawn closer together. According to Eq. 18.2, the forces increase as the distance r decreases. The closer the particles move together, the greater the gravitational force becomes. The whole mass would continue to contract until the original volume of mass becomes a fantastically massive point called a **singularity**.

The singularity is surrounded by an invisible spherical boundary known as the **event horizon**. Any matter or radiation within the event horizon cannot escape the influence of the singularity. See Fig. 18.18 for a sketch of a singularity and its event horizon for a nonrotating black hole. The value R (called the **Schwarzschild radius**), the radial distance the event horizon is located from the singularity, can be determined by equating the escape velocity equation ($v^2 = 2GM/R$) to the speed of light. We obtain

$$R = \frac{2GM}{c^2} \quad (18.3)$$

where G = universal gravitational constant,

M = mass of the collapsed star,

c = speed of light.

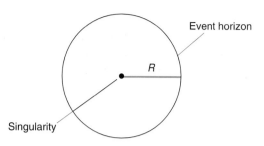

Figure 18.18 Configuration of a nonrotating black hole.
The black dot represents a singularity surrounded by the event horizon at a distance *R*. Anything located within the event horizon cannot escape. Thus, space inward from the event horizon is known as a black hole.

Anything located within the event horizon cannot escape. The event horizon is a one-way boundary because matter and radiation can enter but cannot leave. Thus space inward from the event horizon is a **black hole**.

EXAMPLE 18.2 Calculating the Radius of a Black Hole

Calculate the distance between the singularity and the event horizon for a nonrotating black hole having a mass of 7.0×10^{31} kg.

Solution

Use Eq. 18.2.

$$R = \frac{2GM}{c^2}$$

$$= \frac{2\left(6.67 \times 10^{-11} \text{ kg} \times \frac{m}{s^2} \times \frac{m^2}{kg^2}\right) 7.0 \times 10^{31} \text{ kg}}{(3.0 \times 10^8)^2 \frac{m^2}{s^2}}$$

$$= \frac{2 \times 6.67 \times 10^{-11} \times 7.0 \times 10^{31} \times 10^{-16}}{9.0} \text{ m}$$

$$= 1.0 \times 10^5 \text{ m}$$

Because most stars are rotating, a black hole will probably also be rotating. And because angular momentum is conserved, the rotation rate will increase as the star gets smaller, so the rotation of the black hole will be extremely rapid. Thus rather than being spherical, a black hole is likely to have the shape of an oblate spheroid.

Because nothing can escape the influence of a black hole, how is it detected? One method is the detection of X-rays coming from the vicinity of a black hole. The detection of X-rays by Earth satellites from a double-star system in which one companion may be a black hole has provided astronomers with data indicating that black holes probably do exist.

For example, the supergiant star HDE 226868,* which is located some 8000 ly from Earth in the constellation Cygnus, has a companion called Cygnus X-1 that cannot be detected directly, indicating that it may be a black hole. The detected X-rays coming from the double-star system are generated by gases captured by Cygnus X-1 from the supergiant star. The captured gases, accelerating to extremely high speeds, go into orbit around Cygnus X-1, forming a spiraling flat disk of matter called an *accretion disk*. These gases become extremely hot because of internal friction and emit high-energy X-rays (Fig. 18.19).

Calculations from experimental data indicate that Cygnus X-1 is smaller than Earth and more than seven times as massive as the Sun. This object is too massive to be a white dwarf or a neutron star, leaving the possibility that Cygnus X-1 is a black hole.

Another candidate for a black hole is the dark companion of A0620-00, an orange dwarf star, spectral type K, located in our galaxy in the constellation Monoceros. The A in the name refers to Ariel V, the X-ray satellite that first detected high energy X-rays coming from the binary system. The numbers give the right ascension and declination. A0620-00 is a main sequence star with a mass estimated at 0.7 solar mass located 3200 ly from Earth. The orange dwarf star circles the dark companion (considered a black hole of 8 solar mass) every 7.75 hours.

The most recent candidate for a black hole is at the center of the giant elliptical galaxy M87 in the Virgo cluster of galaxies, which is 15 million

---

*Henry Draper catalog. A nine-volume catalog of stars completed in 1924; it gives the positions, magnitudes, and spectral classes of 225,300 stars.

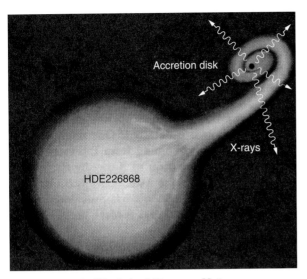

Figure 18.19 Model of Cygnus X-1.
The blue supergiant star HDE 226868 ejects material, some of which is captured by its unseen companion (assumed to be a black hole), forming an accretion disk. X-rays generated at the inner edge of the disk have been detected by astronomers.

parsecs from Earth. The Hubble Space Telescope's Faint Object Spectrograph has provided data that indicate a speed of 550 km/s of the gaseous disk some 60 ly from the center of the galaxy. Calculations indicate, for a speed of this magnitude, the galaxy's center must possess a mass equal to about three billion suns. Only a fraction of this enormous mass is contributed by visible stars in the central volume. Thus, the enormous mass of the extremely fast rotating disk must be a black hole.

CONFIDENCE QUESTION 18.4

Calculate the radial distance between the singularity and the event horizon of a black hole formed by the gravitational collapse of a star having a mass 30 times that of the Sun.

18.5 Galaxies

Learning Goals:

To identify the classes of galaxies.

To state Hubble's law and explain its use to calculate distances to remote celestial objects.

The Sun and its satellites occupy a very small volume of space in a very large system of stars known as the Milky Way galaxy (Greek *galaxias*, "milky way"). A **galaxy** is an extremely large collection of stars bound together by gravitational attraction and occupying an enormously large volume of space. It is the fundamental component for the structure of the universe. A galaxy is classified as irregular, spiral (normal or barred), or elliptical, depending on how it appears when photographed.

Astronomer Edwin P. Hubble established a system of classifying galaxies according to their appearance in photographs. The system starts with spherical ellipticals, spreads out with increasing flatness, then branches off into a normal spiral sequence and a barred spiral sequence, with a scattering of irregulars outside the two branches (Fig. 18.20).

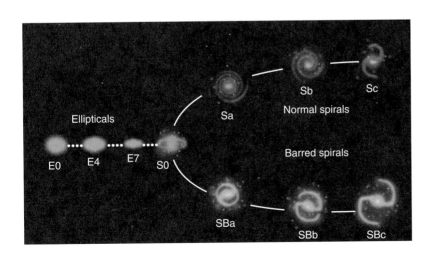

Figure 18.20 Hubble's classification for galaxies.
This diagram illustrates elliptical, normal spiral, and barred spiral galaxies. It does not show the evolutionary sequence of galaxies.

Figure 18.21 The Sombrero galaxy, M104, type-Sab spiral.
The dark band across the galaxy's center is composed of dust and gas.

Figure 18.23 Barred galaxy Type SBb.

Figure 18.22 Spiral galaxy NGC 2997, type Sc.

The type Sa spirals have their spiral arms closely wound to the central region (Fig. 18.21); whereas the Sb spirals have arms that spread out more from the center and the Sc spirals have very loose spiral arms (Fig. 18.22). The S0 type links the normal spirals to the smooth ellipticals. The classification of barred spirals, which are distinguished by a broad bar that extends outward from opposite sides of the central region, follows similar unwinding of the spiral arms (Fig. 18.23).

The irregular-type galaxies have no regular geometric shape. Examples of this type are the Small and Large Magellanic Clouds that can be easily seen with the unaided eye from the Southern Hemisphere (Figs. 18.24 and 18.27). They are named in honor of Magellan, who reported seeing them on his famous voyage around the world.

In a given volume of space there are more elliptical galaxies than spiral galaxies, and the irregular types account for only about 3% of all galaxies. The elliptical galaxies are made up of older stars and are usually dimmer than the spirals. The spiral galaxies account for over 75% of the brighter galaxies observed.

Carl Seyfert, an American astronomer at the Mount Wilson Observatory in California, reported in 1944 that a few spiral galaxies exhibit very bright centers and their spectra show broad emission lines that indicate that hot gas is present and expanding at very rapid rates. This evidence indicates that violent activity is taking place in the central core of the galaxies. Large amounts of energy are being released from these central cores, and the best theory of the source of this energy is the gravitational collapse of an enormous amount of matter. Over 100 of these galaxies have now been reported and are known as *Seyfert galaxies* (Figure 18.25).

Our local galaxy, the **Milky Way,** is believed to be a Hubble-type Sb spiral. It contains some 100

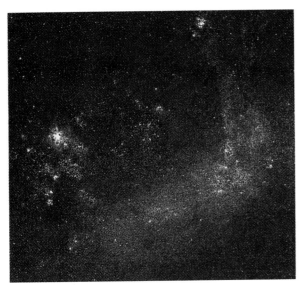

Figure 18.24 The Large Magellanic Cloud.
This irregular galaxy (the closest galaxy to our Milky Way) is only 160,000 ly away. The huge bright region at the left, which is 800 ly in diameter, is the Tarantula nebula.

Figure 18.25 The Seyfert galaxy NGC 4151.
Notice the very bright nucleus. If the galaxy were at a very extreme distance, only the bright nucleus would be visible. Perhaps quasars are the tremendous energetic sources in the nuclei of Seyfert galaxies.

billion (10^{11}) stars and is believed to have an appearance similar to that of the Great Galaxy in Andromeda (Fig. 18.26). Our galaxy (Fig. 18.27) is about 10^5 ly in diameter and has a thickness in the Sun's region of approximately 2×10^3 ly. It is rotating eastward or counterclockwise as viewed from the North Celestial Pole. The period of rotation, which is not the same for all regions of the galaxy, is more than 2×10^8 y for the region that contains our solar system. Our solar system is located in the plane of rotation between the Perseus and Sagittarius spiral arms of the galaxy. It is about 3.0×10^4 ly from the galactic center and is moving at a rate of approximately 240 km/s (150 mi/s) in the direction of the constellation Lyra.

The galactic equator, a great circle positioned halfway between the galactic poles, is inclined about 62° from the celestial equator (Fig. 18.28). The North Galactic Pole has a right ascension of 12 h 40 min and a declination of +28°. The South Galactic Pole has a right ascension of 0 h 40 min and a declination of −28°.

In the immediate neighborhood of the Milky Way and confined to an ellipsoidal volume of space—some 9×10^5 pc for the major axis and about 8×10^5 pc for the minor axis—is located a small group of galaxies known as the **Local Group.** This group has at least 21 known members, and others are believed to exist that have not been detected because of their low magnitude. Our Milky Way is a member located near one end of the major axis. Messier 31 (Great Galaxy in Andromeda, see Fig. 18.26) is also a member and is positioned near the opposite end of the major axis. Messier 31 (M31) can be seen with the unaided eye from the United States. The Andromeda constellation is about 20° south of the constellation Cassiopeia. See Appendix XI for star charts of the night sky.

Figure 18.26 The spiral galaxy M31 in Andromeda.
M31 is a type Sb galaxy 2.25 million ly from Earth. Two elliptical galaxies, NGC 205 (lower right) and NGC 221 (M32) are also shown.

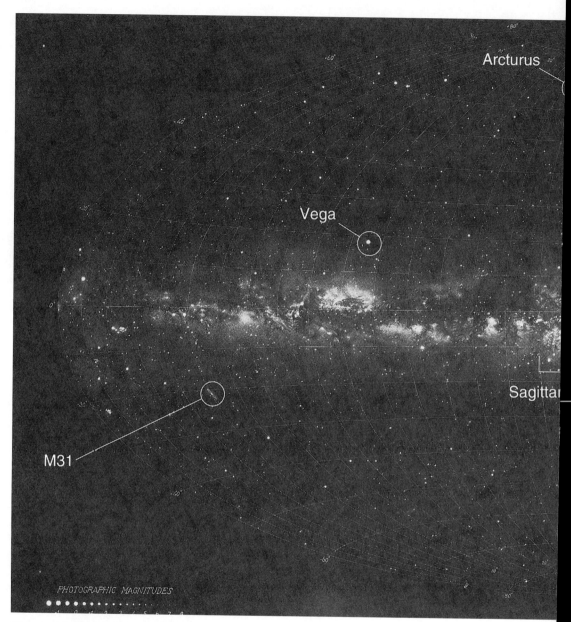

Figure 18.27 Knut Lundmark's map of the Milky Way.
This panorama of the Milky Way is a composite made from many photographs. It illustrates the Milky Way on an Aitoff projection, and accurately shows 7000 stars with known coordinates.

The galaxies of the Local Group seem to be moving with random motions. The two Magellanic Clouds are moving away from our galaxy, and several others, including M31, are moving toward our galaxy.

The galaxies astronomers photograph throughout the vast volume of the universe are lumped together in *clusters* that vary in size and number. The clusters, classified as regular or irregular, range in size from 3 to 15 million ly in diameter. Some contain a few galaxies such as our Local Group, and others contain thousands. As Hubble found regarding the distribution of galaxies, astronomers have discovered that the distribution of

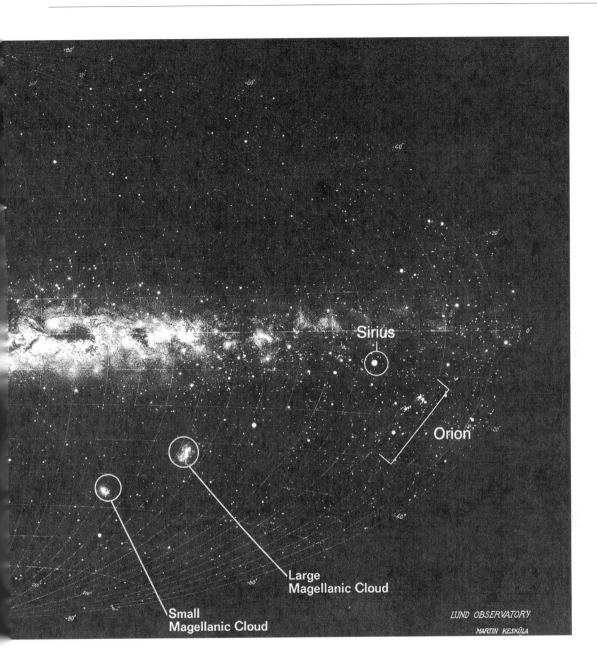

clusters is isotropic and homogeneous—the same in all directions and at all distances.

The galactic clusters lump together into what are called *superclusters;* that is, clusters of clusters. These superclusters have diameters as large as 300 million ly and masses equal to or greater than 10^{15} solar masses. Our Local Group and adjoining groups and clusters such as the Virgo cluster form what is called the Local Supercluster.

What is the origin of galaxies? Did they form from huge clouds of gas and dust? Any theory must take into account the three experimental facts discussed in Section 18.7 on cosmology. Cosmologists do not agree in what way the galaxies evolved. How

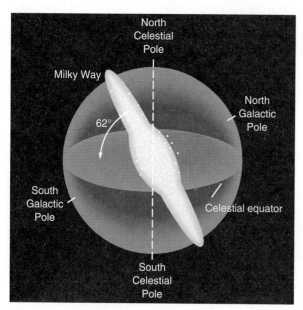

Figure 18.28 Diagram illustrating the inclination of the plane of revolution of the Milky Way galaxy to the celestial equator.

did the universe, which seems to have been extremely uniform and smooth (energy and mass were evenly distributed) in the beginning, evolve into the galaxies, clusters, and superclusters we observe today?

There are two theories concerning the direction taken by the early universe in the formation of galaxies. One is called the *bottom-up theory*, the other is called the *top-down theory*. Did the universe start from small particles of mass that came together to form galaxies, which evolved into clusters, then into superclusters, or did vast clouds of matter, shaped like pancakes, provide the seeds for the formation of galaxies? In December 1991, radio astronomers using the Very Large Array in New Mexico reported the detection of a huge mass of primordial hydrogen near the edge of the observable universe that has the structure of a flat disk or pancake. The detected neutral hydrogen structure has an estimated mass as large as 500 trillion solar masses and measures some 5 million ly across. Astronomers believe galaxies would form within the large hydrogen structure and in time break away. Although this new discovery tends to support the top-down theory, the question of the origin of galaxies is still unanswered.

Many astronomers believe that most of the matter that makes up the universe is not detected by any part of the electromagnetic spectrum. The unobserved matter has been incorrectly referred to as *missing matter*, but that term has been replaced by the term dark matter. No one knows the composition of *dark matter*. Astronomers originated the concept to explain why a cluster of galaxies exists as a gravitationally bound system. The observed mass of a typical cluster is not sufficient to hold the cluster together as a unit. Therefore, matter must be present throughout the cluster that is undetected. The dark matter seems not to be typical particles of matter or dust, but some unknown exotic form of matter. Other astronomers have doubts about the existence of this so-called dark matter.

When Edwin Hubble began looking at Doppler shifts of galaxies, he found nothing surprising at first. In the Local Group, the Doppler shifts were small; some were blue and some were red. But as he looked at galaxies farther and farther away, he found only red shifts. In fact, the farther away the galaxy, the larger the red shift. From these measurements the distances to remote galaxies can be determined by converting the observed red shift of the galaxy to radial velocity and then plotting the logarithm of the velocity against the apparent magnitude (Fig. 18.29). Hubble's discovery, now known as **Hubble's law,** can be written as

$$v = Hd \qquad (18.4)$$

where v = recessional velocity of the galaxy,

d = distance away from the galaxy,

H = Hubble's constant = $\dfrac{50 \text{ km/s}}{10^6 \text{ pc}}$.

The Hubble constant is believed to have a value of 50 to 100 km/s per million pc. If H is 50 km/s per million pc, the observed galaxy is moving away from our Sun at a speed of 50 km/s for every one million pc the galaxy is from our Sun. Hubble's law is extremely important; it gives astronomers vital information about the structure of the universe.

Astronomers and other scientists determine the structure of the universe by detecting and analyzing electromagnetic waves that come from the stars and galaxies. From the collected data, they contemplate the way in which these objects are

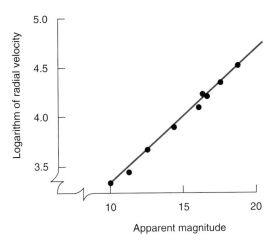

Figure 18.29 Hubble's law.
This graph shows the relationship (Hubble's law) between the logarithm of the radial velocity of a galaxy and its distance from the Milky Way. Recall that apparent magnitude and distance are related.

distributed throughout the vast volume of the universe. Scientists estimate that at least 10^9 (one billion) galaxies that can be photographed are within range of astronomers' optical telescopes.

Photographs of distant stars, galaxies, or galactic clusters show the way the objects appeared at the moment the light radiated from them. A galaxy that is observed (photographed) at a distance of 1 billion ly is being seen the way it appeared 1 billion years ago. In other words, we are taking an image of the galaxy's past, not its present. When we photograph the Sun, we get a picture of its surface as it existed eight minutes ago. That is the time it takes for the light to reach Earth.

Even if they were inclined to do so, astronomers would not have the time to photograph the tremendous volume of the universe to make images of all galaxies. Instead, the astronomers' model of the structure of the universe is based on a sampling of different regions. Using the 60-in. and 100-in. reflecting telescopes on Mount Wilson, Hubble obtained photographs of more than 1200 sample regions of space, counted some 44,000 galaxies, and estimated that it would be possible to photograph 100 million galaxies. After correcting his data for such things as the obstruction presented by the Milky Way and interstellar dust, he concluded that when observation of the universe is made over a large volume, the distribution of galaxies is isotropic and homogeneous. That is, when we observe a large volume of space, we observe as many galaxies in one direction as in any other, and the observations are the same at all distances. This concept of the uniformity of the universe is known as the **cosmological principle** and is the basic assumption for most theories of *cosmology*.*

But recent data concerning the structuring of galaxies indicate that the universe may be quite different from what we presently believe, and the cosmological principle may not be true. Astronomers, using Hubble's law as basic fact, have measured the recessional velocity of approximately 6000 nearby galaxies. They used the measured recessional velocities to calculate the distance to each galaxy. All of the galaxies were within 500 million ly of Earth. The distance data plus the right ascension and declination for each galaxy were put into a computer that produced an illustration showing the location and relative positions of the 6000 galaxies.

The illustration showed the galaxies structured in volumes about 400 million ly long, 200 million ly high, and up to 15 million ly thick. This grouping of galaxies, the largest known structure observed in the universe thus far, was called the "Great Wall" of galaxies by the research astronomers (Fig. 18.30). Nearly every galaxy in the collected data belongs to either a two-dimensional sheet or a one-dimensional thread that is millions of light-years in length. The data also revealed large volumes of space, millions of light-years across, that contained very few galaxies. That is, the two-dimensional sheets were separated by large volumes of almost empty regions of space.

The 60-in. telescope at the Mount Hopkins Observatory near Tucson, Arizona, was used to collect the data. The data were taken on only four pie-shaped volumes of space. Each layer or volume, starting just above the horizon, was 6° high and slightly less than 180° wide. The other three volumes were taken above the first layer. The rotation of Earth on its axis was used to take the 60-in. telescope through the width angle.

---

*Cosmology is the study of the structure and evolution of the universe.

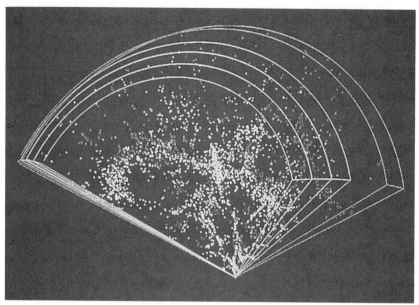

Figure 18.30 "Great Wall" of galaxies.
Almost 6000 galaxies are illustrated in this map. The galaxies extend horizontally some 400 million ly, vertically some 200 million ly, and 15 million ly in depth. At the bottom center is the Coma cluster, located at an average distance of 350 million ly from us. The galaxies are in the general direction of right ascension 12 h and declination +15°.

The data embrace an extremely small volume of space compared to the entire universe. Therefore, additional data on galaxies that are located at greater distances and in other directions of the universe will have to be obtained before conclusive decisions can be made concerning the structure of the universe.

CONFIDENCE QUESTION 18.5
(a) Name the four classifications of galaxies.
(b) Which classification is most abundant in the known universe?

18.6 Quasars

Learning Goal:
To list the physical characteristics of quasars.

In about 1960, radio astronomers used their newly developed radio telescopes with high-resolution (having the ability to distinguish the separation of two points) and began detecting extremely strong radio signals from sources having small angular dimensions. The sources were named **quasars**—a shortened term for "quasistellar radio sources."

Hundreds of quasars have now been detected (● Fig. 18.31), and they have two important characteristics. First, quasars have extremely large red shifts. Therefore, from Hubble's law, they must be very far away. When their brightness was considered, however, scientists showed that if they were indeed extremely distant, quasars must be enormously powerful sources of energy. The mystery of quasars then becomes a question of energy. Quasars have an energy output equivalent to about 10,000 times that of a typical spiral galaxy.

Quasars are considered to be the most distant objects in the universe. The quasar OH 471 has an enormous red shift that, according to Hubble's law, places it 18 billion ly from our solar system. In addition to having enormous red shifts, quasars emit tremendous amounts of electromagnetic radiation (energy) at all wavelengths, and most of the radiation varies in intensity (some regular, some random) over a range of a few days to years. Quasars

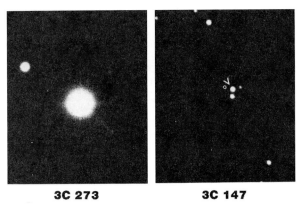

Figure 18.31 Quasistellar radio sources (quasars).
Note the jetlike prong of 3C 273. This quasar has an apparent visual magnitude of 12.8 and can be seen with a good 8-in. telescope.

are very small, appear to be blue, and have absolute magnitudes as high as −25. Although not conclusive, the best evidence points to the theory that the "powerhouse" of quasar energy is a supermassive black hole at their center.

Recent evidence indicates that quasars are somehow related to the centers of galaxies. As mentioned in Section 18.5, Seyfert galaxies have very bright centers, and the broad emission lines present in their spectra indicate that they contain gas that is extremely hot. Because quasars are very distant objects, perhaps they are the bright nuclei of spiral galaxies. The spiral arms of a galaxy would not be visible at a great distance from us.

CONFIDENCE QUESTION 18.6

Give three physical characteristics of quasars.

18.7 Cosmology

Learning Goals:

To describe the present known structure of the universe.

To identify theories concerning the evolution of the universe.

Cosmology is the study of the structure and evolution of the entire universe. We humans observe the universe, and our minds synthesize the observed phenomena. Concepts are then conceived and developed based on the observations. When we construct such concepts, we must remember that any concept of the universe that originates in the human mind is the way the universe is *conceived*, which is not necessarily the way it actually *is*.

The concept presently accepted by most astronomers is called the Big Bang. It has received broad acceptance because experimental evidence supports the concept in three major areas:

1. Astronomers observe galaxies that show a shift in their spectrum lines toward the low-frequency (red) end of the electromagnetic spectrum. This red shift is known as the **cosmological red shift**. (The cosmological red shift is not a Doppler shift, but a lengthening in the wavelength due to the expansion of the universe. The red shift is a result of the increasing size of the universe and is not related to velocity, although astronomers refer to the cosmological red shift with reference to velocity.)

2. Scientists detect a cosmic background radiation coming from space in all directions, commonly referred to as the 3-K cosmic microwave background.

3. Astronomers observe a mass ratio of hydrogen to helium of 3 to 1 in stars and in interstellar matter.

The red shift indicates an expanding universe.* To explain this observation, a theory was formulated that the universe began with a violent event called the **Big Bang** and we are now seeing the universe expanding from that event.

Because the event came before the expansion, we can calculate the time that has elapsed since the event. Astronomers observe galaxies receding from each other and assume they have been receding from each other since the event. To determine the maximum time (t_{max}), let d represent the distance to the most remote galaxy that we observe, and assume that the rate of recession has always been

---

*The geometry and mathematics of an expanding universe were first conceived and formulated by Alexander Friedman, a Russian mathematician, in 1922. Five years later, in 1927, Georges Lemaitre, a Belgian Catholic priest and cosmologist, developed a new cosmology theory of an expanding universe and postulated an explosive beginning from a Primeval Atom. Lemaitre, having heard Hubble lecture at Harvard University, knew of Hubble's evidence concerning the red shift of galaxies.

congruent with what we presently observe. The recessional velocity (v_r) is given by Hubble's law. That is, $v_r = Hd$, where H is Hubble's constant. We know, by definition, that velocity is distance divided by time ($v = d/t$).

EXAMPLE 18.3 Calculating the Age of the Universe

Calculate the maximum age of the universe in years. Use the minimum value for Hubble's constant:

$$H = \frac{50 \text{ km/s}}{10^6 \text{ pc}}$$

Solution

STEP 1

Given: Hubble's law $v = Hd$ and Hubble's constant $\frac{50 \text{ km/s}}{10^6 \text{ pc}}$.

STEP 2

Wanted: The age of the universe in years. Since Hubble's law does not include time, substitute a definition of velocity ($v = d/t$) for v.

$$v = Hd = \frac{d}{t}$$

STEP 3

Cancel the distance d, and solve the equation for t.

$$H = \frac{1}{t}$$

or

$$t = \frac{1}{H}$$

STEP 4

Substitute in the minimum value for H.

$$t = \frac{1}{\frac{50 \text{ km/s}}{10^6 \text{ pc}}}$$

STEP 5

Convert 10^6 pc to kilometers in order to cancel kilometers.

$$1 \text{ pc} = 3.086 \times 10^{13} \text{ km}$$

$$10^6 \text{ pc} = 3.086 \times 10^{19} \text{ km}$$

STEP 6

Substitute 3.086×10^{19} km for 10^6 pc and cancel kilometers.

$$t = \frac{1}{\frac{50 \text{ km/s}}{3.086 \times 10^{19} \text{ km}}}$$

STEP 7

Solve for t by inverting the denominator and multiplying.

$$t = \frac{3.086 \times 10^{19} \text{ s}}{50} = 6.18 \times 10^{17} \text{ s}$$

STEP 8

Convert 6.18×10^{17} s to years.

$$1 \text{ y} = 3.15 \times 10^7 \text{ s}$$

$$t = \frac{6.18 \times 10^{17} \text{ s}}{1} \times \frac{1 \text{ y}}{3.15 \times 10^7 \text{ s}}$$

$$= 1.95 \times 10^{10} \text{ y}$$

$$= 19.5 \text{ billion y}$$

Since the minimum value for Hubble's constant was used in the above calculation, the answer represents the maximum age of the universe. Exercise 12 at the end of the chapter asks you to calculate the minimum age of the universe.

What is the origin of all matter and radiation in the universe? One possible answer is that they came from nothing. When astronomers take a realistic view of the universe, space and time change. At the beginning of the Big Bang, which is considered the origin of the universe, space and time did not exist. Thus, no space, no time, and nothing expanded violently creating space, time, and something. It seems meaningless to conceive of nothing, but out of nothing (a so-called *false vacuum* with immeasurable energy) comes something when we theoretically allow space to expand and time to flow. Thus the possibility exists for the transformation from no time, no space, and nothing to space, time, and something.

Presently, what the something might have been at the beginning of the Big Bang is only speculation. General relativity fails to provide answers for a universe of infinite density and curvature. Scientists theorize that all kinds of particles existed in equilibrium with radiation during the first fractional part of a second that the Big Bang took place. Particles and antiparticles were being produced in pairs from photons, and annihilating and reconverting to photons. But by the time the explosion was one second old and the temperature about 10^{10} K, matter as we know it today (electrons, protons, neutrons, and neutrinos) existed. About one million years after the Big Bang, the temperature had dropped below 3000 K, and the particles could combine to form hydrogen, helium, and a trace of deuterium. The ratio of hydrogen to helium was then 3 to 1 by mass, which is the value presently observed in stars and interstellar matter.

People often ask, "Where did the Big Bang take place?" This question has no meaning, because at the moment of the explosion the makeup of the Big Bang was the entire universe. The explosion did not throw matter off into space. Space expands with time and forms the universe. Thus time and space do not exist outside the universe.

To comprehend an expanding three-dimensional universe of galaxies, observe the expanding two-dimensional surface of a toy rubber balloon speckled with small dots (representing galaxies) as the balloon is being inflated. Think of yourself as standing where a dot is located on the surface of the balloon. As the balloon expands, all dots are observed to recede from you. It makes no difference which dot you choose to observe from; the result is the same. There is no central point on the surface for the dots. Likewise, there is no central point for the galaxies in our three-dimensional universe. Note also that the surface of the balloon has no edge: You could examine every point on the surface of the balloon and find no edge. Similarly, there is no edge to our three-dimensional universe.

As early as 1948, scientists predicted that residual radiation from the early universe should be present here and now, and the radiation should be at radio wavelengths. The radiation should also resemble the radiation from a black body at a temperature of a few degrees above absolute zero. Because the radiation from the early universe was everywhere at once, it should fill the entire universe and be isotropic. That is, the radiation should be the same in any direction that we observe.

This radiation was first detected in 1965 by Arno Penzias and Robert Wilson at Bell Telephone Laboratories in New Jersey. They were testing a new microwave antenna used to make measurements of the absolute intensity of microwave radiation coming from certain regions of the Milky Way. After debugging and accounting for known static sources in their equipment, an unknown source of static remained. The static came from every direction, and was received equally from space at all times from any direction. After consulting with scientists at Princeton University who were designing and building equipment to detect the dying glow of the Big Bang, they concluded that the static (radiation) was indeed the greatly red-shifted radiation of the extremely hot universe that existed about one million years after the Big Bang. Today, the dying glow is called the **3-K cosmic background radiation,** and its presence is considered evidence of an ancient, extremely hot universe.

Will the universe continue to expand forever? Astronomers do not know the answer to this question, but the key to our understanding lies in determining the average density of matter in the universe. If the average density of matter is great enough, then space has a positive curvature and we live in a closed universe. That is, gravity in the universe is great enough to stop the expansion and eventually all matter will collapse in what is called the *Big Crunch*. If the average density of matter is too small, then space has a negative curvature and we live in an open universe that will continue to expand forever.

Present data for the value of the average density of matter are not precise enough to determine whether space has a positive or a negative curvature, but seems to indicate that we are living in a nearly flat universe.

The Big Bang theory of cosmology has problems—three will be mentioned here. First, the theory is based on the cosmological principle, which states that the universe is homogeneous and isotropic. Thus, proponents of the Big Bang base their theory on a universe that is uniform or smooth throughout. To support this view, contrary to the observation of great sheets of galaxies some 200 Mpc across, 50 Mpc high, and 5 Mpc thick, they assume a volume of space (a few hundred Mpc on a side) anywhere in the universe would have approximately the same composition. Theoretical proponents assume the galaxies, the cluster of galaxies, and the huge superclusters as points in their calculations. They also assume that no structures exist larger than those presently observed. Obviously, opponents of the Big Bang theory believe these views are false.

Second is the horizon problem. As mentioned above, the cosmic background radiation, measured by NASA's Cosmic Background Explorer at 2.735 K, is the same from all directions of the universe. The horizon distance is the maximum distance light can have traveled since the Big Bang. Since nothing can travel faster than the speed of light, objects separated by a greater distance cannot be in contact with one another. Thus, how can objects that we observe on opposite sides of the universe have exactly the same temperature? Proponents of the Big Bang do not have an acceptable answer.

Third is the flatness problem. As mentioned above, the observed density of the universe has a value that does not support an open or closed universe. Instead, it supports one near a value that indicates the universe is flat—*flat* meaning a universe that does not expand forever or collapses back to a singularity. The Big Bang theory provides no acceptable reason why the density of the universe is so close to a value that indicates that the universe will exist forever rather than one that indicates an open or a closed universe.

Another problem that challenges astronomers is the age paradox. Cosmologists report the age of the universe between 8 and 15 billion years. Stellar astronomers report that the oldest stars in the known universe are between 16 and 19 billion years old. Both groups cannot be correct. The difference in values is related directly to reliable distance indicators, which are difficult to obtain. Cosmologists and stellar astronomers use different distance indicators, which give different values for distance measurements. Until the problem is resolved, how the universe formed and evolved will remain a mystery.

Today, astronomers are applying the laws of physics and using the techniques of modern computer technology to probe the near and distant volumes of space, seeking knowledge of its matter, radiation, composition, motion, and structure. They are using data collected by sophisticated land-based optical telescopes and others that operate from space, plus an array of new devices operating in the X-ray, gamma ray, microwave, radio, ultraviolet, and infrared regions of the electromagnetic spectrum. Space probes have given astronomers a great deal of information about the solar system: 12 astronauts have gone to the Moon; robots have chemically analyzed the Martian surface; fly-by missions have examined the clouds and surface of Venus; and spacecraft on fly-by missions have photographed and transmitted data on Jupiter, Saturn, Uranus, and Neptune.

The Hubble Space Telescope (Fig. 18.32) is providing new information about the solar system and the universe beyond. The first object it observed was the planet Pluto. Figure 15.37 shows, for the first time, Pluto and its satellite Charon as two distinct objects. After the telescope was placed in orbit on April 24, 1990, astronomers and engineers discovered that images would not focus precisely to give excellent photographs. The primary mirror was not ground perfectly, causing spherical aberration. The defect was corrected in December 1993 when a servicing crew of astronauts repaired the telescope's blurry vision. At the same time the astronauts replaced the solar arrays and the guidance system.

The search for knowledge about the universe continues on several fronts. The world's largest optical telescope—the $94 million Keck, on Mauna Kea in Hawaii—was placed in operation in 1991, and its twin, also in Hawaii, will be in operation in 1996.

The 10-meter-diameter mirror of the Keck Telescope (Fig. 18.33) is made up of 36 hexagonal segments, each with its own support system, as well as sensors that detect misalignment between adjacent segments. The segments operate together as a single unit, and they are computer-controlled for perfect alignment. The huge mirror has four times the light-gathering power of the 5-meter (200-in.) reflector on Palomar Mountain in California.

18.7 Cosmology 491

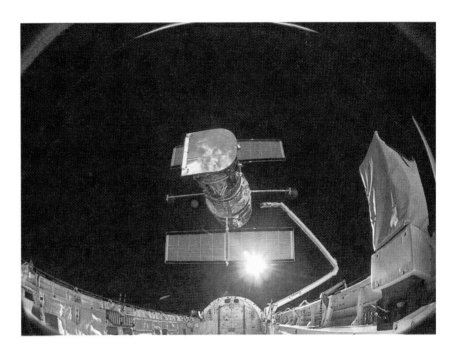

Figure 18.32 Hubble Space Telescope.
This photo shows the Hubble Space Telescope being released from the payload bay of the space shuttle *Discovery* on April 15, 1990. The telescope's 2.4 m (7.8 ft) mirror is covered. The solar array panels are shown fully deployed.

Figure 18.33 The 10-meter-diameter Keck Telescope under construction.
The photo shows the outer circle of 18 hexagonal mirror segments. The completed telescope has two more circles of mirrors; one inner circle of 6 mirrors and 12 more placed between the inner 6 and the outer 18. An opening at the center of the 36 mirror segments, the size of one segment, is for the light collected by the mirrors to be reflected to the recording instruments. This arrangement of 36 mirrors acting as one huge paraboloid is possible because computer technology keeps the segments in alignment. See the text for more details.

HIGHLIGHT

Other Cosmologies

In the late 1940s Fred Hoyle, British astrophysicist and cosmologist, along with two collaborators, Thomas Gold and Hermann Bondi, formulated the *steady-state model* of the universe. The model is based on the generalization of the cosmological principle called the *perfect cosmological principle*—on a large scale the universe is the same for all observers at all places and for all time. The steady state universe has a constant density, is infinite, had no beginning, and exists forever.

The major flaws in the model are its failure to provide an adequate explanation for a steady state (constant density) universe that appears to be expanding, and an acceptable explanation of the microwave background radiation. Recently one of Hoyle's collaborators has indicated that there are other explanations for the observed red shift of galaxies, which indicate an expanding universe. Incidentally, it was Hoyle, during an interview on a British radio broadcast, who coined the name Big Bang for the cataclysmic explosive event that other cosmologists put forth as the birth of the universe.

Plasma cosmology is a theoretical concept proposed by Hannes Alfvén, Swedish Nobel Laureate, in the 1970s. According to Alfvén, the universe has always existed, and its structure and evolution is controlled not by gravity alone as stated in the Big Bang model, but just as much by electrical currents and magnetic fields. Supporters of plasma cosmology using computer simulations, which neglect gravitational forces, show that the rotation of plasma in a magnetic field can initiate star formation. When electrical currents are extremely large (10^{16} amperes) the formation of galaxies, the basic structural unit for the universe, will take place.

Any model or theory of cosmology must address the evidence listed by number in Section 18.7. Plasma cosmologists agree that the observed abundance of helium (about 25%) in the universe and the cosmic background radiation, can be explained by the birth of massive stars in the formation of galaxies. Supporters of Alfvén cosmology indicate that 25 percent helium is produced in thermonuclear reactions of massive stars when part of the stars' hydrogen is transformed into helium. The stars eventually explode into supernovae, releasing the helium, which in turn is available for the formation of smaller stars.

The energy radiated by massive stars is absorbed by interstellar dust, which in turn emits the microwave background. This microwave background is smoothed out by the absorption and re-emitting of the radiation due to intergalactic magnetic fields provided by powerful jets emitted from galactic nuclei. The radiation is scattered in all directions and exists as a radiation fog.

Alfvén and his collaborators suggest one possible explanation for the cosmological red shift. What astronomers observe is an expansion that began some 10 to 20 billion years ago in our part of the universe, but not a Big Bang that created matter, space, and time.

Another thought concerning the red shift is the idea of "tired light." Some scientists hypothesize that light loses energy as it travels through space, causing the red shift.

Paul A. M. Dirac, British theoretical physicist and Nobel Laureate, suggested in 1938 that all space (objects and the space between) is expanding. No real expansion is occurring since the density remains constant. Thus, distant galaxies only appear to be moving farther apart.

The Big Bang, steady-state, and plasma cosmologies are rival concepts or models of the universe. Each competing model must be evaluated as to how well it is supported by experimental evidence and valid predictions. Unless these two tests are satisfied, a concept may be useful but cannot be fully accepted. At present there is not a satisfactory model of the universe acceptable to all cosmologists.

A radio telescope under construction in Green Bank, West Virginia, will be the world's largest fully steerable radio telescope. The antenna, with a reflecting surface of 2.3 acres, plus electronic and recording equipment, is scheduled to be operational in 1996.

The Very Large Array of radio telescopes in central New Mexico (Fig. 18.16) is laid out with three arms, each 13 mi long. The Very Large Array consists of 27 radio antennas, 9 on each arm, that, with the aid of a computer, act as a single antenna and can produce a high-resolution radio map of the radiation coming from a celestial object.

Other regions of the electromagnetic spectrum are also being used to collect data from celestial objects. The Gamma Ray Observatory was placed in orbit in 1991, and the Advanced X-ray Astrophysics Facility is scheduled to be launched in 1997, followed by the launching of the Space Infrared Telescope Facility sometime before the year 2000.

Scientists supported by many nations will continue to send their improved instruments into orbit around Earth, and they will send more advanced probes throughout the solar system and beyond to collect data that will increase our knowledge of the universe in which we humans exist.

CONFIDENCE QUESTION 18.7

(a) How does the calculated age of the universe vary as the value of Hubble's constant increases? (b) How does the gravitational force between galaxies affect the calculated age of the universe?

Important Terms

These important terms are for review. After reading and studying the chapter, you should be able to define and/or explain each of them.

Sun
photosphere
plasma
chromosphere
solar wind
sunspots
neutrino
celestial sphere
declination
right ascension
celestial prime meridian
astronomical unit

light-year
parsec
H-R diagram
main sequence
red giants
cepheid variables
supernova
nucleosynthesis
neutron star
pulsars
gravitational collapse
singularity

event horizon
Schwarzschild radius
black hole
galaxy
Milky Way
Local Group
Hubble's law
cosmological principle
quasars
cosmological red shift
Big Bang
3-K cosmic background radiation

Important Equations

Distance to a star: $d = \dfrac{1}{p}$

Newton's law of gravitation: $F = \dfrac{G m_1 m_2}{r^2}$

Hubble's law: $v = Hd$

Questions

18.1 The Sun

1. Which of the following is a physical characteristic of the Sun?
   (a) chromosphere
   (b) plasma
   (c) prominences
   (d) corona
   (e) all of the above
2. The Sun's energy comes from
   (a) the fission of hydrogen to form helium.
   (b) the fusion of protons to form nuclei of helium.
   (c) the fusion of hydrogen to form carbon nuclei.
   (d) none of the above.
3. What is plasma?
4. Name the two most abundant elements in the Sun.
5. What is the temperature of the Sun (a) at its center and (b) at its surface?
6. (a) What are sunspots? (b) Are they cyclical? Explain.
7. Name the visible surface of the Sun.
8. What is the solar wind?
9. When is the Sun's corona visible?
10. What are neutrinos and photons, and how are they different?

18.2 The Celestial Sphere

11. Which of the following does not refer to the celestial sphere?
    (a) declination
    (b) parsec
    (c) right ascension
    (d) vernal equinox
12. Which of the following is not a unit of distance measurement to a celestial object?
    (a) astronomical unit
    (b) light-year
    (c) seconds of arc
    (d) parsec
13. What is the celestial sphere?
14. Define *declination* and give an example.
15. Define *right ascension* and give an example.
16. What is the celestial prime meridian?
17. Define each term: (a) light-year, (b) parsec, and (c) astronomical unit.
18. Which is a larger unit of distance, the light-year or the parsec? How much larger is one compared to the other?
19. (a) What are the units used to measure the annual parallax of a star?
    (b) State the relationship between the annual parallax and the distance to a star in parsecs.
20. The vernal equinox is a point on the celestial sphere that is between what two constellations?
21. What are asterisms? Give an example.

18.3 Stars

22. Which of the following does *not* refer to a type of star?
    (a) red giant
    (b) cepheid variable
    (c) white dwarf
    (d) binary
    (e) nova
23. The Hertzsprung-Russell (H-R) diagram is a plot of the absolute magnitude of stars against their
    (a) apparent magnitudes.
    (b) surface temperatures.
    (c) distances.
    (d) brightness.
24. What is a star?
25. Except for the Sun, Sirius is the brightest star in the sky. Give Sirius' apparent brightness and distance from the Sun in light-years.
26. (a) What two major elements make up a star?
    (b) Give the relative abundance of each element.
27. What is an H-R diagram?
28. On an H-R diagram, give the position of each of the following: (a) the Sun, (b) the main sequence, (c) red giants, (d) cepheid variables, (e) white dwarfs, and (f) red supergiants.
29. (a) What is the closest star to the Sun?
    (b) Give its distance from the Sun in light-years.
    (c) Can the star be seen from the United States? Explain.
30. Which has the higher surface temperature, a white or yellow star?
31. What is a binary star system? Do most stars appear as a binary system?
32. What are cepheid variables? Why are they important to astronomers?
33. State in what order the following possible stages of a star occur: hydrogen burning, cepheid variable, white dwarf, gravitational accretion, and red giant.

34. The line spectrum of a star shows a shift toward the red end of the spectrum. What does this indicate concerning the motion of the star?
35. What is (a) a neutron star and (b) a pulsar?

18.4 Gravitational Collapse and Black Holes

36. Which of the following does not refer to a black hole?
    (a) singularity
    (b) Schwartzchild radius
    (c) event horizon
    (d) nucleosynthesis
37. The singularity of a nonrotating black hole is surrounded by an invisible spherical boundary known as the
    (a) Schwartzchild radius.
    (b) gravitational boundary.
    (c) event horizon.
    (d) black hole boundary.
38. Describe the gravitational collapse of a star.
39. What is a singularity?
40. What factors determine the radial distance of the event horizon?
41. What was the first candidate to be a black hole?
42. Describe a method for detecting black holes.
43. How does the radial distance of the event horizon of a black hole vary with respect to (a) the mass of a collapsing star and (b) the speed of light?
44. (a) Name three major physical characteristics of a black hole.
    (b) Name the most recent black hole candidate that has these three characteristics.

18.5 Galaxies

45. Galaxies are
    (a) extremely large collections of stars.
    (b) classified as elliptical, normal spiral, barred spiral, or irregular.
    (c) found in clusters.
    (d) found in clusters of clusters called *superclusters*.
    (e) all of the above.
46. Hubble's law
    (a) relates the relationship between the recessional velocity of a galaxy and its distance from us.
    (b) indicates that the universe is expanding.
    (c) is sometimes referred to as the law of red shifts.
    (d) all of the above
47. What is a galaxy? Name three.
48. (a) What physical property is used to classify galaxies?
    (b) Name three major classifications of galaxies.
49. State the structure and dimensions of the Milky Way.
50. (a) Name the small group of galaxies of which the Milky Way is a member.
    (b) How many known members are in the group?
    (c) Which member of the group can be seen with the unaided eye from the United States?
51. Distinguish between isotropic and homogeneous distribution of galaxies.
52. What is dark matter? What is its significance?
53. What is the "Great Wall" of galaxies? Give its dimensions.
54. (a) State the cosmological principle.
    (b) Is there any reason to believe that it may not be true?

18.6 Quasars

55. Quasars
    (a) exhibit very large red shifts.
    (b) are believed to be the most distant objects from us.
    (c) exhibit extremely powerful sources of energy.
    (d) are believed to be galaxies with active galactic nuclei.
    (e) all of the above
56. Quasars
    (a) are very large objects.
    (b) were first detected by radio telescopes.
    (c) exhibit blue shifts in their spectrum.
    (d) all of the above
57. What are quasars, and what are two important characteristics of quasars?
58. (a) What is the apparent color of quasars?
    (b) What is the maximum value of their absolute magnitude?
59. (a) What information is there to support the belief that quasars are the most distant objects observed by astronomers?
    (b) In what respect do quasars resemble stars?
    (c) In what respect do they differ from stars?

18.7 Cosmology

60. Which of the following are true concerning the Big Bang model of cosmology?
    (a) The universe began as a singularity of high density and temperature.

(b) The model has received broad acceptance because of the observed cosmological red shift, the 3-K cosmic microwave background, and the 3 to 1 ratio of hydrogen to helium in stars.
(c) The model predicts an open universe.
(d) The model provides information about the universe back to the very beginning.
(e) Answers (a) and (b) are true.

61. The cosmological principle
  (a) is based on the assumption that the universe is homogeneous.
  (b) implies that the universe has no edge.
  (c) is based on the assumption that the universe is isotropic.
  (d) implies that the universe has no center.
  (e) all of the above
62. Describe the Big Bang model of the universe.
63. State three experimental facts that support the Big Bang model.
64. What property of the universe can be determined by taking the reciprocal of Hubble's constant?
65. How old is (a) the Sun and (b) the universe?
66. Why is the universe thought to be expanding?

Food for Thought

1. Where on Earth's surface must an observer be in order to see all visible stars over a period of one year?
2. For an Earth observer, is it conceivable that the universe is both finite and infinite? (Does Hubble's law set a limit on time?)
3. Do you think our universe is one of a kind, or is it possible that others exist? Why or why not.
4. What is your mental concept of the universe? Has your personal view changed since entering college?

Exercises

18.2 The Celestial Sphere

1. Find the distance in light-years to a star with a parallax of 0.20 s.
   Answer: 16 ly
2. Find the distance in parsecs to a star with a parallax of 0.20 s.
3. Calculate the number of miles in a light-year, using 186,000 mi/s as the speed of light.
   Answer: 5.87×10^{12} mi
4. Calculate the number of meters in a light-year, using 3.00×10^8 m/s as the speed of light.
5. How long in years does it take light to reach us from Alpha Centauri, which is about 4.3 ly away?
   Answer: about 4.3 ly
6. How many parsecs away is Alpha Centauri, which is at a distance of about 4.3 ly?

18.3 Stars

7. The Crab Nebula is expanding at a rate of 70 million mi/day. If it is the remnant of a supernova of A.D. 1054, how many miles in diameter will it be in A.D. 2000?

18.4 Gravitational Collapse and Black Holes

8. Determine the radial distance of the event horizon of a black hole formed from the gravitational collapse of a star having a mass of 15×10^{30} kg.
   Answer: 22 km
9. Calculate the distance between the singularity and the event horizon for a gravitationally collapsed star having a mass of 2.0×10^{32} kg.

18.5 Galaxies

10. Suppose our universe contained 100 billion galaxies with 100 billion stars in each.
    (a) How many stars would there be?
    (b) Is this number greater or less than Avogadro's number (6×10^{23})?

    Answer: (b) less

11. It takes the Sun 2×10^8 y to complete one revolution around the center of the galaxy. How many times has it (and our solar system) gone around the galaxy during its life of 5 billion years?

    Answer: 25 times

18.7 Cosmology

12. Determine the minimum age of the universe, using a value for Hubble's constant of 100 km/s per 10^6 pc.

    Answer: 9.78×10^9 y

13. Hubble's constant is estimated to be between 50 and 100 km/s per 10^6 pc. Use the average of these two values and determine the age of the universe.

Answers to Multiple-Choice Questions

1. e   11. b   22. e   36. d   45. e   55. e   60. e
2. b   12. c   23. b   37. c   46. d   56. b   61. e

Solutions to Confidence Questions

18.1 The Sun is a star, spherical in shape with a diameter of 1.39×10^6 km that radiates electromagnetic energy.

18.2 The given latitude in degrees. When an observer is at 90°N, all Northern Hemisphere stars move in concentric circles above the horizon. When an observer's latitude is 0°, all stars rise and set daily.

18.3 Gravitational accretion, hydrogen burning, red giant, cepheid variable, white dwarf.

18.4
$$R = \frac{2GM}{c^2}$$
$$= \frac{2\left(6.67 \times 10^{-11} \text{ kg} \times \frac{\text{m}}{\text{s}^2} \times \frac{\text{m}^2}{\text{kg}^2}\right) 30 \times 2 \times 10^{30} \text{ kg}}{(3.0 \times 10^8)^2 \frac{\text{m}^2}{\text{s}^2}}$$
$$= 9 \times 10^4 \text{ m}$$

18.5 (a) Elliptical, normal spiral, barred spiral, and irregular. (b) Ellipticals.

18.6 Quasars have extremely large red shifts, are enormously powerful sources of energy, are very small, appear to be blue, and have absolute magnitudes as high as -25.

18.7 (a) The calculated age decreases as H increases.
    (b) The effect of gravitational forces on galaxies will decrease the recessional velocity of the galaxies. Hubble's constant will be smaller, because the constant is proportional to the velocity. This change will give an increase in the calculated age of the universe.

Chapter 19

The Atmosphere, Weather, and Climate

19.1 Atmospheric Composition and Structure

19.2 Atmospheric Energy Content and Measurements

Highlights: Why the Sky Is Blue and Sunsets Are Red

The Greenhouse Effect

19.3 Air Motion and Clouds

19.4 Air Masses and Storms

19.5 Atmospheric Pollution and Climate

Highlight: The Ozone Hole

*E*arth science is a collective term that includes the study of our planet—land, sea, and air, and even its history. In this chapter we will be concerned with the composition and workings of the atmosphere. The final two chapters of the textbook will be devoted to geological features and processes of our planet.

Our **atmosphere** (from the Greek *atmos*, "vapor," and *sphaira*, "sphere") is the gaseous shell, or envelope, of air that surrounds Earth. Just as certain sea creatures live at the bottom of the ocean, we humans live at the bottom of this vast atmospheric sea of gases.

In recent years, the study of the atmosphere has expanded because of advances in technology. Every aspect of the atmosphere, from the ground to outer space is now investigated in what is called *atmospheric science*. An older term, *meteorology* (from the Greek *meteora*, "the air"), is now more commonly applied to the study of the lower atmosphere. The continuously changing conditions of the lower atmosphere is what we call *weather*. Lower atmospheric conditions are monitored daily, and changing patterns are studied to help predict future conditions. "Meteorologists" give you weather forecasts on radio and TV.

The air we now breathe may have been far out over the Pacific Ocean a week ago. As air moves into a region, it brings the temperature and humidity that are mementos of its past travel, such as cold, dry arctic air. Moving air transports the physical characteristics that influence the weather and weather changes. The movement of large air masses depends a great deal on Earth's air-circulation structure and seasonal variations. When air masses meet, variations of their properties may trigger storms, which may be violent and sometimes destructive. Such storms remind us of the vast amount of energy contained in the atmosphere and also of its capability to affect our lives.

An unfortunate topic of atmospheric science is that of pollution. We hear concerns about acid rain, the greenhouse effect, greenhouse gases, CFCs, ozone holes, and global warming. We will first look at the normal conditions in our study of "this most excellent canopy, the air . . ." (Shakespeare, *Hamlet*), and then turn to atmospheric pollution and how it affects our environment and potentially our climate.

19.1 Atmospheric Composition and Structure

Learning Goals:

To identify the composition of air.

To distinguish how the atmosphere is divided into regions.

Table 19.1 Composition of Air

Nitrogen	N_2	78% (by volume)
Oxygen	O_2	21%
Argon	Ar	0.9%
Carbon dioxide	CO_2	0.03%
Others (traces)		*Others (variable)*
Neon	Ne	Water vapor (H_2O) 0–4%
Helium	He	Carbon monoxide (CO)
Methane	CH_4	Ammonia (NH_3)
Nitrous oxide	N_2O	Solid particles—dust,
Hydrogen	H_2	pollen, etc.

The air of the atmosphere is a mixture of gases. In addition, the air holds many suspended liquid droplets and solid particles. However, only two gases comprise about 99% of the volume of air near Earth. From Table 19.1, we see that this air is primarily composed of nitrogen (78%) and oxygen (21%), with nitrogen being about four times as abundant as oxygen. The other main constituents are argon (0.9%) and carbon dioxide (0.03%).

Minute quantities of many other gases are found in the atmosphere, along with particulate matter. Some of these gases, especially water vapor and carbon monoxide (CO), vary in concentrations, depending on conditions and locality. The amount of water vapor in the air depends to a great extent on temperature, as will be discussed later. Carbon monoxide is a product of incomplete combustion (Section 13.4).

CONFIDENCE QUESTION 19.1

What percentage of air is made up by each of its four main constituents, N_2, O_2, Ar, and CO_2?

In general, the relative amounts of the major constituents of the atmosphere remain fairly constant. Nitrogen, oxygen, and carbon dioxide are involved in the life processes of plants and animals. Nitrogen is taken in by some plants and released in organic decay. Animals breathe in oxygen and exhale carbon dioxide, while plants convert carbon dioxide to oxygen.

Plants produce oxygen by **photosynthesis**, the process by which CO_2 and H_2O are converted into sugars (needed for plant life) and O_2, using energy from the Sun (Fig. 19.1). The key to photosynthesis is the ability of chlorophyll, the green pigments in plants, to convert sunlight into chemical energy. Over half of the photosynthesis takes place in the oceans, which contain many forms of green plants.

Because of gravitational attraction, the density of air is greatest near Earth's surface, and decreases with increasing altitude. Over half of the mass of the atmosphere lies below an altitude of 11 km (7 mi), and almost 99% lies below an altitude of 30 km (19 mi). There is no clearly defined upper

Figure 19.1 Photosynthesis.
Energy from the Sun is necessary in the photosynthesis process whereby plants produce sugars and oxygen from water and carbon dioxide.

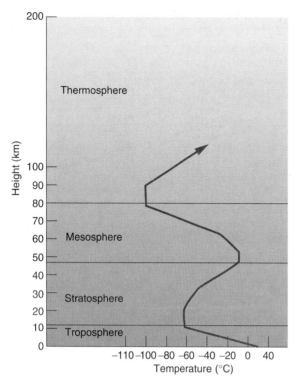

Figure 19.2 Vertical structure of the atmosphere.
Divisions of the atmosphere are based on variation in physical properties such as temperature, as shown here.

limit of Earth's atmosphere. It simply becomes more and more tenuous.

To distinguish different regions of the atmosphere, we look for distinct physical variations that occur with altitude. A couple of atmospheric properties that show vertical variations are (1) temperature and (2) ozone and ion concentrations.

Temperature

In measuring the temperature of the atmosphere versus altitude, we find that distinctions do occur. Near Earth's surface, the temperature of the atmosphere decreases with increasing altitude at an average rate of $6\frac{1}{2}$ C°/km (or $3\frac{1}{2}$ F°/1000 ft) up to about 16 km (10 mi). This region is called the **troposphere** (from the Greek *tropism*, "to change," Fig. 19.2).

Above the troposphere the temperature of the atmosphere increases nonuniformly up to an altitude of about 50 km (30 mi). See Fig. 19.2. This region of the atmosphere is called the **stratosphere** (from the Greek *stratum*, "covering layer").

The temperature of the atmosphere then decreases rather uniformly with altitude up to an altitude of about 80 km (50 mi). This region is called the **mesosphere** (from the Greek *meso*, "middle").

Above the mesosphere, the thin atmosphere is heated intensely by the Sun's rays. This region, extending to the outer reaches of the atmosphere, is called the **thermosphere** (from the Greek *therme*, "heat"). The temperature of the thermosphere varies considerably with solar activity.

Ozone and Ion Concentrations

The atmosphere may also be divided into two parts based on regions of concentrations of ozone and ions, with the ozone region lying below the ion region. **Ozone** (O_3) is formed by the dissociation of molecular oxygen and the combining of atomic oxygen with molecular oxygen ($O + O_2 \rightarrow O_3$). Energetic ultraviolet (uv) radiation from the Sun provides the energy to dissociate the molecular oxygen, and the major production and concentration of ozone depends on the appropriate balance of uv radiation and oxygen molecules. This occurs at an altitude of about 30 km (20 mi), and in this region the major concentration of ozone is found, as illustrated in ● Fig. 19.3. The ozone concentration becomes less with increasing altitude (because of less oxygen), up to an altitude of about 70 km (45 mi). The region of the atmosphere below this is referred to as the **ozonosphere.**

Because the ozone layer absorbs energetic uv radiation, one can expect an increase in temperature in the ozonosphere. A comparison of Figs. 19.2 and 19.3 shows the ozone layer lies in the stratosphere. Hence, the ozone absorption of uv radiation provides an explanation for the temperature increase in the stratosphere. Ozone is very unstable in the presence of sunlight, and it readily dissociates back into atomic and molecular oxygen. When an oxygen atom combines with an ozone molecule, two ordinary oxygen molecules may be formed ($O + O_3 \rightarrow O_2 + O_2$), thus destroying the ozone. This process and the formation of ozone go on simultaneously and at the same rate, so there is a balance in the concentration of the ozone layer.

Little ozone is naturally present near Earth's surface, but you may have experienced the ozone

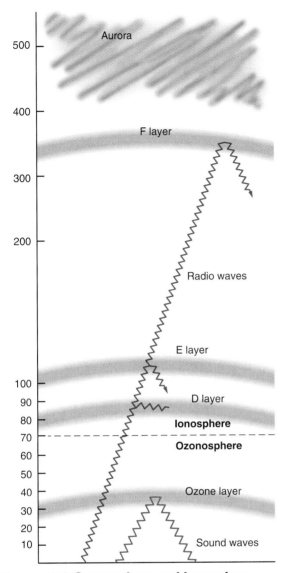

Figure 19.3 Ozonosphere and ionosphere. These atmospheric regions are based on ozone and ion concentrations. A warm-air layer occurs in the ozonosphere because ozone absorbs ultraviolet radiation. This warm-air layer reflects sound waves and was first investigated in this manner. The upper ion layers reflect radio waves.

formed by sparking electrical discharges. The gas is easily detected by its distinct pungent smell from which it derives its name (Greek *ozein*, "to smell"). The ozone layer in the stratosphere acts as an umbrella that shields life against harmful ultraviolet radiation from the Sun by absorbing most of the short wavelengths of this radiation. The portion of the uv radiation that gets through the ozone layer burns and tans our skin (and may cause skin cancer). Were it not for the ozone absorption, we would be badly burned and find sunlight intolerable.

In the upper atmosphere, energetic particles from the Sun cause the ionization of gas molecules. For example, $N_2 + energy \rightarrow N_2^+ + e^-$. The electrically charged ions and electrons are trapped in Earth's magnetic field and form ionic layers in the upper region of the atmosphere that is referred to as the **ionosphere.** Variations in the ion density with altitude give rise to three distinct regions or layers. See Fig. 19.3.

The D layer generally absorbs radio waves, but these waves are reflected E and F layers. At night, without the influx of solar radiation, most of the electrons and ions in the D and E layers recombine, and these layers virtually disappear. However, the F layer exhibits only a small reduction in concentration, and hence provides for global radio communications by reflecting waves to other parts of the world that are blocked by Earth's curvature.

Solar disturbances may cause fluctuations in the ion density and upset the reflection process. However, such solar disturbances, which produce a shower of incoming energetic particles, are associated with the beautiful display of lights in the upper atmosphere of regions of higher latitudes, particularly in the polar regions. In the Northern Hemisphere, these are called *northern lights* or **aurora borealis** (Fig. 19.4). The Southern Hemisphere tends to be forgotten by people living north of the equator. However, light displays of similar beauty occur in the southern polar atmosphere and are called *aurora australis.*

19.2 Atmospheric Energy Content and Measurements

Learning Goals:

To identify how insolation is distributed in the atmosphere.

To explain why the sky is blue and how the greenhouse effect occurs.

To describe some of the common measurements of atmospheric characteristics.

Figure 19.4 Aurora.
The aurora borealis, or northern lights, as seen in Alberta, Canada.

The Sun is by far the most important source of energy for Earth and its atmosphere. The sunlight or radiation incident on Earth's atmosphere is called **insolation** (standing for *in*coming *sol*ar rad*iation*). The solar radiation (energy) output fluctuates, but Earth receives a relatively constant average intensity at the top of the atmosphere. However, only 50% or less of the insolation reaches Earth's surface, depending on atmospheric condition.

In considering the energy content of the atmosphere, one might think that it comes directly from isolation. However, surprisingly enough, most of the direct heating of the atmosphere comes not from the Sun, but from Earth. To understand this, we need to examine the distribution and disposal of the incoming solar radiation (see Fig. 19.5).

About 33% of the insolation is returned to space as a result of reflection by clouds, scattering by particles in the atmosphere, and reflection from terrestrial surfaces, such as water and ice. (Recall from Chapter 15 that Earth's *albedo* is 0.33 or 33%, whereas the Moon's albedo is only 0.07 or 7%. Why?) Scattering of insolation occurs in the atmosphere from gas molecules of the air, dust particles, water droplets, and so on. (*Scattering* is the absorption of incident light and its reradiation in all directions.) As shown in Fig. 19.5, some of the scattered radiation is scattered back into space, and some toward Earth's surface. An important type of scattering involving a common atmospheric phenomenon is **Rayleigh scattering,** named after Lord Rayleigh (1842–1919), the British physicist who developed the theory. Lord Rayleigh showed that the amount of scattering by particles of a given molecular size was proportional to $1/\lambda^4$, where λ is the wavelength of the incident light. That is, the longer the wavelength, the less the scattering (the denominator λ^4 becomes very large). It is this scattering that gives rise to the color of the sky. See the chapter Highlight on page 504.

After reflection, scattering, and direct absorption, about 50% of the total insolation reaches Earth's surface. This radiation goes into terrestrial surface heating, primarily through the absorption of visible radiation. The atmosphere—in particular the troposphere—derives most of its energy directly from Earth. This absorption is accomplished in three main ways. In order of decreasing contribution, they are (1) absorption of terrestrial radiation, (2) latent heat of condensation, and (3) conduction from Earth's surface.

Absorption of Terrestrial Radiation

Earth, like any warm body, radiates energy that may be subsequently absorbed by the atmosphere. The wavelength of the radiated energy emitted by Earth depends on Earth's temperature. From the wavelength relationship of the energy of a photon and the temperature of the emitting source, it can be shown that $\lambda \propto 1/T$. That is, the wavelength of radiation emitted by a source is inversely proportional to its temperature.

Earth's temperature is such that it radiates energy primarily in the long-wavelength infrared region. Water vapor and carbon dioxide (CO_2) are the primary absorbers of infrared radiation in the atmosphere, with water vapor being more important. We refer to these gases as being selective absorbers, because they absorb certain wavelengths and transmit others. This gives rise to the so-called *greenhouse effect,* about which you now hear a great deal. See the chapter Highlight on page 506.

The other two main ways the atmosphere derives energy are easily understood. Approximately 70% of Earth's surface is covered by water. Consequently, a great deal of evaporation occurs because

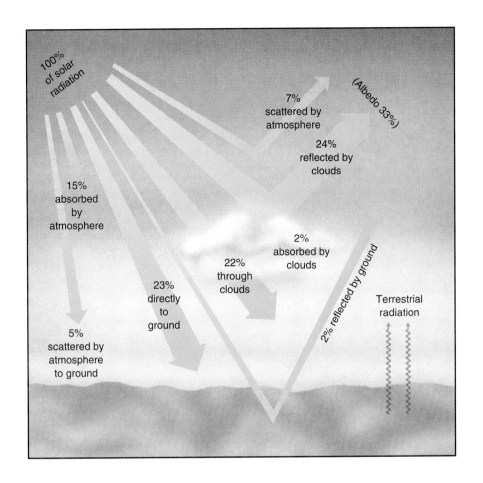

Figure 19.5 Insolation distribution.
An illustration of how the incoming solar radiation is distributed. The percentages vary somewhat, depending on atmospheric conditions.

of the insolation reaching the surface. Recall from Chapter 5 that it takes a relatively large amount of (latent) heat for a liquid-vapor phase change of water. This is 540 kcal/kg at the boiling temperature, but this is also a good approximation for evaporation at normal temperatures. Then, with condensation, a large amount of energy is transferred to the atmosphere in the form of latent heat. Energy is released in the atmosphere in the formation of clouds, fog, rain, dew, and so on.

A comparatively smaller, but significant, amount of heat energy is transferred to the atmosphere by conduction from Earth's surface. Because air is a relatively poor conductor of heat, this process is restricted to the layer of air in direct contact with Earth's surface. The heated air is then transferred aloft by convection. As a result, the temperature of air tends to be greater near the surface of Earth and decreases gradually with altitude (note the troposphere in Fig. 19.2).

The energy-transfer processes in the lower troposphere give rise to a great deal of change in atmospheric conditions. The troposphere contains over 80% of the atmospheric mass and virtually all the clouds and water vapor. The atmospheric conditions of the lower troposphere are referred to collectively as **weather.** Changes in the weather reflect the local variations of the atmosphere near Earth's surface.

Measurements of the atmosphere's properties and characteristics are very important in its study. We watch and listen to daily atmospheric readings to obtain a qualitative picture of the conditions for the day in a region and around the country. Fundamental measurements made in the atmosphere include (1) temperature, (2) pressure, (3) humidity, (4) wind speed and direction, and (5) precipitation. Let's look at these briefly.

Having discussed temperature measurements in Chapter 5, we will not dwell on this property.

HIGHLIGHT

Why the Sky Is Blue and Sunsets Are Red

The gas molecules of the air account for most of the scattering in the visible region of the spectrum. In the visible spectrum the wavelength increases from violet to red. The blue end of the spectrum is therefore scattered more than the red end. (The colors of the visible spectrum—and the rainbow—may be remembered with the help of the name of ROY G. BIV—red, orange, yellow, green, blue, indigo, and violet.)

As the sunlight passes through the atmosphere, the blue end of the spectrum is preferentially scattered. Some of this scattered light reaches the ground, where we see it as blue skylight. (See Fig. 1.) Keep in mind that all colors are present in skylight, but the dominant wavelength or color lies in the blue. You may have noticed that the skylight is more blue directly overhead or high in the sky and less blue toward the horizon, becoming white just above the horizon. You see these effects because there are fewer scatterers along a path through the atmosphere overhead than toward the horizon, and multiple scattering along the horizon path mixes the colors to give the white appearance. If Earth did not have an atmosphere, the sky would appear black except for the area in the immediate vicinity of the Sun.

Because Rayleigh scattering is greater the shorter the wavelength, you might be wondering why the sky isn't violet, since this color has the shortest wavelength in the visible spectrum. Violet light is scattered, but the eye is more sensitive to blue light than to violet light; also, sunlight contains more blue light than violet light. The predominant color component is yellow-green, and the distribution generally decreases toward both the violet and red ends of the spectrum.

The scattering of sunlight by the atmospheric gases *and* small particles gives rise to red sunsets. One might think that because the sunlight travels a greater distance through the atmosphere to an observer at sunset, most of the shorter wavelengths would be scattered from the sunlight and only light in the red end of the spectrum would reach the observer. However, the dominant color of this light, were it due solely to molecular scattering, would be orange. Hence there must be additional scattering by small particles in the atmosphere that shifts the light from the setting (or rising) Sun toward the red. Foreign particles (natural or pollutants) in the atmosphere are not necessary to give a blue sky and even detract from it. Yet such particles are necessary for deep red sunsets and sunrises.

The beauty of red sunrises and sunsets is often made more spectacular by layers of pink-colored clouds. The cloud color results from the reflection of red light.

Larger particles of dust, smoke, haze, and those from air pollution in the atmosphere may preferentially scatter long wavelengths. These scattered wavelengths, along with the scattered blue light resulting from Rayleigh scattering, can cause the sky to take on a milky blue appearance—white being the presence of all colors. Hence the intensity of blue of the sky gives an indication of atmospheric purity.

Keep in mind that it is the air temperature that is being measured. An air temperature measurement should not be made in sunlight, which may result in a higher temperature. A truer air temperature in the summer is often expressed as being so many degrees "in the shade," implying that in making a temperature measurement, the thermometer should not be exposed to the direct rays of the Sun.

Pressure is defined as the force per unit area ($p = F/A$). At the bottom of the atmosphere, we experience the resultant weight of the gases above us. Because we experience this weight before and after birth as a part of our natural environment, little thought is given to the fact that every square inch of our bodies sustain an average weight of 14.7 lb at sea level, or a pressure of 14.7 lb/in^2. We refer to 14.7 lb/in^2 as being *one standard atmosphere of pressure.*

Atmospheric pressure measurements are generally made by a device called a **barometer.** The principle of a *mercury* barometer is illustrated in Fig. 19.6a. A filled tube of mercury is inverted into a pool of mercury. Some runs out, but with standard atmospheric conditions, a column of mercury 76 cm (30 in.) is left in the tube. Because the column of mercury has weight, some force

Figure 1 Rayleigh scattering.
Left: The preferential scattering by molecules of the air causes the sky to appear blue, which in turn, along with scattering by small particles, produces red sunsets. *Above:* Rayleigh scattering in action—blue sky and red sunset.

must hold up the column. The only available force is that of the atmospheric pressure on the surface of the mercury pool. So, the greater the height, the greater the pressure, or $p \propto h$. Other liquids could be used in the barometer, but mercury, with a relatively large density, gives a smaller and more manageable column. (If water were used, the atmosphere would support a column of about 10 m!)

Because mercury is toxic, the liquid is in a sealed arrangement. A modern version of a mercury barometer is shown in Fig. 19.6b. Another type of barometer is the *aneroid* (without fluid) *barometer*. This is a mechanical device having a metal diaphragm that is sensitive to pressure, much like a drum head. Aneroid barometers with dial faces are common around the home. Altimeters used by airplane pilots and sky divers are really aneroid barometers. The pressure reading decreases rather uniformly with decreasing height in the troposphere. When a barometer dial face is replaced with one calculated inversely in height, the barometer becomes an altimeter.

The standard units of pressure ($p = F/A$) are N/m^2 or lb/in^2. However, these units are not used for barometric readings given on radio and TV weather reports. Instead, the readings are ex-

HIGHLIGHT

The Greenhouse Effect

The absorption of terrestrial radiation, primarily by water vapor and CO_2, adds to the energy content of the atmosphere. This heat-retaining process of such gasses is referred to as the **greenhouse effect,** because of a similar effect that occurs in greenhouses. The absorption and transmission properties of regular glass are similar to those of atmospheric gases—in general, visible radiation is transmitted, and infrared radiation is absorbed (Fig. 1).

We have all observed the warming effect of sunlight passing through glass; for example, in a closed car on a sunny, but cold, day. In a greenhouse, the objects inside become warm and reradiate long-wavelength infrared radiation, which is absorbed and reradiated by the glass. Thus, the air inside a greenhouse heats up and is quite warm on a sunny day, even in the winter. Actually, in this case, the maintained warmth is primarily due to the prevention of the escape of warm air by the glass enclosure.

The greenhouse effect in the atmosphere is quite noticeable at night, particularly on cloudy nights. With a cloud and water vapor cover to absorb the terrestrial radiation, the night air is relatively warm. Without this insulating effect, the night is usually "cold and clear," because the energy from the daytime insolation is quickly lost.

In spite of the daily and seasonal gain and loss of heat, the average temperature of Earth has remained fairly constant. Thus, Earth must lose or reradiate as much energy as it receives. If it did not, the continuous gain of energy would cause Earth's average temperature to rise. It is the selective absorption of atmospheric gases that provides a thermostatic or heat-regulating process for the planet.

To illustrate, suppose that Earth's temperature was such that it emitted radiation with wavelengths that were absorbed by the atmospheric gases. As a result of this absorption, the lower atmosphere would become warmer and effectively hold in the heat, thus insulating Earth. With additional insolation, Earth would become warmer and its temperature would rise. But, according to the $\lambda \propto 1/T$ relationship, the greater the temperature, the shorter the wavelength of the emitted radiation. And so, the wavelength of the terrestrial radiation would be shifted to a shorter wavelength.

The wavelength would eventually be shifted to a "window" in the absorption spectrum where little or no absorption takes place and the terrestrial radiation would pass through the atmosphere into space. Thus, Earth would lose energy and its temperature would decrease. But with a temperature decrease, the terrestrial radiation would return to a longer wavelength, which would be absorbed by the atmosphere. So, we have a turning on and off, so to speak, similar to the action of a thermostat. Averaged over the total spectrum, the greenhouse selective absorption of gases plays an important role in maintaining Earth's average temperature.

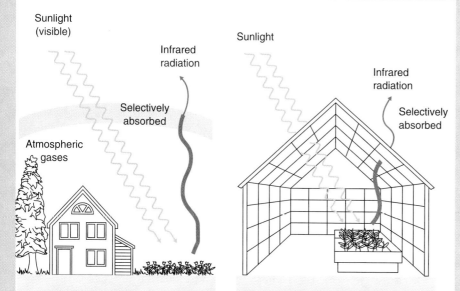

Figure 1 The greenhouse effect.
The gases of the lower atmosphere transmit most of the visible portion of the sunlight, as does the glass of a greenhouse. The warmed Earth emits infrared radiation, which is selectively absorbed by atmospheric gases, whose absorption spectrum is similar to that of glass. This absorbed energy heats the atmosphere and helps maintain Earth's average temperature.

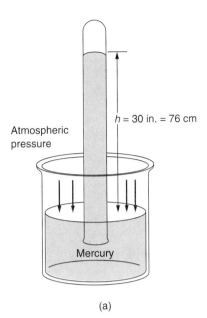

(a)

Figure 19.6 Mercury barometer.
(a) An illustration of the principle of the barometer. The external air pressure on the surface of a pool of mercury supports the mercury column in the inverted tube. The height of the column depends on the atmospheric pressure and provides a means for measuring it. (b) A mercury barometer mounted on a wall in a laboratory.

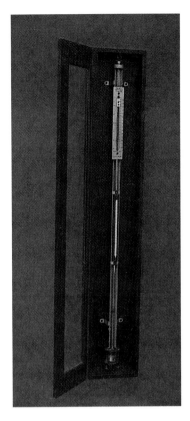

(b)

pressed in length units, that is, so many "inches" (of mercury). For weather circumstances, we are primarily interested in changes of pressure, and since $p \propto h$, the variation in column heights will give this directly. Another unit you may hear used occasionally is the *millibar* (mb), and one atmosphere (atm) is about 1000 mb.

CONFIDENCE QUESTION 19.2

How would a pressure cooker help with cooking at high altitudes?

Humidity is a measure of the moisture, or water vapor, in the air. It affects our comfort and indirectly our ambition and state of mind. In the summer, it may be hot and humid, and we run dehumidifiers. In the winter, the air in heated homes may be dry, and we run humidifiers so as to place moisture in the air. On weather reports we hear what the *relative* humidity is in terms of a percent; for example, the relative humidity is 70%. This is a relative measure of how much moisture the air contains compared to the maximum amount it could contain (at a particular air temperature). The amount of humidity in the air is of prime importance in condensation and precipitation. Therefore, we will defer a detailed discussion of humidity until the next section, which describes how clouds are formed.

Wind speed and direction is ordinarily included in weather reports. Wind speed is measured with an **anemometer.** This instrument consists of three or four cups attached to a rod that is free to rotate, like a pin wheel. The cups catch the wind, and the greater the wind speed, the faster the anemometer rotates (Fig. 19.7).

A **wind vane** indicates the direction from which the wind is blowing. This instrument is simply a freely rotating indicator that, because of its shape, lines up with the wind and points the wind

Figure 19.7 Anemometer and wind vane.
An array of two anemometers and two wind vanes is shown here. The shapes of the wind vanes cause them to point in the direction *from which* the wind is blowing.

direction (Fig. 19.7). Keep in mind that wind direction is reported as the direction *from which* the wind is coming. For example, a north wind is coming *from* the north (and blowing south).

The major forms of precipitation are rain and snow. Rainfall is measured with a **rain gauge.** This device is simply an open container with its height calibrated in inches that is placed outdoors. After a rainfall the rain gauge is read, and the amount of precipitation is reported in inches. The assumption is that this much rainfall is distributed relatively evenly over the surrounding area.

If precipitation is in the form of snow, the depth of the snow (where not drifted) is reported in inches. The actual amount of water received depends on the density of snow. To obtain this, a rain gauge is sprayed with a chemical that melts the snow, and the actual amount of water is recorded.

Other features commonly seen on TV weather reports are radar scans of your area and satellite photos. **Radar** (*ra*dio *d*etecting *a*nd *r*anging) is used to detect and monitor clouds and precipitation. It operates on the principle of reflected electromagnetic waves. Continuous radar scans are commonly seen on TV reports.

Probably the greatest progress in general weather observation has come with the advent of the *weather satellite*. Before satellites, weather observations were unavailable for more than 80% of the globe. Weather satellites were first operational in the 1960s and only monitored a limited area under their low orbital paths. Today, a fleet of satellites, which orbit at very high altitudes (22,000 mi) have the same orbital period of Earth's rotation and hence are "stationary" over a particular location. At such an altitude, satellites can send back pictures of large portions of Earth's surface. Geographic boundaries and grids are prepared by computer and electronically combined with the picture signal so areas of particular weather disturbances can be easily identified (Fig. 19.8).

19.3 Air Motion and Clouds

Learning Goals:

To describe what causes air motion, and analyze some local winds and Earth's general circulation pattern.

To explain how clouds are formed and classified.

If the air in the atmosphere were static, there would be little change in the local atmospheric conditions that constitute weather. Air motion is important in many processes, even biological—for example, in carrying scents and the distribution of pollen. As air moves into a region, it brings with it the temperature and humidity that are mementos of its travels.

Wind is the horizontal movement of air or air motion along Earth's surface. Vertical air motions are referred to as updrafts and downdrafts, or collectively as **air currents.** As in all dynamic situations, forces are necessary to produce motion and changes in motion. The gases of the atmosphere are subject to two primary forces: *gravity* and *pressure differences due to temperature variations*. Once the air is in motion, friction may retard and cause direction changes. Also, Earth's rotation has an effect on the motion of air.

The force of gravity is vertically downward and acts on each gas molecule. Although this force is often overruled by forces in other directions, the downward gravity component is ever-present and

Figure 19.8 Satellite image. A weather picture for the southeastern United States. Most of the states shown have clear skies. The notable exceptions are South Carolina and Georgia, where the photo shows hurricane Hugo making landfall on September 22, 1989.

accounts for the greater air density near Earth's surface.

Because air is a mixture of gases, its behavior is governed by the gas laws of Chapter 5 and other physical principles. The pressure of a gas is directly proportional to its temperature ($p \propto T$), so if there is a temperature variation, there will be a pressure difference ($\Delta p \propto \Delta T$). Recall that pressure is the force per unit area, so a pressure difference corresponds to an unbalanced force. When there is a pressure difference, the air moves from a high-pressure to a low-pressure region.

Recall from Chapter 5 that the pressure and volume of a gas is related to its temperature ($pV \propto T$). A change in temperature, then, causes a change in the pressure and/or volume of a gas. With a change in volume, there is also a change in density ($\rho = m/V$). For example, if the air is heated and expands, the air density decreases. As a result of this relationship, localized heating sets up air motion in a **convection cycle,** or gives rise to thermal circulation (Fig. 19.9).

Convection cycles due to geological features give rise to local winds. Land areas heat up more quickly during the day than do water areas; and the warm, buoyant (less dense) air over the land rises. As air flows horizontally into this region, the rising air cools and falls, and a convection cycle is set up. As a result, during the day when the land is warmer than the water, a lake or *sea breeze* is experienced, as shown in Fig. 19.9a. You may have noticed these daytime sea breezes at an ocean beach. Notice that a wind is named after the direction *from which* it comes. For example, a wind blowing in from the sea is called a *sea breeze*. A wind blowing from north to south is called a *north wind*.

At night, the land loses its heat more quickly than the water, and the air over the ocean is warmer. The convection cycle is then reversed, and at night a *land breeze* blows (Fig. 19.9b). Sea and land breezes are sometimes referred to respectively as *onshore and offshore winds*.

Because of Earth's rotation, projectiles or particles moving in the Northern Hemisphere are apparently deflected to the right as observed in the direction of motion. Similarly, particles moving in the Southern Hemisphere are deflected to the left. This effect (called the *Coriolis force*) varies with latitude, being zero at the equator and increasing toward the poles.

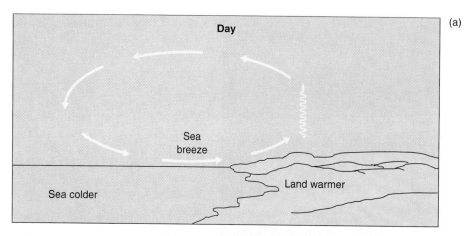

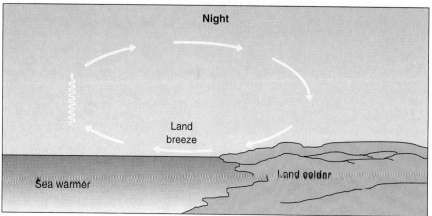

Figure 19.9 Daily convection cycles over land and water.
(a) During the day, the land surface heats up more quickly than a large body of water, and this sets up a convection cycle in which the surface winds are from the water—a sea breeze. (b) At night, the land cools more quickly than the water, and the cycle reverses, with the surface winds coming from the land—a land breeze.

Consider this effect on wind motion. Initially, air moves toward a low-pressure region (a "low") and away from a high-pressure region (a "high"). Because of the apparent rotational deflection, in the Northern Hemisphere the wind tends to rotate counterclockwise around a low and clockwise around a high, as viewed from above (Fig. 19.10). These disturbances are referred to as *cyclones* and *anticyclones*, respectively. Water motion or currents in the oceans are also affected by this effect.

CONFIDENCE QUESTION 19.3

What are the circulation patterns around cyclones and anticyclones in the Southern Hemisphere?

Air motion changes locally with altitude, geographical features, and the seasons. However, the air near Earth's surface does possess a general circulation pattern. Because of rotational effects, land and sea variations, and other complicated reasons, the hemispheric circulation is broken up into six general convection cycles, or pressure cells, as shown in Fig. 19.11. Many local variations occur within the cells, which shift seasonally in latitude because of variations in insolation. However, the prevailing winds of this semipermanent circulation structure are important in influencing general weather movements around the world.

The conterminous United States lies generally between the latitudes of 30°N and 50°N.[*] Note in Fig. 19.11 that this is in the westerly wind zone. As

---

[*]*Conterminous* means contiguous or adjoining, so the conterminous United States refers to the original 48 states. This is different from the *continental* United States. Why?

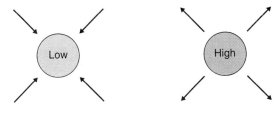

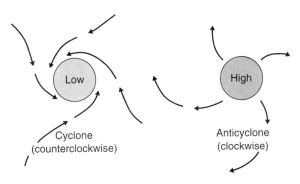

Figure 19.10 Effects of the Coriolis force on air motion.
In the Northern Hemisphere, the Coriolis deflection to right produces counterclockwise air motion around a low and clockwise rotation around a high (as viewed from above).

a result, our weather conditions generally move from west to east across the country.

You now commonly hear about another, high-altitude wind on TV weather reports, particularly in the winter. In the upper troposphere there are fast-moving "rivers" of air called **jet streams.** Several jet streams meander like rivers around each hemisphere. The behavior of jet streams is variable and not well understood. The so-called polar jet stream moves from west to east across the United States. It varies in altitude and latitude with the seasons, reaching lower latitudes in the winter. This jet stream is believed to influence the severity of our winters.

Clouds

Clouds are both a common sight and an important atmospheric consideration. **Clouds** are buoyant masses of visible water droplets or ice crystals. Their size, shape, and behavior are useful keys to the weather.

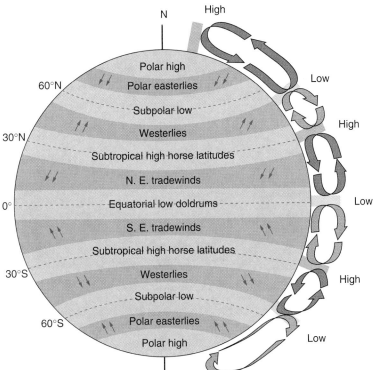

Figure 19.11 Earth's general circulation structure.
For rather complicated reasons, Earth's general circulation structure has six large convection cycles or cells. The conterminous United States lies in the westerly wind zone.

Table 19.2 Cloud Families and Types

High clouds (above 6 km). All composed of ice crystals.

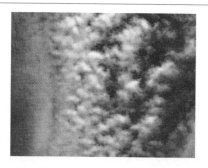

(a)
Cirrus Wispy and curling. Known as "artist's brushes" or "mares' tails."
Cirrocumulus Layered patches. Known as "mackerel scales."

(b)
Cirrostratus Thin veil of ice crystals. Scattering from crystals gives rise to solar and lunar halos.

Middle clouds (1.8–6 km). All names have the prefix *alto-*.

(c)
Altostratus Layered forms of varying thicknesses. May hide the Sun or Moon and cast a shadow.

(d)
Altocumulus Woolly patches or rolled, flattened layers.

Low clouds (ground level–1.8 km)

(e)
Stratocumulus Long layers of cottonlike masses, sometimes with wavy appearance.

(f)
Stratus Thin layers of water droplets. May appear dark, common in winter. Fog may be thought of as low-lying stratus clouds.

(g)
Advection fog Forms when moisture in air moving over a colder surface condenses. Advection fogs "roll in."

Low clouds (**ground level–1.8 km**) *(continued)*

(h)
Radiation fog Condensation in stationary air overlying a surface that cools. Occurs typically in valleys, and called a "valley fog."

(i)
Nimbostratus Dark, low clouds given to precipitation.

Clouds of vertical development (**5–18 km**). **Formed by updrafts.**

(j)
Cumulus Billowy; commonly seen on a clear day.

(k)
Cumulonimbus Darkened cumulus clouds, referred to as "thunderheads."

Clouds are classified according to their shape, appearance, and altitude. There are four basic root names: *cirrus*, meaning "curl" and referring to wispy, fibrous forms; *cumulus*, meaning "heap" and referring to billowy, round forms; *stratus*, meaning "layer" and referring to stratified or layered forms; and *nimbus*, referring to a cloud from which precipitation is occurring or threatens to occur. These root forms are then combined to describe types of clouds and precipitation potential.

When classified according to altitude or height, clouds are separated into four families: (1) **high clouds,** (2) **middle clouds,** (3) **low clouds,** and (4) **clouds of vertical development.** These families are listed in Table 19.2 along with their approximate heights and cloud types belonging to each family. Brief descriptions or common names are also given, along with a collection of illustrative photos in Table 19.2. Study these so you will be able to identify the clouds you see by name.

Humidity and Cloud Formation

Humidity refers to the invisible water vapor in the air, whereas clouds are visible droplets of water. Clouds form in the air, and so their formation must be related in some way to the humidity. That is, when does the invisible water vapor form into visible droplets and clouds? To understand this, let's first take a closer look at humidity and how it is measured.

Humidity is a measure of the moisture content of the air and can be expressed in several ways. *Absolute humidity* is simply the amount of water vapor in a given volume of air. In the United States, this is commonly expressed in units of grains per cubic foot. The grain (gr) is a small weight unit, with 1 lb = 7000 gr. Medicines are sometimes measured in grains, for example, a 5-gr aspirin tablet. An average value of humidity is around 4.5 gr/ft^3.

The most common method of expressing the water vapor content of the air is in terms of relative humidity. **Relative humidity** is the ratio of the actual moisture content and the maximum moisture capacity of a volume of air at a given temperature. This ratio is commonly expressed as a percentage:

$$(\%)\ RH = \frac{AC}{MC}\ (\times 100\%) \qquad (19.1)$$

where *RH* is the relative humidity,

AC is the actual moisture content of a volume of air, and

MC is the maximum moisture capacity of the volume of air.

The actual moisture content (*AC*) is the absolute humidity or the amount of water vapor in a given volume of air. The maximum moisture capacity (*MC*) is the maximum amount of water vapor that a given volume of air can hold *at a given temperature.*

Relative humidity is essentially a measure of how "full" of moisture a volume of air is at a given temperature. For example, if the relative humidity is 0.50 or 50%, then a volume of air is "half full," or contains half as much water as it is capable of holding at that temperature.

To better understand how the water vapor content of air varies with temperature, consider an analogy of a saltwater solution. Just as a given amount of water at a certain temperature can dissolve only so much salt, a volume of air at a given temperature can hold only so much water vapor. When the maximum amount of salt is dissolved in solution, we say the solution is *saturated* (Section 11.1). This condition is analogous to a volume of air having its maximum moisture capacity or all the water vapor it can hold at a particular temperature.

The addition of more salt to a saturated solution results in salt on the bottom of the container. However, more salt may be put into solution if the water is heated and the temperature is raised. Similarly, when air is heated, it can hold more water vapor at a higher temperature. That is, warm air has a greater capacity for water vapor than does colder air.

Conversely, if the temperature of a nearly saturated salt solution is lowered, at a certain temperature the solution will become saturated. Any additional lowering of temperature will cause the salt to crystallize and come out of solution, because it will be oversaturated. In an analogous fashion, if the temperature of a sample of air is lowered, it will become saturated at some temperature.

The temperature to which a volume of air must be cooled to become saturated is called the **dew point** (temperature). Hence, at the dew point, the relative humidity is 100%. (Why?) Cooling below this point causes oversaturation and may result in condensation to form the water droplets of clouds.

Humidity may be measured by several means, but the most common method uses the **psychrometer.** This instrument consists of two thermometers, one of which measures the air temperature, while the other has its bulb surrounded by a cloth wick that is kept wet. These thermometers are referred to as the *dry bulb* and *wet bulb,* respectively. They may be simply mounted, as in Fig. 19.12.

The dry bulb measures the air temperature, while the wet bulb has a lower reading that is a function of the amount of moisture in the air. This reading occurs because of evaporation of water from the wick around the wet bulb. The evaporation removes (latent) heat from the thermometer bulb. If the humidity is high and the air contains a lot of water vapor, little water evaporates and the wet bulb is only slightly cooled. Consequently, the temperate of the wet bulb is only slightly lower than the temperature of the dry bulb, or the wet-bulb reading is only slightly "depressed."

If, however, the humidity is low, a great deal of evaporation occurs, accompanied by considerable cooling, and the wet-bulb reading will be considerably depressed. Hence, the temperature difference of the thermometers, or depression of the wet-bulb reading, is a measure of relative humidity.

Using the air (dry-bulb) temperature and the wet-bulb depression, the relative humidity, the maximum moisture capacity, and the dew point can be read directly from Tables A.1 and A.2 in Appendix X. An example of how this is done follows.

EXAMPLE 19.1

The dry bulb and wet bulb of a psychrometer have respective readings of 80°F and 73°F. Find the following for the air: (a) relative humidity, (b) maximum moisture capacity, (c) actual moisture content, and (d) dew point.

Solution

Given: 80°F (dry-bulb reading)
73°F (wet-bulb reading)

Find: (a) *RH* (relative humidity)
(b) *MC* (maximum capacity)
(c) *AC* (actual content)
(d) dew point (temperature)

First, we find the wet-bulb depression,

$$\Delta T = 80°F - 73°F = 7°F$$

(a) Then, using Table A.1 in Appendix X to determine the relative humidity, we find the dry-bulb temperature in the first column and then locate the wet-bulb depression in the top row of the table. Move down the column under the wet-bulb depression to the row that corresponds to the dry-bulb temperature reading. The intersection of the row and column gives the value of the percent relative humidity. (It may be helpful to move a finger down and another one across to help find the intersection.) The relative humidity is 72% in this case.

(b) The maximum moisture capacity (*MC*) is read directly from the table. Find the dry-bulb temperature in the first column, and the *MC* for that temperature is given in the adjacent column. In this case, *MC* = 10.9 gr/ft³.

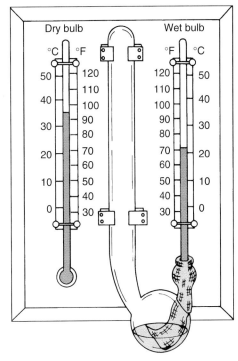

Figure 19.12 Psychrometer and relative humidity.
The dry bulb of a psychrometer records the air temperature, which is greater than that of the wet bulb (because of evaporation). The lower the humidity, the greater the evaporation and the greater the difference in the temperature readings. The temperature difference between the two thermometers is then inversely proportional to the humidity. Thus, the psychrometer provides a means for measuring relative humidity.

(c) Knowing *RH* and *MC*, we can find the actual moisture content (*AC*) from Eq. 19.1, *RH* = *AC*/*MC*. Rearranging.

$$AC = RH \times MC = 0.72 \times 10.9 \text{ gr/ft}^3 = 7.8 \text{ gr/ft}^3$$

(d) The dew point is found using Table A.2 in the same manner as the relative humidity was found in part (a). The intersection value of the appropriate row and column is 70°F.

Hence, if the air is cooled to 70°F, it will be saturated and the relative humidity 100%, with *AC* = *MC*. Note from Table A.1 that the *MC* for 70°F is 7.8 gr/ft³, which is the value found in part (c). The value of the actual content at 80°F should correspond to the maximum capacity at the 70°F dew point.

CONFIDENCE QUESTION 19.4

On a particular day, the dry-bulb and wet-bulb readings of a psychrometer are 70°F and 66°F, respectively. (a) What is the relative humidity, and (b) how many degrees would the air temperature have to be lowered for the air to be saturated?

Cloud Formation

To be visible as droplets, the water vapor in the air must condense. Condensation requires a certain temperature—namely, the dew point temperature. Hence, if moist air is cooled to the dew point, the water vapor contained therein will generally condense into fine droplets and form a cloud.

The air is continuously in motion, and when an air mass moves into a cooler region, cloud formation may take the place. Because the temperature of the troposphere decreases with height, cloud formation is associated with the vertical movement of air. In general, clouds are formed in vertical air movement or currents, and are shaped and moved about by horizontal air movement or winds.

Air may rise as a result of wind motion along an elevating surface, such as up the side of a mountain or boundary of a front (Section 19.4), or as a result of heating. We will consider here cloud formation resulting from the rising of warm, buoyant air (clouds of vertical development). As the warm air ascends, it becomes cooler because it expands and its internal energy is used to do the work of expansion against the surrounding stationary air—with less energy it becomes cooler. The temperature of the air in the troposphere decreases with altitude, and the rate of this temperature decrease with height is called the **lapse rate.** The normal lapse rate in stationary air in the troposphere is about $6\frac{1}{2}$ C°/km (or $3\frac{1}{2}$ F°/1000 ft).

Because energy is used in the expansion of a warm air mass, the rising air has a greater lapse rate than does the surrounding air. Thus, rising air cools more quickly. When the rising air mass cools to the same temperature as the surrounding stationary air, their densities become equal. The rising air mass then loses its buoyancy and is said to be in a stable condition. A heated air mass rises until stability is reached, and this portion of the atmosphere is referred to as a stable layer.

Figure 19.13 Vertical cloud development.
The cloud begins to form at the elevation where water vapor in rising air condenses into small droplets and can be seen. When the rising air is cooled to the same temperature as the surrounding air, stability is reached and this level defines the top of the cloud. Winds shape the clouds and may break them up into smaller forms. Note the anvil shape of this cumulus cloud as its upper part is moved along the stable layer.

Clouds are formed when water vapor in the rising air condenses into droplets and can be seen. If the rising air reaches its dew point before becoming stable, condensation occurs at that height; and the rising air carries the condensed droplets upward, forming a cloud (Fig. 19.13).

CONFIDENCE QUESTION 19.5

Once a cloud is formed, is it permanent, or are clouds destroyed in some manner? Explain.

It was stated that condensation occurs when the dew point is reached. However, it is quite possible for an air mass containing water vapor to be cooled below the dew point without condensation occurring. In this state, the air is said to be *supersaturated* or *supercooled*. How, then, are visible droplets formed? The probability of water molecules coming together is quite remote. Instead, water droplets form around microscopic foreign particles, called *hygroscopic nuclei*, already present in the air in the forms of dust, smoke, and so on. Because foreign particles initiate the formation of droplets that may eventually fall as precipitation, condensation provides a mechanism for cleansing the air.

Larger droplets of precipitation are believed formed by what is known as the **Bergeron process,** named after the Swedish meteorologist who suggested it. This process involves clouds that contain ice crystals in their upper portions and have some supercooled vapor in their lower portions (Fig. 19.14). Mixing or agitation within such a cloud allows the ice crystals to come into contact with the supercooled vapor. Acting as nuclei, the ice crystals grow larger from the vapor condensing on them. The ice crystals melt into large droplets in the warmer lower portion of the cloud, coalesce or grow greater by collision, and then fall as precipitation. Air currents are the normal mixing agents.

Closer to home, results of cooled vapor may be observed. *Dew* is formed by atmospheric water vapor condensing on various surfaces. The land cools quickly at night, and the temperature may fall below the dew point. Water vapor then condenses on available surfaces such as blades of grass, giving rise to the "early morning dew."

If the dew point is below freezing, the water vapor condenses in the form of ice crystals as *frost*. Frost is not frozen dew, but results from the direct change of water vapor to ice (the reverse process of sublimation, called *deposition*).

19.4 Air Masses and Storms

Learning Goals:

To define air masses and how they are classified.

To identify various types of local and tropical storms.

As we know, the weather changes with time. However, we often experience several days of relatively uniform weather conditions. Our general weather depends in large part on vast air masses that move across the country.

When a large body of air takes on physical characteristics that distinguish it from the surrounding air, it is referred to as an **air mass.** The main distinguishing characteristics are tempera-

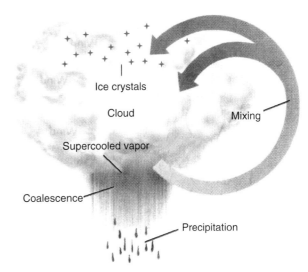

Figure 19.14 The Bergeron process.
The essence of the Bergeron process is the mixing of ice crystals and supercooled vapor, which produces water droplets and initiates precipitation.

ture and moisture content. A mass of air remaining for some time over a particular region, such as a large body of land or water, takes on the physical characteristics of the surface of the region.

The region from which an air mass derives its characteristics is called its **source region.** The time required for an air mass to take on its source region's characteristics depends on the surface conditions.

An air mass eventually moves from its source region, bringing its characteristics to regions in its path and causing changes in the weather. As an air mass travels, its properties may become modified because of local variations. For example, if Canadian polar air masses did not become warmer as they move southward, Florida would experience some extremely cold temperatures.

Whether an air mass is termed cold or warm is relative to the surface over which it moves. Quite logically, if an air mass is warmer than the land surface, it is referred to as a *warm air mass*. If colder than the surface, it is called a *cold air mass*. Remember, though, that these terms are relative. The warm and cold prefixes do not always imply warm and cold weather. A "warm" air mass in winter may not raise the temperature above freezing.

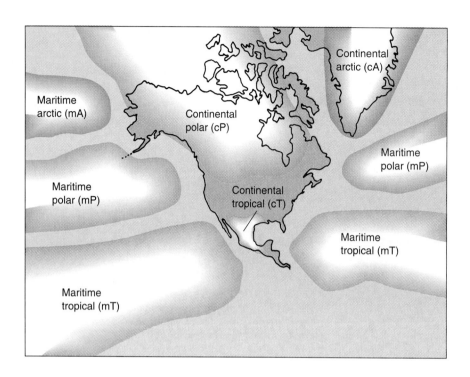

Figure 19.15 Air-mass source regions.
The map shows the source regions for the air masses of North America.

Air masses are classified according to the surface and general latitude of their source regions:

Surface	Latitude
Maritime (m)	Arctic (A)
Continental (c)	Polar (P)
	Tropical (T)
	Equatorial (E)

The surface of the source region, abbreviated by a small letter, gives an indication of the moisture content of an air mass. Forming over a body of water (maritime), an air mass would naturally be expected to have a greater moisture content than one forming over land (continental).

The general latitude of a source region, abbreviated by a capital letter, gives an indication of the temperature of an air mass. For example, mT designates a maritime tropical air mass, which would be expected to be a warm, moist one. A list of the air masses that affect the weather in the conterminous United States is given in Table 19.3, along with their source regions, and are illustrated in Fig. 19.15.

CONFIDENCE QUESTION 19.6

What type of air mass would be formed over the southern Pacific (between Australia and South America)?

Table 19.3 Air Masses That Affect the Weather of the United States

Classification	Symbol	Source Region
Maritime arctic	mA	Arctic regions
Continental arctic	cA	Greenland
Maritime polar	mP	Northern Atlantic and Pacific oceans
Continental polar	cP	Alaska and Canada
Maritime tropical	mT	Caribbean Sea, Gulf of Mexico, and Pacific Ocean
Continental tropical	cT	Northern Mexico, southwestern United States

The boundary between two air masses is called a **front**. A *warm front* is the boundary of an advancing warm air mass over a colder surface, and a *cold front* is the boundary of a cold air mass moving over a warmer surface. (The graphic symbols of the fronts are often seen on weather maps. The scallops and triangles indicate the direction the front is moving.) It is along fronts, which divide air masses of different physical characteristics, that drastic changes in weather occur. Turbulent weather and storms usually characterize a front.

Horizontal views of two advancing fronts are illustrated in Fig. 19.16, along with the general widths of the frontal zones in which changes occur. An advancing cold front moving into a warmer region causes the lighter, warm air to be displaced upward over the front. Cold fronts have sharper vertical boundaries than warm fronts. As a result, cold fronts are accompanied by more violent or sudden changes in weather. The sudden decrease in temperature is often described as a "cold snap." The sudden cooling and rising warm air may set off rainstorm or snowstorm activity along the front.

A warm front may also be characterized by precipitation and storms. Because of the warm air rising over the colder air, the approach of a warm front is more gradual. It is usually heralded by a period of lowering clouds as illustrated in Fig. 19.16. Also, warm fronts travel from 15 to 25 km/h (10 to 15 mi/h), about half the speed of a cold front.

As a faster-moving front advances, it may overtake another air mass and push it upward. The boundary between these two air masses is called an *occluded front* and is indicated graphically by . The symbol indicates that there is one front on top of another, having been *occluded* from the surface. Sometimes fronts traveling in opposite directions meet. The opposing front may balance each other so that no movement occurs. This case is referred to as a *stationary front* ( ).

Air masses and fronts move across the country bringing changes in the weather. Dynamic situations give rise to cyclonic disturbances around low-pressure and high-pressure regions, as learned earlier; and fronts tend to move around these regions. As a low or cyclone moves, it carries with it rising air currents, clouds, and possible precipita-

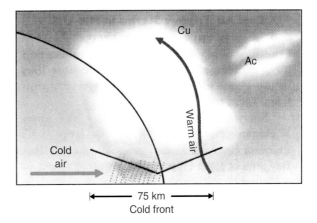

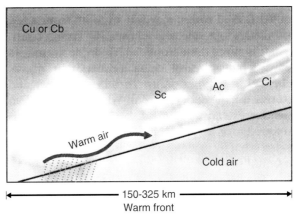

Figure 19.16 Side views of cold and warm fronts.
Notice in the upper diagram the sharp, steep boundary that is characteristic of cold fronts. The boundary of a warm front, as shown in the lower diagram, is less steep. As a result, different cloud types are associated with the approach of the two types of fronts.

tion and generally bad weather. Highs or anticyclones, on the other hand, are generally associated with good weather. The lack of rising air and cloud formation in highs gives clear skies and fair weather.

There are a variety of storms in the lower troposphere. Let's look at a few. Locally, we have *rainstorms*, with which everyone is familiar. More impressive is a *thunderstorm*, which is a rainstorm distinguished by lightning and thunder. In the turmoil of a thundercloud or "thunderhead," there is a separation of charge associated with the breaking up and movement of water droplets. This charge separation gives rise to an electric potential (voltage). When this is of sufficient magnitude, **lightning** occurs. Lightning can take place within

Figure 19.17 Lightning.
Lightning discharges can occur between a cloud and Earth, between clouds, and within a cloud.

a cloud (intracloud or cloud discharges), between two clouds (cloud-to-cloud discharges), or between a cloud and Earth (cloud-to-ground discharges), or between a cloud and the surrounding air (air discharges). See Fig. 19.17.

Lightning has even been reported to occur in clear air, apparently giving rise to the expression "a bolt from the blue." When lightning occurs below the horizon or behind clouds, it often illuminates the clouds with flickering flashes. This commonly occurs on a still summer night and is known as *heat lightning*.

Lightning causes a great deal of damage and even death. To know what to do in case of a thunderstorm, see the boxed feature on lightning safety.

A lightning stroke's sudden release of energy explosively heats the air, producing compressions we hear as **thunder**. When one is relatively near the lightning stroke, thunder consists of one loud bang or "clap." When heard at a distance of 1 km (0.62 mi) from the discharge channel, thunder generally consists of a rumbling sound punctuated by several large claps. In general, thunder cannot be heard at distance of more than 25 km (16 mi) from the discharge channel.

Because lightning strokes generally occur near the storm center, the resultant thunder provides a method of approximating the distance to the storm. Light travels at approximately 297,600 km/s

Lightning Safety

If you are outside during a thunderstorm and feel an electrical charge, as evidenced by hair standing on end or skin tingling, fall to the ground fast! Lightning may be about to strike.

Statistics show that lightning kills, on the average, 200 people a year in the United States and injures another 550. Most deaths and injuries occur at home. Indoor casualties occur most frequently when people are talking on the telephone, working in the kitchen, doing laundry, or watching TV. During severe lightning activity, the following safety rules are recommended:

- Stay indoors away from open windows, fireplaces, and electrical conductors such as sinks and stove.
- Avoid using the telephone. Lightning may strike the telephone lines outside.
- Do not use electrical plug-in equipment such as radios, TVs, and lamps.
- Should you be caught outside, seek shelter in a building. If no buildings are available, seek protection in a ditch or ravine, *not* under a tree. Getting wet is a lot better than being struck by lightning.

A person in the vicinity of a lightning strike may experience an electric shock that causes breathing to fail. In such a case, mouth-to-mouth resuscitation or some other form of artificial respiration should be given immediately, and the person kept warm as a treatment for shock.

(186,000 mi/s), so a lightning flash is seen instantaneously. Sound, however, travels at approximately $\frac{1}{3}$ km/s (or $\frac{1}{5}$ mi/s), so a time lapse occurs between seeing the lightning flash and hearing the thunder. This phenomenon is also observed if you are watching someone at a distance fire a gun or hit a baseball. The report of the gun or the "crack" of the bat is heard after the smoke or flash of the gun is observed or the baseball is well on its way.

By counting the seconds between seeing the lightning and hearing the thunder (by saying one-thousand-one, one-thousand-two, etc.), you can estimate your distance from the lightning stroke or storm. For example, if 5 s elapsed, then the distance would be approximately 1.6 km (1 mi), taking the speed of sound to be $\frac{1}{3}$ km/s or $\frac{1}{5}$ mi/s. That is, $\frac{1}{3}$ km/s $\times$ 5 s = 1.6 km (or $\frac{1}{5}$ mi/s $\times$ 5 s = 1 mi).

CONFIDENCE QUESTION 19.7

If thunder is heard 3 s after seeing the flash of a lightning stoke, approximately how far away is the lightning?

Large pellets of ice, or *hail*, are also associated with thunderstorms. The pellets result from successive vertical descents and ascents in vigorous convection cycles into supercooled regions that are below freezing. This may produce layered-structure hailstones of golf-ball and baseball sizes. When cut in two, the layers of ice can be observed, much like the rings in tree growth.

There are some common storms associated with cold weather. If the dew point is below 0°C, the water vapor freezes on condensing, and ice crystals fall as *snow*. When a snowstorm is accompanied by high winds and low temperatures, it is commonly referred to as *blizzard*. The wind whips the snow into blinding swirls, and when blowing across level terrain, huge drifts are formed against some obstructing objects. Drifting is common on the flat prairies of the western United States.

Frozen rain, or pellets of ice in the form of *sleet*, occurs when rain falls through a cold surface layer of air and freezes or, more likely, when the ice pellets fall directly from a cloud without melting before striking the ground. If the temperature of Earth's surface is below 0°C and raindrops do not freeze before striking the ground, the rain will freeze on striking cold surface objects. The process of building up the resultant glaze is called an *ice storm*. The ice layer may build up to over half an inch in thickness, depending on the magnitude of the rainfall.

The most violent of storms is the **tornado.** The concentration of energy in a relatively small region gives the tornado this violent distinction. Charac-

(a)

(b)

Figure 19.18 Tornado or "twister."
(a) A tornado funnel touched down. Destruction lies in its path. (b) The high-speed winds of a tornado can push in the windward wall of a house, lift off the roof, and push the other wall outward. Notice all of the debris that was flying around when the tornado passed.

terized by a whirling, funnel-shaped cloud that hangs from a dark cloud mass, the tornado is commonly known as a *twister* (Fig. 19.18).

Tornadoes occur around the world but are most prevalent in the United States and Australia.

In the United States, most tornadoes occur in the Deep South and in a broad, relatively flat basin between the Rockies and Appalachians. But no state is immune. The peak months of tornado activity are April, May, and June, with southern states usually hit hardest in the winter and spring, and northern states in spring and summer. However, tornadoes have occurred in every month at all times of the day and night. A typical time of occurrence is on an unseasonably warm, sultry spring afternoon between 3 and 7 P.M.

Most tornadoes travel from southwest to northeast, but the direction of travel can be erratic and may change suddenly. They usually travel at an average speed of about 48 km/h (30 mi/h), and the wind speed of a major tornado may vary from 160 to 480 km/h (100 to 300 mi/h).

The complete mechanism of tornado formation is not known. One essential component, however, is rising air, which occurs in thunderstorm formation and in the collision of cold and warm air masses. As the ascending air cools, clouds are formed that are swept to the outer portions of the cyclonic motion and outline its funnel form. Under the right conditions, a full-fledged tornado develops. When the funnel is fully developed, it may "touch down" or be seen extending up from the ground as a result of dust and debris picked up by the swirling winds.

The alerting system for tornadoes has two phases. A **tornado watch** is issued when atmospheric conditions indicate that tornadoes may form. A **tornado warning** is issued when a tornado has actually been sighted or indicated on radar. The similarity between the terms *watch* and *warning* is sometimes confusing. Remember that you should watch for a tornado when the conditions are right; and when you are given a warning, the situation is dangerous and critical—no more watching. (See boxed feature on tornado safety.)

We mentioned that a tornado can be detected on radar. Radio waves are reflected by the water droplets in clouds and storms and so can be detected. Conventional radar, however, can detect the hooked signature of a tornado only after the storm is well developed, and provides just over a 2 minute-warning (Fig. 19.19a). A more advanced radar system has been developed, and a network of new weather radars is being deployed. The new system is called **Doppler radar**. Like conventional radar, Doppler radar measures the distribution and intensity of precipitation over a broad area. However, Doppler radar has the additional ability of measuring wind speeds. This is based on the Doppler effect (Chapter 6), the same principle used in police radar to measure the speeds of automobiles.

Radar waves are reflected from raindrops in storms. The direction of a storm's wind-driven rain, and hence a wind "field" of a storm region, can be mapped. This map provides strong clues, or signatures, of developing tornadoes. Hence, Doppler radar can penetrate a storm and monitor its wind speeds. Forecasters are able to predict tornadoes some 20 min before they touch down, as compared with the just over 2 min for conventional radar.

Doppler radars at major airports have another important use. Several airplane crashes or near-crashes have been attributed to dangerous downward wind bursts known as *wind shears*. The winds generally result from high-speed downdrafts in the turbulence of thunderstorms but can occur in clear air when rain evaporates high above the ground. In such wind shears, an airplane attempting to land can quickly lose altitude and possibly crash when

---

Tornado Safety

Knowing what to do in the event of a tornado is critically important. If a tornado is sighted, if the ominous roar of one is heard at night, or if a tornado warning is issued for your particular locality, *seek shelter fast!*

The basement of a home or building is one of the safest places to seek shelter. Avoid chimneys and windows. There is a great danger from flying glass and debris. Get under a sturdy piece of furniture, such as an overturned couch, or in a stairwell or closet, and cover your head.

In a home or building without a basement, seek the lowest level in the central portion of the structure and the shelter of a closet or hallway.

If you live in a mobile home, evacuate it. Seek shelter elsewhere.

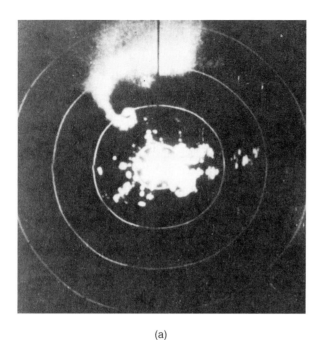 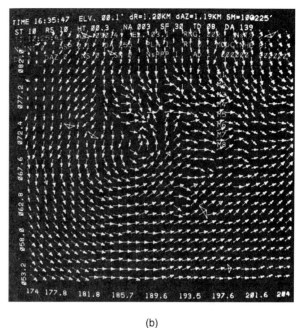

(a) (b)

Figure 19.19 Radar.
(a) Conventional radar scan showing the characteristic hooked signature of a tornado. (b) Doppler radar display showing wind-field pattern.

it is near the ground. Since Doppler radar can detect the wind speed and direction of raindrops in a cloud, as well as the motion of dust and other objects floating in the air, it can provide an early warning of wind shear conditions.

The last storm we will discuss is a tropical storm. *Tropical storms* refer to the massive disturbances that form over tropical oceanic regions. A tropical storm becomes a **hurricane** when its wind speed exceeds 118 km/h (74 mi/h). The hurricane is known by different names in different parts of the world. For example, in southeast Asia it is called a *typhoon,* and in the Indian Ocean, a *cyclone* (Fig. 19.20).

Regardless of the name, this storm is characterized by high-speed rotating winds, whose energy is spread over a large area. A hurricane may be 480 to 960 km (300 to 600 mi) in diameter and have wind speeds of 118 to 320 km/h (74 to 200 mi/h).

Hurricanes form over tropical oceanic regions where the Sun heats large masses of moist air, resulting in an ascending spiral motion. When the moisture condenses, the latent heat provides additional energy and more air rises up the column.

This latent heat is the chief source of a hurricane's energy and is readily available from the condensation of the evaporated moisture from its source region.

Unlike the tornado, a hurricane gains energy from its source region. As more and more air rises, the hurricane grows, with clouds and increasing winds that blow in a large spiral around a relatively calm, low-pressure center—the *eye* of the hurricane (Fig. 19.21).

Hurricane winds cause much damage, but oddly enough, drowning is the greatest cause of hurricane deaths. As the eye of a hurricane comes ashore or *"makes landfall,"* a great dome of water called a **storm surge,** often 80 km (50 mi) wide, comes sweeping across the coast line. It brings huge waves and storm tides that may reach 5 m (17 ft) or more above normal (Fig. 19.22). The storm surge comes suddenly, often flooding coastal low-lands. Nine out of ten casualties are caused by the storm surge.

The torrential rains that accompany the hurricane commonly produce flooding as the storm moves inland. As the storm winds diminish, rainfall floods become a hurricane's greatest threat.

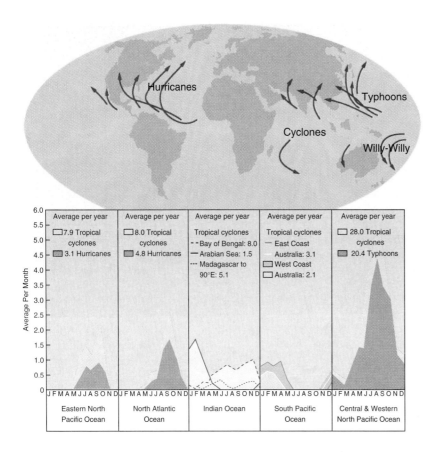

Figure 19.20 Tropical storm regions of the world.
Tropical storms are known by different names in different parts of the world. The average tropical storm activities by month are shown in the graphs.

Once cut off from the warm ocean, the storm dies, starved for moisture and heat energy, and dragged apart by friction as it moves over land.

The breeding grounds of hurricanes affecting the United States are in the Atlantic Ocean southeast of the Caribbean Sea. They are constantly monitored by radar-equipped airplanes or "hurricane hunters," and by satellite. The hurricane alerting system, like that for a tornado, has two phases. A *hurricane watch* is issued for coastal areas when there is a threat of hurricane conditions within 24 to 36 hours. A *hurricane warning* indicates that hurricane conditions are expected within 24 hours.

Tropical storms and hurricanes are named using the first names of people. A six-year list of names for Atlantic storms is given in Table 19.4. A similar list is available for Pacific storms. The names have an international flavor, names beginning with the letters Q, U, X, Y, and Z are omitted because of their scarcity. The lists are recycled. For example, the 1995 list will be used again in 2001.

In 1992 hurricane Andrew was a particularly destructive storm. It made landfall in Florida, causing millions of dollars worth of damage. Notice that Andrew is not in the 1998 list. If a hurricane is particularly destructive, its name is replaced, so as not to cause confusion with another possibly destructive storm of the same name in the next six-year cycle. Thus, the first tropical storm in 1998 will be called Alex rather than Andrew. Similarly, the 1989 hurricane Hugo that hit the South Carolina coast had its name replaced. (See Fig. 19.8.)

CONFIDENCE QUESTION 19.8

(a) What name replaced the destructive 1989 hurricane Hugo? (b) What will be the name given to the third Atlantic tropical storm in 2022?

19.4 Air Masses and Storms 525

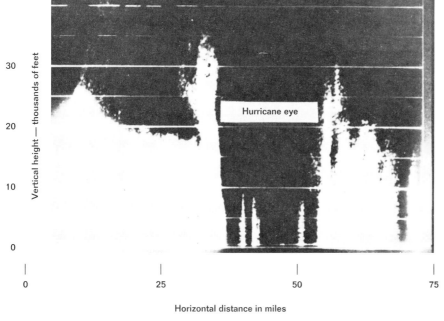

Figure 19.21 The eye of the hurricane, as seen by satellite and radar. The lower photo is a radar profile of a hurricane. Note the generally clear eye about 35 to 55 miles from the radar station and the vertical build-up of clouds (over 30,000 ft) near the sides of the eye.

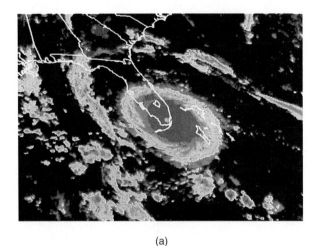

(a)

Figure 19.22 Hurricane storm surge and damage.
(a) This color-enhanced infrared image of hurricane Andrew was taken on August 24, 1992. The eye of the hurricane (yellow) is seen as it passed over the Florida coast. Winds in the storm reached sustained speeds of 230 km/h (140 mph). (b) With landfall comes the storm surge. The drawing illustrates a storm surge coming ashore at high tide, which is more dangerous. Why? (c) An aerial view of housing destruction in the wake of a hurricane.

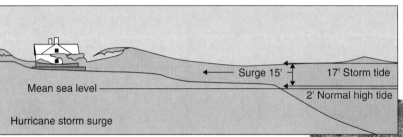

(b)

19.5 Atmospheric Pollution and Climate

Learning Goals:

To identify major atmospheric pollutants.

To explain some pollutant problems, such as smog and acid rain.

To define climate and explain some possible pollutant effects.

(c)

The chapter introduction includes a quote from Shakespeare about the atmosphere: "this most excellent canopy, the air . . . ." The quotation continues: "this most excellent canopy, the air, look you, this brave o'erhanging firmament, this majestical roof fretted with golden fire, why, it appears no other thing to me but a foul and pestilent congregation of vapours." (Shakespeare, *Hamlet*)

Even Shakespeare in his day made reference to atmosphere pollution—an unfortunate topic of earth science. By **pollution,** we mean any atypical contributions to the environment resulting from the activities of human beings. Of course, gases and particulate matter are spewn into the air from

Table 19.4 The Six-Year List of Names for Atlantic Storms*

1995	1996	1997	1998	1999	2000
Allison	Arthur	Ana	Alex	Arlene	Alberto
Barry	Bertha	Bob	Bonnie	Bret	Beryl
Chantal	Cesar	Claudette	Charley	Cindy	Chris
Dean	Diana	Danny	Danielle	Dennis	Debby
Erin	Edouard	Erika	Earl	Emily	Ernesto
Felix	Fran	Fabian	Frances	Floyd	Florence
Gabrielle	Gustav	Grace	Georges	Gert	Gordon
Humberto	Hortense	Henri	Hermine	Harvey	Helene
Iris	Isidore	Isabel	Ivan	Irene	Isaac
Jerry	Josephine	Juan	Jeanne	Jose	Joyce
Karen	Klaus	Kate	Karl	Katrina	Keith
Luis	Lili	Larry	Lisa	Lenny	Leslie
Marilyn	Marco	Mindy	Mitch	Maria	Michael
Noel	Nana	Nicholas	Nicole	Nate	Nadine
Opal	Omar	Odette	Otto	Ophelia	Oscar
Pablo	Paloma	Peter	Paula	Philippe	Patty
Roxanne	Rene	Rose	Richard	Rita	Rafael
Sebastien	Sally	Sam	Shary	Stan	Sandy
Tanya	Teddy	Teresa	Tomas	Tammy	Tony
Van	Vicky	Victor	Virginia	Vince	Valerie
Wendy	Wilfred	Wanda	Walter	Wilma	William

*Names of particular individuals have not been chosen for inclusion in the list of hurricane names.

volcanic eruptions and lightning-initiated forest fires, but these are natural phenomena over which we have little control.

Air pollution results primarily from the products of combustion and industrial processes released into the atmosphere. It has long been a common practice to vent these wastes, and the resulting problems are not new, particularly in areas of population concentrations. Smoke and soot from the burning of coal plagued England over seven hundred years ago. The 1700s saw the beginning of the Industrial Revolution, more burning, increased population, and more domestic heating.

As a result of such air pollution, London has experienced several disasters involving the loss of life. Thick fogs are quite common in this island nation, and the combination of smoke and fog forms a particularly noxious mixture known by the contraction of *sm*oke-*fog*: **smog.**

The presence of fog indicates that the temperature near the ground is at the dew point; with the release of latent heat on condensation, there is a possibility of what is known as a **temperature inversion.** As discussed earlier in the chapter, atmospheric temperature decreases with increasing altitude in the troposphere (with a lapse rate of about $6\frac{1}{2}$ C°/km or $3\frac{1}{2}$ F°/1000 ft). As a result, hot combustive gases generally rise. However, under certain conditions, such as rapid radiative cooling near the ground surface, the atmospheric temperature may locally increase with increasing altitude. The lapse rate is then said to be inverted, giving rise to a temperature inversion (Fig. 19.23).

Radiation temperature inversions occur daily, particularly when there is a clear night and the land surface and the air near it cool quickly. The air some distance above the surface, however, remains relatively warm, thus causing a temperature inversion. Valley fogs provide common evidence of this cooling effect. High-pressure air masses can also result in temperature inversions. As a high-pressure air mass moves over a region and becomes stationary, the dense air may settle, becoming heated and compressed. If the temperature of the descending air exceeds that of the air below it, then the lapse rate is inverted similar to that shown in Fig. 19.23, and we have a *subsidence temperature inversion.*

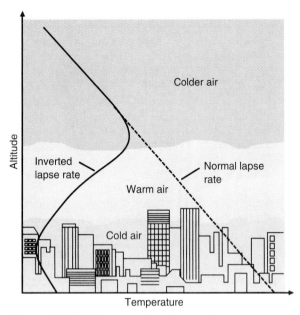

Figure 19.23 Temperature inversion.
Normally, the lapse rate near Earth decreases uniformly with increasing altitude. However, radiative cooling of the ground can cause the lapse rate to become inverted, and the temperature then *increases* with increasing altitude (usually below one mile). A similar condition may occur from the subsidence of a high-pressure air mass.

But even increased concentrations of CO_2 can affect our environment. Carbon dioxide combines with water to yield carbonic acid, a mild acid we all drink in the form of carbonated beverages (carbonated water). Carbonic acid is a natural agent of chemical weathering in geologic processes. But as a product of air pollution, increased concentrations may also aid in the corrosion of metals and react with certain construction materials, causing decomposition. Also, there is concern that an increase in the CO_2 content of the atmosphere may cause a change in global climate through the greenhouse effect. This will be considered later in this section.

Oddly enough, some by-products of complete combustion contribute to air pollution, namely, nitrogen oxides—NO (nitric oxide) and NO_2 (nitrogen dioxide). **Nitrogen oxides** (NO_x) are formed when combustion temperatures are high enough to cause a reaction of the nitrogen and oxygen of

With a temperature inversion, emitted gases and smoke do not rise and are held near the ground. Continued combustion causes the air to become polluted, creating particularly hazardous conditions for people with heart and lung ailments. Smog episodes in various parts of the world have contributed to numerous deaths (● Fig. 19.24).

The major source of air pollution is the combustion of *fossil fuels*—coal, gas, and oil. More accurately, air pollution results from the *incomplete* combustion of *impure* fuels. Technically, combustion (burning) is the chemical combination of certain substances with oxygen. If a fossil fuel is pure and the combustion (oxidation) complete, carbon dioxide (CO_2) is a product. Carbon dioxide is not generally considered a pollutant, because it is a natural part of the atmospheric cycles. However, if the fuel combustion is incomplete, the products include carbon (soot), various hydrocarbons, and carbon monoxide (CO). *Carbon monoxide* results from the incomplete combustion (oxidation) of carbon.

Figure 19.24 Smog.
A scene in Donora, Pennsylvania, taken in 1949. During a five-day smog episode in this Monongahela River Valley town, located 20 miles southeast of Pittsburgh, hundreds of people became ill and at least 20 died.

(a)   (b)

Figure 19.25 Photochemical smog.
(a) On a clear day, the mountains rimming the Los Angeles basin can be seen.
(b) Most days, however, visibility is limited because of photochemical smog.

the air. This reaction typically occurs when combustion is nearly complete, a condition that produces high temperatures, or when combustion takes place at high pressure, for example, in the cylinders of automobiles engines. These oxides can combine with water vapor to form nitric acid (HNO_3), which is very corrosive. This acid also contributes to acid rain, as will be discussed shortly.

Nitrogen oxides are a key substance in the chemical reactions producing smog typical of the city of Los Angeles. This is not the classical London smoke-fog variety, but a smog that results from the chemical reactions of hydrocarbons with oxygen in the air and other pollutants in the presence of sunlight; it is called **photochemical smog.** Since it was first identified in Los Angeles, it is often referred to as *Los Angeles smog.*

Los Angeles has more than its share of temperature inversions, which may occur as frequently as 320 days per year. Its location in a mountain-rimmed basin with a large population, these inversions, a generous amount of air pollution, and an abundance of sunshine set the stage for the production of photochemical smog (Fig. 19.25).

Photochemical smog contains many dangerous contaminants. These include organic compounds, some of which may be *carcinogens,* or substances that cause cancer. One of the best indicators of photochemical reactions, and a pollutant itself is **ozone** (O_3), which is found in relatively large quantities in the photochemically polluted air. In Los Angeles, air pollution warnings of various degrees are given on the basis of ozone and carbon monoxide concentrations in the air.

As noted above, air pollution results chiefly from incomplete combustion of *impure* fuels. Fuel impurities occur in a variety of forms. Probably the most common impurity in fossil fuels and the most critical to air pollution is *sulfur.* When fuels containing sulfur are burned, the sulfur combines with oxygen to form sulfur oxides (SO_x), the most common of which is **sulfur dioxide** (SO_2). A majority of SO_2 emissions come from burning coal and an appreciable amount from burning fuel oils.

Sulfur dioxide in the presence of oxygen and water can react chemically to produce sulfurous and sulfuric acids (H_2SO_3 and H_2SO_4). Sulfurous acid is mildly corrosive and is used as an industrial bleaching agent. Sulfuric acid is a very corrosive acid and widely used as an industrial chemical. Anyone familiar with sulfuric acid can appreciate its undesirability as an air pollutant. Sulfuric acid is the electrolyte used in car batteries.

The sulfur pollution problem has received considerable attention because of the occurrence of **acid rain.** Rain is normally acidic as a result of carbon dioxide combining with water vapor to form carbonic acid. However, sulfur oxide and nitrogen oxide pollutants cause precipitation from

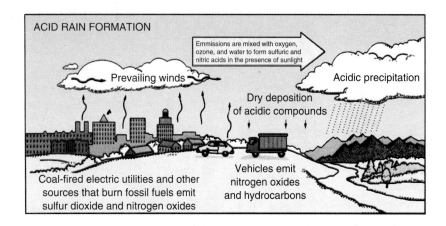

Figure 19.26 Acid rain formation.
Sulfur dioxide and nitrogen oxide emissions react with water vapor in the atmosphere to form acid compounds. The acids are deposited in rain or snow, and may also join dry airborne particles and fall to Earth as dry deposition.

contaminated clouds to be even more acidic, giving rise to acid rain (and also acid snow, sleet, fog, and hail; ● Fig. 19.26).

The problem is most serious in the northeastern United States and Canada, where pollution emissions from the industrialized areas in the midwestern United States are carried by general weather patterns. Other areas are not immune. Acid rain now occurs in the Southeast, and acid fogs are observed on the West Coast. Regulations on the level of sulfur emissions are now in place. Before these, rainfall with a pH of 1.4 had been recorded in New England. This surpasses the pH of lemon juice (pH 2.2), with a yearly average pH of rain in northeastern regions being about 4.2 to 4.4. (Recall from Chapter 13 the pH of pure water is 7.0.) In addition to acid rain, there are acid snows. Over the course of a winter, acid precipitations build up in the form of snowpacks. During the spring thaw and resulting runoff, the sudden release of these acids gives an "acid shock" to streams and lakes.

Acid precipitations lower the pH of lakes, which threatens aquatic plant and animal life. As a result, many lakes in the northeast are "dead" or are in jeopardy of dying. Natural buffers in area soils tend to neutralize the acidity, so waterways and lakes in an area don't necessarily match the pH of the rain. However, the neutralizing capacity in some regions is being taxed, and the effects of acid rain still pose a problem.

● Figure 19.27 shows the major air pollutants and their sources. Study this graph and think about it.

CONFIDENCE QUESTION 19.9

In Fig. 19.27, what is meant by "Fuel combustion in stationary sources"?

Climate

Although not immediately obvious, there is a general belief that there are (and will be) changes in global climate brought about by atmospheric pollution. **Climate** is the long-term average weather conditions of a region. Some regions are identified by their climates. For example, when someone mentions Florida or California, one usually thinks of a warm climate, and Arizona is known for its dryness and low humidity.

Dramatic climate changes have occurred throughout Earth's history. Probably most familiar is that of the Ice Age, when glacial ice sheets advanced southward over the world's northern continents. The most recent ice age ended some 10,000 years ago, after glaciers came as far south as the northern conterminous United States. Also, climatic fluctuations continually occur on a smaller scale than during an ice age. From 1880 to 1940, the average annual temperature of Earth's surface increased slightly. Since 1940, the average temperature has decreased. Associated with this lowering temperature has been a shift in the frost and ice boundaries, a weakening of zonal wind circulations, and marked variations in the world's rainfall pattern. However, this pattern appears to be changing, possibly as a result of the depletion of the ozone layer, as will be discussed shortly.

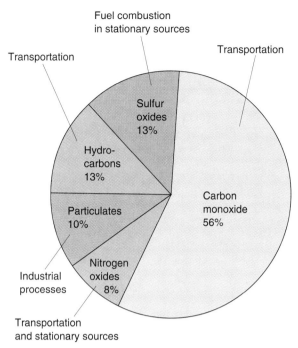

Figure 19.27 Air pollutants and their major sources.

Measurements indicate that Pinatubo was probably the largest volcanic eruption of the century, belching out tons of debris and sulfur dioxide (SO_2). The particulate matter caused beautiful sunrises and sunsets during the following year. However, there is concern about the long-term effects of the millions of tons of sulfur dioxide that was put into the atmosphere from the eruption. The gas reacts with oxygen and water vapor to form tiny droplets, or aerosols, of sulfuric acid, which may stay aloft for years before falling back to Earth. The aerosols are carried around the globe by prevailing winds (Fig. 19.29). Computer models suggest that such acid aerosols could cause a thinning of Earth's protective ozone layer, allowing more ultraviolet radiation to reach the ground. Other environmental concerns about the ozone layer are discussed in the following chapter Highlight.

Global climate is sensitive to atmospheric contributions that affect the radiation balance of the atmosphere. These contributions include the concentration of CO_2 and other "greenhouse" gases, the particulate concentration, and the extent of cloud cover, all of which affect Earth's albedo.

Particulate pollution could contribute to changes in Earth's thermal balance by decreasing the transparency of the atmosphere to insolation. We know particulate matter is emitted during volcanic eruptions, and these particles cause changes in the albedo (the percentage of insolation reflected back into space). For example, the 1815 eruption of the volcano Tambora, located on an island just east of Java, vented cubic miles of debris into the air. The fine volcanic dust was circulated around the globe by global wind patterns, and the winter of 1816 was unseasonably cold because of an increase in the albedo.

Of more recent memory is the eruption of Mount Pinatubo in the Philippines in 1991. Debris was sent over 15 mi into the atmosphere and volcanic ash piled up over a foot deep in the regions surrounding the volcano (Fig. 19.28).

Figure 19.28 Clark Air Force Base.
A thick mushroom cloud of ash and steam from Mount Pinatubo hovers over a portion of the base in Angeles City in the northern Philippines on June 12, 1991.

HIGHLIGHT

The Ozone Hole

In 1974 scientists in California warned that chlorofluorocarbon (CFC) gases might seriously damage the ozone layer through depletion. Observations generally supported this prediction, and in 1978 the United States put a ban on the use of these gases as propellants in aerosol spray cans. Even so, millions of tons of CFCs continue to leach into the atmosphere each year, primarily from refrigerants and spray propellants manufactured in other countries. The release of CFCs from car air conditioners is the single largest source of emissions. CFCs are also used in the manufacture of plastic foams.

When the major CFCs (there are several types of gas) are released, the gases rise slowly, finally reaching the stratosphere—a process that takes 20 to 30 years. In the stratosphere the CFC molecules are broken apart by ultraviolet radiation, with the release of chlorine (Cl) atoms. It is these atoms that react with and destroy ozone molecules in a repeating cycle, such that over the course of a year or two a single Cl atom may destroy as many as 100,000 ozone molecules.

Measurements indicate that the concentrations of CFCs in the atmosphere have more than doubled in the past 10 years. Worldwide ozone levels have declined an estimated 3% to 7% over the past few decades. Part of this decline may be the result of natural fluctuations. But in 1985, scientists announced the discovery of an ozone "hole" over Antarctica (Fig. 1). Measurements have since revealed losses of greater than 50% in the total hole column, and greater than 95% at altitudes of 9 to 12 miles. This polar hole in the ozone layer opens up annually during the southern springtime month of September or October. Satellite measurements taken in October 1991 showed the atmospheric concentration of ozone over the South Pole had dwindled to its lowest recorded level.

The worldwide depletion of the ultraviolet-absorbing ozone layer would have some significant effects. Experts estimate that the cases of skin cancer will increase by 60% and that there will be many additional cases of cataracts. Crops will also be affected. Then too, CFCs are greenhouse gases and can contribute to global warming in this manner. In 1994, the National Weather Service began issuing a "uv-index" forecast for many large cities. On a scale of 0 to 15, the index gives a relative indication of the amount of uv light that will be received at Earth's surface at noontime the next day—the greater the number, the greater the amount of uv. The scale is based on upper-atmosphere ozone levels and clouds.

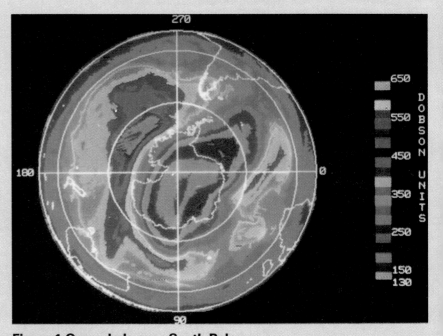

Figure 1 Ozone hole over South Pole.
This satellite spectrometer picture shows the ozone "hole" as a gray, blue, and black oval generally covering Antarctica. The hole is surrounded by regions of high total ozone (yellow, brown, and green in color).

The reduction of the use of CFCs will not be done easily or quickly. In the United States alone, products worth more than $125 billion rely on CFCs. Research on the development of substitutes for these gases is under way. Such replacements will have to be chemically unreactive with ozone, or be destroyed by lower-atmosphere processes before reaching the ozone layer. The sale of CFCs has been restricted, and their production in the United States will be banned in 1996, so substitutes will have to be used in air conditioners and refrigerators when the supply runs out. For refrigerants, initial substitutes have not been as efficient as CFCs. As a result, more energy will have to be used to produce the same amount of cooling.

International efforts are also underway to reduce the use of CFCs. In 1989 some 24 nations signed the Montreal Protocol on Substances that Deplete the Ozone Layer. The Montreal Protocol was amended in 1990, and more than 90 nations have agreed to phase out CFCs by the year 2000. This would drastically reduce the use of CFCs, as Fig. 2 shows. Even so, there have been calls for advancing the scheduled reductions to prevent further depletion of the ozone layer.

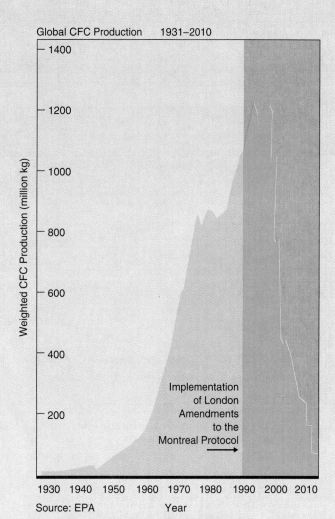

Figure 2 Global CFC production.
The graph shows the predicted decrease in CFC production from the time of international agreement into the twenty-first century. Production has decreased somewhat, and in 1996 CFC substitutes will have to be used in air conditioners and refrigerators in the United States.

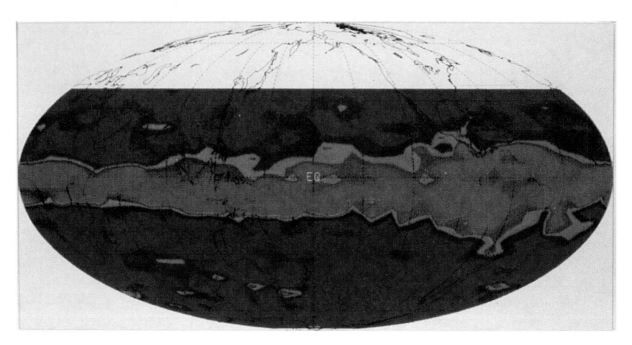

Figure 19.29 Mount Pinatubo aftermath.
A band of sulfur dioxide from the eruption of Mount Pinatubo girdles the globe, as observed by NASA's upper atmosphere research satellite.

Pollution may also affect Earth's temperature. Vast amounts of CO_2 are expelled into the atmosphere as a result of the combustion of fossil fuels. As discussed in Section 19.2, CO_2 and water vapor play important roles in Earth's energy balance, because of the greenhouse effect. An increase in the atmospheric CO_2 concentration could alter the amount of radiation absorbed from the planet's surface and produce an increase in temperature.

Possible effects of increasing temperature or global warming include the melting of the polar ice caps, which would cause a rise in the sea level and flood coastal areas. Also, there would be drier summer months in the middle latitudes. In the United States, there would be dry farmland in the South and a longer growing season in the North. Many other possibilities are not well understood, including the accelerated release of greenhouse gases, such as methane (CH_4) from swamps and bogs because of a warming climate.

Thus, air pollution may contribute to a variety of atmospheric effects—both local and global. The effects of pollution on local climatic conditions are somewhat clear. For example, cities receive more rainfall than the surrounding countryside. However, there are too few data to understand and accurately predict global climatic effects. We are inclined to believe that increased atmospheric CO_2 concentrations and ozone-layer depletions would give rise to an increase in global temperature. Such a temperature increase would cause more water evaporation and an increase in relative humidity. Particulate pollution could then give rise to increased cloud formation and coverage, which would increase the albedo and cause a decrease in global temperature.

Certainly, little comfort can be found in this speculative, counterbalancing pollution cycle. There are too many unanswered questions related to the effects of pollution, which has occurred over a relatively short time; and to the natural interacting cycles of the atmosphere and the biosphere, which has taken millions of years to become established.

Important Terms

These important terms are for review. After reading and studying the chapter, you should be able to define and/or explain each of them.

atmosphere	radar	front
photosynthesis	wind	lightning
troposphere	air currents	thunder
stratosphere	convection cycle	tornado
mesosphere	jet stream	tornado watch
thermosphere	cloud	tornado warning
ozone	high clouds	Doppler radar
ozonosphere	middle clouds	hurricane
ionosphere	low clouds	storm surge
aurora borealis	clouds of vertical development	pollution
insolation	humidity	smog
Rayleigh scattering	relative humidity	temperature inversion
greenhouse effect	dew point	nitrogen oxides
weather	psychrometer	photochemical smog
barometer	lapse rate	ozone
anemometer	Bergeron process	sulfur dioxide
wind vane	air mass	acid rain
rain gauge	source region	climate

Important Equations

Relative humidity: $(\%)\ RH = \dfrac{AC}{MC} (\times 100\%)$

Questions

19.1 Atmospheric Composition and Structure

1. The third most abundant gas in the atmosphere is (a) oxygen, (b) carbon dioxide, (c) nitrogen, (d) argon.
2. The ozone layer lies in the (a) thermosphere, (b) troposphere, (c) stratosphere, (d) mesosphere.
3. What is the difference between atmospheric science and meteorology?
4. What is the composition of the air you breathe?
5. Humans inhale oxygen and exhale carbon dioxide. With our world population, wouldn't this reduce the atmospheric oxygen level over a period of time? Explain.
6. Why is the plant pigment chlorophyll so important?
7. Describe how the temperature of the atmosphere varies in each of the following regions: (a) the mesosphere, (b) the stratosphere, (c) the troposphere, and (d) the thermosphere.
8. Of what importance is the atmospheric ozone layer?
9. What is believed to cause the displays of lights called *auroras*?

19.2 Atmospheric Energy Content and Measurements

10. Earth's average temperature is regulated by (a) Rayleigh scattering, (b) the greenhouse effect, (c) atmospheric pressure, (d) photosynthesis.

11. Approximately what percentage of insolation directly reaches Earth's surface: (a) 33%, (b) 40%, (c) 50%, or (d) 75%?
12. Humidity is measured with (a) an anemometer, (b) a barometer, (c) a wind vane, (d) a psychrometer.
13. What does the word *insolation* stand for?
14. From what source does the atmosphere receive most of its direct heating, and how is the overall heating accomplished?
15. The maximum insolation is received daily around noon. Why, then, is the hottest day part of the day around 2 P.M. or 3 P.M.?
16. (a) Why is the sky blue? (b) In terms of Rayleigh scattering, explain why it is advantageous to have amber fog lights and red tail lights on cars.
17. (a) Explain what is meant by the "greenhouse effect." (b) How does the selective absorption of the atmospheric gases provide a thermostatic effect for Earth?
18. What is the principle of the liquid barometer, and what is the height of a mercury barometer column for one atmosphere of pressure?
19. Explain the operation of a skydiver's altimeter.
20. Which way, relative to the wind direction, does a wind vane point and why?
21. At small airports, a wind sock (a tapered bag pivoted on a pole) acts as a wind vane and gives some indication of the wind speed. Explain its operation.

19.3 Air Motion and Clouds

22. Near a large body of water, the predominant wind during the day is (a) a sea breeze, (b) a north wind, (c) an up draft, (d) a jet stream.
23. The root name meaning "heap" is (a) stratus, (b) cirrus, (c) nimbus, (d) cumulus.
24. The altostratus cloud is a member of which of the following families: (a) high clouds, (b) middle clouds, (c) low clouds, (d) clouds of vertical development.
25. What are the primary forces of air motion?
26. Explain how convection cycles are set up.
27. (a) What is the general wind direction for the conterminous United States and why?
    (b) Generally speaking, on which side of town would it be best to build a house in the United States so as to avoid smoke and other air pollutants generated in the town?
    (c) Should the prevailing wind direction be of any consideration in the heating plan and insulation of a house?
28. Name the cloud family for each of the following: (a) nimbostratus, (b) cirrostratus, (c) altostratus, (d) stratus, (e) cumulonimbus.
29. Name the cloud type associated with each of the following: (a) mackerel sky, (b) solar or lunar halo, (c) the hazy shade of winter, (d) thunderhead.
30. What conditions are necessary for cloud formation?
31. What happens if the dew point of a rising air mass is not reached before stability?
32. Why does water condense on the outside of a glass containing an iced drink?
33. Explain the principle of the psychrometer.

19.4 Air Masses and Storms

34. Which of the following air masses would be expected to be cold and dry: (a) cP, (b) mA, (c) cT, (d) mE?
35. The critical alert for a tornado is a (a) tornado alert, (b) tornado warning, (c) tornado watch, (d) tornado prediction.
36. How are air masses classified? Explain the relationship between air-mass characteristics and source regions.
37. What air masses affect the weather in the conterminous United States?
38. What is a front? List the meteorological symbols for four types of fronts and explain what they mean.
39. Describe the characteristics and weather associated with warm and cold fronts. What is the significance of the sharpness of their vertical boundaries?
40. An ice storm is likely to result along what type of front? Explain.
41. What is the most violent of storms and why?
42. What is the major source of energy for a tropical storm? When does a tropical storm become a hurricane?
43. Distinguish between a hurricane watch and a hurricane warning.

19.5 Atmospheric Pollution and Climate

44. A major source of air pollution is (a) nuclear electrical generation, (b) incomplete combustion, (c) temperature inversions, (d) acid rain.
45. A change in Earth's albedo could result from (a) nitrogen oxides, (b) acid rain, (c) photochemical smog, (d) particulate matter.
46. Define air pollution.
47. What are the products of complete combustion? Of incomplete combustion?

48. Are nitrogen oxides products of complete or incomplete combustion? Explain.
49. Distinguish between classical smog and photochemical smog. What is one of the best indicators of the latter?
50. What are the causes and effects of acid rain? In which areas is acid rain a major problem and why?
51. What are the possible effects of increased atmospheric CO_2 concentrations?
52. What are possible explanations for changes in Earth's average temperature?
53. What effects could CFCs have on Earth's climate? What is being done to prevent this?

Food for Thought

1. At the time of formation, planets and their satellites are generally believed to have atmospheres. Why did the planet Mercury and our Moon lose their atmospheres?
2. If Earth were as small as the Moon, how many times brighter would it appear to an astronaut equidistant from each body?
3. If the surface of Earth were all land or all water, what would be the direction(s) of the atmospheric general circulation pattern?
4. What air masses affect the weather of (a) Alaska and (b) Hawaii?
5. Why do household barometers often have descriptive adjectives such as *stormy* and *fair* on their faces, along with direct pressure readings?
6. How could CO_2 pollution be decreased while supplying our energy needs?

Exercises

19.1 Atmospheric Composition and Structure

1. On a vertical scale of altitude in km and mi above sea level, locate the heights of the following (the heights not listed may be found in the chapter):
    (a) the top of Mt. Everest—29,000 ft
    (b) commercial airline flight—35,000 ft
    (c) supersonic transport flight (SST)—65,000 ft
    (d) communications satellite—400 mi
    (e) the E and F ion layers
    (f) aurora displays
    (g) syncom satellite—23,000 mi (a typical weather satellite with synchronous period to Earth's rotation, so it stays over one location)
2. Express the thickness of the stratosphere, mesosphere, and thermosphere in terms of the thickness of the troposphere. (*Hint:* Use a ratio.) What do these comparisons tell you?

19.3 Air Motion and Clouds

3. On a day when the air temperature is 75°F, the wet-bulb reading of a psychrometer is 68°F. Find each of the following: (a) relative humidity, (b) dew point, (c) maximum moisture capacity of the air, (d) actual moisture content of the air.
   Answers: (a) 70%, (b) 64°F, (c) 9.4 gr/ft^3, (d) 6.6 gr/ft^3

4. A psychrometer has a dry-bulb reading of 95°F and a wet-bulb reading of 90°F. Find each of the quantities asked for in Exercise 3.

5. On a very hot day with an air temperature of 105°F, the wet-bulb thermometer of a psychrometer reads 102°F. (a) What is the actual moisture content of the air? (b) How many degrees would the air temperature have to be lowered for the relative humidity to be 100%?
   Answer: (a) 21.1 gr/ft^3, (b) 4 F°

6. On a winter day a psychrometer has a dry-bulb reading of 35°F and a wet-bulb reading of 29°F. (a) What is the actual moisture content of the air? (b) Would the water in the wick of the bulb freeze? Explain.
7. The dry-bulb and wet-bulb thermometers of a psychrometer both read 75°F. What are (a) the relative humidity, and (b) the actual moisture content of the air?

Answer: (b) 9.4 gr/ft^3

8. While picnicking on a summer day, you hear thunder 11 s after seeing a lightning flash from an approaching storm. Approximately how far away in miles is the storm?

Answer: 2.2 mi

Answers to Multiple-Choice Questions

1. d	10. b	22. a	34. a	44. b
2. c	11. c	23. d	35. b	45. d
	12. d	24. b		

Solutions to Confidence Questions

19.1 From Table 19.1: 78% (N_2) + 21% (O_2) + 0.9% (Ar) + 0.03% (CO_2) = 99.93%

19.2 The increased pressure in the cooker raises the boiling point of water, and boiled foods cook more quickly at a higher temperature. (See the pressure cooker discussion in Section 5.3.)

19.3 Because the deflection is to the left in the Southern Hemisphere, the rotations are in opposite directions to those in the Northern Hemisphere; that is, rotations are clockwise around a low and counterclockwise around a high, as viewed from above.

19.4 With a depression of $\Delta T = 70°F - 66°F = 4$ F°, from the Appendix X Tables, (a) $RH = 81\%$, and (b) with a dew point = 64°F, the air temperature (70°F) would have to be lowered 6 F° for the relative humidity to be 100% and for precipitation to be likely.

19.5 Clouds are formed in rising air that is cooled to its dew point. The temperature of downward moving air (in downdrafts) is increased, the temperature is raised above the dew point, and the clouds dissipate (evaporate). As an example, we say the Sun "burns off" a fog—that is, it raises the temperature above the dew point so the fog dissipates.

19.6 Over water, so maritime, and tropically warm, so mT.

19.7 $d = vt = \frac{1}{3}$ km/s $\times$ 3 s = 1 km

$= \frac{1}{5}$ mi/s $\times$ 3 s = $\frac{3}{5}$ mi = 0.6 mi

19.8 (a) The 1989 list in the six-year cycle becomes the 1995 list, and from Table 19.4, we see that Humberto replaced Hugo. (b) As may be seen from Table 19.4, the 1996 list will be the 2002 list, and the third name is Cesar (provided it does not get deleted).

Chapter 20

Minerals, Rocks, and Geologic Events

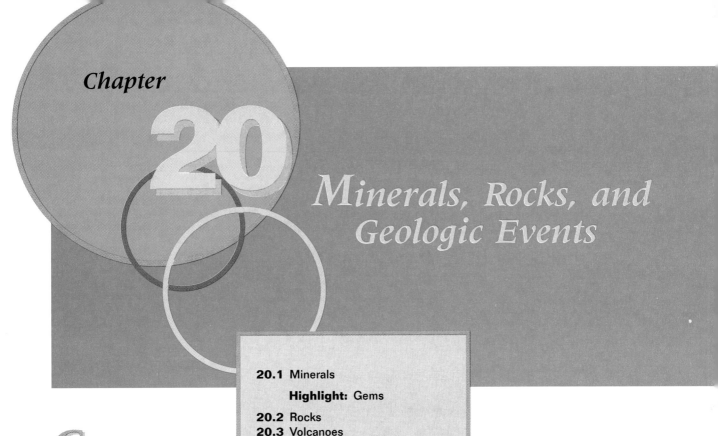

- **20.1** Minerals
  - **Highlight:** Gems
- **20.2** Rocks
- **20.3** Volcanoes
- **20.4** Earthquakes

Geology is an earth science. Historically, *geology* refers to the study of planet Earth, its composition, structure, and history. Today, this definition is expanded to include the study of the Moon, as well as other planets and their moons.

Earth is unique in having abundant water, rich soil, and an atmosphere with the right amount of oxygen. These resources, plus the planet's position and orientation with respect to the Sun, provide energy and temperatures that support life. With the recent increasing emphasis on natural resources obtained from Earth, geology has taken on new importance. Except for water and soil, mineral resources such as coal, gas, and petroleum are nonrenewable. The geologist must discover new mineral deposits to provide the energy-producing materials on which our civilization is based. Understanding the various geologic processes is critical in locating and developing mineral deposits and energy resources.

To illustrate how our way of life depends on Earth's energy resources, consider what would happen if liquid fuels became unavailable. Our reserves of processed fuels would be consumed within a brief period of time, and every heat engine operating on gasoline or similar fuel would stop. Thus trucks, cars, airplanes, trains, farm tractors, and other machines, including industrial equipment, would cease to operate. The result would be chaos in our present way of life.

In this chapter and the following chapter, we introduce and discuss the concepts of geology necessary to comprehend the physical nature of the planet on which we live. The study begins with the outer layer of the planet. This outer layer, called the crust, is a thin shell with a thickness of only 4.8 to 48 km (3 to 30 mi). The crust is composed of minerals and rocks.

This first chapter on geology begins with a discussion of minerals followed by a discussion of rocks and the activity that forms them. Then, we concentrate on the structural geology of the planet, the interactions of its crust and internal processes, and the methods of geological dating and interpreting Earth's history.

20.1 Minerals

Learning Goals:
To define minerals, identify their physical properties, and list their general uses.
To describe methods of mineral identification.

A **mineral** is a naturally occurring, crystalline, inorganic substance (element or compound) that possesses a fairly definite chemical composition and a distinctive set of physical properties. Minerals are composed, for the most part, of eight elements. These elements, along with their relative abundance in Earth's crust, are listed in Table 20.1. Note that two elements, oxygen and silicon, make up about 75 percent of the crust. Over 2000 minerals have been found in Earth's crust; approximately 20 of them are common, and fewer than 10 account for over 90 percent of Earth's crust by weight.

The minerals are arranged naturally in groups to form a consolidated mixture called *rock*. Rocks are composed primarily (90 percent or more, by volume) of oxygen atoms. Therefore, the oxygen atom (or ion) is the dominating influence controlling the number of possible element combinations in the formation of minerals.

Most rock-forming minerals are composed mainly of oxygen and silicon. The fundamental silicon-oxygen compound is silicon dioxide, **silica,** which has the formula SiO_2. Because carbon and silicon are in the same chemical group, one might think that SiO_2 would be a gas similar to carbon dioxide, CO_2, but the bonding is very different in the two compounds. The silicon-oxygen structure of silica, SiO_2, is based on a network of SiO_4 tetrahedra (Fig. 20.1) with shared (covalent bond) oxygen atoms rather than SiO_2 molecules (Fig. 20.2a). Quartz is an example (Fig. 20.2). It is a hard and brittle solid.

In silica the oxygen-to-silicon ratio is 2 to 1; however, the oxygen-to-silicon ratios in the *silicates* are greater than 2 to 1 and can vary greatly. The ratio varies greatly because the silicon-oxygen tetrahedra may exist as separate independent units or may share oxygen atoms at corners, edges, or sometimes faces in many different ways. Thus the structures of the silicate minerals are determined by the way the SiO_4 tetrahedra are arranged.

The rock-forming minerals are composed mainly of oxygen and silicon plus, in most cases, aluminum and at least one more element from Table 20.1. Feldspars are the most abundant minerals found in Earth's crust. There are two main

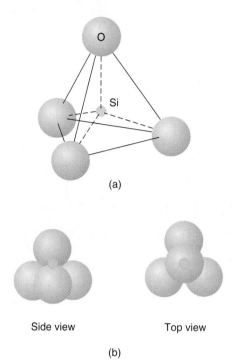

Figure 20.1 The tetrahedral structure of the SiO_4^- ion.
(a) Silicate is a polyatomic ion in which covalent bonds hold the Si atom and the O atoms together. The entire structure has a −4 charge. (b) Close-packed side and top views.

Table 20.1 Relative Percentages of Elements in Earth's Crust

Element	Approx. Percentage (Weight)
Oxygen (O)	46.5
Silicon (Si)	27.5
Aluminum (Al)	8.1
Iron (Fe)	5.3
Calcium (Ca)	4.0
Magnesium (Mg)	2.7
Sodium (Na)	2.4
Potassium (K)	1.9
All Others	1.6
Total	100.0

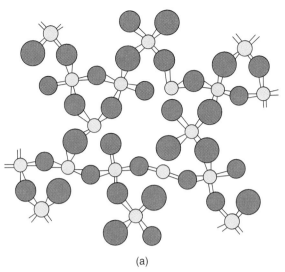

(a) (b)

Figure 20.2 Quartz.
(a) The structure of quartz, SiO_2. The structure is based on interlocking SiO_4 tetrahedra in which each oxygen atom (blue) is shared with two silicon atoms (buff). (b) Quartz is one of the most common minerals. Under favorable conditions, it occurs in hexagonal crystals, as shown here, and is harder than glass.

types: (1) plagioclase feldspar, which contains oxygen, silicon, aluminum, and calcium or sodium (see Table 20.2 for the chemical formula of each); (2) potassium feldspar, composed of oxygen, silicon, aluminum, and potassium. This feldspar is sometimes called orthoclase feldspar. The properties of each type of feldspar are given in Table 20.2.

Table 20.2 also lists several nonsilicate materials and their general or economic use. The nonsilicate minerals include the carbonates (such as calcite), sulfides (galena), sulfates (gypsum), halides (halite), oxides (hematite), and native elements such as gold, silver, sulfur, and diamond.

The main rock-forming materials are all silicates. Some typical silicate minerals are shown in Fig. 20.3. Olivine is a mineral with a single independent tetrahedron that is combined with either Mg^{2+} or Fe^{2+}. These metal ions can substitute for each other, depending on the conditions of mineral formation.

Everywhere around us we see minerals or the products of minerals. Some are quite valuable, while others are essentially worthless. For example, precious stones such as diamonds and rubies are valuable minerals. Also, we speak of a nation's mineral wealth when referring to natural raw materials such as ores or minerals containing iron, gold, silver, or copper. However, the minerals of common rock, such as sandstone, have little monetary value.

The term *mineral* has also taken on popular meanings. For example, foods are said to contain vitamins and "minerals." In this case the term *mineral* refers to compounds in food that contain elements needed in small quantities by the human body, such as Fe, Na, I, Mn, Mg, and Cu. The chemical compositions of these compounds may be the same as those of naturally occurring minerals that make up Earth's crust, or they may actually be the naturally occurring minerals themselves. The names of minerals, like those of chemical elements, have historical connotations and may reflect the names of localities.

Mineral classification is advantageous because it is based on the physical and chemical properties of substances, which distinguish between different forms of minerals composed of the same element or compound. For example, graphite—a soft, black, slippery substance commonly used as a lubricant—and diamond are both composed of carbon. But because of different crystalline structures, their properties are different. (Graphite mixed with other substances to obtain various degrees of hardness is the "lead" in lead pencils.)

Table 20.2 Some Common Minerals and Their Properties

Minerals	Chemical Composition	Color	Luster	Hardness (Mohs' scale)	Specific Gravity* (approx.)	Cleavage	General Use or Occurrence
Asbestos (chrysotile)	Hydrous magnesium silicate, $Mg_3(Si_2O_5)(OH)_4$	Green (different shades)	Silky	Low	2.5	Yes	Fireproofing; insulation
Calcite	Calcium carbonate, $CaCO_3$	White or colorless	Vitreous to earthy	3	2.7	Yes	Cement manufacture; optical application; lime
Clay	Hydrous aluminum silicates containing Na, K, Fe, Mg, etc.	Light	Earthy	Very low	2.6	Fracture	Chinaware; pottery
Dolomite	Calcium (Mg, Fe) carbonate, e.g., $CaMg(CO_3)_2$	White-gray	Pearly	3.5–4	2.9	Yes	Cement manufacture; building materials; lime
Feldspar							
Plagioclase	Sodium and calcium aluminum silicate, $NaAlSi_3O_8$, $CaAl_2Si_2O_8$	White-gray	Vitreous	6	2.6–2.8	Yes	Ceramic glazes
Orthoclase	Potassium aluminum silicate, $KAlSi_3O_8$	White-pink	Vitreous	6	2.5	Yes	Elongated crystals in igneous rock
Fluorite	Calcium fluoride, CaF_2	Colorless and range of colors	Vitreous	4	3.2	Yes	Commonly found in metal ores
Galena	Lead sulfide, PbS	Gray	Metallic	2.5	7.6	Yes	Lead ore
Garnet group	Orthosilicates	Commonly red	Vitreous	6.5–7.5	3.1–4.3	Fracture	Gemstones; abrasives
Gypsum	Hydrous calcium sulfate, $CaSO_4 \cdot 2H_2O$	Colorless or white	Silky-dull	2	2.3	Yes	Plaster of Paris; wallboard
Halite	Sodium chloride, NaCl	Colorless or white	Vitreous	2.5	2.2	Yes	Table salt
Hematite	Iron oxide, Fe_2O_3	Gray-reddish	Metallic, dull	5.5–6.5	5.1	Fracture	Iron ore
Hornblende	Hydrous Ca, Na, Mg, Fe, Al silicates	Dark green to black	Vitreous	5–6	2.9–3.4	Yes	Rock-forming mineral
Magnetite	Iron oxide, Fe_3O_4	Black	Metallic, dull	6	5.2	Fracture	Magnetic iron ore
Mica	Group of hydrous silicates						
Biotite		Black-green	Pearly	2.5–3	3.0	Yes	"Isinglass" heat-proof windows; electrical insulator
Muscovite		Reddish brown	Pearly	2.5–3	2.9	Yes	

Mineral	Composition	Color	Hardness	Specific Gravity*	Luster	Cleavage	Uses
Olivine	Metallic silicate (Mg, Fe)$_2$SiO$_4$	Green	6.5–7	3.2–4.3	Vitreous	Fracture	Igneous rock mineral
Pyrite	Iron sulfide, FeS$_2$ (fool's gold)	Brass yellow	6–6.5	5.0	Metallic	Fracture	Source of iron; sulfur for sulfuric acid
Pyroxene	Metallic aluminum silicate (Ca, Mg, Fe, Na)	Green to black or brown	5–6	3.4	Vitreous	Yes	Igneous rock mineral
Quartz	Silicon dioxide, SiO$_2$	Colorless when pure	7	2.65	Vitreous	Fracture	Optical applications
Serpentine	Mg$_3$Si$_2$O$_5$(OH)$_4$	Greenish; brownish	2.5–5	2.2–2.6	Vitreous or waxy	None	Source of Mg; decorative stone
Sphalerite	Zinc sulfide, ZnS	Black, yellow-brown	3.5–4	4.0	Resinous	Yes	Zinc ore
Talc	Hydrous magnesium silicate, Mg$_3$Si$_4$O$_{10}$(OH)$_2$	Green-white, silvery	1	2.7	Pearly, greasy	Yes	Cosmetics; ceramics

*The specific gravity of a mineral is the ratio of the weight of a given volume of the mineral to the weight of an equal volume of water.

Minerals can be identified by chemical analysis, but most of these methods are detailed and costly and are not available to the average person. More commonly, the distinctive physical properties are used as the key to mineral identification. See Table 20.2 for a list of several minerals with their chemical composition and some of their more distinctive properties. These properties are well known to all serious rock and mineral collectors. Some of the physical properties used in mineral identification are described in the following paragraphs.

Crystalline structure refers to the way the atoms or molecules that make up the mineral are arranged internally. This arrangement is a function of the size and shape of the molecules and the forces that bind them.

All crystalline substances crystallize in one of seven major geometrical patterns. When a mineral grows in unrestricted space, the mineral develops the external shape of its crystal form. However, during the growth of most crystals the space is restricted, resulting in an intergrown mass of crystals that does not exhibit its crystal form. A branch of physical science called *crystallography* deals with the external shapes of crystals and with the geometrical relations among the atomic planes of crystals. The crystalline forms are studied in detail by means of X-ray analysis.

Hardness is a comparative property and so is indicated by a harder mineral being able to scratch a softer one. The varying degrees of hardness are represented on **Mohs' scale of hardness,** which runs from 1 to 10, soft to hard. This arbitrary scale is expressed by the 10 minerals listed in Table 20.3. Talc is the softest and diamond is the hardest. A particular mineral on the scale is harder than (can scratch) all those with lower numbers. Using these minerals as standards, one finds the following on the hardness scale: fingernail, 2–3; a penny, 3; window glass, 5–6.

Cleavage refers to the tendency of some minerals to break along definite smooth planes. The mineral may exhibit distinct cleavage along one or more planes, or it may exhibit indistinct cleavage or no cleavage. The degree of cleavage that a mineral exhibits is a clue to the identification of the mineral.

Fracture refers to the way a mineral breaks. The mineral may break into splinters, ragged or

Silicate structure	Arrangement of tetrahedron (top view)	Typical mineral
Single independent tetrahedron		Olivine
Single chain		Pyroxene
Double chain		Hornblende
Continuous sheet		Mica
Three-dimensional network	Too complex to be shown by simple two-dimensional sketch	Quartz and feldspar

(a)

(b)

(c)

(d)

(e)

Figure 20.3 Molecular structure of several common silicate minerals.
(a) olivine, (b) pyroxene, (c) hornblende, (d) mica, and (e) feldspar.

Figure 20.4a Fluorescent minerals in daylight.

Figure 20.4b Mineral fluorescence in full-spectrum ultraviolet light.
Front, left to right: calcite/red, willemite/green; calcite/red; optical calcite/blue. *Back, left to right:* powellite/yellow; scheelite/blue; willemite/green; opal patch/green.

rough irregularly surfaced pieces, or shell-shaped forms known as conchoidal fractures.

Color refers to the property of reflecting light of one or more wavelengths. Although the color of a mineral may be impressive, it is not a reliable property for identifying the mineral, because the presence of small amounts of impurities may cause drastic changes in the color of some minerals.

Streak refers to the color of the powder of the mineral. A mineral may exhibit an appearance of several colors, but it will always show the same streak. A mineral rubbed (streaked) across the surface of an unglazed porcelain tile will thereby be powdered and will show its true color.

Luster refers to the appearance of the mineral's surface in reflected light. Mineral surfaces appear to have a metallic or nonmetallic luster. A metallic luster has the appearance of polished metal; a nonmetallic appearance may be of varying lusters as described by such terms as greasy, pearly, silky, and vitreous (glassy).

In addition to the physical properties already mentioned, there are two others associated with mineral identification: *magnetism* and *fluorescence*. Some metallic minerals are magnetic, having formed in the magnetic field of Earth. When some are irradiated with ultraviolet light, they fluoresce or reemit visible light (Fig. 20.4).

CONFIDENCE QUESTION 20.1

Which of the following naturally occurring substances are minerals: Asbestos, calcite, diamond, iron oxide, or quartz?

20.2 Rocks

Learning Goals:

To define rock and describe the rock cycle.

To explain the three classifications of rocks.

Table 20.3 Mohs' Scale of Hardness

1. Talc	6. Feldspar (orthoclase)
2. Gypsum	7. Quartz
3. Calcite	8. Topaz
4. Fluorite	9. Corundum
5. Apatite	10. Diamond

A **rock** is defined as a cohesive aggregate of one or more minerals; rock is a natural and substantial part of Earth's crust. When we look at a mountain cliff, we see rock rather than individual minerals. The majestic mountains of our western states are made of rock. The Colorado River has carved the Grand Canyon through layers of rock, and the continents and ocean basins are composed of rock.

HIGHLIGHT

Gems

A gem is any mineral or other precious or semiprecious stone valued for its beauty. Most gems have a crystalline structure, and when they are shaped and polished, their beauty is enhanced immensely. Their beauty depends on the special characteristics of brilliancy, color, prismatic fire, luster, optical effects, and durability.

The brilliancy of a gem is a function of its index of refraction. Diamond, which is composed only of carbon atoms bonded into a single giant molecule by single bonds, has one of the highest refractive indexes (zircon, $ZrSiO_4$ is slightly higher) of the well-known gemstones. Brilliancy also depends on the gem's transparency, the angles of the facets, and the polish (degree of smoothness) of its surfaces.

Gems that are chemically pure are colorless. The crystalline form of aluminum oxide (Al_2O_3) occurs in nature as the mineral corundum. Pure aluminum oxide is colorless, but minute amounts of chemical impurities create a variety of colors. For example, in sapphires (a variety of corundum) a trace of chromium creates pink, iron creates green and yellow, chromium and iron create orange, and titanium with iron creates blue. Corundum is called a ruby when sufficient amounts of chromium are present to create a deep red. To give another example, the green of emerald (a variety of the mineral beryl, $Be_3Al_2Si_6O_{18}$) is due to the presence of chromium oxide.

Radiation in the visible region of the electromagnetic spectrum is composed of many wavelengths. We see the radiation as white light. When white light passes into a gemstone, all wavelengths are bent at different angles.

Figure 1
The Logan Sapphire, a 423-carat, blue sapphire, the largest one on public display.

These changes produce flashes of different colors emerging from the gemstone. This property of several transparent gems is called dispersion, or prismatic fire (see Section 7.2).

Special effects are created in rubies, sapphires, and a few other gems due to the property of double refraction. These gems resolve a single beam of light into two beams, which are absorbed unequally and emerge as different colors. The six-pointed star effect seen in some rubies and sapphires is due to the reflection of light from microscopic, needle-shaped rutile (titanium oxide, TiO_2) crystals that intersect at 60° angles. Figure 1 shows a natural sapphire.

A gem must be durable to be used as an ornament. It must possess the ability to resist abrasion, cleavage, and fracture. Thus, it must rank high on the hardness scale. Diamond and jade are very durable gems, but opal (hydrated silicon dioxide) and zircon are rather fragile.

The value of a gem depends on its beauty, rarity, size, fashion (current style), and durability. The most valued gems are the very best quality diamonds, emeralds, and rubies. However, high-quality sapphires, opals, and pearls are more expensive than lower grades of diamonds, emeralds, or rubies.

The size of a gem is determined by its weight, which is a function of its density. The weight is measured in carats (ct). One carat is equivalent to 200 mg or approximately 0.007 oz.

The major sources of diamonds are the Republic of South Africa, southwest Africa, and Tanzania. The finest emeralds, jade, rubies, and sapphires come from Myanmar (formerly Burma). Opal is found in many countries, particularly Australia. Cambodia is a major source of zircon. The finest variety of turquoise is found in the western United States.

Quality synthetic gems can be made with corundum. A method of fusing fine alumina in a very hot flame has been perfected, and by adding the appropriate chemical compound, synthetic rubies and sapphires of large size and fine quality can be made.

There are many precious and semiprecious gemstones. We have included only a few of the most important and well-known here.

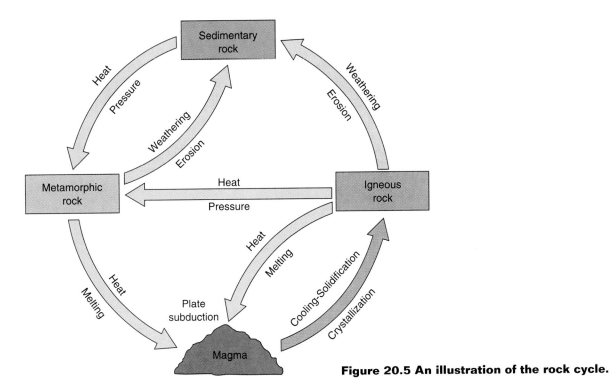

Figure 20.5 An illustration of the rock cycle.

Rocks are classified into three major categories, based on the way they originated.

Igneous rocks are formed by the solidification of magma. **Magma** is molten rock material, including any dissolved gases and crystals, that originates deep beneath Earth's surface. Magma that reaches Earth's surface is called **lava.**

Sedimentary rocks, which are formed at Earth's surface, originate in three ways:

1. the lithification of any preexisting sediment,
2. the precipitation of a mineral from a solution, or
3. the consolidation of plant or animal remains.

Metamorphic rocks are formed by the alteration of preexisting rock due to the effects of pressure, temperature, or the gain or loss of chemical components.

The rock cycle, shown in Fig. 20.5, is a graphic method used to illustrate the interrelationships among the processes that produce the three types of rocks.

The general sequence of mineral formation (crystallization) with decreasing temperature is shown, in a simplified way, in Fig. 20.6. The diagram illustrates what is called the *Bowen reaction series*. A close look at the diagram shows that crystallization takes place in three separate series. The Bowen reaction series can be applied only to certain basaltic magmas, but it shows the sequence in which minerals crystallize from a magma.

Basaltic magma is a term applied to the hot melt that solidifies into a huge mass of rock called basalt. The word *basalt* is believed to have had an ancient Ethiopian origin, meaning "black stone." Basalt is a fine-grained rock composed primarily of pyroxene and calcium-rich feldspar. Its color varies from dark green to black, and it is the most common igneous rock formed in Earth's crust.

The diagram also illustrates that minerals crystallize at different temperatures. The first material to crystallize from the hot melt is olivine, and in most cases, at approximately the same temperature, calcic plagioclase begins to crystallize. The olivine-through-biotite series is discontinuous; that is, the formation of each mineral takes place in discrete steps. Plagioclase-feldspar formation is a continuous series that begins with calcium-rich plagioclase, which, as the temperature falls, gradually becomes more abundant in sodium ions. Several of these materials may exist simultaneously at

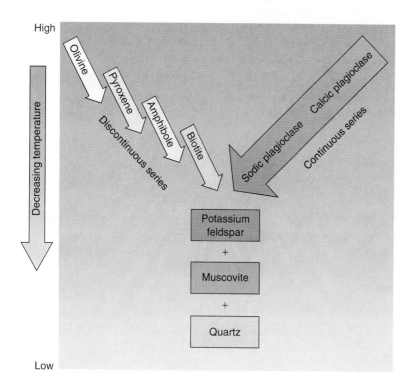

Figure 20.6 The Bowen reaction series.
The diagram illustrates the general sequence of crystallization of minerals with decreasing temperature in a basaltic melt. The minerals at the top crystallize first, the ones at the bottom last.

any one time. The third series, which is not a real series, takes place after most of the magma has solidified. The melt that remains forms the minerals potassium feldspar, muscovite, and quartz. The whole process is more complex than the diagram indicates.

Igneous Rock

As mentioned above, igneous rock forms when molten material from somewhere deep beneath Earth's surface cools and solidifies. Accumulations of solid particles thrown from the throat of a volcano are also considered igneous rocks. The molten material is known as magma as long as it is beneath Earth's surface but becomes lava if it flows on the surface. The term *lava* may be a bit confusing, because it can be used to mean either the hot molten material or the resulting igneous rock.

Assuming that Earth was originally molten, the very first rocks of the continents and ocean basins must have been igneous. Geologic processes, however, long ago obscured any recognizable remnant of these most ancient materials of Earth. Nevertheless, igneous activity has continued so actively throughout the history of Earth that igneous rocks are by far the most abundant type; they are estimated to constitute as much as 80 percent of Earth's crust.

Most people, of course, know more about lava than magma. The eruption of a volcano is a spectacular geologic phenomenon and may be reported around the world. Any igneous rock that cools from molten lava is described as *extrusive rock*. Few people realize that the vast majority of magma (molten rock) never finds its way to the surface but cools to solid rock somewhere within Earth's interior as *intrusive rock*. We see intrusive rock only where erosion has stripped away the burden of overlying rock or where movement of Earth's crust has brought it to the surface.

Anyone who has seen a volcano erupt must wonder about Earth's interior, from which such hot, molten material comes. When people began to sink mines and deep wells in search of buried natural resources, they discovered that Earth grows 1°F hotter for each 150 ft of depth. If the increase in temperature continued at this same rate to the center of Earth, the temperature would be so high that rock more than a few tens of miles below Earth's surface would have to be molten regardless of pressure. This view was held by early geologists, who thought of the thin, solid outer layer as Earth's crust. We now know from several lines of evidence that outside the deep molten core Earth is solid and that its interior, though hot, is not nearly as hot as scientists once thought.

Even so, what makes Earth hot inside? It was once theorized that the internal materials of Earth were poor thermal conductors and the interior had retained its molten origin. However, if this were correct, Earth would have cooled down by now and all geologic processes would be extinct. The current view is that Earth's interior hotness is caused by radioactivity. The decay of radioactive elements produces heat. We know from the distribution and abundance of these radioactive elements in Earth's interior that they supply at least enough heat to keep Earth at its present interior temperature.

Texture, which is grain size, is an important characteristic of both intrusive and extrusive rocks. Grain size is determined primarily by the rate at which the molten rock cools. An igneous body must cool slowly if its mineral grains are to grow large. Below Earth's surface, magma cools more slowly, so intrusive igneous rocks generally have a coarse, or large-grain, texture. Rapid cooling almost invariably yields small grains or perhaps none at all (glassy). Extrusive rocks, which have cooled relatively quickly on Earth's surface, generally have a fine-grain texture.

All objects must be classified and have names if they are to be discussed intelligently. A classification of igneous rocks must account for their widely differing physical and chemical characteristics, and the classification must be logical and useful. Geologists classify igneous rocks according to their mineral composition and their texture, as indicated in Table 20.4.

Igneous rocks can be roughly divided into those rich in silica (SiO_2) and those relatively low in silica. Rocks rich in silica contain minerals with abundant silicon, sodium, and potassium. These minerals are mostly light in color. Rocks low in silica are rich in minerals containing iron, magnesium, and calcium and are much darker. The color of an igneous rock is therefore a convenient guide to its chemical composition. Figure 20.7 shows the appearance of three igneous rocks.

Large bodies of intrusive igneous rock are called **plutons,** after Pluto, the Greek god of the underworld. These plutonic bodies are classified according to their size, shape, and orientation in surrounding rock. Intrusive rock bodies are referred to as being *concordant* when they lie more or less parallel to older formations, and as being *discordant* when they cut across older formations. The four major plutonic forms are illustrated in Fig. 20.8.

Large discordant bodies are called **batholiths.** Batholiths are enormous in size, and to be so classified, must occupy an area of at least 104 km^2 (40 mi^2), but many are vastly larger than that. For example, the Coast Range Batholith in western Canada is more than 1600 km (1000 mi) long and in some places more than 160 km (100 mi) wide. All surface exposures indicate that batholiths grow larger with depth, but the nature of their bottoms

Table 20.4 Classification of Common Igneous Rocks

Texture	High in Silica		Low in Silica	
	Orthoclase and Quartz Dominant; Some Hornblende, Biotite, and Muscovite	Light (Na) Plagioclase Dominant; Abundant Hornblende	Dark (Ca) Plagioclase and Pyroxene Dominant	Pyroxene and Olivine Dominant
Coarse grained	Granite	Diorite	Gabbro	Peridotite
Fine grained	Felsite	Andesite	Basalt	
Glassy	Obsidian			
Glassy and porous	Pumice	Pumice		
Pyroclastic	{ Volcanic tuff { Volcanic breccia	{ Volcanic tuff { Volcanic breccia	Scoria	

Figure 20.7 Igneous rocks.
Left to right: Granite, basalt, and obsidian.

remains uncertain, because no canyons or mines have penetrated that deep.

Another discordant plutonic body is a **dike**, which is formed when magma fills a nearly vertical fracture in rock layers. A dike is tabular, with its width or thickness much smaller than its other dimensions, as shown in Fig. 20.8. Although originally formed below Earth's surface, some intrusive bodies are now visible, due to erosion and other geologic processes.

Sills and laccoliths are intrusive concordant bodies. **Sills** are similar to dikes, but form between and parallel to existing rock beds (Fig. 20.8). A **laccolith** is a blisterlike intrusion that has pushed up the overlying rock layers.

Sedimentary Rock

In general rocks near the surface of Earth are *sedimentary rocks*. These are rocks formed from sediment composed of particles of older rocks and other material. Rocks on Earth's surface are continually being both physically and chemically worn down by environmental processes. The resulting rock particles and dissolved minerals make up the sediment that is transported by streams and rivers and eventually deposited in seas and oceans. The loose sediment eventually settles to form layers, and the consolidation of these sedimentary layers forms sedimentary rock.

Sediments and sedimentary rocks make up only 5 percent of Earth's crust, but their impor-

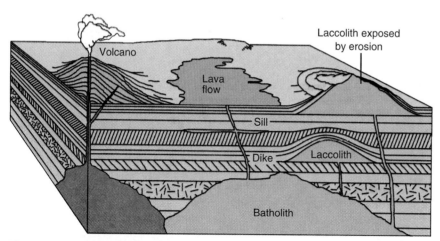

Figure 20.8 Plutonic bodies.
Magma solidifying within Earth forms intrusive igneous bodies called plutons. The batholith is the largest. Sills and laccoliths are concordant bodies that lie parallel to existing rock layers. The discordant dike cuts across existing rock formations. A volcanic stock is shown rising from the batholith.

tance is entirely out of proportion to their limited abundance. One reason is that they cover about 75 percent of the continents and even more of the ocean basins. In a sense, they are a bit like clothing; they cover only the surface, but they are conspicuous. The North American continental interior is composed of a foundation of igneous and metamorphic rocks with a sedimentary veneer only a few thousand feet thick. Residents of many of the interior states of the United States may never have seen any rocks other than sedimentary ones. Our modern way of life owes much to sedimentary rocks, because they contain abundant petroleum, coal, and metal deposits and many of the materials essential to the construction industry.

Sedimentary rocks, with their many varieties, commonly form more interesting landscapes than do igneous and metamorphic rocks in the same setting. A comparison between the Grand Canyon of the Colorado River and the Snake River Canyon illustrates this point. Both are magnificent gorges in terms of sheer size, but the beauty and variety of the Grand Canyon with its staircase of brightly colored slopes and cliffs of sedimentary rock is famous around the world, while the Snake River Canyon, carved through a more monotonous sequence of drab gray to black volcanic rocks, is little known and seldom visited.

The rock and mineral fragments carried to the oceans eventually settle out as **clastic sediments.** The dissolved mineral matter may be extracted from seawater by plants or animals and accumulate on the seafloor as **organic sediments,** or it may be physically precipitated on the floor as **chemical sediments.** Sedimentary rocks, therefore, fit into three major categories—clastic, organic, and chemical. The most important types are listed in Table 20.5.

Geologists gain much information from sedimentary rocks because their characteristics tell so much about their origin and history. Color, rounding, sorting, bedding, fossil content, ripple marks, mud cracks, footprints, and even raindrop prints are common characteristics of sedimentary rocks that indicate the conditions under which the rock was formed. Figure 20.9 shows the appearance of three sedimentary rocks.

Bedding, or stratification, is the layering that develops at the time the sediment is deposited. The bedding may be conspicuous, as shown in Fig.

Figure 20.9 Sedimentary rocks.
Clockwise from top: Fossiliferous limestone, dolomite, and shale.

Figure 20.10 Bedding
Bedding, or stratification, is evident in this limestone formation.

20.10, or vague. It may be in either thick or very thin layers. Because most sediment comes to rest on a level surface, most bedding is horizontal. There are many environments, however, where sediments accumulate in tilted layers known as

Table 20.5 Types of Sedimentary Rocks

Clastic Sedimentary Rocks

Sediment	Grain Size (mm)	Rock Name	Characteristics
Gravel	More than 2 mm	Conglomerate	Rounded pebbles
		Breccia	Angular pebbles
Sand	$\frac{1}{16}$ to 2 mm	Sandstone	
		Quartzose	Dominantly quartz
		Arkose	Quartz and angular feldspar
		Graywacke	Quartz, feldspar, chlorite, clay particles, volcanic debris, variable sizes
Mud	Less than $\frac{1}{16}$ mm	Shale	Silt and clay particles, rich in quartz
Shell fragments and calcite grains	Variable	Limestone	
		Coquina	Porous aggregate of shell fragments
		Oolitic limestone	Small, rounded grains

Organic Sedimentary Rocks

Rock Name	Characteristics
Bituminous coal	Compacted plant remains. Breaks into rectangular lumps.
Limestone	Compacted or cemented calcareous plant or animal remains. May be crystalline.
Fossiliferous limestone	Fossils present.

Chemical Sedimentary Rocks

Rock Name	Characteristics
Limestone	Compacted, cemented, calcareous material. May be crystalline.
Dolomite	Composed of dolomite instead of calcite.
Rock gypsum	Evaporite. Fine grained.
Anhydrite	Evaporite. Fine to coarsely granular.
Rock salt	Evaporite. Composed of halite.
Chert	Siliceous precipitate. Not crystalline.

cross-bedding (Fig. 20.11). Much research and many pages have been devoted to the geometry and origin of the many kinds of cross-bedding. In every case the surface is sloping where the sediments come to rest. Cross-bedding is especially common where a river empties into a lake, where sediments fill in depressions scoured by floods along river channels, or where wind drapes sand down the flanks of a dune. By recognizing the type of cross-bedding, the geologist makes an observation that helps in unraveling the origin and history of the rock.

The most distinctive and most interesting characteristic of a sedimentary rock is the fossils it sometimes contains (Fig. 20.12). Although a mystery to those who lived several centuries ago, a **fossil** is now known to be the remains of an organism that lived in the past. The fossil organism need not be extinct, though many are, but it must have died in prehistoric times.

Those who have looked along the bed of a briskly flowing creek or have seen the sandy bottom of a shallow pond may have noticed that movement or agitation of the water has developed *ripple marks* in the sediment, which look somewhat like waves. Current ripple marks form on stream bottoms in moving water, and oscillation ripple marks (Fig. 20.13) form in shallow, standing water. Because these features are sometimes buried and preserved in sedimentary rock, they are

Figure 20.11 Cross-bedding
Sediments accumulating in tilted layers give rise to cross-bedding, as seen in this sandstone formation.

Figure 20.13 Ripple marks.
Oscillation ripple marks in sandstone.

Figure 20.12 Fossils.
Fossilized shellfish (brachiopods) in rock.

still another clue in deciphering the history of ancient rocks.

Mud deposited in shallow water, or along a valley bottom, may be quickly exposed to the atmosphere when the water recedes. The mud may then be marked by the footprints of animals that walk across it or possibly pocked by the impact of raindrops; or as it dries, it may shrink to form *mud cracks* like those in Fig. 20.14. The vast majority of these features do not even survive the season, but those rare ones that are buried and preserved become a part of the rock record and may later be exposed to the inquiring eye of a geologist.

The transformation of sediment into a clastic sedimentary rock is a process called **lithification.** During this process the loose, solid particles (sediment) are compacted by the weight of overlying material and eventually cemented together. Common cementing agents are silica (SiO_2), calcium carbonate ($CaCO_3$), and iron oxides, which are dissolved in groundwater that permeates the sediment. An example is the most common clastic sedimentary rock, shale. When fine-grained mud is subjected to pressure from overlying rock material, water is driven off and the clay minerals begin to compact (consolidate). As groundwater moves through the compacted sediment, materials dissolved in the water precipitate around the individual mud-size particles, and cementation of the particles occurs. A clastic sedimentary rock results. Sandstones and conglomerates occur in the same manner but with differing sediment sizes.

Figure 20.14 Mud cracks.
Mud cracks along the Rio Grande in Texas.

Figure 20.15 Dripstone.
Dripstone in a limestone cavern. Stalactites extend downward from the ceiling and stalagmites extend upward from the floor. They sometimes meet to form a column (right).

Chemical sedimentary rocks are formed from the precipitation of a material from a solution, usually water. There are two subtypes of chemical rocks: indirect and direct chemical. Indirect chemical rocks are formed by biochemical reactions during the activities of plants and animals; that is, certain organisms, such as coral, extract calcium carbonate from seawater to build skeletal material. When coral dies, the collected skeletal deposits subsequently form biochemical limestone.

Direct chemical rocks are formed when the evaporation of water with dissolved materials leave behind a residue of chemical sediment, such as sodium chloride (rock salt) and calcium sulfate dihydrate (gypsum). Another example of directly formed chemical sedimentary rock is cave dripstone, which is formed primarily by calcium carbonate precipitated from dripping water. Dripstone takes on a variety of forms, but most common are the icicle-shaped *stalactites* extending down from the cave ceilings, and cone-shaped *stalagmites* formed on the cave floor (Fig. 20.15).

Metamorphic Rock

About 15 percent of Earth's crust is metamorphic. Metamorphic rocks are those that have been changed under the influence of great temperature and/or pressure deep beneath Earth's surface. These changes that transform the parent material into metamorphic rocks result from a very fundamental property of virtually all minerals. Minerals (and therefore rocks) are stable only in the environment in which they form. The minerals remain stable and endure as long as the environment remains constant but become unstable and break down if the environment changes.

Sedimentary rocks, which form under surface conditions, are especially susceptible to change by heat and pressure, but even igneous rocks may be affected. The conditions far beneath Earth's surface that produce metamorphic rock are not so fundamentally different from those that form magmas. Both intrusive igneous and metamorphic rocks are the products of processes that act at great depth and are, therefore, found in close association once they have been uncovered by erosion.

Two major kinds of change occur during metamorphism. The first is mechanical change and includes fracturing, crushing, and shaping caused by intense pressure. The second type of change is chemical and consists of recrystallization and loss of water. Recrystallization, which is the growth of minerals, may involve reorganization and growth

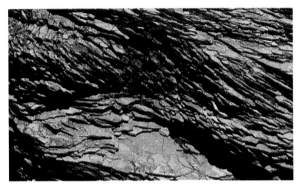

Figure 20.16 Slate bed.
Slate is a fine-grained metamorphic rock that possesses a type of foliation known as slaty cleavage.

Figure 20.17 Gneiss foliation.
This metamorphic rock formation shows rough foliation due to high temperature and pressure. Feldspar and quartz are the chief minerals in gneiss.

of minerals originally present; or it may, by more profound chemical changes, produce an entirely new set of minerals. The new minerals will be those that are stable in the changed environment.

A metamorphic rock itself is classified according to its texture, mineral composition, and foliation. *Foliation* is the ability of a metamorphic rock to split along a smooth plane. Foliated rocks have this property because they contain a large number of flat, platy, or sheetlike minerals, all oriented in the same direction. Minerals such as mica, chlorite, and hornblende are essential ingredients of foliated metamorphic rocks.

Geologists are fairly certain that foliation develops in response to pressure. It is tempting to give credit to the weight of the enormous overburden that bears down upon a rock undergoing metamorphism. This confining pressure, however, is exerted equally in all directions and could not possibly force the development of platy minerals with any one particular orientation. Foliation can develop under confining pressure only if there is an additional directed pressure, the sort that accompanies mountain building, when the rocks are squeezed under tremendous horizontal pressure.

The progressive metamorphism of shale is a good illustration of changes that occur as a sedimentary rock is subjected to more and more intense metamorphism. Shale, after relatively mild metamorphism, is transformed to slate (Fig. 20.16), a fine-grained metamorphic rock similar in many respects to shale but differing fundamentally in its excellent slaty cleavage, a variety of foliation. If the shale is subjected to more intense heat and

Figure 20.18 Metamorphic rocks.
Clockwise from top: Marble, quartzite, and mica schist.

pressure, it will change to schist, a metamorphic rock whose grains are visible to the unaided eye but whose foliation, though good, is less perfect than slate's. Very intense metamorphism produces gneiss (Fig. 20.17), an even coarser-grained rock, with a rough foliation characterized by distinct banding. The higher grades of metamorphism, therefore, produce larger grains but rougher foliation. Fig. 20.18 shows three other types of metamorphic rock.

Metamorphic rocks that lack foliation are said to be massive, and they include such well-known

Table 20.6 Metamorphic Rocks

Foliated

Rock Name	Original Rock Name	Principal Minerals and Characteristics
Slate	Shale, tuff	Mica, quartz—excellent foliation, very fine grained
Chlorite schist	Basalt, felsite, tuff	Chlorite, plagioclase—good foliation, coarse grained
Mica schist	Shale, felsite, tuff	Muscovite, quartz, biotite—very good foliation, coarse grained
Hornblende schist	Basalt, gabbro, felsite	Hornblende, plagioclase—good foliation, coarse grained
Gneiss	Granite, felsite, shale, mica schist	Feldspar, quartz, plagioclase, hornblende—rough foliation, coarse grained, banded

Massive

Rock Name	Original Rock Name	Principal Minerals and Characteristics
Hornfels	Any fine-grained rock, especially shale	Variable composition—fine grained, hard
Quartzite	Sandstone	Quartz—fine to coarse grained
Marble	Limestone, dolomite	Calcite—coarse grained
Anthracite	Bituminous coal	Carbon—conchoidal fracture

examples as marble, which is formed from limestone, and quartzite, which is metamorphosed sandstone. A simplified classification of metamorphic rocks is listed in Table 20.6.

CONFIDENCE QUESTION 20.2

Describe the cyclic nature of the rock cycle.

20.3 Volcanoes

Learning Goals:

To describe the different types of volcanoes.

To identify the eruption characteristics of volcanoes.

A geological process that provides evidence of Earth's interior hotness is the eruption of a volcano. Volcanic eruptions also provide a means to observe and study materials from Earth's mantle. But, what comes out of a volcano?

1. The expulsion of gas is the most widespread general characteristic of all volcanoes. A volcano may burp gas in its very earliest infancy, at the height of its activity, and during its final dying gasps, when all other signs of life are gone. Steam is by far the most abundant and may comprise as much as 90 percent of the gases.

2. Volcanoes may erupt lava in variable quantities and conditions. Some, but not all volcanoes, produce vast amounts of very fluid lava, which flows easily and quietly with little explosive violence. If much gas escapes at the same time, there may be minor explosions, which hurl incandescent lava into the air. See Fig. 20.19.

   These lava fountains, impressive enough by day, form magnificent fireworks at night. Clots of lava sometimes harden in midair and strike the ground as spindle-shaped volcanic bombs of various sizes. Other volcanoes erupt relatively small volumes of lava so stiff and viscous that it can barely flow at all.

3. Many volcanoes spew enormous volumes of solids, which can range in size from the finest dust to huge boulders. Such particles, known collectively as *pyroclastic debris* or *tephra*, include not only fragments of rock but also gas-laden material that the volcano hurls out in molten form but that hits the ground as a solid. Such debris is blasted into the air by violently explosive volcanoes.

Figure 20.19 Volcanic eruption.
The force of gas escaping from the underground reservoir of magma throws molten lava high into the air over the crater of Hawaii's Kilauea volcano. Notice the lava flows in the foreground.

There have been few volcanoes in written history that can match the 1815 eruption of Tambora in violence and in the volume of solids hurled into the atmosphere. Tambora, on the island of Soembawa just east of Java, erupted with a thunderous roar from April 10 to 20. By the time the eruption was finished, there was a hole where before there had been a high mountain.

The 145 km^3 (35 mi^3) of solid debris thrown into the air so darkened the sky that for several days total darkness reigned for hundreds of miles around. Enough fine volcanic dust was circulated around the Earth that the Sun's rays were partially blocked, and the year 1816 was uncommonly cold. The cold weather that accompanied this and a few other similar eruptions have led some geologists to suggest that the recent ice age, whose great glaciers once blanketed much of Earth's surface, was caused by an unusually large number of violently erupting volcanoes. We must, however, learn much more about changes in climate before we can really know why the ice age began.

The behavior of a volcano depends to a considerable extent upon the viscosity of the lava that fills its throat. *Viscosity* is defined as the property of a fluid that resists the force tending to cause the fluid to flow. A liquid with a high viscosity is, therefore, one that is thick and stiff and does not flow readily—like "molasses in January."

Chemical composition and temperature are probably the most important of the several factors that determine the viscosity of a magma or a lava. As you learned in Section 20.2, magmas (and lavas) can vary in their content of silica (SiO_2). Low-silica magmas are much more fluid and can flow more easily than high-silica magmas, as long as other factors, such as gas content and temperature, are the same.

Temperatures are not always the same, however. Very hot lavas flow more easily than those just above their melting temperature. Viscous, silica-rich lavas tend to clog their vents and are removed only when enough pressure builds up that the vent is cleared by violent explosions.

Let us examine now the various types of eruptions. Some lava reaches the surface through long fractures in the surface rocks. More commonly, however, the lavas pour out of central vents or volcanic cones, which previous eruptions have constructed.

Fissure eruptions, which issue from long fractures, have been very rare in human history. Until very recently, the latest such volcanic event had affected Iceland in 1783. On January 23, 1973, however, a fissure eruption began that threatened to inundate the small island of Heimaey just off the south shore of Iceland. This eruption was of great scientific interest, but it was also a matter of grave concern to residents of Iceland, whose economy depends heavily on the fishing industry centered in the town of Vestmannaeyjar on Heimaey.

The basaltic lavas originating from fissure eruptions are so extremely fluid that they flow many miles over Earth's surface without constructing volcanoes. Individual lava flows are not thick, however, because they extend for many miles over the surface. As one lava flow follows another over millions of years, the entire landscape may be eventually drowned in a sea of solid basalt, with perhaps here and there a high peak forming an island in the black and desolate ocean of rock. The Blue Mountains of Oregon are examples of such islands within the Columbia Plateau.

Figure 20.20 Lava flows.
Rivers of lava flow from Hawaii's Mauna Loa volcano. The Hawaiian Islands are the exposed tips of volcanic mountains.

Figure 20.21 Volcano Parícutin.
This cinder-cone volcano erupted in a farmer's cornfield in Mexico in 1943.

Volcanoes may erupt in a variety of ways. Some perform quietly, almost unobtrusively, and yet manage in their unspectacular way to pour great volumes of lava onto the surface. These eruptions may cause some destruction of property but little loss of life. With only a few exceptions, basaltic volcanoes erupt quietly.

One of the most noted examples of such eruptions is in the chain of volcanoes that forms the foundation of the Hawaiian Islands (Fig. 20.20). Eruptions of huge quantities of lava from cracks in the floor of the Pacific Ocean have caused the formation of one of the most extensive mountain chains on Earth, most of which is beneath the ocean. The Hawaiian chain is nearly 2600 km (1600 mi) long, and the volcanic formations on the ocean's floor, nearly 5 km (3 mi) below sea level, rise to form islands that project far above sea level.

Other volcanoes erupt with explosive violence. The felsic lava that may accompany these eruptions is usually subordinate to the much greater volume of pyroclastic debris. An explosive volcano may lie dormant for years, centuries, or even thousands of years and then finally erupt with such violence that it blasts apart the volcanic cone it had previously constructed, destroying the surrounding countryside. Viscous felsite is largely responsible for this behavior because it clogs and seals the volcano's vents and causes pressure to build up that can be relieved only by powerful blasts.

In 1902, geologists were reminded of the unusually destructive behavior of certain explosive volcanoes. After several months of violent, threatening activity, Mt. Pelée, a volcano at the northern end of Martinque in the Caribbean Sea, blasted out incandescent, cloudlike mixtures of superheated gas and pyroclastic debris, which swept down the side of the volcano and over the nearby city of St. Pierre. In a matter of seconds virtually all life and property were destroyed.

From more recent times are the eruptions of Parícutin, Surtsey, Mount St. Helens, and El Chichón. In 1943, Parícutin erupted in a Mexican farmer's cornfield. In the first year the volcanic cone rose to a height of over 300 m (1000 ft) (Fig. 20.21). After about 10 years of subsiding activity, Parícutin is now a dormant volcano with a cone height of over 400 m (1300 ft).

In 1963, volcano Surtsey (after *Surtr*, the subterranean god of fire in Icelandic mythology) boiled up from the ocean floor off the coast of Iceland. Having expelled sufficient lava to form a barrier against the sea, Surtsey is now a permanent volcanic island that may be found on the world map (Fig. 20.22). Geologists were able to study Surtsey from its "birth."

Many people remember the 1980 eruption of Mount St. Helens. This volcano is located in southwestern Washington about 64 km (40 mi) from Portland, Oregon. Prior to 1980, Mount St. Helens

Figure 20.22 Volcano Surtsey.
A 1963 suboceanic eruption off the southern coast of Iceland formed the volcanic island of Surtsey.

Figure 20.23 Mount St. Helens.
A view of the volcano after the May 18, 1980, eruption. An area larger than 384 km² (150 mi²) was devastated.

was a placid, snow-capped, dormant volcano, one of many such volcanoes in the Cascade Mountains of Oregon and Washington. Its last period of activity had been between 1800 and 1857.

In March 1980, a series of minor earthquakes occurred near Mount St. Helens, and the first eruptions took place. On May 18, 1980, a massive eruption occurred, devastating an area of more than 384 km² (150 mi²) and leaving more than 60 people dead or missing (Fig. 20.23). A column of ash rose to an altitude of more than 19 km (12 mi). Nearby cities were blanketed with ash, and a light dusting of ash fell as far away as 1450 km (900 mi) to the east.

Huge mudflows from the ash caused flooding and silting in the rivers near the volcano, destroying and damaging many homes. Mount St. Helens became relatively quiet in 1981. However, it has shown minor activity in the years since, and geologists expect that intermittent activity may continue for years, and even decades, if the behavior of the volcano follows the pattern of its eruptions in the 1800s. Volcano-triggered mudflows also had a devastating effect in the 1985 eruption of a volcano in Colombia, South America. Over 20,000 people were killed.

Another significant volcanic eruption in that decade was that of El Chichón in Mexico. In late March 1982 El Chichón roared into life with a tremendous explosion that sent a column of ash and gases 16 km (10 mi high) within an hour. Although the eruption of El Chichón was not very impressive compared with its earlier Mexican neighbor Paricutin, its significance arises from the projection of debris some 26 mi into the atmosphere. This debris in the stratosphere may have long-term climate effects (see Section 19.5).

The occurrence of volcanic activity is for the most part unpredictable. New volcanoes may be formed unexpectedly, while existing volcanoes lying dormant may suddenly erupt with practically no warning. However, the locations of eruptions and potential eruptions are known. From the theory of plate tectonics (Chapter 21), volcanic activity takes place predominantly at plate boundaries above subduction zones where one plate is deflected downward into Earth's interior (see Fig. 21.14). For example, the area of the Pacific Ocean is one of widespread volcanic activity. The outer rim of the Pacific Ocean is marked by a ring of volcanoes known as the **"Ring of Fire"** (Fig. 20.24). Comparing Figs. 20.24 and 21.10 clearly shows the marked correlation with plate boundaries.

Recent activity on the Ring of Fire occurred in 1991. Mount Unzen in Japan erupted, and Mount Pinatubo in the Philippines came to life after being dormant for 600 years. There was loss of life, and a great deal of damage was suffered with both. Mount Pinatubo sent volcanic ash 21 miles into the atmosphere. Ash "rained" down on the surrounding region and piled up more than a foot deep. Evacuations, including that of nearby American military bases, were hampered by winds and rains

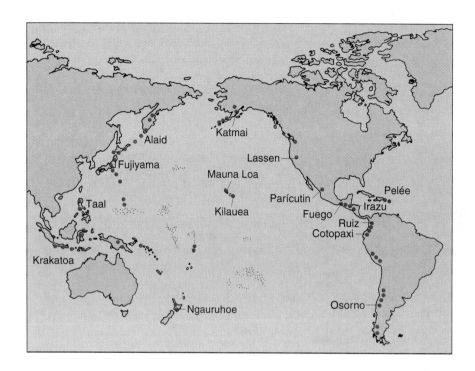

**Figure 20.24
The "Ring of Fire."**
The outer rim of the Pacific Ocean is circumscribed by a ring of volcanoes that tend to have violent eruptions, hence the name "Ring of Fire."

from a typhoon that caused flooding and produced slippery volcanic mud.

The formation of the Hawaiian Islands in the central Pacific Ocean is explained by the **hot-spot theory.** This theory hypothesizes that a hot spot exists within the mantle that ejects magma onto the overlying seafloor. As the Pacific Plate moved over the fixed hot spot, a succession of volcanoes were generated. Each volcano moves with the plate, leaving room for the next.

According to the hot-spot theory, volcanic island chains trace out a plate's movement over a hot spot. The Hawaiian Islands are cited as a classic example of this process, and studies (involving age dating of seafloor samples) have yielded data in support of the theory. However, not every island chain appears to fit the theory, so geologists have more work to do in explaining Earth's internal processes.

As pointed out previously, many volcanoes do not erupt violently but relatively calmly, ejecting liquid magma from fissures in Earth's surface. If the lava has a relatively low viscosity so that it flows easily, a gently sloping, low-profile **shield volcano** is formed by frequently repeated lava flows.

The classic example of a shield volcano is Mauna Loa in the Hawaiian Island chain. In fact, Mauna Loa is a huge volcanic mountain—the largest single mountain on Earth in sheer bulk. Although not as tall as Mt. Everest (slightly over 8800 m—29,000 ft—above sea level), Mauna Loa rises 4600 m (15,000 ft) from the ocean floor to sea level and protrudes an additional 4150 m (13,600 ft) above sea level for a total height of about 8750 m (28,600 ft). Its bulk comes from the fact that this partially submerged mountain has a base almost 160 km (100 mi) in diameter.

Volcanic eruptions of both lava and pyroclastic debris form a more steeply sloping, layered cone that is called a **stratovolcano** (also called a **composite volcano**). The lava of stratovolcanoes has a relatively high viscosity, and eruptions are more violent and generally less frequent than those of shield volcanoes. Many stratovolcanoes have an accumulation of material 1800–2400 m (6000–8000 ft) above their base and have a characteristic symmetrical profile. Mount St. Helens (Fig. 20.23) is a stratovolcano. Dormant stratovolcanoes include Mt. Fuji in Japan (Fig. 20.25) and Mts. Shasta, Hood, and Rainier in the Cascade Mountains of Washington and Oregon.

Figure 20.25 Mount Fuji in Japan.
Mount Fuji is a dormant stratovolcano that was formed from eruptions of both lava and pyroclastic debris.

Figure 20.26 Volcano Izalco.
A cinder-cone volcano in El Salvador.

A volcanic eruption may also consist primarily of pyroclastic debris. In this case, steeply sloped **cinder cones** are formed, which rarely exceed 300 m (1000 ft) in height. A cinder cone is shown in Fig. 20.26 rising above a previous lava flow. Volcano Izalco in El Salvador is an example of these small, steep, symmetric cones of pyroclastic debris.

Volcanic activity is usually not confined to the region of the central vent of a volcano. Fractures may split the cone, with volcanic material being emitted along the flanks of the cone. Also, material and gases may emerge from small auxiliary vents, forming small cones on the slope of the main central vent. A funnel-shaped depression called a *volcanic crater* exists near the summit of most volcanoes from which material and gases are ejected.

Many volcanoes are marked by a much larger depression called a **caldera.** These roughly circular, steep-walled depressions may be up to several miles in diameter. Calderas result primarily from the collapse of the chamber at the volcano's summit from which lava and ash were emitted. The weight of the ejected material on the partially empty chamber causes its roof to collapse, much like the collapse of a snow-ladened roof of a building.

Crater Lake on top of Mount Mazama in Oregon occupies the caldera formed by the collapse of

Figure 20.27 Crater Lake.
Mount Mazama in Oregon was once an active stratovolcano. The water-filled caldera, which is about 10 km (6 mi) in diameter, is known as Crater Lake.

the volcanic chamber of this once-active stratovolcano (Fig. 20.27). In some instances calderas are formed by a violent explosion rather than a collapse. Felsitic lava may clog the volcanic vent, and the resulting buildup of pressure causes the volcano to "blow its top," so to speak.

562 CHAPTER 20 Minerals, Rocks, and Geologic Events

CONFIDENCE QUESTION 20.3
(a) What is a volcano? (b) What are the two most important factors that determine the nature of a volcanic eruption?

20.4 Earthquakes

Learning Goals:

To explain earthquakes and the science of seismology.

To define the Richter scale and the modified Mercalli scale.

An earthquake, as those who have experienced a substantial one know, is manifested by the vibrating and sometimes violent movement of Earth's surface. But waves are also propagated through Earth. It is these waves that provide one of the most useful ways of studying Earth's interior. **Seismology** is the branch of geophysics that studies these waves and uses them as probes to "see" below Earth's surface to determine its internal structure.

Earthquakes, which result from shock waves traveling through Earth, rattle our globe perhaps a million times each year, but the vast majority are so mild that they can be detected only with delicate, sensitive instruments. A very powerful quake, however, can destroy a large area.

Earthquakes may be caused by explosive volcanic eruptions or even explosions caused by humans, but most earthquakes appear to be associated with movements in Earth's crust. These movements in some instances form large fractures in Earth called **faults.** According to the theory of plate tectonics (see Section 21.3), the optimum place for such movement is at the plate boundaries, and the major earthquake belts of the world are observed in these regions (see ● Fig. 20.28).

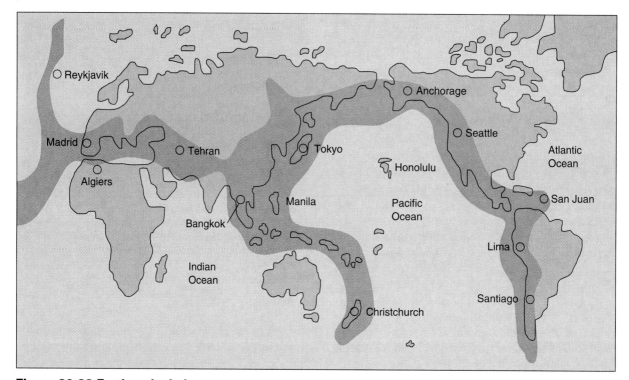

Figure 20.28 Earthquake belts.
The shaded regions show the earthquake belts of the world. Notice that the belt circumscribes the Pacific Ocean, as does the volcanic "Ring of Fire" in Fig. 20.24.

Notice how the earthquake region around the Pacific Ocean is similar to that of the volcanic "Ring of Fire" (Fig. 20.24).

Movements of Earth's plates would certainly put strains on rock formations in Earth's crust near the plate boundaries. Rocks possess elastic properties; energy is stored in the elastic deformations of the lithosphere. If the forces causing this deformation are great enough to overcome the force of friction along a plate boundary or some nearby fault, then the fault walls move suddenly and the stored energy in the rocks is released, causing an earthquake.

If the elastic limit of a stressed rock formation is not exceeded, the rock can rebound elastically and snap back to its original shape after a shift along a fault has relieved the stress—much like a spring. However, if the elastic limit of the rock is exceeded, the rock cannot fully recover when the strain is removed. If the elastic limit is greatly exceeded by the deformation force, the rock formation can even rupture.

Relative horizontal and parallel displacement of rock on each side of a fracture is called strike-slip faulting. Many large strike-slip faults are associated with plate boundaries and are called transform faults. An example of a transform fault is the famous San Andreas Fault in California. It is the master fault of an intricate network of faults that runs along the coastal regions of California (●Fig. 20.29). This huge fracture in Earth's crust is more than 960 km (600 mi) long and at least 32 km (20 mi) deep. Over much of its length a linear trough of narrow ridges reveals the fault's presence.

As you can see in Fig. 20.29, the San Andreas fault system lies on the boundary of the Pacific

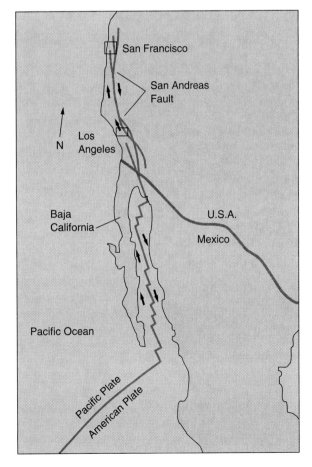

Figure 20.29 The San Andreas Fault.
(a) The photo shows an aerial view of the exposed fault line on the Carizzo Plain in California. (b) The map shows how the fault marks the boundary between the Pacific and American Plates, which are moving in opposite directions. The San Andreas Fault is a classic example of a transform fault.

**Figure 20.30
Earthquake devastation.**
The photo shows the devastation caused by an earthquake measuring 7.2 on the Richter scale on January 19, 1995, in Kobe, Japan.

Plate and the American Plate (see Fig. 21.10). Movement along the transform fault arises from the relative motion between these plates. The Pacific Plate is moving northward relative to the American Plate at a rate of several centimeters per year. At this current rate and direction, in about 10 million years Los Angeles will have moved northward to the same latitude as San Francisco. In another 50 million years the segment of continental crust on the Pacific Plate will have become completely separated from the continental land mass of North America.

The horizontal movement along the San Andreas Fault is a cause for considerable concern for the populated San Francisco Bay area through which it runs. The famous San Francisco earthquake of April 18, 1906, which measured 8.3 on the Richter scale (earthquake scales will be discussed shortly), resulted in the loss of approximately 700 lives and in millions of dollars of property damage. Many other milder earthquakes have occurred since then along the San Andreas and other branch faults.

On October 17, 1989, an earthquake shook the San Francisco Bay area. Measuring 7.1 on the Richter scale, this major earthquake damaged property, bridge structures, and highways. More recently, on January 17, 1995, a violent earthquake measuring 7.2 on the Richter scale jolted western Japan. It devastated the city of Kobe, killing 4,100 persons and injuring 21,600. Over 30,000 buildings were damaged (Fig. 20.30).

When an earthquake does occur, the point or region of the initial energy release or slippage is called the *focus* of the quake. The vast majority of earthquakes originate in the crust or upper mantle, so an earthquake's focus generally lies at some depth—from a few to several hundred miles. Consequently, geologists designate the location on Earth's surface directly above the focus. This point is called the **epicenter** of the earthquake.

The energy released at the focus of an earthquake propagates outwardly as **seismic waves.** Two general types of seismic waves are produced by earthquake vibrations: *surface waves*, which travel along Earth's surface or boundary within it; and *body waves*, which travel through Earth. Surface waves cause most earthquake damage because they move along Earth's surface. Body waves propagate through Earth's interior and so do less damage.

Two types of body waves are P (for primary), or compressional waves, and S (for secondary) or shear waves. The **P waves** are longitudinal compressional waves that are propagated by particles in the propagating material moving longitudinally back and forth in the same direction as the wave is traveling (see Chapter 6). Sound waves are an ex-

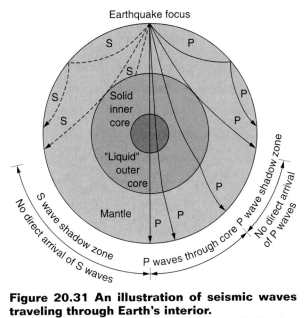

Figure 20.31 An illustration of seismic waves traveling through Earth's interior.
Because of the S and P wave shadow zones, Earth is believed to have a liquid outer core. See text for description.

with depth. As a result, the waves are curved or refracted. Also, the waves are refracted when they cross the boundary or discontinuity between different media—in the same manner that light waves are refracted (Chapter 7). The refraction of seismic waves and the fact that S waves cannot travel through liquid media provide our present view of Earth's structure, as shown in Fig. 20.31. Because of the so-called *shadow zones* of the S and P waves, the outer core of Earth is believed to be a highly viscous liquid.

The seismic waves of an earthquake are monitored by an instrument called a **seismograph,** the principle of which is illustrated in Fig. 20.32a. The recorded seismogram gives the time delay of the arrival of different types of waves, as well as an indication of their energy. (The greater the energy, the greater the amplitude of the traces on the seismogram.) Seismologists are aided by computers, which enhance the interpretation of geoseismic data as shown in Fig. 20.32b.

The severity of earthquakes is represented on different scales. The two most common are the **Richter scale,** which gives an absolute measure of the energy released by calculating the energy of seismic waves at a standard distance, and the modified **Mercalli scale,** which describes the severity of an earthquake by its observed effects (see Table 20.7). A comparison of these two scales is shown in Table 20.8. Notice that the 1985 Mexico earthquake was less severe on the Richter scale, but the death and destruction it caused gave it a high ranking on the Mercalli scale. The epicenter of the quake was off the Pacific coast of Mexico; however, it had devastating effects in Mexico City 400 km (250 mi) away. This city, the second most populous on Earth, suffered thousands of deaths and injuries and billions of dollars of damage. The first earthquake was followed by an aftershock 36 h later that measured 7.5 on the Richter scale.

Of these two scales, the Richter scale (developed in 1935 by Charles Richter of the California Institute of Technology) is used more often, with magnitudes expressed in whole numbers and decimals, usually between 3 and 9. However, the scale is a logarithmic function—that is, each whole-number step represents about 31 times more energy than the preceding whole-number step. For example, an earthquake that registers a magnitude of 5.5 on the Richter scale indicates the measuring

ample of longitudinal compressional waves. In **S waves** the particles move at right angles to the direction of wave travel and hence are transverse waves. An example of transverse shear waves is the vibration in a plucked guitar string.

There are two important differences between P and S waves. S waves can travel only through solids. Liquids and gases cannot support a shear stress; that is, they have no elasticity in this direction and therefore their particles will not oscillate in the direction of a shearing force. For example, little or no resistance is felt when one shears a knife through a liquid or gas. The compressional P waves, on the other hand, can travel through any kind of material—solid, liquid, or gas.

The other important difference is the speed of the waves. P waves derive their name "primary" because they always travel faster than the "secondary" S waves in any particular solid material and hence arrive earlier at a seismic station. It is these differences between P and S waves that allow seismologists to locate the focus of an earthquake and learn about Earth's internal structure.

The speeds of the body waves depend on the density of the material, which generally increases

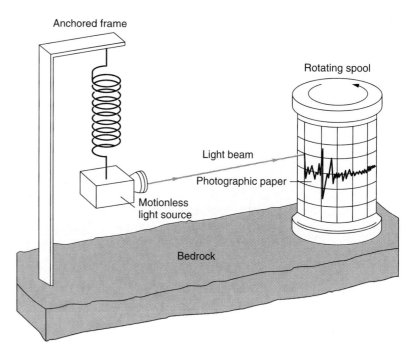

Figure 20.32a Seismograph principle.
The rotating spool (anchored in bedrock) vibrates during an earthquake. A light beam from the relatively motionless source on a spring traces out a record or seismogram of the quake's energy on the light-sensitive photographic paper.

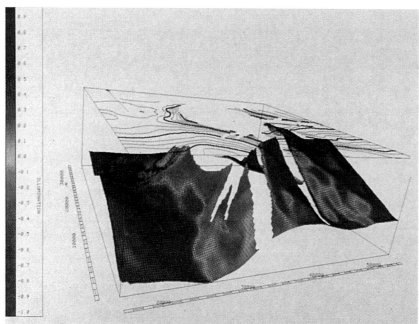

Figure 20.32b Contour map.
Computer systems provide visualization tools needed to interpret geoseismic data. Shown here is a three-dimensional perspective with a color-coded seismic amplitude of a fault horizon and a projected contour map.

seismograph received about 31 times as much energy as for an earthquake that registers a magnitude of 4.5.

Also, for an increase involving more than one whole-number step, the factor of 31 times more energy doesn't add, but multiplies. For example, an earthquake with a magnitude of 8 does not represent twice as much energy as one with a magnitude of 4 but almost one million times as much! [$31 \times 31 \times 31 \times 31 = (31)^4 \approx 924{,}000$.] As indicated in

Table 20.7 Modified Mercalli Scale

I	Not felt except by a very few under especially favorable circumstances.
II	Felt only by a few persons at rest, especially on upper floors of buildings.
III	Felt quite noticeably indoors, especially on upper floors of buildings, but many people do not recognize as an earthquake.
IV	During the day felt indoors by many, outdoors by few. Sensation like heavy truck striking building.
V	Felt by nearly everyone, many awakened. Disturbances of trees, poles, and other tall objects sometimes noticed.
VI	Felt by all; many frightened and run outdoors. Some heavy furniture moved; few instances of fallen plaster or damaged chimneys. Damage slight.
VII	Everybody runs outdoors. Damage negligible in buildings of good design and construction; slight to moderate in well-built ordinary structures; considerable in poorly built or badly designed structures.
VIII	Damage slight in specially designed structures; considerable in ordinary substantial buildings with partial collapse; great in poorly built structures. (Fall of chimneys, factory stacks, columns, monuments, and other vertically oriented features.)
IX	Damage considerable in specially designed structures. Buildings shifted off foundations. Ground cracked conspicuously.
X	Some well-built wooden structures destroyed. Most masonry and frame structures destroyed with foundations. Ground badly cracked.
XI	Few, if any, (masonry) structures remain standing. Bridges destroyed. Broad fissures in ground.
XII	Damage total. Waves seen on ground surfaces. Objects thrown upward into air.

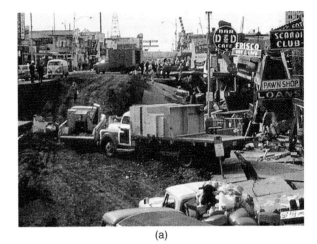

(a)

(b)

Figure 20.33 Subsidence and tsunami damage. (a) Subsidence damage in the 1964 "Good Friday" Alaskan earthquake. (b) Tsunami damage that resulted from the same quake.

Table 20.8, an earthquake with a magnitude of 2 to 3 is the smallest tremor felt by human beings. The largest recorded earthquakes are in the magnitude range of 8.7 to 8.9.

The Richter scale gives no indication of the damage caused by an earthquake, only its potential for damage. Earthquake damage depends not only on the magnitude of the quake but also on the location of its focus and epicenter and the environment of that region—specifically, the local geologic conditions, the density of population, and the construction designs of buildings. The modified Mercalli scale gives a better indication of earthquake effects, because the scale is based on actual observations. The Mercalli scale was developed by Giuseppe Mercalli in the 1890s before the advent of seismographs.

Earthquake damage may result directly from the vibrational tremors or indirectly from landslides and subsidence, as shown in Fig. 20.33a.

Table 20.8 Scales of Earthquake Activity

Richter Scale, Magnitude	Modified Mercalli Scale, Maximum Expected Intensity (at Epicenter)	Description
1–2	I	Usually detected only by instruments
3–4	II–III	Can be slightly felt
4–5	IV–V	Generally felt; slight damage
6–7	VI–VIII	Moderately destructive
7–8	IX–X	Major earthquake
8+	XI–XII	Great earthquake

Specific Earthquakes	Richter	Mercalli
1906 San Francisco	8.3	XI
1964 Alaska	8.4	XI
1985 Mexico	8.1	XI

In populated areas a great deal of the property damage is caused by fires because of a lack of ability to fight them—a result of disrupted water mains and so on.

When the energy release of a quake occurs in the vicinity of or beneath the ocean floor, huge waves called **tsunamis** are sometimes generated. These waves travel across the oceans at speeds up to 960 km/h (600 mi/h). In the open ocean the waves may be 160 km (100 mi) long and only 1 m (3 ft) high. However, as they travel in shallower water, the tsunamis grow in height and may be 15 m (50 ft) high or higher when they smash into shore with immense force (see Fig. 20.33b). Tsunamis coming ashore are commonly and incorrectly referred to as *tidal waves,* although they have no relation to tides. The Hawaiian Islands have experienced many tsunamis, which have raced across the ocean from the Alaska–Bering Strait area or other locations of earthquake activity that circumscribe the Pacific Ocean.

Earthquakes, despite their potential for destruction, can be an aid to science. None of the remarkable scientific advances of the twentieth century has revealed with certainty the composition of the interior of Earth. Our ideas about the interior of our planet must rest upon indirect evidence provided by earthquake body waves, whose speed and direction reflect the type of materials they penetrate, by meteorites whose composition we believe is similar to Earth's, and by laboratory experiments performed on rocks under very high temperature and pressure.

CONFIDENCE QUESTION 20.4

Give the relationship among faults, foci, and epicenters.

Important Terms

These important terms are for review. After reading and studying the chapter, you should be able to define and/or explain each of them.

mineral	rock	magma
silica	gem	lava
Mohs' scale of hardness	igneous rocks	sedimentary rocks

metamorphic rocks
plutons
batholith
dikes
sills
laccolith
clastic sediments
organic sediments
chemical sediments
bedding

fossil
lithification
"Ring of Fire"
hot-spot theory
shield volcano
stratovolcano or composite volcano
cinder cones
caldera
seismology
faults

epicenter
seismic waves
P waves
S waves
seismograph
Richter scale
Mercalli scale
tsunamis

Questions

20.1 Minerals

1. Which of the following is a characteristic property of a mineral?
   (a) naturally occurring
   (b) crystalline
   (c) definite chemical composition
   (d) distinctive set of physical properties
   (e) all of the above
2. The two most abundant elements in Earth's crust are
   (a) oxygen and iron.
   (b) silicon and oxygen.
   (c) aluminum and iron.
   (d) silicon and aluminum.
   (e) silicon and iron.
3. Feldspars
   (a) are the most abundant minerals found in Earth's crust.
   (b) are classified in two main types.
   (c) contain the elements oxygen, silicon, and aluminum.
   (d) all of the above
4. Distinguish between silicon and silicates.
5. State the definition of *mineral*.
6. What two elements make up the majority of Earth's crust and in what percentages?
7. What is the silicon-oxygen tetrahedron? Sketch its structure.
8. What is the chemical composition of (a) asbestos, (b) galena, (c) gypsum, and (d) halite?
9. Name an ore of lead, of iron, and of zinc.
10. State some physical characteristics that are used to identify minerals.

20.2 Rocks

11. Rocks
    (a) are classified as igneous, sedimentary, or metamorphic.
    (b) are composed of minerals.
    (c) are classified based upon the way they originated.
    (d) form the continents and ocean basins.
    (e) all of the above
12. Which of the following is *not* a type of sedimentary rock?
    (a) clastic   (b) organic
    (c) physical   (d) chemical
13. Igneous rocks are
    (a) the most abundant found in Earth's crust.
    (b) described as extrusive or intrusive.
    (c) classified according to their mineral content and their texture.
    (d) all of the above
14. Metamorphic rocks
    (a) are those that have been changed under the influence of great temperature and/or pressure.
    (b) make up about 30 percent of Earth's crust.
    (c) are changed physically but not chemically during metamorphism.
    (d) all of the above
15. Define *rock*.
16. Name the three classifications of rock, and state how each is formed.
17. Define *igneous rock* and name three such rocks.
18. Distinguish between magna and lava.
19. Distinguish between intrusive and extrusive rock. State the physical characteristics of each.

20. What is the source of Earth's internal heat?
21. What two major characteristics are used to classify igneous rock?
22. Distinguish between discordant and concordant igneous bodies.
23. What is the most important discordant igneous body?
24. Distinguish between a dike and a sill.
25. What two factors determine the viscosity of magma and lava?
26. Distinguish between clastic, organic, and chemical sediments.
27. Name some of the characteristics geologists use to give information about the origin of sedimentary rocks.
28. Name three sedimentary rocks, and state the sediment from which they originate.
29. What is meant by *bedding*?
30. Describe the lithification process.
31. Define *metamorphic rock*.
32. Define *foliation*.
33. Name three metamorphic rocks, and state the original rock name.

20.3 Volcanoes

34. Which of the following are true statements?
    (a) Volcanoes are classified as shield, stratovolcano, or cinder cone.
    (b) Volcanic eruptions do not affect the weather.
    (c) Volcanoes erupt lava and pyroclastic debris.
    (d) The viscosity of lava does not affect the behavior of a volcano.
    (e) Both (a) and (c) are correct.
35. Which of the following is a component of the explosive eruption of a volcano?
    (a) water vapor
    (b) carbon dioxide
    (c) rock
    (d) fine ash and dust
    (e) all of the above
36. What is the most abundant gas emitted in the eruption of a volcano?
37. Distinguish between magma and lava.
38. What is the "Ring of Fire"?
39. What is the hot-spot theory?
40. Distinguish between (a) shield volcanoes, (b) stratovolcanoes, and (c) cinder cones.
41. What are caldera, and how are they formed?

20.4 Earthquakes

42. Earthquakes
    (a) have no correlation with the "Ring of Fire."
    (b) generate longitudinal compressional and shear waves.
    (c) are monitored by a seismograph.
    (d) generate waves in Earth's interior called *tsunamis*.
    (e) both (b) and (c)
43. The severity of earthquakes is represented on the _____ scale.
    (a) Richter
    (b) Seismology
    (c) Mercalli
    (d) both (a) and (c)
44. What is an earthquake?
45. What causes an earthquake?
46. What are the focus and epicenter of an earthquake?
47. Give the two general types of seismic waves and the subdivisions of one of these types.
48. What is the difference between S and P waves? How do these waves allow seismologists to locate the focus of an earthquake and to investigate Earth's interior structure?
49. What is the basis of the (a) Richter scale and (b) modified Mercalli scale?
50. Why is *tidal wave* an incorrect and misleading term for a huge ocean wave generated by an earthquake? What is the correct term?

Food for Thought

1. How will our standard of living be affected through the excessive use and loss of our natural mineral resources?
2. Why do most igneous rocks melt over a range of a few hundred degrees rather than at one particular temperature?
3. Name the active or dormant volcano nearest your home.
4. What is a possible method scientists have for controlling the devastating effect of future earthquakes?

Answers to Multiple-Choice Questions

1. e
2. b
3. d

11. e
12. c
13. d
14. a

34. e
35. e

42. e
43. d

Solutions to Confidence Questions

20.1 All of the substances are minerals.

20.2 The rock cycle illustrates how igneous rock forms from magma by means of crystallization, then by weathering, lithification, and metamorphism other rocks are formed. Heat and pressure return formed rock to magma. Bypasses or shortcuts in the cycle also take place.

20.3 (a) A volcano is an opening (vent) in Earth's crust through which molten rock (lava), solid rock fragments, and gases are ejected. A volcanic mountain is formed by the ejected lava and solid rock fragments. (b) Chemical composition and temperature.

20.4 Faults are large fractures in Earth's crust. When a fault boundary suddenly moves and energy is released, an earthquake takes place. The region of the initial energy release is called the *focus* of the earthquake, and the point directly above the focus, on Earth's surface, is called its *epicenter*.

Chapter 21

Structural Geology and Geologic Time

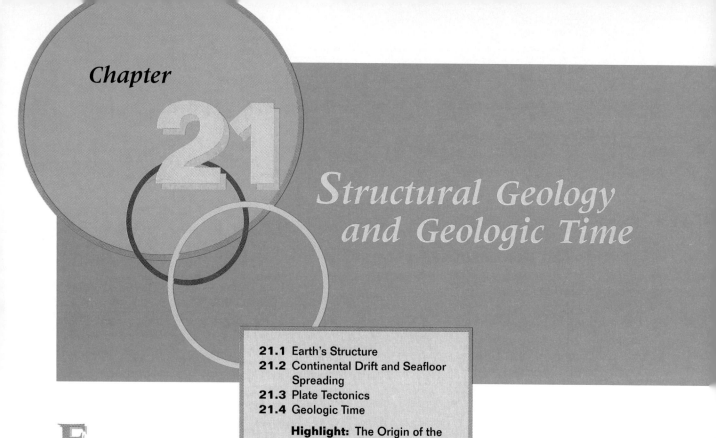

21.1 Earth's Structure
21.2 Continental Drift and Seafloor Spreading
21.3 Plate Tectonics
21.4 Geologic Time

Highlight: The Origin of the Appalachian Mountains

Earth is a dynamic body, both externally and internally. Numerous changes have taken place in Earth's structure during its long history, and changes continue today. Land forms come together then separate, mountains are built then erode, glaciers come and go, oceans inundate the land only to recede and flood again. This chapter deals with processes that bring about changes in Earth's structure, especially those at the surface and in the crust.

The rate of most Earth processes gives geology a unique time scale known as *geologic time*. Rather than an actual time scale, geologic time is more of a concept—that Earth developed gradually over an incredibly long time span. The several thousand years of recorded history of humans is only a "tick" on the geologic clock.

Within the past several years, there has been a revolution in the science of geology that gives us a better insight into the processes occurring below Earth's surface. This change has come about with the development of the theory of plate tectonics, which considers Earth's lithosphere to be made up of movable "plates." The assumptions of the theory of plate tectonics are surprisingly simple, but the consequences are far-reaching. With this set of concepts, a new understanding is gained concerning the formation of mountains and ocean basins, volcanic activity, earthquakes, and the mechanisms of operation of many geologic processes. The internal structure of Earth is also being studied by geophysicists using a technique known as seismic tomography, which employs seismic wave data. Using supercomputers, geophysicists plot three-dimensional pictures of Earth's upper mantle. Then computer graphics reveal the internal structure of the mantle.

Geology is taking on a new importance because of the increasing emphasis on natural resources obtained from planet Earth. An understanding of the various geologic processes is critical in locating and developing these natural resources. Also, such an understanding may allow the prediction of catastrophic events such as volcanic eruptions and earthquakes, which can save lives and property. In addition, geological studies of meteorites and the Moon have helped scientists develop a new theory for the origin of the Moon and to have a better understanding of the solar system.

21.1 Earth's Structure

Learning Goals:

To identify Earth's interior structure and composition.

To describe heat transfer from Earth's interior out to its surface.

Planet Earth is our environment—our home. Since we live on the planet's surface, we can analyze and study directly the soil, the water, the atmosphere, and Earth's crust beneath the surface. To date, we have drilled about 8 km (5 mi) into Earth's crust, and it is unlikely that a hole can be drilled into its deep interior. Materials extruded in volcanic eruptions obviously come from Earth's interior, but these are from relatively shallow depths compared to Earth's radius of approximately 6400 km (4000 mi). How then do geologists study Earth's deep interior structure?

Most of our knowledge of Earth's interior structure comes from the monitoring of shock waves generated by earthquakes. The enormous releases of energy associated with movements along fractures in the solid exterior, which result in earthquakes, generate both transverse and longitudinal waves (Chapter 6) that propagate through Earth. By monitoring these waves at different locations on Earth and by applying their knowledge of wave properties in various types of Earth materials, such as wave speed and refraction, scientists obtain information about Earth's interior structure.

From these indirect observations of Earth's interior, scientists believe that Earth is made up of a series of concentric shells, as illustrated in Fig. 21.1. There are three major shells: (1) the core, which is partly molten, (2) the mantle, and (3) the crust. The different shells are characterized by different composition and physical properties.

The two innermost regions are together called the **core** and have an average density of over 10 g/cm^3. The density suggests a metallic composition, which is believed to be chiefly iron (80%) and nickel. This estimate is based on the behavior of seismic waves passing through the core, the known abundance of iron in meteorites, and the measured proportion of iron in the Sun and stars. The solid inner core has a radius of approximately 1200 km (750 mi), whereas the outer core, some 2250 km (1400 mi) thick, is believed to be composed of molten, highly viscous, "liquid" material.

The magnetic field of Earth is thought to be related to the liquid nature of the outer core. As pointed out in Chapter 8, the magnetic field of Earth resembles that of a huge bar magnet within Earth. However, the interior temperatures of Earth are probably too high to have permanently magnetized ferromagnetic materials with the interior temperatures exceeding the Curie temperatures* of these materials. The slow change of the positions of Earth's magnetic poles suggests that the magnetic field is due to currents. It is thought that Earth's rotation gives rise to motions of the molten metals in the outer core, and to an associated slow change in magnetic field and electric currents.

Around the core is the **mantle,** which averages about 2900 km (1800 mi) thick. The composition of the rocky mantle differs sharply from the metallic core, and their boundary is distinct. The average density of the mantle is about 4.5 g/cm^3, which indicates that its composition is an iron-magnesium, rock-type material.

Around the mantle is the thin, rocky, outer layer upon which we live, called the **crust.** It ranges in thickness from about 5–8 km (3–5 mi) beneath the ocean basins to about 24–48 km (15–30 mi) under the continents. About 65% of Earth's crust is oceanic crust; that is, about 65% of Earth's surface is made up of ocean basins.

An abrupt change in the behavior of body waves as they travel in toward Earth's interior reveals the existence of the Mohorovicic discontinuity, a well-defined boundary that separates the crust from the upper mantle, the next layer down. Both scientists and students prefer the simplified term, Moho, for this important surface. Body waves make another important change in behavior at a depth of about 2900 km (1800 mi) as they cross the Gutenberg discontinuity, the boundary that separates the mantle from the core.

Thus, we see that the internal structure of Earth is somewhat analogous to that of an egg with the yolk, white, and shell corresponding to the

---

*The temperature above which ferromagnetism disappears in a magnetic material. See Chapter 8.

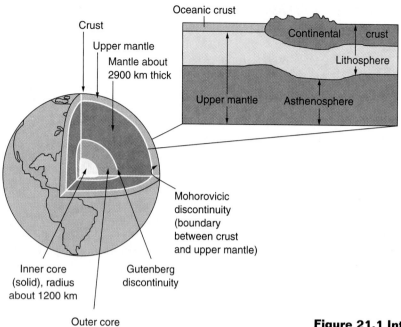

Figure 21.1 Interior structure of Earth.
Not to proportion; the crustal thickness is greatly exaggerated.

core, mantle, and crust, respectively. How did this layered structure of Earth come about? Perhaps when Earth was in an original molten state, the denser material settled to the bottom, in a fashion similar to the gases in the atmosphere. The liquid outer core could not be due to residual heat from Earth's formation, since the interior would have long since had time to cool and solidify. Although the outer core is subjected to enormous pressures that would result in high temperatures, calculations show that these temperatures are not high enough to melt the iron-nickel material of the outer core. Rather, it is believed that the heat is generated from the decay of radioactive materials.

If we view the interior of Earth in terms of its behavior rather than its composition, we can divide it into somewhat different layers. The outer layer, which we call the **lithosphere,** extends to a depth of approximately 80 km (50 mi) and includes all of the crust and the uppermost part of the mantle (Fig. 21.2). The lithosphere is rigid, brittle, and relatively resistant to deformation. Faults and earthquakes are mostly restricted to this layer. Below the lithosphere is the **asthenosphere,** which extends to a depth of roughly 700 km (435 mi) below Earth's surface. This rock plays essential roles in continental drift and seafloor spreading, which are discussed later in this chapter.

The difference in properties and behavior of the lithosphere and asthenosphere may well reflect the different ways in which they transmit heat

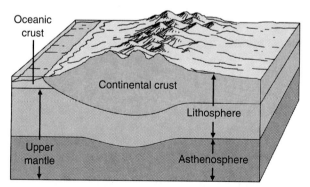

Figure 21.2 Lithosphere and asthenosphere.
Based on behavior rather than on composition, Earth's upper mantle and crust are divided into two layers. One, the lithosphere, extends to a depth of about 80 km (50 mi), and the second, the asthenosphere, extends from there to depth of roughly 700 km (435 mi).

from Earth's interior out to its surface. Heat moves through the lithosphere by conduction. Because this rock is a poor conductor of heat, there is a relatively high temperature gradient from its top to its bottom. The lithosphere, in other words, acts essentially as an insulating blanket, which traps much of the heat brought up to its bottom surface. Rock within the asthenosphere is therefore so hot that it is extremely near its melting temperature. Asthenospheric rock, because of its heat, is plastic and mobile enough to transmit heat mostly by convection.

As you will learn in later sections, convection cells within the asthenosphere are considered by many to be the fundamental forces that have for virtually all of geologic time driven Earth in its restless internal activities. Below the asthenosphere Earth is even hotter, but the melting temperatures under these high pressures are still higher.

CONFIDENCE QUESTION 21.1

How is heat transmitted in (a) the lithosphere, and (b) the asthenosphere?

21.2 Continental Drift and Seafloor Spreading

Learning Goals:

To explain the theory of continental drift and list the geologic evidence offered for its support.

To explain the mechanism and evidence for seafloor spreading and its relationship to continental drift.

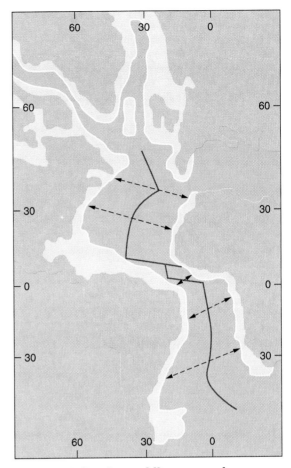

Figure 21.3 Continental jigsaw puzzle.
The coasts of Africa and North and South America appear as though these continents could fit together nicely, as if they were pieces of a jigsaw puzzle. The dashed lines indicate the paths the continents might have followed if they were once together, and then rifted and drifted apart. The heavy line indicating where the continents might once have been joined together marks the present mid-oceanic ridge that runs the length of the Atlantic Ocean.

When looking at a map of the world, we are tempted to speculate that the Atlantic coasts of Africa and North and South America could fit nicely together as though they were pieces of a jigsaw puzzle (Fig. 21.3). This observation has led scientists at various times in history to suggest that these continents, and perhaps the other continents, were once a single, giant supercontinent that broke and drifted apart. However, there was no evidence to support this theory other than the shapes of the continents.

In the early 1900s Alfred Lothar Wegener (1880–1930), a German meteorologist and geophysicist, revived the theory of **continental drift** and brought together various geological evidences for its support. Wegener's theory gave rise to considerable controversy, and only relatively recently has conclusive evidence been found that supports some aspects of the theory of continental drift.

Wegener's assumption was that the continents were once part of a single giant continent, which he called **Pangaea** (from the Greek, pronounced

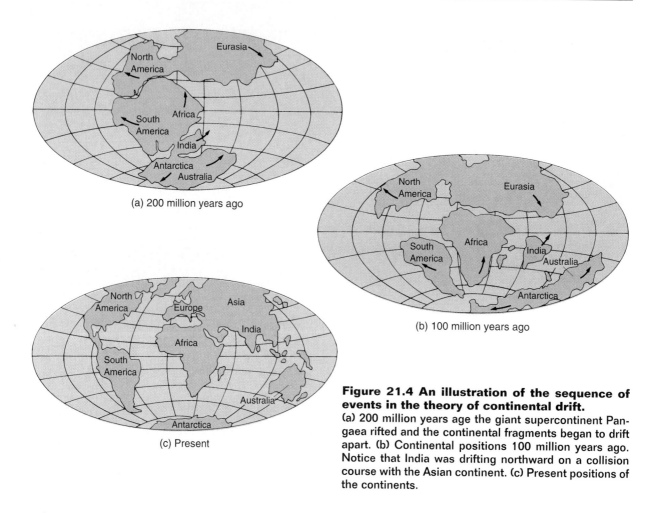

Figure 21.4 An illustration of the sequence of events in the theory of continental drift.
(a) 200 million years age the giant supercontinent Pangaea rifted and the continental fragments began to drift apart. (b) Continental positions 100 million years ago. Notice that India was drifting northward on a collision course with the Asian continent. (c) Present positions of the continents.

"pan-jee-ah" and meaning "all lands"). According to his theory, this hypothetical supercontinent somehow broke apart about 200 million years ago and its fragments ("plates") drifted to their present positions and became today's continents (Fig. 21.4).

The scientific evidence supporting Wegener's theory takes on several different approaches. Some of these approaches are (1) similarities in biological species and fossils found on the various continents; (2) continuity of geologic structures such as mountain ranges, and the distribution of rock types and ages; and (3) glaciation in the Southern Hemisphere. Let's consider each of these briefly.

1. Similarities in biological species and fossils suggest that there was formerly an exchange of these forms between continental regions when they were together as Pangaea. We would not expect that these biological species could traverse the present-day oceans. For example, a certain variety of garden snail is found only in the western part of Europe and the eastern part of North America. Also, a relatively young genus of earthworm is found in the same latitudes of Japan and the Asian and European continents, while it is found in similar latitudes on the east coast but not the west coast of North America. Similarly, fossils of identical reptiles are found in South America and Africa, and identical plant fossils have been found in South America, Africa, India, Australia, and Antarctica.

One suggestion is that the similarities in biological species and fossils can be explained by "land bridges" between the continents that eventu-

Figure 21.5 An illustration of the continuity of continental features.
Continental jigsaw pieces cut from a printed page would show the continuity of printed lines when put back together. If the continents were once together, rifted, and drifted apart, we might expect some similar "printed lines" common to the separated continents in the form of geologic features such as mountains.

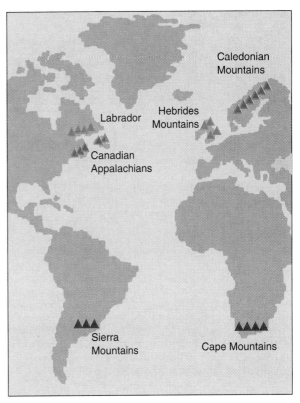

Figure 21.6 Continent continuity.
The continuity of geologic features supports the theory of continental drift. If the continents were fitted together, various mountain ranges of similar structure and rock composition on the different continents would line up analogous to the print in Fig. 21.5

ally sank or were covered by the oceans. However, there is no evidence of such land bridges in the oceans.

2. As has been noted, it was the roughly interlocking shapes of the coastline of the African and American continents that inspired the theory of continental drift. Imagine cutting the continental shapes from a printed page like jigsaw puzzle pieces and attempting to put the separated pieces together again. When we put the page pieces back together, we find that the continuity of the printed lines is common to the fitted pieces, as shown in Fig. 21.5.

If indeed the continents had rifted (split) and drifted apart, we might expect some similar "printed lines" common to the pieces. Such evidence does occur in the form of geologic features. If the continents were put back together, the Cape Mountain Range in southern Africa would line up with the Sierra Range near Buenos Aires, and these mountains are strikingly similar in geologic structure and rock composition (Fig. 21.6). In the Northern Hemisphere, the Hebrides Moun-

tains in northern Scotland match up with similar formations of Labrador, and the Caledonian Mountains in Norway and Sweden have a logical extension in the Canadian Appalachians. Various other similarities in the plateau rock formations of Africa and South America have been found. More recently, with the development of radiometric dating techniques, transatlantic areas of rocks of similar ages have been found.

3. There is solid geological evidence that a glacial ice sheet covered the southern parts of South America, Africa, India, and Australia about 300 million years ago, an ice sheet similar to the one that covers Antarctica today. Hence, a reasonable conclusion is that the southern portions of these continents must have been under the influence of a polar climate at that time. No traces of this ice age exist in Europe and North America. In

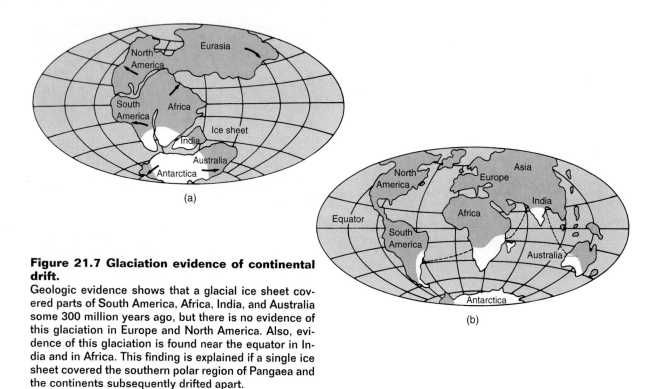

Figure 21.7 Glaciation evidence of continental drift.
Geologic evidence shows that a glacial ice sheet covered parts of South America, Africa, India, and Australia some 300 million years ago, but there is no evidence of this glaciation in Europe and North America. Also, evidence of this glaciation is found near the equator in India and in Africa. This finding is explained if a single ice sheet covered the southern polar region of Pangaea and the continents subsequently drifted apart.

fact, fossil evidence indicates that a tropical climate prevailed in these regions during that period, yet evidence of glaciation is found in India and in Africa near the equator. How could this be? It seems unreasonable that an ice sheet would cover the southern oceans and extend northward to the equator.

Wegener's theory suggests an answer and derives support from these observations. The direction of the glacier flow is easily determined by marks of erosion on rock floors and by moraines (deposits of rock and soil debris transported by glaciers). If the continents were once grouped together as Wegener's theory indicates, then the glaciation area was common to the various continents, as illustrated in ● Fig. 21.7. The glacial movements, as indicated by the arrows in the top drawing, support the idea of a single ice cap and subsequent continental drift.

Although there was evidence supporting Wegener's theory, it was not generally accepted, primarily because the proposed mechanism for continental drift was unsatisfactory. Wegener depicted the continents as giant rafts moving through the oceanic crust owing to Earth's rotation. This mechanism is unacceptable because the force associated with Earth's rotation is not strong enough to overcome the measured strength of the rocks, which would be disintegrated if the continental crust slid over or moved through the oceanic crust.

A more satisfactory theory for the mechanism behind continental drift was suggested in 1960 by H. H. Hess, an American geologist. Geologists know that a mid-oceanic ridge system stretches through the major oceans of the world. In particular, the Mid-Atlantic Ridge runs along the center of the Atlantic Ocean between the continents. This and other mid-oceanic ridges run along large fissures in Earth's crust, as evidenced by volcanic and earthquake activity along the ridges.

Hess suggested a theory of **seafloor spreading,** where the seafloor spreads slowly and moves sideways away from the mid-oceanic ridges. This movement is believed to be accounted for by convection currents of subterranean molten materials that cause the formation of mid-oceanic ridges and surface motions in lateral directions from the fissure, as illustrated in ● Fig. 21.8a.

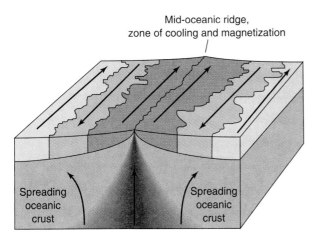

21.8a Seafloor spreading and magnetic anomalies.
Thermal convection currents of molten material rising from the upper mantle are believed to give rise to the formation of mid-oceanic ridges and seafloor spreading. Magnetic anomalies are distributed in symmetrical, parallel bands on both sides of the Mid-Atlantic Ridge. The direction of magnetization is reversed in adjacent parallel regions.

21.8b Mid-Atlantic Ridge.
This feature, along which seafloor spreading is believed to take place, runs along the ocean floor between the continents.

Support for this theory has come from the studies of remanent magnetism and from the determination of the ages of the rocks on each side of the mid-oceanic ridges. **Remanent magnetism** refers to the magnetism of rocks resulting from a group of minerals called ferrites, one of which is ferrous (iron) ferrite, Fe_3O_4, commonly called magnetite or lodestone. When molten material containing ferrites is extruded upward from Earth's mantle (as in a volcanic eruption) and solidifies in Earth's magnetic field, it becomes magnetized. The direction of the magnetization indicates the direction of Earth's magnetic field at the time.

This solidified rock is worn down by erosion. The fragments are carried away by water, and they eventually settle in bodies of water, where they become layers in future rock. In the settling process, the magnetized particles, which are in fact small magnets, become generally aligned with Earth's magnetic field. By studying the remanent magnetization of geologic rock formations and layers, scientists have learned about changes that have taken place in Earth's magnetic field.

Measurements of the remanent magnetism of rock on the ocean floor revealed long, narrow, symmetric bands of **magnetic anomalies** (Fig. 21.8a) on both sides of the Mid-Atlantic Ridge (Fig. 21.8b). That is, the direction of the magnetization was reversed in adjacent parallel regions. Along with data on land rock, the magnetic anomalies indicate that Earth's magnetic field has abruptly reversed fairly frequently and regularly throughout recent geologic time. The most recent reversal occurred about 700,000 years ago. Why this and other reversals should occur is not known. We are evidently living in a period between pole reversals. However, the symmetry of the anomaly bands on either side of the ridge indicates a movement away

Figure 21.9 Seafloor exploration.
The JOIDES *Resolution*, an oceanographic research vessel.

from the ridge at the rate of a few centimeters per year and provides evidence for seafloor spreading.

Remanent magnetism also provides other support for continental drift. The remanent magnetism of a rock remains fixed throughout the history of the rock, even if forces within Earth move the rock. If the rock is moved, its remanent magnetism is no longer aligned with Earth's magnetic field. This situation is similar to the nonalignment of printed lines on moved puzzle pieces, as described in the previous analogy on the continuity of geologic features.

Scientists have studied the remanent magnetism of rock of varying ages and have found evidence of such changes of alignment. Moreover, they have found that rock of the same age on a continent has the same misalignment, but that the misalignment is different for different continents. This observation suggests that the entire continent must have moved as a unit and that each continent moved or "drifted" in its own separate direction.

In recent years investigative drilling into the ocean floor has been done from the oceanographic research vessel JOIDES (Joint Oceanographic Institutions for Deep Earth Sampling) *Resolution* (Fig. 21.9). These drillings have shown that the ocean floor between Africa and South America is covered by relatively young sediment strata. Also, the strata thicknesses increase away from the Mid-Atlantic Ridge, which implies that the older part of the ocean floor is farther away from the ridge. The older parts would be covered by a greater thickness of sediment because there would have been a longer time for the sediment to accumulate.

These observations support the idea of seafloor spreading as a mechanism implementing the theory of continental drift, culminating in the modern theory of plate tectonics. The basics of this theory are described in the next section. However, more current concepts about continental drift and seafloor spreading postulate that crustal plates have been moving over Earth's surface almost as long as there has been solid crustal material to move, long before 200 million years ago.

CONFIDENCE QUESTION 21.2

If Pangaea broke apart 200 million years ago and the current distance between South America and Africa is 8000 km (5000 mi), what is the average annual rate of continental drift? (Hint: $v = d/t$, see Chapter 2.)

21.3 Plate Tectonics

Learning Goals:

To explain the theory of plate tectonics.

To list the general relative motions of plates and the resulting geologic implications.

The view of ocean basins in a process of continual self-renewal has led to the acceptance of the concept of **plate tectonics.** We now view the lithosphere not as one solid rock but as a series of solid sections or segments called plates, which are constantly in very slow motion and interacting with one another. About twenty plates cover the surface of the globe. Some are very large, and some are small. The major plates are illustrated in Fig. 21.10.

The most active, restless parts of Earth's crust are located at the plate boundaries. Along the ridges where one plate is pulling away from another (a **divergent boundary**), new oceanic rock is formed. Where plates are driven together (a **convergent boundary**), rock is consumed. In still other parts of Earth's surface one plate slides along one side of another (a **transform boundary**), and rock is neither produced nor destroyed. These relationships are illustrated in Fig. 21.11.

Figure 21.12 illustrates the structure of the lithosphere and asthenosphere. The interface between these two structures is significant in terms of internal geologic processes. The asthenosphere, which lies beneath the lithosphere, is essentially

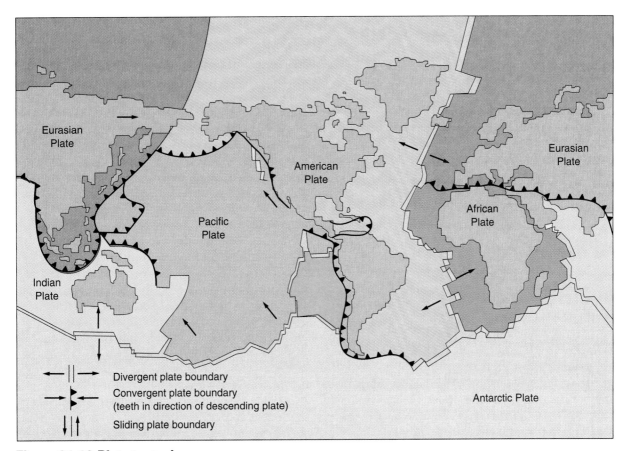

Figure 21.10 Plate tectonics.
The six major plates of the world and several smaller ones. The relative motions along the plate boundaries are indicated.

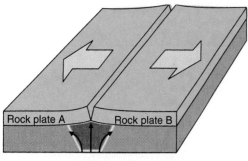
(a) Divergent boundary

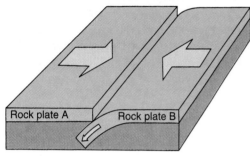
(b) Convergent boundary

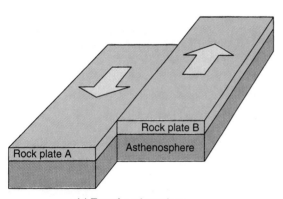
(c) Transform boundary

Figure 21.11 Types of plate movement.
The block diagrams illustrate the relationship between moving rock plates. The interface between the two plates is called the plate boundary. Each boundary type has an identifying name below the diagram.

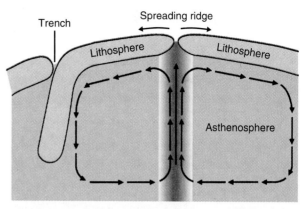

Figure 21.12 Convection cells.
Unequal distribution of temperature within Earth causes the hot, lighter material to rise and the cooler, heavier material to sink. This generates convection currents in the asthenosphere. The lithosphere plates, resting on the asthenosphere, are put into motion by the driving forces of the convection cells.

solid rock, but it is so close to its melting temperature that it contains pockets of molten magma and is relatively plastic. Therefore, it is much more easily deformed than the lithosphere.

The movement of plastic rock during structural adjustments takes place essentially within the asthenosphere. The forces that move the plates one against another, one away from another, or one past another are also found within the asthenosphere. The plates, segments of the lithosphere, are actually passive bodies, driven into motion by the drag of the more active asthenosphere beneath them.

Most earth scientists view this motion in terms of *convection cells*, with the basic source of heat provided by radioactive decay (Fig. 21.12). The role of gravity in the process is to drag the cooler, heavier material deeper into Earth's interior as the hotter, less dense rock rises toward the surface, where it can lose part of its heat.

Let us examine the role of convection cells in plate tectonics by focusing our attention first on the oceanic ridges where the plates are moving away from one another at divergent plate boundaries. (See Fig. 21.13.) Beneath these spreading oceanic ridges, convection currents lift material from the hot mantle (asthenosphere) up into a region of lower pressure, where it begins to melt. The material only partially melts, however, forming magma (molten rock).

As more mantle material wells up from below, the higher mantle material is shouldered to both sides and moves slowly in a horizontal direction beneath the lithospheric plates. It is the drag of the mobile asthenosphere against the bottom of the lithospheric plates that give the plates motion and causes the fracturing at the spreading ridges. The

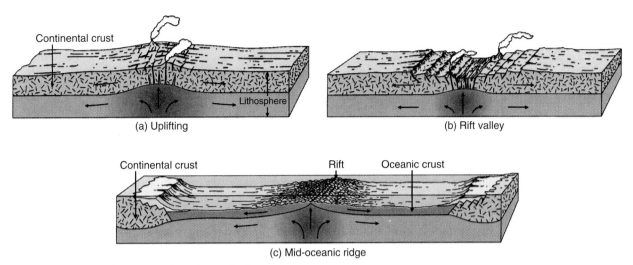

Figure 21.13 Divergent plate boundaries.
The block diagrams illustrate the possible conditions that may occur at divergent plate boundaries. See text for explanation.

fractures are filled, in an area known as a *rift valley*, by the magma produced by the partial melting of rock in the asthenosphere.

In its horizontal journey across the top of the convection cell, beneath the lithospheric plates, some of the asthenosphere cools to well below its melting temperature and becomes part of the lithosphere. The lithospheric plate also cools as it moves slowly away from the hot spreading ridge, and therefore it loses volume. As a result, the top of the plate gradually subsides and causes the oceans to grow progressively deeper away from the spreading ridges. The slope of denser, cooler rock away from the mid-oceanic ridges also contributes to the motion of the plate toward the zones of compression.

The *zones of compression* occur at convergent plate boundaries, and there are three types: (1) oceanic-oceanic, (2) oceanic-continental, and (3) continental-continental.

Initially, when two oceanic plates collide, both begin to be subducted because each is relatively dense (3.0 g/cm^3). Eventually, one plate is subducted more than the other. The place, or zone, where the plate descends into the asthenosphere is called the *subduction zone*. The descending oceanic plate, now in contact with the mantle, begins to melt. The molten material begins to rise, and a series of volcanoes develops in an arc shape on the overriding plate. (See Fig. 21.14.) Deep trenches lie in front of the arc system, marking the places where the plates are being subducted. The deepest trench known is the Marianas Trench in the western Pacific Ocean, which is 11 km (6.8 mi) below sea level. Examples of island arc systems are the Aleutian Islands and the islands of Japan, both of which have deep trenches between the island arcs and the open ocean.

Whenever the oceanic crust collides with lighter-weight continental crust (2.7 g/cm^3), the oceanic crust is always subducted beneath the continental crust. (See Fig. 21.15.) A trench will develop at the boundary where the oceanic plate is being subducted. However, the trench is never as deep as the trench formed in the oceanic-oceanic convergence. The oceanic plate begins to melt as it descends into the asthenosphere. Molten rock material then moves up into the overriding plate, causing large igneous intrusions and often volcanic mountains at the surface. An example is the Andes Mountains of South America.

When two continental plates (both being lightweight, 2.7 g/cm^3) collide, the boundary of rock between the continental edges is pushed and crumpled intensely to form folded-mountain belt systems. (See Fig. 21.16.) In this manner, continents grow in size by suturing themselves together along folded-mountain belt systems. Examples are

584 CHAPTER 21 Structural Geology and Geologic Time

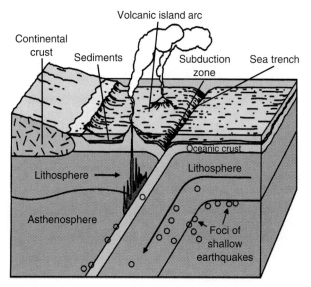

Figure 21.14 Oceanic-oceanic convergent plate boundary.
An illustration of the effects produced by the convergence of two oceanic plates. The subduction zone is the region in which one plate plunges beneath the other. Destructive earthquakes occur most often in subduction zones.

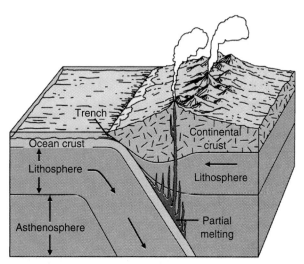

Figure 21.15 Oceanic-continental convergent plate boundary.
An illustration of the effects produced by the convergence of oceanic and continental plates. An example is along the Andes Mountains along the western coast of South America.

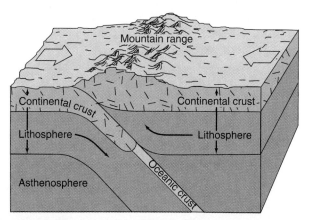

Figure 21.16 Convergent continental plate boundaries.
An illustration of the effects produced by the convergence of two continental plates. The boundary of rock between the continental edges is pushed and crumpled to form folded-mountain systems. The Appalachian mountain range is an example of the effect of converging continental plates.

the Alps, Appalachians, and Himalayas. Geologists believe that the Himalayas formed in this manner when the Indian Plate collided with the Eurasian Plate.

A zone of shear, or transform, fault is a place at which adjacent plates slide past each other without a gain or loss in surface area. This zone occurs along faults, which mark the plate boundaries. Movements and the release of energy along these boundaries give rise to earthquakes. Examples of such fault zones are along the San Andreas Fault in California and the Anatolian Fault in Turkey.

CONFIDENCE QUESTION 21.3
(a) Name the layer of Earth that consists of movable plates. (b) How thick (in km) is the layer?

21.4 Geologic Time

Learning Goals:

To distinguish between relative and absolute geologic time.

To describe the geologic time scale.

The history of Earth is recorded in geologic time and events. Geologists must consider processes that may have taken place over millions of years, a time interval that is very difficult to comprehend. However, if natural processes are constant and the same processes operate today on and within Earth, then the present may be considered the key to the past. This is an important geologic concept called the **principle of uniformity** (or uniformitarianism). Changes occurring on and within Earth today are clues to the long-term total picture.

Relative Geologic Time

From the very birth of geology as a science, geologists have seen the need to measure geologic time. The first and simplest task was to establish the sequence of geologic events within a local area. The individual histories of the many local areas must be related to one another, or geologic history can be little more than a random collection of isolated interpretations.

Using a procedure called **correlation,** geologists have long been busy tying the loose ends together and constructing an organized picture of the geology of large areas and even entire continents. Geologists correlate rock from one area to another by determining whether the separate rock sequences were formed at the same time or different times.

There are many ways to correlate rock, but only a few ways are reliable for rock separated by great distances. One way that has been successful is the use of fossils. A **fossil** is a remnant or trace of an organism preserved from a past geological age, such as a skeleton, footprint, or leaf imprint embedded in a rock.

Fossils are useful in determining the age of rock because life has changed through geologic time. Rocks of different ages contain different fossil assemblages, even if deposited in the same environment. The age of a sedimentary rock can therefore be determined from the fossils contained within it, provided, of course, the age of the fossils is known. However, on a relative basis, one knows only whether a layer of rock is older or younger than another layer. This is determined by the **law of superposition,** which is the simple observation that in a succession of stratified sediment deposits, the younger layers lie on top of the older layers.

The history of life is far from simple, however. The various trends and patterns of organic evolution make a fascinating story of their own. A very thick book would be required to trace the long and involved history of ancient life. The evolutionary changes have differed from one family to another. Certain animals and plants have changed little through time. These creatures in fossil form reveal little of the age of the rock in which they lie buried. Other forms of life, however, have changed in response to their shifting environments and evolved rapidly through geologic history. When these latter creatures become fossilized and are also numerous and widespread, they are very useful in rock-age determinations and are known as **index fossils.**

Using index fossils and all other tools of correlation, geologists through long, patient labor have been able to construct a **relative geologic time scale** similar to our human time scale. Just as human history is divided into units and subdivided into smaller units, so is geologic time. The largest unit of geologic time is the **era,** which is subdivided into smaller time units known as **periods.** Periods may be subdivided into **epochs.** Each era has a name, as has each period and epoch. These are given in Fig. 21.17.

Absolute (Atomic) Geologic Time

Relative time simply identifies an object or event as being younger or older than something else. From the law of superposition, we know that a set of fossils and the associated events in an upper stratum are younger or happened later in time than those in a lower stratum. However, it is difficult to assign times within a stratum. Since no accurate rate of deposition can be determined for most rock strata, the actual length of geologic time represented by a given stratum is at best an educated guess.

As a result, events are assigned to a particular geologic unit. For example, a geologist might say that the last dinosaurs became extinct (as evidenced by fossil remains) at the end of the Mesozoic Era, which precedes the Cenozoic Era. The boundary between these eras is *estimated* to be about 65 million years ago, so the last dinosaurs became extinct about 65 million years ago.

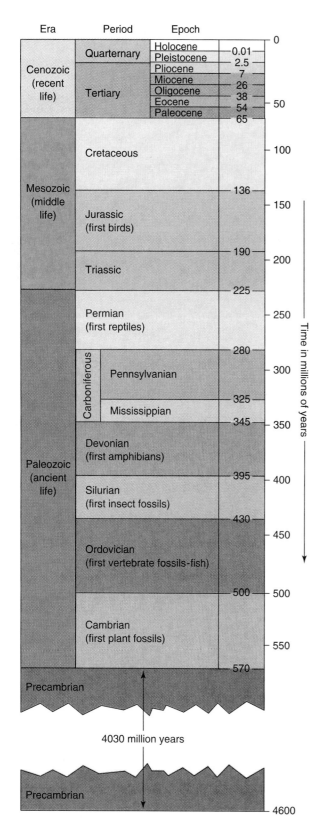

Figure 21.17 The geologic time scale.
The eras, periods, and epochs of the relative geologic time scale are shown on the left. The largest units are eras. The Precambrian Era represents more than four thousand million years, or over 87 percent of the total time scale. The other eras refer to the ancient, middle, and recent periods, respectively, in the history of life on Earth. On the right are the dated years of the absolute time scale.

Modern scientific methods have allowed the development of an **atomic (absolute) time scale** based on natural radioactivity. As presented in Chapter 10, the rate of decay of a radioactive isotope is conveniently expressed in terms of the isotope's half-life, or the time it takes for one-half of the nuclei in a sample to decay. Many isotopes have relatively short half-lives of a few days or years. However, some isotopes decay slowly and several of these are used as "atomic clocks" in measuring the ages of geologic events.

Radiometric dating is simple in theory, as described in Chapter 10 for carbon dating. However, the actual laboratory procedures are complex. The principal difficulty with geologic samples lies in the accurate measurement of very small amounts of radioactive elements and/or their decay products.

Carbon-14 decay is a useful tool in dating relatively recent events. Because of its relatively short half-life, carbon-dating techniques are accurate for dating events that have taken place within the last 50,000 years, a very short interval in geologic time. Fortunately, other radioactive isotopes associated with geologic events have longer half-lives. Radiometric carbon dating is done on once-living material, whereas other methods are used to date rocks. See Section 10.3 for additional information.

The potassium-argon decay scheme is one of the most useful to the geologist because it can be used on rocks as young as a few thousand years as well as on the oldest rocks known. Potassium is a constituent of several common minerals. The half-life of potassium-40 is such that measurable quantities of argon-40 (the daughter product) have accumulated in potassium-bearing rocks of nearly all ages. And, even in very small quantities, the amounts of potassium and argon isotopes can be measured accurately. However, the daughter product, argon, is a gas and can escape from the system, thereby causing the computed age of rock to

HIGHLIGHT

The Origin of the Appalachian Mountains

The Appalachian Mountains extend along the eastern North American continent from central Alabama to Newfoundland, a straight-line distance of 2574 km (1600 mi). The relatively narrow mountain system has a maximum width of slightly over 160 km (100 mi). The highest peak, Mount Mitchell in North Carolina, rises 6684 ft above the Atlantic Ocean.

The Appalachians are very old mountains. The action of weathering and the processes of erosion operating continuously over millions of years have lowered and rounded the once high and rugged peaks. The present structure of the southern Appalachians is shown in Fig. 1. Basic information concerning the composition and structure of the region has been obtained from surface studies of crustal rocks and from seismic-reflection profiling at depths of a few miles.

Although the mountains are complex and vary in structure throughout their length, their formation can be explained in general terms of plate tectonics. The Appalachians were formed when moving crustal plates including North America, Africa, Europe, and Greenland converged to form a huge, single land mass called Pangaea (see Fig. 21.4a). Using plate tectonic theory, the following discussion outlines a hypothesized chain of events that formed the southern Appalachians.

Before the Precambrian Era ended, some 570 million years ago, an ancient supercontinent broke apart, and what is now North America separated from Africa, leaving the ancient Atlantic Ocean between them. During the breakup, a block of continental crust was isolated off the coast of North America and separated from North America by a coastal sea.

The upheaval of Earth's crust that formed the Appalachians began some 550 million years ago, early in the Paleozoic Era, when the ancient Atlantic Ocean began to close. The converging process, which lasted almost 300 million years, caused subduction, together with volcanism, faulting, and folding in Earth's crust. One zone of subduction created a chain of volcanic islands between the block of continental crust off the coast of North America and Africa. Today these remnants are the Carolina Slate Belt.

This convergence continued through the Ordovician Period to early in the Silurian Period, and the coastal sea between the block of continental crust and North America closed when the two masses collided. The collision forced the crust up and over the North American continental sediments. Its remnants are today the Blue Ridge region and part of the Piedmont region of the Appalachian Mountains. West of the Blue Ridge region, thick Paleozoic sediments were folded by compressional forces, creating the valley and ridge region of the Appalachians.

The ancient Atlantic continued to close. Near the end of the Devonian Period, the chain of volcanic islands collided with North America and forced the Blue Ridge and Piedmont regions farther west. The Atlantic closed completely before the end of the Permian Period, when Africa collided with North America and Pangaea was formed. The continuous compressions and the series of collisions over nearly 300 million years deformed and metamorphosed rock throughout the Appalachian region, and forced the Blue Ridge and Piedmont regions more than 241 km (150 mi) westward.

About 200 million years ago, early in the Mesozoic Era, Pangaea began breaking up (see Fig. 21.4b). Africa separated from North America, which initiated the opening of the present Atlantic Ocean. The African continent left a fragment of its land mass attached to North America, which today overlies the coastal plain and continental shelf.

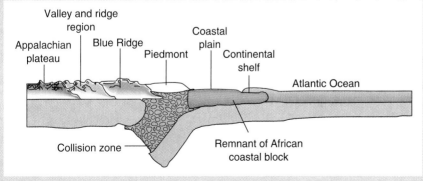

Figure 1 Cross-sectional view of the southern Appalachians. See text for description.

be younger than it actually is. Fortunately, some types of rocks retain the argon quite well.

Each decay scheme dating method finds different geologic applications, but they must be applied carefully. Whenever possible, two or more methods of analysis are used on the same specimen of rock to confirm and verify the results. For example, rubidium-87, with a half-life of 47 billion years, decays in a single step to stable strontium-87. Rubidium-87, which commonly occurs with potassium, is found in more rocks than potassium-40 and has the further advantage of decaying to a daughter element that is not a gas. A disadvantage is the great abundance of strontium-87 that has not formed from radioactive decay and that may have to be corrected for. Geologists often use rubidium-strontium ages to compare with potassium-argon determinations from the same rock. The use of uranium-lead dating is described in section 10.3.

Geologic Time Scale

Relative geologic time and atomic geologic time both are combined to give the **geologic time scale,** which is shown in Fig. 21.17. The time, in millions of years as determined by radioactive methods, has been added to the right side of the relative time units.

Although radiometric dating has provided geologists with many essential dates for the ages of rocks, there are many time intervals where data are questionable. The correlation of the relative geologic time scale deduced from fossil evidence with atomic time poses certain problems, chiefly because rocks solidified from molten material can be dated by radiometric methods more reliably than other kinds of rocks, yet these igneous rocks do not ordinarily contain fossils. Many solidified rocks retain the parent and daughter nuclides within them, thereby allowing accurate determination of the relative amounts of the nuclides.

Rock formed from layered sediments (sedimentary rocks) contain fossils, and may also contain radioactive isotopes. However, the sediments that form these rocks have been weathered from rocks of different ages and are not the same age as the rock layer in which they are found. Correlations must be done carefully. But even with these problems, the "atomic clocks" of radioactive decay processes have helped geologists relate the relative geologic time scale to the atomic scale to give a better measure of geologic time.

CONFIDENCE QUESTION 21.4

Assuming humanoids first appeared about 2 million years ago, designate the era, period, and epoch of this time.

Important Terms

These important terms are for review. After reading and studying the chapter, you should be able to define and/or explain each of them.

core
mantle
crust
lithosphere
asthenosphere
continental drift
Pangaea
seafloor spreading
remanent magnetism
magnetic anomalies
plate tectonics
divergent boundary
convergent boundary

transform boundary
principle of uniformity
correlation
fossil
law of superposition
index fossils
relative geologic time scale
era
periods
epochs
atomic (absolute) time scale
geologic time scale

QUESTIONS

21.1 Earth's Structure

1. Which of the following is *not* a name-labeling part of Earth's interior structure?
   (a) mantle
   (b) Gutenberg discontinuity
   (c) Mohorovicic discontinuity
   (d) igneous discontinuity
   (e) outer core

2. The lithosphere
   (a) is part of Earth's crust.
   (b) is part of Earth's upper mantle.
   (c) is closer to Earth's surface than the asthenosphere.
   (d) extends to a depth of about 80 km (50 mi).
   (e) all of the above

3. State the compositions and structural divisions of Earth's interior. Are they all solid?

4. Explain why there cannot be a permanent magnet within Earth that gives rise to Earth's magnetic field.

5. What is the name of the boundary between the crust and the mantle?

6. How is Earth's internal structure analogous to an egg?

7. What are the lithosphere and the asthenosphere?

21.2 Continental Drift and Seafloor Spreading

8. Pangaea
   (a) means "all lands."
   (b) was named by Alfred Wegener.
   (c) broke apart about 200 million years ago.
   (d) both (a) and (c)

9. Seafloor spreading
   (a) is caused by thermal convection currents.
   (b) is validated by remanent magnetism.
   (c) moves away from the mid-oceanic ridges.
   (d) all of the above

10. Which of the following is geologic evidence supporting continental drift?
    (a) similarities in biological species and fossils found on various continents
    (b) continuity of geologic structures such as mountain ranges
    (c) glaciation in the Southern Hemisphere
    (d) all of the above

11. Describe the geologic evidence that supported Wegener's theory of continental drift.

12. What is the present theory of the mechanism for continental drift? Why was Wegener's proposed mechanism unacceptable?

13. What is meant by remanent magnetism?

14. What evidence supports the theory of seafloor spreading?

15. What is the average rate of seafloor spreading?

21.3 Plate Tectonics

16. Which of the following is *not* a name for a type of plate boundary?
    (a) divergent   (c) subduction
    (b) convergent   (d) transform

17. The basic source of energy provided for the movement of Earth's lithospheric plates is thought to be
    (a) Earth's gravitational potential energy.
    (b) radioactive decay.
    (c) residual heat from Earth's core.
    (d) none of the above

18. What is a "plate" in the context of plate tectonics?

19. Describe the three general types of relative plate motions and the geologic results of each.

20. What is a subduction zone?

21. What are the names of the six major plates?

21.4 Geologic Time

22. The geologic concept called the principle of uniformity is based on the assumption that
    (a) natural processes are not constant.
    (b) relative geologic and atomic time scales are correlated.
    (c) the present may be considered the key to the past.
    (d) none of the above

23. The relative geologic time scale is based upon
    (a) uniformitarianism.
    (b) the law of superposition.
    (c) radioactive decay.
    (d) the law of correlation.

24. Referring to the geologic time scale, which of the following is true?
    (a) The eras are the largest unit of time.
    (b) The Quaternary Period is the youngest period.
    (c) We are presently in the Holocene Epoch.
    (d) The Quaternary Period is part of the Cenozoic Era.
    (e) all of the above

25. What is the principle of uniformity and how is it used in the study of geologic processes?
26. On what is the relative geologic time scale based?
27. What is the atomic time scale and what "atomic clocks" are used in this scale?
28. How are the relative geologic and atomic time scales correlated?
29. Put in order of increasing average length of time: period, epoch, era.

Food for Thought

1. What is the origin and meaning of the term *tectonics*?
2. Suggest a method to measure the absolute motions of individual plates.
3. Distinguish between the ancient Atlantic Ocean and the present Atlantic Ocean.
4. Why do scientists believe the composition of planet Earth resembles that of meteorites?

Answers to Multiple-Choice Questions

1. d	8. d	16. c	22. c
2. e	9. d	17. b	23. b
	10. d		24. e

Solutions to Confidence Questions

21.1 (a) conduction, (b) convection

21.2 Given: separation $d = 8000 \times 10^6$ meters
time $t = 200$ million years
$= 2.0 \times 10^8$ years

Each continent drifted a distance of 4000 km. Therefore, the average rate or speed of drift is

$$v = d/t = \frac{4.0 \times 10^6 \text{ m}}{2.0 \times 10^8 \text{ y}}$$

$$= 2.0 \times 10^{-2} \text{ m/y} = 2.0 \text{ cm/y}$$

21.3 (a) lithosphere, (b) 80 km

21.4 From Fig. 21.17, the Cenozoic Era, Quaternary Period, and Pleistocene Epoch.

Appendixes

Appendix I

The Seven Base Units of the International System of Units (SI)

Meter, m (length)	The meter is defined in reference to the standard unit of time. One meter is the length of the path traveled by light in a vacuum during a time interval of 1/299,792,458 of a second. That is, the speed of light is a universal constant of nature whose value is defined to be 299,792,458 meters per second.
Kilogram, kg (mass)	The kilogram is a cylinder of platinum–iridium alloy kept by the International Bureau of Weights and Measures in Paris. A duplicate in the custody of the National Institute of Standards and Technology serves as the mass standard for the United States. This is the only base unit still defined by an artifact.
Second, s (time)	The second is defined as the duration of 9,192,631,770 cycles of the radiation associated with a specified transition of the cesium-133 atom.
Ampere, A (electric current)	The ampere is defined as that current that, if maintained in each of two long parallel wires separated by one meter in free space, would produce a force between the two wires (due to their magnetic fields) of 2×10^{-7} newtons for each meter of length.
Kelvin, K (temperature)	The kelvin is defined as the fraction 1/273.16 of the thermodynamic temperature of the triple point of water. The temperature 0 K is called *absolute zero*.
Mole, mol (amount of substance)	The mole is the amount of substance of a system that contains as many elementary entities as there are atoms in 0.012 kilograms of carbon-12.
Candela, cd (luminous intensity)	The candela is defined as the luminous intensity of 1/600,000 of a square meter of a black body at the temperature of freezing platinum (2045 K).

Appendix II

Solving Mathematical Problems in Science

Mathematics is fundamental in the physical sciences, and it is difficult to understand or appreciate these sciences without certain basic mathematical abilities. Dealing with the quantitative side of science gives the student a chance to review and gain confidence in the use of basic mathematical skills, to learn to analyze problems and reason them through, and to see the importance and power of a systematic approach to problems. We recommend the following approach to solving mathematical problems in science.

1. Using symbol notation, list what you are given and what is unknown. Include all units—not just the numbers.
2. From that information, decide the type of problem with which you are dealing, and select the appropriate equation.
3. If necessary, rearrange the equation for the unknown. (Appendix III discusses equation rearrangement.)
4. Substitute the known numbers *and their units* into the rearranged equation.

5. See if the units combine to give you the appropriate unit for the unknown. (Appendix IV discusses the analysis of units.)
6. Once the units are adjudged correct, do the math, being sure to express the answer to the proper number of significant figures and to include the unit. (Appendixes V, VI, and VII describe how to use positive and negative numbers, powers-of-10 notation, and significant figures.)
7. Evaluate the answer for reasonableness.

Appendix III

Equation Rearrangement

An important skill for solving mathematical problems in science is the ability to rearrange an equation for the unknown quantity. Basically, we want to get the unknown

- into the numerator
- positive in sign
- to the first power (not squared, cubed, etc.)
- alone on one side of the equals sign ($=$)

Addition or Subtraction to Both Sides of an Equation

The equation $X - 4 = 12$ states that the numbers $X - 4$ and 12 are equal. To solve for X, use the rule, whatever is added to (or subtracted from) one side can be added to (or subtracted from) the other side and equality will be maintained. We want to get X, the unknown in this case, alone on one side of the equation, so we add 4 to both sides (so that $-4 + 4 = 0$ on the left side). Example III.1 illustrates this procedure.

EXAMPLE III.1

$$X - 4 = 12$$
$$X - 4 + 4 = 12 + 4 \quad \text{(4 added to both sides)}$$
$$X = 16$$

EXAMPLE III.2

Suppose in the scientific equation $T_K = T_C + 273$ we wish to solve for T_C. All we need to do is to get T_C alone on one side, so we subtract 273 from each side. The procedure is

$$T_K = T_C + 273$$
$$T_K - 273 = T_C + 273 - 273 \quad \text{(273 is subtracted from both sides)}$$
$$T_K - 273 = T_C \text{ (or } T_C = T_K - 273\text{)}$$

SHORTCUT

After you understand the principle of adding or subtracting the same number or symbol from each side, you may wish to use a shortcut for this type of problem. Namely, to move a number or symbol added to (or subtracted from) the unknown, take it to the other side but change its sign. See how Examples III.1 and III.2 are solved using the shortcut.

Change sign
$$X \overset{\curvearrowright}{- 4} = 12 \quad \text{gives } X = 12 + 4$$

Change sign
$$T_K = T_C \overset{\curvearrowright}{+ 273} \quad \text{gives } T_K - 273 = T_C$$

PRACTICE PROBLEMS

Solve each equation for the unknown. As you proceed, check your answers against those given at the end of this appendix.

(a) $X + 9 = 11; X = ?$ (b) $T_f - T_i = \Delta T; T_f = ?$

(c) $\lambda f = E_i \quad E_f; E_i = ?$ (d) $H = \Delta E_i + W; W = ?$

(e) $\dfrac{\Delta E_p}{mg} = h_2 - h_1; h_2 = ?$ (f) $H = E_p - \dfrac{mv^2}{2}; E_p = ?$

Multiplication or Division to Both Sides of an Equation

For equations such as $\dfrac{X}{4} = 6$, the rule is, multiply (or divide) both sides by the number or symbol that will leave the unknown alone on one side of the numerator. So in the equation $\dfrac{X}{4} = 6$, we multiply both sides by 4, as shown in Example III.3.

EXAMPLE III.3

$$\frac{X}{4} = 6$$

$$\frac{4X}{4} = 6 \times 4 \quad \text{(both sides multiplied by 4)}$$

$$X = 24$$

EXAMPLE III.4

In the scientific equation $F = ma$, solve for a. The unknown is already in the numerator, so all we must do is move the m by dividing both sides by m.

$$F = ma$$

$$\frac{F}{m} = \frac{ma}{m} \quad \text{(both sides divided by } m\text{)}$$

$$\frac{F}{m} = a \left(\text{or } a = \frac{F}{m}\right)$$

EXAMPLE III.5

Suppose the unknown is in the denominator to start with; for example, solving for t in $P = \frac{W}{t}$. Multiply both sides by t to get t in the numerator, then divide both sides by P to move P to the other side. The procedure is

$$P = \frac{W}{t}$$

$$tP = \frac{tW}{t} \quad \text{(both sides multiplied by } t\text{)}$$

$$tP = W$$

$$\frac{tP}{P} = \frac{W}{P} \quad \text{(both sides divided by } P\text{)}$$

$$t = \frac{W}{P}$$

SHORTCUT

After you understand the principle of multiplying or dividing both sides of the equation by the same number or symbol, you may wish to use a shortcut to move the number or symbol. Namely, whatever is multiplying (or dividing) the whole of one side, winds up dividing (or multiplying) the whole other side. See how Examples III.3, III.4, and III.5 are solved using the shortcut.

$$\frac{X}{4} = 6 \quad \text{gives} \quad X = 4 \times 6 \text{ (or 24)}$$

$$F = ma \quad \text{gives} \quad \frac{F}{m} = a$$

$$P = \frac{W}{t} \quad \text{gives} \quad t = \frac{W}{P}$$

(The shortcut saves many steps when an unknown is in the denominator.)

PRACTICE PROBLEMS

Solve for the unknown. The answers are given at the end of this appendix.

(g) $7X = 21; X = ?$ (h) $\frac{X}{3} = 2; X = ?$

(i) $w = mg; m = ?$ (j) $\lambda = \frac{c}{f}; c = ?$

(k) $\lambda = \frac{c}{f}; f = ?$ (l) $\lambda = \frac{h}{mf}; m = ?$

(m) $E_k = \frac{mv^2}{2}; m = ?$ (n) $v^2 = \frac{3kT}{m}; T = ?$

(o) $pV = nRT; R = ?$

Multiple Operations to Both Sides of an Equation

To solve some problems requires the use of both the addition/subtraction and the multiplication/division rules. For example, solving $\frac{X-4}{3} = 8$ and $\frac{X}{3} - 4 = 8$ require both rules. These two equations are similar, but not identical. We will follow a different order of application of the rules as we solve each.

A good way to proceed is to follow the principle: If the *whole side* on which the unknown is found is multiplied and/or divided by a number or symbol, get that number or symbol to the other side first. *Then* move any numbers or symbols that are added to or subtracted from the unknown. (This is the case in the first example.) On the other hand, if only the unknown (and not the whole side) is multiplied or divided by a number or symbol, first move any number or symbol added or subtracted. (This is the case in the second example.) The successive steps for each example are shown.

EXAMPLE III.6

$$\frac{(X-4)}{3} = 8$$

$$\frac{3 \times (X-4)}{3} = 3 \times 8 \quad \text{(both sides multiplied by 3)}$$

$$X - 4 = 24$$

$$X - 4 + 4 = 24 + 4 \quad \text{(4 added to both sides)}$$

$$X = 28$$

EXAMPLE III.7

$$\frac{X}{3} - 4 = 8$$

$$\frac{X}{3} - 4 + 4 = 8 + 4 \quad \text{(4 added to both sides)}$$

$$\frac{X}{3} = 12$$

$$\frac{3X}{3} = 12 \times 3 \quad \text{(both sides multiplied by 3)}$$

$$X = 36$$

Note that in Example III.6, both sides were first multiplied by 3 and then 4 was added to both sides. But in Example III.7, 4 was added to both sides first and then both sides were divided by 3. We used different strategies because in the first example, the 3 was dividing the whole side that X was on, while in the second example it wasn't. Other strategies may be applied to problems like these, but things generally work out better if the recommended principle is used.

The rules for rearranging scientific equations are exactly the same. Before substituting numbers and units for the various letters in a scientific equation, most experts advise that the equation first be rearranged so that the unknown is positive, to the first power, and alone in the numerator on one side. It is easier and more accurate to move several letters than it is to move a bunch of numbers and units, as you would have to do if you substituted first.

EXAMPLE III.8

How would you proceed in solving the equation $a = \dfrac{v_f - v_i}{t}$ for v_f? The v_i and the t need to be moved, and t should be moved first because it is dividing the whole side that the unknown, v_f, is on. The successive steps are

$$a = \frac{v_f - v_i}{t}$$

$$at = \frac{t(v_f - v_i)}{t} \quad \text{(both sides multiplied by } t\text{)}$$

$$at = v_f - v_i$$

$$at + v_i = v_f - v_i + v_i \quad (v_i \text{ added to both sides})$$

$$at + v_i = v_f \quad (\text{or } v_f = at + v_i)$$

Shortcut

Of course, the same shortcuts can be used in these more complicated equations. Just be sure you use them in the correct order. For Examples III.6, III.7, and III.8, the circled numbers show the correct order for the shortcuts:

$$\frac{(X - 4)}{3} = 8 \quad \text{gives} \quad X = 3 \times 8 + 4 = 28$$

$$\frac{X}{3} - 4 = 8 \quad \text{gives} \quad X = 3 \times (8 + 4) = 36$$

$$a = \frac{v_f - v_i}{t} \quad \text{gives} \quad at + v_i = v_f$$

Practice Problems

Solve for the unknown. The answers are given at the end of this appendix.

(p) $3X + 2 = -7; X = ?$

(q) $\dfrac{X + 2}{4} = 6; X = ?$

(r) $a = \dfrac{F_2 - F_1}{m}; F_2 = ?$

(s) $H = \dfrac{B}{U} - M; B = ?$

(t) $T_F = 1.8 T_C + 32; T_C = ?$

(u) $H = mc_w(T_f - T_i); T_f = ?$

(v) $F = \dfrac{m(v - v_o)}{t}; v = ?$

Squared or Negative Unknowns

You will encounter some equations in which the unknown is squared (raised to the second power). For example, solve for v in $a = \dfrac{v^2}{r}$. In such cases, solve for the squared form first; then, either before or after substituting the numbers and units, take the square root of both sides.

EXAMPLE III.9

$$a = \dfrac{v^2}{r}$$

$$ar = \dfrac{rv^2}{r} \quad \text{(both sides multiplied by } r\text{)}$$

$$ar = v^2$$

$$\sqrt{ar} = \sqrt{v^2} \quad \text{(the square root of both sides taken)}$$

$$\sqrt{ar} = v \text{ (or } v = \sqrt{ar}\text{)}$$

Occasionally, you may need to solve for an unknown that is negative at the start. For example, find T_i in $\Delta T = T_f - T_i$. In such cases, multiply both numerators by -1, then proceed as usual.

EXAMPLE III.10

$$\Delta T = T_f - T_i$$

$$-\Delta T = -T_f + T_i \quad \text{(both sides multiplied by } -1\text{)}$$

$$T_f - \Delta T = -T_f + T_i + T_f \quad (T_f \text{ added to both sides)}$$

$$T_f - \Delta T = T_i$$

Practice Problems

Solve for the unknown. The answers are given at the end of this appendix.

(w) $E_K = \dfrac{mv^2}{2}$; $v = ?$ (x) $a = \dfrac{v_f - v_i}{t}$; $v_i = ?$

(y) $F_G = \dfrac{Gm_1m_2}{r^2}$; $r = ?$ (z) $H = mgh - \dfrac{mv^2}{2}$; $v = ?$

Answers

(a) $X = 2$

(b) $T_f = \Delta T + T_i$

(c) $E_i = \lambda f + E_f$

(d) $W = H - \Delta E_i$

(e) $h_2 = \dfrac{\Delta E_p}{mg} + h_i$

(f) $E_p = H + \dfrac{mv^2}{2}$

(g) $X = 3$

(h) $X = 6$

(i) $m = \dfrac{w}{g}$

(j) $c = \lambda f$

(k) $f = \dfrac{c}{\lambda}$

(l) $m = \dfrac{h}{\lambda f}$

(m) $m = \dfrac{2E_k}{v^2}$

(n) $T = \dfrac{v^2 m}{3k}$

(o) $R = \dfrac{pV}{nT}$

(p) $X = -3$

(q) $X = 22$

(r) $F_2 = ma + F_1$

(s) $B = U(H + M)$

(t) $T_C = \dfrac{T_F - 32}{1.8}$

(u) $T_f = \dfrac{H}{mc_w} + T_i$

(v) $v = \dfrac{Ft}{m} + v_o$

(w) $v = \sqrt{\dfrac{2E_k}{m}}$

(x) $v_i = v_f - at$

(y) $r = \sqrt{\dfrac{Gm_1m_2}{F_G}}$

(z) $v = \sqrt{\dfrac{2(mgh - H)}{m}}$

Appendix IV
Analysis of Units

Measured quantities always have dimensions, or *units*; for example, 17 *grams*, 1.4 *meters*, 23°C, and so forth. Analyzing the units is important when dealing with scientific equations because the units can often show you whether you have rearranged the equation and put in your data correctly. For example, if the unknown for which you are solving is a distance, yet the units show you coming out with m/s² (a combination of units characteristic of an acceleration), you have done something wrong (probably rearranged the equation incorrectly). Units follow the same rules as numbers. Only a few basic situations are encountered when analyzing units, and most are illustrated in the problems given in this appendix.

You have often heard that you cannot add apples to oranges; neither can you add, say, grams to meters. When adding and subtracting, the units must be the same.

EXAMPLE IV.1

$$8 \text{ mL} + 2 \text{ mL} = 10 \text{ mL} \quad (\text{that is, mL} + \text{mL gives mL})$$

$$15 \text{ g} - 7 \text{ g} = 8 \text{ g} \quad (\text{that is, g} - \text{g} = \text{g})$$

When multiplying and dividing, a unit in the numerator will cancel its counterpart in the denominator. A unit multiplied by the same unit gives the unit squared.

EXAMPLE IV.2

$$\frac{8 \text{ m}}{2 \text{ m}} = 4 \qquad 5.0 \frac{\text{m}}{\text{s}^2} \times 3.0 \text{ s} = 15 \frac{\text{m}}{\text{s}}$$

$$8 \text{ m} \times 2 \text{ m} = 16 \text{ m}^2$$

What about a problem where $\frac{m}{s}$ is divided by $\frac{m}{s^2}$? Recall how fractions are divided. To divide $\frac{3}{4}$ by $\frac{1}{4}$ you *invert* (turn over) the fraction in the denominator and then multiply by the inverted fraction. Units are handled the same way. Be sure to practice this trickiest part of analyzing units. Do not omit the intermediate step; this is no place to take shortcuts.

EXAMPLE IV.3

$$\frac{\frac{3}{4}}{\frac{1}{4}} = \frac{3}{4} \times \frac{4}{1} = 3 \qquad \frac{\frac{m}{s}}{\frac{m}{s^2}} = \frac{m}{s} \times \frac{s^2}{m} = s$$

PRACTICE PROBLEMS

Analyze these units. As you proceed, check your answers against those given at the end of this appendix.

(a) $\dfrac{g}{mol} \times mol$ 　(b) $\dfrac{\frac{g}{mol}}{mol}$

(c) $\dfrac{cm}{s} + \dfrac{cm}{s}$ 　(d) $\dfrac{g}{\frac{g}{mol}}$

(e) $\dfrac{\frac{g}{mol}}{g}$ 　(f) $\dfrac{m}{s} \times m$

(g) $\dfrac{m}{\frac{s}{m}}$ 　(h) $\dfrac{J}{cal} \times J$

(i) $\dfrac{J}{cal} \times cal$ 　(j) $kg - kg$

(k) $\dfrac{\frac{g}{mL}}{mL}$ 　(l) $kg \times \dfrac{m}{s^2}$

Chapter 1 discusses the use of conversion factors, a procedure in which the proper analysis of units is crucial. Using conversion factors, try your hand on the following practice problems. In (o) through (t), put in the units of the conversion factor first to make sure they will work out correctly, then insert the proper numbers. Check your answers.

PRACTICE PROBLEMS

(m) If 1 L = 1.06 qt, then how many $\frac{qt}{L}$ are there? How many $\frac{L}{qt}$?

(n) If there are 1.61 $\frac{km}{mi}$, then how many $\frac{mi}{km}$ are there?

(o) 15.0 m = ? yd, if 1 m = 1.09 yd

(p) 52.0 L = ? qt, if 1 L = 1.06 qt

(q) 46.0 km = ? mi, if 1 mi = 1.61 km

(r) 93.0 lb = ? kg, if 1 kg = 2.20 lb (at Earth's surface)

(s) 87 kg = ? g, if 1 kg = 1000 g

(t) 49 cm = ? m, if 1 m = 100 cm

Answers

(a) g

(b) $\dfrac{g}{mol^2}$

(c) $\dfrac{cm}{s}$

(d) mol

(e) $\dfrac{1}{mol}$

(f) $\dfrac{m^2}{s}$

(g) $\dfrac{m^2}{s}$

(h) $\dfrac{J^2}{cal}$

(i) J

(j) kg

(k) $\dfrac{g}{mL^2}$

(l) $\dfrac{kg \cdot m}{s^2}$

(m) $1.06\, \dfrac{qt}{L}; \dfrac{1}{1.06}\, \dfrac{L}{qt}$

(n) $\dfrac{1}{1.61}\, \dfrac{mi}{km}$

(o) $15.0\ m\, (1.09)\, \dfrac{yd}{m} = 16.4\ yd$

(p) $52.0\ L (1.06)\, \dfrac{qt}{L} = 55.1\ qt$

(q) $46.0\ km \left(\dfrac{1}{1.61}\right) \dfrac{mi}{km} = 28.6\ mi$

(r) $93.0\ lb \left(\dfrac{1}{2.20}\right) \dfrac{kg}{lb} = 42.3\ kg$

(s) $87\ kg\, (1000)\, \dfrac{g}{kg} = 87 \times 10^3\ g$

(t) $49\ cm \left(\dfrac{1}{100}\right) \dfrac{m}{cm} = 0.49\ m$

Appendix V
Positive and Negative Numbers

Multiplying and Dividing

To multiply and divide positive and negative numbers, follow these simple rules:

> If *both* numbers are positive or *both* numbers are negative, the result is positive.
>
> If one number is positive and the other is negative, the result is negative.

EXAMPLE V.1 (three different cases)

$3 \times 4 = 12$
$-20 \div -5 = 4$
$(-3) \times 4 = -12$

PRACTICE PROBLEMS

Perform the designated operations. The answers are given at the end of this appendix.

(a) $14 \div 2$
(b) $-30 \div 6$
(c) 5×8
(d) $-7 \times (-6)$
(e) $8 \times (-2)$
(f) $-40 \div -5$

Algebraic Addition

Algebraic addition of positive and negative numbers is illustrated in Example V.2 below. The numbers may be grouped and added or subtracted in any sequence without affecting the result.

EXAMPLE V.2 (six different cases)

$4 + 5 = 9$
$4 + (-5) = 4 - 5 = -1$
$4 - 5 = -1$
$4 - (-5) = 4 + 5 = 9$
$-4 - 5 = -9$
$-5 + 4 - 6 + 8 = -11 + 12 = 1$

PRACTICE PROBLEMS

Perform the designated operations. Check your answers.

(g) $7 + 6$
(h) $-8 - 7$
(i) $18 - 10$
(j) $-7 + 3 - 6$
(k) $14 - 5 + 8$
(l) $8 - (-6)$

Answers

(a) 7   (b) −5   (c) 40   (d) 42
(e) −16 (f) 8   (g) 13   (h) −15
(i) 8   (j) −10 (k) 17   (l) 14

Appendix VI
Powers-of-10 Notation

Changing Between Decimal Form and Powers-of-10 Form

Chapter 1 discusses how to switch a number from the decimal form to the powers-of-10 form, and vice versa. Try the following practice problems.

Practice Problems

Put a–d in standard powers-of-10 form, and e–h in decimal form. The answers are given at the end of this appendix.

(a) 2500   (b) 870,000
(c) 0.0000008   (d) 0.0357
(e) 6×10^4 (f) 5.6×10^3
(g) 5.6×10^{-6} (h) 7.9×10^{-2}

Changing Between Powers-of-10 Forms

When changing from one powers-of-10 form to another, the final number must be equal to the number with which you started. So, if the exponential part is made larger, the decimal part *must* become correspondingly smaller, and vice versa.

EXAMPLE VI.1

Change 83×10^5 to the 10^6 form.
Going from 10^5 to 10^6 is an *increase* by a factor of 10. Therefore, the decimal part must *decrease* by a factor of 10. Since 83 divided by 10 is 8.3, the answer is 8.3×10^6.
Change 4.5×10^{-9} to the 10^{-10} form.
Going from 10^{-9} to 10^{-10} is a *decrease* by a factor of 10. Therefore, the decimal part *must increase* by a factor of 10. Since 4.5 multiplied by 10 is 45, the answer is 45×10^{-10}.

Practice Problems

Determine the value required in place of the question mark for the equation to be true. Check your answers.

(i) $3.02 \times 10^7 = ? \times 10^6$ (j) $126 \times 10^{-3} = ? \times 10^{-2}$
(k) $896 \times 10^4 = ? \times 10^6$ (l) $32.7 \times 10^5 = 3.27 \times 10^?$

Addition and Subtraction of Powers-of-10

In addition or subtraction, the exponents of 10 must be the same value.

EXAMPLE VI.2

$$\begin{array}{r} 4.6 \times 10^{-8} \\ + 1.2 \times 10^{-8} \\ \hline 5.8 \times 10^{-8} \end{array} \quad \text{and} \quad \begin{array}{r} 4.8 \times 10^7 \\ - 2.5 \times 10^7 \\ \hline 2.3 \times 10^7 \end{array}$$

Practice Problems

Perform the designated arithmetical operations. Check your answers.

(m) $\begin{array}{r} 4.5 \times 10^5 \\ + 3.2 \times 10^5 \\ \hline ? \end{array}$ (n) $\begin{array}{r} 5.66 \times 10^{-3} \\ - 3.24 \times 10^{-3} \\ \hline ? \end{array}$

Multiplication of Powers-of-10

In multiplication, the exponents are added.

EXAMPLE VI.3

$$(2 \times 10^4)(4 \times 10^3) = 8 \times 10^7 \quad \text{and}$$
$$(1.2 \times 10^{-2})(3 \times 10^6) = 3.6 \times 10^4$$

Practice Problems

Perform the designated arithmetical operations. Check your answers.

(o) $(7 \times 10^5)(3 \times 10^4) = ?$

(p) $(2 \times 10^{-3})(4 \times 10^6) = ?$

Division of Powers-of-10

In division, the exponents are subtracted.

EXAMPLE VI.4

$$\frac{4.8 \times 10^8}{2.4 \times 10^2} = 2.0 \times 10^6 \quad \text{and}$$

$$\frac{3.4 \times 10^{-8}}{1.7 \times 10^{-2}} = 2.0 \times 10^{-6}$$

An alternative method for division is to transfer all powers of 10 from the denominator to the numerator by changing the sign of the exponent. Then, the exponents of the powers of 10 may be added, because they are now multiplying. The decimal parts are not transferred; they are divided in the usual manner. This method requires an additional step, but many students find it leads to the correct answer more consistently. Thus:

$$\frac{4.8 \times 10^8}{2.4 \times 10^2} = \frac{4.8 \times 10^8 \times 10^{-2}}{2.4} = 2.0 \times 10^6$$

PRACTICE PROBLEMS
Perform the designated arithmetical operations. Check your answers.

(q) $\dfrac{18 \times 10^7}{3 \times 10^4} = ?$ (r) $\dfrac{(3 \times 10^{17})(4 \times 10^{-8})}{6 \times 10^{-11}} = ?$

Squaring Powers-of-10

When squaring exponential numbers, multiply the exponent by 2. The decimal part is multiplied by itself.

EXAMPLE VI.5

$$(3 \times 10^4)^2 = 9 \times 10^8$$
$$(4 \times 10^{-7})^2 = 16 \times 10^{-14}$$

PRACTICE PROBLEMS
Perform the designated algebraic operations. Check your answers.

(s) $(8 \times 10^{-5})^2$ (t) $(4 \times 10^3)^2$ (u) $(3 \times 10^{-8})^2$

Finding the Square Root of Powers-of-10

To find the square root of an exponential number, follow the rule, $\sqrt{10^a} = 10^{(\frac{a}{2})}$. Note that the exponent must be an *even* number. If it is not, change to a power-of-10 form that gives an even exponent. Find the square root of the decimal part by determining what number multiplied by itself gives that number.

EXAMPLE VI.6

$$\sqrt{9 \times 10^8} = 3 \times 10^4$$
$$\sqrt{2.5 \times 10^{-17}} = \sqrt{25 \times 10^{-18}} = 5 \times 10^{-9}$$

PRACTICE PROBLEMS
Perform the designated algebraic operations. Check your answers.

(v) $\sqrt{4 \times 10^8}$ (w) $\sqrt{16 \times 10^{-10}}$ (x) $\sqrt{78 \times 10^{-11}}$

Answers

(a) 2.5×10^3 (b) 8.7×10^5 (c) 8×10^{-7} (d) 3.57×10^{-2}
(e) 60,000   (f) 5600   (g) 0.0000056   (h) 0.079
(i) 30.2×10^6 (j) 12.6×10^{-2} (k) 8.96×10^6 (l) 3.27×10^6
(m) 7.7×10^5 (n) 2.42×10^{-3} (o) 21×10^9 (p) 8×10^3
(q) 6×10^3 (r) 2×10^{20} (s) 64×10^{-10} (t) 16×10^6
(u) 9×10^{-16} (v) 2×10^4 (w) 4×10^{-5} (x) 2.8×10^{-5}

Appendix VII
Significant Figures

In scientific work, most numbers are *measured* quantities and thus are not exact. All measured quantities are limited in significant figures (abbreviation SF) by the precision of the instrument used to make the measurement. The measurement must be recorded in such a way as to show the degree of precision to which it was made—no more, no less. Furthermore, calculations based on the measured quantities can have no more (or no less) precision than the measurements themselves. Thus, the answers to the calculations must be recorded to the proper number of significant figures. To do otherwise is misleading and improper.

Counting Significant Figures

MEASURED QUANTITIES

Rule 1: **Non-zero integers** are always significant (e.g., both 23.4 g and 234 g have 3 SF).

Rule 2: **Captive zeros,** those *bounded* on both sides by non-zero integers, are always significant (e.g., 20.05 g has 4 SF; 407 g has 3 SF).

Rule 3: **Leading zeros,** those *not bounded* on the left by non-zero integers, are never significant (e.g., 0.04 g has 1 SF; 0.00035 has 2 SF). Such zeros just set the decimal point; they always disappear if the number is converted to powers-of-10 notation.

Rule 4: **Trailing zeros,** those bounded only on the left by non-zero integers are probably not significant *unless a decimal point is shown,* in which case they are always significant. For example, 45.0 L has 3 SF but 450 L probably has only 2 SF; 21.00 kg has 4 SF but 2100 kg probably has only 2 SF; 55.20 mm has 4 SF; 151.10 cal has 5 SF; 3.0×10^4 J has 2 SF. If you wish to show *for sure* that, say, 150 m is to be interpreted as having 3 SF, change to powers-of-10 notation and show it as 1.50×10^2 m.

EXACT NUMBERS

Rule 5: **Exact numbers** are those not obtained by measurement but by definition or by counting small numbers of objects. They are assumed to have an unlimited number of significant figures. For example, in the equation $C = 2\pi r$, the "2" is a defined quantity, not a measured one, so it has no effect on the number of significant figures to which the answer can be reported. In counting, say, 15 pennies, you can see that the number is exact because you cannot have 14.9 pennies or 15.13 pennies. The $\frac{9}{5}$ or 1.8 found in temperature conversion equations are exact numbers based on definitions.

PRACTICE PROBLEMS

Using Rules 1–5, determine the number of significant figures in the following measurements. As you proceed, check your answers against those given at the end of the appendix.

(a) 4853 g
(b) 36.200 km
(c) 0.088 s
(d) 30.003 J
(e) 6 dogs
(f) 74.0 m
(g) 340 cm
(h) 40 mi
(i) 8.9 L
(j) 1.30×10^2 cal
(k) 0.002710 ft
(l) 4000 mi
(m) 0.0507 mL
(n) 1.6×10^4 N
(o) The 2 in $E_k = \dfrac{mv^2}{2}$

Multiplication and Division Involving Significant Figures

Rule 6: In calculations involving only multiplication and/or division of measured quantities, the answer shall have the same number of significant figures as the fewest possessed by any measured quantity in the calculation.

EXAMPLE VII.1

A calculator gives 4572.768 cm³ when 130.8 cm is multiplied by 15.2 cm and then by 2.3 cm. However, this answer would be rounded and reported as 4.6×10^3 cm³, since 2.3 cm has the fewest significant figures (2). The reasoning behind this is that the measured 2.3 cm could easily be wrong by 0.1 cm. Suppose it were really 2.4 cm; what difference would this make in the answer on the calculator? You would get 4771.584 cm³! Comparing this to what the calculator originally gave, you see that the uncertainty in the answer is in the hundreds place; so the answer is properly reported only to the hundreds place; that is, to 2 SF. The measured quantity with the fewest significant figures will have the greatest effect on the answer because of percentage effects (a miss of 1 out of 23 is more damaging than a miss of 1 out of 1308, for example).

Practice Problems

Rewrite the following calculator-given answers so that the proper number of significant figures is shown in each case. When necessary, use exponential notation to relieve ambiguity in the answer.

(p) $7.7 \frac{m}{s^2} \times 3.222 \text{ s} \times 2.4423 \text{ s} = 60.59199762 \text{ m}$

(q) $0.0075 \text{ cm} \times 0.005 \text{ cm} \times 8211 \text{ cm} = 0.3079125 \text{ cm}^3$

(r) $93.0067 \text{ g} \div 35 \text{ mL} = 2.65733428571 \frac{g}{mL}$

(s) $7.43 \frac{kg}{L} \times 15 \text{ L} = 111.45 \text{ kg}$

(t) $5766 \frac{m}{s} \times 322 \text{ s} = 1{,}856{,}652 \text{ m}$

Addition and Subtraction Involving Significant Figures

Rule 7: In calculations where measured quantities are added or subtracted, the final answer can have only one "uncertain" figure, so it stops at the decimal place on the right that any of the data first stop.

Carry the calculation one place farther, and then round up the answer if justified. (If the last figure is less than 5, drop it; if it is 5 or greater, round the preceding figure up one.)

EXAMPLES VII.2

In each of these examples, the vertical dashed line shows how far over the answer can go. Note that each answer has been calculated to one place farther than will be reported. This is so we can see if rounding is necessary. The final answers are shown below the double underline.

```
  46.6  m          38   cm         5.687 × 10³ g
+  5.72 m        −  7.44 cm      + 11.11  × 10³ g
  ─────           ──────          ──────────────
  52.32 m          30.6 cm         16.797 × 10³ g
  ═════            ════            ══════════════
  52.3  m          31   cm         16.80  × 10³ g
```

Practice Problems

Perform the designated arithmetic operations, being careful to retain the proper number of significant figures.

(u) 0.0012 m
  $+ 1.334$ m
  ?

(v) 879 g
  -79.9 g
  ?

(w) 6.788 cm
  $+ 5.6$ cm
  ?

(x) 67.4 kg
  $- 0.06$ kg
  ?

(y) 54.09×10^4 g
  $+\ 3\ \ \ \ \times 10^4$ g
  ?

Answers

(a) 4 (Rule 1)	(b) 5 (Rule 4)	(c) 2 (Rule 3)	(d) 5 (Rule 2)
(e) unlimited (Rule 5)	(f) 3 (Rule 4)	(g) 2 or 3 (Rule 4)	(h) 1 or 2 (Rule 4)
(i) 2 (Rule 1)	(j) 3 (Rule 4)	(k) 4 (Rules 3, 4)	(l) 1, 2, 3, or 4 (Rule 4)
(m) 3 (Rules 2, 3)	(n) 2 (Rule 1)	(o) unlimited (Rule 5)	(p) 61 m
(q) 0.3 cm³	(r) $2.7 \frac{g}{mL}$	(s) 1.1×10^2 kg	(t) 1.86×10^6 m
(u) 1.335 m	(v) 799 g	(w) 12.4 cm	(x) 67.3 kg
(y) 57×10^4 g			

Appendix VIII

Length Contraction, Time Dilation, and Relativistic Mass Increase

The special theory of relativity is based on two principles:

1. The relativity principle states that the laws of physics are the same in all inertial reference frames. An inertial reference frame is one in

which Newton's law of inertia is valid. Inertial reference frames can move at constant velocity relative to one another. Noninertial reference frames exhibit acceleration.

2. The principle of constancy of the speed of light states that the speed of light in empty space has the same value in all inertial reference frames.

The phenomena called *length contraction, time dilation,* and *relativistic mass increase* are effects that become significant when the relative velocity between the object being measured and the observer is an appreciable fraction of the velocity of light.

In length contraction, the length of a moving object is measured to be less in a direction parallel to the direction of motion than that of a similar object in the rest frame. The equation that relates the lengths of objects in motion to their lengths when at rest is

$$L = L_o \sqrt{1 - \frac{v^2}{c^2}}$$

where L_o = length of object at rest,
L = length of object in motion,
v = velocity of object with respect to observer,
c = velocity of light = 3.0×10^8 m/s.

EXAMPLE VIII.1

What is the length contraction of a meterstick when it is moving in the direction of its length at half the velocity of light?

Solution

Using the preceding equation with $L_o = 1.0$ m, we have:

$$L = (1.0 \text{ m}) \sqrt{1 - \frac{(1.5 \times 10^8 \text{ m/s})^2}{(3.0 \times 10^8 \text{ m/s})^2}}$$

$$= (1.0 \text{ m}) \sqrt{1 - \frac{2.25}{9.0}} = (1.0 \text{ m})(0.866)$$

$$= 0.87 \text{ m}$$

In time dilation (an "increase" of time), moving clocks run slower than clocks at rest with respect to an observer. The equation for relativistic time dilation is

$$t = \frac{t_o}{\sqrt{1 - v^2/c^2}}$$

where t_o = time interval of clock at rest,
t = time interval of clock in relative motion as determined by outside observer,
v = velocity of relative motion,
c = velocity of light.

In the effect of relativistic mass increase, the measured mass of an object at rest is not the same as the measured mass of the object in relative motion. The mass of the moving object is larger, or increased, and is given by the following equation:

$$m = \frac{m_o}{\sqrt{1 - v^2/c^2}}$$

where m_o = rest mass as measured by stationary observer,
m = mass of object moving relative to observer,
v = velocity of relative motion,
c = velocity of light.

EXAMPLE VIII.2

What is the relativistic mass increase of a 1.0-kg mass moving at half the velocity of light?

Solution

With a rest mass of 1.0 kg, we have:

$$m = \frac{1.0 \text{ kg}}{\sqrt{1 - \frac{(1.5 \times 10^8 \text{ m/s})^2}{(3.0 \times 10^8 \text{ m/s})^2}}}$$

$$= \frac{1.0 \text{ kg}}{\sqrt{1 - \frac{2.25 \times 10^{16} \text{ m}^2/\text{s}^2}{9.0 \times 10^{16} \text{ m}^2/\text{s}^2}}}$$

$$= \frac{1.0 \text{ kg}}{\sqrt{1 - 0.25}} = \frac{1.0 \text{ kg}}{0.866} = 1.2 \text{ kg}$$

Appendix IX

Alphabetical List of the Elements

Element	Symbol	Atomic Number	Atomic Mass (u)	Element	Symbol	Atomic Number	Atomic Mass (u)
Actinium	Ac	89	(227.0)	Mercury	Hg	80	200.6
Aluminum	Al	13	27.0	Molybdenum	Mo	42	95.9
Americium	Am	95	(243.1)	Neodymium	Nd	60	144.2
Antimony	Sb	51	121.8	Neon	Ne	10	20.2
Argon	Ar	18	39.9	Neptunium	Np	93	(237.0)
Arsenic	As	33	74.9	Nickel	Ni	28	58.7
Astatine	At	85	(210.0)	Niobium	Nb	41	92.9
Barium	Ba	56	137.3	Nitrogen	N	7	14.0
Berkelium	Bk	97	(247.1)	Nobelium	No	102	(259.1)
Beryllium	Be	4	9.01	Osmium	Os	76	190.2
Bismuth	Bi	83	209.0	Oxygen	O	8	16.0
Boron	B	5	10.8	Palladium	Pd	46	106.4
Bromine	Br	35	79.9	Phosphorus	P	15	31.0
Cadmium	Cd	48	112.4	Platinum	Pt	78	195.1
Calcium	Ca	20	40.1	Plutonium	Pu	94	(244.1)
Californium	Cf	98	(251.1)	Polonium	Po	84	(209.0)
Carbon	C	6	12.0	Potassium	K	19	39.1
Cerium	Ce	58	140.1	Praseodymium	Pr	59	140.9
Cesium	Cs	55	132.9	Promethium	Pm	61	(144.9)
Chlorine	Cl	17	35.5	Protactinium	Pa	91	(231.0)
Chromium	Cr	24	52.0	Radium	Ra	88	(226.0)
Cobalt	Co	27	58.9	Radon	Rn	86	(222.0)
Copper	Cu	29	63.5	Rhenium	Re	75	186.2
Curium	Cm	96	(247.1)	Rhodium	Rh	45	102.9
Dysprosium	Dy	66	162.5	Rubidium	Rb	37	85.5
Einsteinium	Es	99	(252.1)	Ruthenium	Ru	44	101.1
Erbium	Er	68	167.3	Samarium	Sm	62	150.4
Europium	Eu	63	152.0	Scandium	Sc	21	45.0
Fermium	Fm	100	(257.1)	Selenium	Se	34	79.0
Fluorine	F	9	19.0	Silicon	Si	14	28.1
Francium	Fr	87	(223.0)	Silver	Ag	47	107.9
Gadolinium	Gd	64	157.3	Sodium	Na	11	23.0
Gallium	Ga	31	69.7	Strontium	Sr	38	87.6
Germanium	Ge	32	72.6	Sulfur	S	16	32.1
Gold	Au	79	197.0	Tantalum	Ta	73	180.9
Hafnium	Hf	72	178.5	Technetium	Tc	43	(97.9)
Helium	He	2	4.00	Tellurium	Te	52	127.6
Holmium	Ho	67	164.9	Terbium	Tb	65	158.9
Hydrogen	H	1	1.01	Thallium	Tl	81	204.4
Indium	In	49	114.8	Thorium	Th	90	232.0
Iodine	I	53	126.9	Thulium	Tm	69	168.9
Iridium	Ir	77	192.2	Tin	Sn	50	118.7
Iron	Fe	26	55.8	Titanium	Ti	22	47.9
Krypton	Kr	36	83.8	Tungsten	W	74	183.9
Lanthanum	La	57	138.9	Uranium	U	92	238.0
Lawrencium	Lr	103	(262.1)	Vanadium	V	23	50.9
Lead	Pb	82	207.2	Xenon	Xe	54	131.3
Lithium	Li	3	6.94	Ytterbium	Yb	70	173.0
Lutetium	Lu	71	175.0	Yttrium	Y	39	88.9
Magnesium	Mg	12	24.3	Zinc	Zn	30	65.4
Manganese	Mn	25	54.9	Zirconium	Zr	40	91.2
Mendelevium	Md	101	(258.1)				

Note: The atomic masses are based on $^{12}C = 12.0000$ u. They are given to the nearest 0.1 or to at least 3 SF. A value in parentheses denotes the atomic mass of the isotope of longest half-life for a radioactive element. It is recommended that you be able to match the names and symbols of those elements in color.

Appendix X

Psychrometric Tables (Pressure: 30 in. of Hg)

Table A.1 Relative Humidity (%) and Maximum Moisture Capacity

Air Temp. (°F) (Dry Bulb)	Max. Moisture Capacity (gr/ft³)	Degrees Depression of Wet Bulb Thermometer (F°)													
		1	2	3	4	5	6	7	8	9	10	15	20	25	30
25	1.6	87	74	62	49	37	25	13	1						
30	1.9	89	78	67	56	46	36	26	16	6					
35	2.4	91	81	72	63	54	45	36	27	19	10				
40	2.8	92	83	75	68	60	52	45	37	29	22				
45	3.4	93	86	78	71	64	57	51	44	38	31				
50	4.1	93	87	80	74	67	61	55	49	43	38	10			
55	4.8	94	88	82	76	70	65	59	54	49	43	19			
60	5.7	94	89	83	78	73	68	63	58	53	48	26	5		
65	6.8	95	90	85	80	75	70	66	61	56	52	31	12		
70	7.8	95	90	86	81	77	72	68	64	59	55	36	19	3	
75	9.4	96	91	86	82	78	74	70	66	62	58	40	24	9	
80	10.9	96	91	87	83	79	75	72	68	64	61	44	29	15	3
85	12.7	96	92	88	84	80	76	73	69	66	62	46	32	20	8
90	14.8	96	92	89	85	81	78	74	71	68	65	49	36	24	13
95	17.1	96	93	89	85	82	79	75	72	69	66	51	38	27	17
100	19.8	96	93	89	86	83	80	77	73	70	68	54	41	30	21
105	23.4	97	93	90	87	83	80	77	74	71	69	55	43	33	23
110	26.0	97	93	90	87	84	81	78	75	73	70	57	46	36	26

Note: To use the table, determine the air temperature with a dry bulb thermometer and degrees depressed on the wet bulb thermometer. Read the maximum capacity directly. Read the relative humidity (in percent) opposite and below these values.

Table A.2 Dew Point (F°)

Air Temp (°F) (Dry Bulb)	Degrees Depression of Wet Bulb Thermometer (F°)													
	1	2	3	4	5	6	7	8	9	10	15	20	25	30
25	22	19	15	10	5	−3	−15	−51						
30	27	25	21	18	14	8	2	−7	−25					
35	33	30	28	25	21	17	13	7	0	−11				
40	38	35	33	30	28	25	21	18	13	7				
45	43	41	38	36	34	31	28	25	22	18				
50	48	46	44	42	40	37	34	32	29	26	0			
55	53	51	50	48	45	43	41	38	36	33	15			
60	58	57	55	53	51	49	47	45	43	40	25	−8		
65	63	62	60	59	57	55	53	51	49	47	34	14		
70	69	67	65	64	62	61	59	57	55	53	42	26	−11	
75	74	72	71	69	68	66	64	63	61	59	49	36	15	
80	79	77	76	74	73	72	70	68	67	65	56	44	28	−7
85	84	82	81	80	78	77	75	74	72	71	62	52	39	19
90	89	87	86	85	83	82	81	79	78	76	69	59	48	32
95	94	93	91	90	89	87	86	85	83	82	74	66	56	43
100	99	98	96	95	94	93	91	90	89	87	80	72	63	52
105	104	103	101	100	99	98	96	95	94	93	86	78	70	61
110	109	108	106	105	104	103	102	100	99	98	91	84	77	68

Note: To use the table, determine the air temperature with a dry bulb thermometer and degrees depressed on the wet bulb thermometer. Find the dew point opposite and below these values.

Appendix XI
Seasonal Star Charts

Latitude of chart is 34°N, but it is practical throughout the coterminous United States.

To use: Hold chart vertically and turn it so the direction you are facing shows at the bottom.

Chart time (local standard time):

10 P.M. First of month
9 P.M. Middle of month
8 P.M. Last of month

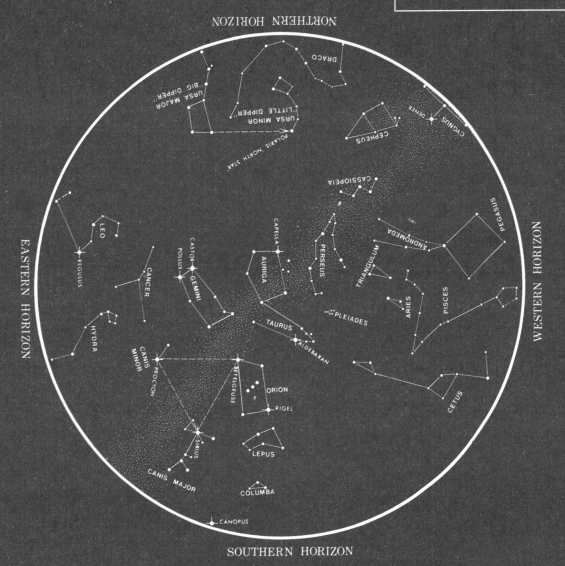

THE NIGHT SKY IN JANUARY

Star Chart from *GRIFFITH OBSERVER*, Griffith Observatory, Los Angeles

A16 Appendix

Latitude of chart is 34°N, but it is practical throughout the coterminous United States.

To use: Hold chart vertically and turn it so the direction you are facing shows at the bottom.

Chart time (local standard time):

10 P.M. First of month
 9 P.M. Middle of month
 8 P.M. Last of month

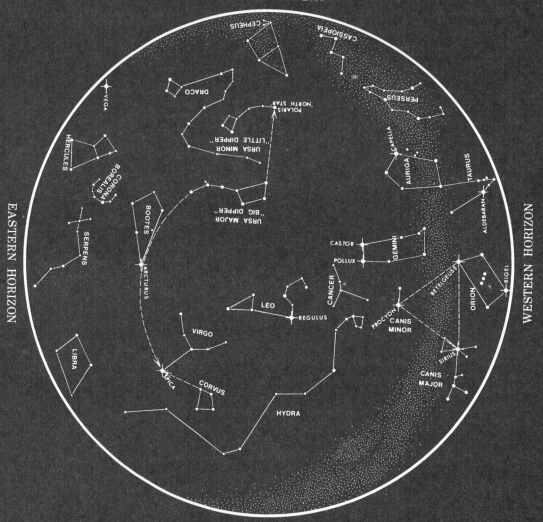

THE NIGHT SKY IN APRIL

Star Chart from *GRIFFITH OBSERVER*, Griffith Observatory, Los Angeles

Latitude of chart is 34°N, but it is practical throughout the coterminous United States.

To use: Hold chart vertically and turn it so the direction you are facing shows at the bottom.

Chart time (local standard time):

10 P.M. First of month
9 P.M. Middle of month
8 P.M. Last of month

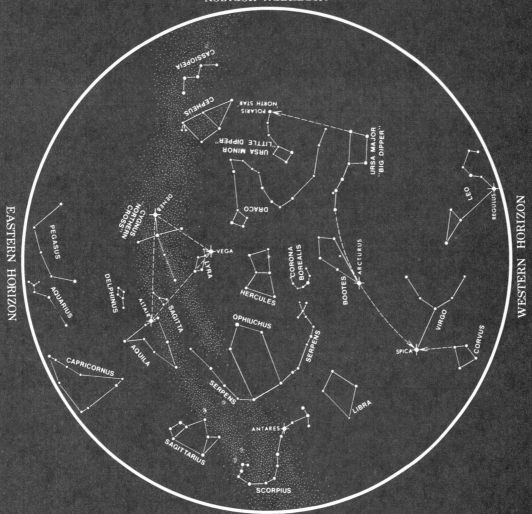

THE NIGHT SKY IN JULY

Star Chart from *GRIFFITH OBSERVER*, Griffith Observatory, Los Angeles

A18 Appendix

Latitude of chart is 34°N, but it is practical throughout the coterminous United States.

To use: Hold chart vertically and turn it so the direction you are facing shows at the bottom.

Chart time (local standard time):

10 P.M. First of month
9 P.M.   Middle of month
8 P.M.   Last of month

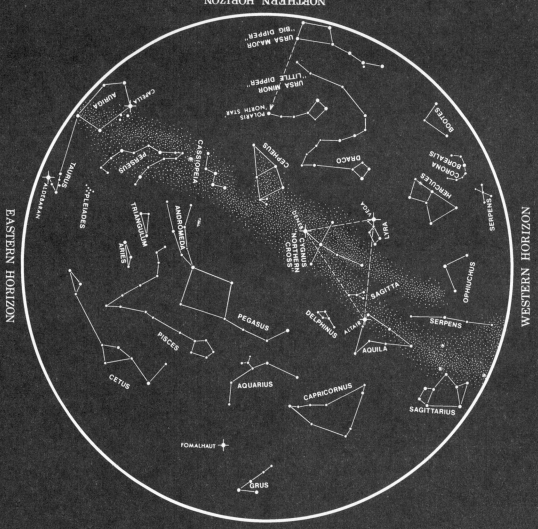

THE NIGHT SKY IN OCTOBER

Star Chart from *GRIFFITH OBSERVER*, Griffith Observatory, Los Angeles

Photo Credits

Chapter 1: p. 4, Gerhard Gscheidle/Peter Arnold, Inc.; p. 8, Mark Helfer/National Institute of Standards & Technology; p. 10 left, Leonard Lessin/Peter Arnold, Inc.; p. 10 right, Don & Pat Valenti/Tom Stack & Associates; p. 13, Custom Medical Stock Photo; p. 16, Peter Arnold, Inc.

Chapter 2: p. 25, David Madison/Duomo; p. 35, The Granger Collection; p. 37, Richard Megna/Fundamental Photographs; p. 39, Focus on Sports.

Chapter 3: p. 46, Ken O'Donoghue; p. 47 left, The Granger Collection; p. 47 right, The Granger Collection; p. 51, John Smith; p. 55, NASA/Johnson Space Center; p. 56, NASA/Johnson Space Center; p. 58, NASA/Johnson Space Center; p. 63 left, Sports Illustrated, Time, Inc.; p. 63 right, Sports Illustrated, Time, Inc.; p. 64, Kairos Latin Stock/Science Photo Library/Photo Researchers; p. 65 left, George Hall/Woodfin Camp & Associates; p. 65 right, Bell Helicopter Textron, Inc.

Chapter 4: p. 72, Paul Sutton/Duomo; p. 74, Daemmrich/The Image Works; p. 84, Jim Richardson/Westlight; p. 85 a, Phil Degginger/Bruce Coleman; p. 85 b, Mark Antman/Phototake.

Chapter 5: p. 91 top, John Smith; p. 91 bottom, Jerry Wilson; p. 93, Bruce Novak; p. 95, Bruce Coleman Inc.; p. 106, Dan McCoy/Rainbow; p. 108, Jim Richardson/Westlight.

Chapter 6: p. 123, John Smith; p. 126 a, United Airlines; p. 126 b, Steven Stone/The Picture Cube; p. 126 c-e, Henry Rachlin; p. 127, Bill Gallery/Stock, Boston; p. 128, Sieman's Lutheran Hospital/Peter Arnold, Inc.; p. 129, Jerry Wilson; p. 131 a, Richard Megna/Fundamental Photographs; p. 131 b, Richard Megna/Fundamental Photographs; p. 131 c, Richard Megna/Fundamental Photographs; p. 132, UPI/Bettmann Newsphotos.

Chapter 7: p. 139, Steve Short/First Light; p. 140, Fundamental Photographs; p. 144 left, Jeff Persons/Stock, Boston; p. 144 right, James H. Karales/Peter Arnold, Inc.; p. 145, E. R. Degginger; p. 146, David Parker/Science Photo Library/Photo Researchers, Inc.; p. 147 b, PASCO Scientific p. 147 c (both), Department of Physics, Imperial College/Science Photo, Library/Photo Researchers; p. 148 left, *PSSC Physics* D. C. Heath; p. 148 right, Peter Aprahamian/Science Photo Library/Photo Researchers; p. 149, *PSSC Physics* D. C. Heath; p. 150, Phil Degginger; p. 151, Polaroid Corporation; p. 152, Polaroid Corporation; p. 153, Lightwave; p. 155, Jerry Wilson; p. 157, Lightwave; p. 159, Ken O'Donoghue; p. 160, Ken O'Donoghue.

Chapter 8: p. 170, Ken O'Donoghue; p. 171, John Smith; p. 161 both, Ken O'Donoghue; p. 175, Yoav/Phototake; p. 180 both, Ken O'Donoghue; p. 182, Paul Silverman/Fundamental Photographs; p. 183 top, Richard Megna/Fundamental Photographs; p. 183 a, Richard Megna/Fundamental Photographs; p. 183 b, Richard Megna/Fundamental Photographs; p. 183 c, Richard Megna/ Fundamental Photographs; p. 188, C. N. K. Baker; p. 192 left, Werner H. Muller/Peter Arnold, Inc.; p. 192 middle, John Zoiner/Peter Arnold, Inc.; p. 192 right, Yoav/Phototake.

Chapter 9: p. 198, Stacy Pick/Stock, Boston; p. 199 left, The Granger Collection; p. 199 right, Tom McHugh/Photo Researchers; p. 201 top, *General College Chemistry*, Fifth Edition (1976), by Charles W. Keenan, Jesse H. Wood, and Donald Kleinfelter; by permission of the authors; p. 201 bottom, The Granger Collection; p. 207, Dagmar Hailer Hamann/Peter Arnold, Inc.; p. 208, Charles Gupton/Stock, Boston; p. 210, Paul Silverman/Fundamental Photographs; p. 211, AIP Neils Bohr Library; p. 213 top, Dr. Jeremy Burgess/Science Photo Library/Photo Researchers; p. 213 middle, Dr. Tony Brain/Science Photo Library/Photo Researchers; p. 213 right, The Granger Collection; p. 215, The Granger Collection.

Chapter 10: p. 231 left, Brown Brothers; p. 231 middle, UPI/Bettmann; p. 231 right, American Institute of Physics; Meggers Gallery of Nobel Laureates; p. 238, UPI/Bettmann; p. 240, Fermi National Laboratory; p. 241, Hank Morgan/Rainbow; p. 244, Sheahan, Gary, *Birth of the Atomic Age*, 1957, Chicago. Courtesy the Chicago Historical Society; p. 245, Photri, Inc.; p. 249, U.S. Air Force; p. 250, Princeton Plasma Physics Lab.

Chapter 11: p. 262, Richard Megna/Fundamental Photographs; p. 263, The Bettmann Archive; p. 266, Photo Researchers; p. 241, Van Pelt Library, University of Pennsylvania; p. 272 left, Chip Clark; p. 272 right, Jerry Wilson; p. 278 left, Tom Pantages; p. 278 right, Tom Pantages; p. 279, Ken O'Donoghue; p. 280 left, Erica & Harold Van Pelt, Photographers; p. 280 right, CNRI/Science Photo Library; p. 281, The Bettmann Archive.

Chapter 12: p. 288, David Jacques Louis, *Antoine Laurent Lavoisier (1743–1794) and His Wife (Marie Anne Pierrette Paulze, 1758–1836)*. The Metropolitan Museum of Art, Purchase, Mr. and Mrs. Charles Wrightsman. Gift, in honor of Everett Fahy, 1977. (1977.10); p. 290, Yoav Levy/Phototake, Inc.; p. 291, The Bettmann Archive; p. 294, E. R. Degginger; p. 295, University Archives/The Bancroft Library; University of California at Berkeley; p. 297, Ken O'Donoghue; p. 300, E. R. Degginger; p. 305 left, Tom Pantages; p. 305 right, Tom Pantages; p. 307, Chip Clark; p. 308 left, E. R. Degginger; p. 308 right, E. R. Degginger.

Chapter 13: p. 316, Chip Clark; p. 317 top, Ken O'Donoghue; p. 317 bottom, Ken O'Donoghue; p. 319 left, Ken O'Donoghue; p. 317 right, Ken O'Donoghue, p. 321, E. R. Degginger; p. 322 left, Don Clegg; p. 322 right, Don Clegg; p. 324 left, Ken O'Donoghue; p. 324 right, Ken O'Donoghue; p. 325, UPI/Bettmann; p. 327 left, E. R. Degginger; p. 327 right, Thomas Eisner/Daniel Aneshansley/Cornell University; p. 327 bottom, UPI/Bettmann; p. 331 top, E. R. Degginger, p. 331 bottom, Dave Buresh/*Denver Post*; p. 333 left, Lawrence Migdale/Science Source/Photo Researchers; p. 333 right, W. H. Breazeale; p. 334, Peticolas/Megna/Fundamental Photographs; p. 335 left, E. R. Degginger; p. 335 right, Chip Clark; p. 336 top, E. R. Degginger; p. 336 bottom, The Granger Collection; p. 337, E. R. Degginger; p. 338, W. H. Breazeale.

Chapter 14: p. 348 all, W. H. Breazeale; p. 352, E. R. Degginger; p. 353, W. H. Breazeale; p. 354, Gamma Liaison; p. 355, Leland C. Clark; p. 356, Bruce Coleman, Inc.; p. 362, B. Masini/Phototake; p. 363 left, E. R. Degginger; p. 363 right, Dr. Jeremy Burgess/Science Photo Library/Photo Researchers; p. 364, David, Jacques Louis *The Death of Socrates*, The Metropolitan Museum of Art, Wolfe Fund, 1931. Catharine Lorillard Wolfe Collection (31.45).

Chapter 15: p. 375, National Museum of American History/Smithsonian Institution; p. 378 left, The Granger Collection; p. 378 right, The Granger Collection; p. 380, The Granger Collection; p. 385, NASA/Johnson Space Center; p. 386, NASA/Johnson Space Center; p. 387, NASA/Johnson Space Center; p. 388 both, NASA/Johnson Space Center; p. 389, NASA/Johnson Space Center; p. 390 top left, USGS; p. 390 bottom, USGS; p. 390 right, NASA/Johnson Space Center; p. 391 top, NASA/Johnson Space Center; p. 391 bottom, NASA/Johnson Space Center; p. 392, JPL/NASA; p. 394 left, NASA/Johnson Space Center; p. 394 right, NASA/Johnson Space Center; p. 395, NASA/Johnson Space Center; p. 396 left, NASA/Johnson Space Center; p. 396 right, NASA/Johnson Space Center; p. 397, NASA/Johnson Space Center; p. 398, NASA/Johnson Space Center; p. 399 left, NASA/Johnson Space Center; p. 399 right, NASA/Johnson Space Center; p. 400, NASA/Johnson Space Center; p. 401, NASA/Johnson Space Center; p. 403, JPL/NASA; p. 404, Meteor Crater Enterprises; p. 405 left, NASA/Johnson Space Center; p. 406, Mount Wilson and Las Campanas Observatories/Carnegie Institution of Washington; p. 407, Photo Researchers; p. 409, JPL/NASA.

Chapter 16: p. 421, National Institute of Standards and Technology, Boulder Laboratories, U.S. Department of Commerce; p. 423, M. Granitsas/The Image Works; p. 435, Art Resource, New York.

Chapter 17: p. 441, NASA/Science Source/Photo Researchers; p. 442, Courtesy of Clementine Science Team/Naval Research Lab/USGS; p. 443, Courtesy of Clementine Science Team/Naval Research Lab/USGS; p. 444, NASA/Science Source/Photo Researchers; pp. 448–449, Lick Observatory, University of California, Santa Cruz; p. 452, Tom Pantages; p. 453, Tom Pantages.

Chapter 18: p. 464, Hale Observatories; p. 465, NASA/Johnson Space Center; p. 466, Brookhaven National Laboratory; p. 468, Anglo-Australian Telescope Board; p. 473, Palomar Observatory/California Institute of Technology; p. 474 left, Hale Observatories; p. 474 right, U.S. Naval Observatory/NASA; p. 476, NRAO/AUI/Science Photo Library; p. 477, National Optical Astronomy Observatories; p. 480 top left, National Optical Astronomy Observatories; p. 480 top right, National Optical Astronomy Observatories; p. 480 bottom, National Optical Astronomy Observatories; p. 481 top left, Palomar Observatory; p. 481 top right, National Optical Astronomy Observatories; p. 481 bottom, Hale Observatories/Photo Researchers; pp. 482–483, Lund Observatory; p. 486, Courtesy of Harvard-Smithsonian Center for Astrophysics; p. 487, Palomar Observatory; p. 491 top, NASA/Johnson Space Center; p. 491 bottom, Peter French/W. M. Keck Observatory.

Chapter 19: p. 499, Hans Blohm/Masterfile; p. 502, Daryl Benson/Masterfile; p. 505, Tom Ives; p. 507, Jerry Wilson; p. 508, Roy Nelson/Phototake; p. 509, NOAA; p. 512 a, E. R. Degginger; p. 512 b, Tom Ives; p. 512 c, E. R. Degginger; p. 512 d, Wayne Decker/Fundamental Photographs; p. 512 e, E. R. Degginger; p. 512 f, E. R. Degginger; p. 512 g, Craig Aurness/Woodfin Camp & Associates; p. 512 h, Tom Bean; p. 512 i, E. R. Degginger; p. 512 j, E. R. Degginger; p. 512 k, Phil Degginger; p. 516, Tom Bean; p. 520, Tom Ives; p. 521 top, E. R. Degginger; p. 521 bottom, E. R. Degginger; p. 523 both, NOAA; p. 525, Photri/NASA; p. 526 top, NOAA/Photo Researchers; p. 526 bottom, Odyssey/Woodfin Camp & Associates; p. 528, EPA-Documerica; p. 529 both, Courtesy of South

Coast Air Quality Management District; p. 531, Reuters/Bettmann; p. 532, NASA/Johnson Space Center; p. 534, NASA/Science Source/Photo Researchers.

Chapter 20: p. 541, E. R. Degginger; p. 544 all, E. R. Degginger; p. 545 both, Paul Silverman/Fundamental Photographs; p. 546, Chip Clark; p. 550 left, Lee Boltin; p. 550 middle, Lee Boltin; p. 550 right, Breck Kent/Earth Scenes; p. 551 top, Paul Silverman/Fundamental Photographs; p. 551 bottom, Gary Withey/Bruce Coleman, Inc.; p. 553 top left, Gary Withey/Bruce Coleman, Inc.; p. 553 top right, Stephen Trimble; p. 553 bottom, W. H. Hodge/Peter Arnold, Inc.; p. 554 left, Stephen J. Kraseman/Peter Arnold, Inc.; p. 554 right, Runk Schoenberger/Grant Heilman Photography; p. 555 top left, E. R. Degginger; p. 555 top right, Noble Proctor/Photo Researchers, Inc.; p. 555 bottom, Paul Silverman/Fundamental Photographs; p. 557, J. D. Griggs/U.S. Geological Survey; p. 558 left, Boone Morrison/Gamma-Liaison; p. 558 right, R. E. Wilcox/U.S. Geological Survey; p. 559 top, Solarfilma; p. 559 bottom, L. Nielson/Peter Arnold, Inc.; p. 561 top left, H. Cazabon/Phototake; p. 561 top right, M. P. L./Bruce Coleman, Inc.; p. 561 bottom, Frank S. Balthis; p. 563, Peter Turnley/Black Star; p. 564, Patrick Robert/Sygma; p. 567 both, U. S. Geological Survey.

Chapter 21: p. 579, Marie Tharp; p. 580, Ocean Drilling Program.

Glossary

Absolute (atomic) time scale the time of past geologic events based on the radioactive decay of certain atomic nuclei. (p. 586)

Absolute humidity the amount of water vapor in a specific volume of air. The amount is measured in grain/ft^3 or grams/m^3. (p. 514)

Absolute zero the lowest possible temperature, about $-273°C$ ($-460°F$), which, according to the third law of thermodynamics, can never be reached. (p. 92)

Abyssal plain large flat areas of the deep ocean floor that are most common near the continents.

Acceleration the change in velocity divided by the change in time: $a = \Delta v/\Delta t$. (p. 31)

Acceleration due to gravity usually given as the symbol g; equal to 32 ft/s^2, 980 cm/s^2, or 9.8 m/s^2. (p. 33)

Acid a substance that gives hydrogen ions (or hydronium ions) in water (Arrhenius definition). (p. 327)

Acid-base reaction the H$^+$ of the acid unites with the OH$^-$ of the base to form water, while the cation of the base combines with the anion of the acid to form a salt. (p. 329)

Acid-carbonate reaction an acid and a carbonate (or hydrogen carbonate) react to give carbon dioxide, water, and a salt. (p. 330)

Acid rain rain that has a relatively low pH (acidic) due to air pollution. (p. 529)

Activation energy the energy necessary to start a chemical reaction; a measure of the minimum kinetic energy colliding molecules must possess in order to react. (p. 321)

Activity series a list of elements in order of relative ability of their atoms to be oxidized in solution. (p. 333)

Addition polymers those formed when molecules of an alkene monomer add to one another. (p. 362)

Air current vertical air movement. (p. 508)

Air mass a mass of air with physical characteristics that distinguish it from other air. (p. 517)

Albedo the reflectivity or average fraction of light a body reflects. (p. 374)

Alchemy a pseudoscience that flourished from approximately A.D. 500 to 1600; its main goals were to change common metals into gold and to find an "elixir of life." (p. 288)

Alcohols organic compounds containing a hydroxyl group, –OH, attached to an alkyl group; general formula, R–OH. (p. 355)

Aliphatic hydrocarbon a carbon-hydrogen compound that contains no benzene rings. (p. 347)

Alkali metals the elements in Group 1A of the periodic table, except for hydrogen. (p. 277)

Alkaline earth metals the elements in Group 2A of the periodic table. (p. 279)

Alkanes hydrocarbons that contain only single bonds; general formula C_nH_{2n+2} (p. 347)

Alkenes hydrocarbons that have a double bond between two carbon atoms; general formula C_nH_{2n}. (p. 351)

Alkyl group a substituent that contains one less hydrogen atom than the corresponding alkane; general symbol, R. (p. 350)

Alkyl halide an alkane derivative in which one or more of the hydrogen atoms has been replaced by a halogen atom; general formula R–X. (p. 354)

Alkynes hydrocarbons that have a triple bond between two carbon atoms; general formula C_nH_{2n-2}. (p. 352)

Allotropes two or more forms of the same element that have different bonding structures in the same physical phase. (p. 265)

Alpha decay the disintegration of a nucleus into an alpha particle and a nucleus of another element. (p. 228)

Alternating current (ac) electric current in which the electrons periodically alternate or change direction. (p. 174)

Altitude the angle measured from the horizon to a celestial object. (p. 429)

Amides nitrogen-containing organic compounds having the general formula RCONHR'. (p. 360)

Amine a basic (alkaline) organic compound that contains nitrogen; general formula, R–NH$_2$. (p. 357)

Amino acid an organic compound that contains both an amino group and a carboxyl group. (p. 361)

Amino group a substituent of general formula –NH$_2$, which, when attached to an alkyl group, forms an amine. (p. 357)

Ampere (A) the unit of electric current defined as that current which, if maintained in each of two long parallel wires separated by one meter in free space, would produce a magnetic force between the two wires of 2×10^{-7} newton for each meter of length. (p. 167)

Anemometer an instrument used to measure wind speed. (p. 507)

Angular momentum mvr for a mass m going at a speed v in a circle of radius r. (p. 61)

Anions negative ions; so-called because they move toward the anode (the positive electrode) of an electrochemical cell. (p. 294)

Annular eclipse an eclipse of the Sun in which the Moon blocks out all of the Sun except for a ring around the outer edge of the Sun. (p. 453)

Antinode a stationary point of zero displacement in a standing wave. (p. 130)

Apogee the point in its orbit at which a satellite is farthest from Earth's center. (p. 446)

Apparent solar day the elapsed time between two successive crossings of the same meridian by the Sun. (p. 421)

Aromatic hydrocarbon a carbon-hydrogen compound that contains one or more benzene rings. (p. 347)

Asteroids small, irregular, rocky and metallic objects that orbit about 2.8 AU from the Sun. Also known as "minor planets." (p. 403)

Asthenosphere the rocky substratum below the lithosphere that is hot enough to be deformed and capable of internal flow. (p. 574)

Astronomical unit (AU) the mean distance between Earth and the Sun—93,000,000 mi. (p. 347)

Astronomy the scientific study of the universe beyond Earth's atmosphere. (p. 373)

Atmospheric science the overall study of the atmosphere from ground to outer space. (p. 498)

Atomic mass the average mass (in *atomic mass units*, u) of an atom of the element in naturally occurring samples. (p. 225)

Atomic number Z, the number of protons in the nucleus of an atom of that element. (p. 224)

Atomic time scale a geologic time scale based on radioactive dating. (p. 586)

Aurora borealis displays of light in the northern upper atmosphere commonly called the "northern lights." (p. 501)

Autumnal equinox the time (near September 21) when the Sun's declination crosses the equator moving south (for the Northern Hemisphere). (p. 428)

Avogadro's number 6.02×10^{23}, symbolized N_A; the number of entities in a mole. (p. 336)

Barometer a device used to measure atmospheric pressure. There are aneroid (without fluid) barometers and mercury barometers. (p. 504)

Base a substance that produces hydroxide ions in water (Arrhenius definition). (p. 328)

Batholith a large, intrusive igneous rock formation that has an area of at least 40 mi^2. (p. 549)

Bedding the stratification of sedimentary rock formations. (p. 551)

Bergeron process the process by which precipitation is formed in clouds. (p. 517)

Beta decay the disintegration of a nucleus into a beta particle and a nucleus of another element. (p. 229)

Big Bang the event that cosmologists consider the beginning of the universe. (p. 487)

Black hole a very dense collapsed star from which no light can escape. (p. 478)

Blizzard a snow accompanied by high winds that whip the snow into blinding swirls and drifts. (p. 521)

British system the system of measurement used in the United States, which uses the foot, pound, second, and coulomb as the standards of length, weight, time, and electric charge, respectively. (p. 5)

British thermal unit (Btu) the amount of heat required to raise one pound of water one degree Fahrenheit at normal atmospheric pressure. (p. 96)

Caldera a roughly circular, steep-walled depression formed as a result of the collapse of a volcanic chamber. (p. 561)

Calorie the amount of heat necessary to raise one gram of pure liquid water one degree Celsius at normal atmospheric pressure. (p. 96)

Carbohydrates organic compounds that contain multiple hydroxyl groups in their molecular structure. A basic component of living matter. (p. 356)

Carbon dating a procedure used to establish the age of ancient organic remains by measuring the relative amount of ^{14}C and comparing it to that of present-day organic remains. (p. 235)

Carboxyl group a substituent of general formula –COOH, which, when attached to an alkyl group, forms a carboxylic acid. (p. 357)

Carboxylic acids a class of organic compounds characterized by the presence of a carboxyl group; general formula, RCOOH. (p. 357)

Carcinogen a cancer-causing agent. (p. 352)

Cartesian coordinate system a two-dimensional coordinate system in which two number lines are drawn perpendicular to each other and the origin assigned at the point of intersection. (p. 418)

Catalyst a substance that increases the rate of reaction but is not itself consumed in the reaction. (p. 325)

Cations positive ions; so-called because they move toward the cathode (the negative electrode) of an electrochemical cell. (p. 294)

Celestial prime meridian an imaginary half-circle running from the North Celestial Pole to the South Celestial Pole and crossing perpendicular to the celestial equator at the point of the vernal equinox. (p. 467)

Celestial sphere the imaginary sphere on which all the stars seem to appear. (p. 466)

Celsius scale a temperature scale with 0°C as the ice point and 100°C as the steam point. (p. 92)

Centi- prefix that means 1/100 or one-hundredth. (p. 17)

Centripetal acceleration the "center-seeking" acceleration of an object in uniform circular motion, $a_c = v^2/r$. (p. 51)

Centripetal force a "center-seeking" force that causes an object to travel in a circle. (p. 52)

Cepheid variables stars that vary in magnitude with a fixed period of between 1 and 100 days. (p. 472)

CFCs chlorofluorocarbons, which are used in air conditioners, refrigerators, heat pumps, and so forth, and which deplete the ozone layer. (p. 354)

Cgs system a metric system, used throughout most of the world, which has the centimeter, gram, second, and coulomb as the standard units of length, mass, time, and electric charge, respectively. (p. 7)

Chain reaction occurs when each fission event causes at least one more fission event. (p. 243)

Characteristic (natural) frequencies the frequencies at which an object will oscillate in resonance. (p. 131)

Chemical properties characteristics that describe the transformation of one substance into another (the chemical reactivity). (p. 316)

Chemical reactions changes that alter the chemical composition of a substance. (p. 316)

Chemical weathering a change in a rock's composition due to a chemical change.

Chemistry the division of physical science that deals with the composition and structure of matter, and the reactions by which substances are changed into other substances. (p. 259)

Chromosphere an outer layer of the Sun, which lies just outside the photosphere. (p. 464)

Cinder cone volcano a volcano with a steeply sloped cinder cone formed by eruptions of pyroclastic debris. (p. 561)

Climate the long-term average weather conditions of a region of the world. (p. 530)

Cloud buoyant masses of visible droplets of water vapor and ice crystals in the lower troposphere. Clouds are classified in families of high clouds, middle clouds, low clouds, and clouds of vertical development. (p. 511)

Coalescence the combining of small droplets of water vapor to make larger drops. (p. 517)

Coefficients the numbers in front of the formulas in chemical equations; they designate the molar ratio in which the substances react. (p. 319)

Cold front the boundary of an advancing cold mass over a warmer surface. (p. 519)

Combination reaction one in which at least two reactants combine to form just one product: $A + B \rightarrow AB$. (p. 320)

Combustion reaction the reaction of a substance with oxygen to produce an oxide, along with heat and light in the form of fire. (p. 322)

Comet a small celestial object composed of ice and dust that revolves about the Sun in a highly elliptical orbit and displays a long tail as it passes near the Sun. (p. 404)

Compound a substance composed of two or more elements chemically combined in a definite, fixed proportion by mass. (p. 260)

Concave (spherical) mirror a mirror shaped like the inside of a small section of a sphere. (p. 152)

Condensation polymers those constructed from molecules that have two or more reactive groups. Each molecule attaches to two others by ester or amide linkages. (p. 363)

Condensation theory a process of solar system formation in which interstellar dust grains act as condensation nuclei. (p. 408)

Conduction (thermal) the transfer of heat energy by molecular collisions. (p. 106)

Conductor (electrical) a material that easily conducts an electric current because some electrons in the material are free to move. (p. 166)

Conservation of angular momentum the angular momentum ($L = mvr$) of an object remains constant if there is no external, unbalanced torque acting on it. (p. 62)

Conservation of linear momentum the total linear momentum of an isolated system remains the same if no external, unbalanced force acts on the system. (p. 59)

Conservation of mass, law of No detectable change in the total mass occurs during a chemical reaction. (p. 286)

Conservation of mechanical energy in a conservative system, the sum of the kinetic and potential energies remains constant. (p. 79)

Conservation of total energy the total energy of an isolated system remains constant. (p. 79)

Continental drift the theory that continents move, drifting apart or together. (p. 575)

Continental shelf the shallowly submerged part of the continental margin that extends from the shoreline to the continental slope.

Convection the transfer of heat through the movement of a substance. (p. 107)

Convection cycle the cyclic movement of matter—e.g., air—due to localized heating and convectional heat transfer. (p. 509)

Convergent boundary a region where moving plates of the lithosphere are driven together, causing one of the plates to be consumed into the mantle as it descends beneath an overriding plate. (p. 581)

Converging (convex) lens a lens that is thicker at its center than at its edge. (p. 157)

Conversion factor an equivalence statement expressed as a ratio. (p. 14)

Convex (diverging) mirror a mirror with a reflecting surface on the outside of a spherical section, commonly called a diverging mirror. (p. 152)

Core the innermost region of Earth, which is composed of two parts—a solid inner core and a molten, highly viscous, "liquid" outer core. (p. 573)

Coriolis force a pseudoforce arising in an accelerated reference frame on the rotating (accelerating) Earth. The apparent deflection of an object is attributed to the Coriolis force. (p. 509)

Correlation establishing the equivalence of rocks in separate regions; correlation by fossils is an example. (p. 585)

Cosmic background radiation the microwave radiation that fills all space and is believed to be the redshifted glow from the Big Bang. (p. 489)

Cosmological principle on the large scale, the universe is both homogeneous and isotropic. (p. 485)

Cosmological red shift the shift toward longer wavelengths caused by the expansion of the universe. (p. 487)

Cosmology the study of the structure and evolution of the universe. (p. 487)

Coulomb (C) the unit of electric charge equal to one ampere-second (A-s). (p. 166)

Coulomb's law the force of attraction or repulsion between two charged bodies is directly proportional to the product of the two charges and inversely proportional to the square of the distance between them. (p. 167)

Covalent bond the force of attraction caused by a pair of electrons being shared by two atoms. (p. 298)

Covalent compounds those in which the atoms share pairs of electrons to form molecules. (p. 298)

Craters (lunar) large- and small-diameter depressions in the surface of the Moon. (p. 422)

Crescent moon the Moon viewed when less than one-quarter of the illuminated surface is facing an observer on Earth. (p. 447)

Critical mass the minimum amount of fissionable material necessary to sustain a chain reaction. (p. 243)

Crust the thin outer layer of Earth. (p. 573)

Curie temperature a high temperature, above which a ferromagnetic material ceases to be magnetic. (p. 185)

Current rate of flow of electric charge. (p. 166)

Cycloalkanes hydrocarbons that have the general molecular formula C_nH_{2n} and possess rings of carbon atoms, with each carbon atom bonded to a total of four carbon or hydrogen atoms. (p. 351)

Daylight Saving Time time advanced one hour from standard time, adopted during the spring and summer months to take advantage of longer evening daylight hours. (p. 426)

Decibel (dB) a unit of sound level intensity. (p. 124)

Declination the angular measure north or south of the celestial equator, measured in degrees. (p. 467)

Decomposition reaction one in which only one reactant is present and breaks into two (or more) products: $AB \rightarrow A + B$. (p. 320)

Definite proportions, law of different samples of a pure compound always contain the same elements in the same proportion by mass. (p. 287)

Delta the accumulation of sediment formed where running water enters a large body of water such as a lake or ocean.

Density a measure of the compactness of matter using a ratio of mass to volume, $\rho = m/V$. (p. 12)

Derived quantities combinations of fundamental quantities. (p. 10)

Destructive interference wave interference in which the combined amplitude is less than that of either wave. (p. 147)

Dew point the temperature at which a sample of air becomes saturated—i.e., has a relative humidity of 100 percent. (p. 514)

Diffraction the bending of waves when an opening or obstacle has a size smaller than or equal to the wavelength. (p. 143)

Diffraction grating an optical instrument, consisting of many narrow, parallel slits spaced very closely, that separates light into a spectrum. (p. 149)

Dike a discordant pluton formation that is formed when magma fills a nearly vertical fracture in rock layers. (p. 550)

Direct current (dc) electric current in which the electrons flow directionally from the negative (−) terminal toward the positive (+) terminal. (p. 174)

Dispersion the "spreading" of light that occurs when different frequencies are refracted at slightly different angles. (p. 142)

Displacement the directional, straight-line distance between two points. (p. 24)

Distance the actual path length between two points. (p. 24)

Divergent boundary a region where plates of the lithosphere are moving away from one another. (p. 581)

Diverging (concave) lens a lens that is thinner at its center than at its edge. (p. 157)

Doppler effect an apparent change in frequency resulting from the relative motion of the source and the observer. (p. 128)

Doppler radar radar that uses the Doppler effect on water droplets in clouds to measure the wind speed and direction. (p. 522)

Doppler red shift the frequency shift, primarily referred to light, from receding objects toward a longer wavelength or the red end of the visible spectrum. (p. 128)

Double-replacement reactions ones that take the form: $AB + CD \rightarrow AD + CB$, in which the positive and negative components of the two compounds "change partners." (p. 331)

Drug a compound that can produce a physiological change in human beings or animals. (p. 364)

Dual nature of light light apparently sometimes acts as a wave and sometimes as a particle in various natural phenomena. (p. 200)

Eclipse an occurrence in which one celestial object is partially or totally blocked from view by another. (p. 452)

Ecliptic the apparent annual path of the Sun on the celestial sphere. (p. 374)

Electric charge a fundamental property of matter that can be either positive or negative and gives rise to electrical forces. (p. 166)

Electric potential energy the potential energy that results from work done in separating electric charges. (p. 171)

Electrochemistry the study of chemical reactions that involve the consumption or production of electric current. (p. 337)

Electrolysis a chemical reaction caused by the direct current from a battery or other source. (p. 337)

Electromagnetic spectrum an ordered arrangement of various frequencies or wavelengths of electromagnetic radiation. (p. 122)

Electromagnetic wave a wave caused by oscillations of electric and magnetic fields. (p. 121)

Electromagnetism the interaction of electric and magnetic effects. (p. 186)

Electron an elementary subatomic particle, with a very small mass of 9.11×10^{-31} kilograms and a negative charge of 1.602×10^{-19} coulombs, that orbits the atomic nucleus. (p. 166)

Electronegativity a measure of the ability of an atom to attract electrons in the presence of another atom. (p. 303)

Electron period a set of energy levels, all of which have approximately the same energy. (p. 219)

Electron shell consists of all electrons with the same principal quantum number (n). (p. 216)

Electron subshell consists of all electrons with the same n and l values. (p. 216)

Element a substance in which all the atoms have the same number of protons, or same atomic number. (p. 224)

Endothermic reaction one that absorbs energy from the surroundings. (p. 321)

Energy the capacity to do work. (p. 75)

Entropy a measure of the disorder of a system. (p. 105)

Epicenter the point on the surface of Earth directly above the focus of an earthquake. (p. 564)

Epoch an interval of geologic time that is a subdivision of a period. (p. 585)

Equilibrium in chemistry, a dynamic process in which the reactants are combining to form the products at the same rate that the products are combining to form the reactants. (p. 317)

Era an interval of geologic time made up of periods and epochs. (p. 585)

Ester an organic compound that has the general formula $RCOOR'$. (p. 358)

Event horizon the position in space at which the escape velocity from a black hole equals the speed of light. (p. 477)

Excess reactant a starting material that is only partially used in a chemical reaction. (p. 290)

Excited state a state of the atom with energies above the ground state; *see* Ground state. (p. 203)

Exothermic reaction one that releases energy to the surroundings. (p. 321)

Fahrenheit scale a temperature scale with 32°F as the ice point and 212°F as the steam point. (p. 92)

Fats esters composed of the trialcohol named glycerol, $C_3H_5(OH)_3$, and long-chain carboxylic acids known as *fatty acids*. A basic component of living matter. (p. 359)

Fault a fracture along which a relative displacement of the sides has occurred. (p. 562)

Ferromagnetic materials materials that are readily magnetized, including iron, nickel, and cobalt. (p. 183)

First harmonic the same as the fundamental frequency ($n=1$) of a standing wave. (p. 131)

First-quarter moon when the Moon is 90° east of the Sun and appears as a quarter moon on the observer's meridian at 6 P.M., local solar time. (p. 447)

Fission a process in which a large nucleus is "split" into two intermediate-size nuclei, with the emission of neutrons and the conversion of mass into energy. (p. 241)

Fluorescence a process in which excited electrons return to the original state in two or more transitions, thereby emitting light photons of lower frequencies than that of the exciting photons. (p. 210)

Focal length the distance from the vertex of a mirror or lens to the focal point. (pp. 151, 158)

Foot-pound (ft-lb) the unit of work in the British system. (p. 70)

Force any quantity capable of producing motion. (p. 44)

Force, unbalanced a net or resultant force of two or more forces. (p. 44)

Formula mass the sum of the atomic masses given in the formula of the compound or element. (p. 287)

Fossils the remains or traces of organisms preserved from the geologic past. (p. 552)

Foucault pendulum a pendulum used to demonstrate the rotation of Earth. (p. 374)

Frequency the number of oscillations of a wave in a given period of time. (p. 119)

Friction (drag) the cause of retardation or deflection of air movements because of the interaction of air molecules among themselves and terrestrial surfaces.

Front the boundary between two air masses. (p. 519)

Full moon occurs for an instant when the Moon is 180° east of the Sun and appears on the observer's meridian at 12 midnight, local solar time. (p. 448)

Fundamental frequency the lowest characteristic or natural frequency of an oscillator. (p. 131)

Fundamental quantities physical quantities that serve as the basis for other physical concepts; length, mass, and time are examples. (p. 3)

Fusion a process in which smaller nuclei are fused (joined) to form larger ones, with the release of energy. (p. 245)

Galaxy a large-scale aggregate of stars plus some gas and dust, held together gravitationally. They have a spiral, elliptical, or irregular structure, and contain, on the average, one hundred billion solar masses. (p. 479)

Gamma decay event in which a nucleus emits a gamma ray and becomes a less energetic form of the same nucleus. (p. 229)

Gas matter that has no definite volume or shape. (p. 112)

Generator a device that converts mechanical work or energy into electrical energy. (p. 190)

Genetic effects defects in the subsequent offspring of recipients of radiation. (p. 251)

Geocentric theory the old false theory of the solar system, which placed Earth at the center of the universe. (p. 378)

Geologic time scale a relative time scale based on the fossil index of rock strata. (p. 588)

Geology the study of Earth, its processes, and its history. (p. 539)

Glacier a thick mass of ice that moves slowly on a land surface.

Gram (g) a unit of mass in the metric cgs system of units. One gram is the mass of one cubic centimeter of pure water at its maximum density (4°C). (p. 8)

Gravitational collapse the collapse of a very massive body because of its attraction for itself. (p. 477)

Great impact theory (Moon's origin) a planet-size object collided with Earth, and the impact ejected enough matter (most of it coming from Earth's mantle) into orbit to form the Moon. (p. 445)

Greenhouse effect the selective absorption process of certain atmospheric gases that regulate Earth's average temperature. (p. 506)

Greenwich meridian the reference meridian of longitude, which passes through the old Royal Greenwich Observatory near London. (p. 419)

Ground state the lowest energy level of an atom. (p. 203)

Groups the vertical columns in the periodic table. (p. 266)

Guyot a submerged seamount with a flat top.

Half-life the time it takes for the decay of half the nuclei in a sample of a given radionuclide. (p. 233)

Halogens the elements in Group 7A of the periodic table. (p. 278)

Hard water water with a high content of dissolved calcium, magnesium, and iron salts.

Heat a form of energy; energy in transit. (p. 96)

Heat engine a device that uses heat energy to perform useful work. (p. 102)

Heat lightning lightning that occurs below the horizon or behind the cloud, which illuminates the cloud with flickering flashes of light. (p. 520)

Heat pump a device used to transfer heat from a low-temperature reservoir to a high-temperature reservoir. (p. 104)

Heisenberg's uncertainty principle it is impossible to know simultaneously the exact velocity and position of a particle. (p. 214)

Heliocentric theory the current theory of the solar system, which places the Sun at the center of the solar system. (p. 378)

Hertz (Hz) one cycle per second. (p. 119)

Horsepower (hp) a unit of power, 550 ft-lb/s. (p. 73)

Hot-spot theory the theory that explains central-plate volcanic chains as being formed as a result of plate movement over a hot spot beneath it. (p. 560)

H-R diagram a plot of the absolute magnitude versus the temperature of stars. (p. 471)

Hubble's law the recessional speed of a distant galaxy is proportional to its distance away. (p. 484)

Humidity a measure of water vapor in the air. (p. 514)

Hurricane a tropical storm with winds of 74 mi/h or greater. (p. 523)

Hydrocarbons organic compounds that contain only carbon and hydrogen. (p. 347)

Hydrogen bond the dipole-dipole force between a hydrogen atom in one molecule and a nearby oxygen, nitrogen, or fluorine atom in the same or a neighboring molecule. (p. 306)

Hydronium ion H_3O^+; a hydrogen ion bonded to a water molecule. (p. 327)

Hydroxyl group a substituent with the formula –OH. When attached to an alkyl group, an alcohol is formed, R–OH. (p. 355)

Hygroscopic nuclei particulate matter that acts as nuclei in the condensation process. (p. 516)

Ice point the temperature of a mixture of ice and air-saturated water at normal atmospheric pressure. (p. 516)

Ice storm a storm with accumulations of ice as a result of the surface temperature being below the freezing point. (p. 521)

Ideal efficiency the ideal or upper limit of heat engine efficiency that can only be approached and never achieved. (p. 103)

Igneous rock rock formed by the cooling and solidification of hot, molten material. (p. 547)

Impulse the product of the force acting on an object and the time during which the force acts. Impulse is equal to the change in momentum. (p. 61)

Index fossil a fossil that is related to a specific span of geologic time. (p. 585)

Index of refraction the ratio of the speed of light in a vacuum and the speed of light in a medium. (p. 141)

Inertia the tendency of an object to resist changes in motion. (p. 45)

Inner transition elements The *lanthanides* and *actinides*, the two rows at the bottom of the periodic table, make up the inner transition elements. (p. 269)

Insolation the solar radiation received by Earth and its atmosphere—*in*coming *sol*ar radi*ation*. (p. 502)

Insulator (electrical) a material that does not readily conduct an electric current or heat because the electrons in the material are not free to move. (p. 167)

Intensity (of sound wave) the rate of energy transfer through a given area, with units of W/m^2. (p. 124)

Interference, constructive a superposition of waves for which the combined wave form has a greater amplitude. (p. 147)

Interference, destructive a superposition of waves for which the combined wave form has a smaller amplitude. (p. 147)

International Date Line (IDL) the meridian that is 180° E or W of the prime meridian. (p. 424)

Interplanetary dust very small particles known as micrometeroids that occupy interplanetary space. (p. 407)

Ion an atom, or chemical combination of atoms, having a net electric charge. (p. 224)

Ionic bonds electrical forces that hold the ions together in the crystal lattice of an ionic compound. (p. 296)

Ionic compounds compounds formed by an electron transfer process, in which one or more atoms lose electrons and one or more other atoms gain them to form ions. (p. 292)

Ionization energy the amount of energy it takes to remove an electron from an atom. (p. 272)

Ionosphere the upper region of the atmosphere containing ionic layers. (p. 501)

Irregular (diffuse) reflection reflection from a rough surface in which the reflected rays are not parallel. (p. 138)

Isobar a line showing the locations of constant pressure.

Isotopes forms of atoms of an element that have the same number of protons but differ in their number of neutrons. (p. 225)

Jet streams fast-moving "rivers" of air in the upper troposphere. (p. 511)

Joule (J) a unit of energy. 1 N-m or $kg\text{-}m^2/s^2$. (p. 70)

Joule heat (I^2R losses) the energy dissipated because of electrical resistance. (p. 173)

Jovian planets the four large outer planets—Saturn, Jupiter, Uranus, and Neptune. All have characteristics resembling Jupiter. (p. 381)

Kelvin (K) the unit on the Kelvin temperature scale. (p. 92)

Kelvin scale the "absolute" temperature scale that takes absolute zero as 0 K. (p. 92)

Kepler's harmonic law The ratio of the square of the period to the cube of the semimajor axis (one-half the larger axis of an ellipse) is the same for all the planets. (p. 379)

Kepler's law of elliptical paths All planets (asteroids, comets, etc.) revolve around the Sun in elliptical orbits. (p. 379)

Kepler's law of equal areas As a planet (or asteroid or comet) revolves around the Sun, an imaginary line joining the planet to the Sun sweeps out equal areas in equal periods of time. (p. 379)

Kilo- prefix that means 10^3 or one thousand. (p. 17)

Kilocalorie the amount of heat necessary to raise 1 kg of water 1°C. (p. 96)

Kilogram (kg) the unit of mass in the mks system; one kilogram has an equivalent weight of 2.2 pounds at Earth's surface. (p. 8)

Kinetic energy energy of motion equal to $\frac{1}{2}mv^2$. (p. 75)

Laccolith a concordant pluton formation formed from a blisterlike intrusion that has pushed up the overlying rock layers. (p. 550)

Land breeze a local wind from land to sea resulting from a convection cycle. (p. 509)

Lapse rate the rate of temperature change with altitude; in the troposphere the normal lapse rate is −3.5 F° per 1000 ft. (p. 516)

Laser an acronym for *l*ight *a*mplification by *s*timulated *e*mission of *r*adiation; it is coherent, monochromatic light. (p. 206)

Last-quarter moon occurs for an instant when the Moon is 270° east of the Sun, and appears on the observer's meridian at 6 A.M. local solar time, with the illuminated side of the Moon toward the east. (p. 448)

Latent heat of fusion the amount of heat required to change a unit mass of a substance from the solid to the liquid phase at the same temperature. (p. 99)

Latent heat of vaporization the amount of heat required to change a unit mass of a substance from the liquid to the gas phase at the same temperature. (p. 99)

Latitude for a point on the surface of Earth, the angular measurement, in degrees, north or south of the equator. (p. 419)

Lava magma that reaches Earth's surface through a volcanic vent. (p. 547)

Law of charges Like electrical charges repel and unlike electrical charges attract. (p. 167)

Law of poles Like magnetic poles repel and unlike magnetic poles attract. (p. 181)

Law of reflection The angle of incidence is equal to the angle of reflection, $\theta_i = \theta_r$. (p. 138)

Le Châtelier's principle Whenever the conditions of a system at equilibrium (temperature, pressure, concentration) are changed, the system will shift in the direction that counteracts the change. (p. 317)

Length the measurement of space in any direction. (p. 3)

Lewis structures Lewis symbols used to show the valence electrons in molecules and ions of compounds. (p. 293)

Lewis symbol the element's symbol represents the nucleus and inner electrons of an atom, and the valence electrons are shown as dots arranged in four groups of one or two dots around the symbol. (p. 293)

Lightning an atmospheric electrical discharge. (p. 519)

Light-year (ly) the distance light travels in one year. (p. 467)

Limiting reactant a starting material that is used completely in a chemical reaction. (p. 290)

Linearly (plane) polarized transverse waves that vibrate in only one direction. (p. 150)

Liquid matter that has a definite volume but no definite shape. (p. 109)

Liter (L) a metric unit of volume or capacity; 1 L = 1000 cm^3. (p. 9)

Lithification the process of forming sedimentary rock from sediment; also called consolidation. (p. 553)

Lithosphere the outermost solid portion of Earth, which includes the crust and part of the upper mantle. (p. 574)

Local group the cluster of galaxies that includes our own Milky Way galaxy. (p. 481)

Longitude for a point on the surface of Earth, an angular measurement, in degrees, east or west of the prime meridian. (p. 419)

Longitudinal wave a wave in which the particle motion is parallel to the direction of the wave velocity. (p. 119)

Longshore currents a flow of water close and parallel to the shore.

Loudness a qualitative measure of sound intensity. (p. 123)

Lunar eclipse an eclipse of the Moon caused by Earth's blocking of the Sun's rays to the Moon. (p. 453)

Magma hot, molten rock material. (p. 547)

Magnetic anomalies adjacent regions of rocks with remanent magnetism of opposite polarities, that is, the directions of the magnetism are reversed. (p. 579)

Magnetic declination the angular variation of a compass from geographic north. (p. 185)

Magnetic domains regions of local atomic alignment in ferromagnetic materials that give rise to magnetism. (p. 183)

Magnetic field a force field characterized by a set of imaginary lines that indicate the direction a small compass needle would point if it were placed at a particular spot. (p. 181)

Main sequence a narrow band on the H-R diagram on which most stars fall. (p. 471)

Mantle the interior region of Earth between the core and the crust. (p. 573)

Mass a quantity of matter and a measurement of inertia. (p. 45)

Mass number the number of protons plus neutrons in a nucleus; the total number of nucleons. (p. 225)

Mass wasting the downslope movement of overburden under the influence of gravity.

Matter (de Broglie) waves the waves produced by moving particles. (p. 211)

Mean solar day the average length of a solar day. One solar day is the elapsed time between two successive crossings of the same meridian by the Sun. (p. 241)

Mechanical energy the sum of the kinetic and potential energies, $E_k + E_p$. (p. 75)

Mercalli scale a scale of earthquake severity based on the physical effects produced by an earthquake. (p. 565)

Meridians imaginary lines along the surface of Earth running from the north geographic pole, perpendicular to the equator, to the south geographic pole. (p. 419)

Mesosphere a region of the atmosphere, based on temperature, that lies between approximately 35 and 60 miles in altitude. (p. 500)

Metal an element whose atoms tend to lose valence electrons during chemical reactions. (p. 270)

Metallic bonding the electrical attractions among the spherical, positively charged, metal ions in the crystal lattice and the "sea" of mobile valence electrons surrounding them. (p. 305)

Metamorphic rock rock that results from a change, or metamorphism, in pre-existing rock because of heat and pressure. (p. 547)

Meteor a small chunk of matter that burns up as it flies through Earth's atmosphere and appears to be a shooting star. (p. 404)

Meteorites chunks of matter from the solar system that fall through the atmosphere and strike Earth's surface. (p. 404)

Meteoroids small, interplanetary objects in space before they encounter Earth. (p. 404)

Meteorology the study of atmospheric phenomena. (p. 498)

Meter (m) the standard of length in the mks system; it is equal to 39.37 inches, or 3.28 feet. (p. 7)

Metric system a decimal measurement system used predominantly throughout the world. (p. 6)

Milky Way the name of our galaxy. (p. 480)

Milli- prefix that means 10^{-3} or one one-thousandth. (p. 17)

Mineral any naturally occurring, inorganic, crystalline substance. (p. 540)

Mixture a type of matter composed of varying proportions of two or more substances that are just physically mixed, *not* chemically combined. (p. 260)

Mks system the metric system that has the meter, kilogram, second, and coulomb as the standard units of length, mass, time, and electric charge, respectively. (p. 7)

Mohs' scale a 10-point scale of hardness based on mineral standards, diamond being the hardest and talc being the softest. (p. 543)

Mole the quantity of a substance that contains 6.02×10^{23} formula units (the number of atoms in exactly 12 g of carbon-12). (p. 336)

Molecule an electrically neutral particle composed of two or more atoms chemically combined. (p. 264)

Momentum, angular the product of mass, velocity, and lever arm, $L = mvr$. (p. 61)

Momentum, linear the product of mass and velocity, $p = mv$. (p. 59)

Monomer a fundamental repeating unit of a polymer. (p. 362)

Moraines ridges of till near the end and sides of a glacier.

Motion the changing of position. (p. 24)

Motor a device that converts electrical energy into mechanical energy. (p. 188)

Neap tide moderate tides with the least variation between high and low. (p. 456)

Neutrino a subatomic particle that has no rest mass or electric charge, but does possess energy and momentum. (p. 465)

Neutron an elementary neutral particle found in the nuclei of atoms. (p. 166)

Neutron number N, the number of neutrons in the nucleus of an atom. (p. 225)

Neutron star an extremely high-density star composed almost entirely of neutrons. (p. 476)

New moon occurs for an instant when Earth, Moon, and Sun are in the same plane with the Moon positioned between Earth and Sun. New moon appears on the observer's meridian at 12 noon, local solar time. (p. 446)

Newton (N) a unit of force, 1 kg-m/s^2. (p. 48)

Newton's first law of motion A body moves at a constant velocity unless acted upon by an external unbalanced force. (p. 45)

Newton's law of universal gravitation $F = Gm_1m_2/r^2$. (p. 54)

Newton's second law of motion The acceleration of an object is equal to the unbalanced force on the object divided by the mass of the object ($a = F/m$). (p. 48)

Newton's third law of motion Whenever one mass exerts a force upon a second mass, the second mass exerts an equal and opposite force upon the first mass. (p. 57)

Nitrogen oxides (NO$_x$) chemical combinations of nitrogen and oxygen, such as NO and NO$_2$. (p. 528)

Noble gases the elements of Group 8A of the periodic table. (p. 276)

Node a position of zero displacement in a standing wave. (p. 130)

Nonmetal an element whose atoms tend to gain (or share) valence electrons during chemical reactions. (p. 270)

Nucleons a collective term for neutrons and protons (particles in the nucleus). (p. 223)

Nucleosynthesis the creation of the nuclei of elements inside stars. (p. 264)

Nucleus the central core of an atom; composed of protons and neutrons. (p. 223)

Nuclide a species of nucleus, characterized by specifying the atomic number and mass number. (p. 228)

Occluded front a frontal boundary where one front has gone under another, causing it to be raised or occluded from the surface. (p. 519)

Octet rule In forming compounds, atoms tend to gain, lose, or share electrons to achieve electron configurations of the noble gases. (p. 292)

Ohm (Ω) a unit of electrical resistance. (p. 171)

Ohm's law a relationship between voltage and current, $V = IR$, which holds for many electrical conductors, metals in particular. (p. 172)

Oort comet cloud the cloud of cometary objects believed to be orbiting the Sun at 50,000 AU, and from which observed comets originate. (p. 406)

Organic chemistry the study of compounds that contain carbon. (p. 345)

Overburden the weathered material that accumulates on base rock.

Oxidation occurs when oxygen combines with another substance (or when an atom or ion loses electrons). (p. 332)

Ozone the compound O_3. It is found naturally in the atmosphere in the ozonosphere and is also a constituent of photochemical smog. (p. 354)

Ozonosphere a region of the atmosphere, between approximately 6 and 43 miles in altitude, characterized by ozone concentration. (p. 500)

Pangaea the single giant supercontinent that is believed to have existed over 200 million years ago. (p. 575)

Parallax the apparent motion, or shift, that occurs between two fixed objects when the observer changes position. (p. 376)

Parallel circuit a circuit in which an entering current divides proportionately between the circuit elements. (p. 176)

Parallels imaginary lines representing degrees of latitude, encircling Earth parallel to the plane of the equator. (p. 419)

Parsec the distance to a star when the star exhibits a parallax on one second. This distance is equal to 3.26 light-years or 206,265 astronomical units. (p. 377)

Pauli exclusion principle no two electrons can have the same set of quantum numbers. (p. 217)

Penumbra a region of partial shadow. During an eclipse, an observer in the penumbra sees only a partial eclipse. (p. 452)

Perigee the point in its orbit at which a satellite is closest to Earth's center. (p. 446)

Period an interval of geologic time that is a subdivision of an era and made up of epochs; the time for a complete cycle of motion; a horizontal row of the periodic table—therefore, elements that have approximately the same energy. (p. 585)

Period (wave) the time for one complete particle oscillation or the time for one wavelength to pass a given point. (p. 120)

Periodic law the properties of elements are periodic functions of their atomic numbers. (p. 266)

pH the measure (on a logarithmic scale) of the concentration of hydrogen ion (or hydronium ion) in a solution. (p. 328)

Phases of matter the physical forms of matter; most commonly, solid, liquid, and gas. (p. 108)

Phosphorescence a process in which excited electrons in a material spend relatively long times in metastable states; emissions occur after the exciting source has been removed—commonly described as "glowing in the dark." (p. 210)

Photochemical smog air-pollution conditions resulting from the photochemical reactions of hydrocarbons with oxygen in the air and other pollutants in the presence of sunlight. (p. 529)

Photoelectric effect the process in which electrons are emitted from certain metallic materials when exposed to light. (p. 119)

Photon a "particle" of electromagnetic radiation. (p. 200)

Photosphere the Sun's outer surface, visible to the eye. (p. 463)

Photosynthesis the chlorophyll-catalyzed process by which plants convert CO_2 and H_2O to sugars with the release of oxygen. (p. 499)

Physical weathering the physical disintegration or fracture of rock, primarily as a result of pressure.

Pitch the highness or lowness of a sound. It is a consequence of the frequency of the sound waves received by the ear. (p. 124)

Plasma a gas of electrons and protons or other nuclei. (p. 112)

Plate tectonics the theory that the outer layer of Earth is made up of rigid plates that are in relative motion with respect to one another. (p. 581)

Polar covalent bond one in which the pair of bonding electrons is unequally shared, leading to the bond having a slightly positive end and a slightly negative end. (p. 303)

Polarization the restriction of the electric vector of a wave to preferred directions. (p. 150)

Polar molecule one that has a positive end and a negative end; i.e., has a dipole. (p. 304)

Pollution any atypical contributions to the environment resulting from the activities of human beings. (p. 526)

Polyatomic ion an electrically charged combination of atoms. Table 11.5 lists the common ones. (p. 273)

Polymer a compound of very high molecular mass whose chainlike molecules are made up of repeating units called *monomers*. (p. 362)

Position refers to the location of an object. (p. 23)

Potential energy the energy a body possesses because of its position. (p. 77)

Power work per unit time. (p. 73)

Precession the slow rotation of the axis of spin of Earth around an axis perpendicular to the ecliptic plane. (p. 432)

Precipitate an insoluble solid that appears when two liquids (usually aqueous solutions) are mixed. (p. 332)

Principal quantum number (n) the numbers (1, 2, 3, . . .) used to designate the various principal energy levels that an electron may occupy in an atom. (p. 202)

Products the substances formed during a chemical reaction. (p. 316)

Prograde motion referred to as forward motion—this motion is west to east (eastward) when applied to the motion of the planets. (p. 381)

Proteins extremely long-chain polyamides formed by the enzyme-catalyzed condensation of amino acids. A basic component of living matter. (p. 361)

Proton a positively charged particle found in the nuclei of atoms. (p. 166)

Psychrometer an instrument used to measure relative humidity. (p. 514)

Pulsar a star that emits radio signal pulses regularly and very rapidly. (p. 476)

Pure substance a type of matter (element or compound) in which all samples have fixed composition and identical properties. (p. 256)

P waves primary (P) waves, so called because they reach a seismic station before the S waves. P waves are longitudinal or compressional waves—i.e., their particle oscillations are in the direction of propagation and are transmitted by solids, liquids, and gases. (p. 564)

Quality (sound) the physiological sound of a tone resulting from different waveforms as determined by the number of waves or overtones that make up the wave. (p. 125)

Quantum a discrete amount. (p. 199)

Quantum mechanics (or wave mechanics) a field of physics used to solve problems in which geometrical sizes are comparable to the wavelengths of particles. Schrödinger's equation forms the basis for wave mechanics. (p. 213)

Quantum physics the branch of physics that explains phenomena in terms of energy being in the form of quanta or packets. (p. 198)

Quasar a shortened term for *quas*i-stell*ar* radio source. (p. 486)

Radar an instrument that sends out electromagnetic (radio) waves, monitors the returning wave that is reflected by some object, and thereby locates the object. Radar stands for *Ra*dio *D*etecting *A*nd *R*anging. Radar is used to detect and monitor precipitation and severe storms. (p. 508)

Radiation the transfer of energy by means of electromagnetic waves. (p. 107)

Radioactivity the spontaneous process of a sample of a radionuclide undergoing a change by the emission of particles or rays. (p. 228)

Radiometric dating a general name for dating rocks and organic remains by measurements utilizing the rate of decay of radionuclides they contain. (p. 234)

Radionuclides species of nuclei that undergo radioactive decay. (p. 228)

Rain gauge an open, calibrated container used to measure amounts of precipitation. (p. 508)

Range (projectile) the horizontal distance a projectile travels. (p. 36)

Ray a straight line that represents the path of light. (p. 138)

Ray (lunar) streaks of light-colored material extending outward from craters. (p. 442)

Rayleigh scattering the preferential scattering of light by air molecules and particles that accounts for the blueness of the sky. The scattering is proportional to $1/\lambda^4$. (p. 502)

Reactants the original substances in a chemical reaction. (p. 36)

Red giant a red star that has a diameter much larger than average. (p. 471)

Redox reaction a reaction in which oxidation and reduction occur. (p. 333)

Red shift a Doppler effect caused when a light source moves away from the observer and shifts the light frequencies lower or toward the red end of the electromagnetic spectrum. (p. 128)

Reduction occurs when oxygen is removed from a compound (or when an atom or ion gains electrons). (p. 332)

Reflection the change in the direction of a wave because of its rebound from a surface or boundary. (p. 137)

Refraction the bending of light waves caused by a velocity change as light goes from one medium to another. (p. 137)

Regular (specular) reflection reflection from a smooth surface in which the rays are parallel. (p. 138)

Relative geologic time scale the relative age of rocks, which are placed in their correct sequence. Only the chronologic sequence of events is established. (p. 585)

Relative humidity the ratio of the actual moisture content of a volume of air to its maximum moisture capacity at a given temperature. (p. 507)

Remanent magnetism the magnetism retained in rocks containing ferrite minerals after solidifying in Earth's magnetic field. (p. 579)

Representative elements the elements of Groups 1A through 8A in the periodic table. (p. 267)

Resistance the opposition to the flow of electric charge. (p. 171)

Resonance a wave effect that occurs when an object has a natural frequency that corresponds to an external frequency, resulting in maximum energy transfer. (p. 132)

Retrograde motion referred to as backward motion; this motion is east to west (westward) when applied to the motion of the planets. (p. 381)

Reversible reaction a reaction that may proceed in either direction; denoted by a double reaction arrow, as in $A + B \leftrightharpoons C + D$. (p. 316)

Revolution the movement of one mass around another. (p. 374)

Richter scale a scale of earthquake severity based on the amplitude or intensity of seismic waves. (p. 565)

Right ascension a coordinate for measuring the east-west positions of celestial objects; the angle is measured eastward from the vernal equinox in hours, minutes, and seconds. (p. 467)

Rille a narrow trench on the Moon's surface where lava once flowed. (p. 443)

Ring of Fire the area generally circumscribing the Pacific Ocean that is characterized by volcanic activity. (p. 559)

Roche limit the distance (approximately 2.5 planetary radii) within which differential gravitational forces (tidal forces) are greater than the binding forces holding the orbiting object together. Also known as the tidal stability limit. (p. 395)

Rock any naturally occurring, solid, mineral mass that makes up part of Earth's lithosphere. (p. 545)

Rotation a spinning motion of a body about an internal axis. (p. 374)

Salt ionic compound that contains any cation except H^+ combined with any anion except OH^-. (p. 329)

Saturated solution one having the maximum amount of solute dissolved in the solvent at a given temperature. (p. 514)

Scalar quantity that has a magnitude, but has no direction associated with it. (p. 24)

Schrödinger wave equation an energy-related equation that uses de Broglie waves or wave functions to describe the characteristics of particles. (p. 212)

Scientific method an investigative process which holds that no theory or model of nature is valid unless the results are in accord with experiment. That is, theory must be substantiated by experiment. (p. 3)

Scientific notation (powers of ten) a mathematical notation in which numbers are expressed in terms of powers of ten. (p. 16)

Sea breeze a local wind from the sea as a result of a convection cycle. (p. 509)

Seafloor spreading the theory that the seafloor slowly spreads and moves sideways away from mid-ocean ridges. The spreading is believed to be due to convection cycles of subterranean molten material that causes the formation of the ridges and a surface motion in a lateral direction from the ridges. (p. 578)

Seamounts an isolated undersea volcanic mountain that may extend to heights over 1.6 km above the seafloor.

Second (s) the standard unit of time. It is now defined in terms of the frequency of a certain transition in the cesium atom. (p. 6)

Sedimentary rock rock formed from the consolidation of layers of sediment. (p. 547)

Seismic waves the waves generated by the energy release of an earthquake. (p. 564)

Seismograph an instrument that records the intensity of seismic waves. (p. 565)

Seismology the geophysical science of earthquakes. (p. 562)

Series circuit a circuit in which an entering current flows individually through all the circuit elements. (p. 174)

Shield volcano a volcano with a low, gently sloping profile formed by a fissure eruption of low-viscosity lava. (p. 560)

SI the abbreviation for the International System of Units, which is a modern form of the metric system. (p. 7)

Sidereal day the rotation period of Earth with respect to the vernal equinox; one sidereal day is 23 h 56 min 4.091 s. (p. 421)

Sidereal month the orbital period of the Moon around Earth with respect to a star other than the Sun. (p. 446)

Sidereal period the orbital or rotation period of any object with respect to the stars. (p. 383)

Sidereal year the time interval for Earth to make one complete revolution around the Sun with respect to any particular star other than the Sun. (p. 427)

Silicate any one of numerous minerals that have the oxygen and silicon tetrahedron as their basic structure. (p. 540)

Sill a pluton formation that lies between and parallel to existing rock layers. (p. 550)

Single-replacement reaction a reaction in which one element replaces another that is in a compound: $A + BC \rightarrow B + AC$ (p. 333)

Singularity the center of a black hole. The point to which the entire mass of a star has contracted. (p. 477)

Slope (of a graph) for a straight line graph, the slope is the ratio of corresponding vertical and horizontal intervals, $\Delta y / \Delta x$. (p. 29)

Smog a contraction of *sm*oke-*fog* used to describe the combination of these conditions. (p. 527)

Snowstorm a storm with a heavy accumulation of snow; referred to as a blizzard when accompanied by high winds. (p. 521)

Soap sodium salts of fatty acids that are formed, along with glycerol, when fats are treated with sodium hydroxide. (p. 359)

Solar eclipse an eclipse of the Sun caused by the Moon blocking the Sun's rays to an observer on Earth. (p. 452)

Solar nebula the large swirling volume of gas and dust from which the solar system is presumed to have formed. (p. 408)

Solar system the Sun, nine planets and their satellites, the asteroids, comets, meteoroids, and interplanetary dust. (p. 374)

Solar wind an outward flow of charged particles from the Sun. (p. 464)

Solid matter that has a definite volume and a definite shape. (p. 109)

Solifluction soil flow, which is the slow downslope movement of weathered material over a solid, impermeable base.

Solution a mixture that looks uniform throughout. Also called a *homogeneous mixture*. (p. 261)

Somatic effects short-term and long-term effects on the health of a recipient of radiation. (p. 251)

Sound a wave phenomenon caused by variations in pressure in a medium such as air. (p. 121)

Sound spectrum an ordered arrangement of various frequencies or wavelengths of sound. The three main regions of the sound spectrum are the infrasonic, the audible, and the ultrasonic. (p. 123)

Source region the region or surface from which an air mass derives its physical characteristics. (p. 517)

Specific heat the amount of heat energy necessary to raise the temperature of a unit mass of a substance one degree Celsius. (p. 97)

Speed, average the distance traveled divided by the time to travel that distance. (p. 26)

Speed, instantaneous the speed, or how fast an object is traveling, at a particular moment or instant. (p. 26)

Speed of light (c) 186,000 mi/s or 3.00×10^{10} cm/s or 3.00×10^8 m/s. (p. 122)

Spherical lenses lenses with spherical surfaces. (p. 157)

Spherical mirror a mirror whose surface is a section of a sphere. (p. 151)

Spring tide the tides of greatest variation between high and low. (p. 456)

Standard unit a fixed and reproducible reference value used for the purpose of taking accurate measurements. (p. 5)

Standing wave A "stationary" wave form arising from the interference of waves traveling in opposite directions. (p. 130)

Stationary front a front composed of two opposing or oppositely traveling fronts that have met and stalled. (p. 519)

Steam point the temperature at which pure water, at normal atmospheric pressure, boils. (p. 92)

Stimulated emission the process in which an excited atom is struck by a photon of the same energy as the allowed excitation transition, and two photons are given off. (p. 206)

Stock system a system of nomenclature for metals that form more than one ion, in which a Roman numeral placed in parentheses directly after the name of the metal denotes its ionic charge. (p. 308)

Storm surge the great dome of water associated with a hurricane when making landfall. (p. 523)

STP standard temperature of 273.16 K (0°C) and pressure of 1 atmosphere (760 mm of Hg).

Stratosphere a region of the atmosphere, based on temperature, approximately 10 to 35 miles in altitude. (p. 500)

Stratovolcano a volcano with a steeply sloping symmetric cone formed by eruption of high-viscosity lava and pyroclastic debris; also called a composite volcano. (p. 560)

Strong nuclear force the short-range force of attraction that acts between two nucleons and holds the nucleus together. (p. 226)

Structural formula a graphic representation of the way the atoms are connected to one another in a molecule. (p. 348)

Structural isomers compounds that have the same *molecular* formula but differ in *structural* formula. (p. 349)

Sublimation the phase change of going from a solid directly to a gas. (p. 99)

Sulfur a major fuel impurity, particularly in coal. (p. 529)

Sulfur dioxides (SO_x) an atmospheric pollutant formed by the oxidation of sulfur, which contributes to acid rain. (p. 529)

Summer solstice the farthest point of the Sun's declination north of the equator (for the Northern Hemisphere). (p. 428)

Sun a star; a self-luminous sphere of gas held together by its own gravity and energized by nuclear reactions in its interior. (p. 463)

Sunspots patches of cooler, darker material on the surface of the Sun. (p. 464)

Supernova an exploding star. (p. 473)

Superposition (law of) In a succession of stratified sediment deposits, the younger layers lie on top of the older layers. (p. 585)

Supersaturated solution a solution that contains more than the normal maximum amount of dissolved solute at a given temperature. (p. 261)

Surface wave (seismic) a seismic wave that travels along Earth's surface or a boundary within it. (p. 564)

S waves secondary (S) seismic waves, so called because they reach a seismic station after the P waves. S waves are transverse or shear waves—i.e., their particle oscillations are at right angles to the direction of propagation, and are transmitted only by solids. (p. 565)

Synodic month the orbital period of the Moon with respect to the Sun—the month of the Moon's phases. (p. 446)

Synodic period the orbital or rotational period of an object as seen by an observer on Earth. (p. 383)

Synthetic a material whose molecule has no duplicate in nature. (p. 361)

Synthetic detergents soap substitutes; their molecules contain a long hydrocarbon chain that is nonpolar (e.g., $C_{12}H_{25}-$ and a polar group such as sodium sulfate ($-OSO_3^- Na^+$). (p. 360)

Temperature a measure of the average kinetic energy of the molecules. (p. 90)

Temperature inversion a condition characterized by an inverted lapse rate. (p. 527)

Terminator the boundary line dividing day and night on the surface of a planet or Moon. (p. 397)

Terrestrial planets the inner four planets—Mercury, Venus, Earth, and Mars. All have earthly characteristics. (p. 381)

Thermal conductivity a measure of the ability of a substance to conduct heat energy. (p. 106)

Thermal efficiency an expression of the successful conversion of heat energy into work by a heat engine; thermal efficiency = work output/heat input. (p. 102)

Thermal expansion the expansion of a material because of a change in temperature. (p. 90)

Thermal insulator a material that is not a good conductor of heat. (p. 106)

Thermodynamics the branch of physics generally concerned with the dynamics of heat, that is, with the production and flow of heat, and the conversion of heat to work. (p. 101)

Thermodynamics, first law of The heat energy added to a system must go into increasing the internal energy of the system, or into any work done by the system, or both. The law also states that heat energy removed from a system must produce a decrease in the internal energy of the system, or any work done on the system, or both. The law is based upon the law of conservation of energy. (p. 101)

Thermodynamics, second law of It is impossible for heat to flow spontaneously from an object having a lower temperature to an object having a higher temperature. (p. 103)

Thermodynamics, third law of A temperature of absolute zero can never be attained. (p. 104)

Thermometer an instrument used to measure temperature. Common thermometers are based on thermal expansion such as that of a bimetallic coil or a liquid-in-glass. (p. 90)

Thermosphere a region of the atmosphere, based on temperature, between approximately 60 and several hundred miles in altitude. (p. 500)

Thunder the sound associated with lightning that arises from the explosive release of electrical energy. (p. 520)

Tidal force a differential gravitational force that tends to elongate a body. (p. 395)

Time the continuous forward-flowing of events. (p. 4)

Titius-Bode law a numerical sequence that gives a good approximate distance of some planets from the Sun in astronomical units. (p. 383)

Tornado a violent storm characterized by a funnel-shaped cloud and high winds. (p. 521)

Tornado warning the alert issued when a tornado has actually been sighted or indicated on radar. (p. 522)

Tornado watch the alert issued when conditions are favorable for tornado formation. (p. 522)

Torque a force about an axis; the product of the magnitude of the force and the perpendicular distance from the line of action of the force to the axis. (p. 62)

Total internal reflection a phenomenon in which light is totally reflected because refraction is impossible. (p. 142)

Transform boundary a region of the lithosphere where a moving plate slides along one side of another without creating or destroying lithosphere. (p. 581)

Transformer a device that increases or decreases the voltage of alternating current. (p. 581)

Transition elements the B-group elements in the periodic table. (p. 269)

Transverse wave a wave in which the particle motion is perpendicular to the direction of the wave velocity. (p. 118)

Tropical year the time interval from one vernal equinox to the next. (p. 427)

Troposphere a region of the atmosphere, based on temperature, between Earth's surface and 10 miles in altitude. (p. 500)

Ultrasound sound with frequency above 20 kHz. (p. 123)

Ultraviolet catastrophe the classical prediction for a thermal oscillator that incorrectly predicted the emission of a large amount of radiation in the high frequency or ultraviolet region. (p. 198)

Umbra a region of total darkness in a shadow. During an eclipse, an observer in the umbra sees a total eclipse. (p. 452)

Universe everything that is—all energy, matter, and space. (p. 373)

Unsaturated solution one in which more solute can be dissolved at the same temperature. (p. 261)

Valence electrons the electrons that are involved in bond formation, usually those in an atom's outer shell. (p. 269)

Valence shell an atom's outer shell, which contains the valence electrons. (p. 269)

Vector a quantity that has not only a magnitude, but also a direction associated with it. (p. 24)

Velocity, average the change in displacement divided by the change in time. (p. 26)

Vernal equinox the point on the celestial sphere at which the Sun crosses the celestial equator from south to north. (p. 428)

Volt (V) the unit of voltage equal to one joule per coulomb. (p. 171)

Voltage the amount of work it would take to move an electric charge between two points, divided by the value of the charge—i.e., work per unit charge. (p. 171)

Waning moon occurs between new and full moon for 14.75 solar days; during this time the illuminated portion of the Moon is decreasing for an Earth observer. (p. 447)

Warm front the boundary of an advancing warm air mass over a colder surface. (p. 519)

Water table the upper boundary of the zone of saturation of ground water.

Watt (W) a unit of power, 1 kg-m^2/s^3 or 1 J/s. (p. 73)

Wave function (ψ) the mathematical solution of a Schrödinger wave equation that can be used to describe particle characteristics. (p. 212)

Wavelength the distance from any point on a wave to an identical point on the adjacent wave. (p. 120)

Wave motion the emanation of energy from the disturbance of matter. (p. 117)

Wave velocity the velocity of a wave, the magnitude of which for a sinusoidal wave is related to the wavelength and frequency, $v = \lambda f$. (p. 118)

Waxing moon occurs between full and new moon for 14.75 solar days; during this time the illuminated portion of the Moon is increasing for an Earth observer. (p. 446)

Weather the atmospheric conditions of the lower troposphere. (p. 503)

Weathering the physical disintegration and chemical decomposition of rock.

Wind horizontal air motion. (p. 508)

Wind vane a pivoted, arrow-shaped device used to indicate the wind direction. (p. 507)

Winter solstice the farthest point of the Sun's declination south of the equator (for the Northern Hemisphere). (p. 428)

Work the product of a force and the parallel distance through which it acts. (p. 70)

Zenith the position directly overhead of an observer on Earth. (p. 429)

Zenith angle the angle between the zenith and the Sun at noon. (p. 429)

Index

Page numbers in boldface type indicate illustrations. The letter n following a page number indicates a footnote.

aberration of starlight, 376
absolute
   humidity, 514
   magnitude, 471
   temperature scale, 92
   (atomic) time scale, 586
   zero, 92
acceleration, 31
   average, 31
   centripetal, 51
   due to gravity, 33
   and force, 47
accelerators, particle, 239, **240**
accretion disk, 478
acetylene, 352, **352**
acid(s), 327
   amino, 361
   common, 275
   fatty, 359
   properties of, 326
   rain, 529
   strong, 327, **328**
   weak, 327, **328**
acid-base
   Arrhenius concept, 325
   reaction, 329, **330**
acid-carbonate reaction, 330, **331**
actinide series, 269
activation energy, 321
activity series, 333
addition polymers, 362
addition reactions, 351–352
air
   composition of, 499
   currents, 508
   mass, 517
      cold, 517
      source region, 517
      warm, 517
   pollution, 526
   resistance, 38
albedo, 374, 502
alchemy, 288

alcohol(s), 355
   ethyl (ethanol), 355
   methyl (methanol), 355
Alfvén, Hannes, 492
aliphatic hydrocarbons, 347
alkali, 335
   metals, 269, **277**
   reaction with water, 335
alkaline earth metals, 269, **279**
alkaloids, 365
alkanes, 347
alkenes, 351
alkyl group, 350
alkyl halide, 354
alkynes, 352
allotropes, 265
alloy, 261
alpha
   decay, 228
   particle, 228
alternating current (ac), 174
altitude (angle), 429
A.M., 422
amide, 360
   formation, 360
amines, 357
amino
   acids, 361
   group, 357
ammonia, 265, 273
   Lewis structure, 299
   properties and uses, 300, **300**
Ampère, André, 166
ampere (unit), 166
amphetamines, 357
amplitude, 119
analgesic, 364
analytical chemistry, 259
anemometer, 507
angular momentum, 61
   conservation of, 62
anion, 294, **297**
annular eclipse, 453
anode, 297, 337
antibiotic, 364
anticyclone, 510
antinode, 130

antipyretic, 364
apogee, lunar, 446
apparent
   brightness, 470
   solar day, 421
argon, 277, **278**
Aristotle, 45, 262, 288
aromatic hydrocarbons, 347, 352
Arrhenius, Svante, 326, **327**
Arrhenius acid-base concept, 326
aspirin, 364
asterism, 469
asteroids, 403
asthenosphere, 574, 582
astronomical unit (AU), 376, 467
astronomy, 373
atmosphere, 498
   divisions of, 500
   standard pressure, 504
atmospheric science, 498
atom, 166
   constituents of, **224**
   diameters of, 271
   hydrogen, Bohr theory of, 200
   multielectron, 215
   relative sizes, 271, **273**
   size of, 224
atomic
   mass, 225
   number, 224
   theory, Dalton's, 290
   (absolute) time scale, 586
aurora
   australis, 501
   borealis, 501
autumnal (fall) equinox, 428
Avogadro, Amedeo, 336
Avogadro's number, 336

Bacon, Francis, 3
Baekeland, Leo, 361
baking soda, 278, 300
Balmer series, 204
barium, 280
barometer, 504
   aneroid, 505
   mercury, 504

basalt, 547
base, 328
   properties of, 326
basins (lunar), 442
batholith, 549
battery, lead storage, 337
Becquerel, Henri, 228, 230
bedding, 551
   cross-, 552
bel (unit), 124
Bell, Alexander Graham, 124
belt of stability, 229
benzene, 347, 352, **353**
Bergeron process, 517
beryl, 279, **280**
beryllium, 279
Berzelius, Jöns Jacob, 263
Bessel, Friedrich W., 376
beta
   decay, 229
   particle, 229
Big Bang, 487
Big Crunch, 489
biochemistry, 259, 345
black hole, 476, **478**
blizzard, 521
Bode, Johann, 381
Bode's law, 381
Bohr, Niels, **201**, 244
Bohr theory of the hydrogen
   atom, 200
boiling point, 92
   of water, 101
bomb
   building of atomic, 244, 247
   hydrogen (fusion), 248
bond
   covalent, 298
      double, 301
      polar, 303
      triple, 301
   hydrogen, 306
   ionic, 296
   metallic, 305
   predicting type, 302
Bondi, Herman, 492
bottom-up theory, 484
boundary, plate
   convergent, 581
   divergent, 581
   transform, 581
Bowen reaction series, 547
Boyle, Robert, 110, 262, **263**, 289
Boyle's law, 110

Brahe, Tycho, 378
branched-chain structure, 349
Brand, Hennig, 263
breeder reactor, 247
breeze
   land, 509
   sea, 509
brilliance (diamond), 142
British system of units, 5
British thermal unit (Btu), 96
bromine, **272**, 279
buckminsterfullerene, 266
buckyball, 266, **267**
butane, 347

calcium, 279
calcium carbonate, 279, 280
caldera, 561
calendar, 433
   Gregorian, 435
   Julian, 434
Calorie (unit), 96
calorie (unit), 96
carbohydrates, 356
carbon
   dating, 235
   tetrachloride, 354
carbon dioxide, **290**, 300
carboxyl group, 357
carboxylic acids, 357
carcinogen, 352, 354, 529
Cartesian coordinate system, 418
Cassini, G. D., 395
Cassini division, 395
catalyst, 325
cathode, 297, 337
cation, 294, **297**
Cavendish, Henry, 54
celestial
   prime meridian, 467
   sphere, 466
cellulose, 356
Celsius, Anders, $92n$
Celsius scale, 92, **92**
center of curvature (mirror), 151
centi-, 7, 17
centigram, 7
centimeter, 7, 8
centrifugal force, 52
centripetal
   acceleration, 51
   force, 52
cepheid variables, 472
CFCs, 354, 532

cgs system, 7
Chadwick, James, 223
chain reaction, 243
characteristic (natural)
   frequencies, 131
chemical
   change, 316
   coefficients, 319
   energy, 320
   equations, balancing, 318
   equilibrium, 316
   subscripts, 272, 319
   energy, 315
   formula, 272
   nomenclature, 272–276
   nomenclature, organic, 350
   nomenclature, Stock system, 307
   properties, 316
   reactions, 316
   reaction types, summary, **335**
   sediments, 551
   symbols, 263
chemiluminescence, 326
chemistry, 259
   analytical, 259
   bio-, 259
   electro-, 337
   inorganic, 259
   organic, 259, 345
   origins of, 288
   physical, 259
Chernobyl, 247
China syndrome, 255
chlorine, 279
chloroform, 301, 354
chromosphere, 464
cinder cone, 561
circuit breaker, 179
circuits, electrical
   parallel, 176
   series, 174
circulation, thermal, 509
cleavage, 543
climate, 530
cloud, 511
   families, 513
   formation, 516
   types, 512–13
cluster, 482
coalescence, 517
cocaine, 365
coefficients, 319
cold front, 519

color
  mineral, 545
  vision, 159
combination reaction, 320
combustion reaction, 322
comet, 404
  Halley's, 405
compound, 260
  covalent, 298
  ionic, 292
  naming of, 272
compression, 123
concave (spherical) mirror, 152
condensation
  polymers, 363
  theory, 408
conduction
  electrical, 166
  thermal, 106
conductivity, thermal, 106
cones, 159
conjunction, 383
conservation of
  angular momentum, 62
  linear momentum, 59
  mass-energy, 84
  mass, law of, 286
  mechanical energy, 79
  total energy, 79
constellation, 469
constructive (complete) interference, 147
continental drift, 575
continuous-chain structure, 349
control rods, 243
convection
  cells, 510, **582**
  cycle, 107, **107**, 509, **510**
  heat transfer by, 107
convergent (plate) boundary, 581
converging (convex) lens, 157
conversion factor, 14
convex (diverging) mirror, 152
Copernicus, Nicolaus, 378
copper, 272
core, 573
Coriolis force, 509
corona, 464
correlation, 585
cosmic
  background radiation (3-K), 489
  rays, 236

cosmological
  principle, 485
  perfect, 492
  red shift, 487
cosmology, 487
Coulomb, Charles, 166
coulomb (unit), 6, 166
Coulomb's law, 166, 167
covalent
  bond, 298
  compound, 298
crater (lunar), 442
  secondary, 443
crescent moon, 447
critical
  angle, 142
  mass, 243
crust, 539, 573, **574**
crystal, 293
crystalline structure, 543
crystallography, 543
Curie, Marie, 185n, 228, 230, **231,** 263
Curie, Pierre, 185n, 228, 230, 263
Curie temperature, 185
current, electric, 166
  alternating, 174
  direct, 174
cycloalkanes, 351
cyclone, 510, 523

Dacron, 363
Dalton, John, 289, **291**
dark matter, 484
Davisson, G., 211
Davy, Humphry, 263
day
  apparent solar, 421
  mean solar, 421
  sidereal, 421
Daylight Saving Time, 426
Dead Sea Scrolls, 238
de Broglie, Louis, 211
de Broglie waves, 211
deceleration, 31
decibel (unit), 124
declination
  coordinate, 467
  magnetic, 185
decomposition reaction, 316, 320
definite proportions, law of, 287
degree (unit), 92
Democritus, 288

density, 12
deposition, 517
derivatives of hydrocarbons, 347
derived quantities, 10
Descartes, René, 418
destructive (complete) interference, 147
detergent, synthetic, 360
deuterium, 225
deuteron, 225
dew, 517
  point temperature, 514
diamond, 265, 266, **266,** 267
  brilliance, 142
  fire, 142
diffraction, 143
  grating, 149
dike, 550
dipole, 304
  -dipole interaction, 306
Dirac, Paul A. M., 492
direct current (dc), 174
dispersion, 142
displacement, 24
distance, 24
divergent (plate) boundary, 581
diverging (concave) lens, 157
Dobereiner, Johann, 268
Doppler, Christian, 128n
Doppler effect, 128
  blue shift, 128
  radar, 522
  red shift, 128
double-replacement reaction, 331, **332,** 333
drugs, 364
dry ice, 290
dual nature of light, 200
dyne (unit), 48n, 70n

Earth, age of, 235
earthquake, 562
echo, 137
eclipse, 452
  lunar, 453
  solar, 452
    annular, 453
    partial, 452
    total, 452
ecliptic, 374
Edison, Thomas, 192
efficiency
  ideal, 103
  thermal, 102

Einstein, Albert, 84, 199, 244, 249, 417
electric(al)
  charge, 3, 5, 166
    law of, 167
  charging
    by friction, 168
    by induction, 170
  circuits, 174, 176
  current, 166
  force, 50, 167
  ground, 177
  potential energy, 171
  power, 172
  resistance, 171
  safety, 178
electrochemistry, 337
electrolysis, 337
electrolyte, 337
electrolytic cell, 337
electromagnet, 184
electromagnetic
  induction, 190
  spectrum, 122
  waves, 121
electromagnetism, 186
electron, 166, 223
  period, 219
  shell, 216
  subshell, 216
  valence, 269
  volt, 203
electronegativity, 303
electrostatics, 168
element(s), 224, 259, **260,** 262
  inner transition, 269
  occurrence of, 264
  representative, 267, **269**
  transition, 269
  transuranium, 240
emerald, 279, **280**
emission
  spontaneous, 206
  stimulated, 206
endoergic reaction, 250
endothermic reaction, 321, **323**
energy, 75
  activation, 321
  barrier, 322
  forms of, 82
  ionization, 272
  kinetic, 75
  mechanical, 75, 79
    conservation of, 79

  potential, 77
    electrical, 171
    gravitational, 77
    spring, 78
    solar, 85
    states (levels), 202
    total, 79
    conservation of, 79
Enola Gay, 245
entropy, 105
enzymes, 326
epicenter, 564
epoch, 585
equator, 418
equilibrium (chemical), 317
equinox
  autumnal (fall), 428
  vernal (spring), 428
equivalence statement, 14
era, 585
Eratosthenes, 377
erg (unit), $70n$
ester, 358
  formation, 358
ethane, 347
ethanol, 356
ethene, 351
ethylene, 311
ethylene glycol, 356
ethyne, 352, **352**
evaporation, 109
event horizon, 477
excess reactant, 290
excited states, 203
exoergic reaction, 250
exothermic reaction, 321, **323**
eye, human, 159

Fahrenheit, Daniel, $92n$
Fahrenheit scale, 92, **92**
false vacuum, 488
Faraday, Michael, 190
fats, 359
fatty acids, 359
fault
  earth, 562
    San Andreas, 563
  lunar, 443
feldspar, 540
Fermi, Enrico, 244
ferromagnetic, 183
fiber optics, 142
fire (diamond), 142
first-quarter moon, 447

fission, nuclear, 82, 241, **242,** 244, **244**
  controlled, 243
  reactor, 243
  uncontrolled, 243
fluorescence, 210
  mineral, 545
fluorine, 278
focal length
  lens, 158
  mirror, 151
focal point
  lens, 158
  mirror, 151
foliation, 555
foot (unit), 5
foot pound (unit), 70
force, 44
  centrifugal, 52
  centripetal, 52
  Coriolis, 509
  electric, 50, 167
  electromagnetic, 50
  electroweak, 50
  external, 44
  gravitational, 50, 54
  internal, 44
  magnetic, 50, 187
  strong nuclear, 50, 226
  tidal, 395, 455
  unbalanced (net), 44
  weak nuclear, 50
formula
  mass, 287
  structural, 348
    condensed, 348
  unit, 293, **294**
fossil, 552, 585
  fuels, 528
  index, 585
Foucault, Jean, 374
Foucault pendulum, 374
fracture, 543
freezing point, of water, 92, 94, 100
frequency, 119
  characteristic (natural), 131
  fundamental, 131
Fresnel, Augustin, 160
Fresnel lens, 160
friction, 72
  charging by, 168
  work against, 72
Friedman, Alexander, $487n$
Frisch, Otto, 244

front, 519
  cold, 519
  occluded, 519
  stationary, 519
  warm, 519
frost, 517
fructose, 356
fruit flies, 241
fuel rods, 243
fuels
  alternative, 83
  fossil, 83
full moon, 448
Fuller, Buckminster, 266
fullerenes, 266
fundamental
  frequency, 131
  quantities, 3
fuse, 178
fusion
  latent heat of, 99
  nuclear, 82, 245, 247, **249**

g, 33
G, 54
galaxy, 479
Galileo, 3, 33, **35**, 43, 45, 380, 394, 422, 464
Galle, John G., 399
gallium, **272**
gamma
  decay, 229
  ray, 122, 229
gas, 112
gasohol, 83, 356
Geiger counter, 233
gegenshein, 407
gem, 546
generator, 190
genetic effects, 251
geocentric theory, 378
geographic poles, 185, 418
geologic time, 572, 584, **586**
  absolute (atomic), 585
  relative, 585
geologic time scale, 588
  absolute (atomic), 586
  relative, 585
geology, 539
germanium, 268
Germer, L. H., 211
gibbous moon, 447
Gilbert, William, 185
glucose, 356

gnomon, 421
Gold, Thomas, 492
gram, 8
graphite, 266, **266**, **267**, 541
graphs, 28
gravitational
  collapse, 477
  constant, universal, 54
  potential energy, 77
gravity, 50
  acceleration due to, 33
great circle, 418
great impact theory, 445
Great Red Spot, 393
greenhouse effect, 502, **506**
Greenwich (prime) meridian, 419
ground (electrical), 177
ground state, 203
group(s), 266
  of elements, 276
Groves, General Leslie, 245
Gutenberg discontinuity, 573

Hahn, Otto, 244
hail, 521
half-life, 233
Halley, Edmond, 405
Halley's comet, 405
halogens, 269, 278
hardness (mineral), 543
  Mohs' scale of, 543
harmonics, 131
heat, 96
  engine, 102
  of fusion, latent, 99
  joule, 173
  latent, 98
  lightning, 520
  mechanical equivalent of, 96
  pump, 104
  specific, 97
  transfer, 106
  of vaporization, latent, 99, 109
heavy water, 225
Heisenberg, Werner, 214
Heisenberg uncertainty principle, 214
heliocentric theory, 378
heliosphere, 464
heliostat, 85
heroin, 365
Hertz, Heinrich, 119
hertz (unit), 119
Hertzsprung, Ejnar, 471

Hess, H. H., 578
Hindenburg syndrome, **281**
Hipparchus, 470
horizon, 429
horsepower (unit), 73
hot-spot theory, 560
Hoyle, Fred, 492
H-R diagram, 471
Hubble, Edwin P., 472, 479
Hubble's law, 484
Hubble Space Telescope, 490
humidity, 507, 514
  absolute, 514
  relative, 507, 514
hurricane, 523
  names, 524
  warning, 524
  watch, 524
hydrocarbon(s), 347
  aliphatic, 347
  aromatic, 347, 352, **353**
  derivatives of, 347, 354
  saturated, 347
  unsaturated, 351
hydrochloric acid, 273, **328**
hydrogen, 265, **281**
  bomb, 245, 248
  bond, 306, **307**
  burning, 475
  chemical classification of, 280
  chloride, 303
  cyanide, 301
  isotopes of, 225
  peroxide, 302
hydrogenation, 359
hydrometer, 13
hydronium ion, 327
hydropower, 83
hydroxyl group, 355
hygroscopic nuclei, 516
hypothesis, 3

ice
  point, 92
  storm, 521
ideal
  efficiency, 103
  (perfect) gas law, 110
igneous rock, 547, 548, **550**
  extrusive, 548
  intrusive, 548
    concordant, 549
    discordant, 549

image
　real
　　lens, 158
　　mirror, 153
　virtual
　　lens, 158
　　mirror, 153
impulse, 61
index
　fossil, 585
　of refraction, 141
inertia, 45, 71, **71**
　law of, 45–46
　work against, 71
inertial confinement, 248n
infrasound, 123
inhibitor, 325n
inner transition elements, 269
inorganic chemistry, 259
intensity, sound, 124
interference, 146
　constructive (complete), 147
　destructive (complete), 147
　double-slit, 149
　thin film, 148
International Date Line (IDL), 424
International System of Units (SI), 7
interplanetary dust, 407
insolation, 502
insulator
　electrical, 167
　thermal, 106, **106**
iodine, 272, 279
ion, 224, 273
　-dipole interaction, 306
　hydronium, 327
　monatomic, 273
　polyatomic, 273, **274,** 320
ionic
　bond, 296
　compound, 292
ionization energy, 272, 274
ionizing radiation, 251
ionosphere, 501
I^{2}R losses (joule heat), 173
irregular (diffuse) reflection, 138
isoelectronic, 295
isomers, structural, 349
isometric process, 110
isothermal process, 110
isotope, 225
IUPAC, 264, 350

jet
　propulsion, 57, 60
　stream, 511
Joliot, Frederick, 231
Joliot, Irene, 231
joule heat (I^{2}R losses), 173
Joule, James Prescott, 70, 89
joule (unit), 70
Jovian planets, 381, 391
Jupiter, 392

karat, 305
Kelsey, Frances O., 364
Kelvin, Lord, 92n
kelvin (unit), 92
Kelvin (absolute) scale, 92, **92**
Kepler, Johannes, 378
Kepler's laws of planetary motion, 379
　of elliptical paths, 379
　of equal areas, 379
　harmonic law, 379
kilo-, 7, 17
kilocalorie, 96
kilogram, 8
kilometer, 6, 8
kilowatt, 74
kilowatt-hour, 74
kinetic energy, 75

laccolith, 550
land breeze, 509
lanthanide series, 269
lapse rate, 516
　inverted, 527
laser, 206
last-quarter moon, 448
latent heat, 98
　of fusion, 99
　of vaporization, 99
latitude, 419
lattice, 109
lava, 547
Lavoisier, Antoine, 286, 289
law(s)
　Bode's, 381
　Boyle's, 110
　of charges, 167
　of conservation of mass, 286
　Coulomb's, 167
　of definite proportions, 287
　Hubble's, 484
　ideal (perfect) gas, 110

　Kepler's
　　of elliptical paths, 379
　　of equal areas, 379
　　harmonic law, 379
　　of multiple proportions, 292
　Newton's
　　of gravitation, 54
　　of motion
　　　first (of inertia), 45
　　　second, 48
　　　third, 57
　Ohm's, 172
　of octaves, 268
　periodic, 266
　of poles, 181
　of reflection, 138
　scientific, 3
　of superposition, 585
　of thermodynamics
　　first, 101
　　second, 103
　　third, 104
　Titius-Bode, 383
LCD, 154
lead storage battery, 337
Leavitt, Henrietta Swan, 472
Le Châtelier, Henri, 317
Le Châtelier's principle, 317, **317, 318**
Lemaitre, Georges, 487n
length, 3
lens, spherical, 157
　converging (convex), 157
　crystalline, 159
　diverging (concave), 157
　Fresnel, 160
Lewis, Gilbert Newton, 293
Lewis
　structures, 293
　symbols, 293
Libby, Willard F., 235
light, 122
　coherent, 208
　dual nature of, 200
　speed of, 122
lightning, 519
　heat, 520
light-year (unit), 467
limiting reactant, 290
linear momentum, 59
　conservation of, 59
linear (plane) polarization, 150
liquid, 109
　crystal display (LCD), 154

liter, 9
lithification, 553
lithosphere, 574, **574**, **582**
litmus, 329
Local Group, 481
local solar time, 422
lone pairs, 293
longitude, 419
longitudinal wave, 119
loudness, 123
lunar eclipse, 453
lye, 278

magma, 547, **547**, 548
   basaltic, 547
magnesium, 279, **321**
magnet
   electro-, 184
   permanent, 184
magnetic
   anomalies, 579
   confinement, 248
   declination, 185
   domains, 183
   field, 181
      Earth's, 185
   force, 187
   poles, 180
      law of, 181
magnetism (mineral), 545
   remanent, 579
magnification (factor)
   lens, 158
   mirror, 156
magnitude (stellar), 470
main sequence, 471
Manhattan Project, 244
mantle, 573
marble, 556
maria, 442
Mars, 389
mass, 3, 4, **5**, 45, 50, 84
   atomic, 225, 287
   conservation of, law, 286
   critical, 243
      subcritical, 243
      supercritical, 243
   formula, 287
   isotopic, 226
   number, 225
   spectrometer, 225
mass-energy, conservation of, 84
matter, 4
   classification of, 259, **260**

   dark, 484
   phases of, 98, **99**, 108
   waves, de Broglie, 211
mean solar day, 421
measurement, 1
mechanical
   energy, 75, 79
      conservation of, 79
   equivalent of heat, 96
mega-, 17
Meitner, Lisa, 244
meltdown, 247
Mendeleev, Dmitri, 266, 268
Mercalli scale, 565
Mercury, 385
mercury, **272**, 316
meridian, 418, **419**
   celestial prime, 467
   Greenwich (prime), 419
mesosphere, 500
metal, 270, **271**
   alkali, 269–70
   alkaline earth, 270
metallic bonding, 305, **306**
metalloids, 270
metamorphic rock, 547, 554, **555**
meteor, 404
meteorites, 404
meteoroids, 404
meteorology, 498
meter, 7
methane, **265**, 273, 301, 347–349
methanol, 356, **356**
metric
   prefixes, 17
   system, 6
Meyer, Julius, 268
micro-, 17
micrometeoroids, 407
microwave(s), 122
   oven, 205
Milky Way, 479, 480, **482**
milli-, 17
millibar, 507
milliliter, 9
millimeter, 8
mineral, 540, **541**
   identification, 543
mirror
   plane, 138
   spherical, 151
      concave (converging), 152
      convex (diverging), 152

mixture, 260, **260**
   heterogeneous, 261
   homogeneous, 261
mks system, 7
moderator, 247
Moho discontinuity, 573
Mohs' scale of hardness, 543
mole, 336
molecule, 264, **265**, 336
   predicting polarity, 304
momentum
   angular, 61
      conservation of, 62
   linear, 59
      conservation of, 59
monomer, 362
month
   sidereal, 446
   synodic, 446
Moon, 440
   composition and origin of, 444
   motions of, 445
   phases of, 446–48
morphine, 365
motion, 24
   prograde, 381
   projectile, 36
   retrograde, 381
   uniform circular, 51
motor, 188
mountain ranges, lunar, 443
mud cracks, 553

nano-, 17
neap tide, 456
nebula, 472
   emission, 474
   Orion, 474
   planetary, 474, 475
neon, 277, **278**
Neptune, 399
neutrino, 465
neutron, 166, 223
   activation analysis, 240
   number, 225
   star, 476
Newlands, John, 268
new moon, 446
Newton, Isaac, 43, **47**, 54, 381, 454
newton (unit), 48
Newton's law(s)
   of motion
      first (of inertia), 45

Newton's law(s) of motion (*cont.*)
   second, 48
   third, 57
   of universal gravitation, 54
nicotine, 365
nitrogen oxides, 528
noble gases, 269, 276, **278**
node
   lunar, 453
   standing wave, 130
nonmetal, 270, **271**
nonpolar molecule, 304
norethynodrel, 365
northern lights, 501
nova, 473
nuclear
   energy, 82
   fission, 241, **242,** 244
   force, 226
   fusion, 82, **245,** 247, **249**
   reaction, 238
   reactor, 243
nucleon, 223
nucleosynthesis, 264, 476
nucleus, 223
   mass of, 224
   size of, 224
nuclide, 228
nylon, 363, **363**

Oak Ridge, 245
occluded front, 519
octet rule, 292
Oersted, Hans, 182
Ohm, Georg, 171
ohm (unit), 171
Ohm's law, 172
oils, 359
olivine, 541
Oppenheimer, J. Robert, 245
Oort, Hendrick, 406
Oort cloud, 406
opposition, 383
organic
   chemistry, 259, 345
   compounds, 345
      bonds in, 345
overtones, 131
oxidation, 332
   -reduction reaction, 333
ozone, 266, 354, 500
   hole, 532
   as pollutant, 354, 529
ozonosphere, 500
P wave, 564

Pangaea, 575
Paracelsus, 289
parallax, 376, 468
parallel circuit, 176
parallels, 419
parsec, 377, 467
particle accelerator, 239
pascal (unit), 110
Pauli, Wolfgang, 217
Pauli exclusion principle, 217
pendulum
   Foucault, 374
   simple, 81
penumbra
   eclipse, 452
   sun spot, 464
Penzias, Arno, 489
percentage composition by mass, 287
perigee, lunar, 446
period
   chemical, 266
   geologic time, 585
   sidereal, 383
   synodic, 383
   wave, 120
periodic
   law, 266
   table, 219, 266
petroleum, 348
PET scan, 240, **241**
pH, 328, 329
phase(s)
   changes of, 98
   in, 147
   of matter, 98, **99,** 108
   of Moon, 446–48
   out of, 147
phenomenon, 3
phlogiston, 289
phosphorescence, 210
photochemical smog, 529
photoelectric effect, 119
photon, 200
photosphere, 463
photosynthesis, 315, 499, **499**
physical
   changes, 316
   chemistry, 259
   properties, 315
physics
   modern, 197
   quantum, 198

Pisa, Leaning Tower of, 35
pitch, 124
plains, lunar, 442
Planck, Max, 198
Planck time, 423
Planck's constant, 199
planet(s)
   Jovian, 381, 391
   terrestrial, 381, 385
   X, 401
plasma, 112, 248, 463
   cosmology, 492
plastics, 361
plate tectonics, 581
Pluto, 392, 400, 401
pluton, 549
plutonium, 245, 247
P.M., 422
polar
   covalent bond, 303
   molecule, 304
Polaris, 432
polarization, 150
   linear (plane), 150
   molecular, 170
   partial, 150
poles
   geographic, 185, 418
   magnetic, 180
pollution, 526
polonium, 230
polyatomic ion, 273, 320
polyethylene, 362, **362**
polymer(s), 362
   addition, 362
   condensation, 363
position, 23
positron, 247
potential energy, 77
   electrical, 171
   gravitational, 77
   spring, 78
pound (unit), 48
power, 73
   electrical, 172
powers-of-ten notation, 16
precession, 432
   of equinoxes, 432
precipitate, 332, 333
pressure, 504
   standard atmosphere of, 504
Priestley, Joseph, 289, **316**
prime (Greenwich) meridian, 419

principal
  axis, 151
  quantum number, 202, 215
principle of uniformity, 585
product (chemical), 316
prograde motion, 381
projectile
  motion, 36
  range, 36
prominences, 465
proteins, 361
protium, 225
proton, 166, 223
proton-proton chain, 465
Proust, Joseph, 287
psychrometer, 514
pulsar, 476
pure substance, 256
pyroclastic debris, 556

quality (timbre), 125
quantum, 199
  mechanics, 213
  number
    magnetic, 216
    orbital, 216
    principal, 202, 215
    spin, 216
  physics, 198
quart, 9
quartz, 540
quasar, 486

radar, 130, 508
  Doppler, 522
radiation
  background, 252
  biological effects, 251
  cosmic background, 489
  electromagnetic, 122
  heat transfer by, 107
  ionizing, 251
  temperature inversion, 527
radioactive decay, 228
radioactivity, 228
  discovery of, 230
  uses of, 241
radioisotopes, 228
radiometric dating, 234
radionuclide, 228
radium, 230, 280
radius of curvature (mirror), 151
radon, 252, 277
  reactions, addition, with alkenes, 351

  reactions, addition, with alkynes, 352
rain gauge, 508
rainbow, 145
rainstorm, 519
range, projectile, 36
rarefaction, 123
ray
  lens
    chief, 158
    parallel, 158
  lunar, 442
  mirror
    chief (radial), 153
    parallel, 153
  optical, 138
Rayleigh, Lord, 502
Rayleigh scattering, 502
reactant, 36
  excess, 290
  limiting, 290
reaction(s)
  acid-base, 329
  acid-carbonate, 330
  addition, 351, 352
  chain, 243
    self-sustaining, 243
  chemical, 316
    endothermic, 321
    exothermic, 321
    forward, 316
    reverse, 317
    reversible, 316
  combination, 320
  combustion, 322
  decomposition, **316**, 320
  double-replacement, 331, **332, 333**
  nuclear, 238
    endoergic, 250
    exoergic, 250
  oxidation-reduction, 333
  rate of, 324–326
  redox, 333
  single-replacement, 332–335
red
  dwarf, 472
  giant, 471
  shift, 128
    cosmological, 487
    Doppler, 128
redox reaction, 333
reduction (chemical), 332
reflection, 137
  irregular (diffuse), 138

  law of, 138
  regular (specular), 138
  total internal, 142
refraction, 140
  angle of, 140
  index of, 141
regolith, 403, 442
relative
  geologic time scale, 585
  humidity, 507
remanent magnetism, 579
representative elements, 267
resistance
  air, 38
  electrical, 171
    in parallel, 176
    in series, 174
resonance, 132
retina, 159
retrograde motion, 381
reverse thrust, 60
reversible reaction, 316
revolution, 374
Richter scale, 565
rift valley, 583
right ascension, 467
rilles (lunar), 443
"Ring of Fire," 559, **560**, 563
ripple marks, 552
Roche limit, 395
rock, 540, 545, **547**
  cycle, 547
  dating of, 235
  igneous, 547, 548, **550**
    concordant, 549
    discordant, 549
    extrusive, 548
    intrusive, 548
  metamorphic, 547, **547**, 554, **555**
  sedimentary, 547, **547**, 550, **551**
    chemical, 551, 554
    clastic, 551
    organic, 551
rods
  control, 243
  (eye), 159
  fuel, 243
Roentgen, Wilhelm, 208, 230
rotation, 374
rubber, 362
Rumford, Count, 89
Russell, Henry, 471
Rutherford, Ernest, 200, 223, 238

S wave, 565
salt (chemical), 329
   bridge, 337
San Andreas fault, 563
saturated
   hydrocarbons, 347
   solution, 261, **261**, 262, 514
Saturn, 395
Segre, Emilio, 244
scalar, 24
scattering, 502
   Rayleigh, 502
Schrödinger, Erwin, 212
Schrödinger equation, 212
Schwarzschild radius, 477
scientific
   method, 3
   notation, 16
sea breeze, 509
seafloor spreading, 578
   convection cells, 582
second (unit), 6, 8
sedimentary rock, 547, **547**, 550, **551**
   chemical, 551, 554
   clastic, 551
   organic, 551
sediments
   chemical, 551
   clastic, 551
   organic, 551
seismic wave(s), 564
   body, 564
   P, 564
   S, 565
   surface, 564
seismograph, 565
seismology, 562
semiconductor, 167
semimetals, 270
senses, human, 2
series circuit, 174
Seyfert, Carl, 480
Seyfert galaxies, 480
shadow zone, 565
shale, 555
shell distribution of electrons, 269, **270**
shield volcano, 560
Shroud of Turin, 238, **238**
SI, 7
sidereal
   day, 421
   month, 446
   period, 383
   year, 427
sievert (unit), 252
single-replacement reaction, 333
singularity, 477
silica, 540
silicates, 540
sill, 550
slate, 555
sleet, 521
slope, 29
slug (unit), 6
smog, 527
   photochemical, 529
smoke detector, 240
snow, 521
snowstorm, 521
soap, 359
sodium, 277, **279**
sodium chloride, 292–294, **293, 294**
sodium hydroxide, 278, 302
solar
   eclipse, 452
   nebula, 408
   system, 374
   wind, 464
solid, 109
solstice
   summer, 428
   winter, 428
solubility, 261
solute, 261
solution, 261
   acidic, 328
   aqueous, 261
   basic, 328
   saturated, 261
   supersaturated, 261
   unsaturated, 261
solvent, 261
somatic effects, 251
sonic boom, 128
sound, 122
   infrasound, 123
   intensity, 124
   spectrum, 123
   speed of, 125
   ultrasound, 123, 125
source region, 517
specific heat, 97
spectroscopy, 143
spectrum
   absorption, 205
   electromagnetic (EM), 122
   emission, 205
   sound, 123
speed
   average, 26
   instantaneous, 26
   of light, 122
   of sound, 125
spherical
   lenses, 157
   mirrors, 151
spontaneous emission, 206
spring tide, 456
stalactite, 554
stalagmite, 554
standard
   time zones, 422
   unit, 5
standing wave, 130
starch, 356
star(s)
   definition, 463
   neutron, 476
   types of, 469–76
stationary front, 519
steady-state model, 492
steam point, 92
stimulated emission, 206
Stock system, 308
storm
   ice, 521
   rain, 519
   surge, 523
   thunder, 519
   tropical, 523
straight-chain structure, 349
Strassman, Fritz, 244
stratosphere, 500
stratovolcano (composite), 560
strong nuclear force, 50, 226
structural
   formula, 348
      condensed, 348
   isomers, 349
subduction zone, 583
sublimation, 99, 517
subscripts, 272, 319
subsidence temperature inversion, 527
substance, pure, 256
sucrose, 356
sulfur, **272**, 529

sulfur dioxide, 529
sugars, 356
summer solstice, 428
Sun, 463
   nuclear reactions in, 248
sunspots, 464
supercluster, 483
superconductivity, 175
supernova, 473
superposition, law of, 585
supersaturated solution, 261
surface
   melting, 100
   wave (seismic), 564
synodic
   month, 446
   period, 383
synthetic, 361
   detergent, 360
system, 79
   isolated, 79
   of units, 5
      British, 5
      metric, 6
         cgs, 7
         mks, 7
         SI, 7
Szilard, Leo, 244

Teflon, 362, **363**
temperature, 90
   dew point, 514
   inversion, 527
      radiation, 527
      subsidence, 527
   scales
      Celsius (centigrade), 92, **92**
      Fahrenheit, 92, **92**
      Kelvin (absolute), 92, **92**
tephra, 556
terminator, 397
terrestrial planets, 381, 385
texture (rock), 549
   bottom-up, 484
   great impact, 445
   hot spot, 560
   top-down, 484
thalidomide, 364
theory, 3
   condensation, 408
   geocentric, 378
   heliocentric, 378
thermal
   circulation, 509
   conduction, 106
   conductivity, 106
   convection, 107, 509
   efficiency, 102
   expansion, 90
   insulator, 106
thermodynamics, 101
   first law of, 101
   second law of, 103
   third law of, 104
thermometer, 90
   constant-volume gas, 111
Thermos bottle, 107
thermosphere, 500
thermostat, 113
Thompson, Benjamin, 89
Thomson, J. J., 223
Three Mile Island, 247
threshold of hearing, 124
thunder, 520
   -storm, 519
tidal
   force, 395, 455
   stability limit, 395
tides, 454, **455**
   neap, 456
   spring, 456
timbre (quality), 125
time, 3, 4
   Daylight Saving, 426
   geologic, 572
   local solar, 422
   Planck, 423
   standard, 422
tincture, 279
Titius, Johann Daniel, 383
Titius-Bode law, 383
TNT, 353
tokamak, 248
top-down theory, 484
tornado, 521
   safety, 522
   warning, 522
   watch, 522
torque, 62
total internal reflection, 142
transform (plate) boundary, 581
transformer, 190
   step-down, 191
   step-up, 191
transition elements, 269
transmutation, 239
transuranium elements, 240
transverse wave, 118

tritium, 225
triton, 225
tropical year, 427
troposphere, 500
typhoon, 523

ultrasound, 123, 125
ultraviolet
   catastrophe, 198
   light, 122
umbra
   eclipse, 452
   sunspot, 464
uniformitarianism, 585
unit(s)
   standard, 5
   system of, 5
universal gravitational constant, 54
universe, 373
unsaturated
   hydrocarbons, 351
   solution, 261
uranium, isotopes of, 243
Uranus, 397

valence
   electron, 269
   shell, 269
vaporization, latent heat of, 99, **99**, 109
vector, 24
Velcro, 363, **363**
velocity
   average, 26
   instantaneous, 27
   wave, 118
Venus, 386
vernal (spring) equinox, 428
viscosity, 557
volcano, 556
   cinder cone, 561
   shield, 560
   strato- (composite), 561
volt (unit), 171
voltage, 171
voltaic cell, 337

waning moon, 447
warm front, 519
water
   formation, **322**
   Lewis structure, 299
   molecule, **265**
   polarity, 303

water hole, 411
Watt, James, 73
watt (unit), 73
wave(s), 117
  amplitude, 119
  electromagnetic, 121
  frequency, 119
  function, 212
  longitudinal, 119
  matter (de Broglie), 211
  period, 120
  radio, 121
  seismic, 564
  sound, 122
  standing, 130
  transverse, 118
  velocity, 118
  water, 119
wavelength, 120

waxing moon, 446
weather, 498, 503
  satellite, 508
Wegener, Alfred, 575
weight, 4, 50, **51**
  on Moon, 4, **5**, 50, 441
weightlessness, 56
white
  dwarf, 472
  light, 142
Wilson, Robert, 489
wind, 508
  direction, 509
  shear, 522
  solar, 464
  vane, 507
wintergreen, oil of, 358
winter solstice, 428
work, 70

  against friction, 72
  against gravity, 71
  against inertia, 71, **71**

X-rays, 122, 208, **280**
  characteristic, 209

year
  sidereal, 427
  tropical, 427
Young, Thomas, 149

zenith, 429
  angle, 429
zero g, 56
Zinn, Walter, 244
zodiac, 433, 469
zodiacal light, 407
zones of compression, 583

Conversion Factors

Mass
1 gram = 10^{-3} kg
1 kg = 10^3 g (equivalent weight = 2.20 lb)
1 u = 1.66×10^{-24} g = 1.66×10^{-27} kg

Length
1 cm = 10^{-2} m = 0.394 in.
1 m = 10^{-3} km = 1.09 yd = 3.28 ft
    = 39.4 in.
1 km = 10^3 m = 0.62 mi
1 in. = 2.54 cm = 2.54×10^{-2} m
1 ft = 12 in. = 30.48 cm = 0.3048 m
1 yd = 3 ft = 0.914 m
1 mi = 5280 ft = 1609 m = 1.609 km
1 Å = 10^{-10} m = 10^{-8} cm
1 pc = 3.26 ly = 2.05×10^5 AU

Volume
1 m^3 = 10^3 L = 264 gal
1 L = 10^{-3} m^3 = 1.06 qt = 0.264 gal
1 ft^3 = 7.48 gal = 0.0283 m^3 = 28.3 L
1 qt = 2 pt = 0.946 L = 946 mL
1 gal = 4 qt = 3.785 L

Time
1 h = 60 min = 3600 s
1 day = 24 h = 1440 min = 8.64×10^4 s
1 year = 365 days = 8.76×10^3 h
    = 5.26×10^5 min = 3.16×10^7 s

Energy
1 joule = 0.738 ft-lb
    = 0.239 cal = 9.48×10^{-4} Btu
    = 6.24×10^{18} eV
1 kcal = 4186 J = 3.97 Btu
    = 0.0016 kWh
1 Btu = 1055 J = 778 ft-lb = 0.252 kcal
1 cal = 4.186 J = 3.97×10^{-3} Btu
    = 3.09 ft-lb
1 ft-lb = 1.36 J = 1.29×10^{-3} Btu
1 eV = 1.60×10^{-19} J
1 kWh = 3.60×10^6 J
    = 3.413×10^3 Btu = 860 kcal

Speed
1 m/s = 3.6 km/h = 3.28 ft/s
    = 2.24 mi/h
1 km/h = 0.278 m/s = 0.621 mi/h
    = 0.911 ft/s
1 ft/s = 0.682 mi/h = 0.305 m/s
    = 1.10 km/h
1 mi/h = 1.467 ft/s = 1.609 km/h
    = 0.447 m/s
60 mi/h = 88 ft/s

Force
1 newton = 0.225 lb
1 lb = 4.45 N
Equivalent weight of 1 kg mass = 2.20 lb
    = 9.80 N

Pressure
1 atm = 14.7 lb/$in.^2$ = 1.013×10^5 N/m^2
    = 30 in. Hg = 76 cm Hg
1 bar = 10^5 Pa
1 millibar = 10^2 Pa
1 Pa = 1 N/m^2 = 10^{-2} millibar

Power
1 watt = 0.738 ft-lb/s = 1.34×10^{-3} hp
    = 3.41 Btu/h
1 ft-lb/s = 1.36 W = 1.82×10^{-3} hp
1 hp = 550 ft-lb/s = 745.7 watt
    = 2545 Btu/h